ANNALS OF THE NEW YORK ACADEMY OF SCIENCES

Volume 834

EDITORIAL STAFF

Executive Editor
BILL M. BOLAND

Managing Editor
JUSTINE CULLINAN

Associate Editor
ANGELA C. FINK

The New York Academy of Sciences
2 East 63rd Street
New York, New York 10021

THE NEW YORK ACADEMY OF SCIENCES
(Founded in 1817)

BOARD OF GOVERNORS, September 1997–September 1998

RICHARD A. RIFKIND, *Chairman of the Board*
ELEANOR BAUM, *Vice Chairman of the Board*
RODNEY W. NICHOLS, *President and CEO* [ex officio]

Honorary Life Governors
WILLIAM T. GOLDEN   JOSHUA LEDERBERG

JOHN T. MORGAN, *Treasurer*

Governors

D. ALLAN BROMLEY	LAWRENCE B. BUTTENWIESER	PRAVEEN CHAUDHARI
RONALD L. GRAHAM	BILL GREEN	HENRY M. GREENBERG
JACQUELINE LEO	WILLIAM J. McDONOUGH	KATHLEEN P. MULLINIX
SANDRA PANEM	CHARLES RAMOND	SARA LEE SCHUPF
JAMES H. SIMONS	WILLIAM C. STEERE, Jr.	TORSTEN WIESEL

MARTIN L. LEIBOWITZ, *Past Chairman of the Board*

HELENE L. KAPLAN, *Counsel* [ex officio]   CRAIG PURINTON, *Secretary* [ex officio]

Na/K-ATPase AND RELATED TRANSPORT ATPases

STRUCTURE, MECHANISM,
AND REGULATION

ANNALS OF THE NEW YORK ACADEMY OF SCIENCES
Volume 834

Na/K-ATPase AND RELATED TRANSPORT ATPases

STRUCTURE, MECHANISM, AND REGULATION

Edited by Luis A. Beaugé, David C. Gadsby, and Patricio J. Garrahan

The New York Academy of Sciences
New York, New York
1997

Copyright © 1997 by the New York Academy of Sciences. All rights reserved. Under the provisions of the United States Copyright Act of 1976, individual readers of the Annals *are permitted to make fair use of the material in them for teaching and research. Permission is granted to quote from the* Annals *provided that the customary acknowledgment is made of the source. Material in the* Annals *may be republished only by permission of the Academy. Address inquiries to the Executive Editor at the New York Academy of Sciences.*

Copying fees: *For each copy of an article made beyond the free copying permitted under Section 107 or 108 of the 1976 Copyright Act, a fee should be paid through the Copyright Clearance Center, Inc., 222 Rosewood Drive, Danvers, MA 01923. The fee for copying an article is $3.00 for nonacademic use; for use in the classroom it is $0.07 per page.*

⊗ *The paper used in this publication meets the minimum requirements of American National Standard for Information Sciences—Permanence of Paper for Printed Library Materials, ANSI Z39.48-1984.*

Cover: *The softcover copy of this volume shows the structure and function of the Ca^{2+}-ATPase from sarcoplasmic reticulum.* LEFT: *Molecular envelope (purple) at 14 Å resolution, determined by cryoelectron microscopy (D.L. Stokes. 1994. FEBS Lett* **346:** *32–38). The transmembrane domain includes 10 alpha helices, colored in 4 groups of 2–3 linearly adjacent helices (model at right), some of which extend upwards into the narrow stalk and thence to the catalytic, cytoplasmic domain (top).* RIGHT: *Linear model (J.P. Andersen et al. :344, this volume) with transmembrane helices shown in the same colors. Results of site-directed mutagenesis suggest that Ca^{2+} ion binding sites in the membrane domain, between helices 4, 5, and 6, are linked to ATP hydrolysis via the stalk helices. Understanding the structural dynamics that couple ATP hydrolysis to Ca^{2+} ion transport represents a major challenge for the immediate future.*

Library of Congress Cataloging-in-Publication Data

NA/K-ATPase and related transport ATPases: structure, mechanism, and regulation/edited by Luis A. Beaugé, David C. Gadsby, Patricio J. Garrahan.
    p. cm. — (Annals of the New York Academy of Sciences, ISSN 0077-8923; v. 834)
    Includes bibliographical references and index.
    ISBN 1-57331-060-3 (cloth: alk. paper). — ISBN 1-57331-061-1 (paper: alk. paper)
    1. Sodium/potassium ATPase—Congresses. 2. Adenosine triphosphatase—Congresses. I. Beaugé, L. II. Gadsby, David C. III. Garrahan, Patricio J. IV. Series.
Q11N5 vol. 834
[Qp609.S63]
500 s—dc21
[571.6′4]
                                                                                   97-35157
                                                                                       CIP

SP
Printed in the United States of America
ISBN 1-57331-060-3 (cloth)
ISBN 1-57331-061-1 (paper)
ISSN 0077-8923

ANNALS OF THE NEW YORK ACADEMY OF SCIENCES

Volume 834
November 3, 1997

Na/K-ATPase AND RELATED TRANSPORT ATPases
STRUCTURE, MECHANISM, AND REGULATION[a]

Editors and Conference Organizers
LUIS A. BEAUGÉ, DAVID C. GADSBY, AND PATRICIO J. GARRAHAN

CONTENTS

Introduction. *By the* EDITORS xv

Part 1. Structure

Emerging Structure of the Neurospora Plasma Membrane H^+-ATPase. *By* GENE A. SCARBOROUGH 1

Two-Dimensional Crystal Formation from Solubilized Membrane Proteins Using Bio-Beads to Remove Detergent. *By* J.-J. LACAPÈRE, D. L. STOKES, G. MOSSER, J.-L. RANCK, G. LEBLANC, and J.-L. RIGAUD 9

Oligomeric Structure of Solubilized Na^+/K^+-ATPase Linked to E_1/E_2 Conformation. *By* Y. HAYASHI, K. KAMEYAMA, T. KOBAYASHI, E. HAGIWARA, N. SHINJI, and T. TAKAGI 19

Organization of the Membrane Domain of the Na/K-Pump. *By* STEVEN J. D. KARLISH 30

Ligand-Induced Conformational Changes in the Na,K-ATPase α Subunit. *By* JACK H. KAPLAN, SVETLANA LUTSENKO, CRAIG GATTO, SYLVIA DAOUD, and LINDA J. KENNEY 45

Plasma Membrane Ca^{2+} Pumps. *By* JOHN T. PENNISTON, AGNES ENYEDI, ANIL K. VERMA, HUGO P. ADAMO, and ADELAIDA G. FILOTEO 56

Structural Aspects of the Gastric H,K ATPase. *By* JAI MOO SHIN, MARIE BESANCON, KRISTER BAMBERG, and GEORGE SACHS 65

P-Type H^+- and Ca^{2+}-ATPases in Plant Cells. *By* B. STANGELAND, A. T. FUGLSANG, S. MALMSTRÖM, K. B. AXELSEN, L. BAUNSGAARD, F. C. LANFERMEIJER, K. VENEMA, F. T. OKKELS, P. ASKERLUND, and M. G. PALMGREN 77

Studies of Na,K-ATPase Structure and Function Using Baculovirus. *By* GUSTAVO BLANCO, WENDY R. HATFIELD, NICOLE T. MINOR, GLADIS SÁNCHEZ, JOSEPH C. KOSTER, ANTHONY W. DETOMASO, and ROBERT W. MERCER 88

[a]This volume is the result of a conference entitled **VIIIth International Conference on the Na/K-ATPase (and Related Transport ATPases)** held in Mar del Plata, Argentina, August 26–30, 1996.

Biochemical Characterization of the Human Renal Na^+,K^+-ATPase. *By* G. CRAMBERT, A. FRANZ, and L. G. LELIEVRE 97

Role of Sugar Residues for Recombinant Gastric H^+,K^+-ATPase. *By* CORNÉ H. W. KLAASSEN, HERMAN G. P. SWARTS, and JAN JOEP H. H. M. DE PONT.. 101

Effects of Adenoviral Mediated Transfer of Na^+,K^+-ATPase Subunit Genes to Alveolar Epithelial Cells. *By* P. FACTOR, C. SENNE, K. RIDGE, H. A. JAFFE, G. BLANCO, R. W. MERCER, and J. I. SZNAJDER.. 104

Abundance of $\alpha 3$, $\alpha 2$, and $\alpha 1$ Isoforms of Na,K-ATPase in Rat Kidney as Estimated by Competitive RT-PCR and [^{3}H]Ouabain Binding. *By* K. LÜCKING and J. M. NIELSEN... 107

Cellular and Developmental Distribution of the Na,K-ATPase β Subunit Isoforms of Neural Tissues. *By* PABLO MARTÍN-VASALLO, EMILIA LECUONA, SONIA LUQUÍN, DIEGO ALVAREZ DE LA ROSA, JULIO AVILA, TERESA ALONSO, and LUIS MIGUEL GARCÍA-SEGURA................................ 110

Characterization of an Isoform of Na^+/K^+-ATPase with High Affinity for [^{3}H]Ouabain in the Rat Vas Deferens. *By* FRANÇOIS NOËL, LUIS EDUARDO M. QUINTAS, AFONSO CARICATI-NETO, SIMONE SETTE L. LAFAYETTE, and ARON JURKIEWICZ 115

Differences in Uncoupled Sodium Efflux between Red Blood Cells and Kidney Na,K-ATPase Are Not Based on Differences in the cDNA for the α Subunit. *By* MARTIN STENGELIN and JOSEPH F. HOFFMAN 119

Isolation of Six Putative P-type ATPase β Subunit PCR Fragments from the Brain of the European Eel (*Anguilla anguilla*). *By* C. P. CUTLER, I. L. SANDERS, and G. CRAMB 123

Structural Characterization of the Glycation Process of the Plasma Membrane Calcium Pump. *By* F. LUIS GONZÁLEZ FLECHA, PABLO R. CASTELLO, JUAN J. GAGLIARDINO, and JUAN P. F. C. ROSSI 126

Cloning of the Eel Electroplax Na^+,K^+-ATPase α Subunit. *By* SHUNJI KAYA, AKIHIKO YOKOYAMA, TOSHIAKI IMAGAWA, KAZUYA TANIGUCHI, J. P. FROEHLICH, and R. W. ALBERS.. 129

pH-Dependent Change in the Oligomeric Structure of the Solubilized Na^+/K^+-ATPase. *By* T. KOBAYASHI, E. HAGIWARA, N. SHINJI, and Y. HAYASHI.. 132

Characterization of Na,K-ATPase α/α Oligomerization. *By* JOSEPH C. KOSTER, WENDY R. HATFIELD, GUSTAVO BLANCO, and ROBERT W. MERCER............. 135

Purification of Heterologously Expressed Plant Plasma Membrane H^+-ATPase by Ni^{2+}-Affinity Chromatography. *By* FRANK C. LANFERMEIJER, KEES VENEMA, and MICHAEL G. PALMGREN 139

Probing of Membrane Topology and Stability of Sarcoplasmic Reticulum Ca^{2+}-ATPase and Na^+, K^+-ATPase with Sequence-Specific Antibodies. *By* J. V. MØLLER, B. JUUL, P. FALSON, and M. LE MAIRE............................. 142

FTIR Studies on Proteolysed Na/K-ATPase. Effects of Occluded Ions. *By* VLAD BRUMFELD, ALLA SHAINSKAYA, and STEVEN J. D. KARLISH............ 146

Atomic Force Microscopy of *Escherichia coli* FoF1-ATPase in Reconstituted Membranes. *By* KUNIO TAKEYASU, HIROSHI OMOTE, SAJU NETTIKADAN, FUYUKI TOKUMASU, ATSUKO IWAMOTO-KIHARA, and MASAMITSU FUTAI........ 149

Isolation and Structural and Functional Analysis of New Types of Na,K-ATPase Isozymes. *By* NATALIA M. VLADIMIROVA, EKATERINA A. PLATOSHKINA, RIAD E. EFENDIYEV, and NATALIA A. POTAPENKO 153

Heterologous Expression of the Metal-Binding Domains of Human Copper-Transporting ATPases (P_1-ATPases). *By* SVETLANA LUTSENKO, KONSTANTIN PETRUKHIN, T. CONRAD GILLIAM, and JACK H. KAPLAN 155

P-Type ATPases in *Tetrahymena. By* SHUSHENG WANG, DONGHONG GAO, JEFFREY PENNY, SANJEEV KRISHNA, and KUNIO TAKEYASU 158

Part 2. Mechanism

Transport-Linked Conformational Changes in Na,K-ATPase. Structure-Function Relationships of Ligand Binding and E_1-E_2 Conformational Transitions. *By* PETER L. JORGENSEN, JAKOB H. RASMUSSEN, JESPER M. NIELSEN, and PER AMSTRUP PEDERSEN 161

Structure/Function Analysis of the Ca^{2+} Binding and Translocation Domain of SERCA1 and the Role of Brody Disease of the *ATP2A1* Gene Encoding SERCA1. *By* DAVID H. MACLENNAN, WILLIAM J. RICE, and ALEX ODERMATT 175

Are Pyridoxal and Fluorescein Probes in Lysine Residues of α-Chain in Na^+,K^+-ATPase Sensing ATP Binding? *By* TAKEO TSUDA, SHUNJI KAYA, TAKESHI YOKOYAMA, and KAZUYA TANIGUCHI 186

Cation and Cardiac Glycoside Binding Sites of the Na,K-ATPase. *By* JERRY B LINGREL, JOSÉ M. ARGÜELLO, JAMES VAN HUYSSE, and THERESA A. KUNTZWEILER 194

ATPase Gene Transfer and Mutational Analysis of the Cation Translocation Mechanism. *By* G. INESI, D. LEWIS, C. SUMBILLA, A. NANDI, M. KIRTLEY, and C. P. ORDAHL 207

Kinetic and Energetic Aspects of Na^+/K^+-Transport Cycle Steps. *By* HANS-JÜRGEN APELL 221

Charge Translocation by the Na/K Pump. *By* R. F. RAKOWSKI, FRANCISCO BEZANILLA, P. DE WEER, DAVID C. GADSBY, MIGUEL HOLMGREN, and JONATHAN WAGG 231

The Na,K-Pump as a Channel. A New Approach to the Study of the Structure-Function Relationship of a P-Type ATPase. *By* JEAN-DANIEL HORISBERGER and XINYU WANG 244

Mechanism of Electrogenic Reaction Steps during K^+ Transport by the Na,K-ATPase. *By* JOSHUA R. BERLIN and R. DANIEL PELUFFO 251

Recent Electrical Snapshots of the Cardiac Na,K Pump. *By* DONALD W. HILGEMANN 260

Na^+,K^+-ATPase Pump Currents Activated by an ATP Concentration Jump. Comparison of Studies with Purified Membrane Fragments and Giant Excised Patches. *By* GEORG NAGEL, THOMAS FRIEDRICH, KLAUS FENDLER, and ERNST BAMBERG... 270

Complex Kinetic Behavior in the Na,K- and Ca-ATPases. Evidence for Subunit-Subunit Interactions and Energy Conservation during Catalysis. *By* JEFFREY P. FROEHLICH, KAZUYA TANIGUCHI, KLAUS FENDLER, JAMES E. MAHANEY, DAVID D. THOMAS, and R. WAYNE ALBERS 280

Functional Consequences of Mutations in the Transmembrane Core Region for Cation Translocation and Energy Transduction in the Na$^+$,K$^+$-ATPase and the SR Ca^{2+}-ATPase. *By* BENTE VILSEN, DORTE RAMLOV, and JENS PETER ANDERSEN 297

Eosin as a Probe for Conformational Transitions and Nucleotide Binding in Na,K-ATPase. *By* MIKAEL ESMANN and NATALYA U. FEDOSOVA 310

A Two-Site Model of Interacting ATP Sites. *By* DETLEF THOENGES, HOLGER LINNERTZ, and WILHELM SCHONER 322

Relationship between Ouabain-Sensitive ATPase Activity and Occluded Rb$^+$ at Micromolar ATP Concentrations. *By* R. C. ROSSI, P. J. GARRAHAN, S. B. KAUFMAN, J. G. NØRBY, and P. J. SCHWARZBAUM 327

Site-Directed Mutagenesis Analysis of the Role of the M5S5 Sector of the Sarcoplasmic Reticulum Ca^{2+}-ATPase. *By* JENS PETER ANDERSEN, THOMAS SØRENSEN, and BENTE VILSEN 333

Changes to Na,K-ATPase α-Subunit E779 Separate the Structural Basis for V_M and Ion Dependence of Na,K-Pump Current. *By* R. D. PELUFFO, J. B LINGREL, J. M. ARGÜELLO, and J. R. BERLIN 339

Interaction of Palytoxin and Mercury with the Na,K-ATPase on *Xenopus laevis* Oocytes. *By* XINYU WANG and JEAN-DANIEL HORISBERGER 343

Voltage-Dependent Inhibition of the Na$^+$,K$^+$ Pump by Intracellular Potassium in Rabbit Ventricular Myocytes. *By* PETER S. HANSEN, DAVID F. GRAY, KERRIE A. BUHAGIAR, and HELGE H. RASMUSSEN 347

Effect of pH on Charge Movement by the Na/K Pump in *Xenopus* Oocytes. *By* KEVIN A. KHATER and R. F. RAKOWSKI 350

Sodium Pump of Cultured Guinea Pig Atrial Myocytes. *By* JENS KOCKSKÄMPER and HELFRIED GÜNTHER GLITSCH 354

Dipole Potential Drop due to RH-Dye Adsorption on the Lipid Bilayer and Its Influence on Na$^+$/K$^+$-ATPase Activity. *By* D. YU. MALKOV, K. V. PAVLOV, and V. S. SOKOLOV 357

Electrogenic Reactions of Na$^+$/K$^+$-ATPase Investigated on Solid Supported Membranes. *By* J. PINTSCHOVIUS, K. SEIFERT, and K. FENDLER 361

Study of Electrogenic Transport of Sodium Ions Inside the Na,K-ATPase by Means of Membrane Capacitance Measurements. *By* V. S. SOKOLOV, S. M. STUKOLOV, N. M. GEVONDYAN, and H.-J. APELL 364

Transposing Results from an Artificial Minicell to a Real Cell. Experimental Evidence for a Working Hypothesis Linking Na,K-ATPase Permeability States to Specific Alterations of Cell Life. *By* BEATRICE M. ANNER 367

Transport Activity of a Chimeric Na$^+$,K$^+$-ATPase with Ca^{2+}/Calmodulin Binding Domain from Ca^{2+}-ATPase in *Xenopus* Oocytes. *By* JIANXING ZHAO, LARISA A. VASILETS, QUANBAO GU, TOSHIAKI ISHII, KUNIO TAKEYASU, and WOLFGANG SCHWARZ 372

Solute Effects on the Sodium Pump. An Evaluation of the Osmotic Dehydration Hypothesis. *By* PETER R. ARULANANTHAM, ZACHARY V. EDMONDS, and R. WAYNE ALBERS 376

ADP Dephosphorylation of the E$_1$P Form of the Na$^+$,K$^+$-ATPase. *By* MARTA CAMPOS and LUIS BEAUGÉ 378

K^+ Induces an Acid-Labile Phosphoenzyme (*or* an Occluded P_i Form) in Na,K-ATPase. *By* J. D. CAVIERES, E. BUXBAUM, D. G. WARD, and T. J. H. WALTON 381

Diversity of the E_2P Phosphoforms of Na,K-ATPase. *By* NATALYA U. FEDOSOVA, FLEMMING CORNELIUS, BLISS FORBUSH III, and IRENA KLODOS ... 386

Interaction between Substrate Site and Cation Binding Sites in P_i Phosphorylation of Na,K-ATPase. *By* FLEMMING CORNELIUS, NATALYA U. FEDOSOVA, and IRENA KLODOS 390

Fluorescent Styryl Dyes as Probes for Na,K-ATPase Reaction. Enzyme Source and Fluorescence Response. *By* IRENA KLODOS, NATALYA U. FEDOSOVA, and FLEMMING CORNELIUS 394

Kinetic Properties of the Na,K-ATPase of Goldfish Kidney. *By* MARIELA P. GARCÍA, PABLO J. SCHWARZBAUM, ROLANDO C. ROSSI, and SERGIO B. KAUFMAN 397

Mn as Cosubstrate for the Phosphorylation of the Sarcoplasmic Reticulum Ca-ATPase by P_i. *By* D. A. GONZÁLEZ, G. L. ALONSO, and J.-J. LACAPÈRE ... 400

Comparative Aspects of Ligand Stoichiometry in Na^+,K^+-ATPase from Kidney and in Ca^{2+}-ATPase from Rabbit Sarcoplasmic Reticulum. *By* OTTO HANSEN and JØRGEN JENSEN 404

On the Mechanism of Inhibition of the $PMCa^{2+}$-ATPase by Lanthanum. *By* CLAUDIO J. HERSCHER and ALCIDES F. REGA 407

Nucleotide Binding to Na,K-ATPase. Effect of Ionic Strength and Charge. *By* JENS G. NØRBY and MIKAEL ESMANN 410

Two Unexplained Kinetic Features of Na,K-ATPase May Be Understood as Indicating K^+-Induced Cooperativity between Subunits in a Dimeric Enzyme. *By* IGOR W. PLESNER 412

Models for ATPase-Substrate Activation Kinetics. *By* GRETEL ROBERTS and LUIS BEAUGÉ 416

Properties of the Cytoplasmic Ion Binding Sites. *By* WOLFGANG DOMASZEWICZ, ANNE SCHNEEBERGER, and HANS-JÜRGEN APELL 420

Interactions of Palytoxin with the Na,K-ATPase. Where Are Those Sites? *By* M. T. TOSTESON, D. R. L. SCRIVEN, A. BHARADWAJ, J. ARNADOTTIR, and D. C. TOSTESON 424

Ordered Interaction of Ions with Na/K-Pump May Confound Interpretation of Unidirectional Fluxes. *By* J. WAGG and D. C. GADSBY 426

Nucleotides Trigger the Release of $Co(NH_3)_4ATP$ Tightly Bound to Inactivated Na,K- ATPase. *By* D. G. WARD, W. SCHONER, and J. D. CAVIERES 432

Transient Currents of Na^+/K^+-ATPase in Giant Patches from Guinea Pig Cardiomyocytes Induced by ATP Concentration Jumps or Voltage Pulses. *By* THOMAS FRIEDRICH and GEORG NAGEL 435

Fluorescence Quenching of IAF-Na^+/K^+-ATPase via Energy Transfer to TNP-Labeled Nucleotide. *By* EDWARD H. HELLEN and PROMOD R. PRATAP . 439

Eosin, Energy Transfer, and RH421 Report the Same Conformational Change in Sodium Pump as Fluorescein. *By* SHWU-HWA LIN, IRINA N. SMIRNOVA, VLADIMIR N. KASHO, and LARRY D. FALLER 442

Binding of TNP-ATP to IAF-Labeled Na^+/K^+-ATPase as Examined by Fluorescence Quenching. *By* P. R. PRATAP, E. H. HELLEN, and A. PALIT 445

A Mutant of the Plasma Membrane Ca^{2+} Pump Highly Sensitive to Inhibition by Mg^{2+}. *By* HUGO P. ADAMO, ALCIDES F. REGA, and JOHN T. PENNISTON 449

Amino Acid Residues 18-75 Are Essential for Expression of an Active Plasma Membrane Ca^{2+} Pump. *By* MIRTA E. GRIMALDI, HUGO P. ADAMO, ALCIDES F. REGA, and JOHN T. PENNISTON 452

Increase in Affinity for ATP and Change in E_1-E_2 Conformational Equilibrium after Mutations to the Phosphorylation Site (Asp^{369}) of the α Subunit of Na,K-ATPase. *By* PER AMSTRUP PEDERSEN, JAKOB H. RASMUSSEN, and PETER LETH JØRGENSEN 454

α1T Can Support Na^+,K^+-ATPase: Na^+ Pump Functions in Expression Systems. *By* JULIUS C. ALLEN, XUN ZHAO, TIMOTHY ODEBUNMI, SANDRA JEMELKA, RUSSELL M. MEDFORD, THOMAS A. PRESSLEY, and ROBERT W. MERCER 457

Involvement of Different Sites for Nucleotide Analogs in the Phosphatase Activity of the Red Cell Calcium Pump. *By* CLAUDIA DONNET, ARIEL J. CARIDE, SILVINA A. TALGHAM, and JUAN P. F. C. ROSSI 459

Nonaqueous Nature of the Environment Surrounding the Acyl-phosphate of the Na^+/K^+-ATPase Deduced from Nucleophilic Attack by Hydroxylamine on the Phosphate Group. *By* YOSHIHIRO FUKUSHIMA, YASUO SHINOHARA, and MAKOTO USHIMARU 462

Study of Structure-Functional Organization Features in the Pig Kidney Na,K-ATPase at Different Conformational States. *By* NATALIA M. GEVONDYAN, ASYA V. GRINBERG, and VLADIMIR S. GEVONDYAN 466

High Yield Fermentation of Pig α1β1 Na,K-ATPase in *Saccharomyces cerevisiae*. *By* JAKOB H. RASMUSSEN, PER AMSTRUP PEDERSEN, and PETER L. JORGENSEN 469

Involvement of Glutamic Acid 820 in K^+ and SCH 28080 Binding to Gastric H^+,K^+- ATPase. *By* HERMAN P. G. SWARTS, CORNÉ H. W. KLAASSEN, and JAN JOEP H. H. M. DE PONT 472

Copper-Stimulated Adenosine Triphosphatase from Rat Liver. Isolation and Kinetic Characterization. *By* JULNAR USTA, HANA BARAKEH, HASHEM MAHFOUZ, and NADIM CORTAS 475

Part 3. Regulation

Phosphorylation of Na,K-ATPase by Protein Kinases. Sites, Susceptibility, and Consequences. *By* MARINA S. FESCHENKO, RANDALL K. WETZEL, and KATHLEEN J. SWEADNER 479

Cytoplasmic Regions of the Alpha Subunit of the Sodium Pump Involved in Modulating the Na,K-ATPase Reaction. *By* STEWART E. DALY, LOIS K. LANE, and RHODA BLOSTEIN 489

Subunit Interactions in the Sodium Pump. *By* T. COLONNA, M. KOSTICH, M. HAMRICK, B. HWANG, J. D. RAWN, and D. M. FAMBROUGH 498

Sorting of Ion Pumps in Polarized Epithelial Cells. *By* LISA A. DUNBAR, DENISE L. ROUSH, NATHALIE COURTOIS-COUTRY, THEODORE R. MUTH, C. J. GOTTARDI, VANATHY RAJENDRAN, JOHN GEIBEL, MICHAEL KASHGARIAN, and MICHAEL J. CAPLAN 514

Distinct Distribution of Different Na^+ Pump α Subunit Isoforms in Plasmalemma. Physiological Implications. *By* MAGDALENA JUHASZOVA and MORDECAI P. BLAUSTEIN 524

α and β Subunits of Na,K-ATPase Interact with BiP and Calnexin. *By* AHMED T. BEGGAH and KÄTHI GEERING 537

Structural Domains Implicated in ER Degradation of α Subunits of Na,K-ATPase. *By* PASCAL BEGUIN and KÄTHI GEERING 540

Posttranslational Processing of the Catalytic Subunit from Na^+/K^+-ATPase. *By* THOMAS A. PRESSLEY and SUSAN A. PETROSIAN 543

Functional Control of Na^+,K^+-ATPase by Vasopressin and Aldosterone in the Cortical Collecting Duct. Role of Protein Phosphatase. *By* N. COUTRY, N. FARMAN, J. P. BONVALET, and M. BLOT-CHABAUD 545

Effect of Neurotensin on Synaptosomal Membrane ATPase and p-NPPase Activities. *By* M. G. LÓPEZ ORDIERES and G. RODRÍGUEZ DE LORES ARNAIZ. 548

Na^+/K^+-ATPase Density Is Sexually Dimorphic in the Adult Rat Kidney. *By* LUIS EDUARDO M. QUINTAS, LUCIANE BARREIRO LOPEZ, CADEN SOUCCAR, and FRANÇOIS NOËL 552

Effect of Haloperidol on the Sarcoplasmic Reticulum Ca-ATPase. *By* DELIA TAKARA and GUILLERMO L. ALONSO 555

Changes in Actin Filament Organization Regulate Na^+,K^+-ATPase Activity. Role of Actin Phosphorylation. *By* HORACIO F. CANTIELLO 559

Specific Expression and Regulation of CHIF in Kidney and Colon. *By* CLAUDIA CAPURRO, NATHALIE COUTRY, JEAN-PIERRE BONVALET, BRIGITTE ESCOUBET, HAIM GARTY, and NICOLETTE FARMAN 562

Effects of Extracellular Sodium Concentration on the Activity of Na,K-ATPase in Dogfish Rectal Gland Epithelial Cells. *By* J. EDWARDS, S. MACKENZIE, C. P. CUTLER, and G. CRAMB 565

Phosphorylation Site-Independent Downregulation of Na-Pump Current in A6 Epithelia by Protein Kinase C. Decrease in Na,K-ATPase Cell-Surface Expression. *By* JÖRG BERON and FRANÇOIS VERREY 569

Regulation of the α2β1 and α3β1 Isozymes of the Na,K-ATPase by Ca^{2+}, PKA, and PKC. *By* GUSTAVO BLANCO and ROBERT W. MERCER 572

Roles of PKA and PKC in Regulation of Na^+ Pump Activity in Vascular Smooth Muscle Cells. *By* MIKHAIL L. BORIN 576

Effect of Protein Kinase Modulators on the Sodium Pump Activities of HeLa Cells Transfected with Distinct Alpha Isoforms of Na,K-ATPase. *By* NESTOR B. NESTOR, LOIS K. LANE, and RHODA BLOSTEIN 579

Phosphorylation of Tyr^7, Tyr^{10}, and Ser^{27} of α-Chain in H^+,K^+-ATPase by Intrinsic and Extrinsic Kinases. *By* KATSUHIKO TOGAWA, SHUNJI KAYA, MASANOBU MORI, AKIRA SHIMADA, TOSHIAKI IMAGAWA, KAZUYA TANIGUCHI, SVEN MÅRDH, JACKIE CORBIN, and USHIO KIKKAWA 582

Regulatory Phosphorylation of the Na^+/K^+-ATPase from Mammalian Kidneys and *Xenopus* Oocytes by Protein Kinases. Characterization of the Phosphorylation Site for PKC. *By* LARISA A. VASILETS, HEIKE FOTIS, and EVA-MARIA GÄRTNER 585

Ion-Sensitive Domains of the SERCA- and the Na^+/K^+-ATPases Identified by Chimeric Recombination. *By* SHIGE H. YOSHIMURA, TOSHIAKI ISHII, JIRO C. YASUHARA, MASA H. SATO, and KUNIO TAKEYASU 588

Part 4. Pharmacology

Selective Inhibition of the Gastric H^+,K^+-ATPase by Omeprazole and Related Compounds. *By* P. LORENTZON, A. BAYATI, H. LEE, and K. ANDERSSON 592

Vacuolar H^+-ATPases. Targets for Drug Discovery? *By* D. J. KEELING, M. HERSLÖF, B. RYBERG, S. SJÖGREN, and L. SÖLVELL 600

The Plasma Membrane H^+-ATPase of Fungi. A Candidate Drug Target? *By* DAVID S. PERLIN, DONNA SETO-YOUNG, and BRIAN C. MONK 609

Demonstration of a Specific Transport Protein for Cardiac Glycosides in Bovine Blood. *By* ROBERTO ANTOLOVIC, HOLGER KOST, DIETMAR LINDER, MONICA LINDER, DETLEF THÖNGES, DAVID LICHTSTEIN, and WILHELM SCHONER 618

Functional Characterization of an Endogenous Digitalis-Like Factor in Human Newborn Plasma. Effects on Rat (Na^+/K^+)-ATPase Isoforms and on Binding to Placenta. *By* G. CRAMBERT, S. BALZAN, A. PACI, S. DECOLLOGNE, U. MONTALI, S. GHIONE, and L. G. LELIÈVRE 621

Evidence of an Endogenous Ouabain-Like Immunoreactive Compound with Digitalis-Like Properties in the Human. *By* SILVANA BALZAN, UMBERTO MONTALI, and SERGIO GHIONE 626

Effect of a Low Molecular Weight Factor from Na^+,K^+-ATPase Preparations on Ouabain Binding. *By* CARLOS F. L. FONTES, FABIO E. V. LOPES, HELENA M. SCOFANO, HÉCTOR BARRABIN, and JENS G. NØRBY 631

High Affinity Anti-Digoxin Antibodies as Model Receptors for Cardiac Glycosides. Comparisons with Na^+,K^+-ATPase. *By* R. KASTURI, L. R. MCLEAN, M. N. MARGOLIES, and W. J. BALL, JR 634

Synthetic Candidates for EDLF. Activity on Human Placenta Digitalis Receptors. *By* ANNA PACI, MASAYUKI SAKAKIBARA, PAOLA DEL BENE, and AKI OGAWA UCHIDA 637

Kinetics of Na^+,K^+-ATPase Inhibition by Brain Endobains. *By* G. RODRÍGUEZ DE LORES ARNAIZ, A. REINÉS, T. HERBIN, and C. PEÑA 642

Proscillaridin A Immunoreactivity: A New Endogenous Cardiac Glycoside? *By* SU-QIN LI, CHRISTIAN EIM, RALF SCHNEIDER, BEATE SICH, ULRIKE KIRCH, and WILHELM SCHONER 646

Skeletal Muscle Na,K-ATPase Concentration Changes and Intramuscular and Extrarenal K Homeostasis in Animals and Humans. *By* H. BUNDGAARD, T. A. SCHMIDT, and K. KJELDSEN 648

Alpha-2 Na,K-ATPase Contributes to Lung Liquid Clearance. *By* K. RIDGE, W. OLIVERA, D. H. RUTSCHMAN, R. W. MERCER, B. UHAL, S. HOROWITZ, F. HUGHES, P. FACTOR, M. L. BARNARD, and J. I. SZNAJDER 651

Opposite Expression Pattern of the Human Na,K-ATPase β1 Isoform in Stomach and Colon Adenocarcinomas. *By* JULIO AVILA, EMILIA LECUONA, MANUEL MORALES, ARTURO SORIANO, TERESA ALONSO, and PABLO MARTÍN-VASALLO 653

Immunological Identification of Na,K-ATPase Isoforms in Nonfailing and Failing Myocardium. *By* ODILE BARBEY, ALAIN GERBI, KARINE ROBERT, VINCENT MAYOL, SANDRINE PIERRE, FRANCK PAGANELLI, and JEAN-MICHEL MAIXENT 656

Regulation of Na^+,K^+-ATPase α Subunit Isoforms in Mouse Cortex during Focal Ischemia. *By* I. JAMME, P. TROUVÉ, J. M. MAIXENT, A. GERBI, D. CHARLEMAGNE, and A. NOUVELOT 658

Kinetic Parameters of Na/K-ATPase Modified by Free Radicals *in Vitro* and *in Vivo*. *By* E. KURELLA, M. KUKLEY, O. TYULINA, D. DOBROTA, M. MATEJOVICOVA, V. MEZESOVA, and A. BOLDYREV 661

Na/K-ATPase and Oxidative Stress. *By* ALEXANDER A. BOLDYREV and ELENA R. BULYGINA 666

A Hypothetic to Explain the Non-Michaelis Substrate Dependence Curve of Na/K-ATPase. *By* ALEXANDER BOLDYREV, ANATOLY KOTLOBAY, EKATERINA KURELLA, OLGA LOPINA, and NUNE SARVAZYAN 669

The Plasma Membrane Ca^{2+}-ATPase in Spontaneously Hypertensive Rats. *By* BASIL D. ROUFOGALIS, SHI CHEN, ELEANOR P. W. KABLE, TUAN H. KUO, and G. R. MONTEITH 673

Regulation of Myocardial Na,K-ATPase Concentration in Experimental and Human Heart Disease. *By* T. A. SCHMIDT, H. BUNDGAARD, and K. KJELDSEN 676

Oxygen-Free Radicals Directly Attack the ATP Binding Site of the Cardiac Na^+,K^+-ATPase. *By* KAI Y. XU, JAY L. ZWEIER, and LEWIS C. BECKER 680

Effect of *Leptospira interrogans* Endotoxin and Renal Tubular Na,K-ATPase and H,K-ATPase Activities. *By* M. YOUNES-IBRAHIM, P. BURTH, M. CASTRO-FARIA, L. CHEVAL, B. BUFFIN-MEYER, S. MARSY, and A. DOUCET .. 684

Transgenic Mice Expressing Human α3 Na,K-ATPase Isoform in Heart. *By* RAPHAEL ZAHLER, MARK LUFBURROW, MIRA MANOR, RADHA SHENOY, DIEGO FORNASARI, MARC ROMANA, and WEI SUN 687

***

Index of Contributors 691

Financial assistance was received from:

Sponsor
- BIOMOLECULAR STRUCTURE AND FUNCTION PROGRAM—NATIONAL SCIENCE FOUNDATION
- ASTRA HÄSSLE AB

Supporter
- INTERNATIONAL UNION FOR PURE AND APPLIED BIOPHYSICS (IUPAB)
- INTERNATIONAL UNION OF PHYSIOLOGICAL SCIENCES (IUPS)
- INTERNATIONAL UNION OF BIOCHEMISTRY AND MOLECULAR BIOLOGY (IUBMB)
- CONSEJO NACIONAL DE INVESTIGACIONES CIENTIFICAS Y TECNICAS (CONICET)
- MINISTERIO DE CULTURA Y EDUCACION, ARGENTINA

Contributor
- PAN-AMERICAN ASSOCIATION FOR BIOCHEMISTRY AND MOLECULAR BIOLOGY (PAABMB)
- LATIN AMERICAN NETWORK OF BIOLOGY (RELAB)—UNESCO
- THE BRITISH COUNCIL
- THE CAMPOMAR WEIZMANN COOPERATION AGREEMENT
- FUNDACION ANTORCHAS
- FUNDACION INTERIOR ARGENTINA (FUNINAR)
- FACULTAD DE FARMACIA Y BIOQUÍMICA, UNIVERSIDAD DE BUENOS AIRES
- GLAXO WELLCOME

The New York Academy of Sciences believes it has a responsibility to provide an open forum for discussion of scientific questions. The positions taken by the participants in the reported conferences are their own and not necessarily those of the Academy. The Academy has no intent to influence legislation by providing such forums.

Introduction

LUIS A. BEAUGÉ,[a] DAVID C. GADSBY,[b] AND
PATRICIO J. GARRAHAN[c]

[a]*Instituto M. & M. Ferreyra*
C.C. 389
5000 Córdoba, Argentina

[b]*The Rockefeller University*
1230 York Avenue
New York, New York 10021

[c]*IQUIFIB*
Facultad de Farmacia y Bioquímica
Universidad de Buenos Aires
Junín 956
(1113) Buenos Aires, Argentina

This *Annals* volume comprises the proceedings of the VIIIth International Conference on the Na/K-ATPase (and related transport ATPases), held on August 26–30, 1996, in Mar del Plata, Argentina. This is the latest in the ongoing series of such conferences, held every three years or so in either Europe or America in response to, and as continuing testimony to, the relentless pace of research into the properties of these ubiquitous and essential cation pumps. The first Na/K-ATPase conference was held in New York in 1973, under the aegis of the New York Academy of Sciences, and subsequent meetings have been organized at Aarhus University in Denmark (1978), at Yale University (1981), at Cambridge University (1984), at Aarhus University again (1987), at the Marine Biological Laboratory in Woods Hole (1990), and at Todtmoos in the Black Forest, Germany (1993). Last summer's meeting in Argentina was the first of these conferences to be held in South America and, the relative remoteness of the venue and the organizers' initial apprehension notwithstanding, it was very strongly supported and was attended by well over 200 participants from all over the world. Their latest findings and insights into the structure, mechanism, and regulation of the Na/K-ATPase and related ion-motive ATPases, summarized in this book, will hopefully provide a comprehensive overview of the present state of knowledge and also stimulate new avenues of research, for at least the ensuing three-year period until the next meeting in the series (anticipated to occur in Japan in 1999).

The surface membrane of practically all animal cells is studded with hundreds, thousands, or even millions of copies of the Na/K-ATPase, or sodium pump as it is frequently called. Its seemingly tireless extrusion of cellular Na ions in exchange for extracellular K ions, at the expense of a major fraction of the cell's ATP budget, is responsible for maintaining the steep concentration gradients of those ions, inwardly directed for Na and outwardly for K, that are vital to the life of the cell. These gradients drive numerous co- and countertransporters that supply the cells with glucose, amino acids, and other chemical building blocks, they regulate cell volume, pH, and Ca homeostasis, and they underlie nearly all the electrical activity in the peripheral and central nervous systems and in cardiac and skeletal muscle. Because, during the catalytic and ion transport cycle, a covalent phosphointermediate of the Na/K-ATPase is formed by transfer of the γ phosphate from ATP to an aspartyl residue in the large second intracellular loop, the sodium pump is identified as a member of the large family of P-type ion-motive ATPases. Other prominent members of this family, sharing varying degrees of sequence identity with the Na/K-ATPase, are the gastric H/K-ATPase, the Ca-ATPase of sarcoplasmic and endoplasmic reticulum, the plasma membrane Ca-ATPase, and the plasma membrane H-ATPase of yeast. The small amount of structural information presently available suggests that the overall topology and three-dimensional organization of these prominent P-type ATPases are fairly similar. Thus, they all appear to comprise (as illustrated on the softcover) a transmembrane domain, incorporating 10 α helices, which is connected via a stalk region to a large globular cytoplasmic domain that includes the site of ATP binding and the phosphorylated aspartate. Results of experiments based on techniques such as site-directed mutagenesis, proteolytic digestion, and chemical modification suggest that acidic residues in transmembrane helices 4, 5, and 6 contribute to the cation binding sites and that accessibility of those sites is somehow regulated by ATP hydrolysis-linked conformational changes that are transmitted to the membrane domain through the stalk. But no high resolution structure of a P-type ATPase is available yet to help guide interpretations of present data or to pinpoint appropriate locations where making new mutations ought to be particularly instructive, as we strive to understand the molecular mechanism of ion transport by these important pumps.

There is substantial motivation, apart from sheer intellectual inquisitiveness, for learning how these ubiquitous molecular machines work because each of the human pumps is a target for either potent therapeutic drugs or disease-causing mutations. Still one of the most commonly prescribed cardiac drugs throughout the world today, the cardiotonic steroid digoxin (analogs like ouabain and strophanthidin are more common in the laboratory) has been routinely administered to patients to ward off heart failure since 1785, when William Withering, an English country physician, first reported his use of extract of the foxglove plant, *Digitalis purpurea,* to treat patients with failing hearts. These cardiac steroids bind tightly and selectively to the extracellular surface of one conformation of the Na/K-ATPase α subunit and so prevent it from functioning as a pump; the small increase in cellular Na concentration that accrues from the resulting overall slowing of Na extrusion causes, through operation of the Na/Ca exchanger, a concomitant rise in cellular Ca ion concentration which, in turn, enhances the strength of myocardial contraction. The far more modern synthetic drugs developed to inhibit the gastric H/K-ATPase also have a far more direct therapeutic action. These drugs, exemplified by omeprazole, are activated by the acid environment of the stomach lumen to become effective covalent modifiers of extracellularly accessible sulfhydryl groups on the α subunit of the H/K-ATPase. The so modified H/K pumps are irreversibly inhibited and can no longer generate the stomach acid that gave rise to the heartburn which prompted taking the drug. At least one of these cation pumps has been demonstrated to be the target for genetic disease in humans. Inherited mutations in the gene encoding the sarcoplasmic reticulum Ca-ATPase that result in the production of truncated, and hence dysfunctional, Ca-pump proteins have been shown to underlie Brody disease, a rare functional disorder of skeletal muscle. It is not unreasonable to expect that many additional mutations in this gene, and in those for other cation pumps, will be found responsible for the debilitating symptoms of other genetic diseases.

It was with this background firmly in mind that the felicitous (in retrospect) decision was made to broaden the scope of the conference from its traditional near-exclusive focus on the Na/K-ATPase to incorporate pertinent information on the several homologous P-type ion-motive ATPases. The hope was to stimulate a cross-fertilization of ideas and information that would accelerate progress in understanding how each pump works. And because of the widely held opinion that a three-dimensional structure, at atomic resolution, of any of the ion-motive ATPases will yield invaluable insights into mechanisms that are likely common to all of them, recent advances in solving their molecular structure and in establishing the relationship between primary sequence and the mechanism of ion transport were heavily represented among the topics covered in the nine principal symposia and three evening workshops. Both to make up for the present lack of equivalent high-resolution structure for any P-type ATPase and to provide a beacon of hope, the conference featured inspiring presentations of the beautiful atomic structures of the mitochondrial F_1-ATPase and of protein phosphatases, both of which carry out reactions related to partial reactions of the P-type ATPases. Other topics covered in the symposia and workshops included transcriptional regulation and trafficking and targeting of P-type ATPases, as well as their post-translational regulation by receptor-initiated and kinase-mediated pathways. Acknowledging the evident clinical significance of P-type and other ion-motive ATPases as targets for selective drugs, an entire symposium was devoted to evaluating the roles of transport ATPases in disease, and the efficacy of therapeutic interventions using compounds presently available or those still under development. Papers based on the symposium talks, together with a larger number of short papers based on the contents of posters presented at the meeting, are gathered together here under the four general headings of Structure, Mechanism, Regulation, and Pharmacology. Although in some cases the categorization is somewhat arbitrary, we trust that readers interested in a particular topic will quickly be able to find the relevant papers.

If this expansion of the subject matter of the meeting was a signal success, so was the deliberate emphasis on frequent and protracted opportunities for fruitful interactions among its participants. A number of factors conspired to foster a great deal of both formal and informal discussion, including strict timekeeping during the symposia by experienced senior scientists, the intimate dimensions of the lecture theater, which was adjacent to the large foyer in which all posters were continuously displayed throughout the entire meeting, and the fact that all participants were accommodated in the same large hotel that was also the site of the conference. A further contribution to the success of the meeting was the attendance and active involvement of a relatively large number of younger investigators. Many of these received travel awards, made possible through the generosity of the substantial number of senior scientists who paid all or part of their own travel expenses to attend the conference (waiving financial assistance offered) and of a long list of donors: Biomolecular Structure and Function Program of the National Science Foundation, Astra Hässle, International Union for Pure and Applied Biophysics (IUPAB), International Union of Physiological Sciences (IUPS), International Union of Biochemistry and Molecular Biology (IUBMB), Consejo Nacional de Investigaciones Científicas y Técnicas (CONICET), Ministerio de Cultura y Educación, Pan-American Association for Biochemistry and Molecular Biology (PAABMB), Latin American Network of Biology (RELAB-UNESCO), the British Council, The Campomar-Weizmann Cooperation Agreement, Fundación Antorchas, Fundación Interior Argentina (FUNINAR), Facultad de Farmacia y Bioquímica,

Universidad de Buenos Aires, and GLAXO WELLCOME. Without their financial support, the meeting could not have been organized. The organizers of this meeting also owe a debt of thanks, for sagacious advice and constructive criticism during the early planning stages, to the informal "council" made up of the organizers of the three previous Na/K-ATPase conferences: W. Schoner and E. Bamberg from Germany, J. Kaplan and P. De Weer from the USA, and F. Cornelius, M. Esmann, I. Klodos, J.G. Nørby, J.C. Skou, and A.B. Maunsbach from Denmark. Special thanks are due to G. Sachs for his help in organizing segments of the program related to the H/K-ATPase and to therapeutic aspects of ion-motive ATPases. The difficult job of dealing with registration fees originating from all over the world was handled smoothly and precisely by Susana Pérez in Córdoba, to whom we are most grateful. The organizers are also grateful to EPER TURISMO for their help with travel arrangements for the participants, to the graduate students and other young scientists from the host department, Instituto de Quimica y Fisicoquimica Biológicas, for their general assistance throughout the meeting, and to colleagues A.F. Rega and J.P. Rossi for their constant cooperation and support. Finally, the organizers wish to acknowledge the enormous contribution made by de facto co-organizer Rolando C. Rossi; without his tireless energy and unfailing attention to local organizational detail, the meeting would not have been possible. In the months since the meeting, Kate Egnatz heroically undertook the task of checking and sorting the large number of manuscripts arriving from around the globe. The huge job of uniformly editing those manuscripts, submitted in disparate styles, has been handled with grace, consummate skill, and good humor, by Angela Fink at the Editorial Office of the New York Academy of Sciences, under the experienced guidance of Executive Editor Bill Boland. David L. Stokes graciously helped generate the cover figure. We are extremely grateful to all these people and to all contributors and participants in the conference; we are particularly grateful to the Keynote Lecturer, John E. Walker. We dare hope that, together, the novel structural and mechanistic insights arising from the conference, and encapsulated in this volume, will conspire to engender a flurry of new experiments whose results will coalesce into the next, clearer, picture of how P-type ion-motive ATPases work.

Emerging Structure of the Neurospora Plasma Membrane H$^+$-ATPase

GENE A. SCARBOROUGH[a]

Department of Pharmacology
School of Medicine
University of North Carolina
Chapel Hill, North Carolina 27599

The primary goal of this laboratory is to understand the molecular mechanism of membrane transport catalyzed by the proton-translocating ATPase from the plasma membrane of *Neurospora crassa*, which is an archetype of the so-called P-type ATPase family of ion-translocating ATPases that includes the Na$^+$/K$^+$-ATPase.[1] The field of enzyme catalysis has shown us clearly that to understand how any enzyme works, it is necessary to understand, in detail, its molecular structure. For this reason, most of our recent efforts with the H$^+$-ATPase have been centered on elucidating its structure. In this article, our progress in this regard is briefly described.

EARLIER STRUCTURAL STUDIES

The development of a procedure for isolating large quantities of the H$^+$-ATPase in stable form[2,3] was essential for virtually all of our progress. Subsequent biochemical studies demonstrated that the H$^+$-ATPase is a hexamer as isolated[4] and that the functional properties of this form of the enzyme are the same as those of the ATPase in its monomeric membrane-bound state.[5] The minimum functional unit capable of catalyzing ATP-hydrolysis-driven proton translocation was shown to be a 100-kD monomer.[6] An accurate estimation of the secondary structure composition of the ATPase by circular dichroism was made possible by the unique, largely lipid- and detergent-free nature of the purified hexameric ATPase preparation,[5] and importantly, an essentially identical secondary structure composition was obtained for the ATPase in its membrane-bound state by infrared attenuated total reflection spectroscopy.[7] Thus, the secondary structure elements that the ATPase comprises, about 36% alpha-helix, 20% antiparallel and parallel beta-sheet, 11% beta-turn, and 26% other structure, are reasonably certain.

In an extensive series of experiments, a protein chemistry approach was used to elucidate the transmembrane topography of the H$^+$-ATPase molecule. Protein chemistry procedures for fragmenting the ATPase by trypsinolysis and purifying virtually all of the numerous hydrophilic and hydrophobic fragments produced were first developed.[8,9] Then, utilizing this largely new technology and inside-out H$^+$-ATPase proteoliposomes prepared from purified components, the transmembrane topography of virtually all of the 919 amino acid residues in the H$^+$-ATPase molecule was

---

[a] Address for correspondence: Gene A. Scarborough, Department of Pharmacology, Campus Box #7365 F.L.O.B., University of North Carolina, Chapel Hill, NC 27599 (tel: 919–966–4681; fax: 919–966–5640; e-mail: gas@med.unc.edu).

established.[10–13] The overall model of the topography of the H^+-ATPase arising from this work includes NH_2- and COOH-terminal cytoplasmic strands, two larger internal loops, and four NH_2-terminal and six COOH-terminal membrane-spanning segments.[14] Although the actual number of transmembrane spans is still uncertain, the 10 proposed transmembrane crossings are in line with most models for other members of the P-type ATPase family.[15]

Some information exists as to how the H^+-ATPase molecule folds up into its three-dimensional structure. On the basis of experimental work carried out with the related Ca^{2+}-ATPase of sarcoplasmic reticulum,[16] it is likely that the membrane-embedded segments are alpha-helical. If so, with a total alpha-helix content of about 36%, most of the nonmembrane part of the molecule must be non-alpha-helix. In another series of experiments, we have elucidated the chemical state of the eight cysteine residues present in the ATPase molecule.[17] We have shown that six of the eight cysteines are present as free thiols, and two are present in disulfide linkage. And interestingly, the disulfide bridge is between cys 148 and either cys 840 or cys 869, thus physically linking the NH_2- and COOH-terminal membrane-embedded domains together. The H^+-ATPase structure at the current level of resolution thus comprises two or three probably discrete, largely nonhelical cytoplasmic domains, a cluster of NH_2- and COOH-terminal transmembrane helices, and very little protein mass on the *trans*-ATP side of the membrane.

CONFORMATIONAL DYNAMICS OF THE H^+-ATPase MOLECULE

Substantial information is available regarding the nature of the conformational changes experienced by the H^+-ATPase molecule as it proceeds through its catalytic cycle. Early experiments with the H^+-ATPase in isolated plasma membranes showed that the enzyme is protected against inhibition by mercurials or trypsin by its substrate, MgATP.[18] Subsequently, the differential sensitivity of the ATPase to tryptic degradation in the presence or absence of MgATP served in the identification of the hydrolytic moiety of the ATPase, and during these studies, it was noted that the potent ATPase inhibitor orthovanadate enhances the protective effects of MgATP against tryptic degradation.[19] It was also shown later that the combination of MgATP and vanadate confers stability to the ATPase during detergent solubilization and purification.[20] These various results suggest that the ATPase undergoes significant conformational changes during its catalytic cycle, and because these conformational changes are almost certainly intimately related to the transport mechanism, they were investigated in more detail in a study of the effects of a variety of ATPase ligands and ligand combinations on the sensitivity of the ATPase to degradation by trypsin.[21] The results of these experiments indicated that the ATPase undergoes a clearly definable conformational change on binding its substrate MgATP, which is most cleanly dissociated from subsequent conformational changes by the use of the competitive, nonphosphorylating ligand MgADP. Following the substrate binding conformational change, the ATPase undergoes additional conformational changes as it proceeds through the transition states of the enzyme phosphorylation and dephosphorylation reactions. It then undergoes a final conformational change to its unliganded conformation on product release. The implications of these findings as to the catalytic mechanism of the ATPase, particularly the fundamental importance of transition state binding as the driving force for the overall reaction cycle, have been explained and emphasized on numerous occasions before[21–25] and will not be repeated here.

Although it seemed likely that the conformational changes that the ATPase undergoes during its catalytic cycle are substantial, ligand protection of only a few key amino acid residues could conceivably explain the experimental results. The ligand-induced ATPase conformational changes were thus further explored using attenuated total reflection Fourier transform infrared spectroscopy.[26] The results clearly showed that the secondary structure components of the H^+-ATPase, that is, alpha-helix, beta-sheet, turns, and other structure, do not change when the H^+-ATPase undergoes these ligand-induced conformational changes, findings in agreement with our earlier circular dichroism studies.[5] But importantly, the hydrogen/deuterium (H/D) exchange rates of roughly 175 surface amide linkages in the ATPase polypeptide chain out of a total of about 350 are drastically reduced as the ATPase proceeds from its unliganded conformation to its substrate binding conformation (i.e., bound to MgADP). Similarly, the H/D exchange rates of about 130 residues are reduced in the transition state of the enzyme dephosphorylation reaction (i.e., bound to Mg-vanadate in the presence of ATP). Because virtually all enzymes,[27] particularly those that catalyze phosphoryl-transfer reactions,[28–34] possess structures with at least two, discrete, relatively rigid structural domains separated by a deep cleft that closes during catalysis, these results constitute strong evidence for a similar behavior of the H^+-ATPase during its catalytic cycle. The proposed cleft closure presumably occludes about 175 surface residues in the ATPase molecule from contact with the aqueous environment in the MgADP bound form and about 130 in the Mg-vanadate bound form. Thus, as just suggested, the cytoplasmic loops probably fold in a way that produces at least two domains and an open cleft, and on ligand binding, numerous interdomain surface residues become occluded when the cleft closes.

ENGINEERED H^+-ATPase MOLECULES FOR STRUCTURAL STUDIES

After numerous attempts to develop an expression system for producing site-directed mutants of the H^+-ATPase molecule, the yeast system developed by Serrano and his colleagues[35] proved to be extremely effective. When the essential yeast plasma membrane H^+-ATPase is replaced by the neurospora H^+-ATPase via a high copy number yeast expression vector, the cells survive and produce the neurospora H^+-ATPase in amounts comparable to those produced naturally.[36] This system thus provides us with an excellent tool for producing large quantities of H^+-ATPase molecules useful for structural studies. Along these lines, mutagenesis of the eight individual cysteine residues in the H^+-ATPase molecule to either serine or alanine indicated that none is essential for the function of the ATPase, although cys 376 appears to be nearly essential.[37] Importantly, in certain cases serine replacement was more effective than alanine replacement, and in other cases the converse was true. Exploiting this, an optimized combination of the cys to ala and cys to ser mutations produced a functional single cysteine mutant with the cysteine in the active site near the phosphorylated aspartate asp 378 (unpublished data). This and other mutants like it should prove to be valuable in future studies of the H^+-ATPase structure as will be mentioned again below.

LARGE, SINGLE THREE-DIMENSIONAL CRYSTALS OF THE H^+-ATPase

We have been trying for some time to obtain crystals of the H^+-ATPase that are suitable for structural analysis by X-ray diffraction techniques. Early efforts in this regard were largely unsuccessful, but in the last several years, considerable success in crystallizing the H^+-ATPase has been realized.[38] The key to obtaining large, 3D crystals of the H^+-ATPase was the recognition of the fundamentally chimeric nature of detergent complexes of integral membrane proteins and the development of an experimental approach for finding conditions for crystallizing both parts of the chimera at the same time. The detergent complex of a typical integral membrane protein contains regions of the molecule normally exposed to the aqueous milieu on the cytoplasmic and exocytoplasmic surfaces of the membrane, referred to in the original treatment[38] as the "protein surfaces." The complex contains, in addition, a torus of detergent molecules surrounding and solubilizing the highly hydrophobic region of the molecule normally embedded in the lipid bilayer, termed the "detergent micellar collar." Detergent complexes of integral membrane proteins are thus, in most cases, chimeras comprising surface regions with potentially very different physicochemical properties. It is a simple matter to bring solutions of such detergent complexes of integral membrane proteins to a state of moderate insolubility using conventional protein precipitants as is done for crystallizing soluble proteins, but rarely do membrane protein crystals form when this is tried. It seemed possible that it is the chimeric nature of detergent-membrane protein complexes that precludes their crystallization. That is, whereas it might be possible to empirically establish conditions for crystallizing one of the surfaces of the complex, it is unlikely that those conditions would be suitable for crystallizing the other. It was thus reasoned that at least one key to crystallizing membrane protein-detergent complexes may be the attainment of conditions in which the protein surfaces are moderately supersaturated and the detergent micellar collar is also at or near its solubility limit so that it is not overly or underly willing to coalesce into a solid state.

These considerations suggested that simple attainment of supersaturation using protein precipitants and detergents chosen more or less at random was not likely to be a very successful approach to crystallizing the H^+-ATPase, and numerous early attempts to crystallize it using a variety of protein precipitants lent support to this notion. It was therefore decided to approach the problem more systematically in an attempt to find conditions in which the extent of supersaturation of the protein surfaces of the ATPase and the detergent micellar collar was properly matched. Preliminary experiments had indicated that the ATPase is most stable when the detergent used to maintain its solubility is dodecylmaltoside (DDM). The first step was therefore to screen a variety of commonly used protein precipitants for those that were able to induce the aggregation of pure DDM micelles. The concentration at which any precipitant induced DDM micellar aggregation was hoped to be close to the concentration at which it might induce insolubility of the detergent micellar collar of the DDM-ATPase complex. Of the nine precipitants tried, seven, all polyethylene glycols (PEGs), were able to induce DDM micelle insolubility. The seven PEGs were then tested for their effect on the solubility of the DDM-ATPase complex at a concentration slightly below that necessary to induce DDM micellar aggregation. Three of the PEGs caused extensive precipitation of the ATPase at this concentration and were therefore set aside. The other four PEGs did not induce precipitation at the concentration employed and were subsequently used at this concentration for crystallization trials in which the protein concentration was varied. Encouragingly, crystalline plates of the ATPase were obtained for each of the four PEGs tried, indicating that the overall approach may be valid. Unfortunately, the crystals obtained were visibly flawed, suggesting that the proper balance of protein surface and DDM micelle insolubility had not yet been reached. The ionic strength of the crystallization trials was then raised, which was known from other experiments to render the protein surfaces of the ATPase less soluble while having little effect on the DDM micellar aggregation point. For one of the PEGs, PEG 4000, this brought on a new, well-formed hexagonal crystal habit. Subsequent optimization of the initial conditions has yielded large, single, hexagonal 3D crystals of the H^+-ATPase up to about $0.4 \times 0.4 \times 0.15$ mm in size. X-ray diffraction by the H^+-ATPase crystals to about 8 Å was seen early on, but subsequent efforts led to the disappointing realization that the crystals are extremely fragile, losing their diffracting power after only a few hours in the X-ray beam. This virtually precluded any structural work with the crystals, but recently, in collaboration with Dean Madden, an effective procedure for freezing the crystals at liquid nitrogen temperatures was developed. Such frozen ATPase crystals are essentially immortal in the X-ray beam, which allowed the collection of a complete set of reflections to about 8 Å. Fortunately, the data collected could be indexed, leading to a crystal packing model and determination of the unit cell as rhombohedral of the R3 type (manuscript in preparation). The hexamers are packed side to side in the crystals with a spacing of 167 Å.

TWO-DIMENSIONAL CRYSTALS OF THE H^+-ATPase

Electron microscopy of 3D crystals has often been used to obtain intermediate level resolution images of protein molecules, and this information can sometimes aid in the interpretation of higher resolution X-ray diffraction data.[39] For this reason, exploration of H^+-ATPase crystals by electron microscopic techniques was initiated in collaboration with the laboratory of Werner Kühlbrandt. The first experiments indicated the presence of small 3D crystals in crystallization mixtures similar to those used for obtaining the large, single ATPase crystals, and the aforementioned interhexamer spacing of 167 Å was quickly confirmed. Soon after, however, an even more important discovery was made. Whereas 3D H^+-ATPase crystals grow in the drops, large 2D H^+-ATPase crystals readily grow at or near the surface of such drops.[40] The 2D H^+-ATPase crystals can be picked up by touching the surface of the drops with a carbon-coated electron microscope grid and can then be viewed with or without negative stain at low temperature in a cryoelectron microscope. Hundreds of electron cryomicroscopy images have been taken since the 2D H^+-ATPase surface crystals were first discovered, and analysis of these images by laser diffractometry identified the best ones. Selected images were then digitized in a flatbed densitometer, Fourier transformed, and processed as described.[41] The inverse Fourier transform of one of the first images processed yielded the projection map shown in FIGURE 1. This projection map extends to a resolution of about 22 Å. The ring-like nature of the H^+-ATPase hexamers is clearly evident in the map. The unit cell is hexagonal with a = b = 167 Å. In projection, the individual monomers viewed from the cytoplasmic aspect are shaped like a boot. Their dimensions in this view are similar to those of the Ca^{2+}-ATPase.[42] Further improvements in specimen preparation and merging of the data from several images yielded a projection map that extends to 10.3 Å.[40]

The conventional method for growing 2D crystals of integral membrane proteins involves preparing mixtures containing the protein of interest and a suitable lipid, both solubilized by an appropriate detergent.[41] The detergent is then removed by dialysis, and the 2D crystals form as proteolipid complexes. The precipitant-induced 2D crystals of the DDM complex of the H^+-ATPase thus represent an entirely new approach to obtaining 2D crystals of an integral membrane protein useful for structure analysis. Only time will tell if this approach is useful for obtaining crystals of other integral membrane proteins, but it has already proved to be extremely effective for obtaining quality 2D crystals of the H^+-ATPase useful for structural work in a relatively short period of time.

PATHS TO THE THREE-DIMENSIONAL STRUCTURE

With the past progress and recent advances just outlined, the neurospora H^+-ATPase is one of the more likely of the membrane transport systems currently under investigation to yield its molecular structure and, eventually, its molecular mecha-

FIGURE 1. Projection map of the H^+-ATPase structure at 22 Å resolution.

nism. With the technology already in hand, a structure should be forthcoming. One way to obtain the 3D structure is from a series of images of tilted 2D ATPase crystals obtained by procedures used to produce the projection map shown in FIGURE 1. This is currently underway, and it is likely that a 3D structure with a resolution of roughly 10 Å will be possible from this approach.

A structure is also expected to emerge from the 3D H^+-ATPase crystals. A nearly complete dataset of 3D X-ray reflections to about 8 Å has been obtained. As with all X-ray data, it contains the intensities of the reflections, but not the phases. If we had the phases, we would have the structure. After determining the level of structural resolution obtainable with synchrotron radiation, a standard search for isomorphous replacements to obtain the phases, and hence the structure, will ensue.

But there are other paths as well. We also have a fairly complete set of reflections from the cryoelectron microscope images of the 2D H^+-ATPase crystals to a resolution of about 10 Å. But here, because of the nature of the data collection process, we have the phases as well as the intensities of the reflections, which allows the production of a structure map such as that shown in FIGURE 1. To our good fortune, the correlation coefficient between the 2D and the 3D data is better than 90%, which means that it is possible to use the phases from the 2D data as a starting point for extracting the phases from the 3D data. Attempts to phase the 3D data in this way are currently in progress. Moreover, it may be possible to obtain the phases of the 3D reflections by molecular replacement calculations using the published 14 Å structure of the related Ca^{2+}-ATPase.[42]

Finally, efforts are underway to grow 2D crystals of the single cysteine ATPase mutant with the cysteine at the active site. When that is accomplished, it should be possible to label the cysteine with gold maleimide[43] and identify the position of the active site in the molecule by electron crystallography. Other engineered mutants should allow the identification of other interesting sites in the molecule by similar techniques.

In conclusion, it seems likely that the 3D structure of the H^+-ATPase will emerge in the not-too-distant future. A resolution in the range of 8 Å should be obtainable with current technology. This will not tell us everything that we want to know, but should be extremely valuable nevertheless. With an 8 Å structure, we should be able to define the various domains of the molecule, and because crystals of several of the liganded forms of the molecule are also available, the nature of the substantial interdomain movements may also be understood. It should also be possible to see the transmembrane helices and detect if they move, and we should also be able to clearly define the location of the active site with respect to the membrane, the helices, and the various cytoplasmic domains. With this information alone, it will be exciting to begin to consider how the ATP hydrolysis, the conformational changes, and the proton translocation all act in concert to generate a transmembrane protonic potential difference. And finally, from what has been learned about the ATPase crystals thus far, there is no reason to believe that they cannot be improved, so an even higher resolution structure, with the complete track of the polypeptide chain, may not be too far down the road.

REFERENCES

1. PEDERSEN, P. L. & E. CARAFOLI. 1987. TIBS **12**: 146–150.
2. SMITH, R. & G. A. SCARBOROUGH. 1984. Anal. Biochem. **138**: 156–163.
3. SCARBOROUGH, G. A. 1988. Meth. Enzymol. **157**: 574–579.
4. CHADWICK, C. C., E. GOORMAGHTIGH & G. A. SCARBOROUGH. 1987. Arch. Biochem. Biophys. **252**: 348–356.
5. HENNESSEY, J. P., JR. & G. A. SCARBOROUGH. 1988. J. Biol. Chem. **263**: 3123–3130.
6. GOORMAGHTIGH, E., C. CHADWICK & G. A. SCARBOROUGH. 1986. J. Biol. Chem. **261**: 7466–7471.
7. GOORMAGHTIGH, E., J. -M. RUYSSCHAERT & G. A. SCARBOROUGH. 1988. In The Ion Pumps: Structure, Function, and Regulation. :51–56. Alan R. Liss, Inc. New York.
8. RAO, U. S., J. P. HENNESSEY, JR. & G. A. SCARBOROUGH. 1988. Anal. Biochem. **173**: 251–264.
9. HENNESSEY, J. P., JR. & G. A. SCARBOROUGH. 1989. Anal. Biochem. **176**: 284–289.
10. HENNESSEY, J. P., JR. & G. A. SCARBOROUGH. 1990. J. Biol. Chem. **265**: 532–537.
11. SCARBOROUGH, G. A. & J. P. HENNESSEY, JR. 1990. J. Biol. Chem. **265**: 16145–16149.

12. RAO, U. S., J. P. HENNESSEY, JR. & G. A. SCARBOROUGH. 1991. J. Biol. Chem. **266:** 14740–14746.
13. RAO, U. S., D. D. BAUZON & G. A. SCARBOROUGH. 1992. Biochim. Biophys. Acta **1108:** 153–158.
14. SCARBOROUGH, G. A. 1996. In Handbook of Biological Physics, Volume II. Transport Processes in Membranes. W. N. Konings, H. R. Kaback & J. S. Lolkema, Eds. :75–92. Elsevier Science B. V. Amsterdam.
15. GOFFEAU, A. & N. M. GREEN. 1990. In Monovalent Cations in Biological Systems. C. A. Pasternak, Ed.: 155–169. CRC Press, Inc. Boca Raton.
16. CORBALAN-GARCIA, S., J. A. TERUEL, J. VILLALAIN & J. C. GOMEZ-FERNANDEZ. 1994. Biochemistry **33:** 8247–8254.
17. RAO, U. S. & G. A. SCARBOROUGH. 1990. J. Biol. Chem. **265:** 7227–7235.
18. SCARBOROUGH, G. A. 1977. Arch. Biochem. Biophys. **180:** 384–393.
19. DAME, J. B. & G. A. SCARBOROUGH. 1980. Biochemistry **19:** 2931–2937.
20. ADDISON, R. & G. A. SCARBOROUGH. 1981. J. Biol. Chem. **256:** 13165–13171.
21. ADDISON, R. & G. A. SCARBOROUGH. 1982. J. Biol. Chem. **257:** 10421–10426.
22. SCARBOROUGH, G. A. 1992. In Molecular Aspects of Transport Proteins. de Pont, J. J. H. H. M., Ed. : 117–134. Elsevier Science Publishers B. V. Amsterdam.
23. SCARBOROUGH, G. A. 1982. Ann. N. Y. Acad. Sci. **402:** 99–115.
24. SCARBOROUGH, G. A. 1985. In Environmental Regulation of Microbial Metabolism. I. S. Kulaev, E. A. Dawes & D. W. Tempest, Eds.: 39–51. Academic Press.
25. SCARBOROUGH, G. A. 1985. Microbiol. Rev. **49:** 214–231.
26. GOORMAGHTIGH, E., L. VIGNERON, G. A. SCARBOROUGH & J.-M. RUYSSCHAERT. 1994. J. Biol. Chem. **269:** 27409–27413.
27. SCHULZ, G. E. & R. H. SCHIRMER. 1979. In Principles of Protein Structure. Springer-Verlag. New York.
28. STEITZ, T. A., R. J. FLETTERICK, W. F. ANDERSON & C. M. ANDERSON. 1976. J. Mol. Biol. **104:** 197–222.
29. STUART, D. I., M. LEVINE, H. MUIRHEAD & D. K. STAMMERS. 1979. J. Mol. Biol. **134:** 109–142.
30. SACHSENHEIMER, W. & G. E. SCHULZ. 1977. J. Mol. Biol. **114:** 23–36.
31. EVANS, P. R. & P. J. HUDSON. 1979. Nature **279:** 500–504.
32. BANKS, R. D., C. C. F. BLAKE, P. R. EVANS, R. HASER, D. W. RICE, G. W. HARDY, M. MERRETT & A. W. PHILLIPS. 1979. Nature **279:** 773–777.
33. PICKOVER, C. A., D. B. MCKAY, D. M. ENGELMAN & T. A. STEITZ. 1979. J. Biol. Chem. **254:** 11323–11329.
34. ANDERSON, C. M., F. H. ZUCKER & T. A. STEITZ. 1979. Science **204:** 375–380.
35. VILLALBA, J. M., M. G. PALMGREN, G. E. BERBERIAN, C. FERGUSON & R. SERRANO. 1992. J. Biol. Chem. **267:** 12341–12349.
36. MAHANTY, S. K., U. S. RAO, R. A. NICHOLAS & G. A. SCARBOROUGH. 1994. J. Biol. Chem. **269:** 17705–17712.
37. MAHANTY, S. K. & G. A. SCARBOROUGH. 1996. J. Biol. Chem. **271:** 367–371.
38. SCARBOROUGH, G. A. 1994. Acta Cryst. D **50:** 643–649.
39. MCPHERSON, A. 1982. Preparation and Analysis of Protein Crystals. John Wiley & Sons. New York.
40. CYRKLAFF, M., M. AUER, W. KÜHLBRANDT & G. A. SCARBOROUGH. 1995. EMBO J. **14:** 1854–1857.
41. KÜHLBRANDT, W. 1992. Q. Rev. Biophys. **25:** 1–49.
42. TOYOSHIMA, C., H. SASABE & D. L. STOKES. 1993. Nature **362:** 469–471.
43. MILLIGAN, R. A., M. WHITTAKER & D. SAFER. 1990. Nature **348:** 217–221.

Two-Dimensional Crystal Formation from Solubilized Membrane Proteins Using Bio-Beads to Remove Detergent[a]

J.-J. LACAPÈRE,[b] D. L. STOKES,[c] G. MOSSER,[b]
J.-L. RANCK,[b] G. LEBLANC,[d] AND J.-L. RIGAUD[b]

[b]Institut Curie
Section de recherche
UMR 168 du CNRS and
LRC-CEA 8
11 rue Pierre et Marie Curie
75005 Paris Cedex, France

[c]Skirkball Institut
Medical School
New York University
New York, New York

[d]Laboratoire J. Maetz
URA 1855 du CNRS
CEA Villefranche
BP 68
06230 Villefranche-sur-mer, France

High resolution structural data have been collected from several two-dimensional crystals of membrane proteins, demonstrating that electron crystallography offers a viable alternative to x-ray crystallography. (For reviews see refs. 1 and 2.) However, the ability to solve structures by electron microscopy is limited by the fact that methods for growing two-dimensional crystals have not kept up with the technical capability to produce high resolution density maps of such crystals. Thus, one of the main limitations in reaching high resolution appears related to the quality of two-dimensional crystals. Therefore, various experimental innovations will greatly facilitate future studies on membrane proteins.

One of the most useful procedures for two-dimensional crystallization relies on the general method of reincorporation of membrane proteins into lipid vesicles. (For a review see ref. 3.) This is achieved by removal of the detergent from a lipid-protein-detergent micellar solution. Among all parameters that may influence crystallization and therefore the final quality of the two-dimensional crystals (e.g., buffer composition, temperature, protein and detergent concentrations, and lipid-to-protein ratio), the procedure of detergent removal is an important parameter that has never been explored carefully. Indeed, the three steps of crystal production (i.e., nucleation, growth, and arrest of growth) are probably affected, on the one hand, by the rate of

---

[a]Part of this work was supported by a grant from the E. U. (Biotechnology: No Bio 2 CT 930 078).

detergent removal and, on the other, by the presence of residual detergent in the bilayers.

Dialysis is the most common procedure used to remove detergents in two-dimensional crystallization. Many membrane proteins have been crystallized using this technique which, nevertheless, presents inherent drawbacks for detergents with low critical micelle concentrations, which are difficult to remove or at best require a long period of dialysis.

As an alternative to dialysis, we have explored in two-dimensional crystallization trials the usefulness of detergent removal through hydrophobic adsorption onto polystyrene beads.[4] Kinetic and equilibrium aspects of the removal of different detergents by SM-2 Bio-Beads have been systematically analyzed. The potential of this strategy for producing two-dimensional crystals was evaluated by studying the crystallization of four radically different membrane proteins solubilized in different detergents: (1) Ca^{2+} ATPase from sarcoplasmic reticulum in $C_{12}E_8$; (2) F_0/F_1-ATPase from thermophilic bacillus PS3 in Triton X-100; (3) melibiose permease from *Escherichia coli* in dodecylmaltoside; (4) cytochrome b_6f from *Chlamydomonas reinhardtii* in hecameg.

The foremost outcome of this study is that Bio-Beads-mediated detergent removal enabled us to get two-dimensional crystals of all these membrane proteins, demonstrating the general benefit of this strategy, as an alternative to conventional dialysis, for removing detergent to generate two-dimensional crystals.[5]

DETERGENT REMOVAL BY SM-2 BIO-BEADS

Removal by SM-2 Bio-Beads of different detergents commonly used in membrane protein solubilization and reconstitution has been analyzed in detail. These include Triton X-100, octylglucoside, hecameg, dodecylmaltoside, and $C_{12}E_8$. For each detergent we have examined the binding capacity of Bio-Beads as well as the rate of removal and the effect of temperature.[4-7] The main conclusion of this thorough study is that SM-2 Bio-Beads can entirely adsorb any kind of detergent. The adsorptive capacity relies only on the chemical structures of the hydrophilic/hydrophobic moieties of the detergents. The rates of detergent removal depend on both the amount of beads and the detergent concentration. In addition, hydrophobic adsorption onto the beads relies on the monomeric or micellar form of the detergent. As an example, for $C_{12}E_8$, a detergent with low critical micelle concentration (cmc) used for the purification and crystallization of the Ca^{2+}ATPase, we measured a binding capacity of 0.2 mg $C_{12}E_8$ per milligram of Bio-Beads. The initial rates of detergent removal are shown to increase with the initial $C_{12}E_8$ concentration, exhibiting a slope change around the cmc of this detergent related to a threefold decrease in the rate of micelle adsorption compared to monomer adsorption.

Another important parameter in determining the rate of detergent removal is related to the temperature of the incubation medium; whatever the detergent, rates of detergent adsorption were observed to increase by a factor of 4–5 between 4°C and 37°C.

Importantly for two-dimensional crystallization experiments, phospholipid binding to the Bio-Beads could be minimized or almost avoided by carefully controlling the amount of beads used. By adding an amount of beads two- to threefold greater than that needed for total detergent removal, no more than 20–25% of the lipids initially present were adsorbed under crystallization conditions. Finally, protein adsorption onto Bio-Beads was negligible, whatever the hydrophilic/lipophilic balance of the proteins analyzed during the course of this study.

TWO-DIMENSIONAL CRYSTALLIZATION OF CA^{2+} ATPase

To estimate the enhancement of detergent removal by Bio-Beads in crystallization experiments, we first analyzed the crystallization of Ca^{2+}ATPase as a prototypic membrane protein.

The sarcoplasmic reticulum (SR) Ca^{2+}ATPase is a P-type ion pump and so belongs to a family of proteins that share both amino acid homology and a basic reaction mechanism. Although we have much information on the topology and topography of these proteins through biochemical, biophysical, and molecular biological studies, none of the structures of the enzymes belonging to this family has yet been solved at high resolution. Different crystals forms have been described for some of the proteins in this family, and structural electron microscopic studies have already given low resolution data from two-dimensional crystals.[8] In this context, two different crystals forms have been described for Ca^{2+}ATPase: (1) tubular crystals induced by vanadate, calcium, or lanthanide addition to SR vesicles[9,10]; (2) large multilamellar flat crystals grown from detergent-solubilized SR.[11,12] The only three-dimensional reconstruction of Ca^{2+}ATPase comes from the vanadate-induced tubes,[13] but it is limited to 15 Å resolution. With the large flat crystals grown in the presence of detergent, the diffraction patterns extend to a much higher resolution (4–6 Å), but three-dimensional reconstruction is difficult because of the multilayer structure of the crystals. Several attempts have been made to maximize the crystallization conditions in order to improve the quality of the crystals and make them suitable for higher resolution three-dimensional reconstruction.[14–16]

To this end, we adapted the method of Bio-Beads–mediated detergent removal to produce new two-dimensional crystals of Ca^{2+}ATPase and analyzed the different factors involved in the crystallization process. In particular, we take one advantage of the Bio-Beads strategy to allow control of the rate of detergent removal by simply controlling the amount of Bio-Beads added to the medium, as previously demonstrated for reconstitution.[7] FIGURE 1 depicts the kinetics of $C_{12}E_8$ adsorption onto Bio-Beads during different crystallization conditions. Faster rates of detergent removal correspond to the addition of larger amounts of beads at one time, whereas slow or intermediate rates of detergent removal correspond to many step-by-step additions of smaller increments, but the same total amount of beads. Under the experimental conditions depicted in FIGURE 1, the $C_{12}E_8$ initially present can be removed within 20 minutes, 40 minutes, or 2–3 hours, respectively.

Slow Rate of Detergent Removal

On slow detergent removal, we observed that crystals were only formed within a narrow range of protein-to-lipid ratios. In the absence of added lipids (i.e., at a protein-to-lipid ratio of 5 [w/w], due to the presence of endogenous lipids in the purified ATPase), only some small and not well-ordered crystals could be observed,

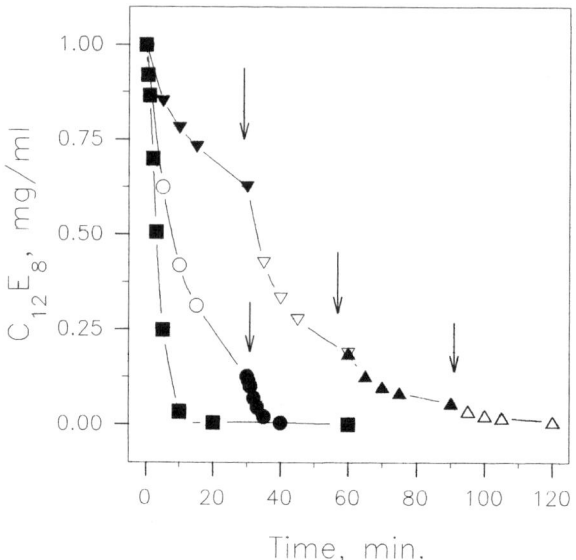

FIGURE 1. Kinetics of detergent removal on crystallization conditions for Ca^{2+}-ATPase. Fast detergent removal (*closed square*) is initiated by one addition of 40 mg SM-2 Bio-Beads; medium rate of detergent removal (*circle*) is initiated by one addition of 10 mg SM-2 Bio-Beads (*open circle*) followed by a second addition of 10 mg Bio-Beads (*arrow*); slow rate of detergent removal (*triangle*) is initiated by one addition of 2.5 mg SM-2 Bio-Beads (*closed triangle*) followed by three additions of 2.5 mg Bio-Beads (*arrows*).

whereas the addition of lipid to give a protein-to-lipid ratio of 1 (w/w) leads to the formation of vesicles with a dense protein packing. At even lower protein-to-lipid ratios, the vesicles tend to have less dense protein packing, and two populations of vesicles (protein-free liposomes and protein-rich proteoliposomes) can be visualized. Interestingly, slow detergent removal from samples at initial lipid-to-protein ratios of between 2.5 and 4 (w/w) induces the formation of large plate crystals. These two-dimensional crystals (FIG. 2A) are composed of a stack of several sheets, as can be seen on the edge and on top of the plates. The optical (or computed) diffraction patterns of these crystals are similar to those previously reported for three-dimensional crystals grown in the presence of detergent.[11,12] Although these crystals diffract up to 20 Å in negative stain and could be used to get higher resolution by electron cryomicroscopy, their multilayered structure will limit three-dimensional reconstruction analysis.

Fast Rate of Detergent Removal

Analysis of the structures formed on fast detergent removal shows that crystals are also only observed within narrow range of protein-to-lipid ratios (2.5–4 w/w), similar to those described for slow detergent removal. No crystals were observed in the absence of added lipids, whereas the addition of lipid in excess of the amount of protein gives rise to the formation of vesicles. However, with this fast rate of detergent removal, we never observed two vesicle populations (liposomes and proteoliposomes). FIGURE 2B shows that fast detergent removal from samples at protein-to-lipid ratios of 2.5–4 induces the formation of wide tubular crystals. They appear to be elongated vesicles (open or closed) with crystalline arrays elongated almost from one end of the vesicle to the other. The diffraction patterns of these crystals are different from those observed on slow detergent removal and reveal two lattices corresponding to the two layers of a flattened tube, each layer similar to that previously observed in slow detergent removal.

Intermediate Rate of Detergent Removal

The different structures formed on detergent removal at intermediate rates have also been studied. Crystal formation follows a dependence on the protein-to-lipid ratio similar to that observed for slow and fast rates of detergent removal with an optimum at a ratio of 2.5–4. The crystals formed are mostly multilamellar, with the coexistence of different shapes from tubes to sheets and even fused tubes and sheets. The multilamellar tubes are longer (more than 1 μ) than those observed with fast rates of detergent removal. They have fewer layers (2–5) than do the large multilamellar sheets obtained with slow detergent removal conditions.

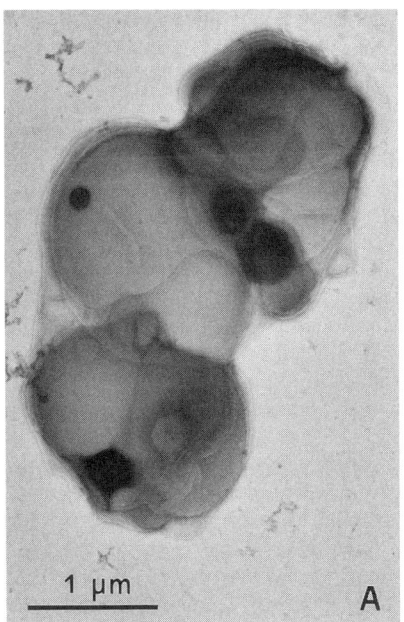
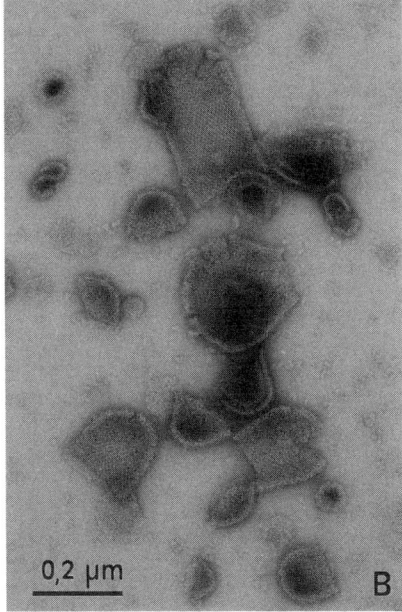

FIGURE 2. Electron micrographs of negatively stained sarcoplasmic reticulum Ca^{2+} ATPase crystals induced by slow (**A**) or fast (**B**) detergent removal.

Freeze-Thaw

Because the monolayered tubular crystals produced on rapid detergent removal are still too small to allow high resolution three-dimensional reconstruction, we worked out conditions to enlarge them. In particular, we treated the different crystals by the freezing-thawing cycles that have been reported to induce membrane vesicles fusion.[17] This process induces enlargement of all types of crystals, but it also induces the formation of either multilamellar tubes or entangled, curved, and rolled multilamellar sheets.

TWO-DIMENSIONAL CRYSTALLIZATION OF OTHER CLASSES OF MEMBRANE PROTEIN

To assess the usefulness of detergent removal by Bio-Beads in crystallization experiments, other membrane proteins, with radically different hydrophilic/lipophilic balances, were examined.

F_1/F_0-ATPases. These are enormous multisubunit complexes containing two major parts, a large hydrophilic F1 part (MW 450 kD) whose structure was recently reported[18] and which contains five types of subunit in an $\alpha_3\beta_3\gamma\delta\epsilon$ stoichiometry and is an approximate sphere 90–100 nm in diameter connected to the membrane part by a stalk about 4.5 nm long, and an intrinsic membrane part, F_0 (MW 100 kD), which, in bacteria, contains three subunits in the ratio $a_1b_3c_{12}$.

The F_1/F_0-ATPase from thermophilic bacillus PS3 is extracted in Triton X-100 and can easily be reconstituted in membrane vesicles using Bio-Beads to adsorb this low cmc detergent.[19] At protein-to-lipid ratios of about 0.5 w/w, the formation of small vesicles with densely packed knob-like structures typical of F_1 particles can be observed after complete detergent removal, regardless of the rate. (FIG. 3A). It is noteworthy that freezing and thawing of these samples induce fusion of the proteoliposomes leading to much larger structures including large vesicular or planar sheet structures. Typical electron micrographs are presented in FIGURE 3B and C, showing densely packed proteins and some examples of small regions of crystallinity.

Melibiose Permease. This is a single polypeptide (MW 53 kD). Predictions and topological studies suggest that this transporter is a very hydrophobic membrane protein, with 12 helical membrane-spanning segments connected by short hydrophilic polar loops and with the NH$_2$- and COOH-terminus on the cytoplasmic surface of the membrane.[20] Fully active recombinant Mel6-His–tagged permease has been solubilized from the membrane and purified to homogeneity in dodecylmaltoside.[21] Two-dimensional crystallization experiments were performed at different temperatures. Detergent removal at 4°C leads to the formation of vesicular structures that tend to fuse and form planar bilayer sheets showing close protein packing and some regions of crystallinity (25 Å resolution). On detergent removal at room temperature (equivalent to a fourfold increase in the rate of detergent removal as compared to that at 4°C), many thin tubular structures were identified showing regular arrangement (FIG. 4). Again in these crystallization experiments, freezing and thawing of samples after detergent removal were demonstrated to improve the growth (number and/or size) of both the planar and the tubular structures.

Cytochrome b_6f Complex. This consists of seven polypeptides (MW 110 kD) with 11–12 transmembrane α helices.[22] This membrane protein is purified and solubilized in hecameg.[23] After detergent removal by Bio-Beads at 4°C and incubation for 1 week,

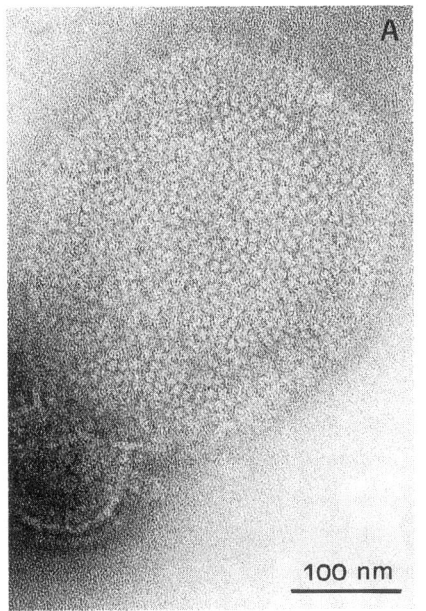
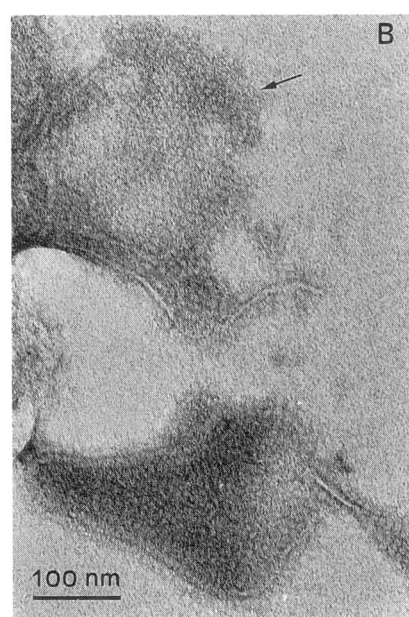
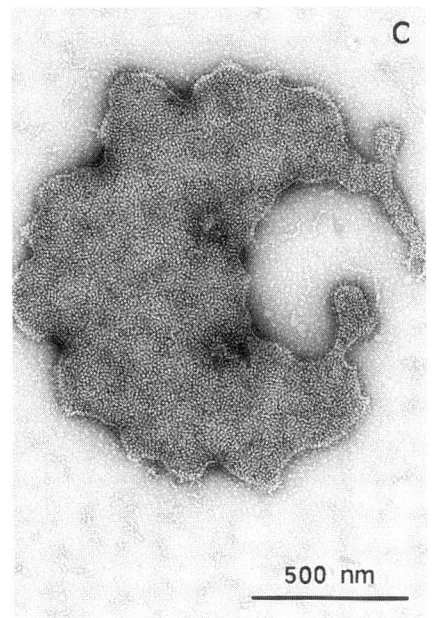

FIGURE 3. Electron micrographs of negatively stained F_1/F_0-ATPase. (**A**) Vesicles with densely packed knob-like structures typical of F_1 particles. (**B**) Some examples of regions of crystallinity. (**C**) Large planar sheets after freezing and thawing.

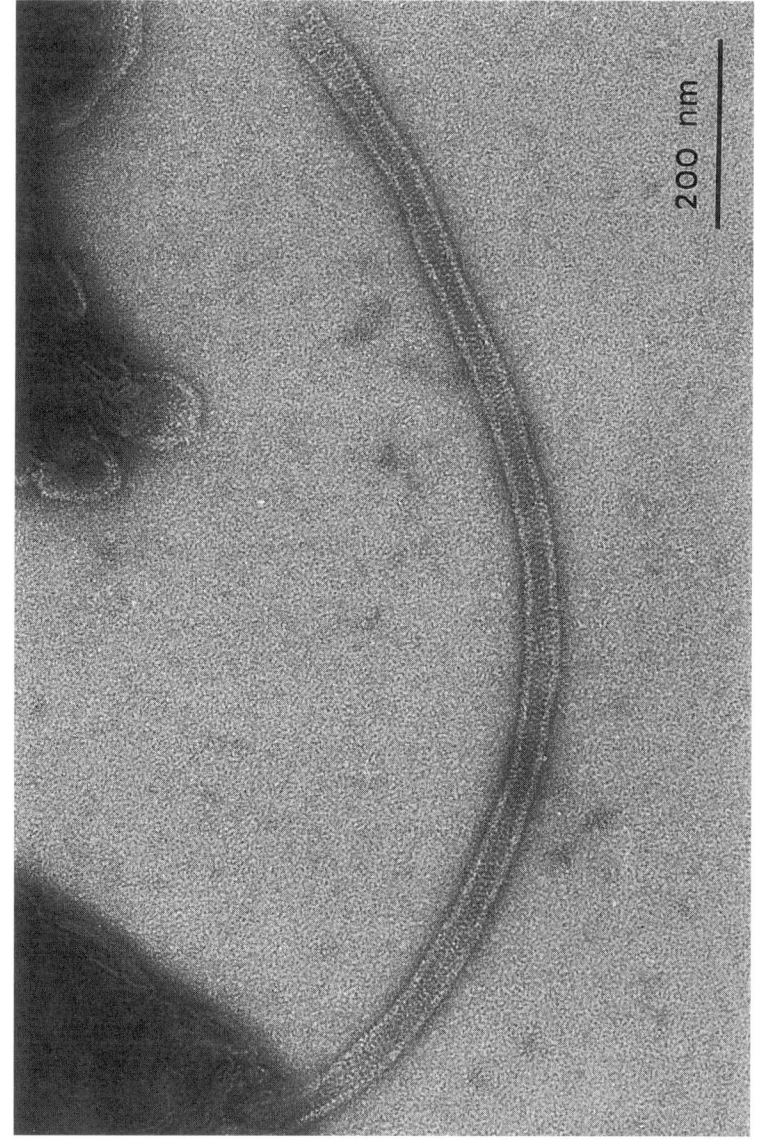

FIGURE 4. Electron micrograph of negatively stained tubular crystals of melibiose permease.

large crystalline sheets could be observed diffracting to about 15–20 Å. Optimization of lipid composition and treatment of the samples by freezing and thawing after detergent removal led to larger plates composed of very few stacked layers and diffracting down to 10 Å in negative stain. A projection map at 8 Å resolution was calculated demonstrating for the first time the potential of Bio-Beads as detergent-removing agents to produce crystals suitable for high resolution structural studies.[24]

CONCLUSION

Methods for growing two-dimensional crystals obviously need to be refined and applied to available membrane proteins in order to make electron crystallography an even more important tool in structure determination of membrane proteins. Some phenomena in two-dimensional crystallization trials are still not well understood. A need obviously exists to outline and evaluate new approaches to coherent two-dimensional crystal formation, thereby providing useful guidelines to enhance the success rate of two-dimensional crystallization experiments.

At the onset of this work it was important to define a new approach to detergent removal that would allow us to perform reliable two-dimensional crystallization experiments. Thus, this work is important in demonstrating that detergent removal by Bio-Beads is successful in producing different types of two-dimensional crystals of different membrane proteins solubilized in different classes of detergents. Our results establish the general usefulness of this strategy as an alternative to dialysis; it is a simple and rapid method for removing all kinds of detergents and, interestingly, for removing detergents with low cmcs, in a short time. In addition, the inherent drawbacks of this method can be avoided, because the detergent can be removed almost entirely with a high recovery of lipids and without protein adsorption.

Another important aspect of this work is that adsorbent beads provide a convenient means for investigating kinetic factors that are important in bilayer formation, protein insertion, and further crystallization. Crystallization experiments at different rates of detergent removal can easily be performed by varying the amount of beads and/or the temperature. For example, our systematic studies of Ca^{2+}ATPase crystallization allowed production of new two-dimensional crystals whose morphologies could be controlled through the rate of detergent removal. Our results led us to hypothesize that when nucleation is rapid, a large number of small tubular crystals is expected, and when nucleation is slow, a smaller number of crystals is expected but they have a chance to grow larger. With Ca^{2+}ATPase crystals, this growth occurs in two directions, in the plane of the lamellae, generating plates with a large diameter, but also perpendicular to the lamellae, generating appreciable stacking.

Finally, an important finding of our studies on different membrane proteins relates to freezing and thawing of the samples after detergent removal. Indeed, treating the samples through freeze-thaw cycles improves the crystallization process. A reasonable explanation is that such treatment not only induces fusion of proteoliposomes but also creates some defects in the bilayer which allow efficient growth of two-dimensional crystals.

In conclusion, it can be stressed that although this paper is not concerned with high resolution analysis per se, the methods developed here will later be important in determining the conditions needed to get large and well-organized two-dimensional crystals suitable for electron crystallography and further three-dimensional reconstruction.

ACKNOWLEDGMENTS

We would like to thank C. Breyton and J. -L. Popot (Paris) for the gift of the cytochrome b_6f and for further collaborations on the high resolution analysis of this complex.

REFERENCES

1. JAP, B. K., M. ZULAUF, T. SCHEYBANI, A. HEFTI, W. BAUMEISTER, U. AEBI & A. ENGEL. 1992. Ultramicroscopy **46:** 45–84.
2. KÜHLBRANDT, W. 1992. Q. Rev. Biophys. **25:** 1–49.
3. RIGAUD, J. L., B. PITARD & D. LEVY. 1995. Biochim. Biophys. Acta **1231:** 223–246.
4. LEVY, D., A. BLUZAT, M. SEIGNEURET & J. L. RIGAUD. 1991. Biochim. Biophys. Acta **1025:** 179–190.
5. RIGAUD, J. -L., G. MOSSER, J. -J. LACAPÈRE, A. OLOFSSON, D. LEVY & J. -L. RANCK. 1997. J. Struct. Biol. **118:** 226–235.
6. LEVY, D., A. GULIK, M. SEIGNEURET & J. -L. RIGAUD. 1990. Biochemistry **29:** 9480–9488.
7. LEVY, D., A. GULIK, A. BLUZAT & J. L. RIGAUD. 1992. Biochim. Biophys. Acta **1107:** 283–298.
8. STOKES, D. L. 1991. Current Opin. Structural Biol. **1:** 555–561.
9. DUX, L. & A. MARTONOSI. 1983. J. Biol. Chem. **258:** 2599–2603.
10. DUX, L., K. TAYLOR, H. P. TING-BEALL & A. MARTONOSI. 1985. J. Biol. Chem. **260:** 11730–11743.
11. DUX, L., S. PIKULA, N. MULLNER & A. MARTONOSI. 1987. J. Biol. Chem. **262:** 6439–6442.
12. STOKES, D. L. & N. M. GREEN. 1990. Biophys. J. **57:** 1–14.
13. TOYOSHIMA, C., H. SASABE & D. L. STOKES. 1993. Nature **362:** 469–471.
14. MISRA, M., D. TAYLOR, T. OLIVIER & K. TAYLOR. 1991. Biochim. Biophys. Acta **1077:** 107–118.
15. VARGA, S., K. TAYLOR & A. MARTONOSI. 1991. Biochim. Biophys. Acta **1070:** 374–386.
16. SHI, D., H. -H. HSIUNG, R. C. PACE & D. L. STOKES. 1995. Biophys. J. **68:** 1152–1162.
17. PICK, U. 1981. Arch. Biochem. Biophys. **212:** 186–194.
18. ABRAHAMS, J. P., A. G. W. LESLIE, R. LUTTER & J. E. WALKRE. 1994. Nature **370:** 621–628.
19. PITARD, B., P. RICHARD, M. DUNACH, G. GIRAULT & J. -L. RIGAUD. 1996. Eur. J. Biochem. **235:** 769–778.
20. POOLMAN, B., J. KNOL, C. VAN DER DOES, P. J. F. HENDERSON, W. J. LIANG, G. LEBLANC, T. POURCHER & L. MUS VETEAU. 1996. Structure-activity relationship of sugar transport proteins. Mol. Microbiol. **19:** 911–922.
21. POURCHER, T., S. LECLERCQ, G. K. BRANDOLIN & G. LEBLANC. 1995. Biochemistry **34:** 4412–4420.
22. CRAMER, W. A., G. M. SORIANO, M. PONOMAREV, D. HUANG, H. ZHANG, S. E. MARTINEZ & J. L. SMITH. 1996. Annu. Rev. plant. Physiol. Plant. Mol. Biol. **47:** 477–508.
23. PIERRE, Y., C. BREYTON, D. KRAMER & J. L. POPOT. 1995. J. Biol. Chem. **270:** 29342–29349.
24. MOSSER, G., C. BREYTON, A. OLOFFSON, J. L. POPOT & J. L. RIGAUD. 1997. J. Biol. Chem. **272:** in press.

Oligomeric Structure of Solubilized Na$^+$/K$^+$-ATPase Linked to E$_1$/E$_2$ Conformation

Y. HAYASHI,[a,c] K. KAMEYAMA,[b] T. KOBAYASHI,[a]
E. HAGIWARA,[a] N. SHINJI,[a] AND T. TAKAGI[b]

[a]*First Department of Biochemistry*
Kyorin University School of Medicine
Mitaka, Tokyo 181, Japan

[b]*Institute for Protein Research*
Osaka University
Suita, Osaka 565, Japan

The quaternary structure of Na$^+$/K$^+$-ATPase has been studied by solubilizing the membrane-bound enzyme, purified from mammalian kidneys, with nonionic surfactants such as C$_{12}$E$_8$ and Lubrol.[1–6] These studies have shown consistently that the two polypeptides, α and β, are noncovalently combined in a minimum structural unit, an αβ-protomer. The enzyme solution obtained thus, however, contained many other oligomers, such as (αβ)$_2$, (αβ)$_3$, and (αβ)$_4$, as shown by chemical cross-linking of the solubilized enzyme.[3] Conflicting conclusions about the structure necessary for Na$^+$/K$^+$-ATPase enzymatic activity have been obtained using the sedimentation equilibrium technique, so that whether the αβ-protomer (P) or the (αβ)$_2$-diprotomer (D) protein unit is the minimum active unit remains to be established.[1–5] We have devised a low-angle laser light scattering photometry coupled with a high-performance gel chromatography (HPGC/LALLS) method to study the minimum functional unit of the quaternary structure.[7] This method has shown so far that P and D are major protein components of the solubilized enzyme and that they are in an equilibrium of association-dissociation (2P ⇌ D) at a moderately high temperature.[8] This suggests that the two oligomeric forms of D and P occur simultaneously during exhibition of ATPase activity, regardless of whether P or D is the minimum active unit. Therefore, to clarify the relation between the structure and function of Na$^+$/K$^+$-ATPase, it is necessary to determine if a specific oligomer of P and/or D works without converting its structure to any other oligomeric form or if such a conversion is essential for its function.

In this paper, by measuring the dependence of M_r of the solubilized enzyme on the protein concentration using a conventional, but precise, light-scattering photometer at 20°C, it was established further that the solubilized enzyme is in equilibrium with association constants (K_a) which are 100-fold different from the E$_1$ and E$_2$ conformational states. We also show that P is converted into D on producing phosphoenzyme intermediates at the E$_2$ state (E$_2$-P) by the addition of ATP,[9] and that the E$_1$ and E$_2$ conformational states therefore correspond to the protomeric and diprotomeric struc-

---

[c]To whom correspondence should be addressed. Tel: +81-422-76-7651; fax: +81-422-76-7650.

tures, respectively, by simultaneous measurement of fluorescence intensity (conformational state) and M_r (oligomeric structure) of solubilized fluorescein 5′-isothiocyanate (FITC)-labeled enzyme.

STATIC OLIGOMERIC STRUCTURE OF SOLUBILIZED Na^+/K^+-ATPase

The membrane-bound enzyme was purified from the outer medulla of frozen dog kidney to a specific activity of 40–48 μmol P_i/min/mg protein at 37°C by the zonal rotor method of Jørgensen.[8,10] The membrane-bound enzyme was solubilized using octaethyleneglycol dodecyl ether ($C_{12}E_8$) in a 1:3 weight ratio of surfactant to protein in the presence of either 0.1 M KCl or NaCl at 0°C.[8] The M_r and the oligomeric structure of the solubilized enzyme were determined using the HPGC/LALLS method.[7,8] The HPGC/LALLS system is composed of a main column of TSKgel G3000SW$_{XL}$ and the following three kinds of detectors which are connected in series to the column: a LALLS photometer (TSK model LS-8000), a differential refractometer (RI), and an ultraviolet spectrophotometer (UV). The column and a flow cell of the LALLS photometer were always cooled to around 0°C by a circulating medium. This cooling was essential for the success of this study. The column was equilibrated using an elution buffer of 0.2 mg · ml^{-1} $C_{12}E_8$ containing 0.05 M NaCl/0.05 M KCl/4 mM $MgCl_2$/1 mM EDTA/10 mM imidazole/13 mM Hepes, at 0°C and at pH 7.0. The solubilized enzyme was charged onto the column and eluted with the same elution buffer as just described. The solubilized enzyme was separated into a minor component (H) and two major components of D and P. The M_r values of the protein moiety itself of the two protein components were estimated to be 302,000 ± 10,000 and 156,000 ± 4,000, respectively, by the HPGC/LALLS method.[7,8] The two major components isolated were composed of the α and β subunits in a molar ratio of 1:1 by sodium dodecyl sulfate-polyacrylamide gel electrophoresis (SDS-PAGE).[7,8] We have also shown that the M_r values of the α and β subunits are 118,000 ± 3,000 and 39,400 ± 900, respectively, by the HPGC/LALLS method in the presence of SDS.[11] Therefore, the major protein components of D and P were identified unambiguously as $(\alpha\beta)_2$-diprotomer and αβ-protomer, respectively. The minor component of H was thought to be an associate and/or aggregate because its M_r values were much higher than those of the major components and it had a nonstoichiometric molar ratio of α:β. The elution pattern with the three peaks of H, D, and P were obtained when chromatography was performed at around 0°C, although the relative contents of D to P were altered depending on the kind of ligand included in the elution buffer at neutral pH and on the pH value of the elution buffer.[12] The D and P that emerged from the column contained about 15 mol of phospholipid in common, and 150 and 200 mol of $C_{12}E_8$ formed part of D and P, respectively, per αβ-protomeric unit.[7,8] The content of phospholipid decreased with increasing temperature of the chromatography column, but it did not alter when the ligands in the elution buffer were changed between Na^+ and K^+.[8]

DISSOCIATION-ASSOCIATION EQUILIBRIUM OF 2P ⇌ D

Computer Simulation of Chromatography Behavior

The two protein peaks corresponding to D and P, separated by HPGC at 0°C, merged into a single major protein peak when the temperature of the chromatography was increased to 20°C, regardless of whether K^+ or Na^+ was included in the elution buffer.[8] The M_r of the resultant main component, revealed when the elution buffer contained either 0.1 M KCl or NaCl at 20°C, was estimated to be 300,000 or 255,000, respectively, by the HPGC/LALLS method. Under different conditions, when the elution buffer contained ligands other than Na^+ and K^+, the single main peak could be reproduced by performing chromatography at 20°C, and the M_r values were estimated to be 247,000–258,000 or 298,000–295,000, with the elution buffer containing the ligands favorable for the E_1 or E_2 conformation, respectively (TABLE 1).[8,13] The M_r of around 297,000 coincided with that of D. That of around 253,000 coincided with neither the M_r of D nor that of P, but it was intermediate between them. SDS-PAGE still showed that the molar ratio of α:β was approximately 1:1 for the merged protein component, indicating that this component was composed of an αβ-protomeric unit.

The behavior of the solubilized enzyme thus observed was analyzed using a computer simulation technique developed for a reversibly associating protein by Stevens and Schiffer.[14] As a result, the elution pattern that produced the merged protein peak could be simulated by assuming that D and P were in an equilibrium of dissociation-association with a very fast rate compared with the elution time. By assuming the association constant (K_a) of the equilibrium to be either $2 \cdot 10^6$ M^{-1} or more than $1 \cdot 10^8$ M^{-1}, the best fit could be obtained for the elution pattern observed under conditions favorable for the E_1 or E_2, respectively, conformational states of the solubilized enzyme at 20°C.[8,13] The ratio of the weight concentration of D:P was revealed at the top of the peak by computer simulation, and it allowed us to calculate a weight-averaged M_r ($\overline{M}_w$) at the same top of the peak as that adopted for the experimental estimation of M_r by the HPGC/LALLS method.[8] As shown in TABLE 1, the calculated $\overline{M}_w$ value (248,000) was in agreement with the values obtained experimentally. Thus, the different M_r of the solubilized enzyme obtained under the conditions favorable for either the E_1 or E_2 state could be attributed to the difference in the K_a of the equilibrium of $2P \rightleftharpoons D$.

Confirmation of Equilibrium by Direct Measurement of the Dependence of M_r on Protein Concentration

The solubilized enzyme was run, at 20°C, through the same column as that just described to exclude aggregates larger than the diprotomer, and the protein component that eluted as a single peak exclusively consisting of D and P was isolated as the purified solubilized enzyme without concentrating. After being diluted to various protein concentrations using the effluent eluted through the column before its void volume, the solubilized enzyme thus purified was applied to a conventional, but precise, light-scattering photometer (Ohtsuka Electronics Co., DLS-700) equipped with a cylindrical batch-cell 12 mm in diameter and an Ar laser-light source (15 mW output) to estimate a $\overline{M}_w$ value for the enzyme. Measurements were taken at scattering angles between 30° and 150° and extrapolated to 0° to estimate $\overline{M}_w$. The $\overline{M}_w$s of the solubilized enzyme obtained by chromatography using the two kinds of elution buffer, containing either 0.1 M NaCl or KCl, are shown in FIGURE 1. By simply assuming the equilibrium of $2P \rightleftharpoons D$ for the purified solubilized enzyme, the dependence of M_r on protein concentration at values higher than 0.15 mg/ml could be well fitted by association constants of about $2.5 \cdot 10^6$ and $3.2 \cdot 10^8$ M^{-1} at concentrations of 0.1 M NaCl and KCl, respectively. The K_as thus obtained were fairly consistent with the values obtained from computer simulation. The discrepancy between the experimental

TABLE 1. M_rs and Association Constants (K_a) of Solubilized Na$^+$/K$^+$-ATPase in the Two Conformational States of E_1 and E_2 Revealed by the HPGC/LALLS Method and the Conventional Light-Scattering Method Using a Batch-Cell at 20°C[a]

Ligands in Elution Buffer	Conformational State (expected)	HPGG/LALLS Methods			Conventional Light-Scattering Method[b]	
		M_r (experiment)	M_r (simulation)	K_a (simulation)	M_r (experiment)	K_a (simulation)
0.1 M NaCl		255,000			263,000–290,000	2.5×10^6 M^{-1}
0.1 M NaCl + 9.8 µg/ml oligomycin	E_1	247,000	248,000[c]	2×10^6 M^{-1}	—	—
0.1 M NaCl + 4 mM MgCl$_2$		258,000			—	—
0.1 M KCl		300,000			320,000–327,000	3.2×10^8 M^{-1}
0.05 M KCl + 0.05 M NaCl	E_2	298,000	—	$\geqq 1 \times 10^8$ M^{-1}	—	—
0.05 M KCl + 0.05 M NaCl + 4 mM MgCl$_2$		295,000				

[a]The solubilized enzyme was subjected to gel chromatography on a TSKgel G3000SW$_{XL}$ column with an elution buffer containing the ligands indicated in the table as well as 0.2 mg/ml C$_{12}$E$_8$ at 20°C. The M_r of the main protein component eluted was measured using the HPGC/LALLS method, and the K_a for this component was obtained by computer simulation, assuming that the enzyme was in an equilibrium of 2 Protomer $\rightleftharpoons$ Diprotomer. The main protein component was isolated and then diluted with the effluent containing no protein. The M_r of the protein component at various protein concentrations was measured directly by the conventional, light-scattering method with a novel precision apparatus equipped with batch-cell and Ar-laser light source. The K_a was obtained from the dependence of M_r on the protein concentration obtained (FIG. 1) by assuming the same equilibrium as above.

[b]Data obtained at protein concentrations higher than 0.15 mg/ml were used.

[c]Calculated using the contents of D and P obtained by computer simulation.[8]

(solid line in FIG. 1) and the calculated (dotted line) curve shown at protein concentrations of less than 0.1 mg/ml might be attributable to a release of phospholipids from the enzyme, because phospholipids would make the protein components of the solubilized enzyme associate.[13]

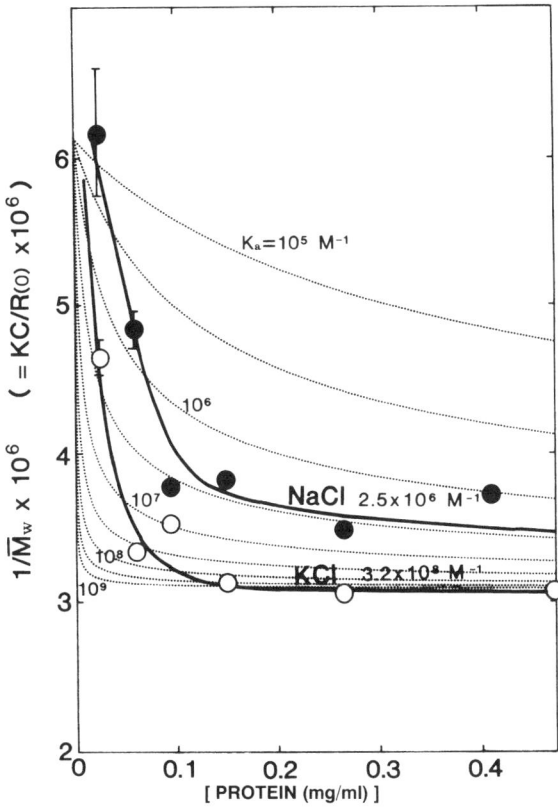

FIGURE 1. Dependence of reciprocal weight-averaged molecular weight ($\overline{M}_w$) of the solubilized enzyme on protein concentration, estimated using an elution buffer containing 0.1 M NaCl (●) or KCl (○) at 20°C. The $\overline{M}_w$s were measured as described in the text. *Dotted lines* show the dependence calculated by assuming that the enzyme is in the equilibrium of $2P \rightleftharpoons D$ with various values of association constant (K_a).

CONVERSION OF P TO D BY ATP-INDUCED PHOSPHORYLATION

The solubilized enzyme was incubated with various concentrations of ATP under conditions expected to produce a phosphoenzyme intermediate of E_2-P. The resultant enzyme was charged onto a column equilibrated with an elution buffer containing 0.1

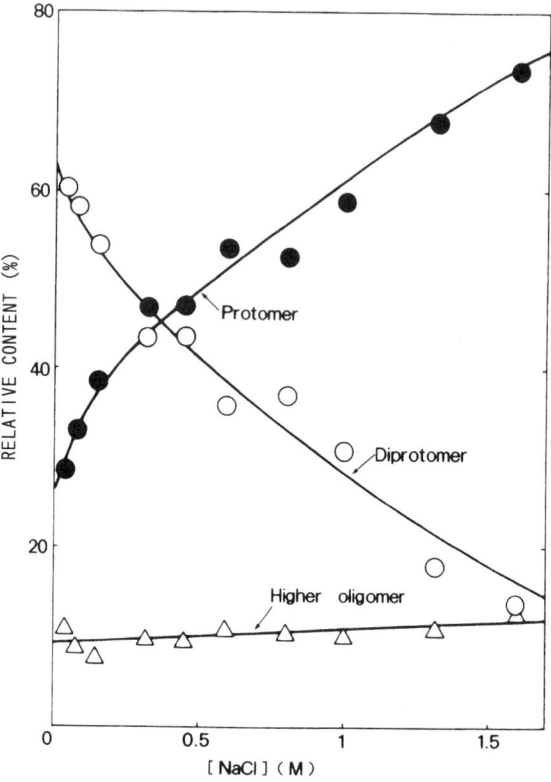

FIGURE 2. Change in the relative contents of $(\alpha\beta)_2$-diprotomer (D), $\alpha\beta$-protomer (P), and higher oligomer (H) involved in the solubilized enzyme incubated with ATP under conditions producing phosphoenzyme intermediates in the presence of the various concentrations of NaCl indicated in the horizontal axis. Solubilized enzyme was incubated with 1 mM ATP to produce a phosphoenzyme intermediate at 0°C and was subjected to gel chromatography using an elution buffer containing 0.1 M NaCl at 0°C. Contents of the protein components were estimated by calculating the area under the peaks corresponding to the respective protein components. The relative contents of D, P, and H compared to those of all the protein eluted were plotted against the concentration of NaCl used in the incubation.

M NaCl at 0°C and eluted with the same elution buffer. Chromatography allowed us to estimate the relative contents of D and P of the solubilized enzyme from the areas under the two resulting protein peaks of D and P.[12] With increasing ATP concentrations from 0 to more than 1 mM, the content of P decreased from 56 to 26%, whereas that of D increased from 37 to 64%. Thus, it was concluded that ATP induced the conversion of P into D.[9] ADP, however, did not have this effect. In another series of experiments, incubation of the solubilized enzyme with ATP was followed by incubation with a stoichiometric amount of [^{3}H]ouabain for 5 minutes at 0°C, and then the enzyme was subjected to the same chromatography as that just described.[9] The conversion of P to D occurred in the same way as in the case without [^{3}H]ouabain, as already mentioned. The amount of [^{3}H]ouabain bound to D increased with increasing

ATP concentration in parallel with the conversion of P to D. The $K_{0.5}$ values of [ATP] for an increase in D and a decrease in P were consistently 0.17 mM. The $K_{0.5}$ values of [ATP] for [^{3}H] ouabain binding to D were 0.19 mM. Thus, both $K_{0.5}$ values were consistent with each other. Therefore, it was concluded unambiguously that the phosphoenzyme intermediate of E_2-P would have been produced first, that it induced the conversion of P to D, and that ouabain bound to the resultant D.

The solubilized enzyme was incubated with 1 mM ATP in the presence of various concentrations of NaCl ranging between 43 mM and 1.6 M to produce a phosphoenzyme intermediate of E_1-P as well as E_2-P and then subjected to the same chromatography as that just described. As shown in FIGURE 2, with increasing concentration of NaCl, the content of D decreased and that of P increased. But the content of the higher oligomer (H) did not change. The ADP sensitivity of the phosphoenzyme intermediates produced under the same conditions was investigated. The intermediate produced in the presence of 83 mM NaCl was not sensitive to ADP, showing that it was E_2-P. With increasing NaCl concentration, the phosphoenzyme became more sensitive to ADP, and at 1.3 M NaCl the sensitivity reached 93%, showing that the intermediate of E_1-P had been produced. Thus, conformation of the phosphoenzyme was converted from E_2-P to E_1-P with increasing NaCl concentration from 83 mM to 1.3 M in parallel with the curve of decreasing D or increasing P versus NaCl concentration. Ouabain is known to bind to the membrane-bound enzyme in its E_2-P form.[15-17] Our data strongly suggest the following relation between conformation of the phosphoenzyme and the oligomeric structure: E_1-P had a protomeric structure, and it was dimerized to D on changing the conformation to E_2-P, and ouabain bound to the E_2-P form and kept the structure diprotomeric.

SIMULTANEOUS DETERMINATION OF QUATERNARY STRUCTURE AND CONFORMATIONAL STATE USING A SOLUBILIZED FITC-ENZYME

FITC acts as a selective label for the ATP binding site, and the E_1 and E_2 conformational states are easily distinguished by large changes in the fluorescein fluorescence emission.[18] To establish whether D or P can exhibit a fluorescence signal that allows the distinction between E_1 and E_2 conformation, the membrane-bound enzyme was first labeled with FITC by the method reported by Carilli et al.,[19] and then the FITC-labeled enzyme was solubilized with $C_{12}E_8$ in the same way as just described. As shown in the upper portion of FIGURE 3, the solubilized FITC-enzyme as well as the original membrane-bound FITC-enzyme could distinguish the E_1 from the E_2 conformation by emitting a 20–30% higher intensity of fluorescence in 0.1 M NaCl than in 0.1 M KCl. The reversibility of the conformational change was also confirmed with both types of the FITC-enzyme (the lower portion of FIG. 3). The solubilized FITC-enzyme was subjected to the LALLS system additionally equipped with a spectrofluorometric detector. FIGURE 4 shows the four kinds of elution patterns of LS, the fluorometer (FL), UV, and RI obtained with an elution buffer containing 0.1 M NaCl. The two major protein components were identified as D and P according to their M_r value. As shown in the inset of FIGURE 4, the patterns of FL and UV were superimposed on each other, thus making the heights of the peaks of D in the two patterns the same. For P then, the height of the new peak of FL was 1.22-fold higher than that of the UV peak, whereas for D the FL peak was 1.00-fold higher than the UV

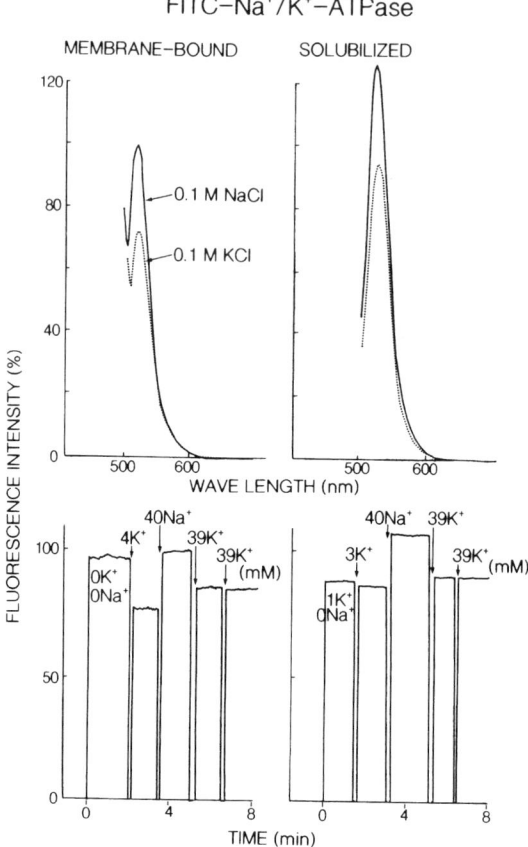

FIGURE 3. Preservation of the capacity to report the conformational change between E_1 and E_2 states after solubilization of membrane-bound FITC-enzyme. Membrane-bound enzyme was bound by FITC and then solubilized to produce the solubilized FITC-enzyme. *Upper portion:* Fluorescence emission spectra were scanned for the two kinds of FITC-enzyme in 0.1 M NaCl (—) or KCl (·····) at pH 7.0 with a fixed excitation wavelength of 485 nm. *Lower portion:* Fluorescence intensity at a fixed emission wavelength of 520 nm was estimated when NaCl or KCl was added to the two kinds of FITC-enzyme. Numerical values within the bars represent the concentration (mM) of cation involved in the initial protein solution and those above the bars represent the concentration of cation added to the solution.

peak. In other words, when the ratio of fluorescence intensity to absorbance at 280 nm, (output)$_{FL}$/(output)$_{UV}$, for D was normalized to 1.00, that for P was 1.22. The ratio thus normalized is equivalent to the difference in fluorescence intensity per unit of protein concentration between D and P. Ratios were 1.00 and 1.31 for D and P, respectively, with the elution buffer containing 0.1 M KCl (FIG. 5, upper portion). Almost the same values were obtained with NaCl (lower portion). If each D and P can take the conformation of E_2 with KCl and that of E_1 with NaCl, the fluorescence intensity of the two protein components with NaCl should become 20–30% higher than that with

KCl. This was the case neither for D nor for P. A difference of about 30%, however, was found between D and P regardless of whether NaCl or KCl was included in the elution buffer (FIG. 5). This implies that D itself is E_2 and, in reverse, that P itself is E_1. The contents of D and P changed according to which monovalent cation of KCl and NaCl was included in the elution buffer (compare the upper UV pattern with the lower one in FIG. 5). We presented a paper on this subject elsewhere in this volume.[12] Therefore, it was concluded that KCl and NaCl increased the content of D and P, respectively, and that this alteration was followed by a change in fluorescence intensity that was proportional to the content in the solubilized enzyme.

CONCLUSIONS

Solubilized Na^+/K^+-ATPase was in an association-dissociation equilibrium of $2P \rightleftharpoons D$ at 20°C. The K_a under the ionic environment favorable for the E_2 state was about 100 times higher than that for the E_1 state. NaCl and KCl shifted the equilibrium to P and D, respectively. ATP-induced formation of E_2-P dimerized P into D and then it promoted ouabain binding to each P of the D produced. With the solubilized FITC-enzyme, P emitted 30% higher fluorescence intensity than did D, in 0.1 M KCl

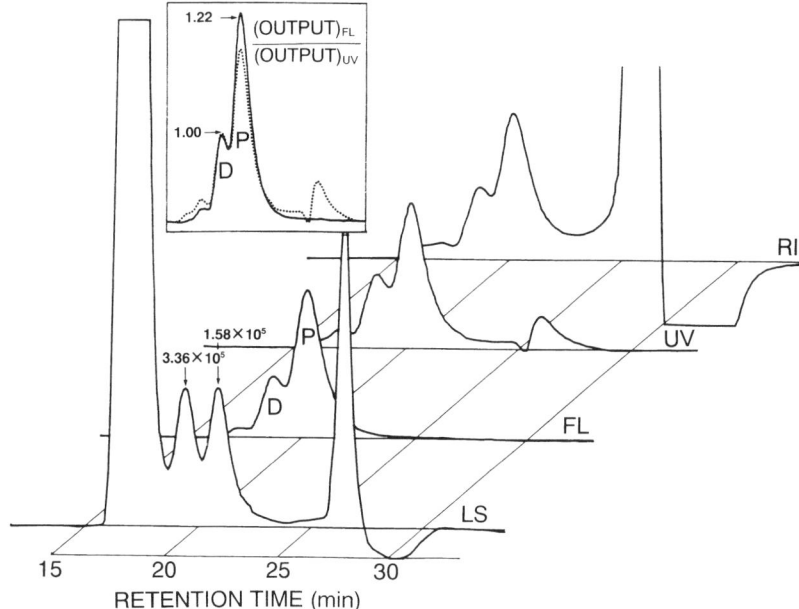

FIGURE 4. Elution patterns of solubilized FITC-enzyme obtained with an elution buffer containing 0.1 M NaCl using the HPGC/LALLS method. A spectrofluorometric monitor was added to the usual system of the HPGC/LALLS method to estimate the conformational state simultaneously with the molecular weight of the protein component separated. **Inset** shows the elution pattern of FL (—) and UV (·····) which are superimposed on each other, making the heights of the peaks of D in the two patterns the same.

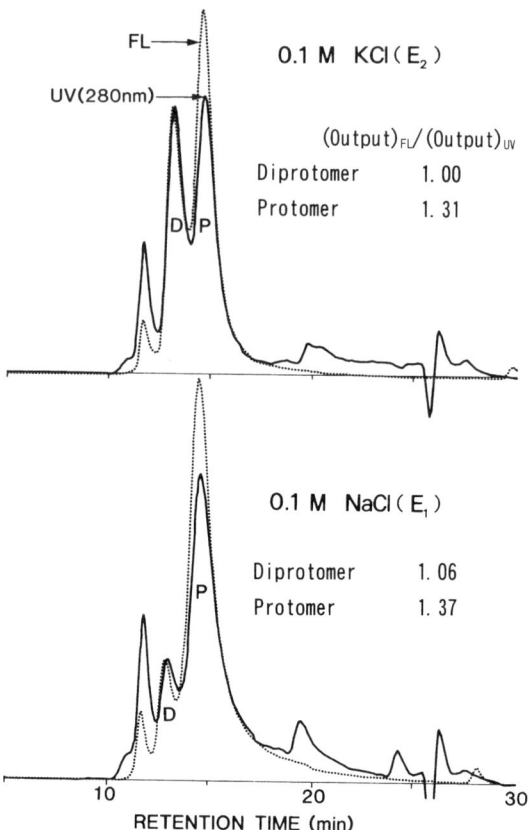

FIGURE 5. Ratios of fluorescence intensity, (output)$_{FL}$, to absorbance at 280 nm, (output)$_{UV}$, for D and P in an elution buffer containing 0.1 M KCl or NaCl at pH 7.0. *Solid* and *dotted lines* were obtained as described in the legend for the inset of FIGURE 4. The ratio of (output)$_{FL}$/(output)$_{UV}$ for D in the elution buffer containing 0.1 M KCl was normalized to 1.00; the other ratios are represented as relative to it.

as well as in 0.1 M NaCl, indicating that P and D structures correspond to E_1 and E_2 conformational states, respectively. Therefore, the interconversion of oligomeric structure between D and P is essential for enzymatic function. The molecular structure of Na$^+$/K$^+$-ATPase in the membrane is consequently thought to be as follows: the E_1 and E_2 conformations correspond to a loosely associated diprotomer and a tightly associated diprotomer, respectively; two-dimensional crystals of the membrane-bound enzyme grown at pH 4.8 in sodium citrate buffer[20] can be expected to reveal a three-dimensional structure of the enzyme in E_2-conformation for reasons described elsewhere.[12]

REFERENCES

1. HASTINGS, D. F. & J. A. REYNOLDS. 1979. Biochemistry **18:** 817–821.

2. ESMANN, M., C. CHRISTIANSEN, K. A. KARLSSON, G. C. HANSSON & J. C. SKOU. 1980. Biochim. Biophys. Acta **603**: 1–12.
3. CRAIG, W. S. 1982. Biochemistry **21**: 2667–2674.
4. CRAIG, W. S. 1982. Biochemistry **21**: 5707–5717.
5. BROTHERUS, J. R., L. JACOBSEN & P. L. JØRGENSEN. 1983. Biochim. Biophys. Acta **731**: 290–303.
6. HAYASHI, Y., T. TAKAGI, S. MAEZAWA & H. MATSUI. 1983. Biochim. Biophys. Acta **748**: 153–167.
7. HAYASHI, Y., H. MATSUI & T. TAKAGI. 1989. Methods Enzymol. **172**: 514–528.
8. HAYASHI, Y., K. MIMURA, H. MATSUI & T. TAKAGI. 1989. Biochim. Biophys. Acta **983**: 217–229.
9. HAYASHI, Y., T. KOBAYASHI, T. NAKAJIMA & H. MATSUI. 1994. In The Sodium Pump, Structure Mechanism, Hormonal Control and its Role in Disease. E. Bamberg & W. Schoner, Eds.: 453–456. Steinkopff. Darmstadt.
10. JØRGENSEN, P. L. 1988. Methods Enzymol. **156**: 29–43.
11. TAKAGI, T., S. MAEZAWA & Y. HAYASHI. 1987. J. Biochem. **101**: 805–811.
12. KOBAYASHI, T., E. HAGIWARA, N. SHINJI & Y. HAYASHI. 1997. Ann. N.Y. Acad. Sci., this volume.
13. MIMURA, K., H. MATSUI, T. TAKAGI & Y. HAYASHI. 1993. Biochim. Biophys. Acta **1145**: 63–74.
14. STEVENS, F. J. & M. SCHIFFER. 1981. Biochem. J. **195**: 213–219.
15. CHARNOCK, J. S. & R. L. POST. 1963. Nature **199**: 910–911.
16. MATSUI, H. & A. SCHWARTZ. 1968. Biochim. Biophys. Acta **151**: 655–663.
17. ASAMI, M., T. SEKIHARA, T. HANAOKA, T. GOYA, H. MATSUI & Y. HAYASHI. 1995. Biochim. Biophys. Acta **1240**: 55–64.
18. KARLISH, S. J. D. 1980. J. Bioenerg. Biomembr. **12**: 111–136.
19. CARILLI, C. T., R. A. FARLEY, D. M. PERLMAN & L. C. CANTLEY. 1982. J. Biol. Chem. **257**: 5601–5606.
20. TAHARA, Y., S. OHNISHI, Y. FUJIYOSHI, Y. KIMURA & Y. HAYASHI. 1993. FEBS Lett. **320**: 17–22.

Organization of the Membrane Domain of the Na/K-Pump[a]

STEVEN J. D. KARLISH[b]

Biochemistry Department
Weizmann Institute of Science
Rehovoth, 76100 Israel

This article summarizes aspects of our recent work on transmembrane topology of the α subunit of the Na/K-ATPase, the cation occlusion domain, and K(Rb)- or Na-induced interactions between transmembrane segments, development of affinity labels for cation sites, and initial attempts to identify interacting transmembrane segments to produce a model of helix packing.

TRANSMEMBRANE TOPOLOGY

The necessity for experimental determination of topology of α subunits of P-type pumps arose because of ambiguities in hydropathy analysis. Four NH_2-terminal segments are clear and hydrophobic and have also been demonstrated clearly experimentally, as have the cytoplasmic location of NH_2- and COOH-terminals.[1] However, several candidate segments in the COOH-terminal region are not particularly hydrophobic. Therefore, the real challenge has been to determine the number of segments in this domain.

We studied topology by proteolytic cleavages of the α subunit in well-defined vesicle systems including renal microsomes or reconstituted proteoliposomes. The work provided evidence for six segments in the COOH-terminal domain, or 10 segments altogether, and factors stabilizing transmembrane segments.[2,3]

FIGURE 1 presents models of topological organization based on this work. Consider first cytoplasmic cleavages (FIG. 1B). Initially we demonstrated the cytoplasmic location of Asn831.[2] This indicated the existence of a pair M5 and M6 rather than one segment as inferred originally from hydropathy. Several antibody-binding studies confirm the result,[4,5] but paradoxically one monoclonal antibody against the M6/M7 loop binds at the outside.[6] The extracellular loop between M5 and M6 has not been demonstrated for Na/K-ATPase, but it has been demonstrated for H/K-ATPase by omeprazole labeling.[7] Recently, we identified a fragment containing the M7/M8 pair, cleaved at cytoplasmic NH_2- and COOH-terminals.[3] The extracellular location of the M7/M8 loop was shown also by antibody binding,[4,6] interaction with the β subunit,[8] and proteolysis.[3] Cytoplasmic cleavage also revealed a transient fragment containing the M9/M10 pair.[3] This finding and that on the fragment containing M7/M8 indicate that the loop between M8 and M9 is cytoplasmic, in agreement with pyridoxal phosphate labeling of Lys943[9] and phosphorylation of Ser936[10] by protein kinase A in

---

[a]This work was supported by grants from the U.S.-Israel Binational Science Foundation and the Minerva Foundation, München, Germany.

[b]Tel: 972 8 934 2278; fax: 972 934 418; e-mail: BCKARLIS@WEIZMANN.WEIZMANN.AC.IL

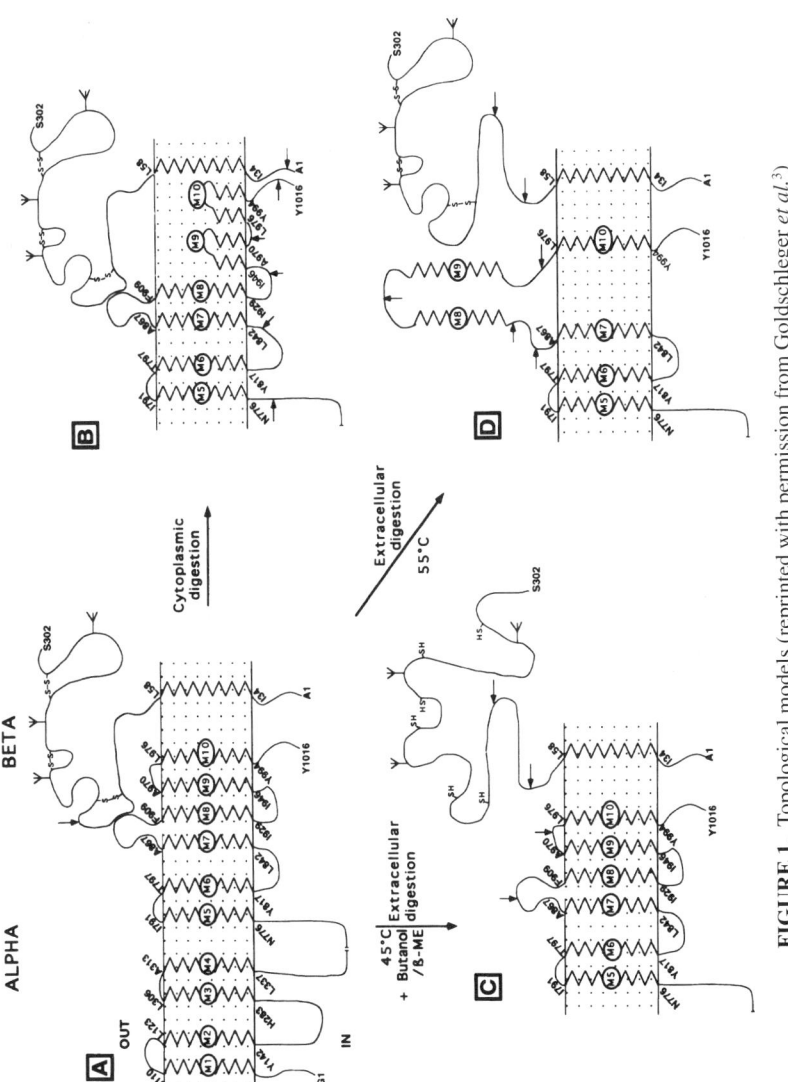

FIGURE 1. Topological models (reprinted with permission from Goldschleger *et al.*[3])

whole cells. An unexpected cleavage was observed between M9/M10 at the cytoplasmic surface.[2] A similar anomaly was reported by Sachs *et al.* in digestion of H/K[7]- and Ca-ATPases.[11] In principle, this could indicate that M9 and M10 are not true transmembrane segments, as depicted in FIGURE 1B, or are intrinsically mobile. In fact, we believe that M9 and M10 are true transmembrane segments, but they become destabilized by initial cleavages at the cytoplasmic surface, exposing the M9 and M10 connecting loop to trypsin. This interpretation is favored by the kinetics of cleavage at the cytoplasmic surface and by proteolytic cleavages at the extracellular surface, which show that the M9 to M10 connecting loop can also be cut at that surface. Although normally the extracellular surface is inaccessible to proteases, digestion can be induced after prior heating of renal microsomes (FIG. 1C and D).[3] In relatively mild conditions (FIG. 1C), we observed the splits between M7/M8 and M9/M10 expected for the 10-segment model. After heating at 55°C (FIG. 1D), a change in topology occurred, causing exposure of the M8 and M9 loop to the exterior, where it also was cut. A similar effect was reported by Arystarkhova *et al.*[12] Rb ions completely protected against thermal destabilization of the COOH-terminal domain, demonstrating the existence of Rb-induced stabilizing interactions between the α and β subunits, far from the Rb occlusion site. *In vitro* expression experiments[13] also support the idea that M9 and M10 are transmembrane segments.

Apart from the information on the number of segments, an important conclusion from this work is that protein-protein interactions outside the membrane, both between α and β subunits and within α subunits, stabilize topological organization of the M7/M10 domain. Similar conclusions appear to hold for the M5/M6 hairpin, as shown by Lutsenko *et al.*[14] for Na/K-ATPase and Shin and Sachs[15] for H/K-ATPase. For proteins such as Na/K and H/K-ATPase with an appreciable fraction of the mass located outside the membrane, hydrophobicity is not necessarily the major factor determining stability of transmembrane segments. By contrast, bacterial membrane proteins, such as bacteriorhodopsin, with little of the protein outside the membrane, are stabilized primarily by interactions between side chains of transmembrane segments.[16] Stabilization by extramembrane protein-protein interactions could explain why nonhydrophobic sequences in the COOH-terminal domain can be transmembrane segments in the native P-type pumps and are expelled from the membrane domain in denatured protein. Segments dependent on such protein-protein interactions could be conformationally sensitive and functionally important, whereas highly hydrophobic segments such as M1-4 are likely to be static structures.

CATION OCCLUSION SITES

An essential tool for the remainder of the work described in this article is 19 kD membranes[17,18] (FIG. 2). 19-kD membranes are produced by extensive tryptic digestion of renal Na/K-ATPase in the presence of Rb ions and the absence of Ca ions, and they consist of 19 kD and smaller fragments (8–11.7 kD) of the α subunit, containing M7-M10 and the pairs M5/M6, M3/M4, and M1/M2. The preparation includes transmembrane segments and extracellular loops and relatively short cytoplasmic tails. The latter may correspond to the stalk region described for Ca-ATPase.[19] The major cytoplasmic loops are removed. The β subunit is partially split into a 16-kD

NH$_2$-terminal, and a glycosylated ≈50 kDa, fragment. In 19-kD membranes, K(Rb) and Na occlusion and ouabain binding are intact, whereas ATP binding is absent.[17,20]

One major conclusion is that cation sites are located within transmembrane segments and communicate with ATP sites by conformational interactions via the stalk. What else can be inferred on cation sites?

Current evidence based on mutations[21-27] and biochemical studies[14] implies that M4, M5, and M6 play a central role in cation occlusion and transport. Mutations of charged or oxygen-rich residues in M4, (Glu327),[21-23] M5 (Ser, 775, Glu 779),[21,24-26] and especially M6 (Asp804, Asp807)[21,27] suggest that these are important for interactions with the cations; however, whether they directly ligate the cations is uncertain. For example, Glu779, which is chemically modified by the carboxyl reagent[28]

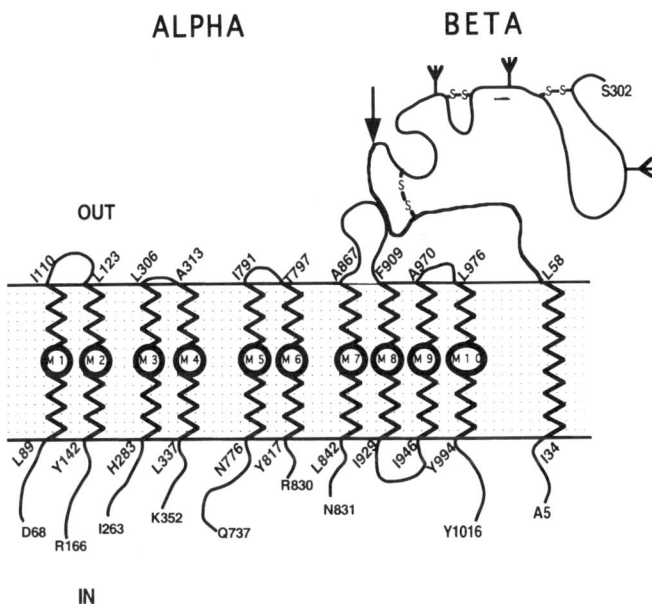

FIGURE 2. Organization of tryptic fragments in 19-kD membranes.

4-(diazomethyl)-7-(diethylamino)-coumarin (DEAC) may not be directly involved.[25] We reported that the carboxyl reagent N,N'-dicyclohexylcarbodiimide (DCCD) modifies Glu953 in M9 in an Rb-sensitive fashion, and another residue, probably Glu327, in M4.[29] However, DCCD may have inhibited occlusion by a conformational effect, for mutations to Glu953 and Glu954 do not produce striking functional effects.[30] Nevertheless, M9 must be a mobile or conformationally sensitive segment, because chemical modification of Glu953 by DCCD is prevented by Rb ions.[29] The fluorescent probe of conformation N-[p-(2-benzimidazolyl)phenyl]maleimide (BIPM), which labels Cys964, also demonstrates mobility of M9.[31,32] Assuming that M4, M5, and M6 are crucial elements, the question arises as to the role of the COOH-terminal domain. One proposal is that the cation binding and transport path lies entirely within

M4, M5, and M6 and the role of the COOH-terminal is structural, to stabilize the M5/6 hairpin.[1,33] The basis for this proposal is (1) mutations of Ca-[34,35] and Na/K-ATPase, and (2) the fact that heavy metal pumps (for Cu and Cd) contain six transmembrane segments analogous to M1-M6 of the Na/K-, H/K-, and Ca-pumps (with two extra segments at the NH$_2$-terminal domain). By contrast, evidence to be discussed below suggests that probably the COOH-terminal of Na/K-ATPase is also involved in cation binding. Important differences between heavy metal pumps and pumps that exchange cations (3Na/2K, 2H/2K, and 2Ca/2H) include (1) the requirement for cation specificity in catalysis, for example, Na-dependent phosphorylation versus K-dependent dephosphorylation, and (2) the fact that the latter transport more than one cation per cycle. One possibility is that the COOH-terminal plays a role in Na/K specificity. This

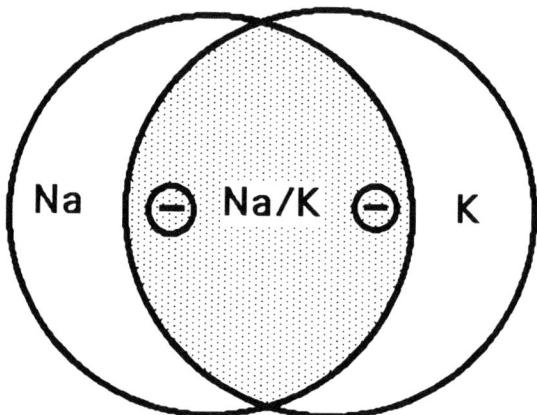

FIGURE 3. Overlapping sites for Na and K ions?

notion is illustrated by a model with overlapping cation sites (FIG. 3). Two Na and K ions could bind to the same charged residues in a common domain, say M4, M5, or M6, while additional oxygen-rich ligating groups for K ions could be provided by more COOH-terminal segments (e.g., serine or threonine hydroxyls, backbone carbonyls). The third Na ion would be bound at an extra site containing only noncharged ligating groups.

The concept of K/Na selectivity proposed in FIGURE 3 is analogous to that established for an enzyme of known structure, dialkylglycine decarboxylase, with two monovalent cation binding sites.[36] In one site, a carboxylate, three carbonyl oxygens, and a serine hydroxyl, from different locations in the primary sequence, ligate the K ion (metal-ligand distance $\approx$2.73 Å), but the serine moves out of the site when the smaller Na ion is bound (metal-ligand distance $\approx$2.33 Å). The minor initial movement is translated at a distance to the active site and accounts for the specificity of K as an activator and Na as an inhibitor of enzyme activity. A second Na site, without a functional role, is formed by a tight turn of the polypeptide chain and consists of neutral oxygen ligands, but no carboxylate.

AFFINITY LABELS FOR CATION SITES

We are developing affinity labels of cation sites as a complementary approach to mutational analysis. A high affinity cation analogue, as a precursor of a reactive derivative, is a necessary requirement, because a low affinity reagent would probably label the protein unselectively. We recently described synthesis and properties of a new family of aromatic, positively charged isothiouronium compounds that act as reversible competitive Na-like cation antagonists.[37,38] Compounds such as Br_2-TITU, shown in FIGURE 4, act similarly to a previous series of aromatic bis-guanidinium and amiloride derivatives[39,40] in competing for occlusion with Rb or Na ions, stabilizing the E_1 conformation, and blocking the ATPase activity, at cytoplasmic sites. However, Br_2-TITU (and a Br-TITU derivative) has a much higher affinity, $K_D \approx 0.2$ μM. At high concentrations ($K_D \approx 10$ μM), the brominated TITU derivatives can also stabilize the E_2 form, suggesting the possibility of simultaneous binding at both high and low affinity sites.

Analysis of their mode of interaction with the protein shows that competitive Na antagonists do not protect the 19-kD fragment against proteases, or 19-kD membranes against thermal inactivation of Rb occlusion, unlike occluded cations K(Rb) or Na, but like the occluded cations, they do protect against inactivation of occlusion by DCCD.[41] Furthermore, analysis of effects of diffusion potentials in reconstituted proteoliposomes shows that inhibition by Na antagonists or by K(Rb ions) at the cytoplasmic surface is unaffected by voltage, whereas Na binding is affected by voltage.[42] These findings support a model of cation sites with two negatively charged sites to which two K or two Na ions bind and a third neutral site for a Na ion[43,44] and suggest that the neutral site lies in a "cation well" so that binding of the one Na ion is affected by voltage. The mechanism of action of Na antagonists appears to involve competition at negatively charged cytoplasmic sites with either two Na or two K ions, but not at the third, neutral, Na site. Occlusion occurs in two steps, an initial recognition step followed by a conformational change to the occluded state. The Na antagonists compete with K and Na in the first step, but they are not occluded because of steric hindrance, and so they block active transport and ATPase activity.

Photoactivated derivatives are now being synthesized. If these are successful, we can expect to label the entrance to occlusion sites. This could also provide information on the proximity of different transmembrane segments.

INTERACTIONS NEAR THE MEMBRANE INDUCED BY OCCLUDED CATIONS AND OUABAIN

As described originally,[17,18] stabilization of 19-kD and smaller fragments against proteases occurs in the presence, but not in the absence, of an occluded cation. In the pig kidney enzyme K, Rb and Na ions were all effective, although Na was slightly less so. Interestingly, in shark rectal gland[45] and dog kidney,[46] Na is ineffective at protecting the 19-kD fragment, and occlusion is not maintained. The latter result points to K selectivity in the COOH-terminal in these cases. In all cases, maintenance of Rb occlusion is correlated with intactness of the 19-kD fragment. The absence of occluded cations, or displacement of occluded Rb by Ca, leads to loss of occlusion. The mechanism involves first thermal inactivation of occlusion and then digestion of

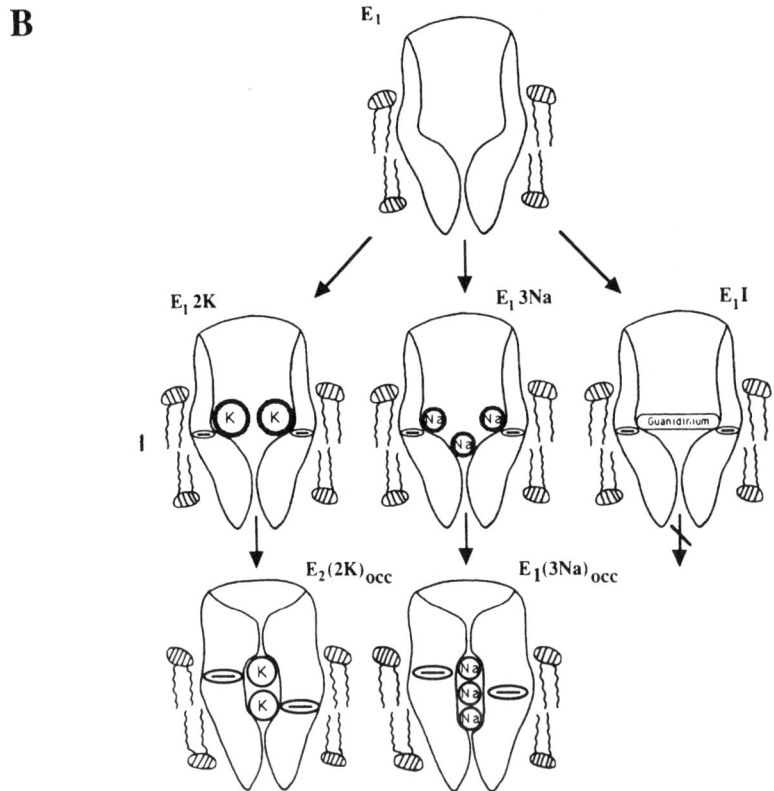

FIGURE 4. Competitive Na-like antagonists. (**A**) 1,3-dibromo-2,4,6-tris(methyleneisothiouronium)benzene tribromide (Br_2-TITU). (**B**) Mechanism of inhibition of cation occlusion and transport.

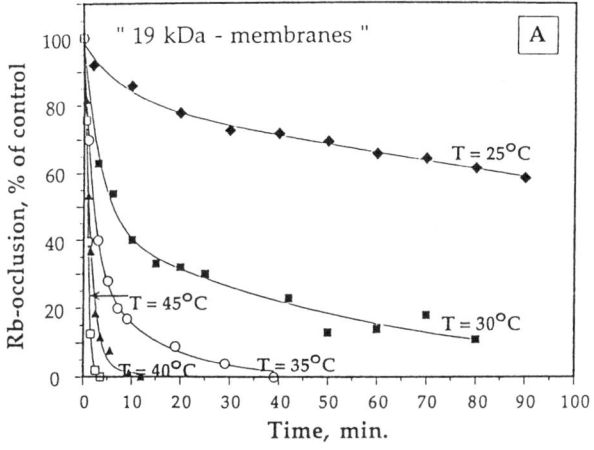

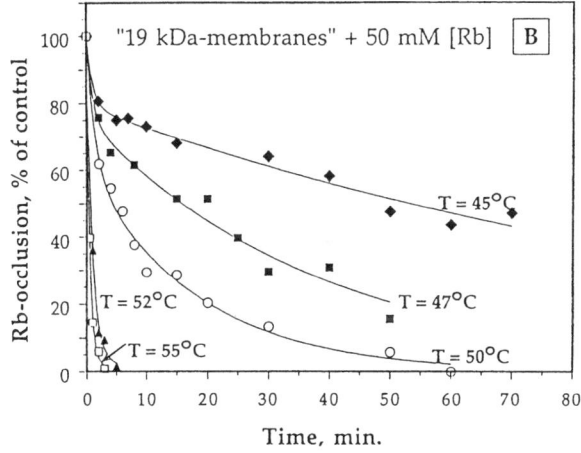

FIGURE 5. Thermal inactivation of Rb occlusion in 19-kD membranes in the absence (**A**) or presence (**B**) of Rb ions.

the 19 kD and other fragments.[47] We now studied thermal inactivation in detail and the effects of occluded cations (Shainskaya, Nesatyi, and Karlish, manuscript in preparation). FIGURE 5 demonstrates the kinetics in the absence (A) or presence (B) of Rb ions. Rb ions protect strongly, as seen from the difference in temperature, and other occluded cations, including Na, are also effective in our system. The kinetics are biphasic and were analyzed by the Lumry-Eyring[48] two-step model:

$$N \underset{k2}{\overset{k1}{\rightleftarrows}} U \overset{k3}{\rightarrow} I$$

N, U, and I represent Native, Unfolded, and Irreversibly denatured forms. Only N is able to occlude Rb ions. TABLE 1 presents activation parameters, based on the fitted values of rate constants k1, k2, and k3. There are two interesting features. First, the $\Delta H\#$ and $\Delta S\#$ of k1 and k3 have positive values, compatible with disorganization and disruption of multiple weak interactions. By contrast, the values for k2 are negative, suggestive of a reorganization process. Second, Rb ions increase all parameters by large factors, consistent with induction of a less flexible structure and multiple weak interactions. Thus, Rb ions appear to induce global stabilizing interactions.

We have identified the limit peptide products of tryptic digestion after thermal inactivation[47] (FIG. 6). The difference between intact and digested 19-kD membranes indicates which peptide segments became exposed or, conversely, which are protected by occluded Rb ions. As can be seen, all fragments of the α or β subunit were truncated. One implication is that Rb ions bind to several transmembrane segments and stabilize a complex of interacting fragments. Another is that interacting tails or loops outside the membrane stabilize the complex or form barriers to dissociation of occluded ions, because residues within transmembrane segments thought to ligate cations (e.g., Glu327, Asp804, and Asp807) are still present in the shaved preparation. Note, specifically, that a long tail was removed from the M5/6 hairpin and that the 19-kD fragment was digested to single transmembrane segments. Thus, Rb ions protect all the loops and tails of the 19-kD fragment. This extensive protection as well as the specificity for K(Rb) in some systems, is one reason for proposing that Rb ions bind within the M7/M10 domain as well as to the M5/6 pair.

Lutsenko and Kaplan[46] reported that ouabain protects the 19-kD fragment against trypsin. We have found that ouabain, like Rb ions, also protects against thermal inactivation, and the combined effects of ouabain and Rb ions are largely complementary (Shainskaya, Nesatyi, and Karlish, manuscript in preparation). In addition, we have investigated the quantitative relation between thermal inactivation of Rb occlusion and selective release from the membrane of the M5/6 hairpin described by Lutsenko et al.,[14] using an antibody raised against a peptide corresponding to the sequence Leu815-Gln828.[4] In the pig kidney enzyme, about half the total amount of M5/M6 is solubilized and it follows thermal inactivation (Shainskaya, Nesatyi, and Karlish, manuscript in preparation). Thus, it can be concluded that release of the M5/M6 fragment is the consequence of thermal inactivation and is not the direct cause of inactivation of Rb occlusion, a distinction that could not be made previously.[14] A common property of occluded cations and ouabain, according to recent mutagenesis work, is that they both interact with the M5/6 hairpin.[49] A reasonable hypothesis could

TABLE 1. Activation Parameters for Thermal Inactivation of Rb Occlusion at 25°C [a]

	Rate Constant	$\Delta G\#$ kCal.mol^{-1}	$\Delta H\#$ kCal.mol^{-1}	$\Delta S\#$ Cal.K^{-1}.mol^{-1}
−Rb ions	k1	21.9	30.7	29.43
	k2	21.1	−0.61	−72.91
	k3	22.2	28.7	21.81
+Rb, 50 mM	k1	23.8	46.8	77.0
	k2	16.4	−55.3	−240.6
	k3	24.5	44.9	68.2

[a]Values (−Rb) were derived directly from experimental data. Values (+Rb) were extrapolated from data obtained at 45–55°C.

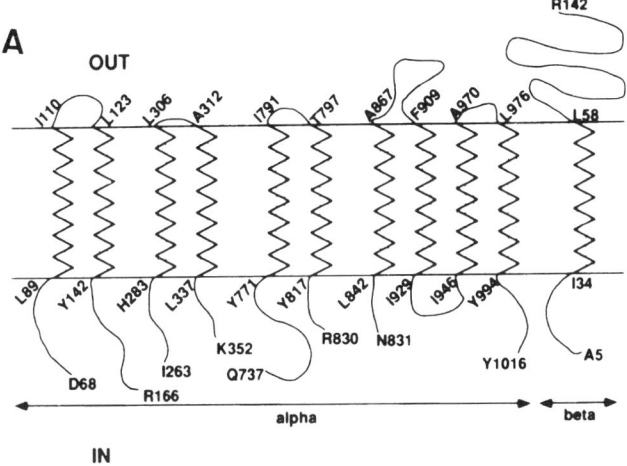

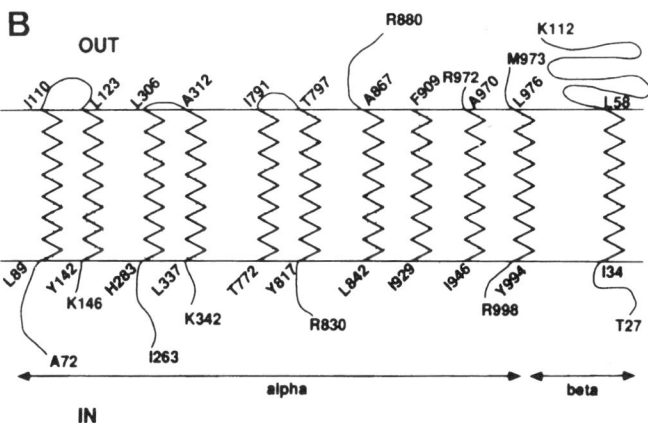

FIGURE 6. (**A**) Intact 19-kD membranes. (**B**) Limit tryptic peptides (reprinted from ref. 47).

be that thermal inactivation involves a disruption of links between the M5/6 hairpin and the M7/10 domain, and then the M5/6 pair is released. Direct evidence for such an interaction has been obtained (see below).

IDENTIFICATION OF INTERACTING FRAGMENTS

By identifying interacting transmembrane segments or polypeptide segments near the membrane surface, we hope eventually to construct a model for organization of transmembrane helices in the plane of the membrane. For model building, evidence on α/β subunit interactions is particularly useful, because the β subunit has only a single transmembrane segment.[50,51]

Chymotryptic Cleavage of the Cytoplasmic Domain of the β Subunit

An indirect approach to identifying interacting segments supposes that truncation in one domain of the protein exposes interacting domains but not others. Structural and functional interactions between α and β subunits have been studied intensively, particularly at the extracellular surface.[52] However, additional interactions are likely

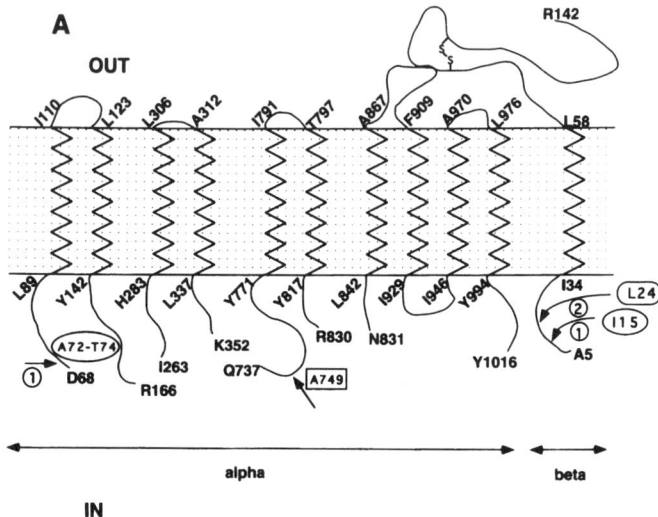

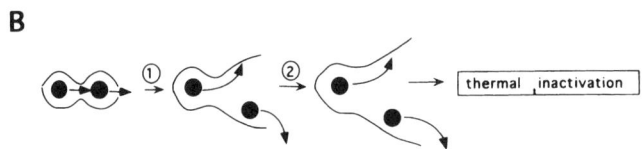

FIGURE 7. Model depicting chymotryptic splits in fragments of 19-kD membranes and accompanying effects on Rb deocclusion (reprinted from ref. 53).

to exist. In this regard, we recently described specific chymotryptic cleavage of the cytoplasmic tail of the β subunit in 19-kD membranes associated with destruction of Rb occlusion.[53] Cleavage occurs in several stages (FIG. 7A). Initially, chymotrypsin removes 10 residues from the β subunit. At 20°C, cleavage stops at this point and a stable intermediate can be isolated. Rb occlusion is still intact, but the affinity for Rb ions is much reduced at 37°C. This is detected as a fast, disordered dissociation of 2 Rb ions, compared to the characteristic ordered dissociation in 19-kD membranes[17] (FIG. 7B). In the second stage an additional 10 residues are removed, resulting in a further decrease in Rb affinity, thermal inactivation, and finally cleavage of all fragments. The observations imply that the cytoplasmic domain of β affects access of Rb ions to and from their sites, presumably via interactions with the α subunit. Our findings suggest that the NH_2-terminal of the α subunit could interact with A5-F14 in the NH_2-terminal domain of the β subunit. First, the initial cleavage of β is associated with cleavage of four or six residues off the M1/2 fragment (FIG. 7A). Second, the first cleavage is associated with a large reduction in ouabain binding affinity. Third, the first cleavage does not affect exposure of the COOH-terminal domain detected in various ways (experiments not shown). Geering et al.[54] have also reported findings that indicate a role of the cytoplasmic domain of the β subunit.

Solubilization of 19-kD Membranes and Cross-Linking of Fragments

A direct approach could be to covalently cross-link fragments in 19-kD membranes. The possibility of artefactual intermolecular cross-links in the membrane has led us to devise a procedure for detergent solubilization of 19-kD membranes containing occluded Rb ions.[55] In a soluble complex, artefactual intermolecular cross-linking should be minimal because the protein is greatly diluted.

Initial attempts to solubilize occluded Rb with $C_{12}E_{10}$ were only partially successful. Subsequently it was found that inclusion of ouabain together with the Rb allowed essentially full solubilization of the occluded Rb and bound ouabain at 0°C and an optimal detergent:protein ratio. Ouabain reduces the rate of dissociation of occluded Rb, thus maintaining Rb-dependent stabilizing interactions that are weakened on solubilization, and ouabain may also stabilize by itself.

Comparison of the protein components of the soluble supernatant and unsolubilized precipitate, at increasing detergent concentrations, showed cosolubilization of all fragments of the α and β subunits of the pump, and also the γ subunit, whereas contaminant proteins were not solubilized. Cosolubilization of the pump fragments suggested that they form a complex. This idea was tested by centrifuging on a sucrose gradient. A single peak of Rb occlusion was found to overlap a protein peak, and the peak fractions indeed contain all the fragments derived from the pump. Thus, the experiment demonstrated directly the existence of the complex of fragments, containing occluded Rb ions and bound ouabain. Based on HPLC separations and specific activity, the complex appears to contain one copy of each fragment.

Interestingly, when this experiment was repeated in the absence of Rb and ouabain, the complex dissociated. The 19-kD and beta fragments remained associated, but fragments corresponding to M5/M6 and M3/M4 dissociated and sedimented together in lighter fractions. M1/M2 occupied an intermediate position. The experiment demonstrates directly a strong 19-kD/β interaction even in the absence of Rb and ouabain. (See also refs. 15 and 52.) Evidently, Rb and ouabain preserve the intact complex, by maintaining interactions between the M5/6 and M3/4 fragments and the 19-kD/β pair. The simplest explanation is that Rb ions and ouabain interact with both M3/M4 and M5/6 and also with the M7/M10 fragment and so keep them together.

Using the soluble complex, a variety of short-length bifunctional agents were screened (Or and Karlish, unpublished data). One agent is Cu/phenanthroline which catalyzes disulfide bridge formation. A specific cross-link between the β subunit, 19-kD fragment, and M1/M2 fragment was observed. This finding is similar to that reported by Sarvazyan et al.[56] The interest is that it shows that the protein is folded so that NH_2- and COOH-termini of the α subunit make contact.

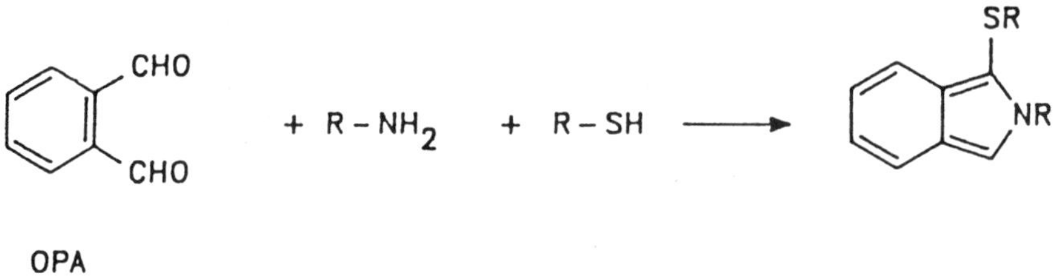

FIGURE 8. A mechanism of cross-linking by o-phthallic dialdehyde.

Another cross-linker used was ortho-phthallic dialdehyde (OPA) which can cross-link proximal cysteine and lysine residues (FIG. 8). This compound induced formation of two cross-linked products of 31 and 25 kD. The products were screened with different antibodies. The 31-kD product contained the 19-kD and the 16-kD fragments of the β subunit, consistent with the solubilization result and with prior information in the literature that α and β subunits interact within a segment of 26 residues in the extracellular loop between M7 and M8.[57] The 25-kD fragment contained the 19-kD and M5/M6 fragments, demonstrating directly what had been surmized on the basis of the thermal inactivation data. An intensive effort is now being invested to establish the positions of all these cross-links.

REFERENCES

1. MOELLER, J. V., B. JUUL & M. LE MAIRE. 1996. Biochim. Biophys. Acta **1286:** 1–51.
2. KARLISH, S. J. D., R. GOLDSHLEGER & P. L. JØRGENSEN. 1993. J. Biol. Chem. **268:** 3471–3478.
3. GOLDSHLEGER, R., D. M. TAL & S. J. D. KARLISH. 1995. Biochemistry **34:** 8668–8679.
4. NING, G., A. B. MAUNSBACH, Y. -J. LEE & J. V. MØLLER. 1993. FEBS Lett. **336:** 521–524.
5. YOON, K. L. & G. GUIDOTTI. 1994. J. Biol. Chem. **269:** 28249–28258.
6. MOHRAZ, M., E. ARYSTARKHOVA & K. J. SWEADNER. 1994. J. Biol. Chem. **269:** 2929–2936.
7. BESANÇON, M., J. M. SHIN, F. MERCIER, K. MUNSON, M. MILLER, S. HERSEY & G. SACHS. 1993. Biochemistry **32:** 2345–2355.
8. FAMBROUGH, D. M., M. V. LEMAS, M. KAMRICK, M. EMERICK, K. J. RENAUD, E. M. INMAN, B. HWANG & K. TAKEYASU. 1994. Am. J. Physiol. **266:** C579–C589.
9. ANDERBERG, S. 1995. Biochemistry **34:** 9508–9516.
10. BEGUIN, P., A. T. BEGGAH, A. V. CHIBALIN, P. BURGENER-KAIRUZ, F. JAISSER, P. M. MATHEWS, B. C. ROSSIER, S. COTECCHIA & K. GEERING. 1994. J. Biol. Chem. **269:** 24437–24445.
11. SHIN, J. M., M. KAJIMURA, J. M. ARGUELLO, J. H. KAPLAN & G. SACHS. 1994. J. Biol. Chem. **269:** 22533–22537.
12. ARYSTARKHOVA, E., D. L. GIBBONS & K. J. SWEADNER. 1995. J. Biol. Chem. **270:** 8785–8796.
13. BAMBERG, K. & G. SACHS. 1994. J. Biol. Chem. **269:** 16909–16919.
14. LUTSENKO, S., R. ANDERKO & J. H. KAPLAN. 1995. Proc. Natl. Acad. Sci. USA **92:** 7936–7940.
15. SHIN, J. M. & G. SACHS. 1994. J. Biol. Chem. **269:** 8642–8646.
16. LEMMON, M. A. & D. M. ENGELMAN. 1994. Q. Rev. Biophys. **27:** 157–218.
17. KARLISH, S. J. D., R. GOLDSHLEGER & W. D. STEIN. 1990. Proc. Natl. Acad. Sci. USA **87:** 4566–4570.

18. CAPASSO, J. M., S. HOVING, D. M. TAL, R. GOLDSHLEGER & S. J. D. KARLISH. 1992. J. Biol. Chem. **267:** 1150–1158.
19. TOYOSHIMA, C., H. SASABE & D. L. STOKES. 1993. Nature **362:** 469–471.
20. SCHWAPPACH, B., W. STÜRMER, H.-J. APELL & S. J. D. KARLISH. 1994. J. Biol. Chem. **269:** 21620–21626.
21. JEWELL-MOTZ, E. A. & J. B LINGREL. 1993. Biochemistry **32:** 13523–13530.
22. VILSEN, B. 1993. Biochemistry **32:** 13340–13349.
23. KUNTZWEILER, T. A., E. T. WALLICK, C. L. JOHNSON & J. B LINGREL. 1995. J. Biol. Chem. **270:** 2993–3000.
24. ARGÜELLO, J. M. & J. B LINGREL. 1995. J. Biol. Chem. **269:** 22764–22771.
25. VILSEN, B. 1995. Biochemistry **34:** 1455–1463.
26. KOSTER, J. C., G. BLANCO, P. B. MILLS & R. W. MERCER. 1996. J. Biol. Chem. **271:** 2413–2421.
27. JORGENSEN, P. L., J. H. RASMUSSEN & P. A. PEDERSEN. 1997. Ann. N.Y. Acad. Sci. This volume.
28. ARGUELLO, J. M. & J. H. KAPLAN. 1994. J. Biol. Chem. **269:** 6892–6899.
29. GOLDSHLEGER, R., D. M. TAL, J. MOORMAN, W. D. STEIN & S. J. D. KARLISH. 1992. Proc. Natl. Acad. Sci. USA **89:** 6911–6915.
30. VAN HUYSSE, J. W., E. JEWELL & J. B LINGREL. 1993. Biochemistry **32:** 819–826.
31. TANIGUCHI, K., K. SUSUKI & S. IIDA. 1982. J. Biol. Chem. **257:** 10659–10667.
32. NAGAI, M., K. TANIGUCHI, K. KANAGAWA, H. MATSUO, S. NAKAMURA & S. IIDA. 1986. J. Biol. Chem. **261:** 13197–13202.
33. LUTSENKO, S. & J. H. KAPLAN. 1995. Biochemistry **34:** 15607–15613.
34. CLARKE, D. M., T. W. LOO, G. INESI & D. H. MACLENNAN. 1989. Nature **339:** 476–478.
35. ANDERSEN, J. P. & B. VILSEN. 1995. FEBS Lett. **359:** 101–106.
36. TONEY, M. D., E. HOHENESTER, S. W. COWAN & J. N. JANSONIUS. 1993. Science **261:** 756–759.
37. TAL, D. M. & S. J. D. KARLISH. 1995. Tetrahedron **51:** 3823–3830.
38. HOVING, S., M. BAR-SHIMON, J. J. TIJMES, D. M. TAL & S. J. D. KARLISH. 1995. J. Biol. Chem. **270:** 29788–29793.
39. DAVID, P., H. MAYAN, E. J. CRAGOE, JR. & S. J. KARLISH. 1993. Biochim. Biophys. Acta **1146:** 59–64.
40. DAVID, P., H. MAYAN, H. COHEN, D. M. TAL & S. J. D. KARLISH. 1992. J. Biol. Chem. **267:** 1141–1149.
41. OR, E., P. DAVID, A. SHAINSKAYA, D. M. TAL & S. J. D. KARLISH. 1993. J. Biol. Chem. **268:** 16929–16937.
42. OR, E., R. GOLDSHLEGER & S. J. D. KARLISH. 1996. J. Biol. Chem. **271:** 2470–2477.
43. GOLDSHEGER, R., S. J. D. KARLISH, A. REPHAELI & W. D. STEIN. 1987. J. Physiol. **387:** 331–355.
44. GLYNN, I. M. & S. J. D. KARLISH. 1990. Annu. Rev. Biochem. **59:** 171–205.
45. ESMANN, M., S. J. D. KARLISH, L. SOTTRUP-JENSEN & D. MARSH. 1994. Biochemistry **33:** 8044–8050.
46. LUTSENKO, S. & J. H. KAPLAN. 1994. J. Biol. Chem. **269:** 4555–4564.
47. SHAINSKAYA, A. & S. J. D. KARLISH. 1994. J. Biol. Chem. **269:** 10780–10789.
48. LUMRY, R. & H. EYRING. 1954. J. Phys. Chem. **58:** 110–120.
49. PALASIS, M., T. A. KUNTZWEILER, J. M. ARGUELLO & J. B. LINGREL. 1996. J. Biol. Chem. **271:** 14176–14182.
50. GEERING, K. 1991. FEBS Lett. **285:** 189–193.
51. CHOW, D. C. & J. G. FORTE. 1995. J. Exp. Biol. **198:** 1–17.
52. FAMBOROUGH, D. M., M. V. LEMAS, M. HAMRICK, M. EMERICK, K. J. RENAUD, E. M. INMAN, B. HWANG & K. TAKEYASU. 1994. Am. J. Physiol. **266:** 579–589.
53. SHAINSKAYA, A. & S. J. D. KARLISH. 1996. J. Biol. Chem. **271:** 10309–10316.
54. GEERING, K., A. BEGGAH, P. GOOD, S. GIRARDET, S. ROY, D. SCHAER & P. JAUNIN. 1996. J. Cell Biol. **133:** 1193–1204.

55. OR, E., R. GOLDSHLEGER, D. M. TAL & S. J. D. KARLISH. 1996. Biochemistry **21:** 6853–6864.
56. SARVAZYAN, N. A., N. N. MODYANOV & A. ASKARI. 1995. J. Biol. Chem. **270:** 26528–26532.
57. LEMAS, M. V., M. HAMRICK, K. TAKEYASU & D. M. FAMBROUGH. 1994. J. Biol. Chem. **269:** 8255–8259.

Ligand-Induced Conformational Changes in the Na,K-ATPase α Subunit[a]

JACK H. KAPLAN,[b,d] SVETLANA LUTSENKO,[b]
CRAIG GATTO,[b] SYLVIA DAOUD,[b] AND LINDA J. KENNEY[c]

[b]Department of Biochemistry and Molecular Biology and
[c]Molecular Microbiology and Immunology
Oregon Health Sciences University
3181 S.W. Sam Jackson Park Road
Portland, Oregon 97201–3098

The coupled activities of cation transport and ATP hydrolysis carried out by the Na,K-ATPase involve interactions between distinct regions of the protein specialized to perform these independent functions. ATP binding and hydrolysis occur in one region of the protein, while transport and cation occlusion occur at spatially distinct sites on the protein. The way in which these domains interact to produce the tightly coupled and stoichiometric active transport of Na and K ions lies at the heart of the problems we are trying to address.

In recent years it has become apparent that most of the structural determinants for ATP binding resides in the major cytoplasmic loop of the Na,K-ATPase α subunit, which is not part of the membrane-contained segments.[1] It is equally apparent that many of the requirements for cation binding are provided by amino acid residues of the intramembrane segments of the α subunit.[2]

This article summarizes our recent work which examines each of these functional "domains" separately and begins to address the issues of how they might interact under the influence of physiological ligands.

CATION BINDING AND THE INTRAMEMBRANE SEGMENTS

Recent studies using a positively charged carboxyl-directed reagent, 4-(diazomethyl)-7-(diethylamino)-coumarin (DEAC), identified E^{779} as a specific amino acid residue of the α subunit important for cation occlusion and transport.[3] According to most topological models of the α subunit, E779 is in M5, the fifth transmembrane segment (FIG. 1). These results served to focus attention on this segment of the intramembrane part of the Na,K-ATPase as having an important role in cation binding and occlusion. Subsequent studies from other laboratories (particularly those of Lingrel *et al.*) using site-directed mutagenesis have confirmed the importance of this region of the molecule for cation occlusion and transport.[4]

---

[a]This work was supported by National Institutes of Health grants GM39500 and HL30315 to J.H.K. and by grants from the American Heart Association, Oregon Affiliate, to L.J.K. and C.G.

[d]Address for correspondence: Professor Jack H. Kaplan, Department of Biochemistry & Molecular Biology, L224, Oregon Health Sciences University, 3181 S.W. Sam Jackson Park Road, Portland, OR 97201-3098 (tel: 503-494-1001; fax: 503-494-1002; e-mail: kaplanj@ohsu.edu).

In 1990 Karlish and co-workers[5] made the important observation that extensive digestion of the Na pump in the presence of K (or Rb) ions resulted in a membrane preparation that was composed of several fragments within the membrane, was devoid of most of the extramembrane protein, and yet was still able to occlude K ions. These results suggested that if indeed occlusion in this preparation was very similar to occlusion in native enzyme, then the cation-occlusion sites were contained almost exclusively in the intramembrane segments. In subsequent studies we were able to show that removal of K ions from the preparation led to loss of the ability to occlude K which was associated with loss of the M5M6 hairpin from the membrane to the aqueous phase.[6] This redistribution from the membrane phase to the aqueous phase was prevented by K ions (not by Na or other pump ligands). These observations

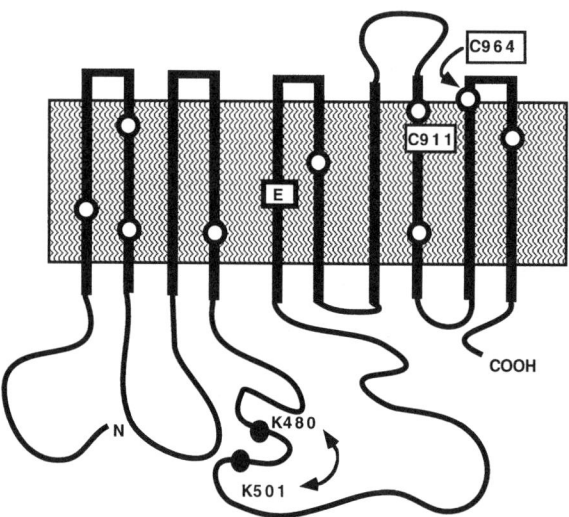

FIGURE 1. Topological model of the α subunit of Na,K-ATPase. Cysteine residues in the transmembrane portion of the molecule are marked by *circles,* mobile residues at the extracellular surface of the α subunits (Cys964 and Cys911) are identified by *letters and numbers,* and Lys480 and Lys501 in the ATP-binding domain are targets for the H_2DIDS cross-linking and are marked with *filled circles.*

further emphasized the importance of the M5M6 region in cation occlusion and transport.

Several aspects of these observations led us to speculate about the possible functional roles for this region of the protein. The M5 segment of the α subunit is one of the two direct connections between the intramembrane segments (and cation occlusion sites) and the major cytoplasmic loop containing the ATP binding and phosphorylation site (the M4 segment providing the other connection).[7] Thus, the link between conformational changes in the ATP binding domain and cation occupancy may be provided directly (indeed perhaps mechanically) via M5, which we now know is essentially involved in cation occlusion. Furthermore, if in the proteolyzed preparation the M5M6 hairpin escapes from the membrane perhaps, such behavior is an exaggerated or amplified version of movements that occur in the intact protein during its reaction cycle. Bearing in mind that the assignment of protein sequences as transmembrane segments is based on thermodynamic considerations, it is striking that M5M6 which "should" be more stable in the membrane redistributes to the aqueous environment. It is possible that charge interactions between aspartate residues and glutamate residues in M5 and M6 are reduced by the presence of monovalent cations; in the absence of such cations these interactions render M5 and M6 less stable in the membrane phase. We believe it possible that during the reaction cycle of the Na,K-ATPase, as K ions enter and leave their occlusion sites, conformational changes in the protein result in movements of the M5M6 region perpendicular to the plane of the membrane. If indeed such changes do occur, to what extent are they associated with similar movements in other intramembrane segments of the protein?

In an initial approach to answering such questions and also examining some topological issues, we recently began a series of studies to determine the accessibility of cysteine residues in transmembrane segments of the α subunit to reagents in the extracellular medium. With the use of membrane-permeable and membrane-impermeable cysteine-reactive reagents we are trying to map which of the cysteine residues of the TM segments are exposed to the outside and to determine if any of these alter their location as the pump takes up various conformations.

The experimental strategy employs right-side-out vesicles from dog renal medulla, and the approach is outlined in FIGURE 2.

Using this approach we recently identified two cysteine residues at the extracellular portion of the α subunit of the Na,K-ATPase.[8] These are Cys964 in transmembrane segment M9 and Cys904 in M8 (FIG. 1). Both Cys residues became exposed at the extracellular surface of the membrane and available for modification with the membrane-impermeable Cys-directed reagent SDSM (4-acetamido-4′-maleimidylstilbene-2,2′ disulfonic acid) when the Na pump is stabilized in the phosphorylated state (in the presence of Mg and phosphate). Modification of these cysteine residues resulted in partial inactivation of the ATPase activity and Rb occlusion, whereas ATP binding remained unaffected. Upon binding of K, both Cys residues (Cys964 and Cys911) seemed to undergo movement and became buried in the membrane, as evidenced by their availability for modification only with the hydrophobic membrane-permeable Cys-directed probe CPM (7-diethylamino-3-(4′-maleimidyl)-4-methyl coumarin). Because the presence of K protects Cys964 and Cys911 against modification with membrane-impermeable SDSM, but not with the hydrophobic CPM, we concluded that the cysteines in M9 and M8 transmembrane segments were not directly involved in the coordination of K.

Cys911 is less readily modified with CPM than is Cys964, suggesting that steric hindrance may preclude its reactivity. It is possible that involvement of this segment of the α subunit in the interaction with the β subunit[9] is responsible for such a decrease in reactivity. We find this possibility particularly appealing because such conformational mobility of M8 agrees well with the conformational transitions in the β subunit we observed earlier.[10]

Previous studies on labeling of the purified Na,K-ATPase with BIPM (a hydrophobic, sulfhydryl-directed chromophore, N-[p-(2-benzimidazolyl)phenyl]maleimide) showed little, if any, effect of modification of Cys964 on Na,K-ATPase activity.[11] This supports our conclusion that although the accessibility/reactivity of Cys964 is altered by K occlusion, the residue is probably not essentially involved in the K occlusion cavity. Our observation that Cys964 and Cys911 have different reactivities in different enzyme conformations is reminiscent of the finding that the fluorescence emission of

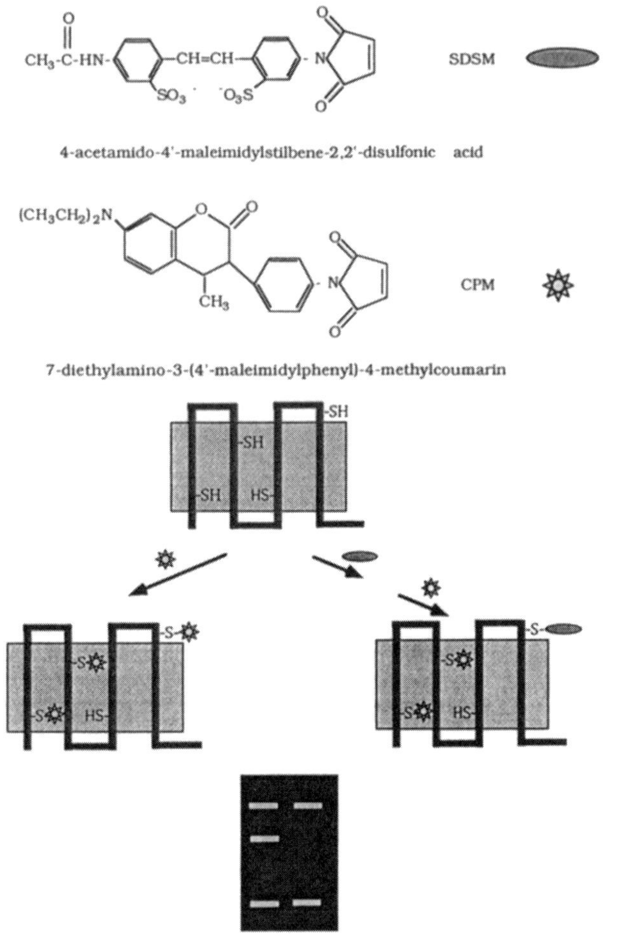

FIGURE 2. Schematic presentation of the experimental procedure used for differential labeling of Cys residues located at the extracellular surface of the α subunit of Na,K-ATPase. Two Cys-directed reagents were used to modify Na,K-ATPase in the right-side-out oriented vesicles. Both reagents have the same reactive group (maleimide), but differ in their hydrophobic and fluorescent properties. Hydrophobic membrane-permeable CPM is a highly fluorescent probe, while hydrophilic membrane-impermeable SDSM has very low fluorescence. Prior modification of Na,K-ATPase with SDSM blocks the extracellular Cys residues and prevents their labeling with CPM. This can be monitored as a selective loss of the fluorescent band following separation of the fragments of the CPM or SDSM/CPM-labeled α subunit by gel electrophoresis.

BIPM-labeled Cys964 changes depending on Na,K-ATPase conformation.[12] Although not critical for cation binding and occlusion, both cysteine residues 964 and 911 are involved in ligand-induced movements, which are likely coupled to rearrangements within the cation-binding and translocation moiety. It is also interesting that modification of Cys964 with the hydrophobic reagent (BIPM) probably does not affect relocation of Cys964 during the ATPase cycle, whereas attachment of a negatively charged hydrophilic moiety (SDSM) results in a significant effect on Rb (K) occlusion (a loss of about 50%). This loss in Rb occlusion probably results from destabilization of the $E_2(K_2)$ form when SDSM is attached to Cys964 and Cys911.

The differential labeling approach described in FIGURE 2 was also used to identify Cys residues located at the intracellular border of the membrane. To do so we used the post-tryptic membrane preparation of Na,K-ATPase obtained by extensive trypsin digestion of Na pump in the presence of K/Rb.[12] These posttryptic membranes, obtained and stored in the presence of K, retain an Rb occlusion capacity that is similar to the K occlusion capacity of the native enzyme. Labeling of the posttryptic membrane preparations with 1 mM CPM demonstrates that prior incubation with SDSM does not significantly diminish the CPM fluorescence of any of these membrane-associated fragments, indicating that in the presence of occluded K none of the 11 cysteines are aqueous exposed and/or accessible to SDSM.

Removal of Rb or K from this posttryptic preparation is accompanied by release of the M5M6 transmembrane hairpin from the membrane and loss of Rb-occlusion capacity (see above). We noted that under these conditions the COOH-terminal 21-kD fragment became more exposed to CPM labeling, indicating that residues that were not accessible to CPM in the presence of K, became exposed after the M5M6 hairpin was released from the membrane. Additional experiments demonstrated that when Rb (and consequently M5M6) is removed, not only do some cysteine residues become more accessible to CPM, but also a fraction of these are relocated outside the membrane.[8] FIGURE 3 shows that such reorganization and associated increase of CPM labeling is occurring largely in the COOH-terminal 21-kD fragment, whereas CPM labeling of Cys residues in the M1M2 and M3M4 transmembrane hairpins remain essentially unchanged. Thus, Cys residues in the M1M2 and M3M4 segments are unlikely to be involved in direct interactions with the M5M6 hairpin. Our further experiments led to the identification of Cys983 in M10 as a residue in the COOH-terminal portion of the α subunit which becomes exposed to the aqueous phase when

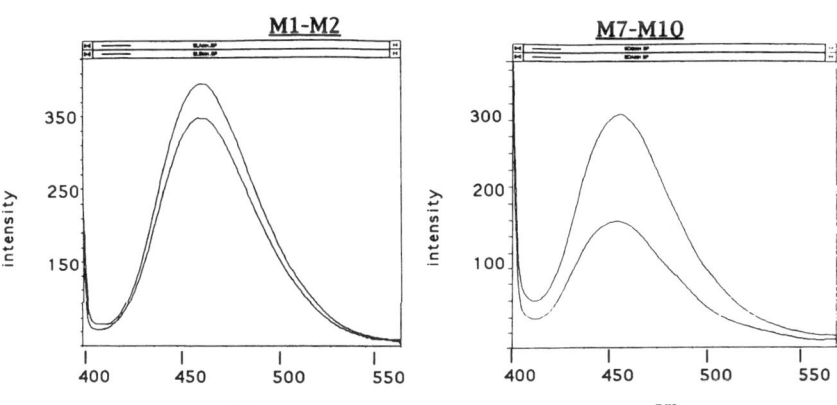

FIGURE 3. Release of the M5M6 transmembrane hairpin from the membrane selectively alters labeling of the COOH-terminal 21-kD segment. Fluorescence spectra of the isolated transmembrane fragments modified in the membrane-bound form with CPM in the presence of Rb (*lower line* in both cases) and after replacement of Rb with Tris and release of the M5M6 hairpin (*upper line* in both cases). **Left panel:** NH_2-terminal hairpin M1M2; **right panel:** COOH-terminal 21-kD fragment, including transmembrane segments M7-M10.

the Rb occlusion capacity is lost and the M5M6 leaves the membrane.[8] Prior to the loss of the M5M6 segment this residue is not accessible to SDSM.

These results provide new suggestive evidence for interactions between specific segments of the α subunit of the Na,K-ATPase. It seems likely that the M9M10 hairpin in the COOH-terminal portion of the α subunit is tightly associated with segments that are involved in cation occlusion and energy-coupling functions. Our observation that Cys983 in M10 has an increased susceptibility to modification with SDSM following M5M6 release from the membrane indicates a tight interaction between M5M6 and M10. The M5M6 hairpin is a rather hydrophilic and proline-rich transmembrane hairpin (FIG. 4), and stabilization via interactions with M10 may play

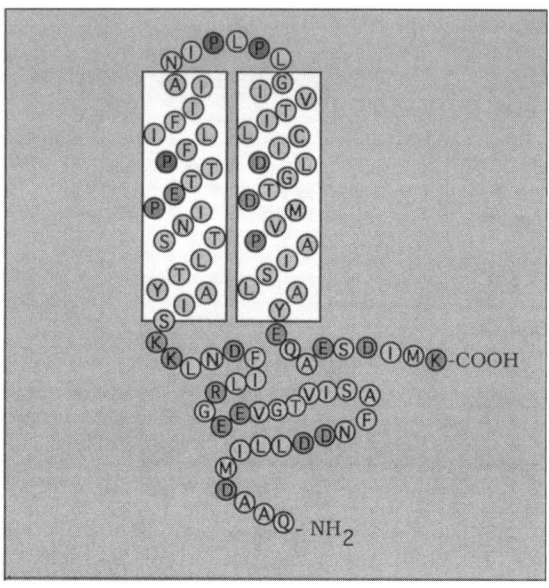

FIGURE 4. Amino acid composition of transmembrane segment M5M6. The structure corresponds to the peptide which is retained in the membrane after proteolytic degradation of Na,K-ATPase in the presence of potassium and is released from the membrane after replacement of K with Tris. Note a long hydrophylic NH_2-terminal portion which is resistant to proteolytic degradation in the presence of K and large amounts of Pro residues.

an important role in the intramolecular interactions with the membrane-bound domain of the α subunit.

It is clear that the M9M10 hairpin undergoes changes in its location in the membrane as different physiological ligands of the Na pump are bound to the α subunit. Such changes in the M9M10 segment are probably coupled to changes in the transmembrane location of the M5M6 segment. The extent to which such structural changes in M9M10 and M5M6 during pump turnover are transmitted to some or all of the other α-subunit transmembrane segments remains to be determined. Piston-like movements of M5M6 can be associated with similar movements of, say, M9 and M10 or, alternatively, changes in depths of M5M6 and M9 and M10 in the membrane may result from coupled rotations (or screw movements) between these tightly associated transmembrane helices.

ATP BINDING AND THE CYTOPLASMIC LOOP

Most of the amino acid residues that have been identified as playing a role in or being associated with ATP binding reside in the major cytoplasmic loop between transmembrane segments M4 and M5 of the Na,K-ATPase. The identification of most of these residues has resulted from the use of either ATP affinity reagents or group-specific reagents whose inactivation effects were prevented by the simultaneous presence of ATP. More recently, mutagenesis studies on the Na,K-ATPase or other closely related P_2-ATPases have identified other important residues in the region.[2]

The segment of this loop located between D369 and G502 is particularly interesting. It has been observed that a variety of aryl isothiocyanates, including NIPI,[13] SITS,[14] and FITC,[17,18] covalently label K501 in the absence of ATP but not in its presence, consistent with K501 residing in or close to the ATP site. Also, the photoaffinity ATP analogs, 2-azido-ATP and 8-azido-ATP, have been shown to modify K480 and G502, respectively; these modifications are eliminated by the simultaneous presence of ATP.[18] Another interesting observation is that the reactivity of K501 with isothiocyanates is altered by cation binding in much the same way that ATP binding is sensitive to cation binding. For example, the binding of Na^+ dramatically facilitates the reaction, whereas the binding of K has a biphasic effect on the reaction. That is, at low $[K^+]$ (<2 mM), the reaction between K501 and an isothiocyanate is significantly diminished, as opposed to high $[K^+]$ (>25 mM) where the protective effect of potassium is lost (e.g., see the reaction with FITC in FIG. 5).

This biphasic K^+ effect has also been observed for SITS and H_2DIDS inactivation.[21,22] Moreover, Rb^+ and Cs^+ (but not Na^+ or choline) have a similar biphasic dependence in protecting the Na^+ pump against H_2DIDS (see FIG. 5, ref. 22). Compared to either K or Rb, the decreased ability of Cs^+ to protect against H_2DIDS inactivation is consistent with the cation selectivity (Rb > K ≫ Cs) reported for the extracellular K^+ transport site.[21] It is possible that at low $[K^+]$ the binding of only one K^+ ion to the enzyme leads to protection of K501 from modification. (The protection can be achieved either by blocking the access to K501 or by effectively raising its pK, thus decreasing its nucleophilicity.) Then, as the $[K^+]$ increases, a second molecule of K^+ binds and causes a conformational change that once again allows K501 to react with H_2DIDS.

Although H_2DIDS reacts with K501 like all of the aryl isothiocyanates just mentioned, it is unique in that it has two reactive isothiocyanate functionalities. In our recent experiments on modification of the Na pump with H_2DIDS we obtained two peptides that were radioactively labeled with H_2DIDS in approximately equal amounts. These were 496HLLVMxGAPER and 470IVEIPFNSTNxYQL which identified K480 and K501 (corresponding to x in each of the sequences) as the H_2DIDS-modified residues. These two peptides electrophoresed as a single band with the apparent mass of 12 kD even after the most exhaustive tryptic degradation, suggesting that these two fragments could be cross-linked by H_2DIDS.

Because the lack of tryptic cleavage could also be a result of poor accessibility for protease, we chose to avoid such complication using chemical cleavage with the small molecule, CNBr. In this work, CNBr was especially useful because of the strategic

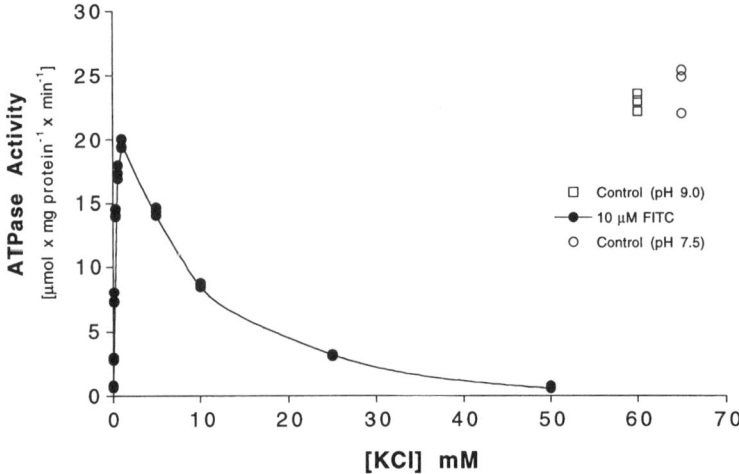

FIGURE 5. FITC inactivation of purified dog Na,K-ATPase in the presence of varying KCl concentrations. The enzyme was treated with 10 μM FITC in 50 mM Tris, 2 mM EDTA (pH 9.0) for 30 minutes at 37°C. The reaction was stopped by diluting the reaction mixture 10-fold with ice-cold stopping solution containing: 50 mM imidazole, 50 mM β-mercaptoethanol, and 0.6 mg/ml BSA. Incubation of the enzyme at pH 9.0 alone did not decrease activity; c.f., *open squares* and *open circles*.

location of methionine residues at M463, M500, and M584. CNBr treatment (which cleaves on the COOH-terminal side of methionine residues) of the H$_2$DIDS-modified α subunit gave a single labeled fragment of about 15 kD (the predicted molecular weight for the cross-linked product). Sequencing of the labeled peptide proved that it contained only the two predicted sequences. Thus, H$_2$DIDS indeed cross-links K501 and K480 of the α subunit.[20]

Formation of a cross-link by H$_2$DIDS (and presumably DIDS) had been the major source of the difficulty in determining its site of modification in earlier experiments. Previous work from our laboratory,[14] demonstrated that the single isothiocyano-substituted stilbene, SITS, was an irreversible inhibitor of the Na,K-ATPase. Like DIDS, SITS inactivation was completely preventable by the presence of either ATP or low K$^+$. Subsequently, the lysine residue modified by SITS was identified as K501 following tryptic digestion and amino acid sequencing. The fact that SITS labeled only K501 suggests that the additional isothiocyano group on H$_2$DIDS is important for reaction with K480. Lysine-480 is not highly reactive (or SITS would also modify it), and only becomes modified after reaction with K501. Presumably K501 is more reactive/accessible than K480 to such aromatic isothiocyanates, and the initial anchoring at K501 brings the second isothiocyanate of H$_2$DIDS in close proximity to K480.[19] A diagram of the reaction between H$_2$DIDS and the Na pump is shown in FIGURE 6. The labeled lysine residues K480 and K501 thus appear to be about 14 Å apart (i.e., the spacing between the two-NCS moieties of H$_2$DIDS).

The information derived from affinity labeling and site-directed mutagenesis studies on the whole Na,K-ATPase has been useful in predicting amino acid residues that are likely to be involved with nucleotide binding. Thus far, all of the implicated residues reside in the large cytoplasmic loop between transmembrane segments M4 and M5 (for a review, see ref. 21.) However, detailed structural information about the secondary and tertiary structure involved in nucleotide coordination remains a mystery. To gather structural information about the nucleotide binding site more directly, we overexpressed the large cytoplasmic loop in *Escherichia coli* as a glutathione S-transferase fusion protein (FIG. 7).

The appropriate DNA oligonucleotide primers were used to synthesize a 1.2-kb fragment via polymerase chain reaction. *Eco*R1 sites were engineered on both ends of the ATP-loop construct, so ligating with the pGEX-1λT vector (Pharmacia Biotech) was achieved in a single step. The construct encoded the M4M5 loop (K354–K774) of the rat α1 subunit. *E. coli* (strain DH5α) cells were transformed with the plasmid and then selected by growth on LB agar with 125 µg/ml ampicillin. The entire 1262 base insert was sequenced to ensure that no random mutagenesis occurred during the PCR reaction.

Using this approach we can successfully isolate 3–5 mg of the GST-loop fusion protein from a liter of cell culture. Similar approaches have been employed by other

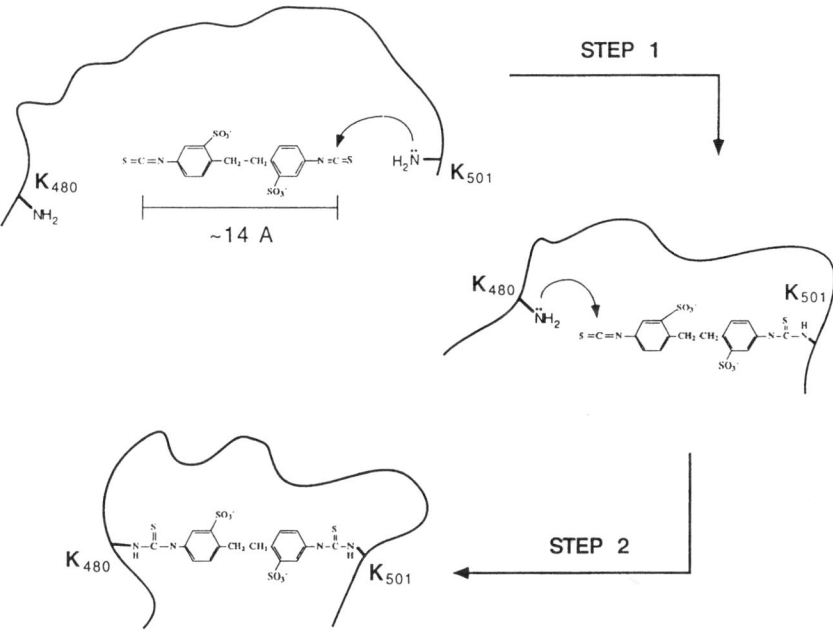

FIGURE 6. A scheme for the cross-linking of K480 and K501 of the Na,K-ATPase by H_2DIDS. The initial reaction of the protein with H_2DIDS occurs when the highly nucleophilic epsilon amino group of K501 attacks one of the two available isothiocyanate groups on H_2DIDS (step 1). Once a molecule of H_2DIDS is tethered to K501, the second isothiocyanate group is "fixed" in close proximity to K480, thus facilitating the reaction with the less reactive K480 (step 2). This two-step reaction sequence is supported by the fact that the protective effects of low K^+ and ATP are the same for H_2DIDS as they are for isothiocyanates that modify K501 only. Reproduced with permission from Gatto *et al.*[20]

laboratories to isolate the ATP-binding domains of the Na^+ pump,[22] the SR Ca^{2+} pump,[23] and the yeast H^+ pump.[24]

To determine if the GST-loop fusion protein binds ATP, we tested whether ATP would protect the expressed protein against modification with FITC. We incubated 50 µg of GST-loop with 20 µM FITC for 20 minutes (pH 9.0) in the absence of nucleotide or in the presence of 3 mM ATP or 3 mM AMP. The presence of ATP, but not AMP, fully prevented FITC incorporation into the GST-loop fusion protein. This experiment was simultaneously performed on purified dog Na,K-ATPase; it is also clear that in the native enzyme only ATP protected against FITC labeling. The important experiments characterizing the nucleotide binding properties of the M4M5 loop now await large scale purification of the isolated loop following cleavage from GST.

It seems then that significant advances in our understanding of how the Na,K-ATPase may perform its coupled functions will accrue from studies of isolated

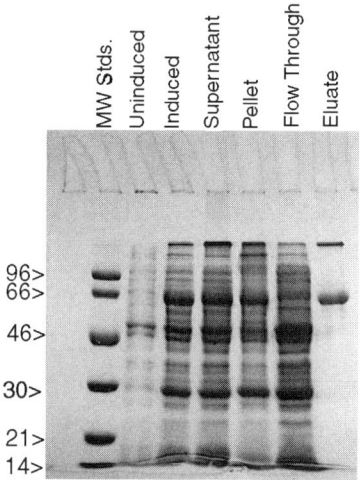

FIGURE 7. A 10% Laemmli gel showing GST-fusion protein production in *E. coli* DH5α cells. A single colony was grown overnight and used to inoculate 1 L of LB_{amp} (125 µg/ml). After an OD_{600} of 0.8 was reached, protein synthesis was introduced with 0.1 mM IPTG, and the cells were grown for an additional 6 hours at 25°C. There was clearly production of a protein at ~65 kD after induction (*lane 3*). After cell lysis, a significant amount of the protein remained soluble (routinely 30–40%; *lane 4*). The GST-fusion protein was purified from the supernatant via a glutathione sepharose column (*lane 7*). Yield is ~3–5 mg GST-M4M5 loop/L cell culture.

domains. However, it is important to remember that evidence is already in hand that the β subunit may also play a role in active cation transport.[25,26]

To understand the important details of the mechanism of active transport by the Na pump it will be necessary to synthesize the information we have on how the separate parts of the α subunit work. It is equally necessary also to account for the important interactions that occur between the α and β subunits in order to resolve the essential role of β.

REFERENCES

1. KAPLAN, J. H. 1991. *In* The Na Pump: Structure, Mechanism, and Regulation. J. H. Kaplan & P. DeWeer, Eds.: 118–128. Rockefeller University Press. New York.
2. LINGREL, J. B. & T. A. KUNTZWEILER. 1994. J. Biol. Chem. **269:** 19659–19662.
3. ARGUELLO, J. M. & J. H. KAPLAN. 1994. J. Biol. Chem. **269:** 6892–6899.
4. ARGUELLO, J. M. & L. B. LINGREL. 1995. J. Biol. Chem. **270:** 22764–22771.

5. KARLISH, S. J. D., R. GOLGSHLEGER & W. D. STEIN. 1990. Proc. Natl. Acad. Sci. USA **87:** 4566–4570.
6. LUTSENKO, S., R. ANDERKO & J. H. KAPLAN. 1995. Proc. Natl. Acad. Sci. USA **92:** 7936–7940.
7. KAPLAN, J. H., J. M. ARGUELLO, G. C. R. ELLIS-DAVIES & S. LUTSENKO. 1994. The Sodium Pump: Structure, Mechanism, and Hormonal Control, and its Role in Disease. E. Bamberg & W. Schoner, Eds.: 321–331. Springer-Verlag. New York.
8. LUTSENKO, S., S. DAOUD & J. H. KAPLAN. 1997. J. Biol. Chem. **272:** 5249–5255.
9. LEMAS, M. V., M. HAMRICK, K. TAKEYASU & D. M. FAMBROUGH. 1994. J. Biol. Chem. **269:** 8255–8259.
10. LUTSENKO, S. & J. H. KAPLAN. 1993. Biochemistry **32:** 6737–6743.
11. NAGAI, M., K. TANIGUCHI, K. KANGAWA, H. MATSUO, S. NAKAMURA & S. IIDA. 1986. J. Biol. Chem. **261:** 13197–31202.
12. TANIGUCHI, K., H. TOSA, K. SUZUKI & Y. KAMO. 1988. J. Biol. Chem. **263:** 12943–12947.
13. ELLIS-DAVIES, G. C. R. & J. H. KAPLAN. 1993. J. Biol. Chem. **268:** 11622–11627.
14. PEDEMONTE, C. H., T. L. KIRLEY, M. H. TREUHEIT & J. H. KAPLAN. 1992. FEBS Lett. **314:** 97–100.
15. FARLEY, R. A., M. C. TRAN, C. T. CARILLI, D. HAWKE & J. E. SHIVELY. 1984. J. Biol. Chem. **259:** 9532–9535.
16. KIRLEY, T. L., E. T. WALLICK & L. K. LANE. 1984. Biochem. Biophys. Res. Commun. **125:** 767–773.
17. TRAN, C. M., E. E. HUSTON & R. A. FARLEY. 1994. J. Biol. Chem. **269:** 6558–6565.
18. TRAN, C. M., G. SCHEINER-BOBIS, W. SCHONER & R. A. FARLEY. 1994. Biochemistry **33:** 4140–4147.
19. PEDEMONTE, C. H. & J. H. KAPLAN. 1988. Biochemistry **27:** 7966–7973.
20. GATTO, C., S. LUTSENKO & J. H. KAPLAN. 1997. Arch. Biochem. Biophys. **340:** 90–100.
21. SACHS, J. R. & G. L. WELT. 1967. J. Clin. Invest. **46:** 1433–1441.
22. TRAN, C. M., E. HROUDA, C. M. GRISHAM & R. A. FARLEY. 1996. Biophys. J. **70:** A329.
23. MOUTIN, M.-J., M. CUILLEL, C. RAPIN, R. MIRAS, M. ANGER, A.-M. LOMPRE & Y. DUPONT. 1993. J. Biol. Chem. **269:** 1–8.
24. CAPIEAUX, E., C. RAPIN, D. THINES, Y. DUPONT & A. GOFFEAU. 1993. J. Biol. Chem. **268:** 21895–21900.
25. LUTSENKO, S. & J. H. KAPLAN. 1993. Biochemistry **32:** 6737–6743.
26. KAWAMURA, S., S. NOGUCHI, S. ENO, M. KUSABA & K. TAKEDA. 1994. The Sodium Pump: Structure, Mechanism and Hormonal Control, and its Role in Disease. E. Bamberg & W. Schoner, Eds.: 287–296. Springer-Verlag. New York.

Plasma Membrane Ca^{2+} Pumps[a]

JOHN T. PENNISTON,[b] AGNES ENYEDI,[b,c]
ANIL K. VERMA,[b] HUGO P. ADAMO,[b,d]
AND ADELAIDA G. FILOTEO[b]

[b]*Department of Biochemistry and Molecular Biology*
Mayo Foundation
Rochester, Minnesota 55902

[c]*Department of Cell Metabolism*
National Institute of Haematology and Immunology
Daroczi Ut 24, 1113 Budapest, Hungary

[d]*Instituto de Quimica y Fisicoquimica Biologicas (UBA-CONICET)*
Facultad de Farmacia y Bioquimica
Buenos Aires, Argentina 1113

Plasma membrane Ca^{2+} pumps are P-type ATPases and therefore are members of the same superfamily as the Na$^+$,K$^+$ ATPase and the sarcoplasmic reticulum Ca^{2+} pump. Like them, this pump moves a biologically essential ion up its concentration gradient at the expense of hydrolysis of ATP. The pump is important in shaping the Ca^{2+} signal that controls the biological responses of cells. When cells are simulated by agonists, Ca^{2+} flows into the cytosol through channels from either intracellular stores or outside the cell. Activation of these channels determines the rising phase of the Ca^{2+} signal, but the falling phase is controlled by the two Ca^{2+} pumps, those of the plasma membrane and the sarco/endoplasmic reticulum, and in some cells by the Na$^+$/Ca^{2+} exchanger.

ROLE OF PLASMA MEMBRANE Ca^{2+} PUMPS

The relative roles of the proteins that remove Ca^{2+} from cytosol vary depending on the kind of cell. For example, in skeletal muscle the sarcoplasmic reticulum plays a dominant role, with the Na$^+$/Ca^{2+} exchanger also moving a large share of the Ca^{2+}. A similar situation exists in heart muscle, but in smooth muscle the sarco/endoplasmic reticulum is much less abundant, and the plasma membrane Ca^{2+} pump plays a coordinate role with the sarco/endoplasmic reticulum pump.

In other kinds of cells, the plasma membrane Ca^{2+} pump plays a more dominant role. The most extreme example is the mature red cell, in which this pump is the only mechanism for controlling intracellular Ca^{2+}. Pancreatic acinar cells lack the Na$^+$/Ca^{2+} exchanger,[1] and the plasma membrane Ca^{2+} pump is believed to play a major role in removal of Ca^{2+} from the cytosol. The Na$^+$/Ca^{2+} exchanger has long been thought to play a major role in excitable cells, where it is abundant. However, a recent

---

[a]This work was supported in part by grant GM28835 from the US National Institutes of Health and by an International Research Scholarship from the Howard Hughes Medical Institute.

study of the soma of dorsal root ganglion cells concluded that the plasma membrane Ca^{2+} pump is the major enzyme involved in Ca^{2+} efflux from the soma, with the other two efflux systems playing a negligible role.[2] This is a case in which the cell is differentiated strongly between regions, and the Na^+/Ca^{2+} exchanger may be much more important in the synaptic region.

Another important role of the plasma membrane Ca^{2+} pump is in transcellular Ca^{2+} movement by epithelia. In the intestinal epithelium, in the epithelium of the kidney tubule, and in the placenta, this pump is situated on the side towards which the Ca^{2+} is being moved. This indicates that it is involved in the ejection from the epithelial cell of the Ca^{2+} which is on its way to its biological target.

REGULATION

Unlike the Na^+,K^+ pump and the sarco/endoplasmic reticulum Ca^{2+} pump, the plasma membrane Ca^{2+} pump has a regulatory region attached to the molecule. The Na^+,K^+ pump requires a β subunit for activity. The role of this subunit is still not clearly defined, but it may be important in correct folding and orientation and perhaps in regulation of activity. The sarco/endoplasmic reticulum Ca^{2+} pump does not have a regulatory protein in skeletal muscle, but in cardiac muscle it is regulated by phospholamban. The degree of inhibition of the pump by phospholamban varies, with highly phosphorylated phospholamban inhibiting the least. In the plasma membrane Ca^{2+} pump, a similar regulatory function is assumed by a unique carboxyl terminal region which is an integral part of the molecule. When P-type ion pumps are aligned, the plasma membrane Ca^{2+} pump extends beyond the carboxyl terminus of the other pumps. This regulatory region (about 120 amino acid residues in isoform 4b) serves as an autoinhibitor of the pump's activity as observed in other calmodulin-regulated enzymes. The intact plasma membrane Ca^{2+} pump with a fully functional regulatory region is relatively inactive and has a low affinity for Ca^{2+}. The enzyme can be activated in several different ways, all of which partially or completely remove the interaction between the regulatory carboxyl terminus and the catalytic core of the enzyme. The fullest activation of the pump is caused by acidic phospholipids, which interact not only with the regulatory carboxyl terminus, but also with a lipid binding region elsewhere in the molecule.[3-5] The most acidic lipids, such as phosphatidylinositol bisphosphate, are the most potent. Because the concentration of this lipid can change drastically as part of the Ca^{2+} cycle, it may be an important regulator of the pump.[6] Activation nearly as great as that caused by lipid is accomplished by binding calmodulin to a basic calmodulin-binding region in the regulatory carboxyl terminus or by removing this region (by proteolysis or other means). The function and structure of this region are discussed more fully below. Another means of fully activating the pump is via dimerization in which the calmodulin-binding domain may interact with an acidic domain in the regulatory carboxyl terminus of an adjacent molecule, activating both molecules.[7,8]

A final means of activating this enzyme is via the action of protein kinases phosphorylating the regulatory carboxyl terminus. The best studied kinase has been protein kinase C, and its action was first localized to the carboxyl terminus by Wang *et al.*[9] The degree of activation caused by this phosphorylation has been the subject of controversy. We discuss new evidence from our laboratory below.

ISOFORMS OF THE PLASMA MEMBRANE CA^{2+} PUMP

The plasma membrane Ca^{2+} pump has isoforms coded by four different genes, and in addition alternative splices are known to occur in at least two locations.[10,11] These combinations can produce at least 20 different forms of the pump, each of which may have a significant physiological role in certain cell types. The four gene products show their greatest variation in the amino terminal and carboxyl terminal regions, with lesser differences scattered along the central part of the molecule. The form of the plasma membrane Ca^{2+} pump that has been studied most extensively is that in the human erythrocyte, which consists primarily of isoform 4b, with a small amount of 1b.[12] The existence of such a pump was first discovered in human red cells by Schatzmann[13] and the easy preparation of relatively large amounts of plasma membranes from human red cells has made possible extensive studies on this version of the pump. Because of this, we also initiated studies on the pump using isoform 4b, and this paper considers primarily isoforms 4b and 4a (4a is derived from 4b by an alternative splice).

It is expected that differences in regulation between the sarco/endoplasmic reticulum and plasma membrane Ca^{2+} pumps result in differences in the manner in which they influence the shape of the intracellular Ca^{2+} spike in living cells. The sarco/endoplasmic reticulum Ca^{2+} pump requires no external protein activator, so that it would be expected to respond promptly to a rise in intracellular Ca^{2+} and begin to remove Ca^{2+} from the cytosol concurrently with a rise in the cytosolic Ca^{2+} concentration. This immediate response to Ca^{2+} would continue throughout the Ca^{2+} spike unless the reticular lumen becomes saturated with Ca^{2+}. An increase in the free Ca^{2+} concentration inside the lumen will cause an inhibition of the sarco/endoplasmic reticulum Ca^{2+} pump and slow its action. The activatory pattern shown by the plasma membrane Ca^{2+} pump could be different, because its stimulation by calmodulin would not be instantaneous. Under conditions in which calmodulin activates the pump only slowly, the action of this pump might be significantly delayed and only reach full operation late in the Ca^{2+} spike. One study of the erythrocyte pump has indicated that the time required for binding of calmodulin might be as long as 30 seconds.[14] Such a slow interaction between calmodulin and the pump would have a strong influence on the shape of the intracellular Ca^{2+} spike.

STRUCTURE OF THE REGULATORY CARBOXYL TERMINUS

The discovery that calmodulin is an activator of the erythrocyte Ca^{2+} pump[15,16] led to its purification using a calmodulin affinity column.[17] The success of this purification made it evident that calmodulin bound rather tightly to the pump. That the calmodulin-binding domain is not too far from one end of the pump became evident when activation similar to that due to calmodulin was caused by proteolysis, which removed about one quarter of the enzyme.[18] A subsequent detailed study of proteolysis of the pump showed that the calmodulin-binding domain was within 15–20 kD of one end of the molecule.[19] With the cloning and sequencing of cDNAs for the pump[10,20] it became evident that only one stretch of amino acids of the pump had the properties of a good calmodulin-binding domain. This region of 28 residues had the mixture of basic and hydrophobic residues characteristic of calmodulin-binding domains. That this portion of the sequence indeed represented the calmodulin-binding domain was confirmed by a cross-linking study.[21] This 28-residue calmodulin-binding domain defines the beginning of the carboxyl terminal regulatory region.

The calmodulin-binding domain inhibits the enzyme by interacting with its catalytic core. This was first demonstrated by mixing a peptide representing the calmodulin-binding domain with the pump from which the calmodulin-binding domain had been removed proteolytically.[22] The addition of a calmodulin-binding domain inhibited the pump and brought its activity back down near that displayed by the full-length pump without added calmodulin. The regions in the catalytic core with which the calmodulin-binding domain interacts have been identified by cross-linking experiments.[23,24]

The cloning of the pump also disclosed that the carboxyl terminus contained a very high percentage of serine and threonine residues. This percentage ranges between 17 and 31%, depending on the isoform. Isoform 4b, which has been the most studied, has the lowest percentage of serine and threonine. Additional attention was directed towards this region by the study of Wang et al.,[9] which showed that protein kinase C phosphorylates the enzyme in this region. In this study, which was carried out on purified pump in the presence of detergent and at a high Ca^{2+} concentration (1.5 mM), phosphorylation was found in at least two locations, one of them being a downstream serine and the other being a threonine in the middle of the calmodulin-binding domain.

STUDIES WITH TRUNCATED MUTANTS

The studies mentioned so far serve as background to a new initiative in our laboratory to explore the regulatory domains using truncated mutants of the plasma membrane Ca^{2+} pump. We began these studies by transient overexpression of isoform 4b in COS-1 cells.[25] The amount of Ca^{2+} uptake demonstrated in this first experiment was low, but subsequent adjustments in the assay technique allowed us to achieve a specific activity approximately equal to that found in the native erythrocyte pump,[26] which set the stage for a series of studies using the truncated mutants. These truncated mutants are named according to the number of residues eliminated from the carboxyl terminus. Thus, the first mutant that we made was called ct120 (*c*arboxyl *t*runcated *120* residues) or, when it was necessary to be more specific, 4b-ct120. This mutant lacked the calmodulin-binding domain and everything downstream from that. We found that ct120 had an activity indistinguishable from that obtained by adding calmodulin to the full-length 4b. In a subsequent study we made ct111, ct100, and ct92[27]; we found that the shorter mutants ct120, ct111, and ct100 had little or no affinity for calmodulin, whereas ct92 bound calmodulin very tightly. The apparent affinity of ct92 for calmodulin was the same as the full-length 4b. This allowed us to establish that the 28 residues added back to ct120 to yield ct92 were sufficient to account for all of the affinity of the enzyme for calmodulin. This shows that no other portion of the molecule makes a significant contribution to the high-affinity binding of the enzyme to calmodulin, although calmodulin may bind with lower affinity (and no biochemical effect) to another part of the molecule.[4,19] Another result of this study was that ct92 still showed a higher basal activity in the absence of calmodulin than did full-length 4b. About three quarters of the inhibitory effect of the carboxyl terminus was accounted for by the 28-residue calmodulin-binding domain, but the remainder was apparently due to another inhibitory domain further downstream.

In our most recent study,[28] we investigated longer truncated mutants, ct71 and ct48. In this study, ct48 showed inhibition just as great as that of full-length 4b, indicating that the inhibitory domain was in the 45 residues immediately downstream of the calmodulin-binding domain. However, its exact position within this region has still not been determined. We also used these mutants to explore the effects of phosphorylation of isoform 4b by protein kinase C. Using our assay of Ca^{2+} uptake by the intact enzyme in the natural membrane environment of COS cell membranes, we established phosphorylation by protein kinase C and activation of the enzyme under almost identical conditions. This had not been possible in previous studies in other laboratories, where changes in activity due to phosphorylation were measured under rather different conditions from those used for measuring the incorporation of phosphate. In our assays, the only difference in the conditions used for the two types of assays was the necessary incorporation of vanadate in the phosphorylation experiment. This was needed to prevent interference by the acylphosphorylation that is a part of the enzyme's reaction cycle.

We found that the site that was most susceptible to phosphorylation by protein kinase C was not in the calmodulin-binding domain, but in a region of about 20 residues immediately downstream of the calmodulin-binding domain. Phosphorylation at this site occurred readily; a low concentration of C kinase (0.2 nM), a low concentration of Ca^{2+} (10 μM), and a short incubation time (10 minutes) sufficed to give substantial phosphorylation. Phosphorylation at this site gave partial activation of the enzyme causing about 25% of the activity increase caused by binding of calmodulin. The activity increase due to phosphorylation and that due to calmodulin were not additive, because both together produced only the same amount of activity as was present with calmodulin alone. The location of the phosphorylation site and the amount of activation led us to a model of the structure of the autoregulatory carboxyl terminus of isoform 4b, which is summarized in FIGURE 1. The phosphorylation due to protein kinase C occurs in the downstream inhibitory domain and relieves the enzyme of the inhibition due to that region. This increases the activity of the enzyme by about 25% of the total stimulation possible when calmodulin binds. The binding of calmodulin to the phosphorylated enzyme relieves the remaining inhibition due to the calmodulin-binding domain and produces a "fully activated" enzyme. Calmodulin relieves both inhibitions, presumably because it wags the downstream inhibitory domain, which is attached as a tail to the calmodulin-binding domain. Movement of the calmodulin-binding domain away from its target would be expected to remove the downstream inhibitory domain from its target.

Isoform 4a of the plasma membrane Ca^{2+} pump is produced from the same gene as is 4b, but a 154 base pair insert produces an almost entirely different carboxyl terminal regulatory region, with differences beginning in the middle of the calmodulin-binding domain. FIGURE 2 shows some of these differences. A comparison of the calmodulin-binding domain of 4b with the corresponding residues from 4a shows a striking difference in the carboxyl terminal third of the domain. We expected, on the basis of this difference in the sequence and on our previous comparison of peptides representing the a and b splices,[29] that these two domains would have a different affinity for calmodulin. When the full-length isoforms 4a and 4b were expressed in COS-1 cells, this expectation was borne out.[30] The $K_{1/2}$ value for stimulation of 4b by calmodulin was about 18 nM, whereas that for stimulation of 4a was about 130 nM. Perhaps the most interesting result of this study came from our comparison of the

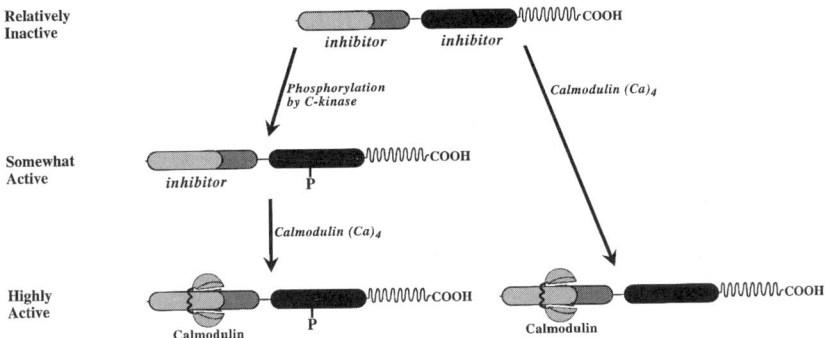

FIGURE 1. Regulation of isoform 4b by calmodulin and C-kinase. C-kinase phosphorylates in a region downstream of the calmodulin-binding domain, and the phosphorylated enzyme is partially activated even in the absence of calmodulin. A more highly activated form occurs when calmodulin binds, even when no phosphorylation has occurred.

Ca^{2+} dependence of the activities of 4b and 4a in the presence of a high concentration of calmodulin. This approximated the conditions believed to exist in a living cell, where calmodulin is usually present in a very high concentration. Under these circumstances, the $K_{1/2}$ values for Ca^{2+} activation of the enzyme were 0.25 µM for 4b and 0.54 µM for 4a. This change in the effective affinity for Ca^{2+} of the enzyme occurs in the presumed physiological range of Ca^{2+} concentrations so that expression of 4a rather than 4b would be expected to have significant consequences for cellular function. One might expect that cells in which higher calcium concentrations were achieved for longer periods would benefit from isoform 4a as compared with 4b. In

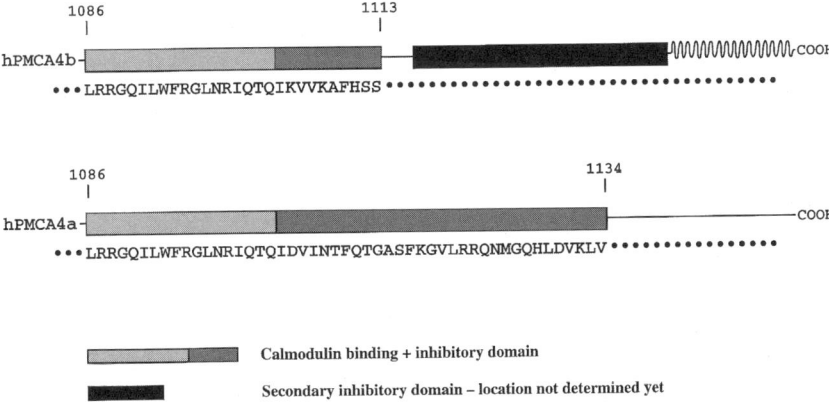

FIGURE 2. Structural difference between regulatory regions of hPMCA4a and hPMCA4b. Isoform 4b has a compact calmodulin-binding domain, which also accounts for most of the autoinhibition of the pump; the rest of the inhibition is contributed by a downstream domain that does not interact with calmodulin. Isoform 4a has an extended domain that binds calmodulin less tightly and that also accounts for all of the inhibition observed in this isoform.

fact, 4a is found primarily in excitable tissues, which tends to support this characterization of it.

This difference in Ca^{2+} affinity is achieved by alterations only in the regulatory region, which is not required for Ca^{2+} transport. In fact, it appears that the lower affinity of 4a for calmodulin leads directly to the lower effective Ca^{2+} affinity. Because 4a is less highly loaded with calmodulin at the low intracellular Ca^{2+} concentrations, the Ca^{2+} curve is displaced toward higher Ca^{2+} concentrations.

In a subsequent study, three truncated mutants of isoform 4a were expressed; they were 4a-ct56, 4a-ct44, and 4a-ct35.[31] The study of these truncated mutants showed us that additional residues, downstream of the 28 residue domain, were needed to account for the affinity of isoform 4a for calmodulin. We found that the calmodulin-binding region of 4a covered approximately 49 residues and that a portion of these 49 residues was probably not involved in the binding. The study of the sequences of the calmodulin-binding domains of the two isoforms (FIG. 2) suggests why these differences in structure exist. In isoform 4b the entire 28 residues contain a net plus charge of 6 and a high proportion of hydrophobic residues, which tend to make it a good calmodulin-binding domain. However, in isoform 4a, the calmodulin-binding domain is interrupted at its twentieth residue by negatively charged aspartic acid, and the next nine residues, down to the next lysine, have no positively charged amino acids, so that these nine residues have a net charge of -1. The next region downstream from the glycine appeared much more favorable for binding of calmodulin. Three positively charged residues and a substantial number of hydrophobic residues are present. This region contains motifs similar to those observed at the beginning of the calmodulin-binding domain. The LRR motif is repeated here from the first three residues of the calmodulin-binding domain and the next four residues at the amino terminus of the calmodulin-binding domain, GQIL, are echoed by GQHL in this downstream region. This structure suggests that the calmodulin-binding domain of 4a consists of two calmodulin-binding regions consisting of hydrophobic and positively charged residues separated by the short hairpin of a negatively charged region that probably does not react with calmodulin.

A study of the activity of the truncated mutants in the absence of calmodulin showed that the autoinhibitory portion of the regulatory region is included within this extended calmodulin-binding region and that no residues downstream of the end of the calmodulin-binding region are needed to account for further inhibition. FIGURE 2 compares the structure of the regulatory regions of isoforms 4a and 4b, showing the second downstream inhibitory region of 4b and the self-contained calmodulin-binding and autoinhibitory region of 4a. It should be remembered that both the inhibition and the affinity for calmodulin are less in 4a than in 4b.

It is clear from these studies that the variations that are exhibited in the carboxyl terminal regions of these pumps have a strong influence on the biological properties of the pump. Alterations in the affinity of the domains for calmodulin may also be accompanied by differences in the rate at which calmodulin binds to the enzyme, which will alter the contribution of the pump to a Ca^{2+} spike, and the effective affinity for Ca^{2+} is affected by these changes. All four of the gene products of calmodulin have a and b splices that would be expected to show differences similar to those displayed by 4a and 4b. Therefore, the results of these studies have strong implications not only for isoform 4 but also for other gene products.

REFERENCES

1. MUALLEM, S., T. BEEKER & S.-J. PANDOL. 1988. Role of Na^+/Ca^{2+} exchange and the plasma membrane Ca^{2+} pump in hormone-mediated Ca^{2+} efflux from pancreatic acini. J. Membr. Biol. **102:** 153–162.
2. WERTH, J. L., Y. M. USACHEV & S. A. THAYER. 1996. Modulation of calcium efflux from cultured rat dorsal root ganglion neurons. J. Neurosci. **16:** 1008–1015.
3. ENYEDI, A., M. FLURA, B. SARKADI, G. GARDOS & E. CARAFOLI. 1987. The maximal velocity and the calcium affinity of the red cell calcium pump may be regulated independently. J. Biol. Chem. **262:** 6425–6430.
4. FILOTEO, A. G., A. ENYEDI & J. T. PENNISTON. 1992. The lipid-binding peptide from the plasma membrane Ca^{2+} pump binds calmodulin, and the primary calmodulin-binding domain interacts with lipid. J. Biol. Chem. **267:** 11800–11805.
5. BRODIN, P., R. FALCHETTO, T. VORHERR & E. CARAFOLI. 1992. Identification of two domains which mediate the binding of activating phospholipids to the plasma membrane Ca^{2+} pump. Eur. J. Biochem. **204:** 939–946.
6. PENNISTON, J. T. 1982. Plasma membrane Ca^{2+}-pumping ATPases. Ann. N.Y. Acad. Sci. **402:** 296–303.
7. KOSK-KOSICKA, D., T. BZDEGA & A. WAWRZYNOW. 1989. Fluorescence energy transfer studies of purified erythrocyte Ca^{2+}-ATPase: Ca^{2+} regulated activation by oligomerization. J. Biol. Chem. **264:** 19495–19499.
8. VORHERR, T., T. KESSLER, F. HOFMANN & E. CARAFOLI. 1991. The calmodulin binding domain mediates the self-association of the plasma membrane Ca^{2+} pump. J. Biol. Chem. **266:** 22–27.
9. WANG, K. K. W., L. C. WRIGHT, C. L. MACHAN, B. G. ALLEN, A. D. CONIGRAVE & B. D. ROUFOGALIS. 1991. Protein kinase C phosphorylates the carboxyl terminus of the plasma membrane Ca^{2+}-ATPase from human erythrocytes. J. Biol. Chem. **266:** 9078–9085.
10. SHULL, G. E. & J. GREEB. 1988. Molecular cloning of two isoforms of the plasma membrane Ca^{2+}-transporting ATPase from rat brain. Structural and functional domains exhibit similarity to Na^+,K^+- and other cation transport ATPases. J. Biol. Chem. **263:** 8646–8657.
11. STREHLER, E. E. 1991. Recent advances in the molecular characterization of plasma membrane Ca^{2+} pumps. J. Membr. Biol. **120:** 1–15.
12. STREHLER, E. E., P. JAMES, R. FISCHER, R. HEIM, T. VORHERR, A. G. FILOTEO, J. T. PENNISTON & E. CARAFOLI. 1990. Peptide sequence analysis and molecular cloning reveal two calcium pump isoforms in the human erythrocyte membrane. J. Biol. Chem. **265:** 2835–2842.
13. SCHATZMANN, H. J. 1966. ATP-dependent Ca^{2+}-extrusion from human red cells. Experientia **22:** 364–365.
14. SCHARFF, O. & B. FODER. 1982. Rate constants for calmodulin binding to Ca^{2+} ATPase in erythrocyte membranes. Biochim. Biophys. Acta **691:** 133–143.
15. GOPINATH, R. M. & F. F. VINCENZI. 1977. Phosphodiesterase protein activator mimics red blood cell cytoplasmic activator of Ca^{2+} Mg^{2+} ATPase. Biochem. Biophys. Res. Commun. **77:** 1203–1209.
16. JARRETT, H. W. & J. T. PENNISTON. 1977. Partial purification of the Ca^{2+}-Mg^{2+} ATPase activator from human erythrocytes: Its similarity to the activator of $3':5'$-cyclic nucleotide phosphodiesterase. Biochem. Biophys. Res. Commun. **77:** 1210–1216.
17. NIGGLI, V., J. T. PENNISTON & E. CARAFOLI. 1979. Purification of the Ca^{2+}-Mg^{2+} ATPase from human erythrocyte membranes using a calmodulin affinity column. J. Biol. Chem. **254:** 9955–9958.
18. ENYEDI, A., B. SARKADI, I. SZASZ, G. BOT & G. GARDOS. 1980. Molecular properties of the red cell calcium pump. II. Effects of calmodulin, proteolytic digestion and drugs on the

calcium-induced membrane phosphorylation by ATP in inside-out red cell membrane vesicles. Cell Calcium **1**: 299–310.

19. ZURINI, M., J. KREBS, J. T. PENNISTON & E. CARAFOLI. 1984. Controlled proteolysis of the purified Ca^{2+} ATPase of the erythrocyte membrane. A correlation between the structure and the function of the enzyme. J. Biol. Chem. **259**: 618–627.

20. VERMA, A. K., A. G. FILOTEO, D. R. STANFORD, E. D. WIEBEN, J. T. PENNISTON, E. E. STREHLER, R. FISCHER, R. HEIM, G. VOGEL, S. MATTHEWS, M. STREHLER-PAGE, P. JAMES, T. VORHERR, J. KREBS & E. CARAFOLI. 1988. Complete primary structure of a human plasma membrane Ca^{2+} pump. J. Biol. Chem. **263**: 14152–14159.

21. JAMES, P., M. MAEDA, R. FISCHER, A. K. VERMA, J. KREBS, J. T. PENNISTON & E. CARAFOLI. 1988. Identification and primary structure of a calmodulin-binding domain of the Ca^{2+} pump of human erythrocytes. J. Biol. Chem. **263**: 2905–2910.

22. ENYEDI, A., T. VORHERR, P. JAMES, D. J. MCCORMICK, A. G. FILOTEO, E. CARAFOLI & J. T. PENNISTON. 1989. The calmodulin binding domain of the plasma membrane Ca^{2+} pump interacts both with calmodulin and with another part of the pump. J. Biol. Chem. **264**: 12313–12321.

23. FALCHETTO, R., T. VORHERR, J. BRUNNER & E. CARAFOLI. 1991. The plasma membrane Ca^{2+} pump contains a site that interacts with its calmodulin binding domain. J. Biol. Chem. **266**: 2930–2936.

24. FALCHETTO, R., T. VORHERR & E. CARAFOLI. 1992. The calmodulin-binding site of the plasma membrane Ca^{2+} pump interacts with the transduction domain of the enzyme. Protein Sci. **1**: 1613–1621.

25. ADAMO, H. P., A. K. VERMA, M. A. SANDERS, R. HEIM, J. L. SALISBURY, E. D. WIEBEN & J. T. PENNISTON. 1992. Overexpression of the erythrocyte plasma membrane Ca^{2+} pump in COS-1 cells. Biochem. J. **285**: 791–797.

26. ENYEDI, A., A. K. VERMA, A. G. FILOTEO & J. T. PENNISTON. 1993. A highly active truncated mutant of the plasma membrane Ca^{2+} pump. J. Biol. Chem. **268**: 10621–10626.

27. VERMA, A. K., A. ENYEDI, A. G. FILOTEO & J. T. PENNISTON. 1994. Regulatory region of plasma membrane Ca^{2+} pump. 28 residues suffice to bind calmodulin but more are needed for full auto-inhibition of the activity. J. Biol. Chem. **269**: 1687–1691.

28. ENYEDI, A., A. K. VERMA, A. G. FILOTEO & J. T. PENNISTON. 1996. Protein kinase C activates the plasma membrane Ca^{2+} pump isoform 4b by phosphorylation of an inhibitory region downstream of the calmodulin-binding domain. J. Biol. Chem. **271**: 32461–32467.

29. ENYEDI, A., A. G. FILOTEO, G. GARDOS & J. T. PENNISTON. 1991. Calmodulin-binding domains from isozymes of the plasma membrane Ca^{2+} pump have different regulatory properties. J. Biol. Chem. **266**: 8952–8956.

30. ENYEDI, A., A. K. VERMA, R. HEIM, H. P. ADAMO, A. G. FILOTEO, E. E. STREHLER & J. T. PENNISTON. 1994. The Ca^{2+} affinity of the plasma membrane Ca^{2+} pump is controlled by alternative splicing. J. Biol. Chem. **269**: 41–43.

31. VERMA, A. K., A. ENYEDI, A. G. FILOTEO, E. E. STREHLER & J. T. PENNISTON. 1996. Plasma membrane calcium pump isoform 4a has a longer calmodulin-binding domain than 4b. J. Biol. Chem. **271**: 3714–3718.

Structural Aspects of the Gastric H,K ATPase[a]

JAI MOO SHIN,[b] MARIE BESANCON, KRISTER BAMBERG, AND GEORGE SACHS

Department of Physiology and Medicine
UCLA and Wadsworth VA Hospital
Los Angeles, California 90073

In the absence of resolved crystals of any P-type ATPase allowing determination of the number and orientation of the transmembrane segments, evidence for these has necessarily been based on application of molecular biological and biochemical methods. The importance of this domain for ion transport was first established by the demonstration of an occluded form of K^+ in the Na,K ATPase almost a quarter of a century ago, long before any cDNA-based sequence analysis had been thought of.[1,2] The occluded form was conceived as being bound or trapped within the enzyme, probably in the membrane domain of the enzyme.

As the first cDNA sequences became available for the Na,K and sr Ca ATPases, hydropathy plots were interpreted as showing the presence of 8 and 10 transmembrane helices, respectively.[3,4] The hydropathy profile of these two enzymes was clear in predicting four NH_2-terminal transmembrane domains, but it was difficult to interpret in the region of the fifth through seventh hydrophobic segments, which resulted in varying interpretations. Experimental methods had to be developed to establish the number of transmembrane segments. Even today, some of these methods have resulted in models inconsistent with most experimental data.[5,6] Several have been applied to the gastric H,K ATPase, because no single method proved adequate for all segments that appeared intuitively to be present.

TOPOGRAPHICAL ANALYSIS

Epitope Localization

Antibody epitope localization appears capable of giving clear data. However, almost immediately, the problem of predictable cell permeabilization resulted in placement of the COOH-terminus of the Na,K ATPase as being extracytoplasmic.[7] On the other hand, a comparison of intact and broken sarcoplasmic vesicles using two separate antibodies generated against the loop between the seventh and eighth transmembrane segments provided direct evidence for this sector of the sr Ca ATPase, establishing the likelihood for 10 transmembrane segments in this enzyme.[8,9] Results of mutagenesis of hydrophilic amino acids within the predicted transmembrane domain also could most easily be interpreted as reflecting the presence of 10 transmembrane segments.

Recently, epitope localization was attempted on heat-treated Na,K ATPase with data inconsistent with a stable 10-membrane segment model.[5] The addition of antibody-reactive sequences also resulted in a controversial placement of the extracytoplasmic domain in the COOH-terminal region of the enzyme.[10]

---

[a]This work was supported by US Veterans Administration SMI and by National Institutes of Health grants DK40615, 41301, and 17294.
[b]Tel: 310-268-4672; fax: 310-312-9478.

With gastric H,K ATPase, inappropriate localization of a monoclonal antibody against the α subunit[11] also resulted in misinterpretation of the topography of this member of the eukaryotic P ATPases. Interpretation of an α and β subunit antibody (mAb 146) generated by injection of rat parietal cells as being directed only against the α subunit[12] also resulted in confusing data. In fact, this particular mAb 146 antibody showed the presence of a shared epitope recognizing a sequence in the M7/M8 membrane interface (AGFTD) as defined by octamer walking and also recognized a sequence in the β subunit of rat and rabbit but not hog ATPase.[13] These data suggested simultaneous presentation of these two fragments due to strong interaction in the region of the shared epitopes. Reactivity against the rat β sequence, lower reactivity against the rabbit β sequence, and virtually no recognition of the hog β sequence allowed assignment of the epitope for this antibody to a specific region of the β subunit where significant amino acid differences were observed among the three species. This is in the region around positions 165 and 166.

Evidently, epitope localization has advantages and disadvantages. Data from such methods have provided controversial models and should be viewed as fragments of a picture, not the entire picture.

Tryptic Digestion

A particularly useful method is tryptic or other protease cleavage of cytoplasmic side-out oriented vesicles of the gastric H,K or sr Ca ATPase.[14,15] After exhaustive cleavage of the cytoplasmic domain and sequencing of peptides remaining in the membrane, four pairs of transmembrane sequences were detected in the gastric H,K ATPase and four pairs and an additional H9 sequence were detected in the sr Ca ATPase. Detection of membrane sequences in this method depended on reaction of free SH groups with the fluorescent reagent fluorescein-5-maleimide. Absence of reactive SH groups in the tenth hydrophobic domain of the sr Ca ATPase accounts for inability to detect all 10 transmembrane segments. The lack of reactivity of the ninth hydrophobic sequence of the gastric H,K ATPase to FMI remains mysterious. Neither disulfide-reducing agents nor thioester cleavage renders this segment visible on separating gels (Shin, unpublished data). Cleavage by trypsin at the arg residue between M9 and M10 of the sr Ca ATPase was also unexpected.

In vitro *Translation*

Inspection of the hydropathy plot of the H,K ATPase shows adequate hydrophobicity for predicting membrane-spanning properties of the first four and the eighth through tenth hydrophobic region. The H5/6 sequence detected as containing a membrane-spanning pair by trypsinolysis and labeling is predicted to be a single membrane-spanning domain in this enzyme, and the seventh segment is apparently insufficiently hydrophobic to be predicted by standard algorithms. On the other hand, the ninth and tenth sequences should form a pair of transmembrane domains based on almost all predictive algorithms.

A method of *in vitro* translation was developed to analyze all possible membrane-spanning domains of the gastric H,K ATPase, and this method was applied later to the SERCA 2 Ca ATPase.[21,22] Membrane insertion of polytopic integral membrane proteins is thought to proceed via insertion of alternating signal anchor and stop transfer sequences,[23] and any cDNA sequence can be analyzed for the presence of such sequences using the vector system to be described.

Two plasmid vectors were constructed in a modified pGEM plasmid for transcription-translation in the presence or absence of microsomal membranes using the TNT reticulocyte lysate system. The first contained the cDNA encoding the first 101 amino acids of the α subunit of the H,K ATPase, a linker sequence to allow insertion of putative transmembrane segments and the cDNA encoding for the last 177 amino acids of the β subunit with five potential N-linked glycosylation sites. This was the M0 vector, able to define membrane insertion sequences capable of insertion on their own, namely, signal anchor sequences. The vector designed to detect stop transfer activity was similar to the M0 vector, but it began with the first 139 triplets of the α subunit, which contains the first transmembrane-coding sequence. After incubation with appropriate medium-containing amino acids and energy source in the absence and presence of microsomes, the radioactive products of the vector translation experiment are separated on SDS gels to determine if the N-linked glycosylation sites are glycosylated. This depends on insertion of an odd number of transmembrane sequences to see glycosylation in the M0 vector or on the insertion of an even number in the M1 vector preventing glycosylation. This method of *in vitro* translation scanning can be applied to any putative transmembrane sequence. FIGURE 1 shows construction of the vectors.

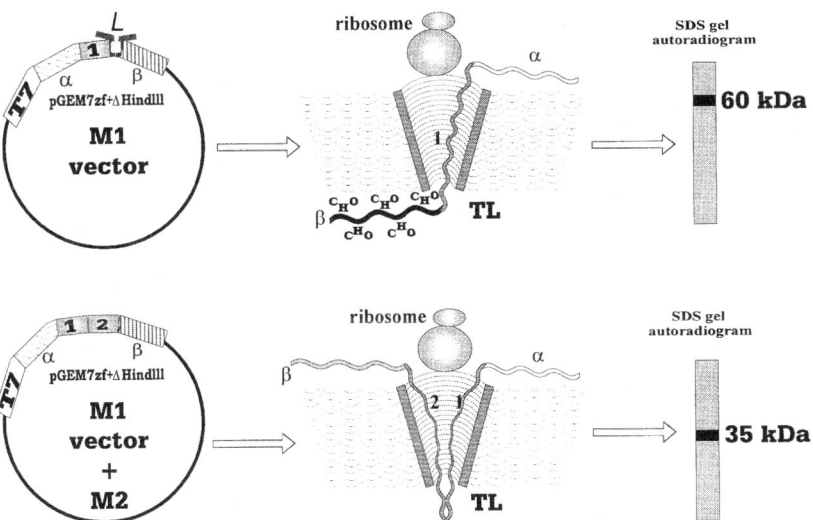

FIGURE 1. The construction of vectors able to analyze the presence of membrane signal sequences in the α subunit of P-type ATPases. This method depends on translation in the absence and presence of microsomes and detects glycosylation of the COOH-terminal sequence.

The first four segments behaved as membrane insertion signals. Thus, H1, H2, and H3 acted as both signal anchor and stop transfer sequences, whereas H4 showed only stop transfer ability. However, this was consistent with its location in the mature protein. Surprisingly, H5, H6, and H7 were without signal insertion properties in this translation system, whereas H8 had stop transfer. When H9, H10, or H9 + H10 were translated in these vectors, clear evidence was obtained for signal anchor properties of H9, stop transfer properties of H10, and membrane pair insertion of H9 + H10. This finding of *in vitro* translation is interpreted as showing the presence of this membrane pair in the mature, assembled protein and substitutes for evidence that should have been present using more classical methods.

The translation method, therefore, does not read the properties of three of the known transmembrane domains of the gastric ATPase.[21] However, evidence that these M5, M6, and M7 segments exist on their own in strong and translation data should not result in rejection of data from another technique. This method applied to the SERCA 2 Ca ATPase, predicted to have 11 transmembrane segments,[24] provided appropriate signal insertion properties for 9 of the 11 segments, H5 and H6 again giving anomalous results.

The hydropathy profile of this H5/H6 region in these two enzymes is similar and is interpreted by all algorithms as a single transmembrane sequence. An explanation for the anomalous results of translation may be that factors involved in folding of the proteins in this region are missing in the TNT reticulocyte-dog microsome system, that posttranslational insertion occurs in this region, or that this pair is not α helical. Most algorithms predict a mixture of β sheet and random coil for the TM5/6 domain. *In vitro* translation of the sequences in this domain, if not α helical, would therefore not detect these as membrane insertion sequences.

Analysis of the Mg ATPase of *Salmonella typhimurium* using insertion of either β lactamase or β galactosidase as external and internal markers, respectively, resulted in data suggesting a 10-membrane segment-spanning model that seems now to be the consensus topography for these alkali cation ATPases.[25] By contrast, bacterial transition metal P-type ATPases have only eight transmembrane sequences as shown by the *in vitro* translation method just described.[26]

Translation and digestion studies focusing on the β subunit provide evidence for a single membrane-spanning segment.

Extracytoplasmic Reagents

General Considerations

The availability of extracytoplasmic inhibitors of the Na,K and H,K ATPases has proved particularly useful for these two enzymes. For example, comparison of the ouabain-insensitive rat sequence with the more sensitive human or hog sequence showed that predicted boundary amino acids in the first and second transmembrane domain were responsible for this difference.[16,17] Saturation mutagenesis provided additional evidence that the loop joining TM5 and TM6 was also involved in ouabain binding[18] which correlated with the presence of ion binding sites in this region.

With H,K ATPase, reagents such as thiophilic acid-activated antiulcer drugs, such as omeprazole, lansoprazole, and pantoprazole, bound to the second, third, and fourth predicted extracytoplasmic loops of a 10-transmembrane segment model.[14,15] A variant of the extracytoplasmic K^+ competitive, 1,2α-imidazopyridine, SCH 28080,[19] ^{3}H Me-DAZIP, bound to the first and second transmembrane segments.[20]

Reaction with the Substituted Pyridyl-2-Methylsulfinyl-Benzimidazoles

There are two classes of H,K ATPase inhibitors, as already discussed. The K^+ competitive type, as exemplified by the imidazopyridines and arylquinolines, bind noncovalently from the extracytoplasmic side and must have two photoactivatable groups to be able to act as cross-linking reagents. These compounds were ineffective cross-linking reagents. Radioactive compounds bound mainly to the lipid and not to the membrane protein. We therefore focus in the rest of this brief review on the covalent inhibitors of the gastric H,K ATPase.

The substituted pyridyl methylsulfinyl benzimidazoles, as just stated, are weak bases that accumulate in acidic spaces. Their general structure and mechanism of inhibition are shown in FIGURE 2. After accumulation in an acidic space, they undergo an acid-catalyzed rearrangement to form tetracyclic sulfenamides that are permanent cations.[27–29] The mechanism of rearrangement depends on a nucleophilic attack of the unprotonated pyridine N on the 2C of the benzimidazole. Therefore, an increase in pK_a of this N will accelerate the rearrangement. However, the increased protonation in the acid space of the parietal cell will decrease reactivity. Hence, the influence of various substituents in either the pyridine or the benzimidazole ring is sometimes difficult to predict. An additional complication is introduced by the structure of the enzyme itself.

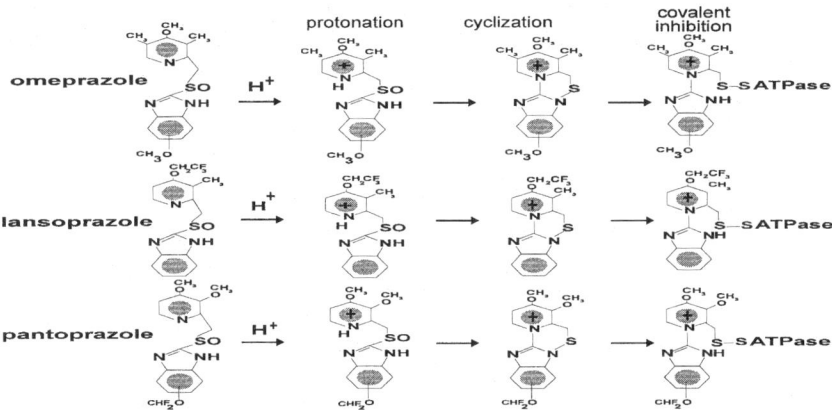

FIGURE 2. The structure and reaction mechanism of the three proton pump inhibitor drugs currently available commercially. The compounds accumulate in an acidic space of pH less than 4.0, undergo acid activation to the cationic sulfenamide, and react with accessible SH groups from the luminal surface of the H,K ATPase.

The rate of inhibition of the ATPase under acid transport conditions by omeprazole and lansoprazole is equal to but faster than the more acid stable compound pantoprazole. The rate of inhibition of acid transport by the vesicles followed the same pattern, the rate of inhibition being faster with omeprazole and lansoprazole than with pantoprazole. It is notable that when ATP is added in the presence of these drugs, the initial rate of acidification is the same as that in the control in the absence of drug. These particular compounds at 10 µM inhibit only after acidification.

Measurement of the sites of covalent labeling by lansoprazole, omeprazole, and pantoprazole was performed by reacting the radioactive compounds with the isolated vesicular gastric H,K ATPase under acid-transporting conditions. When the enzyme was digested and the membrane pellet separated on SDS tricine gradient gels, despite their similar mechanism of acid activation and disulfide bond formation, distinct labeling patterns were found. From microsequence analysis, lansoprazole labels the enzyme at three positions, cysteine 321 in the TM3/4 domain, cys 813/822 in the TM5/6 domain, and cys 892 in the TM7/8 domain. Omeprazole labels the enzyme at two positions, again cys 813/822 in the TM5/6 domain and cys 892 in the TM7/8 domain. The other cysteine present in the TM7/8 domain was shown not to be labeled by analysis of fragments resulting from V8 protease digestion.[14] Pantoprazole labels only the cysteines in the TM5/6 domain. FIGURE 3 illustrates the result found with pantoprazole labeling.

CROSS-LINKING STUDIES

An unresolved problem for any P-type ATPase is the arrangement of the transmembrane segments. Two general cross-linking approaches can be taken, either to use cross-linking sites within the amino acid sequence such as natural or inserted cysteines (MacLennan, this volume) or to use chemical cross-linking reagents. Thus far we have been unable to express the H,K ATPase in sufficient quantity for cross-linking studies using internal mutations.

The membrane domain, although composed of five pairs of transmembrane segments and five extracytoplasmic loops with, in addition, the transmembrane domain of the β subunit and its extracytoplasmic domain, is likely to be extremely compact to be able to transport ions but not to create an ion leak. Furthermore, there are no obvious groups on the lipid-facing surfaces of the enzyme that would react quantitatively with bifunctional reagents.

We first used a series of commercially available reagents, then synthesized some of our own. None worked satisfactorily in that no obvious quantitative changes in any of the membrane segments were detected after digestion by either staining or FMI labeling. We decided that not enough reactive groups were accessible to these reagents in the membrane domain.

The method illustrated here, which gave adequate cross-linking to allow conclusions as to the nearest loops to the loop between TM5/TM6, was to synthesize an azido derivative similar to pantoprazole, to give similar acid stability and similar covalent binding specificity.

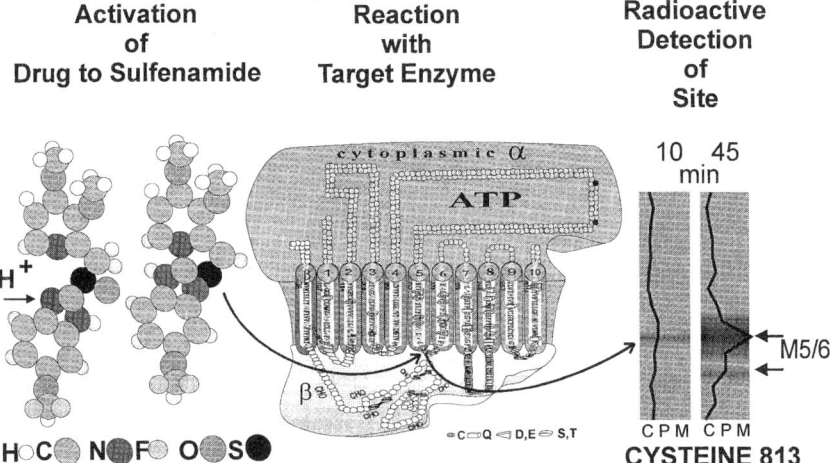

FIGURE 3. Autoradiograms of the membrane segments left after tryptic digestion of the enzyme after inhibition by radioactive pantoprazole, showing the activation, a two-dimensional map of the ATPase, and the autoradiographic bands with counts on the gel superimposed.

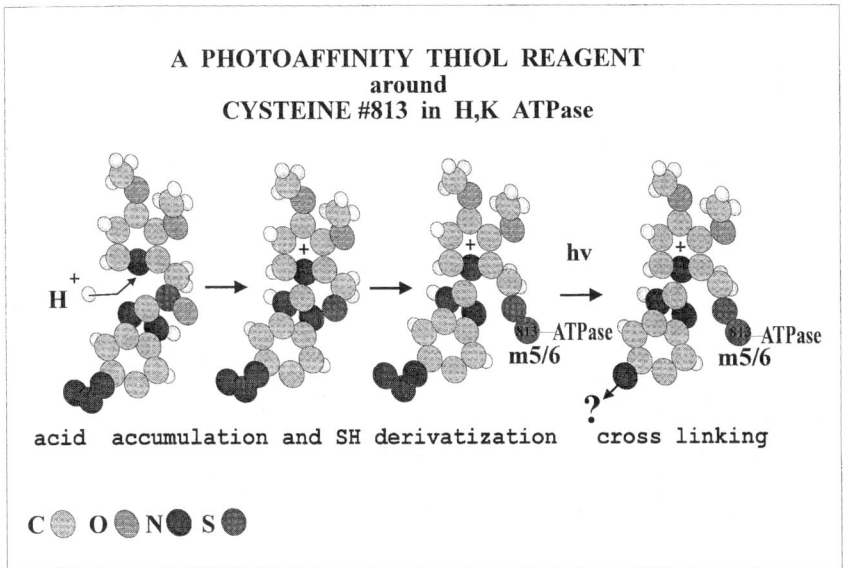

FIGURE 4. Steps involved in generation of cross-linking using an azido derivative of pantoprazole.

Each compound binds to characteristic cysteines. Pantoprazole was relatively more selective in binding to the loop between TM5 and TM6 and was therefore chosen as the lead structure for making the first photoaffinity analog. The substitution of a large group such as a benzophenone resulted in an inactive compound, probably because the substituent was too large. The azido group, however, is about the same size as the resident difluoromethoxy group and was substituted in the same position in the ring to give 2[(3,4-methoxy)pyridin-2-yl-methylsulfinyl]-1H-(5-azido)-benzimidazole. This azido derivative was able to inhibit the H,K ATPase under acid-transporting conditions, although somewhat more stable than the parent difluoromethoxy derivative, the 5 azido substituent being more electrophilic. The compound and the steps used to generate a covalent cross-link are illustrated in FIGURE 4.

The derivatization of the enzyme is carried out under acid-transporting conditions, so that there is selective covalent labeling of the TM5/TM6 extracytoplasmic loop. The enzyme is then exposed to ultraviolet light. This is followed by tryptic hydrolysis, solubilization in SDS, and labeling with fluorescein maleimide to detect the transmembrane segments. Controls are run without the benzimidazole and without UV irradiation. Cross-linking was found between the covalent binding site on the loop between TM5 and TM6 and TM3 and TM4 and some to the TM7 and TM8 loop. The size of the bound derivative suggests that these loops are within 8–10 Å of the central TM5/TM6 loop, as illustrated in FIGURE 5.

DISCUSSION

The consensus that has developed for the alkali cation P-type ATPases suggests the presence of 10 transmembrane domains. Various methods applied to the gastric H,K

ATPase provide direct evidence for eight segments, and extrapolation from *in vitro* translation also puts the number of such segments at 10. FIGURE 6 illustrates a model of such an arrangement, with association of the β subunit with the loop between TM7 and TM8 as well as a point of close approach to another α subunit at cys 565 and/or 615.

There are certainly still caveats for an uncritical acceptance of this model. These relate to the position of TM9 and TM10. The weight of evidence for extracytoplasmic exposure of the loops between TM1 and TM2, TM3 and TM4, TM5 and TM6 as well as TM7 and TM8 is overwhelming for the gastric H,K, the SERCA Ca ATPases and the Na,K ATPase.[4,14,15,31] However, it is not clear why TM9 and TM10 are not visualized by FMI in the gastric H,K ATPase, and cleavage between H9 and H10 in the sr Ca ATPase also points to a readier access of trypsin to this potential cleavage site as compared to an earlier site between TM3 and TM4. The finding that this *in vitro* translation and oocyte expression of partial constructs of the Na,K ATPase[32] demonstrates the presence of TM9 and TM10 is encouraging for the model. Evidently crystal structure will be conclusive, but it is not available at sufficiently high resolution as yet.[33]

The status of the two transmembrane segments M5 and M6 is another vexing problem. Whereas digestion or labeling provides unequivocal evidence that these segments exist, the absence of membrane insertion properties and the destabilization of segments extending beyond TM5 in the frog oocyte point to a peculiarity in this region of the enzyme in terms of either structure and/or mechanism of membrane assembly. Apparently TM7 is also at least somewhat deficient in its signal anchor properties in the Ca ATPase and certainly so in the H,K ATPase. Oocyte expression data also point to an anomaly in the assembly of this region of the Na,K ATPase.[34]

The order of arrangement of the transmembrane helices for any P-type ATPase is

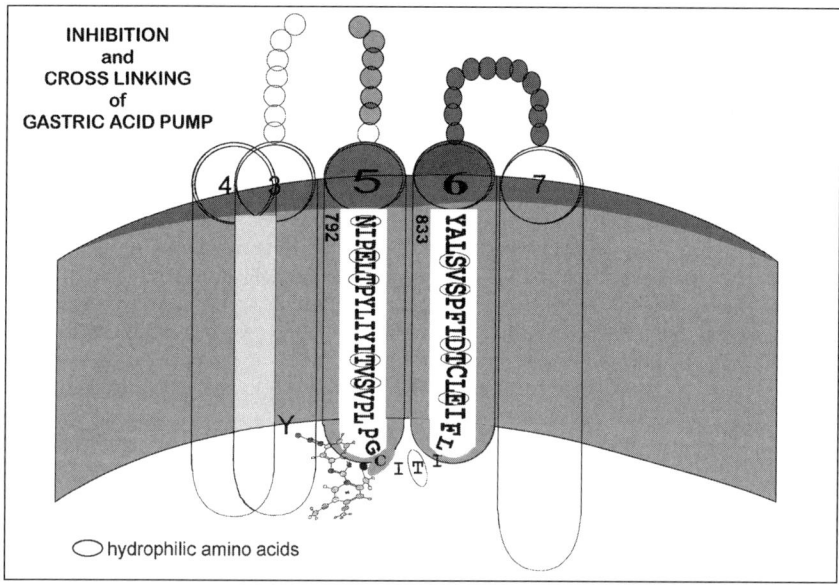

FIGURE 5. A model of the reaction of the azido derivative after SH reaction under acid-transporting conditions and photolytic activation. A possible site of cross-linking is the tyrosine in TM3.

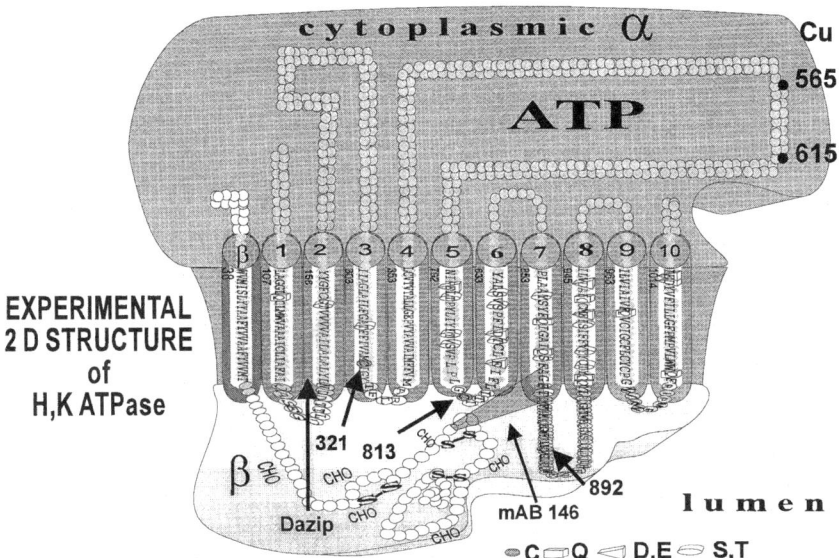

FIGURE 6. A model summarizing the data obtained for the topography of the gastric H,K ATPase. This shows the 10-membrane segments obtained from trypsinolysis and *in vitro* translation, sites of reaction with the benzimidazoles and 3H-Me-DAZIP, sites of Cu cross-linking, and association between the α and β subunits from WGA chromatography and cross-reactivity of mAb 146 in the β sequence and in the boundary region of TM7.

unknown. The use of cross-linking is at the moment the necessary biochemical approach for these ATPases because they are not as yet expressed in sufficient quantity to enable the cysteine energy transfer technique that has been applied to lac permease.[35]

Whereas Cu-dependent cross-linking was successfully applied to define cysteine proximity between two α subunits,[36] it was unable to cross-link any of the transmembrane segment pairs. Either there are no cysteines in close proximity or Cu was unable to penetrate the membrane domain. A variety of commercially available cross-linking reagents were also tried without success, as were other chemicals synthesized in our laboratory intended to mimic K^+ competitive reagents.

The use of the benzimidazoles was thought perhaps to be a fruitful approach, because the generation of a photoaffinity group linked to the aryl group in the benzimidazole would provide a means of first quantitatively forming a covalent disulfide at the benzimidazole binding site(s) within the membrane or extracytoplasmic domain and then using nitrene or 7-membered ring generation after irradiation to detect reactive groups (perhaps tyrosines) within about 10 Å of the cysteine(s) to which the benzimidazole is bound after acid transport activation. The presence of a covalent disulfide link renders the long half-life of the potential nitrene or 7-membered ring generated after irradiation useful rather than confounding. This approach appears to yield quantitative cross-linking between at least two membrane pairs in the H,K ATPase.

The data obtained provided evidence for proximity between the derivatized TM5/6 domain and the TM3/4 domain. The efficacy of this reagent suggests that the use of similar derivatives of either omeprazole or lansoprazole will be able to define the nearest neighbors of TM7/8 and TM3/5, respectively, so that we can order the pairs of transmembrane sequences. For deciding which member of the pair is coupled, we will need to know the particular amino acid that is covalently modified. A likely target amino acid for the photoaffinity part of the reagent is tyrosine which is present close to the extracytoplasmic surface in all of the predicted transmembrane pairs. Another acid-transporting P-type ATPase is the H ATPase of yeast or neurospora. This type of reagent, coupled to cysteine mutagenesis in the extracytoplasmic loops, might allow mapping of all the transmembrane segments of these P-type ATPases.

SUMMARY

The gastric H,K ATPase is an α β heterodimeric member of the eukaryotic alkali-cation P-type ion-motive ATPase family. The α subunit is composed of 1033 amino acids and the β subunit of 291 amino acids with 6 or 7 potential N-linked glycosylation sites. Much effort has been expended to define the membrane domain of P-type ATPases. A membrane domain of the large subunit consisting of 10 membrane-spanning sequences is suggested by a combination of methods such as (1) tryptic digestion, separation, and sequencing of membrane peptides, (2) labeling with extracytoplasmic reagents, and (3) *in vitro* translation of hydrophobic segments. The β subunit has a single transmembrane segment with strong hydrophobic interactions with the α subunit. Blue native gel electrophoresis shows that the enzyme is an $(\alpha\text{-}\beta)_2$ dimer.

Cross-linking with Cu-phenanthroline provides evidence that association is between the α subunits, and the potential SH groups that are Cu sensitive are at cysteine 565 and cysteine 615, in the region of the large cytoplasmic loop between the fourth and fifth transmembrane segments. No cross-linking is observed in the membrane domain. ATP prevents cross-linking because of a conformational change at the surface of the protein induced by ATP or by direct binding of the nucleotide at the site of cross-linking. The WGA binding properties of the β subunit allow investigation of the region of interaction with the α subunit. Thus, digestion of the enzyme by trypsin followed by SDS solubilization and selective elution from a WGA column resulted in coelution of the membrane fragment containing TM7 and TM8. This result demonstrates major hydrophobic interaction between the seventh and eighth transmembrane segments and the β subunit. An antibody generated against rat parietal cells also recognized shared epitopes in the same region of both the α and β subunits. Biochemical investigation of the arrangement of the transmembrane segments has been hindered by the lack of effective cross-linking reagents probably because of the compact arrangement of this domain, preventing even Cu access.

A series of antiulcer drugs has been developed that have a unique chemistry related to their inhibition of the gastric H,K ATPase. They are 2-(substituted pyridyl methylsulfinyl) benzimidazoles, weak bases with a pK_a of 4.0. After accumulation in the acidic space generated by the H,K ATPase either *in vivo* or *in vitro*, they undergo acid-catalyzed conversion to a tetracyclic sulfenamide which reacts with luminally accessible SH residues to form stable disulfide derivatives. In the particular case of pantoprazole, 2-(3,4-dimethoxy-2-pyridyl-methylsulfinyl)-5-difluoromethoxy benzimidazole, reaction is confined largely to cysteine 813, placed between the fifth and sixth transmembrane segments. The 5 azido analog of pantoprazole provided acid transport-dependent inhibition of the isolated transporting ATPase by this photoactivatable covalent SH reagent. The inhibited enzyme was then photolyzed, cleaved with trypsin, and the membrane fragments compared before and after photolysis. Disappearance of the segment corresponding to TM3,4 and a relative loss of the segment corresponding to TM7,8 suggests close proximity of these two membrane pairs to the loop joining the fifth and sixth transmembrane segments, in particular TM3,4. Use of this type of covalent, photoactivatable site-specific reagent to determine loop proximity can be extended to other acid transporters.

REFERENCES

1. POST, R. L. 1989. Seeds of sodium, potassium ATPase. Annu. Rev. Physiol. **51:** 1–15.
2. GLYNN, I. M. & S. J. D. KARLISH. 1990. Occluded cations in active transport. Annu. Rev. Biochem. **59:** 171–205.
3. SHULL, G. E., A. SCHWARTZ & J. B LINGREL. 1985. Amino-acid sequence of the catalytic subunit of the (Na^+-K^+)ATPase deduced from a complementary DNA. Nature **316:** 691–695.
4. MACLENNAN, D. H., C. J. BRANDL, B. KORCZAK & N. M. GREEN. 1985. Amino-acid sequence of Ca^{2+}-Mg^{2+}-dependent ATPase from rabbit muscle sarcoplasmic reticulum, deduced from its complementary DNA sequence. Nature **316:** 696–700.
5. ARYSTARKHOVA, E., D. L. GIBBONS & K. J. SWEADNER. 1995. Topology of the Na,K-ATPase. Evidence for externalization of a labile transmembrane structure during heating. J. Biol. Chem. **270:** 8785–8796.
6. XIE, Y., S. A. LANGHANS-RAJASEKARAN, D. BELLOVINO & T. MORIMOTO. 1996. Only the first and the last hydrophobic segments in the COOH-terminal third of Na,K-ATPase α subunit initiate and halt, respectively, membrane translocation of the newly synthesized polypeptide. Implications for the membrane topology. J. Biol. Chem. **271:** 2563–2573.
7. OVCHINNIKOV, Y. A., N. M. LUNEVA, E. A. ARYSTARKHOVA, N. M. GEVONDYAN, N. M. ARZAMAZOVA, A. T. KOZHICH, V. A. NESMEYANOV & N. N. MODYANOV. 1988. Topology of Na+,K+-ATPase. Identification of the extra- and intracellular hydrophilic loops of the catalytic subunit by specific antibodies. FEBS Letts. **227:** 230–234.
8. CLARKE, D. M., T. W. LOO & D. H. MACLENNAN. 1990. The epitope for monoclonal antibody A20 (amino acids 870–890) is located on the luminal surface of the Ca-ATPase of sarcoplasmic reticulum. J. Biol. Chem. **265:** 17405–17408.
9. MATTHEWS, I., R. P. SHARMA, A. G. LEE & J. M. EAST. 1990. Transmembranous organization of (Ca-Mg)-ATPase from sarcoplasmic reticulum. Evidence for luminal location of residues 877–888. J. Biol. Chem. **265:** 18737–18740.
10. SHYJAN, A. W. & R. LEVENSON. 1989. Antisera specific for $\alpha 1$, $\alpha 2$, $\alpha 3$, and β subunits of the Na,K-ATPase: Differential expression of α and β subunits in rat tissue membranes. Biochemistry **28:** 4531–4535.
11. SMOLKA, A., L. ALVERSON, R. FRITZ, K. SWIGER & R. SWIGER. 1991. Gastric H,K-ATPase topography: Amino acids 888–907 are cytoplasmic. Biochem. Biophys. Res. Commun. **180:** 1356–1364.
12. MERCIER, F., H. REGGIO, G. DEVILLIERS, D. BATAILLE & P. MANGEAT. 1989. Membrane-cytoskeleton dynamics in rat parietal cells: Mobilization of actin and spectrin upon stimulation of gastric acid secretion. J. Cell Biol. **108:** 441–453.
13. MERCIER, F., D. BAYLE, M. BESANCON, T. JOYS, J. M. SHIN, M. J. M. LEWIN, C. PRINZ, A. M. REUBEN, A. SOUMARMON, H. WONG, J. H. WALSH & G. SACHS. 1993. Antibody epitope mapping of the gastric H/K-ATPase. Biochim. Biophys. Acta **1149:** 151–165.
14. BESANCON, M., J. M. SHIN, F. MERCIER, K. MUNSON, M. MILLER, S. HERSFY & G. SACHS. 1993. Membrane topology and omeprazole labeling of the gastric H,K-adenosinetriphosphatase. Biochemistry **32:** 2345–2355.
15. SHIN, J. M., M. BESANCON, A. SIMON & G. SACHS. 1993. The site of action of pantoprazole in the gastric H/K-ATPase. Biochim. Biophys. Acta **1148:** 223–233.
16. PRICE, E. M. & J. B LINGREL. 1988. Structure-function relationship in the Na,K-ATPase α subunit: Site-directed mutagenesis of glutamine-111 to arginine and asparagine-122 to aspartic acid generates a ouabain-resistant enzyme. Biochemistry **27:** 8400–8408.

17. PRICE, E. M., D. A. RICE & J. B. LINGREL. 1990. Structure-function studies of Na,K-ATPase. Site-directed mutagenesis of the border residues from the H1-H2 extracellular domain of the α subunit. J. Biol. Chem. **265:** 6638–6641.
18. PALASIS, M., T. A. KUNTZWEILER, J. M. ARGUIELLO & J. B. LINGREL. 1996. Ouabain interactions with the H5-H6 hairpin of the Na,K-ATPase reveal a possible inhibition mechanism via the cation binding domain. J. Biol. Chem. **271:** 14176–14182.
19. WALLMARK, B., C. BRIVING, J. FRYKLUND, K. MUNSON, R. JACKSON, J. MENDLEIN, E. RABON & G. SACHS. 1987. Inhibition of gastric H,K-ATPase and acid secretion by SCH 28080, a substituted pyridyl[1,2α]imidazole. J. Biol. Chem. **262:** 2077–2084.
20. MUNSON, K. B., C. GUTIERREZ, V. N. BALAJI, K. RAMNARAYAN & G. SACHS. 1991. Identification of an extra-cytoplasmic region of H,K-ATPase labeled by a K^+-competitive photoaffinity inhibitor. J. Biol. Chem. **266:** 18976–18988.
21. BAMBERG, K. & G. SACHS. 1994. Topological analysis of the H,K-ATPase using *in vitro* translation. J. Biol. Chem. **269:** 16909–16919.
22. BAYLE, D., D. WEEKS & G. SACHS. 1995. The membrane topology of the rat SERCA ATPases by *in vitro* translation scanning. J. Biol. Chem. **270:** 25678–25684.
23. FRIEDLANDER, M. & G. BLOBEL. 1985. Bovine opsin has more than one signal sequence. Nature **318:** 338–343.
24. KEETON, T. P., S. E. BURK & G. E. SHULL. 1993. Alternative splicing of exons encoding the calmodulin-binding domains and C termini of plasma membrane Ca^{2+}-ATPase isoforms 1, 2, 3, and 4. J. Biol. chem. **268:** 2740–2748.
25. SMITH, D. L., T. TAO & M. E. MAGUIRE. 1993. Membrane topology of a P-type ATPase. The MgtB magnesium transport protein of *Salmonella typhimurium*. J. Biol. Chem. **268:** 22469–22479.
26. MELCHERS, K., T. WEITZENEGGER, A. BUHMANN, W. STEINHILBER, G. SACHS & K. P. SCHAFER. 1996. Cloning and membrane topology of a P type ATPase from *Helicobacter pylori*. J. Biol. Chem. **271:** 446–457.
27. LINDBERG, P., P. NORDBERG, T. ALMINGER, A. BRANDSTROM & B. WALLMARK. 1986. The mechanism of action of the gastric acid secretion inhibitor, omeprazole. J. Med. Chem. **29:** 1327–1329.
28. SENN-BILLFINGER, J., U. KRUGER, E. STURM, V. FIGALA, K. KLEMM, B. KOHL, G. RAINER, H. SCHAEFER, T. J. BLAKE, D. W. DARKIN, R. J. IFE, C. A. LEACH, R. C. MITCHELL, E. S. PEPPER, C. J. SLTER, N. J. VINE, G. HUTTNER & L. ZSOLNAI. 1987. J. Org. Chem. **52:** 4582–4592.
29. SINGH, P., R. C. SHARMA & T. N. OJHA. 1991. Drug Design & Delivery **7:** 131–138.
30. RABON, E., G. SACHS, S. BASSILIAN, C. LEACH & D. KEELING. 1991. K^+-Competitive fluorescent inhibitor of the H,K-ATPase. J. Biol. Chem. **266:** 12395–12401.
31. KARLISH, S. J. D., R. GOLDSHLEGER & P. L. JORGENSEN. 1993. Location of asn831 of the α chain of Na/K-ATPase at the cytoplasmic surface. Implication for topological models. J. Biol. Chem. **268:** 3471–3478.
32. MATHEWS, P. M., D. CLAEYS, F. JAISSER, K. GEERING, J. D. HORISBERGER, J. P. KRAEHENBUHL & B. C. ROSSIER. 1995. Primary structure and functional expression of the mouse and frog α-subunit of the gastric H^+-K^+-ATPase. Am. J. Physiol. **268:** C1207–C1214.
33. TOYOSHIMA, C., H. SASABE & D. STOKES. 1993. Three-dimensional cryo-electron microscopy of the calcium ion pump in the sarcoplasmic reticulum. Nature **362:** 469–471.
34. JAUNIN, P., J. D. HORISBERGER, K. RICHTER, P. J. GOOD, B. C. ROSSIER & K. GEERING. 1992. Processing, intracellular transport, and functional expression of endogenous and exogenous a-b3 Na,K-ATPase complexes in xenopus oocytes. J. Biol. Chem. **267:** 577–585.
35. FRILLINGOS, S. & H. R. KABACK. 1996. Probing the conformation of the lactose permease of *Escherichia coli* by *in situ* site-directed sulfhydryl modification. Biochemistry **35:** 3950–3956.
36. SHIN, J. M. & G. SACHS. 1996. Dimerization of the gastric H,K-ATPase. J. Biol. Chem. **271:** 1904–1908.

P-Type H$^+$- and Ca^{2+}-ATPases in Plant Cells[a]

B. STANGELAND,[b] A. T. FUGLSANG,[b,c] S. MALMSTRÖM,[b,d]
K. B. AXELSEN,[b,c] L. BAUNSGAARD,[b,c]
F. C. LANFERMEIJER,[b] K. VENEMA,[b] F. T. OKKELS,[e,f]
P. ASKERLUND,[d] AND M. G. PALMGREN[b,g]

[b]*Molecular Biology Institute*
Copenhagen University
Øster Farimagsgade 2A
1353 Copenhagen K, Denmark

[c]*Department of Plant Biology*
Royal Veterinary and Agricultural University
Thorvaldsensvej 40
1871 Frederiksberg C, Denmark

[d]*Department of Plant Biochemistry*
Lund University
P.O. Box 117
S-221 00 Lund, Sweden

[e]*DANISCO Biotechnology*
Langebrogade 1
Postboks 17
1101 Copenhagen K, Denmark

PLASMA MEMBRANE H$^+$-ATPASE

Plasma membrane H$^+$-ATPase and Na$^+$/K$^+$-ATPase belong to the same subfamily of ion pumps that has been termed P$_2$-ATPases.[1] A phylogenetic tree showing the evolutionary relation between members of the P-type ATPase superfamily is presented in FIGURE 1. Plasma membrane proton pumps appear to constitute a monophyletic group of P-type ATPases and are present in plants, fungi, algae, protozoa, and archaea. This wide distribution among species suggests that P-type H$^+$-ATPases appeared very early in evolution. On the contrary, Na$^+$/K$^+$-ATPases probably appeared late in evolution because they are restricted to animals, who evolved in the sea and were dependent on systems to extrude sodium. However, in plants and animals respectively, H$^+$-ATPase and Na$^+$/K$^+$-ATPase serve analogous functions. Thus, in plants (and

---

[a]The work in the authors' laboratories was funded by the Danish Veterinary and Agricultural Research Council, the Danish Natural Science Research Council, the Swedish Natural Science Research Council, NOVO Nordisk Fonden, Carlsbergfonden, and the European Communities' BIOTECH Programme, as part of the Project of Technological Priority.

[f]Present address: DLF Trifolium, Hojerupvej 31, DK-4660 Store Heddinge, Denmark.

[g]To whom correspondence should be addressed. Tel: +45 35322132; fax: +45 3532 2128; palmgren@biobase.dk

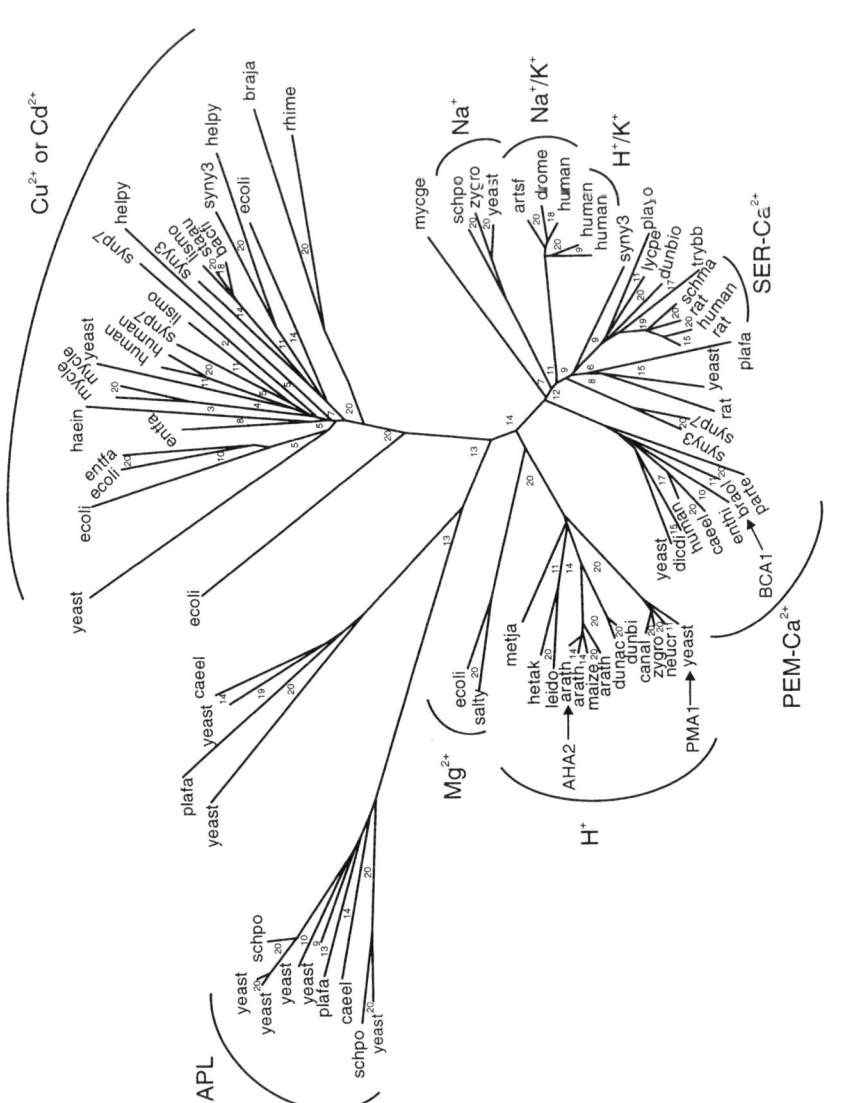

FIGURE 1. See legend on facing page.

algae, fungi, protozoa, and many bacteria as well), secondary transport is driven by an H^+ gradient and not by a Na^+ gradient as in animals. This H^+ gradient is generated by the plasma membrane H^+-ATPase, a single subunit enzyme of approximately 950 amino acids.[2] This pump is a powerful electrogenic enzyme, and membrane potentials around -200 mV generated by this enzyme are routinely measured across plant and fungal plasma membranes.[3]

OVEREXPRESSION OF PLASMA MEMBRANE H^+-ATPase IN TRANSGENIC TOBACCO PLANTS

It has been suggested that the plasma membrane H^+-ATPase is rate limiting for growth.[4] To test this hypothesis we attempted to overexpress plasma membrane H^+-ATPase in transgenic tobacco plants. The *AHA2* gene, encoding an isoform of H^+-ATPase from the plant *Arabidopsis thaliana*, was put under control of the strong cauliflower mosaic virus promoter and introduced into a transferred DNA (T-DNA) sequence containing the hygromycin phosphotransferase gene that in plants confers resistance to hygromycin B. By *Agrobacterium tumaefaciens* mediated transformation, the T-DNA was introduced into the genome of tobacco leaf cells. Thirty-three independent transgenic seedlings were regenerated from hygromycin B-resistant cells

←
FIGURE 1. Phylogenetic tree of eight conserved regions (totalling 265 amino acids) of all cloned full-length P-type ATPases. The ions transported by the various ATPases are given when known. The conserved regions of a subset of the P-type ATPases were identified by inspection of multiple alignments. The conserved regions were extracted from all the ATPases, and phylogenetic analysis was performed on these sequences. The tree was constructed using the Protdist, Fitch, and Drawtree programs of the Phylip package.[35] Bootstrapping was performed with the bootstrapping facility of the program package. Twenty datasets were created with seqboot. These were analyzed with Protdist and Fitch and the consensus was found using Consense. *Arrows* indicate the position of specific ATPases discussed in the text. Accession numbers, gene names, and a more detailed characterization can be found at http://biobase.dk/~palmgren/TRAP.html. Abbreviations: APL = aminophospholipids; arath = *Arabidopsis thaliana;* artsf = *Artemia franciscana;* bacfi = *Bacillus firmus;* bovin = *Bos taurus;* braja = *Bradyrhizobium japonicum;* braol = *Brassica oleracea;* bufma = *Bufo marinus;* caeel = *Caenorhabditis elegans;* canal = *Candida albicans;* cavco = *Cavia cobaya;* crigr = *Cricetulus griseus;* ctefe = *Ctenocephalides felis;* dicdi = *Dictyostelium discoidum;* drome = *Drosophila melanogaster;* dunac = *Dunaliella acidophila;* dunbi = *Dunaliella bioculata;* ecoli = *Eschericia coli;* entfa = *Enterococcus hirae;* enthi = *Entamoeba histolytica;* haein = *Haemophilus influenzae;* helpy = *Helicobacter pylori;* hetak = *Heterosigma akashiwo;* hisca = *Histoplasma capsulatum;* human = *Homo sapiens;* klula = *Kluyveromyces lactis;* leido = *Leishmania donovani;* lismo = *Listeria monocytogenes;* lycpe = *Lycopersicon esculentum;* maize = *Zea mays;* metja = *Methanococcus jannaschii;* mouse = *Mus musculus;* mycge = *Mycoplasma genitalium;* mycle = *Mycobacterium leprae;* myctu = *Mycobacterium tuberculosis;* neucr = *Neurospora crassa;* nicpl = *Nicotiana plumbaginifolia;* orysa = *Oryza sativa;* parte = *Paramecium tetraurelia;* phavu = *Phaseolus vulgaris;* plafa = *Plasmodium falciparum;* playo = *Plasmodium yoelii;* PEM = Plasma(endo)membrane; pneca = *Pneumocystis carinii;* rat = *Rattus norvegicus;* rhime = *Rhizobium meliloti;* salty = *Salmonella typhimurium;* schma = *Schistosoma mansoni;* schpo = *Schizosaccharomyces pombe;* SER = Sarco(endo)plasmatic reticulum; staau = *Staphylococcus aureus;* synp7 = *Synechococcus* PCC7942; syny3 = *Synechocystis* PCC6803; trybb = *Trypanosoma brucei;* vicfa = *Vicia faba;* xenla = *Xenopus laevis;* yeast = *Saccharomyces cerevisiae;* zosma = *Zostera marina;* zygro = *Zygosaccharomyces rouxii*.

and were transferred to soil where they gave seeds. Seeds were sown on medium containing hygromycin where the germination was observed. Two lines, which were growing poorly on hygromycin (15 mg/ml; FIG. 2), were found by genomic polymerase chain reaction (PCR) analysis to contain both the *AHA2* and the hygromycin B phosphotransferase gene. The *AHA2* gene was expressed as shown by RT-PCR

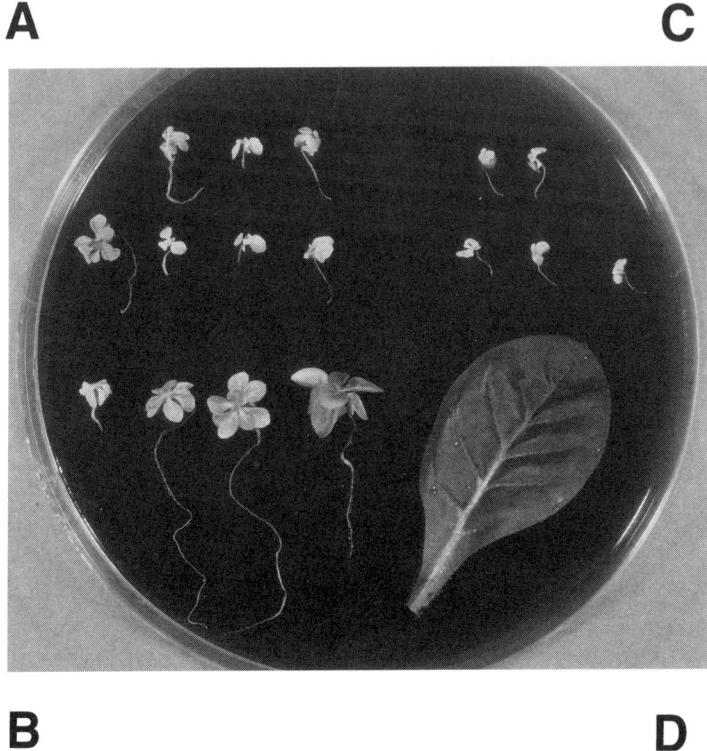

FIGURE 2. Transgenic tobacco (*Nicotiana tabacum*) plants overexpressing plant plasma membrane H^+-ATPase (*AHA2*). Seeds obtained from homozygous T_1 generation plants were sown on solid LS medium containing hygromycin (15 mg/ml), and growth was recorded after 40 days. The T_0 generation plants were regenerated from transgenic callus tissue. (**A**) Line 345,3-19 and (**B**) line 345,4-06 plants expressing *AHA2* and containing the hygromycin B phosphotransferase gene (*HYG*) conferring hygromycin resistance. (**C**) Wild-type tobacco (hygromycin-sensitive) without *AHA2* and *HYG*. (**D**) Positive control (line 315,5-18) containing *HYG* but not *AHA2*. Only a single leaf is shown.

analysis. On the contrary, 50 lines, which were capable of growing at 50 mg/ml hygromycin, contained only the hygromycin resistance gene. The high frequency of *AHA2* gene abortion suggests that under the experimental conditions, the AHA2 product is toxic. AHA2 is expected to influence the size of the membrane potential, and because hygromycin B is a lipophilic cation which is taken up according to the membrane potential,[5] it is likely that overexpression of AHA2 results in increased hygromycin B uptake exceeding the level tolerated even in the presence of hygromycin phosphotransferase.

PRODUCTION OF PLANT PLASMA MEMBRANE H^+-ATPase IN YEAST

A full understanding of the mechanism of plant H^+-ATPase is unlikely to be achieved without knowledge of its three-dimensional structure. Structural techniques, most notably x-ray crystallography, require milligram quantities of protein. The yeast *Saccharomyces cerevisiae* has proven to be an efficient high level expression system for massive production of plant plasma membrane H^+-ATPase.[6–8] Plant H^+-ATPase is correctly targeted to the plasma membrane of transformed yeast cells,[8–10] but in addition it accumulates in the endoplasmic reticulum.[6,11] This organelle was observed to proliferate in the form of stacked membranes surrounding the yeast nucleus to accommodate the large amount of plant H^+-ATPase produced (70 mg from 1 kg yeast).[6] From these membranes, plant H^+-ATPase protein tagged with six consecutive histidine residues in the NH_2-terminus of the enzyme is easily purified to apparent homogeneity by affinity chromatography (Lanfermeijer, Venema, and Palmgren, this volume).

REPLACEMENT OF YEAST AND PLANT H^+-ATPase

Given that fungal and plant P-type H^+-ATPases are carrying out the same function, it should be possible to functionally replace the H^+-ATPase of the yeast *S. cerevisiae* with one of its plant counterparts. We therefore attempted to produce a yeast strain in which secondary transport is energized by an H^+ gradient created by a plant ATPase. Yeast cells express a single plasma membrane H^+-ATPase, *PMA1*. A second gene, *PMA2*, is present but is normally not expressed. PMA1 is an essential enzyme, implying that in the absence of PMA1, yeast cells do not grow at all. An H^+-ATPase isoform from the plant *A. thaliana*, *AHA1*, was expressed in yeast cells in the absence of *PMA1* and was unable to support yeast growth. Therefore, at the beginning of our studies, replacement of *PMA1* by *AHA1* did not seem possible.

In most yeast growth media, glucose is used as the carbon source. Application of glucose to intact yeast cells causes the yeast H^+-ATPase to become posttranscriptionally activated, probably by a protein kinase-mediated phosphorylation event.[12,13] Plant H^+-ATPase, however, is not activated in yeast cells following glucose application, probably because it is not a substrate of the protein kinase regulating the endogenous yeast H^+-ATPase. We therefore hypothesized that plant H^+-ATPase is simply not active enough to replace yeast H^+-ATPase.

Spontaneous mutants arise readily in yeast genes. We therefore decided to wait for spontaneous mutants that allowed the transformed yeast cells to grow in the absence of endogenous ATPase.

In two cases, spontaneous mutants were observed that gave rise to modified plant H^+-ATPases that could actually support yeast growth (TABLE 1). The mutations were both localized to the COOH-terminal end of the pump polypeptide. One was a single conservative amino acid substitution (Trp-874 → Leu). The other was an internal deletion of 72 amino acids (ΔTrp-874 − Lys-935).

The properties of the modified plant ATPases just described have been analyzed in more detail. The amount of plant H^+-ATPase in the yeast plasma membrane is not increased (TABLE 1). Rather it is slightly decreased. Also, no significant change in specific ATPase activity of plasma membrane ATPase activity can be observed (TABLE 1). Although the molecular activity of the ATPase mutants is somehow increased (100–300%), this is compensated for by the reduced expression level. However, when the degree of H^+ extrusion from the transformed yeast cells is measured, there are marked differences. Those cells capable of growing are very efficient in extruding protons to the external medium. Apparently, the rates of H^+ transport by the modified enzymes are increased dramatically, despite the fact that the rates of ATP hydrolysis are less affected. Similar results were recently reported for a tobacco plasma membrane H^+-ATPase expressed in yeast.[14] In addition to 13 mutations in the COOH-terminal hydrophilic domain, these authors observed another six amino acid

TABLE 1. Effect of *AHA1* Mutations on Growth Rate of Transformed Yeast Cells, Expression Level in the Plasma Membrane, Specific ATPase Activity of the Plasma Membrane, and the Rate of H^+ Extrusion from Intact Cells

ATPase Allele	Growth Rate (div/h)	Relative Amount of H^+-ATPase (%)[a]	Specific ATPase Activity[b]	H^+ Extrusion (μmol/g/min)
(no ATPase)	—	—	0.02	2.5
AHA1	—	5.8	0.07	2.6
aha1ΔW874-K935	0.35	1.6	0.09	8.3
aha1W874-L	0.33	3.0	0.08	11.7

[a]Immunoreactive H^+-ATPase was quantified from protein gel blots using a PhosphorImager. The primary antibody was a polyclonal antibody against the central domain of AHA3 and the second antibody was a ^{125}I-labeled anti-rabbit antibody. As ATPase standard, affinity-purified 6 × histidine-tagged AHA2 was used.

[b]ATPase activity given as μmol P_i released from ATP per minute per milligram total protein (pH 7.0; 2 mM ATP).

substitutions in four other domains of the ATPase molecule that similarly gave rise to a pump molecule with increased transport capacity.

It was observed earlier that H^+ pumping and ATP hydrolytic activities by the plasma membrane H^+-ATPase are not necessarily directly coupled to each other. Thus, when the plant H^+-ATPase is activated *in vitro* by proteolytic removal of the COOH-terminal regulatory domain, H^+ pumping is stimulated 3–10 times more than is ATP hydrolysis by the enzyme.[15] Also, when the yeast plasma membrane H^+-ATPase is activated posttranscriptionally after glucose application *in vivo*, H^+ pumping by the enzyme is stimulated about 10 times more than is ATP hydrolysis.[16] The yeast plasma membrane H^+-ATPase has been reconstituted in artificial liposomes before and after glucose application to whole cells. From these studies it is apparent that although it readily splits ATP, the nonactivated yeast H^+-ATPase is incapable of generating a significant membrane potential.[16] This is in contrast to the glucose-activated enzyme which is capable of forming a steep electrical gradient.

We have removed completely the last 92 COOH-terminal amino acids of AHA2, another *Arabidopsis* isoform of the H^+-ATPase, and expressed the truncated enzyme in yeast.[17] The cells transformed with COOH-terminally truncated ATPase had growth rates approximating those of wild-type yeast. Again, compared to cells expressing unmodified AHA2, the cells extruded an increased amount of H^+ into the medium, whereas the specific ATPase activity of the plasma membrane was less affected. We performed deletion analysis of the COOH-terminus of AHA2 and could conclude that the removal of 66 residues is sufficient to produce an enzyme that functionally is able to replace PMA1.[9]

On the basis of biochemical and molecular approaches, the COOH-terminal end of the plant plasma membrane H^+-ATPase has earlier been designated a regulatory domain involved in inhibiting the activity of the pump.[8,11,13] What is consistently observed is an improved coupling of proton transport and ATP hydrolysis following modification of this domain. Improved coupling could indicate that the nonactivated enzyme contains internal H^+ leaks that are repaired in the activated coupled enzyme. This interpretation is problematic, because it would imply that futile cycles of ATP hydrolysis exist *in vivo*. Alternatively, the current-voltage relation of the plasma membrane H^+-ATPase might be dramatically altered after regulatory events that involve the COOH-terminal regulatory domain. This would imply that ATPase and H^+ transport activities exhibit different dependence on membrane voltage and that the two activities must somehow be uncoupled.[18]

REGULATION OF PLANT H^+-ATPase BY 14-3-3 PROTEINS

Fusicoccin is a fungal toxin that, when applied to intact plants or plant segments, activates H^+ extrusion from affected cells.[19] It is known that fusicoccin produces an activated state of the plant H^+-ATPase which resembles that of the enzyme lacking its COOH-terminal regulatory domain.[20–22] However, fusicoccin does not bind directly to the ATPase. The fusicoccin receptor was recently identified and found to belong to the family of 14-3-3 proteins.[23–25] These proteins are thought to function as adaptor molecules that bring together proteins in the cell, typically protein kinases and other signal-transduction proteins.[26] In addition, they are known to influence the activity of at least one plant protein, namely, nitrate reductase which is inhibited by a direct interaction with 14-3-3 proteins.[27]

If the role of the fusicoccin receptor is to guide a regulatory protein to the plant H^+-ATPase or to modify the activity of the ATPase by itself, we speculated that the 14-3-3 protein might interact directly with the ATPase. Using the yeast two-hybrid system we obtained genetic evidence that this is indeed the case. Thus, four tested isoforms of the *A. thaliana* GF14 14-3-3 protein (GF14-v,ψ,χ, and ϕ) all interact strongly with the COOH terminus (amino acids 859-948) of the AHA2 H^+-ATPase (FIG. 3). In addition, these isoforms interact slightly with the NH_2-terminus (amino acids 1-43, data not shown). None of the isoforms interacted with any other part of the molecule tested (FIG. 3). We are currently testing whether this interaction also occurs *in vitro* at the protein level.

The number of hydrophilic residues preceding the first putative transmembrane segment and those following the last membrane-spanning domain are very variable among P-type ATPases. This suggests that the NH_2- and COOH-terminal domains

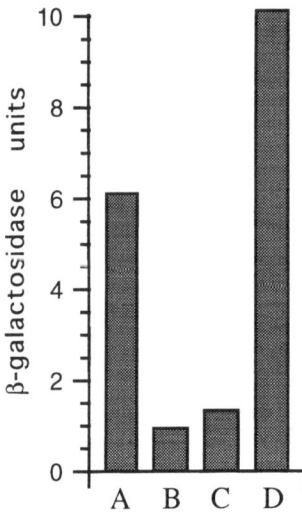

FIGURE 3. Interaction between the COOH-terminus (amino acids 859–948) of plant plasma membrane H$^+$-ATPase (AHA2) and a 14-3-3 protein (GF14-ψ) was observed using the yeast two-hybrid system. Domains of AHA2 were fused to the activation domain of GAL4, and the complete 14-3-3 protein was fused to the DNA binding domain of GAL4 using the vectors described in ref. 36. GAL4 is a transcription factor normally activating galactose-dependent genes. In the yeast strain used, the complete GAL4 molecule, produced and assembled from its two parts, is controlling expression β-galactosidase. Methods were those described in ref. 36. (**A**) Positive control: SNF1 versus SNF4 (plasmids pSE1111 + pSE1112); (**B**) empty vector (plasmid PACT2) versus 14-3-3; (**C**) central domain of AHA2 (amino acids 378–471) versus 14-3-3; (**D**) COOH-terminus of AHA2 (amino acids 852–948) versus 14-3-3.

might be dispensable for the transport function of the ATPases and may have other functions. One can envisage that 14-3-3 binds to either the NH$_2$- or the COOH-terminal domain (FIG. 4). A dimer of 14-3-3 proteins might even bind simultaneously to both termini (FIG. 4). Binding of 14-3-3 to ATPase might directly affect the activity of the enzyme. Thus, contrary to the inhibitory action on nitrate reductase, it is possible that direct binding of 14-3-3 to the ATPase results in activation of proton pumping.

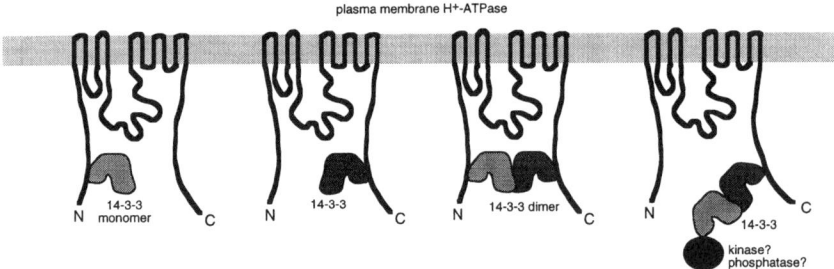

FIGURE 4. Involvement of both the NH$_2$- and the COOH-termini in regulation of plant plasma membrane H$^+$-ATPase. The COOH-terminus of the ATPase is involved in regulating the activity of the pump. Both the NH$_2$ and COOH-termini may be involved in the binding of 14-3-3 proteins. A 14-3-3 monomer possibly binds to either of the termini. In addition, a 14-3-3 dimer might be able to simultaneously bind to both termini. It is not known whether binding of 14-3-3 to the H$^+$-ATPase is sufficient to influence the activity of the pump. Inactivation and/or activation of pump activity might involve yet another regulatory protein. A 14-3-3 dimer bound to any one of the docking sites in the NH$_2$- or the COOH-terminus might guide a regulatory protein (e.g., protein kinase or phosphatase) to the H$^+$-ATPase. After binding of such a Hypothetical protein to the accessible 14-3-3 protein in the dimer, it may interact with the H$^+$-ATPase, resulting in a change in pump activity.

The function of fusicoccin, which activates the enzyme, could be to promote the interaction between 14-3-3 and the plasma membrane H^+-ATPase. Alternatively, binding of 14-3-3 to the H^+-ATPase is not sufficient to modulate pump activity. A 14-3-3 heterodimer bound to any one of the termini might guide yet another regulatory molecule with affinity for 14-3-3 proteins to the H^+-ATPase (FIG. 4). In this model, 14-3-3 proteins are indirectly affecting the activity of the pump.

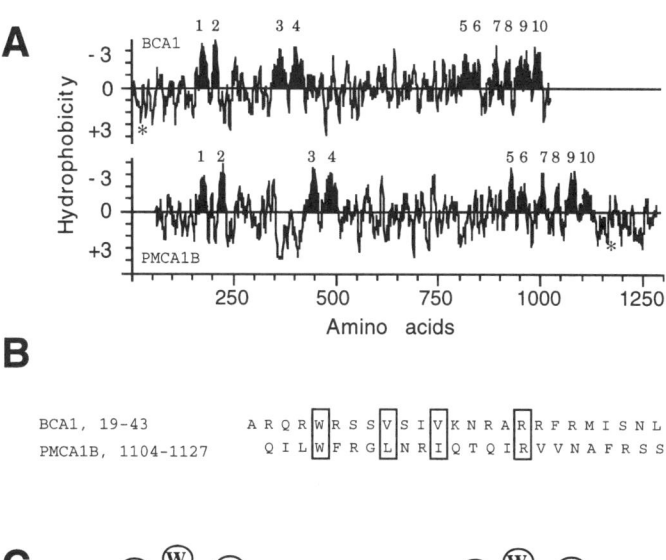

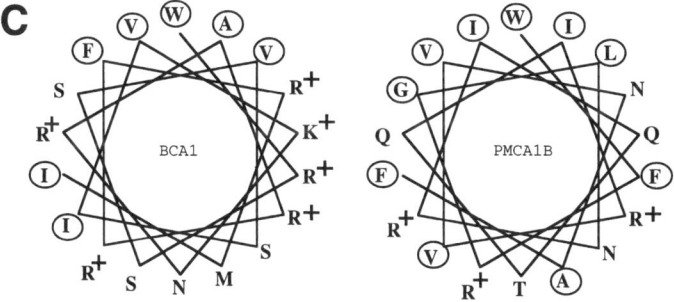

FIGURE 5. Calmodulin-binding domain in the NH_2-terminus of plant BCA1. (**A**) Comparison of the hydropathy profiles for plant BCA1 and plasma membrane Ca^{2+}-ATPase (PMCA1B).[37] Putative transmembrane segments are numbered. *Asterisks* denote the position of putative calmodulin-binding domains. (**B**) Sequence alignment of the putative calmodulin-binding domain of BCA1 (amino acids 19–43) to PMCA1B (amino acids 1104–1127).[37] Hydrophobic and basic residues conserved in a wide range of known and putative calmodulin-binding domains[38] are boxed. (**C**) α-helical wheel diagram of amino acids 23–40 in the calmodulin-binding domain of BCA1 and amino acids 1107–1124 of PMCA1B.[37] Positively charged amino acids are denoted with a + *superscript*. Hydrophobic amino acids are *circled*. In both domains a segregation of basic and polar residues on one side and hydrophobic amino acids on the other is seen.

A NOVEL VACUOLAR Ca^{2+}-ATPASE WITH A PUTATIVE REGULATORY DOMAIN AT ITS NH_2-TERMINUS

In plants, the central vacuole plays an important role as Ca^{2+} reservoir, much like the role of the endoplasmic reticulum of animal cells. Ca^{2+} is released into the cytosol from the vacuole during events of signal transduction.[28–30] Plant CaM-stimulated Ca^{2+}-ATPases are found in the vacuolar membrane and data from several laboratories strongly indicate that CaM-stimulated Ca^{2+}-pumps are present also in the plasma membrane of plant cells.[34] We recently cloned vacuolar Ca^{2+}-ATPase from cauliflower, which probably is responsible for pumping Ca^{2+} back into the vacuole after an increase in cytosolic Ca^{2+} concentration. This Ca^{2+}-ATPase is activated by Ca^{2+}/CaM and was purified from cauliflower low-density membranes using CaM-affinity chromatography. CaM-stimulated Ca^{2+} pumping was demonstrated after reconstitution of the purified ATPase into liposomes.[31,32]

The isolated cDNA clone, BCA1 (database accession # X99972), encodes a 111-kD polypeptide that shows high homology to the CaM-stimulated plasma membrane Ca^{2+}-ATPases of animal cells, but low homology to the sarco(endo)plasmic reticulum-type Ca^{2+}-ATPases. The animal plasma membrane Ca^{2+}-ATPase has a CaM-binding autoinhibitory region in the COOH-terminal end of the molecule.[33] When we obtained the full-length plant cDNA sequence, however, it became evident that the COOH-terminus was very short (22 aa; FIG. 5) and that it did not contain a CaM-binding domain. The NH_2-terminus, on the other hand, was rather long (168 aa; FIG. 5) and contained a putative CaM-binding domain (aa 19–43) which was predicted to fold as an amphipathic α-helix. A synthetic peptide corresponding to this domain bound CaM with high affinity.[39]

In summary, we cloned a novel calmodulin-stimulated Ca^{2+}-ATPase from plant vacuolar membranes with a putative regulatory domain at its NH_2-terminus. This suggests that in P-type ATPases both NH_2- and COOH-termini may serve regulatory functions.

ACKNOWLEDGMENTS

We are indepted to Gertrud Christensen for her excellent technical assistance and Dr. Robert Ferl for the *Arabidopsis* GF-14 cDNA clones.

REFERENCES

1. LUTSENKO, S. & J. H. KAPLAN. 1995. Biochemistry **34:** 15607–15613.
2. SERRANO, R. 1989. Ann. Rev. Plant Physiol. Plant Mol. Biol. **40:** 61–94.
3. SPANSWICK, R. M. 1981. Ann. Rev. Plant Physiol. **32:** 267–289.
4. SERRANO, R. 1990. Bot. Acta **103:** 230–234.
5. PERLIN, D. S., C. L. BROWN & J. E. HABER. 1988. J. Biol. Chem. **263:** 18118–18122.
6. VILLALBA, J. M., M. G. PALMGREN, G. E. BERBERIAN, C. FERGUSON & R. SERRANO. 1992. J. Biol. Chem. **267:** 12341–12349.
7. PALMGREN, M. G. & G. CHRISTENSEN. 1994. J. Biol. Chem. **269:** 3027–3333.
8. DE KERCHOVE D'EXAERDE, A., P. SUPPLY, J.-P. DUFOUR, P. BOGAERTS, D. THINES, A. GOFFEAU & M. BOUTRY. 1995. J. Biol. Chem. **270:** 23828–23837.
9. REGENBERG, B., J. M. VILLALBA, F. C. LANFERMEIJER & M. G. PALMGREN. 1995. Plant Cell **7:** 1655–1666.

10. BAUNSGAARD, L., K. VENEMA, K. AXELSEN, J. M. VILLALBA, A. WELLING, B. WOLLENWEBER & M. G. PALMGREN. 1996. Plant J. **10:** 451–458.
11. DE KERCHOVE D'EXAERDE, A., P. SUPPLY & A. GOFFEAU. 1996. Yeast **12:** 907–916.
12. CHANG, A. & C. W. SLAYMAN. 1991. J. Cell Biol. **115:** 289–295.
13. PORTILLO, F., P. ERASO & R. SERRANO. 1991. FEBS Lett. **287:** 1, 2, 71–74.
14. MORSOMME, P., A. DE KERCHOVE D'EXAERDE, S. DE MEESTER, D. THINES, A. GOFFEAU & M. BOUTRY. 1996. EMBO J. **15:** 5513–5526.
15. PALMGREN, M. G., M. SOMMARIN, R. SERRANO & C. LARSSON. 1991. J. Biol. Chem. **266:** 20470–20475.
16. VENEMA, K. & M. G. PALMGREN. 1995. J. Biol. Chem. **270:** 19659–19667.
17. PALMGREN, M. G. & G. CHRISTENSEN. 1993. FEBS Lett. **317:** 216–222.
18. LÄUGER, P. 1991. Electrogenic Ion Pumps. Sinauer Associates, Inc. Sunderland, MA.
19. MARRÉ, E. 1979. Ann. Rev. Plant Physiol. **30:** 273–288.
20. JOHANSSON, F., M. SOMMARIN & C. LARSSON. 1993. Plant Cell **5:** 321–327.
21. RASI-CALDOGNO, F., M. C. PUGLIARELLO, C. OLIVARI & M. I. DE MICHELIS. 1993. Plant Physiol. **103:** 391–398.
22. LANFERMEIJER, F. C. & H. B. A. PRINS. 1994. Plant Physiol. **104:** 1277–1285.
23. KORTHOUT, H. A. A. J. & A. H. DE BOER. 1994. Plant Cell **6:** 1681–1692.
24. MARRA, M., M. R. FULLONE, V. FOGLIANO, J. PEN, M. MATTEI, S. MASI & P. ADUCCI. 1994. Plant Physiol. **106:** 1497–1501.
25. OECKING, C., C. ECKERSKORN & E. W. WEILER. 1994. FEBS Lett. **352:** 163–166.
26. AITKEN, A. 1996. Trends Cell Biol. **6:** 341–347.
27. BACHMANN, M., J. L. HUBER, P.-C. LIAO, O. A. GAGE & S. C. HUBER. 1996. FEBS Lett. **387:** 127–131.
28. ALEXANDRÉ, J., J. P. LASALLES & R. T. KADO. 1990. Nature **343:** 567–570.
29. CANUT, H., A. CARRASCO, M. ROSSIGNOL & R. RANJEVA. 1993. Plant Sci. **90:** 135–143.
30. ALLEN, G. J., S. R. MUIR & D. SANDERS. 1995. Science **268:** 735–737.
31. ASKERLUND, P. & D. E. EVANS. 1992. Plant Physiol. **100:** 1670–1681.
32. ASKERLUND, P. 1996. Plant Physiol. **110:** 913–922.
33. CARAFOLI, E. 1994. FASEB J. **8:** 993–1002.
34. ASKERLUND, P. & M. SOMMARIN. 1996. *In* Membranes: Specialized Functions in Plants. M. Smallwood, J. P. Knox & D. J. Bowles, Eds.: 281–299. Bios Scientific. Oxford.
35. FELSENSTEIN, J. 1989. Cladistics **5:** 164–166.
36. DURFEE, T., K. BECHERER, P. L. CHEN, S. H. YEH, Y. YANG, A. E. KILBURN, W. H. LEE & S. J. ELLEDGE. 1993. Genes Dev. **7:** 555–569.
37. KATAOKA, M., J. F. HEAD, T. VORHERR, J. KREBS & E. CARAFOLI. 1991. Biochemistry **30:** 6247–6251.
38. CRIVICI, A. & M. IKURA. 1995. Annu. Rev. Biophys. Biomol. Struct. **24:** 85–116.
39. MALMSTRÖM, S., P. ASKERLUND & M. G. PALMGREN. 1997. FEBS Lett. **400:** 324–328.

Studies of Na,K-ATPase Structure and Function Using Baculovirus

GUSTAVO BLANCO, WENDY R. HATFIELD,
NICOLE T. MINOR, GLADIS SÁNCHEZ, JOSEPH C. KOSTER,
ANTHONY W. DeTOMASO, AND ROBERT W. MERCER[a]

*Department of Cell Biology and Physiology
Washington University School of Medicine
St. Louis, Missouri 63110*

Elucidating the role of the individual Na,K-ATPase subunits in directing enzymatic activity and defining the associations of the various isoforms have proven to be difficult tasks. Major advances in understanding Na,K-ATPase structure and function have come from studies using heterologous expression systems. To date, exogenous Na,K-ATPase activity has been expressed in yeast,[1,2] amphibian,[3,4] insect,[5] and mammalian[6] cells. Each of these systems affords distinct advantages and disadvantages. An advantage of using insect cells for the expression of the Na,K-ATPase is that in contrast to most vertebrate cells, certain insect cells have very little or no endogenous Na,K-ATPase activity. The most common expression system for expressing proteins in insect cells uses the baculovirus *Autographica californica*. Using this system, recombinant baculoviruses can be used to direct the production of foreign proteins in cultured insect cells, most commonly a cell line (*Sf*-9) derived from the ovary of the fall armyworm *Spodoptera frugiperda*. *Sf*-9 insect cells provide a useful eukaryotic expression system that, like mammalian cells, will implement most of the posttranslational modifications dictated by the primary amino acid sequence. Generally, *Sf*-9 cells will perform the glycosylation, proteolytic processing, signal peptide cleavage, phosphorylation, palmitoylation, and myristoylation characteristic of their mammalian counterparts. Furthermore, infected *Sf*-9 cells assemble multisubunit proteins and accurately segregate recombinant proteins to their proper cellular domains (reviewed in ref. 7). We used the baculovirus expression system to study the structure and function of the Na,K-ATPase α and β isoforms in *Sf*-9 insect cells. This system allowed us to express and characterize the individual Na,K-ATPase isozymes. In addition, by being able to express the individual subunits in an environment relatively free of endogenous Na,K-ATPase activity, we obtained some surprising and unexpected results on the structure and function of the Na,K-ATPase.

CHARACTERIZATION OF Na,K-ATPase ISOZYMES

Rat α1, α2, α3, β1, and β2 cDNAs were used to produce recombinant baculoviruses for expression of Na,K-ATPase polypeptides in insect cells. Immunocytochem-

---

[a]Address for correspondence: Dr. Robert W. Mercer, Department of Cell Biology and Physiology, Washington University School of Medicine, 660 S. Euclid Avenue, St. Louis, MO 63110 (tel: 314–362–6924; fax: 314–362–7463; e-mail: rmercer@cellbio.wustl.edu).

istry of infected *Sf*-9 insect cells demonstrated that the individual Na,K-ATPase isoforms can be localized to the plasma membrane. Also, the insect cells are able to properly assemble the α and β subunits as defined by immunoprecipitation experiments. Infecting cells with the various recombinant baculoviruses has demonstrated that all α isoforms can stably assemble with either β1 or β2 polypeptides.[8,9] Moreover, all αβ combinations result in catalytically competent Na,K-ATPase molecules. This activity is reflected by (1) Na^+-dependent phosphorylation of the α subunit from ATP that is sensitive to K^+ and inhibited by ouabain; (2) phosphorylation of the α subunit by P_i that is stabilized by ouabain; (3) Na^+- and K^+-dependent ouabain inhibitable hydrolysis of ATP, and (4) specific binding of [^{3}H]ouabain. Specific ATPase activity ranged from 7- to 15-fold that of membranes from uninfected cells, which have little ouabain-sensitive activity (0.05–0.10 μmol P_i/mg protein/h). There was no detectable increase in Na,K-ATPase activity from cells solely expressing the α or β subunit. By analyzing the dose-response curves towards Na^+,K^+, ATP and the inhibitor ouabain, the enzymatic properties of the various isozymes were determined. TABLE 1 summarizes the enzymatic characteristics of the rat Na,K-ATPase isozymes expressed in the insect cells. Comparison of the different Na,K-

TABLE 1. Kinetic Characteristics of the Rat Na,K-ATPase Isozymes Expressed in *Sf*-9 Cells[a]

Isozyme	Na^+ Activation $K_{0.5}$ (mM)	K^+ Activation $K_{0.5}$ (mM)	ATP Activation K_m (mM)	Ouabain Inhibition K_i (M)
α1β1	16.4 ± 0.7	1.9 ± 0.2	0.46 ± 0.10	4.3 ± 1.9 × 10^{-5}
α2β1	12.4 ± 0.5	3.6 ± 0.3	0.11 ± 0.01	1.7 ± 0.1 × 10^{-7}
α2β2	8.8 ± 1.0	4.8 ± 0.4	0.11 ± 0.02	1.5 ± 0.2 × 10^{-7}
α3β1	27.9 ± 1.3	5.3 ± 0.3	0.09 ± 0.01	3.1 ± 0.3 × 10^{-8}
α3β2	17.1 ± 1.0	6.2 ± 0.4	0.07 ± 0.02	4.7 ± 0.4 × 10^{-8}

[a]Apparent affinities ($K_{0.5}$), K_m, and inhibition constant (K_i) parameters were calculated from Na,K-ATPase activity dose-response curves for the indicated effectors. Values represent the mean ± standard error of the mean of three to seven different determinations.

ATPase αβ pairs shows that the major differences in kinetic properties occur among isozymes that differ in their α subunits. Thus, the apparent affinity for Na^+ varies with a rank of order α2β2 > α2β1 > α1β1 ≈ α3β2 > α3β1. Also, the apparent affinity for K^+ differs among the isozymes, following the sequence α1β1 > α2β1 ≈ α2β2 > α3β1 ≈ α3β2. For activation by ATP, the enzymes composed of the α2 and α3 isoforms display equivalent K_m values, which are approximately four times lower than that of the α1β1. Finally, the most conspicuous kinetic difference is in the response to ouabain. The α3β1 and α3β2 isoforms have a high sensitivity, whereas α2β1 and α2β2 an intermediate and α1β1 a low sensitivity to the cardiotonic steroid.

Taken together, these results suggest a functional basis for the existence of Na,K-ATPase isozymes, because their unique enzymatic properties may help establish the ionic milieu required by tissues to meet their physiological requirements. For example, in excitable tissues the lower apparent affinity of α3β1 for Na^+ and K^+ suggests that this isozyme may serve as a reserve pump. Changes in cation concentrations during repetitive firing of action potentials may activate the α3β1 enzyme to help restore the ion gradients. Thus, the unique kinetic properties of the different Na pump isozymes may be at least one of the reasons underlying the diversity in Na,K-ATPase polypeptides.

Whereas the baculovirus expression system has allowed us to characterize the individual isozymes, the ability to independently express the α and β subunits has given us the unique opportunity to characterize the requirements for Na,K-ATPase activity and assembly.

ACTIVITY OF THE INDEPENDENT α SUBUNIT

Although all enzymatic properties of the Na,K-ATPase have been ascribed to the α subunit, it has never been shown if the subunit has activity separate from that of the β subunit. Using baculovirus-produced polypeptides, we showed that the α subunit, in the absence of the β subunit, has catalytic activity.[10] As discussed before, when coinfected with α and β baculoviruses, functional Na,K-ATPase activity is expressed. Activity of the virally produced αβ polypeptides can be demonstrated by the ability of the α subunit to be phosphorylated by ATP. During the reaction cycle, the α subunit of the Na,K-ATPase is phosphorylated by ATP at a β-aspartyl carboxyl group (Asp371).[11,12] This aspartate group is conserved in all Na,K-ATPase α subunits. Phosphorylation by ATP requires the presence of both Mg and Na ions; in the presence of K^+, the phosphate is rapidly released from the subunit as P_i. As shown in FIGURE 1A, the Na,K-ATPase α subunit from αβ coinfected Sf-9 cells can be phosphorylated by ATP. This Na^+-dependent phosphorylation is inhibited by ouabain if ouabain is bound to the enzyme before exposure to ATP. In addition, the α subunit is completely dephosphorylated in the presence of K^+. Because the reaction cycle of the Na,K-ATPase is reversible, the α subunit can also be phosphorylated directly with P_i. Phosphorylation of the α subunit by P_i is stabilized by the presence of ouabain. Although it is more difficult to detect, the Na,K-ATPase from the uninfected Sf-9 cells has an identical pattern of phosphorylation (FIG. 1B). The insect Na,K-ATPase α subunit is phosphorylated only in the presence of Na^+; EGTA does not affect the Na^+-dependent phosphorylation, and Mg^{2+} alone cannot support phosphorylation.

Phosphorylation of the α subunit with different properties is seen when the α subunit is expressed in the absence of the β subunit (FIG. 2). Immunoprecipitation experiments and analysis of sucrose density gradients have shown that the α polypeptides are not associated with other proteins.[5] As shown, phosphorylation of the lone α subunit can occur with only Mg^{2+} present. This Mg^{2+}-dependent phosphorylation is not inhibited by ouabain; however, 150 mM $Tris^+$, K^+, or Na^+ greatly reduce phosphorylation. The lack of inhibition by ouabain is consistent with the finding that [^{3}H]ouabain does not bind to the unassociated α subunit.[9] The β independent phosphorylation of the α subunit is not restricted to the α1 isoform, as both α2 and α3 exhibit identical Mg^{2+}-dependent phosphorylation.

Surprisingly, increasing concentrations of either Na^+ or K^+ inhibit α subunit phosphorylation. However, choline, a cation not thought to influence Na,K-ATPase enzymatic activity, also inhibits independent α subunit phosphorylation. Therefore, although Na^+, K^+, and $Tris^+$ can influence the phosphorylation of the native enzyme, it appears that these ions do not directly affect the phosphorylation of the independent α subunit.[13] Moreover, the inhibition of phosphorylation is not influenced by chloride ions, as substitution of chloride with sulfate has no effect on inhibition. Taken together, these results suggest that phosphorylation of the unassociated α subunit is sensitive to the ionic strength of the medium. At higher ionic strengths, phosphoryla-

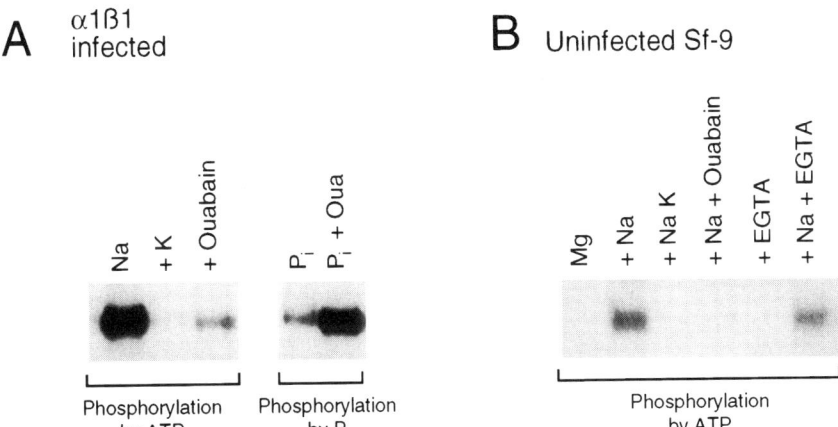

FIGURE 1. (**A**) α1β1 infected *Sf*-9 cells produce functional Na,K-ATPase. 30 mg of cellular membrane proteins from α1β1 coinfected *Sf*-9 cells were phosphorylated with γ[^{32}P]-ATP or [^{32}P$_i$]. (**B**) Uninfected *Sf*-9 cells have low levels of functional Na,K-ATPase. 100 mg of cellular membrane proteins from uninfected *Sf*-9 cells were phosphorylated with γ[^{32}P]-ATP. Exposure times of the gels were over three times those of the gels in **A**. Na,K-ATPase activity of uninfected *Sf*-9 cells is approximately 20-fold less than that of α1β1 coinfected cells.

tion of the independent α subunit is inhibited. However, even at physiological ionic strength, the unassociated α subunit has catalytic activity (see below). The unassociated α subunit can be phosphorylated only by ATP; the subunit is not phosphorylated by P$_i$ in either the presence or the absence of ouabain.

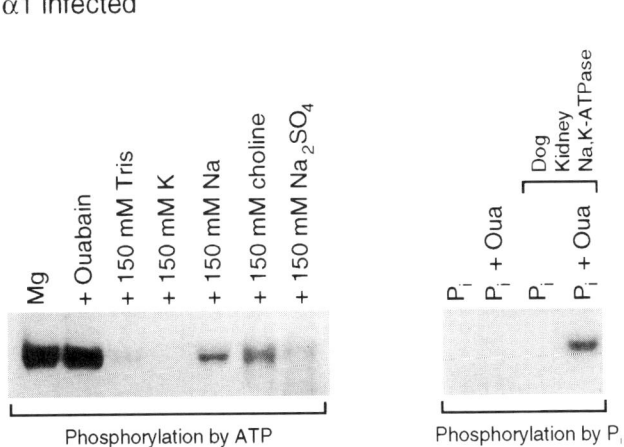

FIGURE 2. The independent α subunit has catalytic activity. 30 mg of cellular membrane proteins from α1 infected *Sf*-9 cells or 2 μg of purified Na,K-ATPase from dog kidney were phosphorylated with γ[^{32}P]-ATP or [^{32}P$_i$].

As found with increases in ionic strength, 1 mM EGTA also reduces Mg^{2+}-dependent phosphorylation of the α subunit. Inhibition of α subunit phosphorylation by EGTA suggests that the phosphorylation may depend on Ca^{2+}. However, inhibition of α subunit phosphorylation by EGTA is not influenced by Ca^{2+}. The addition of 1 mM Ca^{2+} to the Mg^{2+}-EGTA medium does not alter the inhibition by 1 mM EGTA. At these concentrations of Ca^{2+} and EGTA, the free Ca^{2+} concentration is approximately 5 μM higher than would be expected in the Mg^{2+} medium. Consequently, the inhibition by EGTA appears to be mediated through a factor other than Ca^{2+}. EGTA may inhibit activity by chelating another metal ion that is essential for the phosphorylation of the α subunit; however, it is possible that EGTA inhibits activity other than by its metal-binding abilities. In either event, EGTA is useful in defining the activity of the unassociated α subunit as an EGTA-sensitive, ATPase activity. As shown in TABLE 2, membranes from α infected cells exhibit EGTA-sensitive, Mg^{2+}-dependent ATPase activity that is not present in the membranes from uninfected Sf-9 cells. In addition, the Mg^{2+}-dependent ATPase of the α infected cells is reduced by ≈40% under conditions of high ionic strength (130 mM choline). The Mg^{2+}-dependent ATPase activity remaining at the higher ionic strength is completely inhibited by EGTA. These results agree with the phosphorylation studies and demonstrate that phosphorylation of the unassociated α subunit represents the ATPase activity of the enzyme. At this time, it is not known if the transport of ions is coupled to the hydrolysis of ATP.

As mentioned previously, during the catalytic cycle the α subunit is normally phosphorylated at an aspartate residue. If the independent α subunit is phosphorylated at an aspartate residue, then the characteristics of the phosphoenzyme must satisfy certain criteria. The phosphointermediate should be acid stable and alkaline labile.[14] In addition, because the intermediate is an acylphosphate, it should be sensitive to hydroxylamine. This is in contrast to the phosphorylation of serine, threonine, and tyrosine, which results in an esterphosphate that is resistant to hydroxylamine treatment.[14–16] Also, unlike most serine or threonine protein kinases, phosphorylation of the α subunit should be sensitive to inhibition by vanadate. Phosphorylation of the independent α subunit demonstrates that the phosphate bond of the unassociated α subunit phosphointermediate is acid stable and alkaline labile, consistent with phosphorylation at the normal aspartyl residue. The formation of the phosphointermediate is also inhibited by vanadate and is sensitive to hydroxylamine, compatible with it

TABLE 2. ATPase Activity of Membranes from Uninfected and α Infected Sf-9 Cells[a]

	Mg^{2+}-ATPase Activity EGTA-Sensitive	Mg^{2+}-ATPase Activity Ionic Strength-Sensitive
	μmol P_i/mg protein/h	
Sf-9 cells	0.02 ± 0.07	−0.03 ± 0.06
α infected Sf-9 cells	0.34 ± 0.04	0.20 ± 0.03

[a]The ATPase activity of membranes from uninfected and α infected Sf-9 cells was determined in medium containing 2 mM ATP, 0.2 μC γ[^{32}P]-ATP, 25 mM imidazole, pH 7.4, 2 mM $MgCl_2$, with 2 mM EGTA or 130 mM choline. The ouabain-sensitive Na,K-ATPase activity of the uninfected and α infected Sf-9 cells was 0.44 ± 0.09 and 0.30 ± 0.09 μmol P_i/mg protein/h, respectively. Values are the mean of triplicate determinations ± the standard deviation. Similar results were obtained in three separate experiments using different preparations. β infected Sf-9 cells did not exhibit the EGTA-sensitive Mg^{2+}-ATPase and have activity levels similar to those of the uninfected Sf-9 cells (not shown).

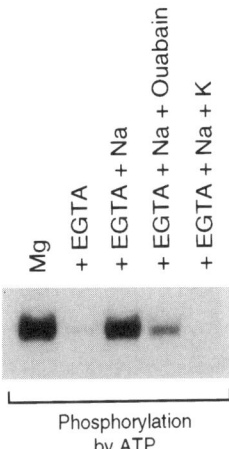

FIGURE 3. Characterization of independent α subunit and αβ phosphorylation in α1β1 coinfected *Sf*-9 cells. Cells were infected with α and β baculoviruses; the amount of α baculovirus added was greater than β. 30 mg of cellular membrane proteins from α1β1 infected *Sf*-9 cells were phosphorylated with $\gamma[^{32}P]$-ATP in medium containing 10 μM ATP, 1 mM $MgCl_2$, 10 mM Tris-HCl, pH 7.2. As indicated, the reactions also contained 1 mM EGTA, 100 mM NaCl, 1 mM ouabain, or 25 mM KCl.

being an acylphosphate and not an esterphosphate. These results provide strong evidence that the independent α subunit is phosphorylated at the normal aspartate residue.

To test if the EGTA-sensitive phosphorylation of the α subunit occurs in the presence of the β subunit, *Sf*-9 cells were infected with both α and β viruses. The multiplicity of infection of the α baculovirus was greater than that of the β baculovirus. Under these circumstances, both EGTA-sensitive and Na^+-sensitive phosphorylation patterns are seen. As shown in FIGURE 3, in the presence of Mg^{2+} the α subunit is phosphorylated, and this phosphorylation is sensitive to EGTA. Presumably this activity is mediated by the unassociated α subunit. When Na^+ is added to the Mg^{2+}-EGTA medium, α subunit phosphorylation is again seen. This Na^+-dependent α subunit phosphorylation is inhibited by ouabain and K^+ and is likely mediated by the αβ enzyme complex. These results suggest that when the α subunit is expressed in excess of the β subunit, the unassociated α subunits exhibit an EGTA-sensitive activity. However, we cannot exclude the possibility that the activities observed are a result of individual cells expressing only the α subunit. This work provides strong evidence that the α subunit, in the absence of the β subunit, can be phosphorylated and mediate ATPase activity. This is a unique activity that generally would not be attributed to the Na,K-ATPase because it is ouabain insensitive. At this time it is not clear if the enzyme is transporting ions or if this activity is physiologically significant.

ASSEMBLY OF α AND β SUBUNITS INTO FUNCTIONAL ENZYME AT THE PLASMA MEMBRANE

We previously showed that both α and β subunits of the Na,K-ATPase are present in the plasma membrane when individually expressed in *Sf*-9 cells.[5] In contrast, the infected *Sf*-9 cell retains murine light and heavy chains when expressed alone, and only secretes functional immunoglobulin when both chains are simultaneously expressed.[5,17] Infected *Sf*-9 cells also retain the α subunit of the GABA receptor intracellularly and only direct it to the plasma membrane when coexpressed with the β subunit.[18] Although heteromeric proteins appear to require assembly for delivery, monomeric plasma membrane proteins are delivered directly to the plasma membrane. Therefore, the baculovirus-infected *Sf*-9 cell, like mammalian cells, can recognize and retain unassembled polypeptides, suggesting that delivery of the unassociated Na,K-ATPase α and β subunits is physiologically relevant. These results imply that α and β assembly may occur, not only in the endoplasmic reticulum (ER), but at the plasma membrane as well. This type of assembly for the Na,K-ATPase has been suggested in insulin-sensitive tissues.[19] To test if the α and β subunits can assemble at the plasma membrane, we decided to take advantage of a unique property of the infected *Sf*-9 cells. The 64K envelope glycoprotein of baculovirus mediates pH-dependent membrane fusion.[20] This fusion activity can be used to fuse infected cells into large syncytia. Infected cells growing in suspension are allowed to settle at a high density onto a culture plate. Fusion is initiated by a 10-minute shift to pH 5.1; this pH shift results in a large syncytium of fused cells. Under these conditions, approximately 50–60% of the infected cells fuse into syncytia. This cell fusion appears to be limited to the plasma membrane. When *Sf*-9 cells are infected separately with murine heavy and light chain immunoglobulins and fused, confocal microscopy demonstrates that the separate, cytoplasmic chains do not associate. To determine if the α and β subunits can assemble at the plasma membrane, cells grown in suspension were infected with either the α1 or the β1 baculovirus. After 72 hours the α and β cells were plated together and allowed to attach for 2 hours; the cells were then metabolically labeled with ^{35}S-methionine, chased for 2 hours, and treated with cycloheximide. This protocol of labeling and treating with cycloheximide before fusion assures that once fusion is initiated only pre-fusion polypeptides are detected and allowed to associate. To start fusion, the medium was shifted to pH 5.1 for 10 minutes. The cells were then incubated for 1 hour, solubilized, and the α subunit immunoprecipitated with an α-specific monoclonal antibody. Immunoprecipitation of the α subunit from fused α and β cells results in the coprecipitation of the β subunit. This association is not present in similarly treated cells not receiving the pH shift. These results demonstrate that the α and β subunits at the plasma membrane can associate and assemble into an enzyme complex.

To determine if the fused α and β subunits are active, the ouabain-sensitive K^+ uptake of the fused cells was assayed. As shown in FIGURE 4, only cells that had been individually infected with α or β baculoviruses and subsequently fused exhibited an increase in ouabain-sensitive uptake. Fusions between cells expressing only one subunit showed no such increase in activity. Also, there was no increase in activity in the mixture of α and β infected cells that have not been fused. Moreover, a hybrid ATPase molecule consisting of a Na,K-ATPase α subunit and a H,K-ATPase β subunit, which efficiently assembles in the ER of coinfected cells, does not assemble at the plasma membrane of fused cells. When cells expressing the Na,K-ATPase α subunit are fused to cells coexpressing the Na,K-ATPase β subunit and the H,K-ATPase β subunit, the Na,K-ATPase α subunit selectively assembles with the Na,K-ATPase β subunit. However, when cells are coinfected and express all three polypeptides, the Na,K-ATPase α subunit assembles with both β subunits in the ER, in what appears to be a random fashion. These results supply strong evidence that the Na,K-ATPase α and β subunits at the plasma membrane can assemble into functional enzyme. These experiments demonstrate that assembly between some polypeptides is restricted to the ER and suggests that the ability of the Na,K-ATPase α and β subunits to leave the ER and assemble at the plasma membrane may represent a novel mechanism of regulation of activity. Under normal conditions it is clear that the majority of α and β assembly occurs in the ER; however, these subunits can assemble into functional Na,K-ATPase at the plasma membrane. The exact functional significance of the plasma membrane assembly of the Na,K-ATPase must await further study. In addition, it will have to be determined if the α and β subunits at the plasma membrane can dissociate and reassociate with other subunits to form functional enzyme. This type of assembly has been suggested to explain the insulin-responsive assembly of α2β1 complexes.[19] Thus, if different combinations of the subunits have distinct functions, then the regulated dissociation and reassociation of the different isoforms may be important in enzyme regulation. The molecular mechanisms leading to this type of assembly remain to be investigated.

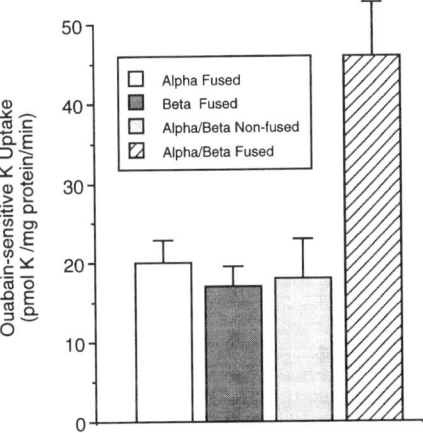

FIGURE 4. Newly formed Na,K-ATPase αβ heterodimers are functional. Fused and nonfused cells were assayed for ouabain-sensitive potassium uptake. Each experimental point is the average of six wells, and the data shown in the histogram are the mean of three separate experiments ± SEM.

CONCLUSIONS

Expression of the Na,K-ATPase polypeptides in insect cells has allowed us to kinetically characterize the individual Na,K-ATPase isozymes and has afforded the opportunity to study the characteristics and assembly of the individual α and β subunits. Some important and novel findings using the baculovirus expression system include the following: (1) The α1β1, α2β1, α3β1, α2β2, and α3β2 enzyme complexes are catalytically active and display unique kinetic properties; (2) the α and β subunits of the Na,K-ATPase can be independently delivered to the plasma membrane; and (3) the α and β subunits of the Na,K-ATPase can assemble at the plasma membrane into functional enzyme.

REFERENCES

1. HOROWITZ, B., K. A. EAKLE, G. SCHEINER-BOBIS, G. R. RANDOLPH, C. Y. CHEN, R. A. HITZEMAN & R. A. FARLEY. 1990. Synthesis and assembly of functional mammalian Na,K-ATPase in yeast. J. Biol. Chem. **265:** 4189–4192.

2. PEDERSEN, A. & P. L. JØRGENSEN. 1992. Expression of Na,K-ATPase in *Saccharomyces cerevisiae*. Ann. N.Y. Acad. Sci. **671:** 452–454.
3. NOGUCHI, S., M. MISHINA, M. KAWAMURA & S. NUMA. 1987. Expression of functional (Na + K)-ATPase from cloned cDNAs. FEBS Lett. **225:** 27–32.
4. GEERING, K., I. THEULAZ, F. VERREY, M. T. HAUPTLE & B. C. ROSSIER. 1989. A role for the β-subunit in the expression of functional Na^+-K^+-ATPase in *Xenopus* oocytes. Am. J. Physiol. **257:** C851–C858.
5. DETOMASO, A. W., Z. J. XIE, G. LIU & R. W. MERCER. 1993. Expression, targeting and assembly of functional Na,K-ATPase polypeptides in baculovirus-infected insect cells. J. Biol. Chem. **268:** 1470–1478.
6. JEWELL, E. A. & J. B. LINGREL. 1991. Comparison of the substrate dependence properties of the rat Na,K-ATPase α1, α2, and α3 isoforms expressed in HeLa cells. J. Biol. Chem. **266:** 16925–16930.
7. LUCKOW, V. A. 1990. *In* Recombinant DNA Technology and Applications, McGraw-Hill. New York.
8. BLANCO, G., J. C. KOSTER, G. SÁNCHEZ & R. W. MERCER. 1995. Kinetic properties of the α2β1 and α2β2 isozymes of the Na,K-ATPase. Biochemistry **34:** 319–325.
9. BLANCO, G., G. SÁNCHEZ & R. W. MERCER. 1995. Comparison of the enzymatic properties of the Na,K-ATPase α3β1 and α3β2 isozymes. Biochemistry **34:** 9897–9903.
10. BLANCO, G., A. W. DETOMASO, J. C. KOSTER, Z. J. XIE & R. W. MERCER. 1994. The α-subunit of the Na,K-ATPase has catalytic activity independent of the β-subunit. J. Biol. Chem. **269:** 23420–23425.
11. POST, R. L. & S. KUME. 1973. Evidence for an aspartyl phosphate residue at the active site of sodium and potassium ion transport adenosine triphosphatase. J. Biol. Chem. **248:** 6993–7000.
12. BASTIDE, F., G. MEISSNER, S. FLEISCHER & R. L. POST. 1973. Similarity of the active site of phosphorylation of the adenosine triphosphatase from transport of sodium and potassium ions in kidney to that for transport of calcium ions in the sarcoplasmic reticulum of muscle. J. Biol. Chem. **248:** 8385–8391.
13. SCHUURMANS STEKHOVEN, F. M. A. H., H. G. P. SWARTS, J. J. H. H. M. DE PONT & S. L. BONTING. 1985. Na^+-like effect of imidazole on the phosphorylation of $(Na^+ + K^+)$-ATPase. Biochim. Biophys. Acta **815:** 16–24.
14. HOKIN, L. E., P. S. SASTRY, P. R. GALSWORTHY & A. YODA. 1965. Evidence that a phosphorylated intermediate in a brain transport adenosine triphosphatade is an acyl phosphate. Proc. Natl. Acad. Sci. USA **54:** 177–184.
15. DUCLOS, B., S. MARCANDIER & A. J. COZZONE. 1991. Chemical properties and separation of phosphoamino acids by thin-layer chromatography and/or electrophoresis. Methods Enzymol. **201:** 10–21.
16. LIPMANN, F. & L. C. TUTTLE. 1945. A specific micromethod for the determination of acyl phosphates. J. Biol. Chem. **159:** 21–28.
17. HASEMANN, C. A. & J. D. CAPRA. 1990. High-level production of a functional immunoglobulin heterodimer in a baculovirus expression system. Proc. Natl. Acad. Sci. USA **87:** 3942–3946.
18. BIRNIR, B., M. L. TIERNEY, S. M. HOWITT, G. B. COX & P. W. GAGE. 1992. A combination of human α1 and β1 subunits is required for formation of detectable GABA-activated chloride channels in Sf-9 cells. Proc. R. Soc. Lond. **250:** 307–312.
19. HUNDAL, H. S., A. MARETTE, Y. MITSUMOTO, T. RAMLAL, R. BLOSTEIN & A. KLIP. 1992. Insulin induces translocation of the α2 and β2 subunits of the Na/K-ATPase from intracellular compartments to the plasma membrane in mammalian skeletal muscle. J. Biol. Chem. **267:** 5040–5043.
20. BLISSARD, G. W. & J. R. WENZ. 1992. Baculovirus gp64 envelope glycoprotein is sufficient to mediate pH-dependent membrane fusion. J. Virol. **66:** 6829–6835.

Biochemical Characterization of the Human Renal Na$^+$,K$^+$-ATPase[a]

G. CRAMBERT,[b] A. FRANZ,[c] AND L. G. LELIEVRE[d]

[b]*Laboratoire de Pharmacologie des Transports
Ioniques Membranaires
Université Paris VII–Hall de Biotechnologie
75251 Paris Cedex 05, France*

The membrane-bound Na$^+$,K$^+$-ATPase involved in the transmembrane electrochemical Na$^+$ plus K$^+$ gradients is also the specific receptor for cardiac glycosides (digitalis) which leads to positive inotropic effects in the heart.

This enzyme consists of a catalytic α subunit and a glycosylated regulatory β subunit. There are three isoforms of the α subunit (α1, α2, and α3) that mainly differ in their respective affinities for cardiac glycosides. Likewise, three isoforms have been described for the β subunit with only two (β1 and β2) expressed in mammalian tissues.

The expression and functional properties of the cardiac Na$^+$,K$^+$-ATPase are well known in most animal models. By contrast, in healthy human heart, controversial information has been published. The three isoforms are expressed in terms of mRNA[1,2] and proteins at the membrane level,[3,4] but only two[4,5] different sensitivities to digitalis could be detected.

The question arises of whether the three α proteins are active at the membrane level and which one represents α1, α2, and α3. Our way to discriminate is to determine the functional characteristics of the human renal Na$^+$,K$^+$-ATPase assuming that, by analogy with animal models, α1 is mainly if not exclusively expressed in kidney and would correspond to the cardiac α1 isoform.

From biopsies ($n = 4$) of healthy human kidneys (cortex and medulla), plasma membranes were purified by differential centrifugation.[6] With these membranes, three approaches were developed: Western blot analysis with specific antibodies, enzymatic assays, and (^{3}H)ouabain binding measurements. Before enzymatic and binding experiments, membrane vesicles are permeabilized to ligands by pretreatment with sodium deoxycholate (0.6 µg/µg proteins for 15 minutes at 23°C, pH 7.0).

---

[a]This work was supported in part by grants from l'Université Paris 7 (3409 R10/11), la DRED (Direction de la Recherche et de l'Encadrement des Doctorants), and Réseau INSERM "Inotropie" n° 493017 (L. G. L.).

[c]A. Franz is an Erasmus student from Humboldt Universität zu Berlin.

[d]To whom correspondence should be addressed. Tel: 33 (0)1 44 27 57 03; fax: 33 (0)1 44 27 69 66; e-mail: lelievre@paris7.jussieu.fr

RESULTS AND DISCUSSION

Immunological Detection

After protein treatment according to Laemmli et al.,[7] samples are resolved in a 7.5% (for α subunits) or 12% (for β subunits) acrylamide gel as described by Charlemagne et al.[8]

Specific polyclonal antibodies directed against a peptidic sequence specific for each α isoform[9] and human β1 and β2 subunits[10] were used. The α2, α3, and β2 subunits could not be detected in the membrane fractions. Anti-α and anti-α1 antibodies revealed a single band at an apparent molecular weight of 103 ± 1 and 104 ± 1 kD, respectively. The glycosylated β1 chain has an apparent molecular weight of 64 ± 1 kD. Its deglycosylation by N glycanase F (0.035 U/μg proteins for 2 hours at 37°C) leads to an apoprotein of 36 ± 1 kD, similar to that found in rat brain and kidney. So, demonstrated here for the first time is that the renal human Na^+,K^+-ATPase is only constituted of α1β1 chains.

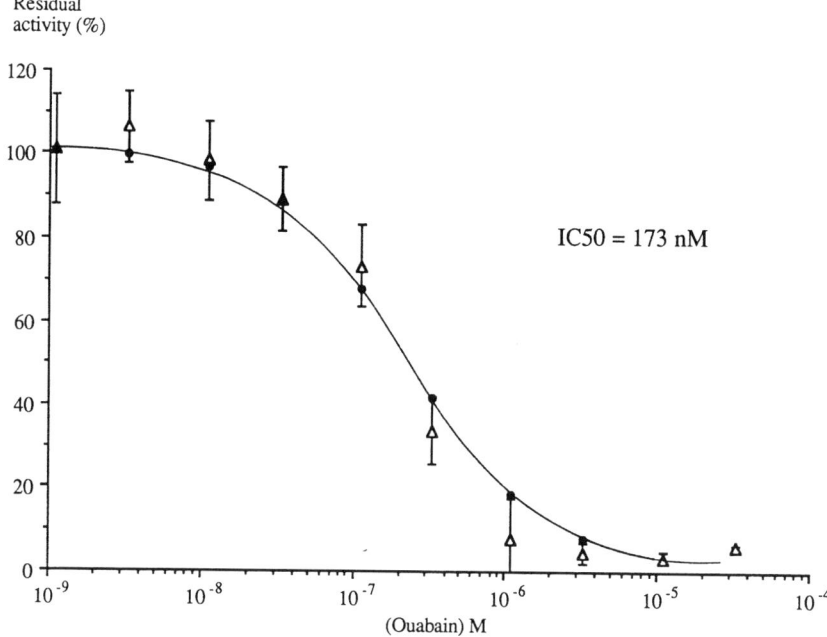

FIGURE 1. Dose-response curve of Na^+,K^+-ATPase activity versus ouabain concentrations in permeabilized plasma membranes from human kidney. Experiments are carried out for 30 minutes at 37°C in the presence of 0.83 μg of membrane proteins per milliliter of an ATP-regenerating medium containing ATP (4 mM), Mg^{2+} (4 mM), Na^+ (100 mM), K^+ (10 mM), imidazole/HCl (30 mM), pH 7.4, and varying ouabain concentrations (from 10^{-9} to 3.10^{-5} M). Experimental values (△) are means (±SD) of at least 12 measurements with 5 different membrane preparations. The curve corresponds to theoretical values (●) calculated assuming *a single population of ouabain sites* with an IC_{50} value of 173 nM.

TABLE 1. Kinetic Parameters of (^{3}H)Ouabain Binding (Association and D Processes Are Carried Out in the Absence of Potassium)

Association	Dissociation		
k1 (M^{-1}·s^{-1})	k − 1 (min^{-1})	Calculated K_d [a] (nM)	Equilibrium K_d [b] (nM)
26,000	0.017 ± 0.002	11	17 ± 3 (no K$^+$)
			46 ± 13 (+5 mM K$^+$)

[a] The K_d value is calculated by k-1/k1 ratio.
[b] K_d values are determined from bindings at equilibrium with or without potassium (Scatchard analysis).

PHARMACOLOGICAL CHARACTERIZATIONS

Sensitivity of Na$^+$,K$^+$-ATPase Activity to Ouabain

Na$^+$,K$^+$-ATPase activity is measured with a spectrophotometric method in an ATP-regenerating medium. Oxidation of NADH (directly correlated to the rate of ATP hydrolysis via a coupled enzyme system) is recorded at 37°C within 30 minutes after the addition of permeabilized membranes.

The specific activity of Na$^+$,K$^+$-ATPase, 35.8 ± 5.5 μmol P$_i$/mg × h ($n > 50$), is the difference between total (no ouabain) and basal (+10^{-4} M ouabain) activities and represents 60% of total activity.

The sensitivity of Na$^+$,K$^+$-ATPase activity to ouabain is assayed in the presence of varying drug concentrations (from 10^{-9} to 3.10^{-5} M). The dose-response curve (FIG. 1) reveals a single population. Half-maximal inhibition (IC$_{50}$ value) occurs with 173 nM. This value is relevant to that found in clinical studies.

(^{3}H)Ouabain-Binding Assays

Experiments were carried out as previously described.[4] It is worthy of note that the nonspecific binding is lower than 10% whatever the (^{3}H)ouabain concentration tested.

Specific (^{3}H)ouabain-binding sites account for 27.1 ± 3.9 pmol/mg. Kinetic parameters obtained from association and dissociation processes and bindings at equilibrium (TABLE 1) show that a single population of (^{3}H)ouabain-binding sites does exist in human kidney.

Turnover Number

The relation between the amount of bound (^{3}H)ouabain and the degree of Na$^+$,K$^+$-ATPase inhibition is linear. The slope corresponds to a turnover number of 140 s^{-1}, which is very close to that found in human red cells (α1 type).[11]

CONCLUSION

Only one population of Na$^+$,K$^+$-ATPase does exist in human healthy kidney. It is functional and of the α1β1 type. Its pharmacological properties, IC$_{50}$ of 173 nM, K_d of

46 nM, K^+/ouabain antagonism (17 vs 46 nM), correspond to those of one of the isoforms found in human cardiac membranes (IC_{50} = 80 nM, Kd = 17 nM with no K^+).[4] In human heart, therefore, the α1β1 form is not only expressed[1-4] but also *functional*.

ACKNOWLEDGMENTS

We wish to thank Dr. Yves Chrétien (Hôpital Necker) for his help in procuring peroperatory biopsies used in this study, Dr. Tom Pressley (Texas Tech University) for his generous gift of anti-α, α1, α2, and α3 antibodies, and Pr. Pablo Martín-Vassalo (Universidad de La Laguna, Tenerife, Spain) for his generous gift of anti-β1 and β2 antibodies.

REFERENCES

1. ZAHLER, R., M. GILMORE-HEBERT, J. C. BALDWIN, K. FRANCO & E. J. BENZ. 1993. Biochim. Biophys. Acta **1149:** 189–194.
2. SHAMRAJ, O. I., D. MELVIN & J. B. LINGREL. 1991. Biochem. Biophys. Res. Comm. **173:** 1434–1440.
3. SWEADNER, K. J., V. L. M. HERRERA, S. AMATO, A. MOELLMANN, D. K. GIBBONS & K. R. H. REPKE. 1994. Circ. Res. **74:** 669–678.
4. LELIEVRE, L. G. & P. A. ALLEN. 1997. Proceeding of the Nobel Conference 1995. Acta Biol. Scand. In press.
5. SHAMRAJ, O. I., I. L. GRUPP, D. M. GRUPP, N. GRADOUX, W. KREMERS, J. B. LINGREL & A. DE POVER. 1993. Cardiovasc. Res. **27:** 2229–2237.
6. BERREBI-BERTRAND, I., J.-M. MAIXENT, G. CHRISTE & L. G. LELIEVRE. 1990. Biochim. Biophys. Acta **1021:** 148–156.
7. LAEMMLI, U. K. 1970. Nature **227:** 680–685.
8. CHARLEMAGNE, D., J.-M. MAIXENT, M. PRETESEILLE & L. G. LELIEVRE. 1986. J. Biol. Chem. **261:** 185–189.
9. PRESSLEY, T. A. 1992. Am. J. Physiol. **262:** C743–C751.
10. GONZALEZ-MARTINEZ, L. M., J. A. MARRERO, E. MARTI, E. LECUONA & P. MARTIN-VASSALO. 1994. *In* The Sodium Pump. E. Bamberg & W. Schoner, Eds.: 218–221. Springer Darmstadt: Steinkopff.
11. ERDMANN, E., K. WERDAN & L. BROWN. 1984. Eur. Heart J. **5:** 297–302.

Role of Sugar Residues for Recombinant Gastric H$^+$,K$^+$-ATPase[a]

CORNÉ H. W. KLAASSEN, HERMAN G. P. SWARTS, AND
JAN JOEP H. H. M. DE PONT[b]

Department of Biochemistry
Institute of Cellular Signalling
University of Nijmegen
P.O. Box 9101
6500 HB Nijmegen, The Netherlands

A major feature of the gastric H$^+$,K$^+$-ATPase β subunit is the presence of six or seven consensus sequences for *N*-linked glycosylation which are all cotranslationally glycosylated. In several reports of Na$^+$,K$^+$-ATPase, it was demonstrated that *N*-glycosylation is not essential for enzymatic activity.[1–3] The present study investigates whether *N*-glycosylation is essential for H$^+$,K$^+$-ATPase activity.

H$^+$,K$^+$-ATPase can be synthesized *in vitro* as an active enzyme using the baculovirus system.[4,5] In contrast to the mammalian enzyme, the β subunit is synthesized in both a nonglycosylated and a core-glycosylated form. Complex glycosylated β subunit is either present or absent in minor amounts.[4] The presence of increasing concentrations of tunicamycin, an inhibitor of *N*-glycosylation, in the culture medium of *Sf*-9 cells resulted in a highly reproducible dose-dependent decrease in the amount of functional H$^+$,K$^+$-ATPase synthesized (FIG. 1). This decrease in H$^+$,K$^+$-ATPase activity is correlated with a simultaneous decrease in the amount of glycosylated β subunits. Tunicamycin treatment had no visible effect on the H$^+$,K$^+$-ATPase α subunit. These results strongly suggest that *N*-glycosylation somehow is essential for H$^+$,K$^+$-ATPase activity.

By using deoxymannojirimycin, a specific inhibitor of α-mannosidase I, trimming of the high-mannose oligosaccharide precursor can be blocked, preventing formation of complex glycosylated forms. Analysis of glycosylated forms of the β subunit indicated that the compound was active. However, no effect on the activity of the recombinant expressed H$^+$,K$^+$-ATPase was measured. Thus, only the presence and not the exact structure of the oligosaccharide moieties is essential for H$^+$,K$^+$-ATPase activity.

Functional H$^+$,K$^+$-ATPase subunits in the standard crude membrane preparation can be separated from nonfunctional H$^+$,K$^+$-ATPase subunits using a discontinuous sucrose density gradient. FIGURE 2 shows that the purified H$^+$,K$^+$-ATPase fraction contained more glycosylated and almost no nonglycosylated β subunits. The nonglycosylated β subunits were more abundant in the pellet fraction, in which only little H$^+$,K$^+$-ATPase activity was found. This supports our conclusion that glycosylation is essential for H$^+$,K$^+$-ATPase activity.

---

[a]This work was supported by the Netherlands Foundation for Scientific Research, Division of Medical Sciences (NWO-GMW), under grant 902-22-086.

[b]To whom correspondence should be addressed. Tel: +31243614260; fax: +31243540525; e-mail: J.dePont@bioch.kun.nl

Confocal microscopy studies show that the α subunit of H^+,K^+-ATPase is found exclusively in intracellular membranous structures. No levels of α subunit are detectable in the plasma membrane. This means that the catalytically active H^+,K^+-ATPase fraction also originates from an intracellular source. The H^+,K^+-ATPase β subunit is partly targeted to the plasma membrane and partly retained in intracellular membranous structures. In the presence of 5 μg/ml tunicamycin, the nonglycosylated β subunit can no longer be found on the plasma membrane (not shown). Apparently, proper processing of the H^+,K^+-ATPase β subunit onto the plasma membrane depends on the presence of N-linked oligosaccharides on this subunit. However, because the H^+,K^+-ATPase α subunit is found exclusively in intracellular membranous structures, processing of the H^+,K^+-ATPase β subunit to the plasma membrane is apparently not essential for synthesis of a functional H^+,K^+-ATPase in insect cells.

In immunoprecipitates from untreated cultures, both glycosylated and nonglycosylated H^+,K^+-ATPase β subunits are precipitated with the anti-α subunit antibody (not shown). This means that both forms of the β subunit must be engaged with the α

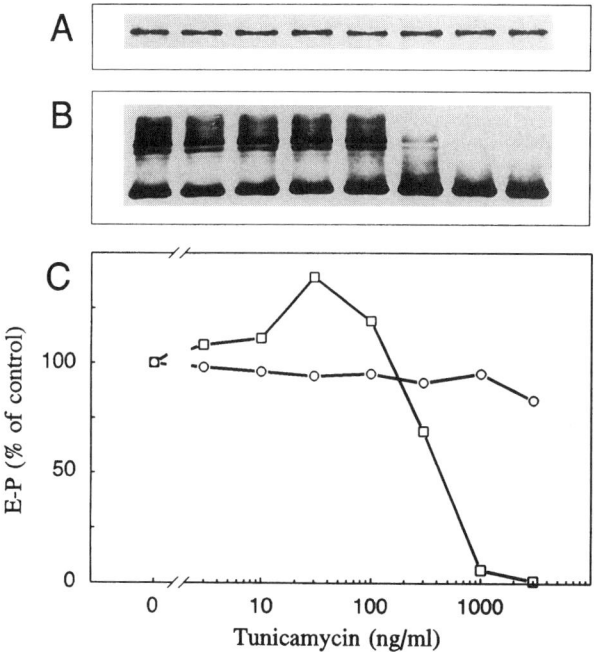

FIGURE 1. Effect of tunicamycin on glycosylation and activity of H^+,K^+-ATPase. The phosphorylation capacity of H^+,K^+-ATPase (*squares*) and endogenous (auto)phosphorylating enzymes (*circles*) in crude membrane fractions is plotted against the tunicamycin concentration in the culture medium. 100% values are 2.16 ± 0.28 pmol/mg (mean ± SEM) for H^+,K^+-ATPase and 0.75 ± 0.04 for endogenous (auto)phosphorylating enzymes from nine experiments. In the *upper panel,* western blot of the α subunit (**A**) and the β subunit (**B**) is shown. Each lane contains 2.0 μg crude membrane protein. H^+,K^+-ATPase subunits were visualized using subunit-specific antibodies. Horizontal position of the lanes in **A** and **B** corresponds to the tunicamycin concentration below.

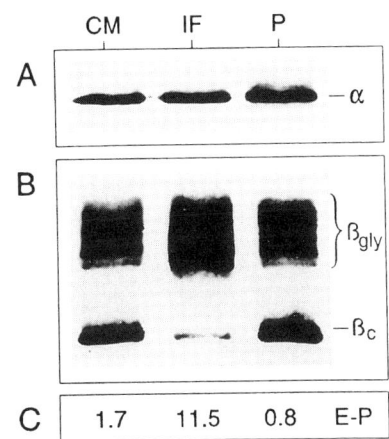

FIGURE 2. Glycosylated β subunits copurify with functional H^+,K^+-ATPase. The H^+,K^+-ATPase α subunit (**A**) or β subunit (**B**) in different Sf-9 membrane fractions was visualized with subunit-specific antibodies. $β_c$ = β subunit core protein (34 kD); $β_{gly}$ = glycosylated β subunits (40–50 kD). The activity of the resulting fraction is given as the steady-state phosphorylation capacity in pmol.mg^{-1} protein and is given below in **C**. CM = crude membranes; IF = 25% (w/v) − 38% (w/v) sucrose interfase; P = 38% (w/v) sucrose pellet. Each lane contains 1.0 µg protein.

subunit in a detergent-resistant complex and hence are tightly associated. This conclusion is supported by the finding that in tunicamycin-treated cultures, where no glycosylated β subunits are produced, the amount of immunoprecipitated nonglycosylated β subunit is increased relative to untreated cultures. Thus, N-glycosylation plays no role in the interaction between α and β subunits.

In conclusion, N-glycosylation is essential for H^+,K^+-ATPase activity and for targeting the β subunit to the plasma membrane. Core glycosylation seems to be sufficient for enzyme activity. Moreover, the presence of sugar residues is not essential for α-β interaction.

REFERENCES

1. TAKEDA, K., S. NOGUCHI, A. SUGINO & M. KAWAMURA. 1988. FEBS Lett. **238:** 201–204.
2. TAMKUN, M. M. & D. M. FAMBROUGH. 1986. J. Biol. Chem. **261:** 1009–1019.
3. ZAMOFING, D., B. C. ROSSIER & K. GEERING. 1989. Am. J. Physiol. **256:** C958–C966.
4. KLAASSEN, C. W. H., T. J. F. VAN UEM, M. P. DE MOEL, G. L. J. DE CALUWE, H. G. P. SWARTS & J. J. H. H. M. DE PONT. 1993. FEBS Lett. **329:** 277–282.
5. KLAASSEN, C. W. H., H. G. P. SWARTS & J. J. H. H. M. DE PONT. 1995. Biochem. Biophys. Res. Commun. **210:** 907–913.
6. FUHRMANN, U., E. BAUSE, G. LEGLER & H. PLOEGH. 1984. Nature **307:** 755–758.

Effects of Adenoviral Mediated Transfer of Na^+,K^+-ATPase Subunit Genes to Alveolar Epithelial Cells[a]

P. FACTOR,[b] C. SENNE, K. RIDGE, H. A. JAFFE,
G. BLANCO, R. W. MERCER, AND J. I. SZNAJDER

Pulmonary and Critical Care Medicine
Michael Reese Hospital
Chicago, Illinois 60616

University of Illinois at Chicago
Chicago, Illinois 60607

Washington University
St. Louis, Missouri 63110

Edema is cleared from the alveolar airspace by active Na^+ transport. Data from studies in rats, rabbits, sheep, and humans indicate that this active transport is due, in part, to the function of alveolar Na^+,K^+-ATPases.[1–6] In the lung, Na^+,K^+-ATPase is principally, but not exclusively, composed of α_1 and β_1 subunits and is primarily located on the basolateral surface of alveolar type 2 (AT2) epithelial cells.[7] This multigene system works in concert with other membrane-bound transport systems, including apical Na^+ channels, to effect the iso-osmotic movement of water out of the airspace clearance.[8]

We previously reported that rats exposed to subacute hyperoxia have increased capacity to clear edema from the alveolus and that a significant fraction of this capacity can be inhibited by the specific Na^+,K^+-ATPase inhibitor ouabain.[1] AT2 cells isolated from these same rats have increased Na^+,K^+-ATPase function and protein expression.[6] In other studies using antisense constructs against Na^+,K^+-ATPase α subunits, we have been able to limit alveolar liquid clearance in rat lungs and Na^+,K^+-ATPase function in isolated AT2 cells.[9] These studies indicate that Na^+,K^+-ATPase contributes to alveolar liquid clearance and plays an important role in keeping the alveolar airspace dry.

We hypothesized that overexpression of Na^+,K^+-ATPase subunits can be used to augment function of this important multigene system in alveolar epithelial cells. We constructed cytomegalovirus promoter-driven replication-deficient human type 5 adenoviruses containing cDNAs for the α_1 and β_1 subunits of rat Na^+,K^+-ATPase and *Escherichia coli lacZ* (AdMRCMVα_1, AdMRCMVβ_1, and AdMRCMVβ-gal, respectively).[10]

---

[a]This work was supported by HL-48129, the American Lung Association, the American Lung Association of Metropolitan Chicago, the American Heart Association of Metropolitan Chicago, and Michael Reese Hospital.

[b]Address for correspondence: Phillip Factor, D.O., FCCP, Pulmonary and Critical Care Medicine, Michael Reese Hospital and Medical Center, 2929 S. Ellis, Room MR255, Chicago, IL 60616 (tel: 312-791-2050; fax: 312-791-2311; e-mail: PFACT@AOL.COM).

Primary alveolar type 2 (AT2) cells were isolated from normal Sprague-Dawley[11] rats and plated for 24 hours prior to infection with adenovirus.[12] Na^+,K^+-ATPase expression and function were evaluated using Western and Northern blot analyses and ouabain inhibitable $^{86}Rb^+$ uptake in these cells 24 hours following infection. In contrast to sham and AdMRCMVβ-Gal-infected controls, AT2 cells infected with multiplicities of infection (MOI, # viruses/plated cell) of 5 or 10 of AdMRCMVβ$_1$ demonstrated significant expression of adenovirus-specific β$_1$ mRNA and protein, no change in α$_1$ expression was noted. Ouabain inhibitable Na^+,K^+-ATPase activity was increased up to 2.5-fold only in AT2 cells infected with AdMRCMVβ$_1$.

Human alveolar carcinoma cells (A549) were plated for 24 hours prior to infection with AdMRCMVα$_1$, AdMRCMVβ$_1$, and AdMRCMVβ-gal. Cells infected with MOI of 25 to 100 of AdMRCMVα$_1$ demonstrated significant increases in α$_1$ mRNA, protein, and ouabain-inhibitable $^{86}Rb^+$ uptake 24 hours after infection. Infection with AdMRCMVβ$_1$ produced no changes in Na^+,K^+-ATPase function, suggesting that the α-subunit may be rate limiting in A549 cells.

The human α$_1$ isoform is 2–3 logs more sensitive to ouabain than is the rat α$_1$ isoform. To demonstrate transgene expression in A549 cells, $^{86}Rb^+$ uptake was measured using 15 different concentrations of ouabain (1×10^{-11} to 1×10^{-3} M). The concentration of ouabain that produced a 50% reduction in $^{86}Rb^+$ uptake (IC_{50}) was determined in cells infected with AdMRCMVα$_1$ (MOI = 25) using a computerized nonlinear regression equation designed to test for the presence of two receptors of differing affinities for the same ligand. AdMRCMVα$_1$-infected cells demonstrated two distinct IC_{50}s ($IC_{50(1)} = 3.51 \times 10^{-8}$, $IC_{50(2)} = 3.17 \times 10^{-5}$). Sham and AdMRCMVβ-Gal infected cells each demonstrated a single IC_{50} that was not different from α$_1$ $IC_{50(1)}$. Identification of a second α$_1$ IC_{50} suggests the presence of two functional α$_1$ isozymes: endogenous (ouabain-sensitive) human α$_1$ and transgenic (ouabain-resistant) rat α$_1$.

These results demonstrate that adenoviral mediated gene transfer can be used to augment Na^+,K^+-ATPase expression and activity and that the response to gene transfer differs between cell types. These gene transfer vehicles allow for new investigations of Na^+,K^+-ATPase expression and function in mammalian cells.

REFERENCES

1. SZNAJDER, J. I., W. G. OLIVERA, K. M. RIDGE & D. H. RUTSCHMAN. 1995. Mechanisms of lung liquid clearance during hyperoxia in isolated rat lungs. Am. J. Respir. Crit. Care Med. **151:** 1519–1525.
2. RUTSCHMAN, D. H., W. OLIVERA & J. I. SZNAJDER. 1993. Active transport and passive liquid movement in isolated perfused rat lungs. J. Appl. Physiol. **75:** 1574–1580.
3. GOODMAN, B. E., K. KIM & E. D. CRANDALL. 1987. Evidence for active sodium transport across alveolar epithelium of isolated rat lung. J. Appl. Physiol. **62:** 2460–2466.
4. MATTHAY, M. & J. WIENER-KRONISH. 1990. Intact epithelial barrier function is critical for the resolution of alveolar edema in humans. Am. Rev. Respir. Dis. **142:** 1250–1257.
5. SUZUKI, S., D. ZUEGE & Y. BERTHIAUME. 1995. Sodium-independent modulation of Na(+)-K(+)-ATPase activity by beta-adrenergic agonist in alveolar type II cells. Am. J. Physiol. **268:** L983–L990.
6. OLIVERA, W., K. RIDGE, L. D. H. WOOD & J. I. SZNAJDER. 1994. Active sodium transport and alveolar epithelial Na-K-ATPase increase during subacute hyperoxia in rats. Am. J. Physiol. **266:** L577–L584.

7. SCHNEEBERGER, E. E. & K. M. MCCARTHY. 1986. Cytochemical localization of Na, K-ATPase in rat type II pneumocytes. J. Appl. Physiol. **20:** 1584–1589.
8. O'BRODOVICH, H. M. 1995. The role of active Na^+ transport by lung epithelium in the clearance of airspace fluid. New Horizons **3:** 240–247.
9. RIDGE, K., J. ILEKIS, P. FACTOR & J. I. SZNAJDER. 1992. Inhibition of the Na,K ATPase by transfection of alveolar type 2 cells with antisense RNA to the Na,K ATPase. Am. Rev. Respir. Dis. **145:** A832.
10. MCGRORY, W. J., D. S. BAUTISTA & F. L. GRAHAM. 1988. A simple technique for the rescue of early region 1 mutations into infectious human adenovirus type 5. Virology **163:** 614–617.
11. DOBBS, L. G., R. GONZALEZ & M. C. WILLIAMS. 1986. An improved method for isolating type 2 cells in high yield and purity. Am. Rev. Respir. Dis. **134:** 141–145.
12. ROSENFELD, M. A., C. CHIN-SHYAN, P. SETH *ET AL.* 1994. Gene transfer to isolated human respiratory epithelial cells *in vitro* using a replication-deficient adenovirus containing the human cystic fibrosis transmembrane conductance regulator cDNA. Hum. Gene Ther. **5:** 331–342.

Abundance of α3, α2, and α1 Isoforms of Na,K-ATPase in Rat Kidney as Estimated by Competitive RT-PCR and [³H]Ouabain Binding

K. LÜCKING[a] AND J. M. NIELSEN

Biomembrane Research Center
August Krogh Institute
Universitetsparken 13
2100 Copenhagen OE, Denmark

Several isoforms of the α and β subunits of Na/K-ATPase are expressed in a tissue-specific manner, but attempts to determine the distribution of α-subunit isoform mRNA in kidney have provided conflicting results. *In situ* hybridization revealed the presence of the α1 isoform in all segments of the rat kidney, whereas the α2- and α3-isoform mRNA could not be detected.[1] In another study, using dot blot analysis, the α1, α2, and α3 isoforms contributed 70, 20, and 10% of the total α-isoform mRNA.[2]

Isozymes of Na/K-ATPase containing the α1 and α2 isoforms have identical kinetic properties, whereas Na/K-ATPase containing the α3 isoform has a lower affinity for cytoplasmic Na^+ and a higher apparent affinity for extracellular K^+.[3] For understanding the regulation of sodium reabsorption, it is therefore important to know if the α2 or α3 isoforms of Na,K-ATPase are expressed in mammalian kidney in addition to the predominant α1β1 isozyme. In this study we applied competitive polymerase chain reaction (PCR) for accurate quantification of mRNA in preparations from the cortex, outer medulla, and papilla of the rat kidney for comparison with data of high affinity [³H]ouabain binding to particulate membrane fractions of the three zones.[4]

Initially the α1, α2, α3 and α1-T isoforms mRNA were demonstrated in all three zones of rat kidney and in brain by conventional RT-PCR. However, the abundance of the α2 and α3 isoforms was so low that 45 cycles in the PCR procedure were necessary to visualize amplification.

In the competitive PCR procedure, an unknown concentration of cDNA is determined by running several PCR samples with a fixed amount of cDNA and a serial dilution of a competitor fragment. DNA competitor fragments are identical to target sequences but synthesized to be ~10% shorter. Estimation of a certain isoform ratio was done from one reverse transcription sample primed with either a generic isoform-specific primer or with a poly(dT) primer. Representative data of a competitive RT-PCR procedure are shown in FIGURE 1. The α3 isoform mRNA was demonstrated to form 0.04–0.05% of the amount of α-isoform mRNA in the cortex, medulla, and papilla of rat kidney. The α2 mRNA constituted 0.03% of the amount of α1 in cortex.

---

[a]Tel: (45) 35 32 16 78; fax: (45) 35 32 15 67; e-mail: klucking@aki.ku.dk

High affinity [^{3}H]ouabain binding was used to estimate the concentration of functional Na-K pumps containing α2 or α3 isoforms in the parenchymal zones of rat kidney. The rationale was that in the rat only, the Na/K-ATPase containing the α2 or α3 isoforms can bind [^{3}H]ouabain with high affinity, whereas the affinity of Na/K-ATPase containing the rat α1 isoform is so low that binding of [^{3}H]ouabain is undetectable. In a modified assay of [^{3}H]ouabain binding, the upper limit for the expression of Na,K-ATPase isozyme protein with high ouabain affinity (K_D 69–141 nM) was estimated to be 0.10–0.14% of the concentration of α1β Na,K-ATPase. Data of competitive PCR and [^{3}H]ouabain binding are summarized in TABLE 1.

Thus, a small but well-defined pool of α2 and α3 isoforms constitutes ≤0.1% of the amount of the α1 isoform at both the mRNA and protein level in rat kidney. This

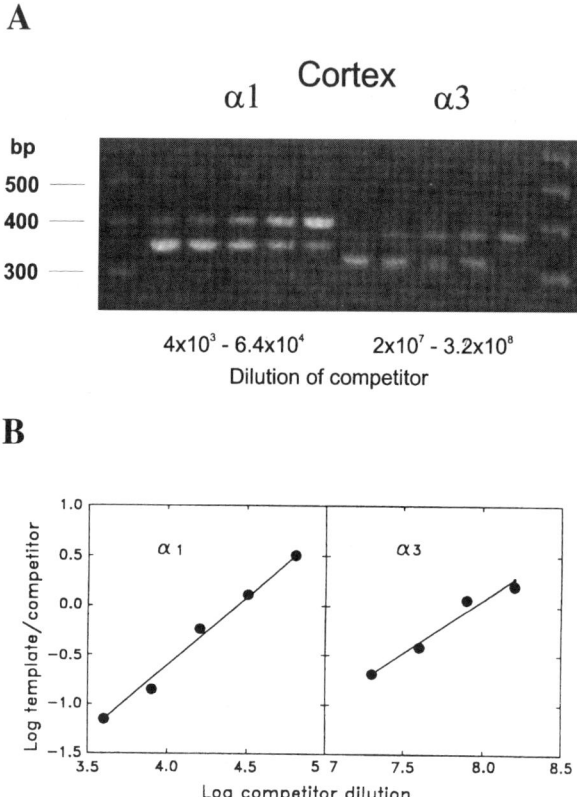

FIGURE 1. Example of the final competitive polymerase chain reaction (PCR) procedure. (**A**) Agarose gel electrophoresis after final competitive PCR on cDNA from α1 and α3 in cortex. Competitor dilution was varied twofold between samples. (**B**) Double logarithmic plots of template-to-competitor ratio versus dilutions of competitor. Data obtained by densitometric scanning of the photograph in **A**. The best fit of the straight line through the data is calculated to determine the dilution of competitor DNA that results in a 1:1 molar ratio between template and competitor DNA.

TABLE 1. Molar Ratios of α-Subunit Isoforms in Rat Kidney Parenchymal Zones and Brain Determined by Competitive Polymerase Chain Reaction and [^{3}H]Ouabain Binding

Parenchymal Zone	mRNA α2/α1	α3/α1	α1-T/α1	Protein (α2 + α3)/α1
Cortex	0.03% (1)	0.05% ± 0.02% (4)	0.8% (2)	0.10%
Outer medulla	ND	0.04% ± 0.03% (3)	0.6% (2)	0.14%
Papilla	ND	0.04% (2)	0.1% (2)	ND
Brain	3.8 (2)	1.6 (2)	0.9% (2)	ND

ND = not determined; number in parentheses is the number of experiments.

Note: Concentration of α1β1 isozyme was calculated from Na,K-ATPase activities estimated to be 86 μmol P$_i$/h · mg protein in the outer medulla and 38 μmol P$_i$/h · mg protein in the cortex. Concentration of α1β1 isozyme was estimated to be 0.205 and 0.090 nmol/mg protein, respectively.

amount of α2 and α3 isoforms agrees with the data of Farman *et al.*[1] and is not expected to be of physiological significance for renal Na$^+$ reabsorption.

REFERENCES

1. FARMAN, N., I. CORTHESY-THEULAZ, J. P. BONVALET & B. C. ROSSIER. 1991. Am. J. Physiol. **260:** C468–C474.
2. CLAPP, W. L., P. BOWMAN, G. S. SHAW, P. PATEL & B. C. KONE. 1994. Kidney Int. **46:** 627–638.
3. MUNZER, J. S., S. E. DALY, E. A. JEWELL MOTZ, J. B. LINGREL & R. BLOSTEIN. 1994. J. Biol. Chem. **269:** 16668–16676.
4. LÜCKING, K., J. M. NIELSEN, P. A. PEDERSEN & P. L. JØRGENSEN. 1996. Am. J. Physiol. **271:** F253–F260.

Cellular and Developmental Distribution of the Na,K-ATPase β Subunit Isoforms of Neural Tissues[a]

PABLO MARTÍN-VASALLO,[b,d] EMILIA LECUONA,[b]
SONIA LUQUÍN,[c] DIEGO ALVAREZ DE LA ROSA,[b]
JULIO AVILA,[b] TERESA ALONSO,[b]
AND LUIS MIGUEL GARCÍA-SEGURA[c]

[b]Laboratorio de Biología del Desarrollo
Departamento de Bioquímica y Biología Molecular
Universidad de La Laguna
Avda Astrofísico Sánchez s/n
38206 La Laguna, Tenerife, Spain

[c]Instituto Cajal, C.S.I.C.
Dr. Arce 37
28002 Madrid, Spain

The Na,K-ATPase of the central nervous system (CNS) maintains the cation gradients necessary for neural impulse conduction in neurons and potassium buffering and neurotransmitter uptake in glial cells. We studied the cellular and subcellular localization of the β1 and β2 isoforms of the Na,K-ATPase in neural cells and their expression patterns throughout development. We performed immunohistochemical studies on preparations of CNS of newborn, 6-, 10-, 15-, and 90-day-old Wistar male rats. β isoform-specific antibodies SpETb1 and SpETb2 (anti-β1 and anti-β2, respectively)[1] and a monoclonal antiserum against glial fibrillary acidic protein (GFAP, Sigma) were used as probes.

Adult Animals. Immunoreactivity for the Na,K-ATPase β1 subunit was detected in neurons scattered all over the CNS of adult rats. Immunoreactivity in neurons was localized to the plasma membrane and the cytoplasm at the level of the perikaryon, although not all of the neuronal somas in a given area were immunoreactive.

---

[a]This work was supported by grants 93/0831 and 96/0453 from FIS (Spain) to P.M.-V. and a grant from DGICYT to L.M.G.-S.

[d]Address for correspondence: Laboratorio de Biología del Desarrollo, Departamento de Bioquímica y Biología Molecular, Universidad de La Laguna, 38206 La Laguna, Tenerife, Spain (tel: 34.22.603728; fax: 34.22.603724; e-mail: PMARTIN@ULL.ES).

Dendrites were not immunoreactive. In contrast, some axon terminals, such as those of the basket cells in the cerebellar cortex, showed intense immunostaining (FIG. 1). The perikarya of some Purkinje cells in the cerebellar cortex showed immunoreactivity for the β1 subunit, but most of them were negative (FIG. 1c). Granule and Golgi neurons of the cerebellar cortex were negative. In contrast, stellate and basket neurons in the molecular layer, axonal baskets around the Purkinje cell somas (FIG. 1c), and some neurons in the central cerebellar nuclei were immunopositive. Immunoreactivity for the Na,K-ATPase β2 subunit showed a different distribution. Cytoplasmic immunoreactivity was not observed in neuronal somas. However, most neurons showed a prominent pericellular immunostaining. This periferal superficial immunostaining of the neuronal perikarya may be due in part to labeling of axonal endings and/or glial processes. For example, some axonal endings, such as basket terminals on Purkinje cells, were immunoreactive. In the granular layer of the cerebellar cortex, mossy fiber glomeruli were also immunoreactive. Although β2 immunoreactive processes were observed in the white matter, double labeling with GFAP revealed that many of these β2-like immunoreactive profiles were astroglial processes. In contrast, astroglial cell somas were negative. However, not all astroglial cell processes were immunostained, as the cell processes of tanycytes in the mediobasal hypothalamus and median eminence and Bergman glia in the cerebellar cortex.

Postnatal Development. Brains of newborn rats (P0) showed faint immunoreactivity for both β1 and β2 Na,K-ATPase subunits. At this age, the intensity of β1 immunostaining was low; its localization was similar to that in the adult. Adult levels of immunostaining intensity were reached by postnatal day 6. The distribution of β2 immunoreactivity in newborn animals was different from that of adults. Both neuronal somas and glial cell somas showed cytoplasmic immunostaining. The intensity of β2 immunostaining increased progressively from P0 to postnatal day 10, but the pattern of immunoreactivity remained constant. By postnatal day 15 the number of neurons showing cytoplasmic immunostaining decreased, with only some neurons in selected areas, such as the primary olfactory cortex and the hippocampus, remaining positive. In contrast, immunoreactivity was detected in astroglial cell processes at this age. In summary, both β1 and β2 (AMOG) isoforms are expressed in the same cellular fashion at the neonatal stage in both neurons and glia. After birth, β1 protein expression increases, reaching the adult level at day 6; after which, the subcellular distribution pattern remains constant in both cell types.[2]

To gain insights into how hormones regulate the expression patterns previously described, we studied the effects of insulin on both β isoforms in mixed rat fetal hypothalamic cultured cells (neurons and glia). Hypothalamic tissue from rat fetus E15 was dispersed, and the cells were cultured for 7 days as described in ref. 3. The most striking finding involved the β2 isoform. The β2-specific mRNA and the β2 protein show increased values after 20 minutes of insulin addition. However, the β2 isoform decreased after 50 minutes of insulin treatment. It also showed cellular redistribution, with progressive disappearance of immunostaining from the neuronal processes (FIG. 2C), and after 2 hours of addition only the somas remain immunostained (FIG. 2D).

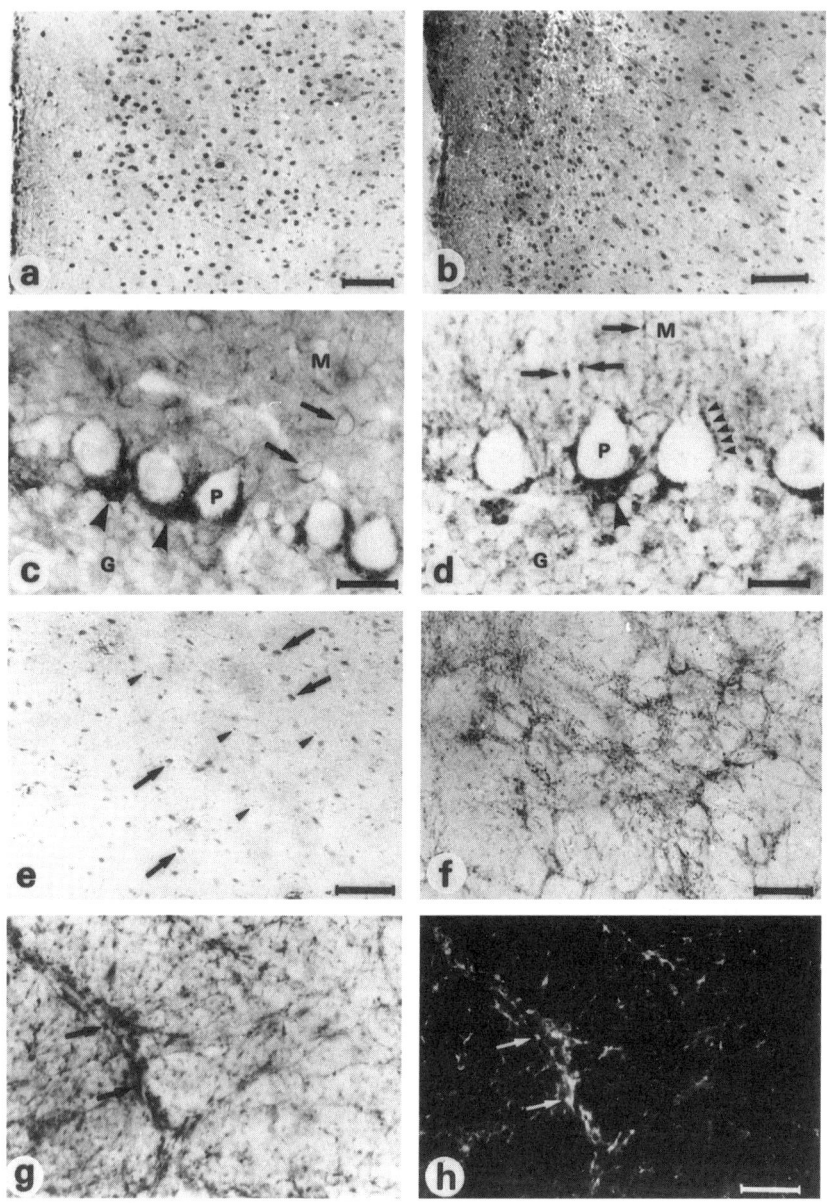

FIGURE 1. *See legend on facing page.*

FIGURE 1. Immunocytochemical localization of the Na,K-ATPase β subunit isoforms (β1 and β2) in the central nervous system of rats. (**a**) Localization in the frontal cortex of the adult rat. Immunoreactive cells are observed in all cortical layers. Cryostat section. Scale bar, 150 μm. (**b**) Immunoreactive cells are observed in all cortical layers in the frontal cerebral cortex of a 10-day-old rat. The staining pattern is similar to that of the β1 subunit in adult rats (**a**). Scale bar, 150 μm. (**c**) β1 localization in the cerebellar cortex of the adult rat. Basket neurons (*long arrows*) and basket axonal terminals (*thick arrows*) around Purkinje cell somas are immunoreactive. M = molecular layer. G = granular layer. P = Purkinje cells. Vibratome section. Scale bar, 20 μm. (**d**) β2 localization in the cerebellar cortex of the adult rat. The pattern of immunoreactivity is similar to that observed for β1. Pericellular basket cell axonal endings on Purkinje cells (*thick arrow*) are immunoreactive. Punctiform labeling is observed on the dorsolateral surface of Purkinje cells (*arrowheads*), probably corresponding to basket cell axon terminals. Punctiform labeling (*long arrows*) in the molecular layer (M) probably corresponds to descending basket cell axons. Scale bar, 20 μm. (**e**) Immunostaining in the caudate/putamen of a 10-day-old rat. Immunoreactive somas of neurons (*arrows*) and glial cells (*arrowheads*) are observed. Scale bar, 100 μm. (**f**) Caudate putamen of a 15-day-old rat. Cell processes, mainly from glial cells, are immunoreactive. Neuronal somas are negative. Scale bar, 100 μm. Colocalization of (**g**) β2 immunoreactivity and (**h**) GFAP immunoreactivity in the spinal trigeminal tract. Most of the cell processes immunoreactive for β2 subunit are also immunoreactive for GFAP (*arrows*). Both figures are at the same magnification. Scale bar, 20 μm.

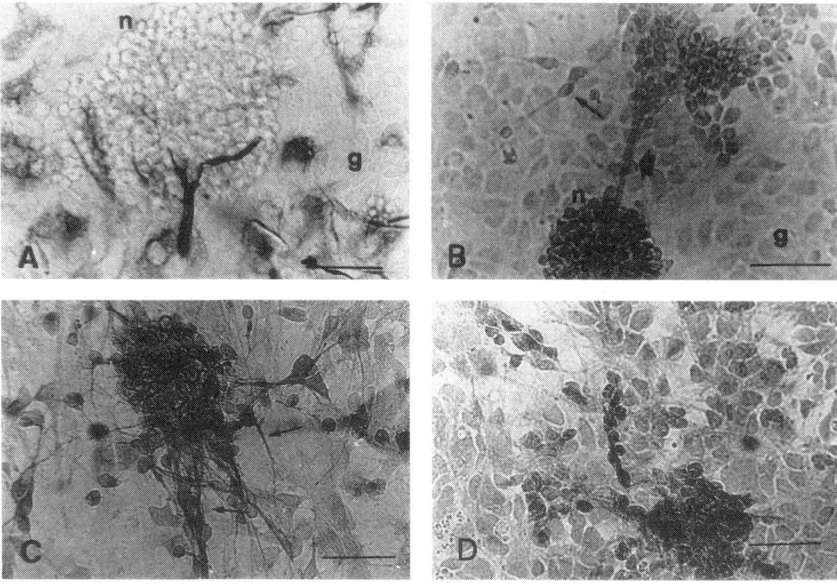

FIGURE 2. Insulin effects in rat fetal hypothalamic cultured cells at 7 days. (**A and B**) Immunostained cells with antibody anti-GFAP and antibody anti-GAP-43, respectively. For immunostaining, the ABC system was used. *Arrow,* neuron; n = neurosphere; *thick arrow,* neuronal processes; g = flat glia; *open arrow,* astrocytes. Bar: 75 μm. Immunolocalization of the Na,K-ATPase β2 isoform using the SpETb2 antibody. (**C**) Control cells. (**D**) Cells treated with 100 nM insulin for 120 minutes, respectively. Neuronal processes (*arrow*) show decreasing immunostaining throughout the 2 hours of treatment with insulin. Bar: 50 μm.

REFERENCES

1. GONZÁLEZ MARTÍNEZ, L. M., J. AVILA, E. MARTÍ, E. LECUONA & P. MARTÍN-VASALLO. 1994. Biol. Cell. **81:** 215–222.
2. LECUONA, E., S. LUQUÍN, J. AVILA, L. M. GARCÍA-SEGURA & P. MARTÍN-VASALLO. 1996. Brain Res. Bull. **40:** 167–174.
3. BRINES, M. L. & R. J. ROBBINS. 1993. Brain Res. **631:** 1–11.

Characterization of an Isoform of Na$^+$/K$^+$-ATPase with High Affinity for [^{3}H]Ouabain in the Rat Vas Deferens[a]

FRANÇOIS NOËL,[b,d] LUIS EDUARDO M. QUINTAS,[b]
AFONSO CARICATI-NETO,[c]
SIMONE SETTE L. LAFAYETTE,[c] AND ARON JURKIEWICZ[c]

[b] *Departamento de Farmacologia Básica e Clínica*
Instituto de Ciências Biomédicas
Universidade Federal do Rio de Janeiro
Ilha do Fundão
21941-590, Rio de Janeiro, Brazil

[c] *Departamento de Farmacologia*
Escola Paulista de Medicina
Universidade Federal do Estado de São Paulo
Caixa Postal 20372
São Paulo 04023-970, Brazil

The α and β subunits of Na$^+$/K$^+$-ATPase present distinct isoforms in different tissues and cells.[1] The α1 isoform was identified in all cells and is considered the "housekeeping" enzyme. The α2 and α3 isoforms have a more limited distribution. The α3 isoform is located predominantly in neural tissue, whereas the α2 isoform is present in neural tissue, heart ventricles, and adipocytes and is the predominant isoform in adult skeletal muscle. Based on mRNA analysis, the α2 isoform seems to be the main isoform in rat adult smooth muscle as well.[1] In smooth muscle of guinea pig vas deferens, there is a specific reduction of the α2 isoform associated with nonspecific postjunctional supersensitivity to various agonists after denervation.[2] In the rat vas deferens, a tissue densely innervated by autonomic nervous system and in particular with no, or little, pump contribution to the resting membrane potential,[3] the denervation also produces a nonspecific postjunctional supersensitivity,[4] but the existence and role of the α2 isoform were not yet demonstrated. Therefore, the present study investigates the presence of the Na$^+$/K$^+$-ATPase α2 isoform in the rat vas deferens.

MATERIALS AND METHODS

Vasa deferentia of about 15–20 adult Wistar rats were cut into small segments and homogenized at 4°C in 25 volumes (v/w) 0.25 M sucrose buffered to pH 7.4 with 5 mM Tris-HCl containing 2 mM dithiothreitol and 0.2 mM PMSF. After filtration under vacuum through four layers of gauze, the homogenate was centrifuged at 100,000 × *g* for 1 hour. Crude preparation was used because it is more suitable for quantitation of

---

[a]This work was supported by FAPERJ, FINEP, FAPESP, and CNPq.
[d]Tel: +55-21-5909522 (ext. 244); fax: +55-21-5901841; e-mail: fnoel@pharma.ufrj.br

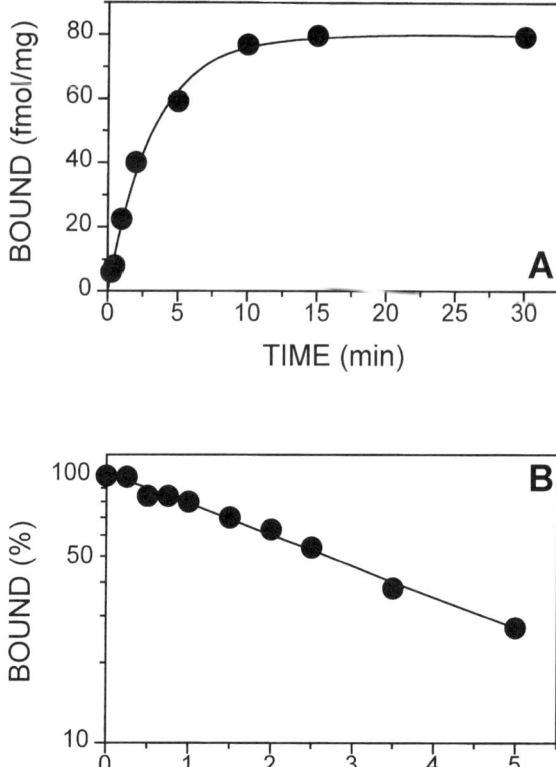

FIGURE 1. (**A**) Time course of [^{3}H]ouabain binding. Each point is the mean value of specific binding of triplicate determinations in a typical experiment. (**B**) Time course of dissociation of [^{3}H]ouabain binding. The crude preparation was incubated for 30 minutes in the presence of 15 nM [^{3}H]ouabain, after which a large excess of nonradioactive ouabain was added to promote the dissociation process. The amount of specifically bound ouabain was expressed as a percentage of specific binding measured just before performing the isotopic dilution. Each point is the mean of triplicate determinations. The curve was drawn using the value of k_{-1} calculated by nonlinear regression analysis of the monoexponential model of decay (TABLE 1).

sites in muscles under conditions of plasticity, because there are fewer problems of variability in yields of required membranes.[5] The binding of [^{3}H]ouabain was performed in a 3-mM Mg-P_i medium, as previously described.[6] SDS-polyacrylamide gel electrophoresis was carried out according to Laemmli[7] on a 6% polyacrylamide gel and transferred to nitrocellulose filter papers. After treatment with nonfat dry milk, nitrocellulose sheets were washed and then incubated overnight at room temperature with anti-rabbit Na$^+$/K$^+$-ATPase α1 isoform (mouse monoclonal IgG), anti-rat Na$^+$/K$^+$-ATPase α2 isoform (rabbit antiserum), or anti-rat Na$^+$/K$^+$-ATPase α3 isoform (rabbit polyclonal IgG) from Upstate Biotechnology Inc., USA. Blots were rinsed and incubated for 1–1.5 hours with rabbit anti-mouse secondary antibody. Immunoreactivity was detected by enhanced chemiluminescence.

RESULTS AND DISCUSSION

The binding of 15 nM [^{3}H]ouabain reached a maximum after about 15 minutes and remained stable thereafter (FIG. 1A). The concentration dependency of [^{3}H]ouabain binding at equilibrium, assessed using concentrations ranging from 10 nM to 5 µM, was well fitted by the model of two classes of specific binding sites. The high-affinity sites were very precisely characterized by their values of K_d, very similar to that reported for the rat α2 isoform in either heart ventricles[6] or transfected cells,[8] and B_{max}; on the other hand, the low-affinity sites, supposed to correspond to the ubiquitous α1 isoform of the Na$^+$/K$^+$-ATPase that exhibits a K_d value of about 15 µM in the rat under similar experimental conditions,[6] were not quantified with precision (TABLE 1). The difficulty in obtaining more precise values for the low-affinity sites can be ascribed to the high concentrations of [^{3}H]ouabain needed for labeling these sites in the rat, added to the use of crude membrane preparations. These data were confirmed in two other experiments. The dissociation curve of [^{3}H]ouabain bound to high-affinity sites was linear in a semi-log scale (FIG. 1B) and well fitted when a monoexponential model of decay was used to calculate the dissociation rate constant (TABLE 1). The actual association rate constant (k_{+1}) based on the values of k_{obs} and k_{-1} calculated from FIGURE 1 is shown in TABLE 1. Similar values for the rate constants, calculated using a different preparation, were also in very good agreement with the values measured for the rat α2 isoform in other preparations under similar experimental conditions.[6,8] Western blot assay indicated a high hybridization with the anti-α1 isoform antibody and a slight one with the anti-α2 isoform antibody, whereas no, or very discrete, reaction occurred with the anti-α3 isoform antibody. Taken together, the present data indicate the existence of a low proportion of high-affinity sites for [^{3}H]ouabain in the rat vas deferens that share all the characteristics of the α2 isoform of Na$^+$/K$^+$-ATPase.

TABLE 1. Equilibrium and Kinetic Constants for Ouabain Binding to the Crude Preparation of Rat Vas Deferens[a]

Ouabain Binding Sites	B_{max} (pmol · mg^{-1})	K_d (µM)	k_{+1} (µM^{-1} · min^{-1})	k_{-1} (min^{-1})	k_{-1}/k_{+1} (µM)
High affinity	0.42 ± 0.04	0.090 ± 0.013	2.87	0.267 ± 0.008	0.093
Low affinity	5.39 ± 2.34	16.4 ± 9.6	—	—	—

[a]Capacity (B_{max}) and dissociation constant (K_d) values were calculated by nonlinear regression analysis using the model of two classes of independent binding sites from a typical experiment. Association and dissociation rate constants (k_{+1} and k_{-1}) were calculated from experiments of FIGURE 1. Note that the parameters were calculated from single experiments performed with the same preparation and are given with their standard deviation as an indication of the "goodness of fit." The observed association rate constant (k_{obs}) was calculated by nonlinear regression using the general rate equation for bimolecular association when performed under pseudo-first order conditions: $B = B_{eq}(1 - e^{-k_{obs} \cdot t})$. The association rate constant (k_{+1}) was calculated as the following: $k_{+1} = (k_{obs} - k_{-1})/[L]$.

REFERENCES

1. MERCER, R. W. 1993. Int. Rev. Cytol. **137C:** 139–168.
2. HERSHMAN, K. M., D. A. TAYLOR & W. W. FLEMING. 1993. Mol. Pharmacol. **43:** 833–837.
3. FLEMING, W. W. 1980. Ann. Rev. Pharmacol. Toxicol. **20:** 129–149.
4. JURKIEWICZ, A., S. S. LAFAYETTE, S. H. NUNES, L. C. MARTINI, L. G. DO CARMO, A. G. WANDERLEY & N. H. JURKIEWICZ. 1994. Eur. J. Pharmacol. **256:** 329–333.
5. BAMBRICK, L. L., S. E. HOWLETT, Z. P. FENG & T. GORDON. 1988. J. Pharmacol. Methods **20:** 313–321.
6. NOËL, F. & T. GODFRAIND. 1984. Biochem. Pharmacol. **33:** 47–53.
7. LAEMMLI, U, K. 1970. Nature **227:** 680–685.
8. O'BRIEN, W. J., J. B. LINGREL & E. T. WALLICK. 1994. Arch. Biochem. Biophys. **310:** 32–39.

Differences in Uncoupled Sodium Efflux between Red Blood Cells and Kidney Na,K-ATPase Are Not Based on Differences in the cDNA for the α Subunit[a]

MARTIN STENGELIN AND JOSEPH F. HOFFMAN

Department of Molecular and Cellular Physiology
Yale University School of Medicine
New Haven, Connecticut 06520-8026

There are differences in the uncoupled sodium efflux mode between red blood cells and kidney Na,K-ATPase reconstituted into vesicles (TABLE 1). These functional differences lead to the question of whether the kidney and red cell enzymes themselves are different. In contrast to the red blood cell enzyme, the amino acid sequence of Na,K-ATPase from kidneys of many species is known. Immunological studies[1] indicated that human red cells probably express the same subunits as the kidney (α1β1), but they did not exclude possibilities such as alternatively spliced isoforms, RNA editing, posttranslational modification, or even the presence of a new isoform. Therefore, we considered it necessary to determine the complete primary structure of the human red cell Na,K-ATPase. Previous attempts to detect Na,K-ATPase mRNA in reticulocytes with Northern blotting were unsuccessful.[2] As a more sensitive alternative method, we examined reticulocytes with reverse transcriptase-polymerase chain reaction (RT-PCR).

TABLE 1. Kinetic Differences between Uncoupled Sodium Efflux in Red Blood Cells and in Other Preparations

	Red Blood Cells	Other Preparations
Electrogenicity	Electroneutral at pH 7.4[5]	Shark: electrogenic[6] Pig kidney: electroneutral at pH 6.5 to 7, electrogenic at higher pH[7]
Ion coefflux	Chloride or sulfate[5] and phosphate[8] are coeffluxed with Na^+. No proton influx at neutral pH[5]	No anion cotransport[9] Proton influx at pH 7 to 8.5[7]
Inhibition by external Na^+	Inhibited by 5 mM external Na^{+10}	No inhibition by 5 mM external Na^{+9}
Magnitude relative to Na,K exchange	≈15%[5,9]	≈2%[7,9]

[a]This work was supported by National Institutes of Health grant R01 HL52720.

METHODS

Blood (100 or 500 ml) from healthy human donors was filtered with a high performance filter (RCZL, Pall) to remove leukocytes and washed six times in saline solution. Messenger RNA was extracted, as previously described.[3] Briefly, red cells were lysed, ribonucleoproteins precipitated, total RNA extracted, and m-RNA isolated (Oligotex, Qiagen). A typical yield was 250 µg of total RNA per 500 ml blood. Messenger RNA was reverse transcribed with Superscript II (Gibco BRL) using either oligo-dT or random hexanucleotide primers or a mixture of both. In some preparations, single-stranded cDNA was treated with RNase H, and primers were removed (QIAquick, Qiagen). Approximately 5 ng of single-stranded cDNA were used per PCR reaction. Typically, the following PCR conditions were used: denaturation for 1 minute at 94°C, annealing for 1 minute at 55–65°C, and extension for 1 minute at 72°C, 40 amplification cycles. TABLE 2 shows the primers used. PCR products were analyzed on agarose gel, appropriate bands excised, and DNA extracted (QIAquick), subcloned (pGEM, Promega), and sequenced (Sequenase II, Amersham).

RESULTS

To detect any α isoform, reticulocyte single-stranded cDNA was used as a template for PCR with primers from highly conserved areas of the α subunit (TABLE 2). Products from more than 10 different PCR reactions were sequenced. Only the α1 isoform or sequences with no homology to the Na,K-ATPase were found. We also used PCR with primers specific for the α2, α3, and α4 isoforms and found no evidence of these isoforms in our preparation. We therefore conclude that only the α1 isoform is expressed in significant amounts in red blood cells.

To test for alternative splicing or RNA editing, we sequenced overlapping PCR products that covered the entire coding region of the red cell α subunit. No difference between the red cell α subunit and α1 was found.

DISCUSSION

This study demonstrates that the α1 isoform is expressed in red cells and that the cDNA for the red cell α1 isoform is identical to the known cDNA from kidney. No evidence for another isoform of the α subunit was found.

TABLE 2. Polymerase Chain Reaction Primers

Sense	Antisense
(a) Isoform Unspecific Primers	
5'-ATGTGGTTCGACAACCAAATCCA-3' (1170–1193)	5'-CATTTGCCCACAGCATTCGGAAC-3' (1817–1795)
5'-CAGCAGAAGCTCATCATTGTGGA-3' (2086–2109)	5'-GTTCTTCATGCCCTGCTGGAAGA-3' (2853–2830)
(b) Isoform Specific Primers	
α2 5'-GGTTCGGCGGACACCCATAGCAATG-3' (828–852)	5'-ACGTGGCTCTGGGGGCTGTCTTCTC-3' (1502–1477)
α3 5'-GGTGCGCAAGACGCCCATCGCCATC-3' (810–833)	5'-TATCGGTTGTCGTTGGGGTCCTCGG-3' (1484–1460)
α4 5'-TAACCTTGGAAGAGCTGAGCA-3' (≈155–175)	5'-TCCTTGTGCAGAGATAAGCC-3' (≈662–643)

We therefore conclude that differences in uncoupled sodium efflux between red blood cells and kidney Na,K-ATPase reconstituted into vesicles are not based on differences in the cDNA for the α subunit. This leaves as possible explanations posttranslational modifications of the enzyme, effects of the membrane environment, or effects of the purification and reconstitution procedure. Therefore, although this study did not solve the problem, it helped to exclude some possible explanations.

There are several reasons to investigate this problem further: first, we need information about properties of uncoupled sodium efflux in intact renal tubular cells or cell lines developed from them. Second, as shown in ref. 4, purified kidney enzyme can be fused into red blood cells. (The limitation of the aforementioned paper is that the signals were too small to study uncoupled efflux.) It would also be interesting to purify the red blood cell enzyme and study its properties after reconstitution into artificial vesicles. Finally, it should be mentioned that preliminary results indicate that in our reticulocyte cDNA preparation, only β2 and β3 isoforms are present. The significance of this result is under investigation.

Note added in proof: Analysis of the α, β, and γ isoform composition of the Na,K-pump in human reticulocytes has now been completed.[11] The results show that human reticulocytes contain only the α1, α3, β2, β3, and γ isoforms.

ACKNOWLEDGMENTS

We thank Drs. Bob Mercer and Cecilia Canessa for helpful discussions and Paul DiPasquale and Olga Potapova for excellent technical assistance.

REFERENCES

1. INABA, M. & Y. MAEDE. 1986. Na,K-ATPase in dog red cells. J. Biol. Chem. **261:** 16099–16105.
2. CHEBAB, F. F., Y. W. KAN, M. L. LAW, J. HARTZ, F. KAO & R. BLOSTEIN. 1987. Human placental Na,K-ATPase α subunit: cDNA cloning, tissue expression, DNA polymorphism, and chromosomal localization. Proc. Natl. Acad. Sci. USA **34:** 7901–7905.
3. GOOSSENS, M. & Y. W. KAN. 1981. DNA analysis in the diagnosis of hemoglobin disorders. Methods Enzymol. **76:** 805–817.
4. MUNZER, J. S., J. R. SILVIUS & R. BLOSTEIN. 1992. Delivery of ion pumps from exogenous membrane-rich sources into mammalian red blood cells. J. Biol. Chem. **267:** 5202–5210.
5. DISSING, S. & J. F. HOFFMAN. 1990. Anion-coupled sodium efflux mediated by the human red blood cell Na/K pump. J. Gen. Physiol. **96:** 167–193.
6. CORNELIUS, F. 1989. Uncoupled sodium efflux on reconstituted shark Na,K-ATPase is electrogenic. Biochem. Biophys. Res. Commun. **160:** 801–807.
7. GOLDSHLEGER, R., Y. SHAHAK & S. J. D. KARLISH. 1990. Electrogenic and electroneutral transport modes of renal Na/K ATPase reconstituted into proteoliposomes. J. Membr. Biol. **113:** 139–154.
8. MARÍN, R. & J. F. HOFFMAN. 1994. Phosphate from the phosphointermediate (EP) of the human red blood cell Na/K pump is coeffluxed with sodium, in the absence of external potassium. J. Gen. Physiol. **104:** 1–32.
9. MARTIN, W. H., D. E. RICHARDS, R. MARÍN, M. JACK-HAYS & J. F. HOFFMAN. 1994. Comparative aspects of Na/K pump-mediated uncoupled sodium efflux in red blood cells and kidney proteoliposomes. Proc. Natl. Acad. Sci. USA **91:** 9881–9885.

10. GARRAHAN, P. J. & I. M. GLYNN. 1967. The behaviour of the sodium pump in red cells in the absence of external potassium. J. Physiol. **192:** 159–174.
11. STENGELIN, M. K. & J. F. HOFFMAN. 1997. Na,K-ATPase subunit isoforms in human reticulocytes: Evidence from reverse transcription-PCR for the presence of $\alpha 1$, $\alpha 3$, $\beta 2$, $\beta 3$, and γ. Proc. Natl. Acad. Sci. USA **94:** 5943–5948.

Isolation of Six Putative P-type ATPase β Subunit PCR Fragments from the Brain of the European Eel (*Anguilla anguilla*)[a]

C. P. CUTLER, I. L. SANDERS, AND G. CRAMB[b]

School of Biological and Medical Sciences
University of St Andrews
St Andrews
Fife, Scotland, UK
KY16 9TS

Over the last decade substantial progress has been made in the identification of isoforms of P-type ATPase subunits. Of the P-type ATPases characterized, only the four α subunit isoforms of Na,K-ATPase[1] or the two α subunit isoforms of the closely related enzyme H,K-ATPase[2] are known to associate with β subunits. To date, studies investigating the existence of P-type ATPase β subunit isoforms have resulted in the identification of five β subunit isoforms (designated β1, β2, β3, H,K β, and βb1) in different species.[3] However, unlike the Na,K-ATPase or H,K-ATPase α subunit isoforms which share relatively high levels of amino acid homology (>60%), the amino acid sequences of β subunit isoforms share much lower levels of homology, sometimes as low as 30%. The low level of homology of β subunit isoforms makes the task of discovering new isoforms considerably more difficult. Here as part of a wider study investigating the role of ion transporters in various osmoregulatory tissues in euryhaline fish, experiments were designed to determine which P-type ATPase β subunit isoforms were expressed in the European eel (*Anguilla anguilla*).

RESULTS AND DISCUSSION

Reverse transcriptase-polymerase chain reaction (PCR) amplifications of RNA samples extracted from eel gill and brain tissues yielded a number of different fragments which were sequenced and subsequently identified as putative P-type ATPase β subunits by virtue of their homology and similarity to existing isoforms.

Experiments using isoform-specific primers and gill and brain RNA resulted in the identification of PCR products that represented eel homologs of the Na,K-ATPase β1[4] and β3[5] isoforms. Despite two separate strategies to identify an eel Na,K-ATPase β2 homolog, no PCR products with significant homology to this isoform were identified.[5] However, in other experiments using brain total RNA, another isoform (β233) was amplified which was highly homologous (amino acid homology 72%; TABLE 1) to the

---

[a]This work was supported by grants from the Wellcome Trust and the Natural Environment Research Council.
[b]Tel: (+44) 1334 463530; fax: (+44) 1334 463600; e-mail: gc@st-and.ac.uk

TABLE 1. Percentage of Amino Acid Homology of Eel P-type ATPase β Subunit Isoforms in Comparison with Other Isoforms

	Eel β233	Rat β2	Eel β3	Eel β179	Eel β85	Eel β185b	Bufo βb1	Rat HKβ
Eel β1	72	18	23	15	19	19	20	20
Eel β233		16	23	15	18	18	20	22
Rat β2			27	31	27	36	35	21
Eel β3				32	27	31	26	23
Eel β179					30	36	24	23
Eel β185						41	25	24
Eel β185b							33	25
Bufo βb1								28

eel Na,K-ATPase β1 isoform.[6] Northern blot analyses showed that this isoform was expressed in a number of epithelial tissues in addition to the brain. Further experiments using brain total RNA identified three other PCR products (β179, β185, and β185b) with amino acid homologies and structural similarities to P-type ATPase β subunits.[6] Although these products were most similar in structure to the Na,K-ATPase β2 or β3 isoforms, the homology shared with all of the other published isoform sequences was low. The three putative β isoforms shared only 15–36% amino acid homology with the known P-type β subunits (β1, β2, β3, H,K β, and βb1) and 30–41% homology with each other. These three PCR products may therefore represent partial fragments of new P-type ATPase β subunits. Northern blot analyses showed that the β179 isoform was expressed in brain and that the β185 isoform was expressed in brain and the eye. Preliminary experiments using the β185b probe failed to yield autoradiographic signals on Northern blots containing 5 μg of total RNA. Further experiments will be required to confirm that the β185b product was isolated from eel RNA (and was not a contaminant from another species) and also to determine the tissue-specific distribution of this isoform.

ACKNOWLEDGMENTS

The authors would like to acknowledge the help of Drs. N. Hazon and M. Tierney with fish dissections.

REFERENCES

1. SHAMRAJ, O. I. & J. B. LINGREL. 1994. A putative fourth Na^+,K^+-ATPase α-subunit gene is expressed in testis. Proc. Natl. Acad. Sci. USA **91:** 12952–12956.
2. GRISHIN, A. V., V. E. SVERDLOV, M. B. KOSTINA & N. N. MODYANOV. 1994. Cloning and characterisation of the entire cDNA encoded by ATP1AL1—a member of the human Na,K/H,K-ATPase gene family. FEBS Lett. **349:** 144–150.
3. JAISSER, F., J. D. HORISBERGER & B. C. ROSSIER. 1993. Primary sequence and functional expression of a novel β subunit of the P-ATPase gene family. Pflugers Arch. **425:** 446–452.
4. CUTLER, C. P., I. L. SANDERS, N. HAZON & G. CRAMB. 1995. Primary sequence, tissue specificity and expression of the Na^+,K^+-ATPase β1 subunit in the European eel (*Anguilla anguilla*). Fish Physiol. Biochem. **14:** 423–429.

5. CUTLER, C. P., I. L. SANDERS & G. CRAMB. 1996. Expression of Na^+,K^+-ATPase β subunit isoforms in the European eel (*Anguilla anguilla*). Fish Physiol. Biochem. In press.
6. CUTLER, C. P., I. L. SANDERS, G. LUKE, N. HAZON & G. CRAMB. 1996. Ion transport in teleosts: Identification and expression of ion transporting proteins in the branchial and intestinal epithelia of the European eel (*Anguilla anguilla*). Soc. Exp. Biol. Semin. Ser. S. J. Ennion & G. Goldspink, Eds. **58:** 43–71. Cambridge University Press. Cambridge.

Structural Characterization of the Glycation Process of the Plasma Membrane Calcium Pump[a]

F. LUIS GONZÁLEZ FLECHA,[b,d] PABLO R. CASTELLO,[b,d]
JUAN J. GAGLIARDINO,[c,d] AND JUAN P. F. C. ROSSI[b,d]

[b]Departamento de Química Biológica–IQUIFIB
Facultad de Farmacia y Bioquímica
Universidad de Buenos Aires
Junín 956
1113-Buenos Aires, Argentina

[c]Centro de Endocrinología Experimental y Aplicada
Universidad Nacional de La Plata
CONICET, 60 y 120
1900-La Plata, Argentina

We previously showed that the Ca^{2+}-ATPase activity of erythrocytes from poorly controlled diabetic patients is significantly lower than that from nondiabetic control subjects. This effect was reproduced *in vitro* by incubation of either intact erythrocytes or erythrocyte membranes with glucose.[1] The degree of Ca^{2+}-ATPase inhibition induced by *in vitro* glycation was modified by changes in the incubation media (e.g., pH increase, presence of phosphates, or replacement of glucose by other reducing sugars), as it generally occurs with the nonenzymatic glycation process.[1,2] The kinetic behavior of the decrease in enzymatic activity was identical to that of incorporation of [6-^{3}H]glucose into the purified Ca^{2+} pump, suggesting that nonenzymatic glycation of this protein caused its inactivation.[2] This work further characterizes the chemical reactions involved in the enzymatic inactivation process.

RESULTS AND DISCUSSION

Glycation of proteins is the result of the nucleophilic addition of their amino moiety to the carbonyl group of glucose. This reaction may occur to any lysine side chain exposed to a hydrophilic milieu.[3] The erythrocyte calcium pump corresponds up to 80% to the hPMCA4b isoform, and it has 80 lysine residues outside the putative transmembrane regions constituting potential targets for the glycation process. To establish whether glucose binds to the Ca^{2+} pump in a specific domain, a purified preparation of this protein was incubated at 37°C for 1 hour with 10 mM [6-^{3}H]glucose and then digested for 2 hours at 25°C with V8 protease (200 ng/ml) in the

---

[a]This work was supported by grants from Universidad de Buenos Aires (UBACYT-FA046), CONICET, and Fundación Antorchas.
[d]Tel: 54 1 964 8289; fax: 54 1 962 5457; e-mail addresses: lgf@qb.ffyb.uba.ar; castello@qb.ffyb.uba.ar; gagliard@isis/unlp.edu.ar; rtjpaul@criba.edu.ar

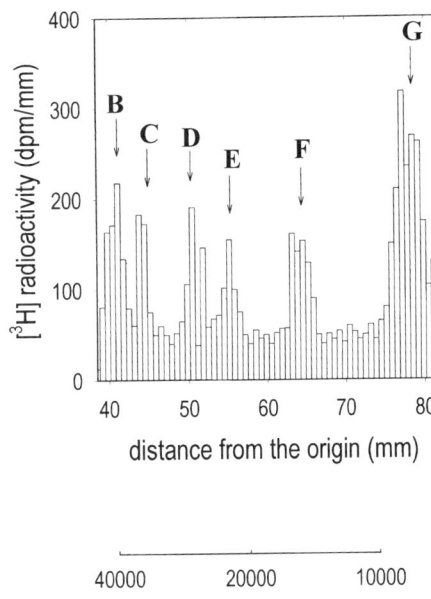

FIGURE 1. SDS-PAGE separation of the V8 protease digest of Ca^{2+} pump labeled with [6-^{3}H]glucose. *Arrows* indicate all proteolytic fragments. (For conditions, see text and ref. 4.)

presence of 0.8% SDS. The resulting peptides were separated by SDS-PAGE. (For experimental procedures see refs. 2 and 4.) Digestion of the pump in this condition produced three complementary fragments of M_r, 58,000 (A), 40,000 (B), and 38,000 (C), spanning the whole sequence of the ATPase. Fragments A and B include the COOH- and NH$_2$-termini, respectively, whereas the C fragment corresponds to a major central portion of this protein. These three fragments were labeled with

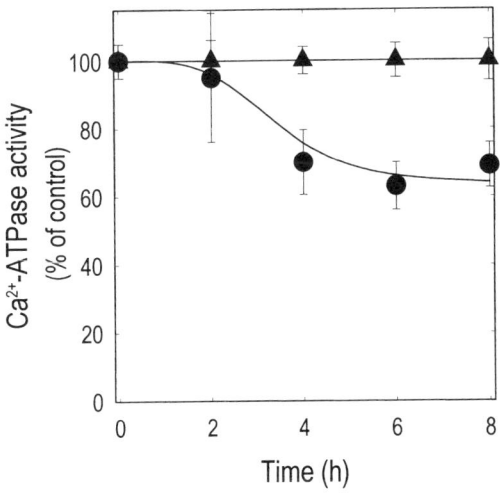

FIGURE 2. Time course of the effect on Ca^{2+}-ATPase activity of the incubation of intact red blood cells with glucose (●) and glucose 6P (▲). (For experimental procedures, see ref. 1.)

[6-^{3}H]glucose, and no selective reaction of [6-^{3}H]glucose with a given peptide was readily apparent. To recognize possible glycation sites in each of the regions just described, a labeled Ca^{2+} pump was incubated 1 hour at 37°C with 10 mM [6-^{3}H]glucose and then digested for 18 hours at 25°C with V8 protease (500 ng/ml) in the presence of 0.8% SDS. FIGURE 1 shows the extent of labeling of the six fragments (B to G) obtained.

Analysis of the specific radioactivity of these peptides suggests a random distribution of labeling rather than a pattern of specific glycation sites. Furthermore, unambiguous assignment of each peptide fragment within the primary structure of the pump showed that glucose was incorporated into no less than five different regions of the protein. The specific incorporation of glucose to the whole pump was, in this condition, about 2 mol of [6-^{3}H]glucose per mole of ATPase, representing a short fraction of the total number of lysine residues of the pump. These results suggest that glucose incorporation into the pump is produced by an unselective and self-limited mechanism as described for the glycation of other proteins.[5]

Experiments with intact cells are useful tools for topological studies in membrane proteins. FIGURE 2 shows the effect on the Ca^{2+}-ATPase activity of the glycation of intact red cells with 10 mM glucose and with 10 mM of the nonpermeant reducing sugar glucose-6-phosphate.[6]

Inasmuch as incubation of membranes, obtained from erythrocytes, with glucose-6-phosphate strongly decreases Ca^{2+}-ATPase activity,[2] this result demonstrates that the glycated residues responsible for the inhibition are located at the cytoplasmic domain of the pump.

REFERENCES

1. GONZÁLEZ FLECHA, F. L., M. C. BERMÚDEZ, N. V. CÉDOLA, J. J. GAGLIARDINO & J. P. F. C. ROSSI. 1990. Diabetes **39:** 707–711.
2. GONZÁLEZ FLECHA, F. L., P. R. CASTELLO, A. J. CARIDE, J. J. GAGLIARDINO, & J. P. F. C. ROSSI. 1993. Biochem. J. **293:** 369–375.
3. WATKINS, N. G., S. R. THORPE & J. W. BAYNES. 1985. J. Biol. Chem. **260:** 10629–10636.
4. CASTELLO, P. R., A. J. CARIDE, F. L. GONZÁLEZ FLECHA, H. N. FERNÁNDEZ, J. P. F. C. ROSSI & J. M. DELFINO. 1994. Biochem. Biophys. Res. Commun. **201:** 194–200.
5. KOWLURU, R. A., D. B. HEIDORN, S. P. EDMONDSON, M. W. BITENSKY, A. KOWLURU, N. W. DOWNER, T. W. WHALEY & J. TREWHELLA. 1989. Biochemistry **28:** 2220–2228.
6. GHIOTTO, G., G. DE SANDRE & S. CORTESI. 1959. Arch. Sci. Med. **108:** 789–799.

Cloning of the Eel Electroplax Na$^+$,K$^+$-ATPase α Subunit[a]

SHUNJI KAYA,[b,e] AKIHIKO YOKOYAMA,[b]
TOSHIAKI IMAGAWA,[b] KAZUYA TANIGUCHI,[b]
J. P. FROEHLICH,[c] AND R. W. ALBERS[d]

[b]*Division of Chemistry*
Graduate School of Science
Hokkaido University
N-10, W-8, Sapporo 060 Japan

[c]*Laboratory of Cardiovascular Science*
National Institute on Aging
National Institutes of Health
Baltimore, Maryland 21224

[d]*Laboratory of Neurochemistry*
National Institute of Neurological and
Communicative Disorders and Stroke
National Institutes of Health
Bethesda, Maryland 20892

Despite many kinetic studies on the enzyme preparation from *Electrophorus electicus*, the primary structure of the α subunit has not been identified. Some functional differences between Electrophorus and mammalian enzymes may be attributed to their primary sequence differences. We have isolated and sequenced the cDNA coding for the α subunit of the eel Na$^+$,K$^+$-ATPase. Chemical modification and phosphorylation by protein kinases were performed and are discussed in terms of the deduced sequence.

CLONING AND SEQUENCING

mRNA from the eel electroplax was purified and converted to double-stranded cDNA by the Superscript II Choice System (BRL). The DNA fragment was ligated to a Zip-lox phage vector and transfected to *Escherichia coli*. From 1×10^6 clones, 12 positive clones were isolated by a hybridization technique using the labeled polymerase chain reaction fragment of the COOH-terminal coding region of the rat Na$^+$,K$^+$-ATPase α subunit (a.a. 700–850) as a probe. In the positive clones, the one clone containing the longest insert (3.5 kbp) was shown to be coding for the Na$^+$,K$^+$-ATPase α subunit by sequence analysis, but it did not cover the full length of the α

---

[a]This work was supported in part by Grants-in-Aid for Scientific Research (06454648 and 07558220) and for International Scientific Research Program (07044049 and 08044047) from the Ministry of Education, Science and Culture of Japan.
[e]E-mail addresses—SK: kayan@S1.hines.hokudai.ac.jp; TI: toshi@S1.hines.hokudai.ac.jp; KT: KTAN@huco.hokudai.ac.jp; JPF: jeffro@vax.grc.nia.nih.gov; RWA: rwalbers@helix.nih.gov

subunit. The 5′-upstream region (500 bp) was further cloned by the 5′-race method (BRL). Both strands of the eel cDNA were sequenced by Sequenase II (USB).

RESULTS AND DISCUSSION

The deduced amino acid sequence from the cloned cDNA has an open reading frame of 1,009 amino acids (FIG. 1). The amino acid sequence of the eel Na^+,K^+-ATPase α subunit has a homology with other species of around 83%. The eel sequence has a conserved sequence for the phosphorylation site (Asp-362), the FITC binding site (Lys-494), and the PLP binding site (Lys-474) as in other species. There are three extra cysteine residues not found in the mammalian sequence (Cys-194, 229 and 395) that may be involved in the inactivation of the enzyme activity by BAL-arsenite and stabilization of the E1 conformation of this enzyme.[1,2] Taking into account the reaction of BAL-arsenite, Cys-194, which has a vicinal cysteine (Cys-197), is most likely the binding site of these cysteine-directed reagents.

In the presence of Triton X-100, the Electrophorus enzyme was phosphorylated by protein kinase A, as in the other Na^+,K^+-ATPase.[3–6] The eel sequence also has a similar phosphorylation site for cAMP-dependent protein kinase (Arg-Arg-Leu-Ser-929). Furthermore, the enzyme preparation was also phosphorylated with protein kinase C. The phosphorylation sites reported in rat sequence and its homologs[6] are not present in the eel sequence. One possible sequence in the NH_2-terminal region (Arg-Gly-Ser-9) could not be phosphorylated, because NH_2-terminal sequencing of the native enzyme showed that 9 amino acid residues are cleaved in the mature enzyme. These results suggest that a new phosphorylation site may be present in the eel sequence. A study of the effect of phosphorylation by protein kinase C on the enzyme activity is now in progress.

```
  1 MGKGFGRGSS SDEKKKDLDE LKKEVALDDH KLSLTDLASR YGVDLNKGLT TKRAAEILER
 61 DGPNALTPPP TTPEWVKFCK QLFGGFSILL WIGAILCFFA YSIQVASEDE PVNDNLYLGV
121 VLAAVVIITG CFSYYQESKS SRIMDSFKNM VPQQAMVIRD GEKRQINAED VVAGDLVEIK
181 GGDRILADVR FISCSGCKVD NSSLTGESEP QSRSPDFTHE NPLETKNICF FSTNCVEGTG
241 RGIVIATGDR TVMGRIATLA SGLEVGQTPI NIEIEHFIHI ITAVAVVLGV AFFFLSLVLG
301 YTWLEAVIFL IGIIVANVPE GLLATVTVCL TLTAKRMARK NCLVKNLEAV ETLGSTSTIC
361 SDKTGTLTQN RMTVAHMWFD NQIQVADTTE DQSGCGAFDK TSPSWKALSR VAGLCNRADF
421 LPGQESVPIL KRDTAGDASE SALLKCIELS CGSVRSLREK NNKVAEIPFN STNKFQLSIH
481 EIEESPTGHI LVMKGAPERI LDRCSSIMIS GQDIPLNDEW TNAFQRAYME LGGLGERVLG
541 FCHLNLPPSQ FPRGFAFDSE DVNFPTEQMC FLGLMSMIDP PRAAVPDAVG KCRSAGIKVI
601 MVTGDHPITA KAIAKGVGII SEGNETVEDI AERLQVPLSQ VNPRDAKACV VHGSDLKDMT
661 SEFLDDLLRN HTEIVFARTS PQQKLIIVEG CQRQGAIVAV TGDGVNDSPA LKRADIGIAM
721 GIAGSDVSKQ AADMILLDDN FASIVTGVEE GRLIFDNLKK SIAYTLTSNI PEISPFLFFI
781 IASVPLPLGT VTILCIDLGT DMVPAISLAY ESAESDIMKR QPRNPKTDKL VNERLISIAY
841 GQIGMIQALA GFFTYFVVLA ENGFLPRGLL GLRVDWDSRD VNDIEDSYGQ QWTYEQRKIV
901 EFTCHTSFFV SIVVVQWADL IICKTRRNSV FQQGMRNKIL IFGLFAETAL AAFLSYCPGM
961 DVALRMYPLK LFWWFCALPY SLLIFVYDEV RKLILRRYPG GWVEKETYY
```

FIGURE 1. Deduced amino acid sequence of the eel electroplax Na^+,K^+-ATPase α subunit.

REFERENCES

1. SIEGEL, G. J. & R. W. ALBERS. 1967. J. Biol. Chem. **242:** 4972–4979.
2. FAHN, S., M. R. HURLEY, G. J. KOVAL & R. W. ALBERS. 1966. J. Biol. Chem. **241:** 1890–1895.
3. BERTORELLO, A. M., A. APERIA, S. I. WALAAS, A. C. NARIN & P. GREENGARD. 1991. Proc. Natl. Acad. Sci. USA **88:** 11359–11362.
4. CHIBALIN, A. V., L. A. VASILETS, H. HENNEKES, D. PRALONG & K. GEERING. 1992. J. Biol. Chem. **267:** 22378–22384.
5. FESCHENKO, M. S. & K. J. SWEADNER. 1994. J. Biol. Chem. **269:** 30436–30444.
6. FESCHENKO, M. S. & K. J. SWEADNER. 1995. J. Biol. Chem. **270:** 14072–14077.

pH-Dependent Change in the Oligomeric Structure of the Solubilized Na^+/K^+-ATPase

T. KOBAYASHI, E. HAGIWARA, N. SHINJI, AND Y. HAYASHI[a]

First Department of Biochemistry
Kyorin University School of Medicine
Mitaka, Tokyo 181, Japan

Solubilized Na^+/K^+-ATPase was clearly separated into $\alpha\beta$-protomer (P) and $(\alpha\beta)_2$-diprotomer (D) using gel chromatography at 0°C, and these components were eluted as a single peak, reflecting a dissociation-association equilibrium of $2P \rightleftharpoons D$ at 20°C.[1,2] A computer simulation technique showed that the association constant of the equilibrium (K_a) for the solubilized enzyme being in the E_2 conformational state was about 50-fold larger than that for it being in the E_1 state.[3] Accordingly, it can be expected that changes in the relative content of D and P depend on the conformational state. In the present study this hypothesis was confirmed by gel chromatography performed at pH 7.0 and at 0°C. Furthermore, the ratio of D to P at various pH values was investigated in the presence of KCl or NaCl, and the equilibrium of $2P \rightleftharpoons D$ was found to shift towards D at acidic pH and towards P at basic pH, regardless of whether NaCl or KCl was present or absent.

DEPENDENCE OF THE RELATIVE CONTENTS OF D AND P ON THE VARIATION IN MONOVALENT CATION CONTENT OF NaCl PLUS KCL MIXTURES AT pH 7.0

The membrane-bound enzyme purified from dog kidney was solubilized in a solution containing variable concentrations of NaCl plus KCl (total concentration = 0.1 M) and then subjected to gel chromatography in a TSKgel G3000SW$_{XL}$ column equilibrated and eluted at 0°C with an elution buffer containing the same composition of monovalent cations as that adopted in the solubilization medium. The solubilized enzyme was separated into the three protein components of D, P, and higher oligomer (H). The relative contents of D, P, and H were estimated from areas under the respective peaks. As shown in FIGURE 1, the content of D increased from 24 to 47% of all the protein eluted, whereas that of P decreased from 57 to 32%, with decreasing NaCl concentration from 0.1 M to 0, although the content of H was approximately constant. The results are consistent with the fact that in the E_2 or E_1 conformational state, the equilibrium of $2P \rightleftharpoons D$ was shifted towards D or P, respectively.[3] The variation in the monovalent cation necessary to produce a half-maximal ($K_{0.5}$) increase in D and a decrease in P was equivalent, that is, 0.012 M KCl

---

[a]Corresponding author. Tel: +81-422-76-7651; fax: +81-422-76-7650.

plus 0.088 M NaCl. Thus obtained, the molar ratio of KCl to NaCl (0.14:1) at the $K_{0.5}$ for the oligomeric conversion was nearly equivalent to that obtained for the maximum ATPase activity of the membrane-bound enzyme (0.020 M KCl plus 0.130 M NaCl) reported by Skou,[4] suggesting that oligomeric conversion is essential for ATP hydrolysis.

ESTIMATION OF THE CONTENTS OF D, P, AND H AT VARIOUS pH VALUES

The solubilized enzyme was treated at pH values ranging between 5.5 and 9.4 for 5 minutes at 0°C in the presence of 0.1 M KCl or NaCl and was then subjected to gel chromatography in a TSKgel G3000SW$_{XL}$ or a Superdex 200HR 10/30 column

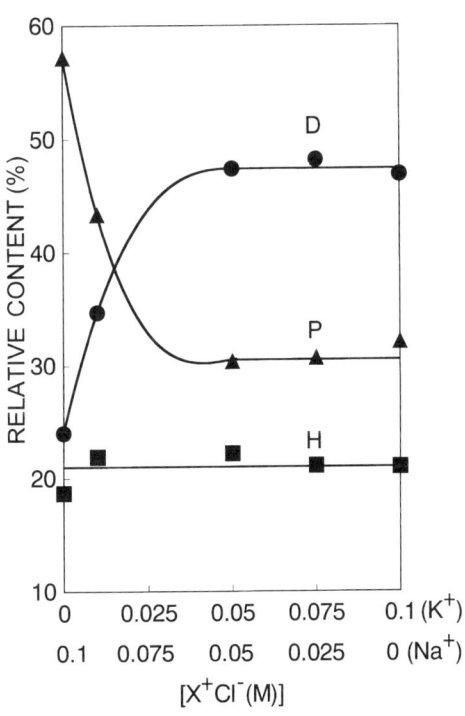

FIGURE 1. Change in the relative content of $(\alpha\beta)_2$-diprotomer (D), $\alpha\beta$-protomer (P), and higher oligomer (H) caused by variation in the concentration of NaCl and KCl. The membrane-bound enzyme was solubilized at pH 7.0 and at 0°C in a solution containing various concentrations of NaCl and KCl, as indicated on the horizontal axis. The resultant solubilized enzyme was subjected to gel chromatography in TSKgel G3000SW$_{XL}$ columns equilibrated and eluted with a buffer containing the same concentrations of NaCl and KCl as adopted in the solubilization solution at pH 7.0 and at 0°C. The relative contents of the protein components were estimated from areas under the resulting respective protein peaks.

equilibrated and eluted at the same pH, and at 0°C, in the presence of the same concentrations of monovalent cation as those adopted in the initial treatment of the enzyme. The solubilized enzyme was separated into D, P, and H as just described, and their relative contents were plotted against the pH (FIG. 2). The content of D increased, whereas that of P decreased, with decreasing pH from 9.4 to 5.5, although levels of H remained almost constant. The pH dependence of each component of the ATPase was similar in the presence of either KCl or NaCl. However, the equivalent ratio of D to P (46% to 46%) was observed at pH 7.2 in the presence of 0.1 M KCl, but at pH 6.1 in

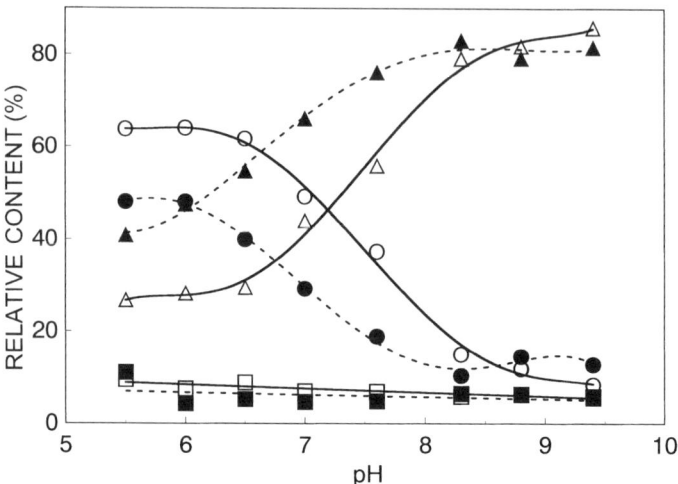

FIGURE 2. pH- and Na$^+$/K$^+$-dependent changes in the oligomeric structure of the solubilized enzyme as revealed by gel chromatography performed at 0°C. The solubilized enzyme was treated at the pH indicated in the figure for 5 minutes at 0°C in the presence of 0.1 M KCl (○, △, □, ——) or 0.1 M NaCl (●, ▲, ■, ------) and then subjected to gel chromatography in a TSKgel G3000SW$_{XL}$ column or a Superdex 200HR 10/30 column equilibrated and eluted with a buffer adjusted to the same pH as that adopted during solubilization, at a flow rate of 0.30 ml/min and at 0°C. The relative contents of D (○, ●), P (△, ▲), and H (□, ■) were estimated as described in the legend for FIGURE 1 and plotted against pH.

the presence of 0.1 M NaCl, consistent with the fact that K$^+$ can produce more D and less P than can Na$^+$. Therefore, it was concluded that the solubilized enzyme took the E$_2$ and E$_1$ conformations at acidic and basic pH, respectively, regardless of whether NaCl or KCl was included. This is not inconsistent with the conclusion that at pH 7.0 the enzyme was in either the E$_2$ or the E$_1$ state, depending on the presence of KCl or NaCl, respectively. The best conditions for the formation of two-dimensional crystals of the membrane-bound enzyme have been found to be at pH 4.8 in sodium citrate buffer.[5] The present results strongly suggest that the enzyme in these crystals is in the E$_2$ state irrespective of the presence of Na$^+$ and the absence of K$^+$.

REFERENCES

1. HAYASHI, Y., K. MIMURA, H. MATSUI & T. TAKAGI. 1989. Biochim. Biophys. Acta **983:** 217–229.
2. HAYASHI, Y., H. MATSUI & T. TAKAGI. 1989. Methods Enzymol. **172:** 514–528.
3. MIMURA, K., H. MATSUI, T. TAKAGI & Y. HAYASHI. 1993. Biochim. Biophys. Acta **1145:** 63–74.
4. SKOU, J. C. 1974. Ann. N.Y. Acad. Sci. **242:** 168–184.
5. TAHARA, Y., S. OHNISHI, Y. FUJIYOSHI, Y. KIMURA & Y. HAYASHI. 1993. FEBS Lett. **320:** 17–22.

Characterization of Na,K-ATPase α/α Oligomerization

JOSEPH C. KOSTER, WENDY R. HATFIELD,
GUSTAVO BLANCO, AND ROBERT W. MERCER[a]

*Department of Cell Biology and Physiology
Washington University School of Medicine
St. Louis, Missouri 63110*

While most structural studies of the Na,K-ATPase support a subunit stoichiometry of one α subunit to one β subunit, the exact quaternary structure of the Na,K-ATPase and its relevance to enzyme function are unknown. Our success in expressing different α isoforms in insect cells and the availability of isoform-specific antibodies have provided the unique opportunity to investigate the existence of specific and stable associations among α subunits. Strong evidence of α/α oligomerization was provided by the specific coimmunoprecipitation of the rat Na,K-ATPase isoforms (α1, α2, and α3) expressed in virally infected *Sf*-9 insect cells.[1,2] When expressed together in the insect cells, the α subunits specifically and stably associate into oligomeric complexes. Moreover, this same association among α subunit isoforms was demonstrated in native enzyme from rat brain. The highly specific nature of the α/α interaction is demonstrated by the inability of the Na,K-ATPase α subunit to stably associate with the homologous H,K-ATPase α subunit in cells coexpressing both polypeptides. To delineate the domain(s) necessary for α/α assembly, we coexpressed a series of Na,K/H,K α subunit chimeras and α subunit deletion mutants with the full-length α subunit in *Sf*-9 insect cells and assayed for detergent-resistant association. Using this method we identified a cytoplasmic region of the Na,K-ATPase α subunit that is necessary for stable and specific α/α association.

METHODS

DNA and Viral Constructions. *Sf*-9 cells grown in 150-mm petri dishes were infected with recombinant baculoviruses as described.[2] Na,K-ATPase and H,K-ATPase α subunit chimeras were constructed as described previously.[2]

Immunoblots and Immunoprecipitations. After 48 hours, uninfected and infected *Sf*-9 cells were lysed in 2% 3-[(3-cholamidopropyl)dimethylammonio]-1-propanesulfonate (CHAPS) in 150 mM NaCl, 25 mM HEPES, pH 7.4 (HBS), or in 0.1% SDS, 1% Triton X-100 in HBS. Proteins from solubilized cells were separated by SDS-

---

[a]Address for correspondence: Dr. Robert W. Mercer, Department of Cell Biology and Physiology, Washington University School of Medicine, 660 S. Euclid Avenue, St. Louis, MO 63110 (tel: 314/362-6924; fax: 314/362-7463; e-mail: rmercer@cellbio.wustl.edu).

polyacrylamide gel electrophoresis (SDS/PAGE), transferred to nitrocellulose paper, and probed with antibodies as described previously.[2] For immunoprecipitations, 150–200 µg of total protein from Sf-9 cells were solubilized in 500 µl of 2% CHAPS in HBS for 15 minutes on ice, and the insoluble material was pelleted in a microfuge (10 minutes; 15,000 × g). The supernatant was removed and the detergent diluted to 1% by the addition of HBS. To precipitate the α polypeptides, 50 µl of the indicated α-specific monoclonal antibody hybridoma supernatant and 70 µl (1 mg/ml) of goat anti-mouse or anti-rabbit coated magnetic beads (BioMag, PerSeptive Diagnostics Inc., Cambridge, Massachusetts) were added. After overnight incubation at 4°C, the magnetic beads were isolated with a magnet. The beads were washed three times with 1 ml of 1% CHAPS in HBS. The precipitated proteins were eluted, separated by SDS/PAGE, transferred to nitrocellulose, and immunoblotted. As a control for the specificity of the immunoprecipitations, 100 µg of protein from individually infected cells were combined, solubilized, and used for immunoprecipitations.

Antibodies. The Na,K-ATPase α1 subunit was identified with a monoclonal antibody (C464-6B) provided by Dr. Michael Caplan (Yale University School of Medicine); the α2 subunit was identified using the monoclonal antibody McB2 provided by Dr. Kathleen Sweadner (Massachusetts General Hospital). The H,K-ATPase α subunit was identified by either a monoclonal antibody (12:18) provided by Dr. George Sachs (University of California, Los Angeles) or a polyclonal antibody (HK9) provided by Dr. Michael Caplan. The 5α monoclonal antibody, which is specific for the Na,K-ATPase α subunit, was provided by Dr. Doug Fambrough (Johns Hopkins University).

RESULTS AND DISCUSSION

A series of H,K-ATPase-Na,K-ATPase chimeras was constructed by combining the NH_2-terminal, cytoplasmic midregion and COOH-terminal segments derived from the Na,K-ATPase α1 (N) and the H,K-ATPase (H) polypeptides (HNN, HNH, NHH, NHN, and HHN). All chimeric α subunits are stably expressed at high levels in the membrane fraction of infected Sf-9 cells and are structurally competent, as judged by their ability to specifically and stably associate with the coexpressed β subunit.[2] Coexpression of these chimeric polypeptides with the Na,K-ATPase α1 subunit demonstrated that the segment important for α/α subunit interaction resides in the cytoplasmic midregion portion of the α1 polypeptide (R350-P785). Consistent with the Na,K-ATPase cytoplasmic domain as being necessary for α/α interactions, the full-length α subunit stably associates with an α NH_2-terminal deletion mutant (ΔG2-L273), but not with an α cytoplasmic deletion mutant (ΔR350-P785). In addition, the naturally occurring COOH-terminal truncated α1 isoform, α1T (ΔG554 to COOH-terminus), does not associate with the α1 subunit in Sf-9 cells coexpressing both polypeptides. Thus, a cytoplasmic region in the α subunit (G554-P785) is necessary for specific α/α association.[2]

To further delineate the domains responsible for α/α oligomerization, additional Na,K-ATPase-H,K-ATPase chimeras were constructed and expressed in insect cells with the α1 subunit. For these chimeras, short segments of the Na,K-ATPase α2 subunit were replaced with the corresponding region of the H,K-ATPase α subunit

(FIG. 1). As shown in FIGURE 2, specific and detergent-resistant association with the native α1 subunit only occurred with a chimera containing the sequence from A725 to P785. These results suggest that the region from P561 to Q709 is required for α/α association. Chimeras containing only part of this region failed to associate into

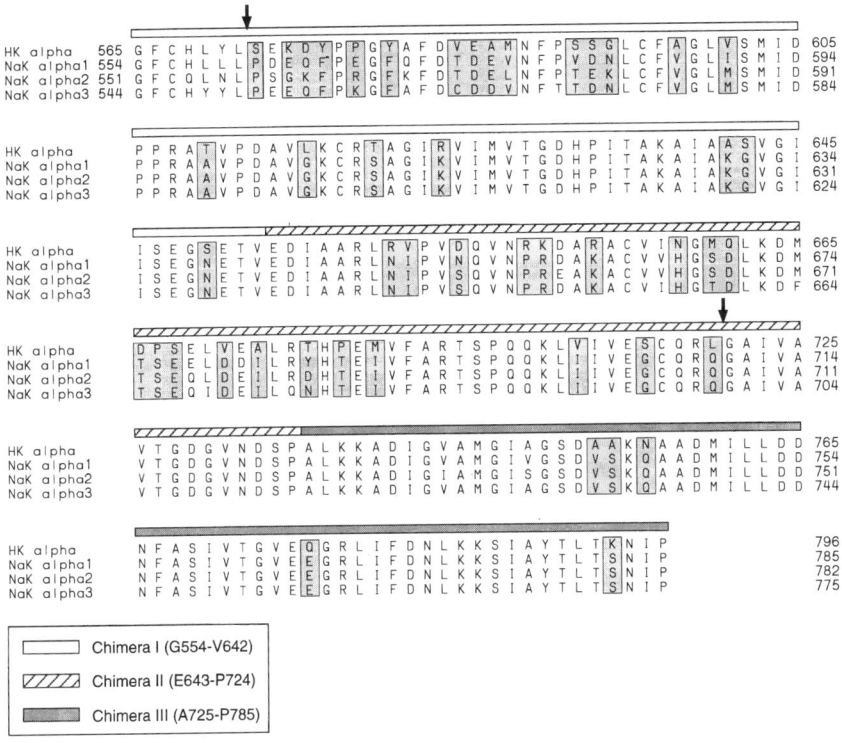

FIGURE 1. Sequence comparison of the Na,K-ATPase α/α association domain and corresponding region of the H,K-ATPase. The amino acid sequences from the cytoplasmic midregion of the α subunit of the rabbit H,K-ATPase and the rat Na,K-ATPase α1, α2, and α3 isoforms are compared. Amino acid dissimilarities between the Na,K-ATPase α isoforms and the H,K-ATPase α subunit are highlighted. Using the cDNAs, segments from the cytoplasmic region of the H,K-ATPase α subunit were interchanged with the corresponding sequence in the full-length Na,K-ATPase α2 subunit to create the α2/H,K α fusion constructs shown. The chimeras are designated based on the position of the chimeric region in the α2 polypeptide: Chimera I (G551-V639), Chimera II (E640-P721), and Chimera III (A722-P782). *Arrows* designate the boundary of the domain necessary for specific Na,K-ATPase α/α association.

oligomers. This region contains a strongly hydrophobic segment that, by analogy with oligomerization of water-soluble proteins, may form the interface of the extramembranous α/α contact site. Additional chimeras are currently being constructed to further delineate the region required for oligomerization and to clarify the functional role of α/α association to the stability and physiological regulation of the enzyme.

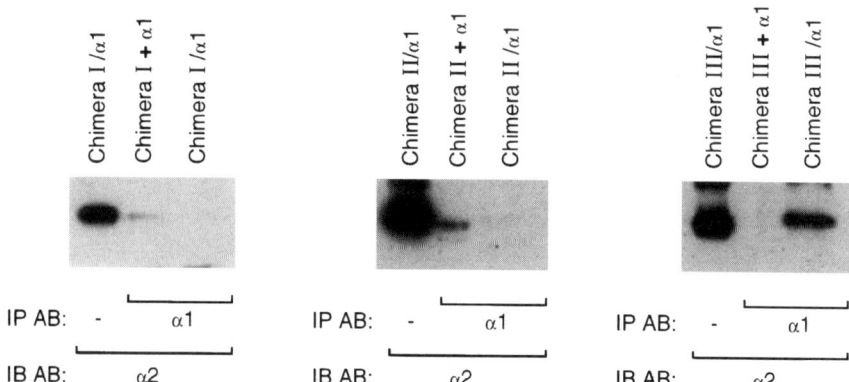

FIGURE 2. Association between the α1 subunit and the NaK-HK Chimera III coexpressed in *Sf*-9 cells. The Na,K-ATPase α1 subunit was coexpressed in *Sf*-9 cells with the NaK-HK α subunit chimeras I, II, or III as indicated. Proteins from coinfected cells were solubilized, immunoprecipitated with an α1-specific monoclonal antibody (C464-6B), and immunoblotted with an antiserum that recognizes all three Na,K-H,K α subunit chimeras. In lanes 2 (Chimera I + α1), 5 (Chimera II + α1), and 8 (Chimera III + α1) proteins from cells individually infected with the indicated recombinant viruses were mixed and immunoprecipitated.

REFERENCES

1. BLANCO, G., J. C. KOSTER & R. W. MERCER. 1994. The α-subunit of the Na,K-ATPase specifically and stably associates into oligomers. Proc. Natl. Acad. Sci. USA **91:** 8642–8546.
2. KOSTER, J. C., G. BLANCO & R. W. MERCER. 1995. A cytoplasmic region of the Na,K-ATPase α-subunit is necessary for specific α/α association. J. Biol. Chem. **270:** 14332–14339.

Purification of Heterologously Expressed Plant Plasma Membrane H^+-ATPase by Ni^{2+}-Affinity Chromatography

FRANK C. LANFERMEIJER,[a] KEES VENEMA,
AND MICHAEL G. PALMGREN[b]

*Department of Plant Physiology
Molecular Biology Institute
Copenhagen University
Øster Farimagsgade 2A
DK-1353, Copenhagen Ø, Denmark*

The ATP-driven plasma membrane proton pump of plants (PM-H^+-ATPase) is a P-type ATPase that pumps protons out of the cytoplasm. The membrane potential and proton gradient generated are used to drive solute transport across the plasma membrane of plant cells. The study on structure-function relationships of this ATPase requires the availability of an effective system for expressing and isolating mutant forms of the ATPase and the ability to grow diffraction-quality crystals of the enzyme.

To facilitate efficient purification of the *Arabidopsis thaliana* PM-H^+-ATPase and its mutant proteins, we constructed a PM-H^+-ATPase with an affinity tag of 6 histidine residues in its NH_2-terminus (6H-AHA2). A $(CAC)_6$-oligonucleotide cassette was inserted into a convenient *Eco*RV site, which coincided with the codons for Asp^6 and Ile^7 (FIG. 1). The chimeric gene was placed under the control of the strong *PMA1* promoter and was subcloned into a yeast multicopy vector that was used to transform the yeast *Saccharomyces cerevisiae*. Production of 6H-AHA2 in the endoplasmic reticulum (ER) of transformed yeast cells was confirmed by SDS-PAGE (FIG. 2A) and Western blotting (FIG. 2B). Production was high and comparable to that of wild-type AHA2. After isolation of ER membranes from the transformed yeast cells, kinetic studies revealed the presence of ATPase activity with characteristics ($K_{m,ATP}$ = 1.2 mM; pH optimum = 6.6) comparable to those of the wild-type plant PM-H^+-ATPase.

These ER membranes were used for purification of the 6H-AHA2 polypeptide. ER proteins were solubilized with 0.3% (w/v) *n*-dodecyl β-D-maltoside (DDM) (detergent/protein ratio = 3). Insoluble material was removed by centrifugation, and the supernatant was exposed to the Ni^{2+}-nitrilotriacetic acid resin. The resin was packed into a column and washed with solutions containing DDM and soybean phosphatidylcholine and increasing concentrations of imidazole.

Purified PM-H^+-ATPase could be detected in the fractions with 150 mM imidazole (FIG. 3). These fractions contained about 4% of the amount of ER protein and close to 27% of the ATPase-activity initially present in the ER. The presence of asolectin in the washing and elution solutions appeared essential. Without asolectin the PM-H^+-

---

[a]Present address: Department of Microbiology, Groningen Biomolecular Sciences and Biotechnology Institute, Kerklaan 30, 9571 NN Haren, The Netherlands (tel: +31-503632170; e-mail: F.C.Lanfermeijer@biol.rug.nl).
[b]Tel: +45-35322132; fax: +45-3532212; e-mail: palmgren@biobase.dk

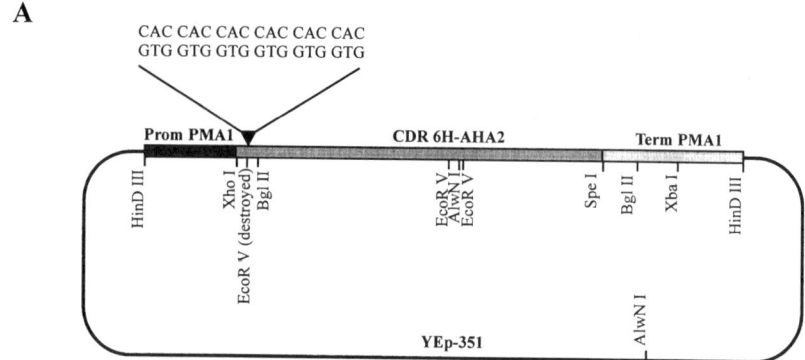

FIGURE 1. (**A**) The yeast multicopy vector containing the coding region (CDR) of the recombinant his-tagged plant PM-H^+-ATPase (6H-AHA2). The gene is placed under control of the strong *PMA1* gene promoter. (**B**) The first 26 NH_2-terminal amino acid residues of the his-tagged plant PM-H^+-ATPase.

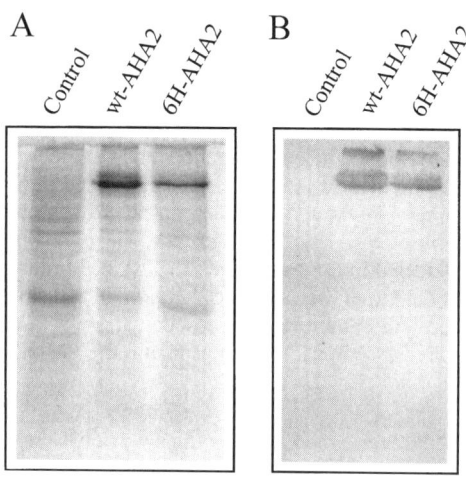

FIGURE 2. Polypeptide composition of ER membranes from yeast strains expressing wild-type and his-tagged AHA2. (**A**) SDS-PAGE of the polypeptides of ER membranes isolated from yeast. Gel was stained with Coomassie brilliant blue. (**B**) Western blot of ER membrane proteins from yeast. Membranes were exposed to antibody raised against the plant PM-H^+-ATPase. For both the Coomassie stained gel and the Western blot, 4 µg of ER protein was loaded on the gels.

FIGURE 3. Polypeptide composition of ER membranes containing 6H-AHA2 and the fraction containing the purified 6H-AHA2. SDS-PAGE of ER membranes and the purified 6H-AHA2 protein. ER protein 4 µg and purified 6H-AHA2 protein 1.5 µg were loaded on the gel. Gel was stained with Coomassie brilliant blue.

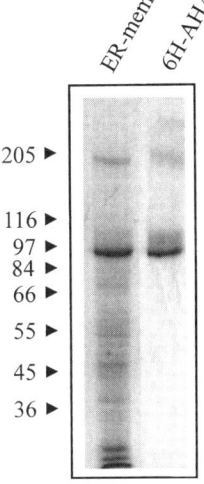

ATPase was irreversibly inactivated. The purified enzyme had characteristics comparable to those of the native wild-type enzyme ($K_{m,ATP}$ = 0.7 mM; pH optimum = 6.5). Moreover, the purified enzyme could be reconstituted in liposomes and demonstrated the ability to translocate protons.

The amino acid composition of the fraction with the purified ATPase confirmed the purity of the enzyme and allowed exact determination of the amount of polypeptide in the fraction. Using this amount we calculated a specific ATPase activity of 32.6 µmol · min^{-1} · mg protein^{-1} at pH 6.5 and 30°C. Using a molecular mass of 104.4 kD, this value could be converted into a molecular activity of 3,400 min^{-1}, which is in accordance with values reported for other P-type ATPases.

In conclusion, we have developed a convenient and rapid method for purification to apparent homogeneity of plant PM-H$^+$-ATPase which is heterologously expressed in yeast. The specific activity and purity of the plant H$^+$-ATPase are so far the highest of those of any starting material. The method should be useful in isolating mutant proteins, among which are inactive mutants, which hitherto have been difficult to separate from endogenous yeast PM-H$^+$-ATPase activity. Finally, the method should also be applicable in other P-type ATPases in other expression systems.

REFERENCES

1. VILLALBA, J. M., M. G. PALMGREN, G. A. BERBERIÁN, C. FERGUSON & R. SERRANO. 1992. J. Biol. Chem. **267:** 12341–12349.
2. PALMGREN, M. G. & G. CHRISTENSEN. 1994. J. Biol. Chem. **269:** 3027–3033.

Probing of Membrane Topology and Stability of Sarcoplasmic Reticulum Ca^{2+}-ATPase and Na^+,K^+-ATPase with Sequence-Specific Antibodies

J. V. MØLLER,[a,c] B. JUUL,[a,d] P. FALSON,[b,e]
AND M. LE MAIRE[b,e]

[a]*Department of Biophysics*
University of Aarhus
Ole Worms Allé 185
DK-8000 Aarhus C, Denmark

[b]*URA 1290*
CNRS
Section de Biophysique des Protéines et des Membranes
Département de Biologie Cellulaire et Moléculaire
Commissariat à l'Energie Atomique
Centre d'Etudes de Saclay
F-91191 Gif-sur-Yvette Cedex, France

Despite efforts to pinpoint the topology of P-type ATPases, membrane folding has not universally been agreed upon for the COOH-terminal domain of these proteins. Models have focused on either six or four membrane traverses in this domain and have taken into consideration the possibility of membrane association other than the classical α-helical traverses. (For a review see ref. 1.) As part of a combined protein-chemical and immunological program to study membrane topology, we prepared a panel of oligopeptide antibodies against Ca^{2+}-ATPase, reacting with amino acid sequences located at the NH_2-terminus, COOH-terminus, and putative loop (L) regions, localized close to the membrane (FIG. 1A). For comparison with Ca^{2+}-ATPase, we also prepared four antibodies against Na^+,K^+-ATPase (1-13, 815-28, 889-903, and 1002-16), homologous to some of the Ca^{2+}-ATPase antibodies. Antisera were raised by immunization of rabbits with oligopeptide coupled to keyhole limpet hemocyanin, and specificity was tested by their ability to react with relevant proteolytic ATPase fragments in Western blots and displacement in ELISA tests from ATPase by peptides corresponding to the epitopic regions against which they had been produced.

The antisera were used in competitive ELISA experiments to examine the reactivity and exposure of Ca^{2+}-ATPase in sarcoplasmic reticulum (SR) vesicles and Na^+,K^+-ATPase in intact (extracellularly exposed) or leaky preparations of Na^+,K^+-ATPase purified from pig kidney renal microsomes. With the exception of NH_2-

---

[c]Tel: +45 8942 2938; fax: +45 8612 9499; e-mail: jvm@biophys.aau.dk
[d]Tel: +45 8942 2934; fax: +45 8612 9499; e-mail: bj@biophys.aau.dk
[e]Tel: +33 1 6908 5362; fax: +33 1 6908 8139; e-mail: falson@dsvidf.cea.fr

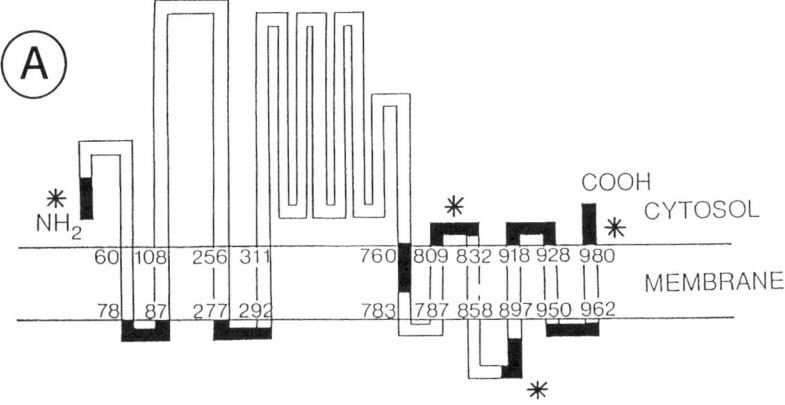

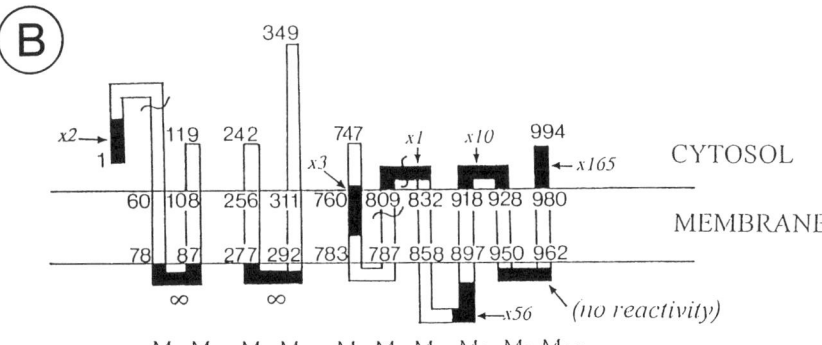

FIGURE 1. Topology of SR Ca^{2+}-ATPase and proteolytic degradation by proteinase K. (**A**) Model of intact ATPase based on 10 transmembrane segments (M1-M10). The filled-out regions of the polypeptide chain indicate the location of epitopic regions for the sequence-directed antibodies. Those produced against homologous regions in the Na$^+$,K$^+$-ATPase are indicated by an *asterisk*. (**B**) Effect of proteinase K proteolysis. Note that all epitope regions remain on the membranes after removal of cytosolic regions. *Squiggles* indicate proteolytic cleavage sites and *italicized numerals* the increase in antibody affinity after solubilization of the membranes with C$_{12}$E$_8$.

terminal Na$^+$,K$^+$-ATPase directed antiserum, we found that the antisera reacted with only low affinity against intact ATPase. After proteolytic treatment (FIG. 1B) to remove cytosolic regions, the reactivity of most antisera was increased (FIG. 2). The only exceptions were the Ca^{2+}-ATPase antisera directed against L1-2 and L3-4 regions and the Na$^+$,K$^+$-ATPase 889-903 antiserum directed against the L7-8 loop. Solubilization with C$_{12}$E$_8$ caused a further increase in reactivity of most epitopes, irrespective of cytosolic or extracytosolic location. The pattern of reactivity is consistent with the 10 transmembrane model for eukaryotic ATPases.[2] However, there was one significant exception; this concerns the L7-8 (877-888) epitope region of Ca^{2+}-ATPase with a putative intravesicular localization, which was found to react with the 877-88 antiserum. Ultrastructural investigation confirmed that the antibody

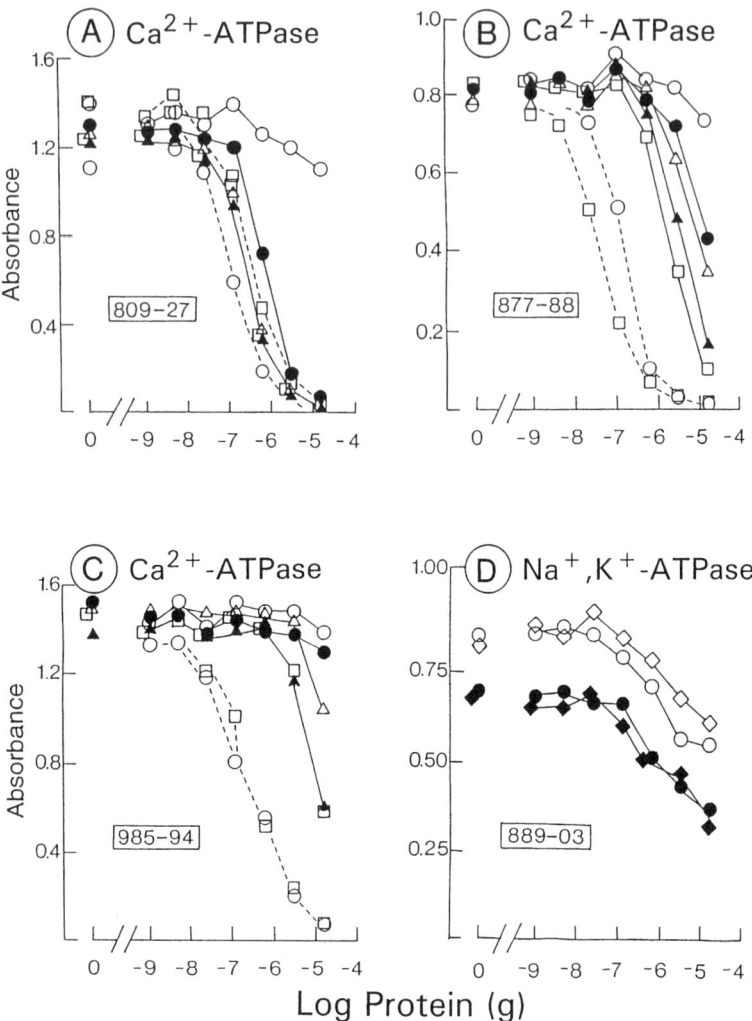

FIGURE 2. Effect of proteolytic digestion and $C_{12}E_8$ treatment on reactivity of the Ca^{2+}-ATPase (**A, B, C**) and Na^+,K^+-ATPase (**D**) with sequence-directed antibodies. Sarcoplasmic reticulum (SR) vesicles were treated with proteinase K (1 hour at 0.03 mg/ml and 23°C) for various periods (○ = 0; ● = 12.5; △ = 25; ▲ = 50; and □ = 75 minutes), followed by a competitive ELISA reaction with the indicated antisera, using denatured Ca^{2+}-ATPase as the stationary phase and detection with peroxidase-conjugated goat-anti-rabbit antibody. *Broken lines* refer to experiments in which the membranous digest was solubilized by the addition of $C_{12}E_8$ before performing the ELISA reaction. **D** shows immunoreactivity resulting from similar experiments with the 889-903 Na^+,K^+-ATPase antibody and purified Na^+,K^+-ATPase (○, ◇) or tight, right-side-out (extracellularly oriented) renal microsomes (●, ◆), treated with trypsin at a 0.3:1 weight ratio for 0 (○, ●) and 60 (◇, ◆) minutes at 37°C.

was bound on the outside of the vesicles. By contrast, the antibody directed against the homologous region in the Na^+,K^+-ATPase (889-903) was found by ultrastructural investigation to be located on the extracellular side of the membranes.[3,4] These findings could indicate either that the membrane topology of the two ATPases is different or that the membrane structure of the Ca^{2+}-ATPase is more labile, so that under certain conditions the 877-888 epitope may appear at an unexpected place. The latter possibility is of interest in relation to the autonomy of the two ATPases. In the Na^+,K^+-ATPase it is possible by proteolytic treatment to produce the so-called "19-kDa" membranes that retain cation binding and occlusion.[5] By contrast, Ca^{2+} binding in the Ca^{2+}-ATPase is exquisitely sensitive to proteolytic cleavage, high affinity binding being abolished by one to two proteolytic cuts in the cytosolic region.[6] However, this study also indicated that the membranous regions could be stabilized in the presence of high (10 mM) concentrations of Ca^{2+}.

We have now investigated if it is possible to stabilize the membranous domain after complexation with CrATP in a Ca^{2+} occluded state. We find that CrATP strongly stabilizes the major part of the Ca^{2+}-ATPase against proteolytic degradation with proteinase K and decreases immunological reactivity, with the exception of that directed against the L7-8 region. The proteolytic fragments that are slowly formed are mainly derived from cleavage in the NH_2-terminal half and middle part of the polypeptide chain, corresponding to the primary tryptic site. The accompanying decline in Ca^{2+} occlusion is slower than the rate of Ca^{2+}-ATPase degradation, but more rapid than the decrease in CrATP complexation; it is consistent with retention of Ca^{2+} occlusion by only those fragments, arising from cuts in the β region and middle of the protein, that are known to retain high affinity Ca^{2+} binding.

In conclusion, our results suggest that stability of the COOH-terminal Ca^{2+}-ATPase domain is dependent on interaction with the cytosolic regions. Absence of such interaction results in labilization of the COOH-terminal domain and exposure of the L7-8 region on the cytosolic side. The fact that this exposure is retained after complexation with CrATP could indicate a functional role of such movements in relation to the conformational changes needed in the translocation of Ca^{2+} across the SR membrane. The L7-8 loop of the Na^+,K^+-ATPase is characterized by a comparatively higher degree of stability which could arise from stabilizing interactions with the β subunit.[7,8]

REFERENCES

1. MØLLER, J. V., B. JUUL & M. LE MAIRE. 1996. Biochim. Biophys. Acta **1286:** 1–51.
2. GREEN, N. M. 1994. *In* The Sodium Pump. Structure, Mechanism, Hormonal Control and Its Role in Disease. E. Bamberg & W. Schoner, Eds.: 110–120. Steinkopff Verlag. Darmstadt.
3. NING, G., A. B. MAUNSBACH, Y. J. LEE & J. V. MØLLER. 1993. FEBS Lett. **336:** 521–524.
4. SKRIVER, E., R. ANTOLOVIC & J. V. MØLLER. 1994. *In* The Sodium Pump. Structure, Mechanism, Hormonal Control and Its Role in Disease. E. Bamberg & W. Schoner, Eds.: 354–357. Steinkopff Verlag. Darmstadt.
5. KARLISH, S. J. D., R. GOLDSHLEGER & W. D. STEIN. 1990. Proc. Natl. Acad. Sci. USA **87:** 4566–4570.
6. JUUL, B., H. TURC, M. L. DURAND, A. G. DE GRACIA, L. DENOROY, J. V. MØLLER, P. CHAMPEIL & M. LE MAIRE. 1995. J. Biol. Chem. **270:** 20123–20134.
7. LEMAS, M. V., M. HAMRICK, K. TAKEYASU & D. M. FAMBROUGH. 1994. J. Biol. Chem. **269:** 8255–8259.
8. SHIN, J. M. & G. SACHS. 1994. J. Biol. Chem. **269:** 8642–8646.

FTIR Studies on Proteolysed Na/K-ATPase

Effects of Occluded Ions

VLAD BRUMFELD, ALLA SHAINSKAYA,
AND STEVEN J. D. KARLISH[a]

Biochemistry Department
Weizmann Institute of Science
Rehovoth, 76100 Israel

This article describes initial FTIR studies which address the question of whether transmembrane segments of Na/K-ATPase are α-helices and whether occlusion of cations K(Rb) or Na is associated with significant changes in secondary structure. The experiments use "19-kD membranes" prepared by extensive tryptic digestion of renal Na/K-ATPase.[1,2] The preparation consists of a ≈19-kD fragment and several 8–11.7-kD fragments containing transmembrane segments M7–M10 and the pairs M5/M6, M3/M4, and M1/M2 with their extracellular connecting loops and segments protruding at the cytoplasmic surface. Most of the cytoplasmic domain of native enzyme is removed. The β subunit is intact or is partially split at the extracellular surface. K(Rb) and Na occlusion and ouabain binding are preserved, but ATP binding is destroyed. An important conclusion is that cation occlusion sites are located in transmembrane segments.

METHODS

19-kD membranes were prepared as described.[2] Before use the membranes were washed twice to remove Rb ions and were resuspended in an ice-cold medium containing imidazole 25 mM, pH 7.5, EDTA, 1 mM with or without RbCl, 2 mM. 19-kD membranes were thermally inactivated by incubation for 15 minutes at 37°C in the absence of Rb ions.[3] 19-kD membranes digested with trypsin in the presence of Ca ions were prepared as described.[4]

FTIR Measurements. Forty microliters of membrane suspensions (7–10 mg/ml) were placed between two CaF_2 windows separated by a 5-μ Mylar spacer. FTIR spectra were taken on a Nicolet spectrophotometer with a spectral resolution of 2 cm^{-1}. Two hundred scans were averaged. Spectral resolution was improved by Fourier self-deconvolution, and the position of these peaks was used to fit the original data using a Voigt band profile to obtain the spectral decomposition.

[a] Tel: 972 8 9342278; fax: 972 8 9344118; e-mail: bckarlis@weizmann.weizmann.ac.il

Spectral Analysis. Published FTIR spectra of proteins and polypeptides having mostly one secondary structure (helix, beta structures, turns, or unordered structures) were decomposed into individual bands, using a Voigt band shape, after enhancement of spectral resolution when necessary. The position of individual bands was collected into groups of 5–10 cm^{-1}. Amplitudes of individual bands were divided by the amplitude of the highest band to create a "base matrix." For experimental spectra, amplitudes and positions were measured using band-decomposed spectra and then amplitudes relative to the highest band were calculated and collected into the same wave number groups as for the base matrix spectra. This procedure generated an experimental vector of the spectrum. The relative amount of secondary structure was estimated by a least squares fit of the experimental vector to the base matrix (Matlab nnls routine).

RESULTS AND DISCUSSION

FIGURE 1 shows FTIR spectra of 19-kD membranes with or without 2 mM Rb, including the amide I and amide II absorptions, and decomposed spectral components.

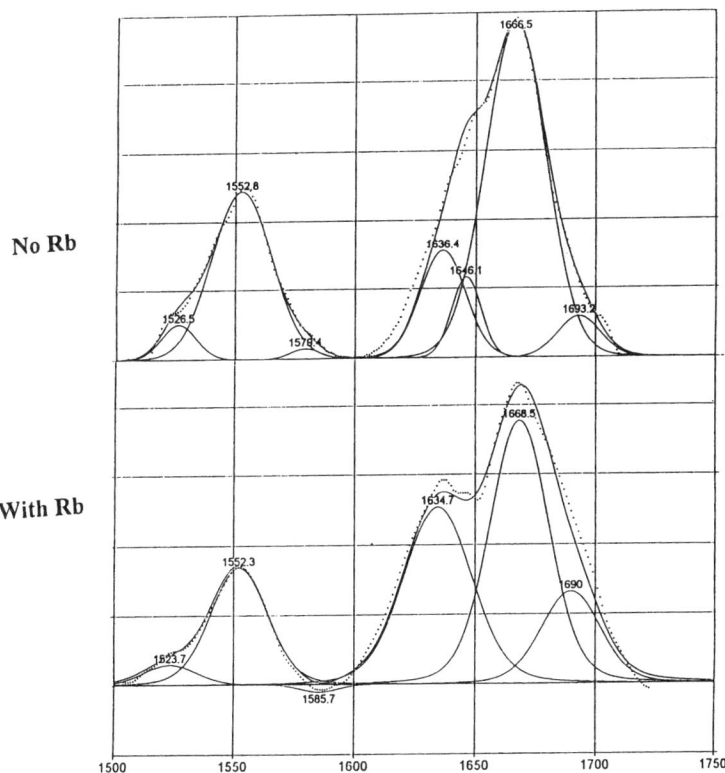

FIGURE 1. FTIR spectra of 19-kD membranes with or without 2 mM Rb, including the amide I and amide II absorptions, and decomposed spectral components.

TABLE 1. Estimated Secondary Structure Components

Condition	α-Helix	β-Sheet	β-Turn	Unordered
Control ($n = 4$)	62 ± 4	10 ± 1	11 ± 5	18 ± 1
+Rb, 2 mM ($n = 4$)	55 ± 4	33 ± 2	12 ± 4	0
Ca/trypsin (average of 2)	76	3	5	17
Thermally inactivated (average of 2)	48	15	9	27

TABLE 1 gives estimates of secondary structure components. Note the preponderance of α-helix and the fact that occluded Rb ions cause a striking change in the shape of the spectrum. This change is estimated to reflect a rise in β-sheet, fall in unordered form, and moderate fall in α-helix. Na, 10 mM, causes a similar but less pronounced change (not shown). 19-kD membranes shaved with Ca plus trypsin to limit peptides, consisting mainly of transmembrane segments,[4] show a distinct increase in α-helix compared to 19-kD membranes without Rb ions. The mechanism of formation of the latter membranes involves, first, thermal inactivation of Rb occlusion and disruption of interaction between the fragments, and then digestion of polypeptide loops and tails outside the membrane.[4] Thermally inactivated 19-kD membranes[3] (without trypsin) are characterized by a distinct fall in α-helix and a rise in unordered structure.

Our tentative inferences are that (1) transmembrane segments are largely α-helix and (2) occlusion of Rb (or Na) ions induces significant changes in secondary structure. The first conclusion is in agreement with previous studies using highly digested Na/K-ATPase and Raman spectroscopy[5] and also with an FTIR study of highly digested sarcoplasmic reticulum Ca-ATPase.[6] Effects of occluded cations could not be looked at in those studies, because the preparations are analogous to our highly shaved preparation (+Ca/trypsin) which cannot occlude cations. Regarding the observed effect of Rb ions, the intriguing question arises of whether the changes occur outside or inside the membrane. The thermal inactivation and proteolysis data indicate that part of these changes may occur in segments lying outside the membrane. However, it is not excluded that changes in secondary structure also occur within transmembrane segments. Conceivably, K(Rb) or Na occlusion is associated with reorganization of the relevant α-helices (M5/M6?), so that backbone carbonyl oxygen atoms, which are otherwise H-bonded, become available to ligate the cations.

REFERENCES

1. KARLISH, S. J. D., R. GOLDSHLEGER & W. D. STEIN. 1990. Proc. Natl. Acad. Sci. USA **87:** 4566–4570.
2. CAPASSO, J. M., S. HOVING, D. M. TAL, R. GOLDSHLEGER & S. J. D. KARLISH. 1992. J. Biol. Chem. **267:** 1150–1158.
3. OR, E., P. DAVID, A. SHAINSKAYA, D. M. TAL & S. J. D. KARLISH. 1993. J. Biol. Chem. **268:** 16929–16937.
4. SHAINSKAYA, A. & S. J. D. KARLISH. 1994. J. Biol. Chem. **269:** 10780–10789.
5. OVCHINNIKOV, Y. A., E. A. ARHYSTARKHOVA, N. M. ARZAMAZOVA, K. N. DZHANDZHUGAZYAN, R. G. EFREMOV, I. R. NABIEV & N. N. MODYANOV. 1988. FEBS Lett. **227:** 235–239.
6. CORBALÁN-GARCÍA, S., J. A. TERUEL, J. VILLALAÍN & J. C. GÓMEZ-FERNÁNDEZ. 1994. Biochemistry **33:** 8247–8254.

Atomic Force Microscopy of *Escherichia coli* FoF1-ATPase in Reconstituted Membranes[a]

KUNIO TAKEYASU,[d] HIROSHI OMOTE,[b]
SAJU NETTIKADAN,[c] FUYUKI TOKUMASU,
ATSUKO IWAMOTO-KIHARA,[b] AND MASAMITSU FUTAI[b]

Department of Natural Environment Science
Faculty of Integrated Human Studies
Kyoto University
Sakyo-ku, Kyoto 606-01, Japan

The structure of *Escherichia coli* FoF1-ATPase (ATP synthase) and its Fo sector reconstituted in lipid membranes[1] was analyzed using atomic force microscopy (AFM), a technique that does not require protein crystals and reveals the spatial organization of membrane proteins. Most FoF1-ATPases (FIG. 1A and C) were visualized as spheres with a calculated diameter of ~90 Å and a height of ~100 Å from the membrane surface.[2] These dimensions are consistent with the information about the F1 sector obtained by electron microscopy[3] and X-ray crystallography.[4,5] In contrast to the height of the F1 sectors, the height of the Fo sectors out of the membrane was very small (<10 Å) (TABLE 1). Two ring-like structures of Fo with a calculated outer diameter of ~130 Å were consistently observed, one with a central hollow of ≥18 Å in depth and a similar one with a central mass (FIG. 1B and D).[2] These images possibly represent two surface views of Fo oriented differently in the reconstituted membrane. Further analysis of these two ring-like structures at higher resolution revealed that in both types of rings, half the ring was about twice as thick as the other half; however, both halves (ridges) of the ring-like structures exhibited similar heights (<10 Å) (TABLE 1). Such asymmetry of the Fo sector was also found in a recent electron microscopic study.[6] It is noteworthy that regardless of the presence of a central mass, a similar ring structure was observed.

The bacterial Fo sector is composed of *a*, *b*, and *c* subunits with a stoichiometry of 1:2:10–12, and all three subunits are necessary to reconstitute a functional sector.[7] Each subunit of Fo traverses the membrane; subunit *a* is very hydrophobic and crosses the membrane several times (probably 6 times), whereas subunit *b* crosses the membrane once. Subunit *c* has a hairpin structure with two transmembrane helical domains with a polar cytoplasmic loop in between as supported by nuclear magnetic-

---

[a]This work was supported by the Japanese Ministry of Education, Science and Culture.
[b]Division of Biological Science, The Institute of Scientific and Industrial Research, Osaka University, Ibaraki, Osaka 567, Japan.
[c]Department of Medical Biochemistry and Neurobiotech Center, The Ohio State University, Columbus, Ohio 43210.
[d]Address for correspondence: K. Takeyasu, Department of Natural Environment Sciences, Faculty of Integrated Human Studies, Kyoto University, Sakyo-ku, Kyoto 606-01, Japan (tel: 81-75-753-6852; fax: 81-75-753-6549; e-mail: takeyasu@gaia.h.kyoto-u.ac.jp).

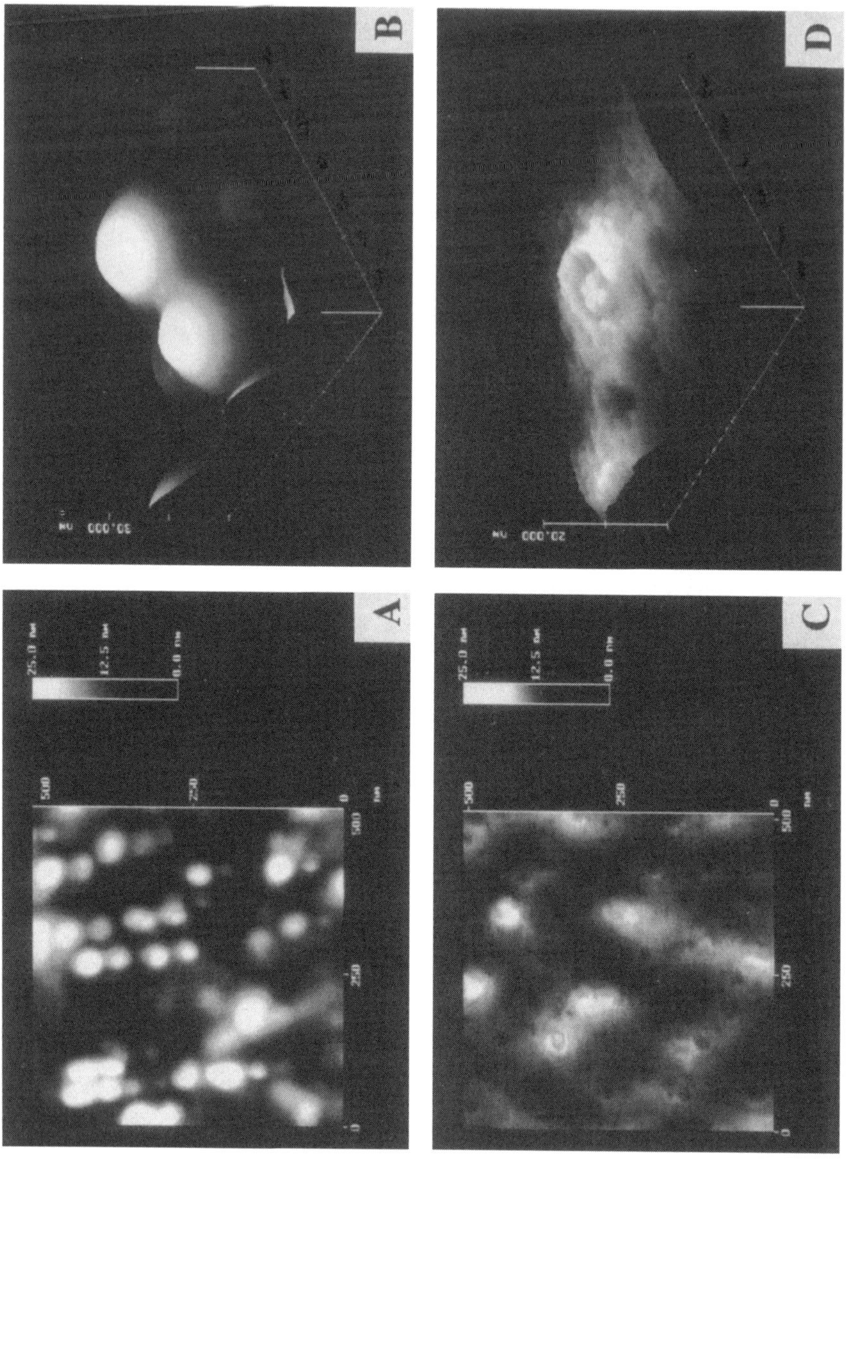

FIGURE 1. *See legend on facing page.*

TABLE 1. Two Types of Fo Images Obtained on Atomic Force Microscopy (AFM)[a]

	Fo with a Central Hollow	Fo with a Central Mass
Outer diameter	175 ± 16Å (~130)[b]	177 ± 12Å (~130)[b]
Inner diameter	43 ± 8	40 ± 3
Depth of hollow	18 ± 2[c]	
Height of central mass		5 ± 1
Major ridge		
Width	87 ± 7	81 ± 4
Height	6 ± 2	7 ± 3
Minor ridge		
Width	48 ± 1	51 ± 4
Height	5 ± 1	6 ± 1

[a]Mean values with standard deviations are shown for the dimensions of the two Fo images. These values were obtained from the AFM images for three separate Fo specimens (15 images for each specimen).
[b]Corrected values for outer diameters.
[c]Minimum possible dimension for depth due to limitations of the AFM technique (see FIG. 2B).

resonance studies.[8,9] The two asymmetric AFM images suggest a possible structure of Fo comprising *a*, *b*, and *c* subunits having 6, 1, and 2 transmembrane helices, respectively. The central mass of the ring-like image may not represent the transmembrane regions of these subunits because other images showed a hollow. Thus, a model with central *a* and *b* subunits surrounded by 10–12 copies of the *c* subunit is unlikely (FIG. 2B). In our model, the *a* and *b* subunits are attached to the ring-like structure formed by the *c* subunits (FIG. 2A). This allows at least 12 *c* subunits to form a symmetrical ring, taking into account the fact that each *c* subunit contains two transmembrane α helices (total 24 α helices for 12 *c* subunits). An asymmetric Fo structure could be formed through the association with *a* and *b* subunits to one side of the symmetric ring formed by the *c* subunits. About 15 residues at the cytoplasmic amino terminus of the *c* subunit may extend beyond one side of the membrane and form the central mass observed in a ring-like structure. Subunit *b* crosses the

---

FIGURE 1. Molecular imaging of FoF1 (**A, C**) and Fo sectors (**B, D**) by atomic force microscopy (AFM). In AFM, the probe (scanning tip) mounted on a cantilever scans the surface of the specimen, and the deflection of the cantilever is used as information about the surface topology. The reconstituted membranes[10,11] were adsorbed onto a mica surface and imaged after staining with 1% uranyl acetate. The images obtained were direct 500-nm scans of the samples under tapping mode. (**A**) Several dozen FoF1-ATPase molecules with F1 sectors facing the scanning tip. (**B**) Several dozen Fo sectors exhibiting ring-like structures, one with a hollow and the other with a central mass in the middle. These two different images may represent different orientations of Fo in the membranes. In the reconstituted FoF1 complex, the Fo sector could not be detected, because it was not possible to image simultaneously two different objects with very different heights of ~100 Å (FoF1) and ≤10 Å (Fo) due to technical limitations. (**C** and **D**) Three-dimensional representation of FoF1 (**C**) and Fo (**D**) in an area (125 × 125 nm) of reconstituted membranes selected from **A** and **B**, respectively. Surface plot program of Nanoscope III™ was used.

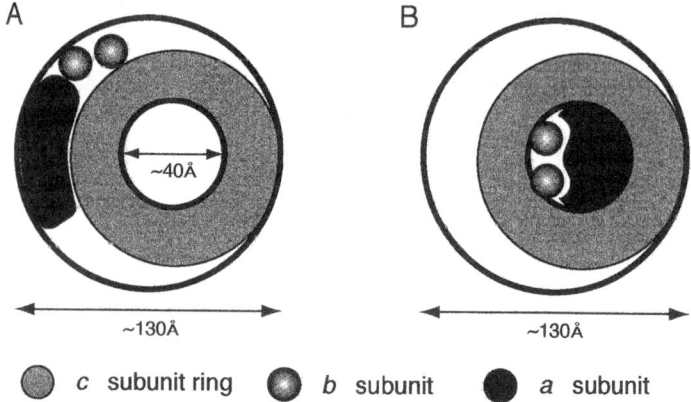

FIGURE 2. Schematic models of the subunit assembly of Fo. *Thick solid lines* represent the boundaries of Fo sectors that are identified by atomic force microscopy (AFM). Model A may be consistent with an assembly having outer and inner diameters of 130 Å and 40 Å, respectively. The a and b subunits directly contribute to the asymmetry of the ring-like structure consisting of the c subunits. Model B suffers from disagreement of the space occupied by the c subunits with the area of the ring-like structure of Fo obtained from AFM images. Using this model, it is hard to explain how a central hollow can be formed.

membrane with a single transmembrane helix, leaving most of the polar domain exposed to the cytoplasm,[7] allowing the formation of the stalk region, along with parts of F1 subunits, that connects F1 and Fo. However, such a structure could not be observed on AFM, although the Fo preparation contained a stoichiometric amount of subunit b. Therefore, it is possible that the b subunit could not be maintained as part of the stalk and thus formed the central mass observed on AFM.

REFERENCES

1. FUTAI, M., T. NOUMI & M. MAEDA. 1989. Annu. Rev. Biochem. **58:** 111–136.
2. TAKEYASU, K., H. OMOTE, S. R. NETTIKADAN, F. TOKUMASU, A. IWAMOTO-KIHARA & M. FUTAI. 1996. FEBS Lett. **392:** 110–113.
3. GOGOL, E. P., E. JOHNSTON, F. AGGELER & R. A. CAPALDI. 1990. Proc. Natl. Acad. Sci. USA **87:** 9585–9589.
4. ABRAHAMS, J. P., A. G. W. LESLIE, R. LUTTER & J. E. WALKER. 1994. Nature **370:** 621–628.
5. BIANCHET, M., X. YSERN, J. HULLIHEN, P. L. PEDERSEN & L. M. AMZEL. 1991. J. Biol. Chem. **266:** 22197–22201.
6. BIRKENHÄGER, R., M. HOPPERT, G. DECKERS-HEBESTREIT, F. MAYER & K. ALTENDORF. 1995. Eur. J. Biochem. **230:** 58–67.
7. SCHNEIDER, E. & K. ALTENDOLF. 1987. Microbiol. Rev. **51:** 477–497.
8. GIRVIN, M. E. & R. H. FILLINGAME. 1994. Biochemistry **33:** 665–674.
9. GIRVIN, M. E. & R. H. FILLINGAME. 1995. Biochemistry **34:** 1635–1645.
10. MORIYAMA, Y., A. IWAMOTO, H. HANADA, M. MAEDA & M. FUTAI. 1991. J. Biol. Chem. **266:** 22141–22146.
11. SONE, N., M. YOSHIDA, H. HIRATA & Y. KAGAWA. 1978. Proc. Natl. Sci. USA **75:** 4219–4223.

Isolation and Structural and Functional Analysis of New Types of Na,K-ATPase Isozymes

NATALIA M. VLADIMIROVA,[a]
EKATERINA A. PLATOSHKINA, RIAD E. EFENDIYEV,
AND NATALIA A. POTAPENKO

Shemyakin and Ovchinnikov Institute of Bioorganic Chemistry
Russian Academy of Sciences
117871 Moscow, Russia

Functionally active enzymes were obtained from microsomes from calf cerebral cortex gray matter, brain stem, and stem axolemma by two different methods involving (1) the selective removal of contaminating proteins, according to Jorgensen,[1] and (2) the selective solubilization of the enzyme with subsequent reformation of the membrane structure, according to Esmann.[2] The protein components of the isolated preparations were separated by polyacrylamide gel electrophoresis, transferred to an immobilon membrane by electroblotting, and subjected to structural analysis. To carry out preliminary structural analysis of the protein components of the enzymes, the preparations were dansylated.[3] Semiquantitative analysis of the proteins was possible using comparison of their NH_2-terminal amino acid residues directly from immobilon, because dansylated proteins with different M_r transferred to immobilon membranes at about the same rate.

Brain gray matter Na,K-ATPase was characterized by biphasic kinetics with respect to ouabain inhibition ($K_i \sim 10^{-6}$, $1.5 \cdot 10^{-8}$ M) and was comprised of a set of isozymes with subunit compositions of $\alpha 1\beta 1$, $\alpha 2\beta m$, and $\alpha 3\beta m$ (where m = 1 and/or 2), with the $\alpha 1\beta 1$ form clearly predominating. Na,K-ATPase from the brain stem and axolemma consisted mainly of the mixture of isozymes $\alpha 2\beta 1$ and $\alpha 3\beta 1$, which had identical ouabain inhibition constants ($K_i \sim 10^{-7}$ M).

Catalytic subunit $\alpha 3$ within the native enzyme complex exhibited increased sensitivity to endogenous proteolysis. Specific proteolysis was localized to the region of the polypeptide chain that is unique to all $\alpha 3$ type isoforms: PNDNR^{492} ↓ (Y^{493}) (according to the numbering of human $\alpha 3$ subunit). As shown for the first time, two other proteins were present in all enzyme preparations containing $\alpha 2$ and $\alpha 3$ isoforms isolated by both methods: $\beta 5$ chain of tubulin and glyceraldehyde-3-phosphate dehydrogenase. The biological meaning of this association is still unclear. However, the influence not only of subunit composition, but also of cytoskeletal structure and other plasma membrane-associated proteins might be taken into account in functional peculiarities of Na,K-ATPase isozymes.

Many diseases and pathological changes in the organism occur directly or indirectly from dysfunction of the sodium pump. The Na,K-ATPase in ischemic

---

[a]Tel: 095 330 6647; fax: 095 330 6456; e-mail: vla@ibch.siobc.ras.ru

tissues and during radiation treatment is a direct target of oxidants, such as oxygen-free radicals and hydrogen peroxide.[4,5]

We thoroughly analyzed the peroxide-mediated reaction—lactoperoxidase (LPO)-catalyzed radioiodination. Being the most stable to oxidation of the Na,K-ATPase form,[6] α1β1 type enzyme from pig kidney was chosen as the target. The residues, which will be iodinated by LPO, must be on the exposed three-dimensional protein surface and must have a proper geometric position.

Tyrosine residues of the α subunit (including COOH-terminal ones)—targets of modification—are shown to be cytoplasmic. Comparative analysis of the number and localization of the iodinated tyrosine residues in both subunits exposed to the enzyme surface in different conformational states of the Na,K-ATPase and evaluation of their reactivity provided information on various changes of both subunits in the enzyme spatial three-dimensional structure on conformational transitions. It was found that LPO-mediated iodination of tyrosine residues in the Na,K-ATPase molecule led to significant inhibition of ATP-hydrolyzing activity. The extent of inhibition of ATP hydrolysis correlated with the extent of modification. ATP protected the enzyme from inactivation. Inactivation was established to occur because of the action of reactive I° (I^+) produced under reaction conditions (3 mM H_2O_2 or a peroxide-generating system). So, a dramatic increase in the inhibitory effect of H_2O_2 when adding an I° (I^+)-generating system was observed. It may be important to understand the reasons for Na-pump damage and the methods of cell protection.

REFERENCES

1. JORGENSEN, P. L. 1974. Biochim. Biophys. Acta **356:** 36–52.
2. ESMANN, M. 1988. Biochim. Biophys. Acta **940:** 71–76.
3. VLADIMIROVA (ARZAMAZOVA), N., N. POTAPENKO, N. LEVINA & N. MODYANOV. 1991. Biomed. Sci. **2:** 68–78.
4. KAKO, K., M. KAKO, T. MATSUOKA & A. MUSTAPHA. 1988. Am. J. Physiol. **254:** C330–C337.
5. HUANG, W.-H., Y. WANG & A. ASKARI. 1992. Int. J. Biochem. **24:** 621–626.
6. HUANG, W.-H., Y. WANG, A. ASKARI, N. ZOLOTARJOVA & M. GANJEZADEH. Biochim. Biophys. Acta **1190:** 108–114.

Heterologous Expression of the Metal-Binding Domains of Human Copper-Transporting ATPases (P_1-ATPases)[a]

SVETLANA LUTSENKO,[c] KONSTANTIN PETRUKHIN,[b]
T. CONRAD GILLIAM,[b] AND JACK H. KAPLAN

Department of Biochemistry and Molecular Biology
Oregon Health Sciences University
Portland, Oregon 97201

[b]*Department of Psychiatry*
Columbia University
New York, New York 10032

Human copper-transporting ATPases (Wilson and Menkes disease proteins) are members of the P_1-subfamily of P-type ATPases.[1] ATPases in this group have unique structural features[1–4] such as (1) a characteristic hydropathy profile with eight putative transmembrane segments, three pairs before the DKTG motif and only one transmembrane hairpin after the ATP-binding domain; (2) heavy-metal binding motif at the NH_2-terminus, which is repeated one to six times depending on the species; (3) CPC/H sequence in the transmembrane segment; and (4) SEHPL motif, which seems to be characteristic of Cu-ATPases[5] and is located after phosphorylation site DKTG.

Based on the symptoms of the associated metabolic diseases, both Wilson (WND) and Menkes (MNK) disease gene products are presumed to be copper-specific proteins. However, the exact cation specificity of these two proteins remains to be determined. The presence of the GMTCxxC motif repeated six times at the NH_2-terminus suggests that the NH_2-terminal domain plays an important role in the selective binding and/or transport of heavy metals by these proteins. Alignment of all six repeats of the MNK and WND proteins[5] revealed that the overall level of identity between repeats varies between 15 and 42%, suggesting that affinity towards heavy metals as well as specificity may vary for the MNK and WND proteins.

Here we report expression of the NH_2-terminal domains of the WND and MNK proteins as fusions with maltose binding protein (MBP) and characterization of their metal-binding properties. Regions corresponding to the NH_2-terminal domain, including all six repeats, were incorporated into pMAL-c2 and pMAL-p2 vectors (New England Biolabs) at the 3′-end of coding sequence for MBP, after the Factor Xa cleavage site. To do so, segments corresponding to 64-1868 bp of WND cDNA and 1-1836 bp of MNK cDNA were amplified by the polymerase chain reaction. Stop codons were engineered at the 3′-end and unique restriction sites (*Hind*II at 5′-end and *Hind*III at 3′-end) were added to create a 5′ blunt end, 3′ sticky end-cloning fragment.

---

[a]This work was supported by National Institutes of Health grant HL 30315 to J. H. K.
[c]Tel: 503 494-6953; fax: 503 494–8393.

Fragments were inserted between Asp700 and *Hind*III polylinker sites of pMal-c2 or pMal-p2 vectors. This procedure added an Asn residue between the Factor Xa cleavage site and the sequence of metal-binding domains in the final fusion product. Inserts have been sequenced to ensure the lack of mutations.

The addition of isopropyl-β-thiogalactoside to the growth media induced expression of fusion proteins in the cells transformed with vectors carrying the insert. Mobility of the expressed proteins corresponded to their calculated molecular weight (112 kD for N-WND/MBP and 116 kD for N-MNK/MBP; FIG. 1). The level of expression from pMal-p2 based constructs was 10–20 times lower than that from pMal-c2-based constructs (not shown). Most of the protein produced in all four cases was found in an insoluble fraction. Thus, a denaturation and refolding protocol was developed. Expressed proteins were dissolved in 6 M urea and then refolded during

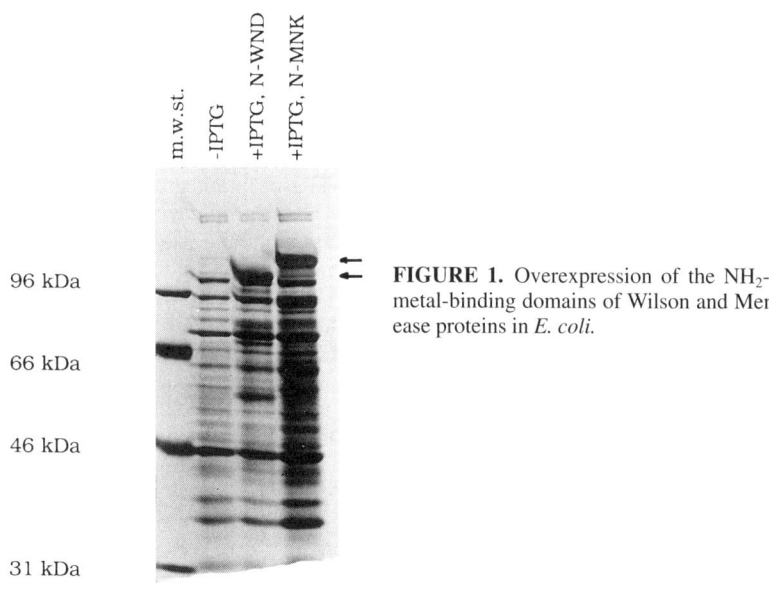

FIGURE 1. Overexpression of the NH_2-terminal metal-binding domains of Wilson and Menkes disease proteins in *E. coli*.

the 48-hour three-step dialysis at 4°C (against Na-phosphate buffer pH 7.5 [Na-P buffer] with 2 M urea and 10 mM β-mercaptoethanol, and then against Na-P with 10 mM β-mercaptoetanol, and the last dialysis step against just a Na-P buffer).

About 10 mg of protein was refolded after denaturation from a liter of cell culture, as indicated by the ability of the MBP portion to bind specifically to a lectin resin. Affinity-purified proteins were used to measure metal-binding properties (FIG. 2) by metal-chelate chromatography on iminodiacetic acid agarose (IAA-columns, Sigma), equilibrated with different heavy metals. FIGURE 2 demonstrates that purified N-WND and N-MNK fusion proteins display a selectivity in binding to IAA-heavy metal columns. MBP by itself did not bind to any of these columns. Moreover, we observed slight differences in the specificity between the N-MNK and N-WND domains

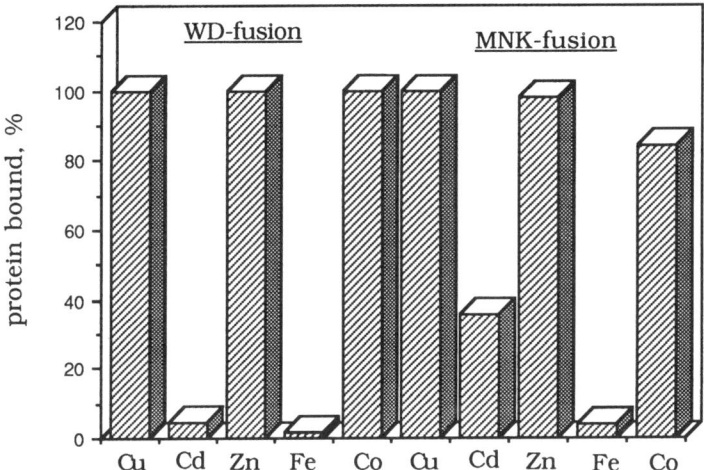

FIGURE 2. Binding of expressed N-WND-fusion and N-MNK-fusion to the IAA-columns, equilibrated with different metal ions. Bound proteins were eluted with EDTA and the amount of bound and nonbound material was analyzed by densitometry of Coomassie-stained bands, following gel electrophoresis.

towards heavy metals (compare the binding of N-MNK and N-WND to cobalt and cadmium-IAA).

In conclusion, NH_2-terminal domains of the human copper-transporting ATPases are directly involved in selective binding of heavy metals. N-MNK and N-WND show different affinities towards heavy metals, copper and zinc being the most tightly bound metals. This property of P_1-ATPases (heavy metal transporting pumps) distinguishes them from nonheavy metal P_2-pumps, in which the NH_2-terminal domain is not directly involved in the coordination of transported cations.[1]

REFERENCES

1. LUTSENKO, S. & J. H. KAPLAN. 1995. Biochemistry **34:** 15607–15613.
2. BULL, P. C. & D. W. COX. 1994. Trends Genet. **10:** 246–252.
3. SILVER, S., G. NUCIFORA & L. T. PHUNG. 1993. Mol. Microbiol. **10:** 7–12.
4. SOLIOZ, M., A. ODERMATT & R. KRAPF. 1994. FEBS Lett. **346:** 44–47.
5. PETRUKHIN, K., S. LUTSENKO, I. CHERNOV, B. M. ROSS, J. H. KAPLAN & T. C. GILLIAM. 1994. Hum. Mol. Genet. **3:** 1647–1656.

# P-Type ATPases in *Tetrahymena*

SHUSHENG WANG,[a,c] DONGHONG GAO,[a]
JEFFREY PENNY,[b] SANJEEV KRISHNA,[b]
AND KUNIO TAKEYASU[a]

[a]*Department of Medical Biochemistry and
Neurobiotechnology Center
The Ohio State University
Columbus, Ohio 43210*

[b]*Division of Infectious Diseases
St. George's Hospital Medical School
London SW17 0RE, UK*

The P-type ATPase gene family of ion pumps is widely distributed in fungi, Protozoa, plants, and animals. However, despite detailed sequence analysis which shows that these ion pumps have highly divergent primary structures, except at a limited number of amino acid residues important for ATP hydrolysis, functional characterization of many pumps is still lacking. Using degenerate oligonucleotides corresponding to the phosphorylation (DKTGTLT) and CIRATP-binding sites (GDGVND), we applied a polymerase chain reaction technique to isolate genes encoding ion pumps in *Tetrahymena*, as a basis for evolutionary and physiological studies. In total, 12 distinct P-type ATPase genes were identified and classified into three families, related closely by sequence (FIG. 1). Family 1 (at least 7 homologous genes) encodes ion pumps similar to those of animal Na^+/K^+- (H^+/K^+-) ATPases, family 2 (1 gene) resembles SERCA ATPase gene, and family 3 gene products are functionally unclassified but share significant homology with PfATPase 1 from *Plasmodium falciparum*.[1]

Southern blot analysis revealed the existence of distinct P-type ATPase genes in the *Tetrahymena* genome; family 1 contains at least 10 genes, whereas genes in the other two families do not seem to possess such multiplicity. Southern blot analysis following pulse-field gel electrophoresis showed that all these genes are located in *Tetrahymena* macronuclei, with members of family 1 widely distributed on at least four different macronuclear chromosomes (FIG. 2). The more recent duplications created two genes closely linked on the same macronuclear chromosomes, whereas genes derived from the older duplications were drifted to the different chromosomes. Phylogenetic analysis (FIG. 1) suggests that *Tetrahymena* family 1 branched off from the phylogenetic lineage of the Na^+/K^+- (H^+/K^+-) ATPases, possibly before separation of the Na^+/K^+- and the H^+/K^+-ATPase. It is interesting that family 1 evolved in this single-cell organism with a high frequency of gene duplication resulting in at least 10 isoforms.

Northern blot analysis under various culture conditions (i.e., different temperatures such as 25°, 30°, 37°, and 42°C, high salt solution [0.2 M NaCl], and starvation [10 mM Tris-HCl, pH 7.5]) constantly revealed a single-size transcript (~3.6 kb) for TPA2 (Na^+/K^+- or H^+/K^+-ATPase), TPA8 (SERCA-ATPase), and TPA9-10 (two of

[c]Address for correspondence: Shusheng Wang, Neurobiotechnology Center, The Ohio State University, Columbus, Ohio 43210 (tel: 1-614-292-8543; fax: 1-614-292-5379).

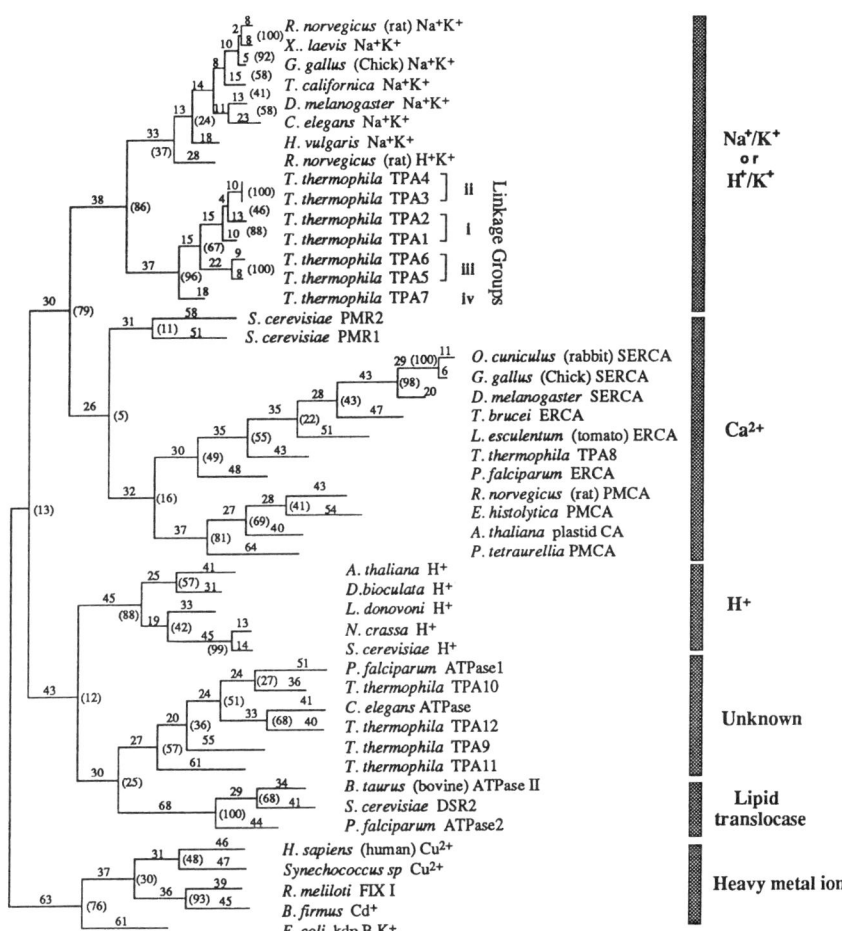

FIGURE 1. Phylogenetic relationships of *Tetrahymena* P-type ATPases with other P-type ATPases. The phylogenetic tree was constructed by the neighbor-joining method based on the amino acid sequences of the ATP binding domains from 12 *Tetrahymena* P-type ATPases and 35 other P-type ATPases obtained from the GenBank database. Branch length values on the tree represent the distances between the sequences. Bootstrap values are shown in *brackets*. Linkage groups within the *Tetrahymena* Na/K- or H/K-ATPase family were indicated as i–iv.

the protozoan ATPases). A high concentration of NaCl did not result in apparent changes in mRNA levels. However, a drastic increase in the level of TPA2 and TPA10 expression was seen at 37°C. Starvation downregulates expression of the TPA2, TPA8, and TPA10 genes as well as the histone H4-I gene, but upregulates TPA9 expression. The TPA9 gene product with a molecular weight of 130 kD was immunolocalized at the oral apparatus, a place undergoing phagocytosis for particulate food uptake, whereas TPA10 was immunolocalized at the area close to basal bodies, specifically at the parasomal sac, a coated pit-like structure where pinocytosis

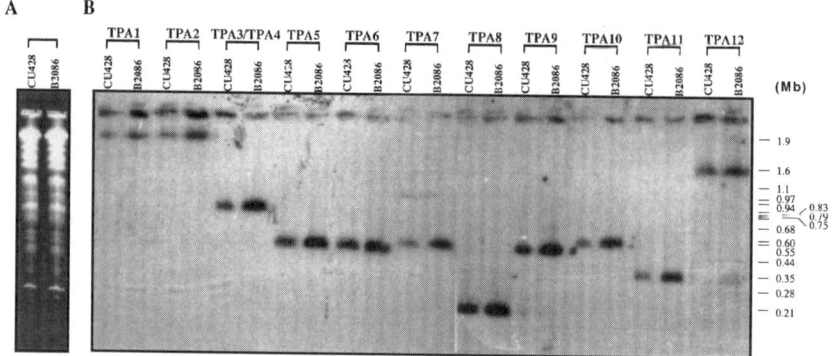

FIGURE 2. Genomic DNA gel-blot analysis of *Tetrahymena* P-type ATPase gene family. (**A**) Etidium-bromide staining of macronuclear DNA from two *Tetrahymena thermophila* strains, CU428 and B2086, separated by pulse-field gel electrophoresis. (**B**) Linkage analysis by high stringency hybridization with the probes made from cloned P-type ATPase gene fragments. Genes within the putative Na/K- or H/K-ATPase family are located on at least four distinct macronuclear chromosomes. Genes of other TPAs are located on different macronuclear chromosomes.

occurs for the uptake of liquid nutrients. The unique localization of TPA9 and TPA10 proteins implies that the function of this novel P-type ATPase family may be involved in endocytosis in general.

A comparison of the full-length sequences between TPA2 and P-type ATPases in higher eukaryotes revealed that several glutamates as well as asparagine and threonine residues within transmembrane domains 4, 5, 6, and 8 are conserved. These results are consistent with the current idea that these charged amino acids are important for cation transport.[2,3] In contrast, these charged amino acids within the transmembrane domains are replaced by amino acids with bulky hydrophobic side chains in TPA9. Analysis of the deduced amino acid sequences of the protozoan ATPases (TPA9–12) identified a sequence similarity to those of the lipid flippases of yeast,[4] *Plasmodium*,[5] *Caenorhabditis elegans*,[6] and bovine.[7] It is interesting to speculate that family 3 of *Tetrahymena* P-type ATPases may translocate aminophospholipids, such as phosphatidylserine, therefore establishing the asymmetric distribution of phospholipids on the two leaflets of the bilayer.

REFERENCES

1. WANG, S. & K. TAKEYASU. 1996. Am. J. Physiol. **272** (Cell Physiol. 41): C715–C728.
2. CLARKE, D. M., T. W. LOO, G. INESI & D. H. MACLENNAN. 1989. Nature **339**: 476–478.
3. ARGUELL, J. M. & J. B LINGREL. 1995. J. Biol. Chem. **270**: 22764–22771.
4. RIPMASTER, T. L., G. P. VAUGHN, & J. L. WOOLFORD. 1993. Mol. Cell. Biol. **13**: 7901–7912.
5. TROTTEIN, F. & A. COWMAN. 1995. Eur. J. Biochem. 214–225.
6. WISON, R. *et al.* 1995. Nature **368**: 32–38.
7. TANG, X., M. S. HALLECK, R. A. SCHLEGEL & P. WILLIAMSON. 1995. Science **272**: 1495–1497.

Transport-Linked Conformational Changes in Na,K-ATPase

Structure-Function Relationships of Ligand Binding and E_1-E_2 Conformational Transitions

PETER L. JORGENSEN,[a] JAKOB H. RASMUSSEN,
JESPER M. NIELSEN, AND PER AMSTRUP PEDERSEN

Biomembrane Research Centre
August Krogh Institute
Copenhagen University
2100 Copenhagen OE, Denmark

Renal Na,K-ATPase is organized in $\alpha 1\beta 1$ units with full activity in the soluble state. The substrate ATP and the inhibitors vanadate and ouabain are bound with a stoichiometry of one ligand per $\alpha\beta$ unit, and cavities for occlusion of $2K^+$ or $3Na^+$ ions are formed within the structure of the $\alpha\beta$ unit. The $\alpha\beta$ unit is also the minimum asymmetric unit in membrane crystals of purified renal Na,K-ATPase. The α subunit may contain 10 transmembrane segments with sites for ATP binding and phosphorylation in the large central cytoplasmic protrusion between the fourth and fifth transmembrane segment.[1-3] Extensive proteolysis has shown that the cation occlusion sites lie in the intramembrane domain.[4] To understand the Na,K-pump mechanism it is necessary to identify the groups coordinating the cations and to know if the pump has independent sites for Na^+ and K^+ or if the cations bind alternately to the same set of sites. It is equally important to identify the segments of the protein involved in the relatively large E_1-E_2 conformational changes in the α subunit that are thought to mediate long-range interactions between the ATP site and the cation occlusion sites in the membrane domain.[5]

Development of a system for expression in high yield of Na,K-ATPase in yeast[6] allows expression of lethal Na,K-ATPase mutations in cells devoid of endogenous Na,K-ATPase activity. The yeast expression system does not distinguish active Na,K-ATPase from inactive mutant $\alpha\beta$ units with respect to biosynthesis and translocation to the cell membranes. It therefore allows targeting of $\alpha\beta$ units with lethal substitutions at the phosphorylation site, Asp[369], or at presumptive cation sites in transmembrane segments IV, V, and VI, Glu[327], Glu[779], Asp[804], and Asp[808], to the cell surface at the same concentration of α subunit and sites for [^{3}H]ouabain binding as for wild-type Na,K-ATPase.[6,7] The absence of endogenous activity allowed the development of assays not previously achieved for recombinant enzyme, such as [^{3}H]ATP binding at equilibrium with demonstration of a large increase in affinity for ATP of D369N as a consequence of reducing the negative charge of Asp[369].[7] This development opens the opportunity for examining several aspects of the transfer of energy from the ATP binding domain to cation translocation. The K^+ nucleotide antagonism (FIG. 1)[8] is a classic feature of Na,K-ATPase that can be used for estimating relative

---

[a]Tel: 45 35 32 16 70; fax: 45 35 32 15 67; e-mail: PLJORGENSE@aki.ku.dk

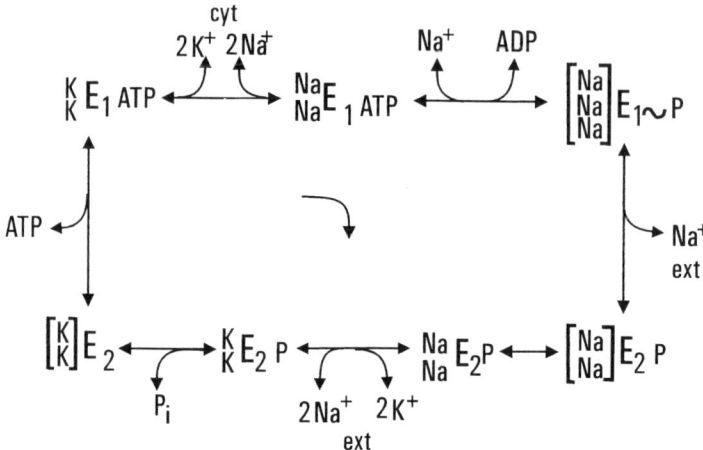

FIGURE 1. E_1-E_2 reaction cycle of the Na,K pump with four major occluded conformations and ping-pong sequential cation translocation. [Na$^+$] or [K$^+$] within brackets are occluded in the $\alpha\beta$ unit and are prevented from exchanging with medium cations. Na$^+$ or K$^+$ without brackets exchange with open forms of the cation binding cavity, in the E_1 form via a cytoplasmic gate and in the E_2 form via an extracellular gate.

affinities for the K$^+$ ions and their effects on the conformational state of the enzyme protein.[9] Exact determinations of intrinsic affinities for binding of ATP and ADP and examination of the K$^+$-ATP antagonism are important for evaluating the role of nucleotide binding as a driving force for translocation of K$^+$. It is also important to examine if the amino acid replacements at the phosphorylation site have long-range effects on the cation binding sites in the transmembrane region of the protein and the extracellular binding site for cardiac glycosides.

CATALYTIC FUNCTIONS OF THE PHOSPHORYLATION SITE, Asp^{369}

Previous studies have shown that Asp^{369} in the conserved CSDKT segment is essential for ATP hydrolysis of cation pumps. Substitutions of Asp^{369} to Asn, Glu, Thr, or Ala abolished ATP hydrolysis[6,10,11]-like mutations to the phosphorylation site of Ca-ATPase from sarcoplasmic reticulum[12] or the H-ATPase from yeast.[13] The observation that the mutations Asp^{369}-Ala and Asp^{369}-Asn could be expressed as [^{3}H]ouabain binding sites at the surface of yeast cells was a surprise, because previous data had shown that mutations to the transiently phosphorylated aspartate residue cause retention in intracellular membranes and failure in targeting of Na,K-ATPase[10,11] or H-ATPase[13] to the plasma membrane.

EXPRESSION OF MUTATIONS TO Asp^{369} IN YEAST

Size exclusion chromatography of the soluble [^{3}H]ouabain complexes showed that the hydrodynamic properties of the mutated Asp^{369}-Ala and Asp^{369}-Asn $\alpha 1 \beta 1$ units are similar to those of recombinant wild-type or pig kidney $\alpha 1\beta 1$-Na,K-ATPase.[6] This together with the observations that high affinity ligand binding sites for nucleotides and ouabain are preserved along with interactions with Na^+ and K^+ and the transitions between E_1 and E_2 conformations, support the notion that unspecific perturbations of tertiary structure are indeed limited after mutations to the phosphorylation site. Both cytoplasmic and extracellular aspects of the mutant proteins appear to be correctly folded and structurally organized in the membrane.

EFFECT OF MUTATIONS TO Asp^{369} ON [^{3}H]ATP AND ADP BINDING AT EQUILIBRIUM

The conditions for demonstrating specific binding of [^{3}H]ATP or [^{3}H]ADP are shown in FIGURE 2. The addition of NaCl up to 10 mM did not alter binding at 13 nM

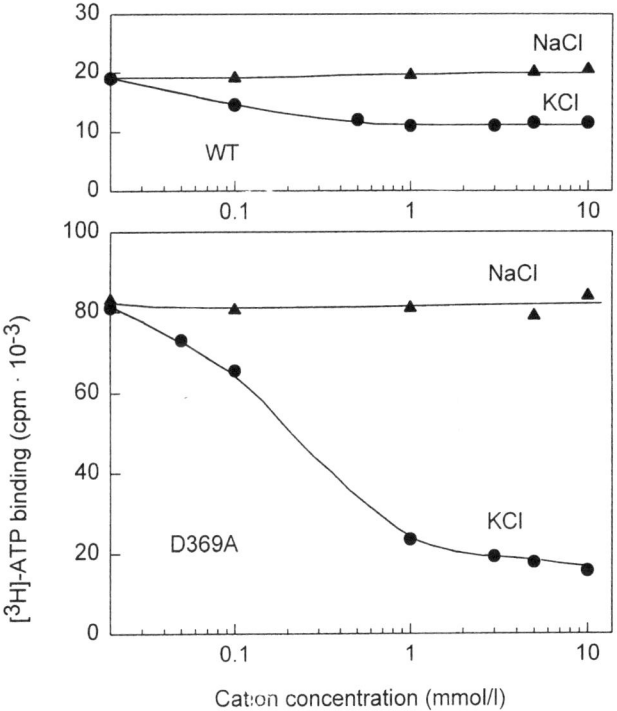

FIGURE 2. Effect of NaCl (▲) or KCl (●) on [^{3}H]ATP binding to wild-type $\alpha_1\beta_1$ Na,K-ATPase (*upper frame*) or to mutation D369A (*lower frame*). Aliquots of about 200 μg SDS-treated gradient membranes were incubated on ice with 10 mM MOPS-Tris pH 7.2, 10 mM EDTA, [^{3}H]ATP (Amersham, specific activity 36 Ci/mmol) at a concentration of 13 nM [^{3}H]ATP and either 0.1–10 mM NaCl or 0.1–10 mM KCl. Bound and unbound [^{3}H]ATP were separated by centrifugation at 265,000 g for 30 minutes at 4°C. The supernatant was discarded and remaining buffer removed with a paper towel. The pellet was resuspended for determination of protein and bound [^{3}H]ATP by scintillation counting.

[^{3}H]ATP, because both the unliganded and the Na$^+$-bound E_1 forms of wild-type and mutant proteins have high affinities for ATP (TABLE 1). In contrast, the addition of 0.1–10 mM KCl displaces [^{3}H]ATP binding, because K$^+$ stabilizes the E_2[2K] form of the protein with affinities for ATP (K_{ap} 200–400 µM) that are more than an order of magnitude lower than those of the E_1 forms.[5,8] Specific binding of ATP or ADP to wild-type and mutant proteins can therefore be determined as the binding at 10 mM NaCl minus that in medium containing 10 mM KCl. Scatchard plots in FIGURE 3 indicate that the mutation Asp^{369}-Ala altered the dissociation constant from 109 ± 11 nM to 2.8 ± 0.5 nM or by 39-fold. A similar plot for Asp^{369}-Asn gave a dissociation constant of 5.9 ± 0.4 nM.

Estimates of the dissociation constants (K_{ADP}) for equilibrium binding between recombinant Na,K-ATPase and ADP were calculated from the curves for equilibrium displacement of [^{3}H]ATP by ADP. The K_{ADP} values were in the same range for wild-type as for the mutations Asp^{369}-Asn and Asp^{369}-Ala. These data allow calculation of the contribution of the γ-phosphate to ATP binding at 0°C. TABLE 1 shows that replacement of the carboxylate group of Asp^{369} with the carboxamide group of Asn (D369N) caused a much larger change in free energy of binding (−7.3 kJ/mol) than does replacement of Asn^{369} with Ala (D369A) (−0.6 kJ/mol). The partial charge of the carbonyl oxygen of Asn is about −0.3 relative to a charge of −1 for the carboxylate group of Asp.[14] The data show that the changes in affinity for ATP observed after mutagenesis correlated with removal of the negative charge of the side chain of residue 369 with subsequent reduction of the electrostatic repulsion between the side chain of Asp and the γ-phosphate of ATP. It will be interesting to learn if the high intrinsic affinity for ATP that has been revealed in our study of the mutations to Asp^{369} can play a role as a driving force for uptake of K$^+$ ions into the cells.

EFFECT OF MUTATIONS TO Asp^{369} ON E_1-E_2 CONFORMATIONAL EQUILIBRIUM

In addition to its function as receptor for the γ-phosphate from ATP and as a major determinant of the ATP affinity, the side chain of residue 369 turned out to be a key residue for regulation of the conformational transition accompanying cation translocation. The evidence for this is estimation of the E_2(K)-E_1(K) conformational equilibrium from K$^+$ titrations of equilibrium ATP binding for the wild type and the two mutants, showing a two- to fourfold increase in the E_2(K) concentration compared to

TABLE 1. Effect of Mutations D369N and D369A on Capacities and Dissociation Constants for [^{3}H]ATP and ADP

	[^{3}H]ATP (K_{ATP}, nM)	[^{3}H]ATP (Capacity pmol/mg pr)	ADP^a (K_{ADP}, nM)	$\Delta G°$(γ-phosphate)^b (kJ/mol)
Wild type	109 ± 11	14.1 ± 0.8	152 ± 10	−0.75
D369N	5.9 ± 0.4	21 ± 1	196 ± 33	−8.0
D369A	2.8 ± 0.5	13.1 ± 1	122 ± 30	−8.6

^{a}Values were estimated from the curves of ADP displacement of [^{3}H]ATP binding[8] using the equation: $K_{ADP} = K_{1/2}$(ADP) $K_{ATP}/(K_{ATP} + [ATP])$.

$^b\Delta G°$(γ-phosphate), the contribution of γ-phosphate to ATP binding at 0°C, was calculated as: $\Delta G°$(γ-phosphate) = $\Delta G°$(ATP binding) − $\Delta G°$(ADP binding).

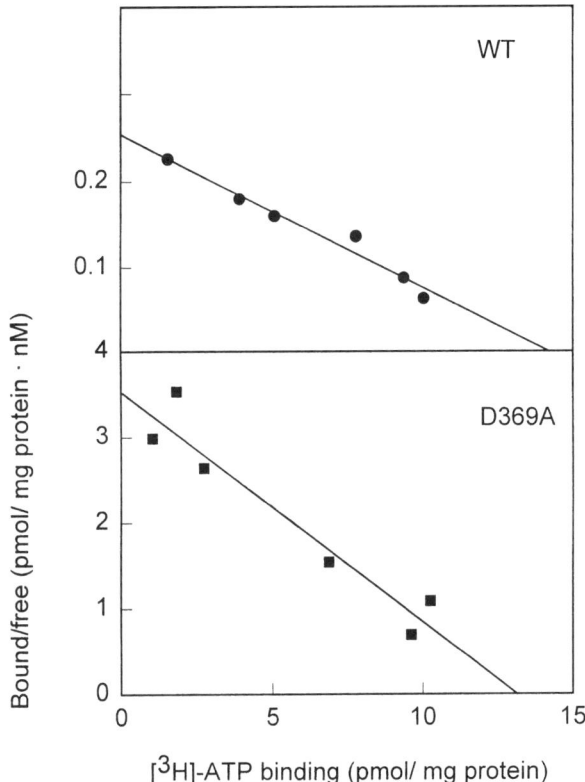

FIGURE 3. Scatchard plot of [³H]ATP binding to SDS-treated gradient membranes of wild-type (●) and mutant $\alpha_1(D369A)\beta_1$ (■) recombinant Na,K-ATPase. Procedures for assay of [³H]ATP binding in 10 mM NaCl or 10 mM KCl medium were as described in FIGURE 2. Specific binding was calculated as binding in the presence of NaCl minus binding in the presence of KCl. Data were fitted by nonlinear least squares regression to the lines with K_D 109 ± 11 nM and a binding capacity of 14.1 ± 0.8 pmol/mg protein for wild type and K_D 3.4 ± 0.5 nM and a binding capacity of 13 ± 1 pmol/mg protein for D369A.

$E_1(K)$ for Asp[369]-Asn and Asp[369] Ala.[7] The K_c value of 762 obtained for the $E_2(K)$-$E_1(K)$ equilibrium for the wild-type enzyme fits very well a K_c value of 1,000 previously reported from studies on renal Na,K-ATPase.[9] Quantitative estimations of E_2-E_1 conformational equilibrium from K^+ titration of ouabain binding also revealed a large increase in concentration of the E_2 compared to the E_1 conformation (TABLE 2). These estimates of the conformational equilibria are consistent with the observation that the affinity for ouabain in Mg^{2+} medium is greatly increased for the Asp[369]-Asn and Asp[369]-Ala mutations (TABLE 3), because high affinity ouabain binding is a recognized property of E_2 conformations.[5]

On a hydropathy scale,[15] Asp (−14.6 kJ/mol) and Asn (−14.6 kJ/mol) are equally hydrophilic, whereas Ala (10.5 kJ/mol) is considerably more hydrophobic. In a hydropathy plot of the peptide sequence of 20 residues around residue 369, it can be

TABLE 2. Effects of Mutations D369N and D369A on K^+-ATP Antagonism and on Equilibrium Constants (K_c) for the $E_1(K)$-$E_2(K)$ Transition as Calculated from $K_{1/2}$ Values Using the Equation: $K_{1/2} = K_p([ATP]/K_{ATP} + 1)/K_c$ [a]

Mutants	[^{3}H]ATP	
	$K_{1/2}$ K^+, mM	K_c
Wild type	0.11 ± 0.02	769
D369N	0.14 ± 0.01	1798
D369A	0.14 ± 0.02	3195

[a] $K_{1/2}$ values were determined from curves like that in Figure 2.

seen that the profile of the plot is unaffected by the substitution of Asp for Asn, whereas the substitution of Asp for Ala causes a pronounced change in the profile with considerable displacement from the hydrophilic to the hydrophobic section of the plot. Although the substitution of Asn[369] for Ala had only a moderate effect on the affinity for ATP, it had a major effect on the conformational equilibrium (TABLE 2) and on the affinity for ouabain (TABLE 3). Our data suggest that the position of the free charge of the carboxylate group of Asp[369] may be such that it contributes to prevent the E_1-E_2 transition in the absence of phosphate and magnesium. Removal of the charge and hydrophobic substitution (Asp[369]-Ala) favor transmission of the conformational transition originating in the phosphorylated segment through the transmembrane segment to the ouabain site at the extracellular surface. Mutation to a hydrophobic residue may in part substitute for the requirements for Mg^{2+} and phosphate, because substitution of Asp[369] or Asn[369] for the hydrophobic Ala allows spontaneous transition from the E_1 to the E_2 form to support high affinity binding of [^{3}H]ouabain in the absence of Mg^{2+} (TABLE 3).

SITE-DIRECTED MUTAGENESIS OF CARBOXYLATE RESIDUES Glu[327], Glu[779], Asp[804], AND Asp[808] IN TRANSMEMBRANE REGIONS

Mutations to oxygen-carrying residues in transmembrane segments of Ca-ATPase of SR (Glu[309], Glu[771], Asn[796], Thr[799], or Asp[800]) are known to block Ca transport and

TABLE 3. Effects of Mutations D369N and D369A on Dissociation Constants (K_D) for Binding of [^{3}H]Ouabain in Media Containing Mg^{2+}, Vanadate plus Mg^{2+}, or EDTA [a]

	Medium			
Mutants	Mg^{2+}, 3 mM Vanadate, 1 mM (K_D, nM)	Capacity (pmol/mg pr)	Mg^{2+}, 3 mM (K_D, nM)	EDTA, 1 mM (K_D, nM)
Wild type	1.8 ± 0.9	13.4 ± 0.4	133 ± 35	[b]
D369N	58 ± 8	5.8 ± 0.4	51 ± 7	[b]
D369A	4.7 ± 1.2	9.4 ± 0.3	4.9 ± 0.9	153 ± 34

[a] K_D values were estimated from Scatchard plots.
[b] In medium containing 1 mM EDTA and no Mg^{2+}, there was no binding of [^{3}H]ouabain to wild-type or D369N mutants.

Ca occlusion.[16] Mutations to homologous counterparts in the intramembrane portion of the α subunit of Na,K-ATPase (Glu^{327}, Glu^{779}, Asp^{804}, and Asp^{808}), using the ouabain selection method for functional expression in HeLa[17] or Cos[18] cells, have provided mixed results. Some are lethal mutations E327(D,A), E779(D,L), and D808(Q,L) in the sense that they do not allow expression of a functional ouabain-resistant Na,K-pump to sustain cell growth in a medium containing 1 μM ouabain. Other mutations, E327(Q,L), E779(A,Q), and D808(E,A), have some activity, and when expressed in ouabain media in HeLa[17] or Cos[18] cells they may display full Na,K-ATPase activity and phosphorylation level, albeit with altered apparent affinities to Na$^+$, K$^+$, and ATP. The expression systems based on ouabain selection have revealed considerable ability to compensate for a reduction in the turnover number per α subunit by expressing three- fivefold higher concentrations of α-subunit protein per cell,[17] thus reaching the same Na,K-ATPase activity and phosphorylation capacity as those of cells expressing wild-type α subunits. On this basis it was concluded that Glu^{327} and Asp^{808} are not essential residues for active Na,K-transport[18,19] and that only Asp^{804} appears to be important for enzyme function, because no functional substitution has been found for this residue.

TABLE 4. Expression in Yeast of Mutations to Intramembrane Charged Residues Glu^{327}, Glu^{779}, Asp^{804}, and Asp^{808} [a]

Mutation	Na,K-ATPase (%)	[^{3}H]Ouabain pmol/mg	[^{3}H]Ouabain K$_d$, nM	[^{3}H]ATP (pmol/mg)	[^{3}H]ATP (K$_d$, nM)
E327Q	37	12	7	16	73
E327D	5	14	8	12	50
E779Q	85	9	16	9	89
E779D	51	8	16	6	69
D804N	17	16	7	12	57
D804E	13	16	10	5	67
D808N	0	17	7	9	54
D808E	0	5	12	9	49
WT	100	13	10	7	52

[a]Mutagenesis, expression in yeast cells, and assay of Na,K-ATPase activity, [^{3}H]ATP, and [^{3}H]ouabain binding at equilibrium were performed as before.[6,7] Data are average values of two or three determinations.

EXPRESSION IN YEAST OF MUTATIONS TO Glu^{327}, Glu^{779}, Asp^{804}, AND Asp^{808}

After expression in yeast, the Na,K-ATPase activity was severely reduced in mutations to Glu^{327}, Asp^{804}, and Asp^{804}, whereas 70–90% of activity was preserved in mutations to Glu^{779}. This was not due to a variation in expression or to disturbance of tertiary structure, because the capacities for binding of [^{3}H]ATP and [^{3}H]ouabain were in the same range as those for wild type (TABLE 4). Western blots of yeast membrane protein[6] show that mutations to Glu^{327}, Glu^{779}, Asp^{804}, and Asp^{808} were expressed at a concentration close to one αβ unit per [^{3}H]ouabain binding site. Chromatography in C$_{12}$E$_8$ on high resolution size fractionation columns (TSK3000SW) shows that the recombinant, wild-type, and mutant Na,K-ATPase are organized in $\alpha\beta$ units with hydrodynamic properties similar to those of native renal Na,K-ATPase.[6]

The affinities for [^{3}H]ouabain binding after mutations of Glu^{327} and Glu^{779} to Asp or Gln and of mutations of Asp^{804} or Asp^{808} to Asn or Glu fell in the same range as for wild-type Na,K-ATPase (TABLE 4). These high affinities for ouabain, with K_D values in the range of 7–16 nM, can only be achieved if the sites for interaction with Mg^{2+} and vanadate (phosphate) are undamaged. In the absence of phosphate or vanadate, the dissociation constant for ouabain in Mg^{2+} medium is close to 150 nM, and ouabain binding is abolished in the absence of Mg^{2+}. Preservation of the high affinity site for ouabain allows the use of ouabain to define the background in determining specific ATP binding to cation site mutants where the use of KCl as a background may not be justified.

TABLE 5. Effect of Mutations to Intramembrane Charged Residues Glu^{327}, Glu^{779}, Asp^{804}, and Asp^{808} on Interactions with Na$^+$ and K$^+$ [a]

Mutation	Na-EP (%)	Na$^+$-Ouabain ($K_{1/2}$ [Na$^+$], mM)	K$^+$-ATP ($K_{1/2}$ [K$^+$], mM)	K$^+$-Ouabain ($K_{1/2}$ [K$^+$], mM)
E327Q	31	54	5.2	>10
E327D	12	81	>10	>10
E779Q	73	—	3.4	4.0
E779D	—	—	>10	>10
D804N	74	5.4	>10	>10
D804E	0	15.5	>10	>10
D808N	10	2.6	>10	6.7
D808E	10	15.3	2.3	3.0
WT	100	6.2	0.04	0.4–1.1

[a]Mutagenesis, expression in yeast cells, and assay of Na-dependent phosphorylation and binding of [^{3}H]ATP or (^{3}H)ouabain at equilibrium in the presence of increasing concentrations of Na$^+$ or K$^+$ were performed as before.[6,7] Data are average values of 2 or 3 determinations.

In assays of [^{3}H]ATP binding to cation site mutants, 2 mM ouabain was used as a background after it was observed that the level of [^{3}H]ATP binding to wild type in the presence of 2 mM ouabain was equal to the level of binding at saturating concentrations of KCl (10 mM). TABLE 4 demonstrates that the capacities and affinities for binding of [^{3}H]ATP at equilibrium also remained unaltered after mutations to acidic side chains. The structure of the ATP binding domain may therefore be preserved in the mutants along with the sites for Mg^{2+} binding and phosphorylation, thus demonstrating the selective nature of the mutations to presumptive cation sites.

EFFECT OF MUTATIONS AT Glu^{327} ON Na$^+$ INTERACTIONS

When expressed in yeast, the Na,K-ATPase activity and phosphorylation level were reduced to 30–37% after the substitution Glu^{327}-Gln and to less than 10% in the mutation Glu^{327}-Asp (TABLE 5). The maximum steady-state levels of phosphorylation from γ[^{32}P]ATP were reduced to 30% of wild-type levels for Glu^{327}-Gln at 5–30 mM Na$^+$, and at higher Na$^+$ concentrations, 100–150 mM, those levels were further depressed. In the presence of oligomycin, 30 µM, the phosphorylation level of

Glu^{327}-Gln at 100–150 mM Na$^+$ approached wild-type levels, but the apparent affinity for Na$^+$ (K$_{1/2}$ = 7 mM) remained four to fivefold lower than that for wild type.

This observation that Glu^{327} in Na,K-ATPase is important for Na$^+$ binding is in accordance with the proposed role of its homologous counterpart Glu^{309} in SR Ca-ATPase.[16] In contrast, when expressed by ouabain selection in Hela[17] or COS[18] cells, the mutations Glu^{327}-Gln or Glu^{327}-Leu yielded a functional enzyme with Na,K-ATPase activities and phosphorylation levels comparable to those of wild type, but with lower apparent affinities for both Na$^+$ and K$^+$. Overexpression of the mutant Glu^{327}-Gln protein has been observed in HeLa cells.[17] The explanation for the discrepancy can therefore be that the ouabain selection system leads to expression of three- to fourfold more α-subunit protein per cell to compensate for the reduction of Na,K-ATPase activity and phosphoenzyme levels.

EFFECT OF MUTATIONS TO Asp^{804} OR Asp^{808} ON Na$^+$ INTERACTIONS

In the Asp^{804}-Glu mutation, no Na-dependent phosphorylation from γ[^{32}P]ATP was observed in the range of 0–150 mM NaCl. The mutation Asp^{804}-Asn incorporated phosphate from ATP to a level of 70% of that of wild type, but the apparent affinity for Na$^+$ was reduced 10-fold from K$_{1/2}$(Na) 0.7–2 mM in wild type to K$_{1/2}$(Na) 10 mM in Asp^{804}-Asn. Substitution of Asp^{808} reduced the maximum steady-state phosphorylation level to 30% of that in wild type, and the apparent affinity for Na$^+$ was reduced (K$_{1/2}$(Na) = 7 mM) about fivefold. In Asp^{808}-Asn the phosphorylation level was about threefold lower than that in Asp^{808}-Glu (TABLE 5).

The reduction in steady-state level of phosphorylation after substitutions to Glu^{327}, Asp^{804}, or Asp^{808} can be due to a reduction in the rate of formation of the E$_1$P[3Na] form, because the mutations have removed some of the coordinating oxygens required to activate the transfer of the γ-phosphate from ATP to Asp^{369}. Another possibility is that the mutations destabilized the E$_2$P[2Na] intermediate, resulting in an increased rate of dephosphorylation in the Na$^+$ medium. The phospho-intermediates E$_1$P[3Na] and E$_2$P[2Na] are the main contributors to the acid-stable phosphorylation level. Reduction of Na$^+$-affinity and the steady-state phosphoenzyme level after expression of the mutants in yeast therefore shows that Glu^{327}, Asp^{804}, and Asp^{804} contribute some of the oxygen groups required for coordination of Na$^+$-ions.

Previously it was observed that the Asp^{808}-Glu mutation was expressed as a functional Na,K-ATPase by HeLa cells in ouabain medium after transfection with ouabain-resistant mutants.[17,20] It is probable that the cell system also in this case compensated for the reduced turnover and phosphorylation level by expression of larger amounts of the mutant α subunits with reduced activity. No functional substitution has been reported for Asp^{804}, and the properties of mutations to this residue have not been determined before.

CONSEQUENCES OF MUTATIONS TO Glu^{327}, Glu^{779}, Asp^{804}, OR Asp^{808} ON INTERACTIONS WITH K$^+$, K$^+$-ATP AND K$^+$-OUABAIN ANTAGONISM

Development of assays of [^{3}H]ATP binding[6] allows titration of K$^+$ displacement of [^{3}H]ATP and stabilization of the alternative conformation, E$_2$[2K], and derivation of values for the affinities of wild type and mutants for K^+.[7] In the wild type, K^+ displaces [^{3}H]ATP with high apparent affinity ($K_{1/2}$ (K^+) = 0.4–1.1 mM). As seen in TABLE 5, substitution at Glu^{327} or Glu^{779} by Gln or Asp strongly reduced the apparent affinity for K^+. The Asp^{808}-Glu substitution also caused a reduction in the apparent affinity for K^+($K_{1/2}$(K) 2.3 mM), whereas the Asp^{808}-Asn substitution almost abolished the K^+-induced displacement of [^{3}H]ATP. Similarly, substitution of Asp^{804} with Asn or Glu abolished the effect of K^+ in the range 0–10 mM of KCl. Additional experiments showed that only partial displacement by K^+ (20–30%) was observed at 200–300 mM KCl using the level of [^{3}H]ATP binding in the presence of 2 mM ouabain as reference.

Previous studies showed that K^+ ions stabilize a conformation with relatively low ouabain affinity.[20,21] Preservation of high affinity ouabain binding in the mutations therefore allows studies of the interaction of K^+ with the ouabain complexes of the E_2 conformation. Substitutions at Asp^{804} abolished the displacement seen after adding increasing concentrations of K^+ to the medium for [^{3}H]ouabain binding to wild type. In substitutions to Asp^{808} to Glu or Asn, KCl was able to displace [^{3}H]ouabain, but with a much lower apparent affinity ($K_{1/2}$(K) 3.0 or 6.7 mM) than for wild type ($K_{1/2}$(K) 0.4–1.1 mM) (TABLE 5).

Titrations with K^+ of [^{3}H]ATP or [^{3}H]ouabain binding thus show that substitutions to Glu^{327}, Glu^{779}, Asp^{804}, or Asp^{808} seriously impaired the ability of the αβ unit to assume the E_2[2K] form. In particular, mutations of Asp^{804} were completely refractory to K^+. Analysis of sequence homologies for this region in FIGURE 4 shows that although the aspartate at no. 808 is a well-conserved feature, a carboxylate-containing side chain was only observed at the position homologous to Asp^{804} in H,K-ATPase,[23,24] whereas Ca-ATPase of plasma membranes[25] and sarcoplasmic reticulum[26] or H-ATPase from yeast[27] or plants[28] have Asn or Ala at this position. This comparison supports the notion that Asp^{804} provides coordinating oxygens for occlusion of K^+ in Na,K-ATPase.

A MODEL FOR ORGANIZATION OF Na$^+$ SITES AND K$^+$ SITES IN Na,K-ATPase

Observations on mutants expressed in yeast cells show that at least three of the carboxylate residues in presumptive TM segments 4 and 6 (Glu^{327}, Asp^{804}, and Asp^{808}) engage in formation of the occlusion cavity for Na$^+$ in phosphoenzyme conformations E_1P[3Na] or E_2P[2Na]. Alternatively, these acid groups (Glu^{327}, Glu^{779}, Asp^{804}, and

Na,K-ATPase	Ref no.	804 808	Homology (%)
Na,K-ATPase	22	PLGTVTILCIDLGTDMVPAISLAYE	100
H,K-ATPase rat	23	PIGTITILfIDLGTDiIPSIALAYE	96
H,K-ATPase pig	24	PLGcITILfIELcTDifPSVSLAYE	80
Ca-ATPase PM	25	PLkaVqmLwVnLimDtfaSLALAtE	52
Ca-ATPase SR	26	aLipVzLLwVnLvtDgLPAtALgfn	48
H-ATPase yeast	27	aLdidlIvfiaIfaDvatlaia-YD	28
H-ATPase ATH	28	dFsafmVLIiaIlnDgtimtis-KD	32

FIGURE 4. Alignment of M6 regions of Na,K, H,K, Ca, and H pumps.

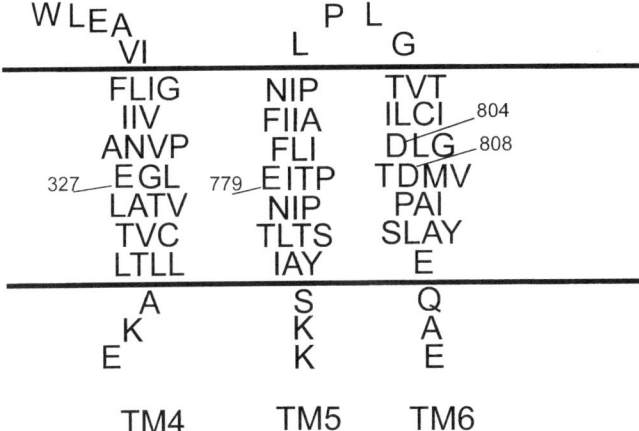

FIGURE 5. Model of cation binding sites in transmembrane segments 4, 5, and 6 of pig kidney Na,K-ATPase based on the topological model of the α subunit proposed in ref. 1.

Asp^{804}) appear to engage in coordination of K$^+$ ions in the occlusion cavity of the E$_2$[2K] form. In particular, Asp^{804} and Asp^{808} are important contributors to coordination of K$^+$ in the occlusion cavity of the E$_2$[2K] conformation, as demonstrated in titrations of K$^+$-[^{3}H]ATP or K$^+$-[^{3}H]ouabain antagonism. The large changes in affinity for K$^+$ after substitution of Glu^{327} or Glu^{779} also argue for contribution of these residues to coordination of K$^+$ in the occlusion cavity.

From the model in FIGURE 5 it is possible that the Glu^{327}, Glu^{779}, Asp^{804}, and Asp^{808} residues in the fourth, fifth, and sixth presumptive transmembrane segments can be organized in amphiphilic helices. The carboxylate oxygens may orient towards a central occlusion cavity, whereas hydrophobic residues may form the back side oriented towards the lipid phase. Inasmuch as five or six oxygen groups may be required to coordinate one dehydrated Na$^+$ or K$^+$ ion,[30] it is obvious that a number of residues in addition to Glu^{327}, Glu^{779}, Asp^{804}, and Asp^{808} may contribute to coordination of Na$^+$ or K$^+$ in the occluded E$_1$P[3Na] or E$_2$[2K] complexes. The model in FIGURE 5 shows that the carboxylate groups in transmembrane segments TM4 (Glu^{329}), TM5 (Glu^{779}), and TM6 (Asp^{804} and Asp^{808}) are surrounded by a number of relatively hydrophilic residues with oxygen- or sulfur-containing side chains (Pro, Asn, Thr, Ser, and Met). The sequence D^{808}MVP in the sixth transmembrane segment of the α subunit is homologous to sites for binding of Ca^{2+} in TM4 (EGLP) and TM6 (DMVP) of SR Ca-ATPase[26,29] and to sites for binding of H$^+$ (DAIP) or Na$^+$ (ESAP) in subunit c of FoF1-ATPase in *Escherichia coli*[31] or *Pseudomonas modestum*.[32] The presence of proline residues near the carboxylate groups may induce instability or kinks in the helix and thus increase the possibility that main chain carbonyls may shift from hydrogen bonding to interaction with Na$^+$ or K$^+$ ions.

The cation sites of Na,K-ATPase bind Na$^+$ or K$^+$ ions in a ping-pong sequence, and hybrid occupancy has not been observed. The change in orientation and specificity of the acid groups in TM segments 4, 5, and 6 can be elicited by binding Na$^+$ or K$^+$ ions in the absence of other ligands, or they can be driven by the scalar reactions in the

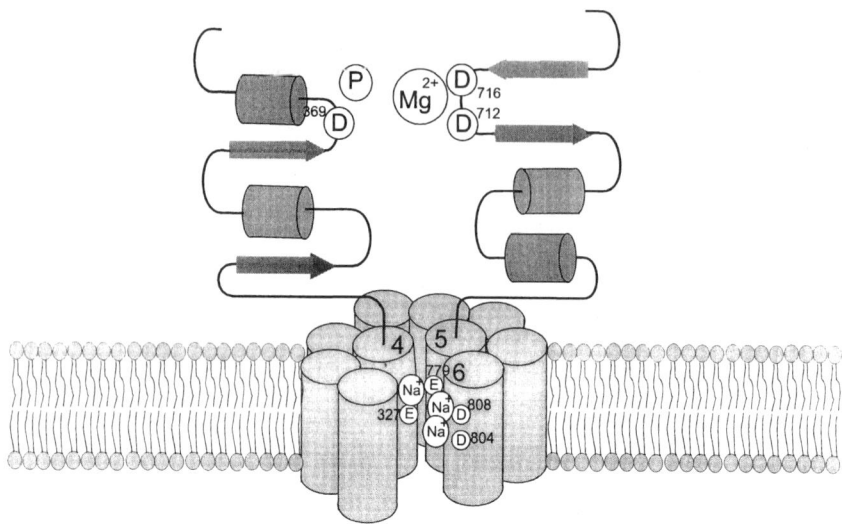

FIGURE 6A. Model of the $E_1P[3Na]$ conformation illustrating the binding of 3 Na^+ ions with ionic diameter 1.9 Å in a cavity with a gate function towards the cytoplasmic side and a lock towards the extracellular surface. Data in this work provide evidence for contributions of carboxylate oxygens from residues Glu^{327}, Asp^{804}, and Asp^{804} to the coordination of Na^+ in this cavity. Covalent attachment of γ-phosphate to Asp^{369} and coordination via Mg^{2+} to Asp^{712} elicits a structural change in TM4 and TM5–6 to close the gate and thus occlude 3 Na^+.

cytoplasmic protrusion, such as ATP binding, phosphorylation, and dephosphorylation.[1,5]

In the E_1Na form, the Na^+ ions have access to the binding site from the cytoplasmic surface. The ATP bound E_1 form binds Na^+ with apparent affinity 1–2 mM ($K_{1/2}$), and the transfer of γ-phosphate from ATP to Asp^{369} is accompanied by transformation of the binding site to an occluded state, the $E_1P[3Na]$ phosphoform, without access to exchange with cation in the medium. Na^+ may fit into the binding site with coordination to five oxygen ligands,[30,33] and the ion selectivity reflects that the size of the ion binding site matches the ionic diameter of the Na^+ ion (1.9 Å). The isomerization between phosphoforms $E_1P[3Na]$ and $E_2P[2Na]$ may involve a change in size of the occlusion cavity and in the number of coordinating oxygens as the change is accompanied by the release of one extracellular Na^+ ion and subsequent exchange of the remaining 2 Na^+ ions with extracellular K^+ to stabilize the $E_2[2K]$ conformation. The ionic diameter of K^+ (2.66 Å) is larger than that of Na^+, and the number of coordinating oxygens may rise from 5 to 6 per ion.[30,33] The nature of these structural changes is not known, but they may be elicited by moderate rotation or tilting of transmembrane helices as proposed before.[16,34]

The model in FIGURE 6 is a working hypothesis for the role of Asp^{369} and Mg^{2+} in the conformational transition coupling ATP hydrolysis to ion translocation. It is based on the observation of the effects of mutations to Asp^{369} on the E_1-E_2 conformational transitions and the interactions with Mg^{2+}.[7] The structural change accompanying

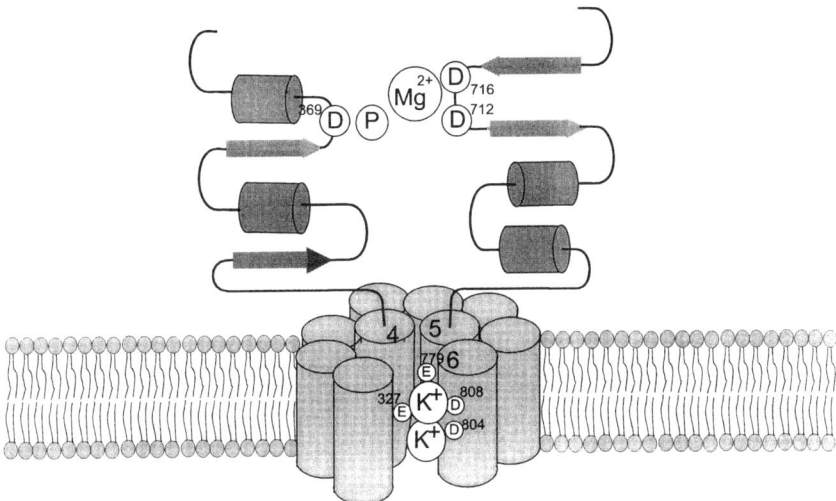

FIGURE 6B. Model of K^+-induced hydrolysis of $E_2P[2Na]$ to $E_2[2K]$ plus inorganic phosphate after exchange of Na^+ for K^+ at the extracellular surface. Residues Glu^{327}, Glu^{779}, Asp^{804}, and Asp^{808} provide carboxylate oxygens for coordination of 2 K^+ ions with ionic diameter of 2.6 Å in a cavity with gate function towards the extracellular side and a lock towards the cytoplasmic portion of the transmembrane segments. The binding of $2K^+$ elicits a structural transition from TM5–6 through the segment to Asp^{712} and the coordinated Mg^{2+} to bring the Asp^{369}-phosphate bond in line for hydrolysis.

isomerization between $E_1P[3Na]$ and $E_2P[2Na]$ is transmitted directly from Asp^{369} to TM4 via the relatively short segment of 42 residues. Through cooperative interactions, the structural transition is transmitted indirectly to TM5–6 via the interaction of Asp^{369}-phosphate with Mg^{2+} which is coordinated to other negatively charged residues in the protein. It can be assumed that Mg^{2+} coordinates to six negatively charged groups, but the relevant residues have not been defined. In FIGURE 6 it is proposed that Mg^{2+} coordinates to Asp^{712} and Asp^{716}, which are separated by only 52 residues from TM5. The segment around Asp^{712} and Asp^{716} residues is the best conserved amino acid sequence among cation pumps.[5] Asp^{712} and Asp^{716} have been identified as sites for covalent insertion of Cl-ATP,[35] but preliminary data from our laboratory suggest that Asp^{712} may not be involved in ATP binding. Moreover, it was previously proposed that this segment is involved in coordination of Mg^{2+}.[36] The force of the structural change around Asp^{369} is thought to be transmitted through the protein, as in a pair of scissors with the segments around Asp^{369} and Asp^{712} forming the handles and the transmembrane helices 4 and 5 forming the blades of the scissors. The movement of the helices may cause structural rearrangement of TM4 and TM5–6, resulting in altered positions and changes in orientation of the acidic side chains. In the previous model by MacLennan et al.[16] for twisting or tilting of transmembrane helices as part of the change in conformation during Ca^{2+} extrusion by Ca-ATPase, there were no suggestions on the nature of energy transduction between the site for binding of ATP and phosphorylation and the cation sites in the intramembrane domain.

REFERENCES

1. JORGENSEN, P. L. 1992. In Molecular Aspects of Transport Proteins. J. J. H. H. M. De Pont, Ed.: 1–26. Elsevier.
2. LINGREL, J. G. & T. KUNTZWEILER. 1994. J. Biol. Chem. **269:** 19659–19662.
3. LUTSENKO, S. & J. H. KAPLAN. 1995. Biochemistry **34:** 15607–15613.
4. CAPASSO, J. M., S. HOVING, D. M. TAL, R. GOLDSHLEGER & S. J. D. KARLISH. 1992. J. Biol. Chem. **267:** 1150–1158.
5. JORGENSEN, P. L. & J. P. ANDERSEN. 1988. J. Membr. Biol. **103:** 95–120.
6. PEDERSEN, P. A., J. H. RASMUSSEN & P. L. JORGENSEN. 1996. J. Biol. Chem. **271:** 2514–2522.
7. PEDERSEN, P. A., J. H. RASMUSSEN & P. L. JORGENSEN. 1996. Biochemistry **35:** 16085–16093.
8. NORBY, J. G. & J. JENSEN. 1988. Methods Enzymol. **156:** 191–201.
9. JORGENSEN, P. L. & S. J. D. KARLISH. 1980. Biochim. Biophys. Acta **597:** 305–317.
10. OHTSUBO K., S. NOGUCHI, K. TAKEDA, M. MOROHASHI & M. KAWAMURA. 1990. Biochim. Biophys. Acta **1021:** 157–160.
11. KUNTZWEILER, T. A., E. T. WALLICK, C. L. JOHNSON & J. B. LINGREL. 1995. J. Biol. Chem. **270:** 2993–3000.
12. MARUYAMA, K. & D. H. MACLENNAN. 1988. Proc. Natl. Acad. Sci. USA **85:** 3314–3318.
13. RAO, R. & C. W. SLAYMAN. 1992. Biophys. J. **62:** 228–234.
14. CREIGHTON, T. E. 1993. Proteins, Structure and Molecular properties. W. H. Freeman & Co. New York.
15. KYTE, J. & R. F. DOOLITTLE. 1982. J. Mol. Biol. **163:** 451–466.
16. MACLENNAN, D. H., D. M. CLARKE, T. W. LOO & I. S. SKERJANC. 1992. Acta Physiol. Scand. **146:** 141–150.
17. JEWELL-MOTZ, E. A. & J. B. LINGREL. 1993. Biochemistry **32:** 13523–13530.
18. VILSEN, B. 1993. Biochemistry **32:** 13340–13349.
19. ARGUELLO, J. M. & J. B. LINGREL. 1995. J. Biol. Chem. **270:** 22764–22771.
20. KUNTZWEILER, T. A., E. T. WALLICK, C. L. JOHNSON & J. B. LINGREL. 1995. J. Biol. Chem. **270:** 16206–16212.
21. HANSEN, O. & J. C. SKOU. 1973. Biochim. Biophys. Acta **311:** 51–66.
22. SHULL, G. E., J. GREEB & J. B. LINGREL. 1986. Biochemistry **25:** 8125–8132.
23. CROWSON, M. S. & G. E. SHULL. 1992. J. Biol. Chem. **267:** 13740–13748.
24. SHULL, G. E. & J. B. LINGREL. 1986. J. Biol. Chem. **261:** 16788–16791.
25. ADEBAYO, A. O., A. ENYEDI, A. K. VERMA, A. G. FILOTEO & J. T. PENNISTON. 1995. J. Biol. Chem. **270:** 27812–27816.
26. BRANDL, C. J., N. M. GREEN, B. KORCZAK & D. H. MACLENNAN. 1986. Cell **44:** 597–607.
27. SERRANO, R., M. C. KIELLAND-BRANDT & G. R. FINK. 1986. Nature **319:** 689–693.
28. HARPER, J. F., T. K. SUROWY & M. R. SUSSMAN. 1989. Proc. Natl. Acad. Sci. USA **86:** 1234–1238.
29. CLARKE, D. M., T. W. LOO & D. H. MACLENNAN. 1990. J. Biol. Chem. **265:** 6262–6267.
30. WEBER, E. & F. VOGTLE. 1981. Topics in Current Chemistry. F. Vogtle, Ed. **98:** 1–42. Springer-Verlag Press. Germany.
31. ZHANG, Y. & R. H. FILLINGAME. 1995. J. Biol. Chem. **270:** 87–93.
32. KLUGE, C. & P. DIMROTH. 1993. Biochemistry **32:** 10378–10386.
33. TONEY, M. D., E. HOHENESTER, S. W. COWAN & J. N. JANSONIUS. 1993. Science **261:** 756–759.
34. TANFORD, C. 1982. Proc. Natl. Acad. Sci. USA **79:** 2881–2884.
35. OVCHINNIKOV, Y. A., K. N. DZHANDZUGAZYAN, S. V. LUTSENKO, A. A. MUSTAYEV & N. N. MODYANOV. 1987. FEBS Lett. **217:** 111–116.
36. GIRARDET, J. L., I. BALLY, G. ARLAUD & Y. DUPONT. 1993. Eur. J. Biochem. **217:** 225–231.

Structure/Function Analysis of the Ca^{2+} Binding and Translocation Domain of SERCA1 and the Role in Brody Disease of the *ATP2A1* Gene Encoding SERCA1[a]

DAVID H. MACLENNAN,[b] WILLIAM J. RICE, AND ALEX ODERMATT

Banting and Best Department of Medical Research
University of Toronto
Charles H. Best Institute
112 College Street
Toronto, Ontario M5G1L6, Canada

The Ca^{2+}-ATPase of fast-twitch skeletal muscle sarcoplasmic reticulum (SERCA1a) has proven to be an excellent model for analysis of structure/function relationships in a P-type ATPase.[1] The protein binds 2 moles of Ca^{2+} per mole of protein and transports 2 moles of Ca^{2+} from cytoplasmic to lumenal spaces for each mole of ATP hydrolyzed. In recent years, site-directed mutagenesis has been used to explore structure/function relationships in SERCA1. Ca^{2+}-ATPase mutants have been divided into several classes: V_{max} mutants, Ca^{2+} binding mutants, apparent Ca^{2+} affinity mutants (possibly reflecting a shift in E_1-E_2 equilibrium), phosphorylation mutants, ATP binding mutants, mutants blocked between E_1P and E_2P, mutants blocked in E_2P dephosphorylation, and uncoupled mutants.[2–10]

Mutagenesis has identified Glu^{309} in transmembrane sequence M4, Glu^{771} in M5, Asn^{796}, Thr^{799}, and Asp^{800} in M6, and Glu^{908} in M8 as probable Ca^{2+} binding amino acids,[2] thereby implicating these four transmembrane helices in the formation of the Ca^{2+} binding and translocation domain. Sites of ATP binding and phosphorylation have been localized in cytoplasmic sequences of the molecule.[11–13] In the Ca^{2+} transport cycle, a series of phosphoprotein intermediates are formed,[14] which probably correspond to a series of transient conformations that occur in the molecule as Ca^{2+} transport proceeds. At least two phosphoprotein intermediates, E_1P, a high energy intermediate,[3] and E_2P, a low energy intermediate,[6] can be stabilized in mutant enzymes. In the process of Ca^{2+} transport, conformational changes initiated in the cytoplasmic nucleotide binding and phosphorylation domains must be transmitted through long-range interactions, involving stalk sequences S4 and S5,[15] to the transmembrane Ca^{2+} binding domain, accounting for the coupling between ATP hydrolysis and the movement of Ca^{2+} ions.[1,16]

---

[a]Original research described in this review was supported by grants (to DHM) from the National Institutes of Health (USA), the Canadian Genetic Diseases Network of Centers of Excellence, and the Muscular Dystrophy Association of Canada. WJR is a predoctoral Fellow of the Medical Research Council of Canada (MRCC), and AO is a postdoctoral Fellow of the MRCC.
[b]Tel: 416 978-5008; fax: 416 978-8528; e-mail: david.maclennan@utoronto.ca

SCANNING MUTAGENESIS REVEALS A SIMILAR PATTERN OF MUTATION SENSITIVITY IN M4, M5, AND M6 BUT NOT M8

To understand the structure/function relationships of the amino acids that form the Ca^{2+} binding and translocation domain, we analyzed the effects of mutagenesis of all residues in transmembrane helices M4, M5, M6, and M8.[17] Our objective was to determine if these helices contain mutation-sensitive amino acids additional to those already reported[2–10] and if the mutation-sensitive amino acids form a pattern or motif in these helices like the "patch" of residues surrounding the Ca^{2+}-binding ligand Glu^{309}, which we described earlier.[7] About 200 novel and prior mutations were included in the study. For mutations that involved large or small hydrophobic amino acids, our strategy was to alter their size. Ala was changed to Val or Gly; Leu and Ile were altered to Val and Ala; and Val was changed to Leu and Ala. Ser was substituted for some large aliphatic residues in tests evaluating the effects of an alteration in hydrophobicity.

Ca^{2+}-ATPase expression levels, based on quantitative ELISA, were measured in all mutants. All mutants were assayed for Ca^{2+} dependence of Ca^{2+} transport, because curve fitting of such data gives a measurement of maximal Ca^{2+} transport activity and apparent Ca^{2+} affinity. Quantitation of Ca^{2+}-ATPase protein permitted direct comparison between Ca^{2+} transport activities for mutant and wild-type protein.

FIGURE 1A summarizes the functional consequences of all mutations in M4, M5, M6, and M8 that were analyzed in the course of several studies.[2–10] The amino acids are shaded according to the severity of the effects of mutations on their function. In order to group a wide range of alterations into a very few categories, the criteria for each category were broad. Loss of Ca^{2+} transport was a clear category, and these mutants are colored black. Mutants with less than 5% of wild-type Ca^{2+} transport activity were recorded as having zero activity, because they could not be distinguished reliably from the background. Ca^{2+} transport-deficient mutants were also tested for possible blocks in the Ca^{2+} transport cycle. To detect blocks in the dephosphorylation of E_1P or E_2P intermediates, the stabilities of the phosphorylated intermediates were measured in the presence of EGTA alone or EGTA plus ADP.[3]

Because most mutations altered the rate of Ca^{2+} transport activity, we did not use lowered V_{max} values alone as a suitable criterion upon which to classify mutants. A reduction in K_{Ca} of 0.5 pCa units (a threefold alteration in Ca^{2+} affinity) was chosen as a cutoff point between major and minor alterations in Ca^{2+} affinity, provided mutant activity was 1 standard deviation away from the cutoff line. These mutants are shaded. Without these broad criteria, truly deleterious mutations were lost in the background of more benign mutations. With a few exceptions, the most deleterious effect among multiple mutations of the same residue was used in shading selection.

Transport Negative Mutants. Transport-deficient mutations were found in all four transmembrane sequences. These mutations involved seven amino acids in M4, six in M5, five in M6, and one in M8. Transport-deficient mutants were tested for their ability to form a phosphoenzyme intermediate when reacted with Ca^{2+} and ATP or with P_i in the presence and absence of Ca^{2+}. Those that did not form a phosphoenzyme intermediate in the presence of Ca^{2+} and ATP, but did form a phosphoenzyme intermediate from P_i in the presence and absence of Ca^{2+} were considered to be Ca^{2+}-binding mutants. These were Glu^{309} in transmembrane sequence M4, Glu^{771} in M5, Asn^{796}, Thr^{799}, and Asp^{800} in M6, and Glu^{908} in M8.

Those transport-deficient mutants that formed a phosphoenzyme intermediate in

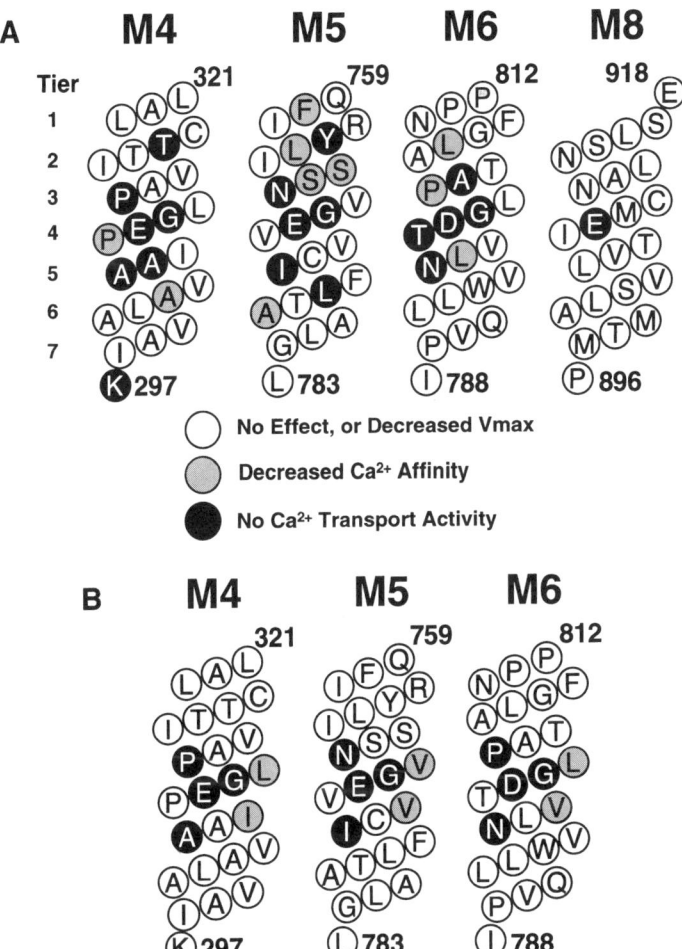

FIGURE 1. Summary of mutational analysis of transmembrane helices M4, M5, M6, and M8. (**A**) Residues are shaded according to the following scheme: *black* = no Ca^{2+} transport activity; *shaded* = reduction of apparent Ca^{2+} affinity by at least 0.5 pCa units, provided that Ca^{2+} affinity was more than 1 standard deviation away from the cutoff; *white* = no effect, reduced V_{max}, or reduction in apparent Ca^{2+} affinity by less than 0.5 pCa units. These results are the combined results from a series of studies.[1-9] Residues are generally shaded according to the most severe mutation found. (**B**) A functional and structural motif found in the central tiers of M4, M5, and M6. Helices were aligned by placing Ca^{2+} binding ligands Glu^{309}, Glu^{771}, and Asp^{800} at the same relative position in each helix. Mutation-sensitive residues in the same relative position in all three helices are colored black. Residues that are hydrophobic in the same relative position in central tiers of all three helices are shaded. (Adapted from Rice & MacLennan,[17] with permission.)

the presence of Ca^{2+} and ATP, which could be dephosphorylated by ADP in the presence of EGTA (representing exchange between a high-energy phosphoenzyme and ADP), but not by EGTA alone (representing the forward reactions of dephosphorylation observed when phosphorylation is prevented by the lack of Ca^{2+}), were considered to be blocked in a high-energy phosphorylated conformation. These mutants, referred to as E_1P to E_2P mutants, were P312G and T317D in M4 and L777A in M5. Those that formed a phosphoenzyme intermediate in the presence of Ca^{2+} and ATP, which could not be dephosphorylated by ADP or EGTA, were considered to be blocked in a low-energy phosphorylated conformation. These mutants are referred to as E_2P dephosphorylation mutants. They were A305V, A306V, and G310V in M4, N768I and I775S in M5, and A804V in M6.

An estimate of the Ca^{2+} affinity of the remaining transport-deficient mutants was obtained by measurement of phosphoenzyme intermediate formation from ATP in the presence of free Ca^{2+} concentrations varying from pCa 4 to pCa 8. Those mutants that formed a phosphoenzyme intermediate in the presence of Ca^{2+} and ATP, but had a drastic decrease in Ca^{2+} affinity, were considered to be Ca^{2+} affinity mutants. They were G310V in M4, F760G in M5, and A804V in M6. Ca^{2+} inhibition of phosphoenzyme formation from P_i was also used to determine Ca^{2+} affinity. Those shown to have large reductions in Ca^{2+} affinity by this assay were G310V in M4 and G770A in M5.

Mutation-Sensitive Amino Acids in M4, M5, and M6. The cluster of large hydrophobic and aromatic residues, Phe[760], Tyr[763], and Leu[764], that are mutation sensitive in tiers 1 and 2 in M5 are of particular interest. The mutation Y763G uncoupled Ca^{2+} transport from ATP hydrolysis.[8] This is a drastic mutation, going from the largest to the smallest residue, but the demonstration that this alteration allows Ca^{2+} to escape from its binding sites in the center of M4, M5, and M6 to the cytoplasmic surface suggests that the presence of seven bulky residues out of seven at the top of M5 plays an important role in "capping" the Ca^{2+} pore. Mutations F760G or of L764G, however, did not affect coupling, but did affect Ca^{2+} affinity.[8]

Another residue of particular interest is Leu[797] in helix M6. Replacement of Leu[797] with any of Ala, Val, or Ile resulted in an approximately threefold reduction in apparent Ca^{2+} affinity. Given the similar total bulk of Leu and Ile, this indicates that Leu[797] requires a near perfect fit in the wild-type enzyme. Ala[305] and Ala[306] in M4 and Ala[804] in M6 are also of unusual interest, because mutation of any of these three to Val results in a block in E_2P dephosphorylation. Accordingly, Ala[304], Ala[305], and Ala[804] may form hydrophobic holes in which larger residues from other helices fit with high precision.

Mutation-Sensitive Patterns in M4, M5, and M6. FIGURE 1A reveals that the overall patterns resulting from mutagenesis are similar for predicted helices M4, M5, and M6, but very different for M8.[17] The mutation-sensitive residues are distributed rather uniformly among the first three helices, with 9 in M4, 11 in M5, and 8 in M6. These patches form similar arrays on the seven tiers of the α-helical net displayed in FIGURE 1A. The patches of mutation-sensitive residues extend from tier 1 to tier 6, with their broadest points in tiers 3 and 4. At their broadest point, they span about 240 degrees of the 360-degree helix (PEG[310] in M4; SSN[768] in M5; TDG[801] in M6), and in tiers 2 and 5 they span about 140 degrees (AA[305] in M4; YL[764] in M5; NL[797] in M6). Except in tiers 1 through 3 in M5, a contiguous column of relatively hydrophobic, mutation-insensitive residues runs down the right-hand side of the seven tiers of each of the three helical nets. This establishes a hydrophobic face in M4, M5, and M6 and indicates their orientation with respect to hydrophobic regions surrounding this cluster.

The pattern of mutation-sensitive residues in the three different helices also highlights an interesting distribution of amino acid sizes. Of the 14 mutation-sensitive residues in tiers 3 and 4 of the helical nets for M4, M5, and M6, only the two Glu residues in M4 and M5 are bulky residues. Of the five mutation-sensitive residues in tiers 1 and 2, four are bulky residues. Of the eight mutation-sensitive residues in tiers 5 and 6, three are bulky. This distribution reflects in part the distribution of large and small residues in the different tiers. Large and bulky residues account for 57% of the total residues in tiers 1 and 2 of the three helices; 9.5% of the total residues in tiers 3 and 4; and 43% of the total residues in tiers 5 and 6. The overall pattern is not uniform, because all seven residues in tiers 1 and 2 of M5 are bulky and all four residues in tier 6 of M6 are bulky. Nevertheless, the overall mass distribution suggests that each of helices M4, M5, and M6 is slightly hour-glass shaped, with a higher concentration of small and polar residues in the central tiers and a higher concentration of bulky and nonpolar residues in the peripheral tiers. The lower mass density in the central tiers might create the space between ligands that is required for Ca^{2+} ion binding and perhaps for a few water molecules. They may also allow for mobility, relative to each other, of those helices that form the Ca^{2+} binding and translocation sites. Mutation sensitivity in the different tiers may result from fundamentally different principles. The small and polar core residues may be mutation sensitive because of their direct involvement with Ca^{2+} binding and translocation. The larger, more hydrophobic, peripheral residues may be mutation sensitive because of space constraints during conformational changes.

Mutation-Sensitive Motif. An important point of similarity between M4 and M6 is the presence in tiers 3 and 4 of a six-residue motif that is highlighted by its mutation sensitivity (FIG. 1B). The motif begins with Ca^{2+} binding amino acids Glu^{309} and Asp^{800} and continues as E/D-G-L-P-A-V/T. Although the sequence similarity does not extend beyond six residues, the similarity in mutation sensitivity actually begins four residues upstream. Thus, sequences $AAIP^{308}$ in M4 and $NLVT^{799}$ in M6 each contain three residues in the same position that are mutation sensitive, and the remaining residue is hydrophobic and mutation insensitive in both sequences (FIG. 1B). There is, however, little correspondence among the mutation classes found at each location. The structural and functional similarity between helices M4 and M6 suggests a possible sequence duplication during evolution. The positions of Glu^{908} and Ala^{912} in M8 are the only indication that M8 might also have evolved divergently from the same helix that gave rise to M4 and M6.

Elements of the motif can also be seen in M5, and these similarities are highlighted in FIGURE 1B. The (reverse) sequence EGV^{769} corresponds to the first three amino acids in the M4/M6 motif (E/D-G-L/V), and the pattern of mutation sensitivity is very similar (FIG. 1B). In M5, convergent evolution may have created a counterpart of the M4/M6 motif, but in the reverse orientation.

In the region of the motif, the acidic residues at position 1 are clearly Ca^{2+} binding residues. Gly, with no side chain, when fitted into position 2, would create an open space on the cytoplasmic side of each of the Ca^{2+} binding ligands, perhaps providing room for rotation among helices or for binding of the Ca^{2+} ion. Leu at position 3 (and Ile or Val at position -2) may represent the faces of the helices that are in contact with lipid. Proline at position 4 might allow the helices to bend away from the Ca^{2+} entry point on the cytoplasmic surface, thus widening the pore. The function of Ala in position 5 or of the relatively isomorphic residues Thr and Val at position 6 is not clear.

Mutation-Sensitive Amino Acids in M8. Although even conservative substitutions for many residues in M4, M5, and M6 resulted in conformational blocks, the only residue in M8 that was found to be mutation sensitive was Glu^{908}, previously proposed to be a Ca^{2+} binding ligand.[2] In earlier studies, replacement of Glu^{908} with Asp led to a transport-negative phenotype, but replacement with Gln had no effect on function. By contrast, substitution of the corresponding amidated residue for Glu^{309}, Glu^{771}, or Asp^{800} resulted in complete loss of Ca^{2+} transport function. This result has been interpreted as indicating that only one oxygen in Glu^{908} is directly involved in the formation of the Ca^{2+} binding site. Andersen and Vilsen,[18] however, showed that mutation E908A retains Ca^{2+} occlusion, suggesting that Glu^{908} may not actually provide an oxygen ligand essential to the Ca^{2+} binding site.

Data from this study expand our knowledge of the size and complexity of the Ca^{2+} binding faces in M4, M5, and M6 and tend to exclude M8 as a participant in the formation of Ca^{2+} binding sites. Some doubt has been cast on the role of Asn^{796} in Ca^{2+} binding,[10] inasmuch as analysis of the Ca^{2+} inhibition of E_2P formation in mutant N796A has shown that it has a higher affinity for Ca^{2+} than do comparable mutants involving the other five Ca^{2+} binding mutants.[2,10] Our data, however, show that it fits well with the mutation-sensitive pattern of the Ca^{2+} binding and translocation motif. Moreover, mutations of Asn^{796} disrupted the ability of the mutant protein to occlude Ca^{2+},[18] making it difficult to remove it from the class of Ca^{2+}-binding amino acids.

Current data do not yet resolve the question of how Ca^{2+} binding faces relate to each other. The residues forming these Ca^{2+} binding sites clearly reside in M4, M5, and M6 and reasonable evidence exists that Glu^{309}, Glu^{771}, Asn^{796}, Thr^{799}, Asp^{800}, and perhaps Glu^{908} contribute different ligands to the two different Ca^{2+} binding sites.[18–20] Accordingly, these residues must face inwards, towards each other. As a corollary, FIGURE 1A shows a contiguous column of residues on the right-hand edge of the helical net for M4, M5, and M6 that locates the lipid binding "face" of these helices. The next step in analysis is to understand how these Ca^{2+} binding faces relate to each other. Site-directed disulfide mapping of amino acids located in the mutation-sensitive patches is being carried out in our laboratory in an effort to understand these relationships.[21] This approach, combined with high resolution analysis of three-dimensional structures,[22] should soon provide interesting structural information.

MUTATIONS IN *ATP2A1*, THE GENE ENCODING SERCA1, ARE ASSOCIATED WITH BRODY DISEASE

Brody disease is a rare inherited disorder of skeletal muscle function. Symptoms include exercise-induced impairment of skeletal muscle relaxation, stiffness, and cramps. Ca^{2+} uptake and Ca^{2+} ATPase activities are reduced in the sarcoplasmic reticulum,[23–29] leading to the prediction that Brody disease results from defects in the *ATP2A1* gene on chromosome 16p12.1–12.2, which encodes SERCA1.[30] In a recent search,[31] however, we did not find any mutations in the *ATP2A1* gene in three Brody patients or in an unpublished study of a fourth Brody patient.

In our more recent study, clinical evaluation identified three individuals in a relatively large family, BD1 (F1), and two individuals in a smaller family, BD2 (F2), with typical signs of Brody disease.[32] In these families, it was possible to carry out haplotype analysis using genetic markers on chromosome 16p. The parents in family 1 shared a common haplotype, and the two patients in family 1 (F1P3 and F1P4) inherited two copies of the common haplotype, one from each parent. The two patients in family 2 (F2P3 and F2P4) inherited the same pair of nonidentical haplotypes from their parents. These data were consistent with autosomal recessive inheritance of Brody disease and suggested that the Brody defect might lie in a region of chromosome 16 containing the *ATP2A1* gene in both families.

We amplified overlapping 2–7 kb DNA sequences spanning the entire *ATP2A1* gene in genomic DNA from F1P3 and F2P3 and used them to sequence all 23 exons, including flanking intron sequences of at least 20 nucleotides, as well as 1 kb of 5' sequence likely to contain the *ATP2A1* promoter. Sequencing of exon 7 in F1P3 and comparison with normal *ATP2A1* sequence revealed the homozygous mutation of C592 to T, resulting in mutation of the Arg^{198} codon CGA to the stop codon TGA.[32] The truncated product, illustrated in FIGURE 2B, would be devoid of the β-strand, phosphorylation, nucleotide binding, and Ca^{2+} binding domains, illustrated in FIGURE 2A.[33] Linkage data for inheritance of the mutation and inheritance of the disease were fully consistent with the recessive inheritance of the C592 to T mutation as the causal factor for Brody disease in family BD1.

Sequencing of amplified genomic DNA revealed two different mutations in *ATP2A1* from F2P3, consistent with inheritance of the two different haplotypes. The mutation of C2025 to A in the TGC codon for Cys^{675} in exon 15 of the paternally inherited chromosome created the stop codon, TGA, predicted to lead to a truncated protein of 674 amino acids (FIG. 2C). The truncated gene product would contain phosphorylation and nucleotide binding domains, but the Ca^{2+} binding domain would be disrupted.[33] The mutation of the invariant GT dinucleotide to CT at the splice donor site of intron 3 in the maternally inherited chromosome was predicted to lead preferentially to skipping of exon 3 and less frequently to partial retention of intron 3.[34] If exon 2 of *ATP2A1* were spliced to exon 4 and transcribed, the product would be truncated, consisting of 45 normal amino acids, followed by 5 novel amino acids. If intron 3 were partially retained and transcribed, the product would also be truncated, consisting of 73 normal amino acids, followed by 49 novel amino acids (FIG. 2D). Thus, both gene products would be missing phosphorylation, nucleotide binding, and Ca^{2+} binding domains.[33]

Linkage data for the inheritance of the two mutations and of Brody disease were fully consistent with the causal nature of the compound heterozygosity for the two *ATP2A1* mutations. These findings, therefore, provide the first genetic evidence that *ATP2A1* is a candidate gene for at least one autosomal recessive form of Brody disease and support the earlier view that the loss of SERCA1 function underlies the manifestation of this form of the disease. The disease is clearly heterogeneous, however, because we have not discovered *ATP2A1* gene defects in four of the six families studied to date.[31,32]

In both family BD1 and BD2, inheritance of *ATP2A1* mutations led to premature stop codons, truncating SERCA1, deleting essential functional domains, and raising the intriguing question: how have these Brody patients partially compensated for the functional "knockout" of a gene product believed to be essential for fast-twitch skeletal muscle relaxation? In a normal muscle contraction-relaxation cycle, Ca^{2+} is released from the sarcoplasmic reticulum to induce contraction and pumped back into the lumen of the sarcoplasmic reticulum by Ca^{2+} ATPases to initiate relaxation.[35]

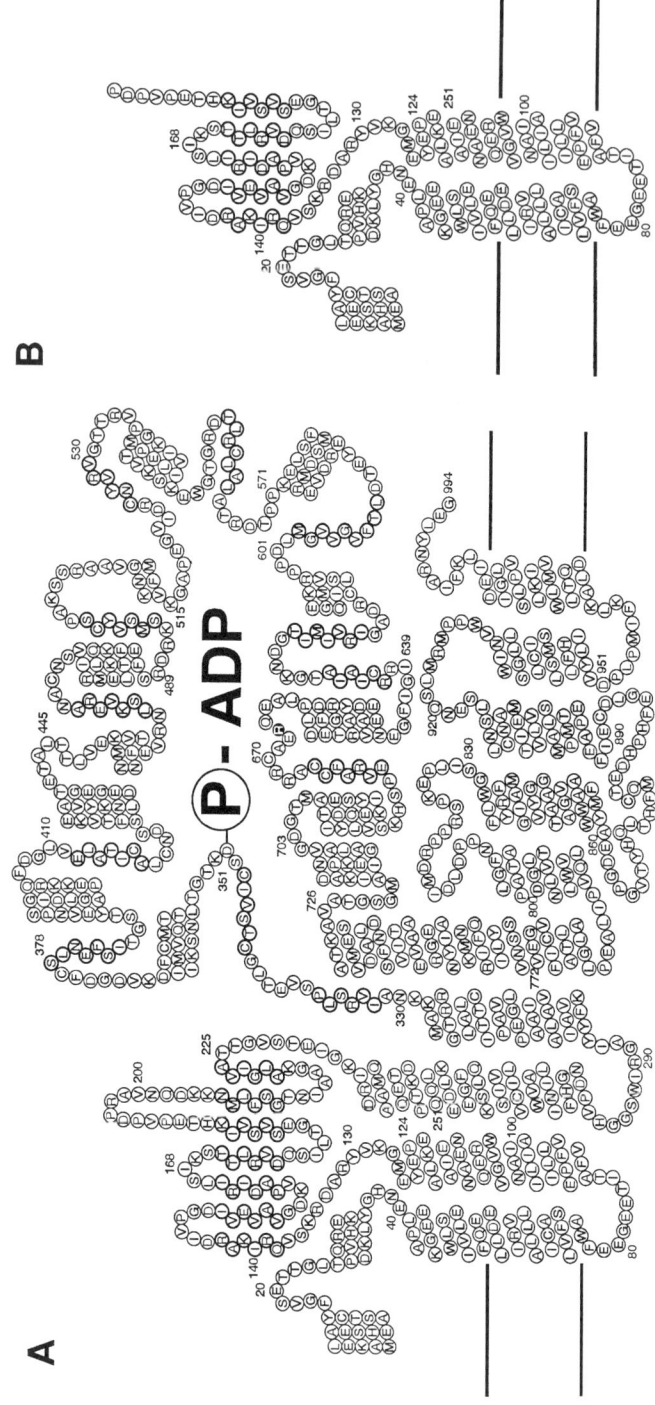

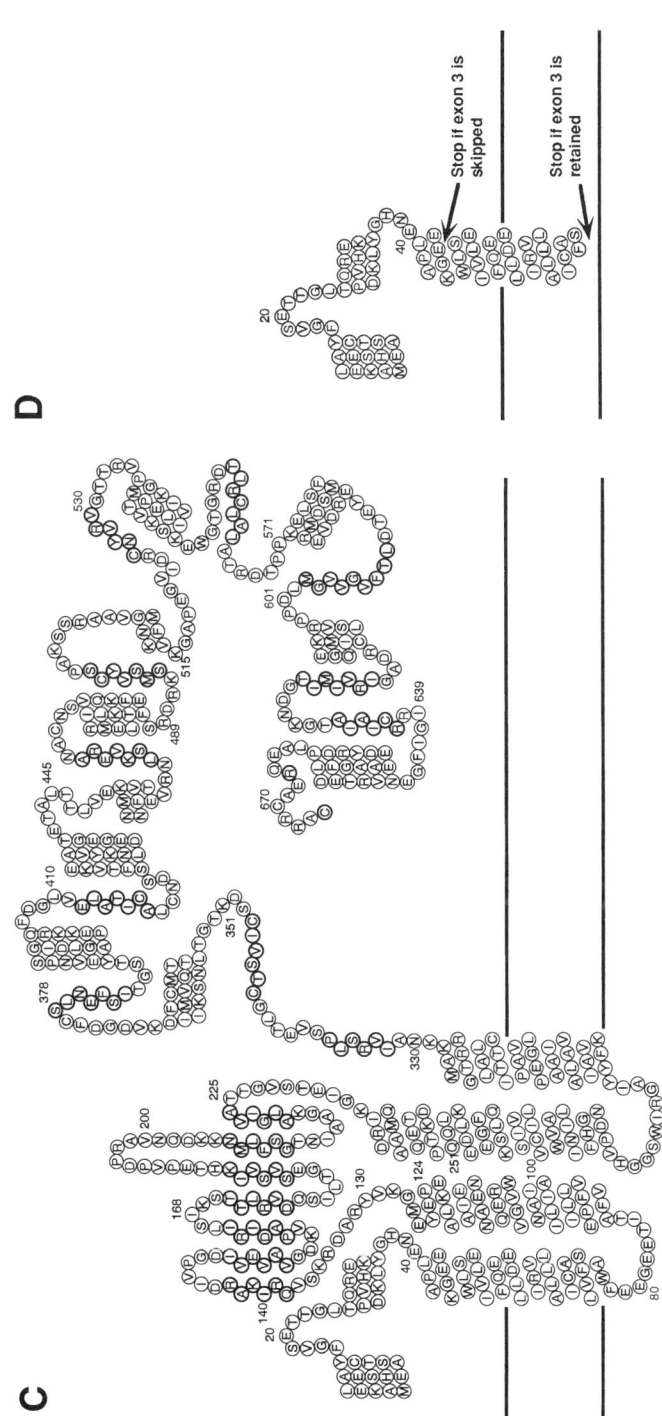

FIGURE 2. A planar circle model of SERCA1a showing the predicted structural features of the wild-type protein (**A**), including 10 transmembrane helices, a cytoplasmic β-strand domain between residues 137 and 225, and cytoplasmic phosphorylation, nucleotide binding, and hinge domains between residues 330 and 725. (**B**) The truncation observed in family BD1 results from the homozygous mutation of C592 to T, leading to replacement of the Arg[198] codon CGA with the stop codon TGA. The two truncations making up compound heterozygosity in Brody patients from family BD2 are: (**C**) mutation of C2025 to A in the TGC codon for Cys[675] in exon 15 of the paternally inherited chromosome creates the stop codon TGA, predicted to lead to a truncated protein of 674 amino acids; and (**D**) mutation of the invariant GT dinucleotide to CT at the splice donor site of intron 3 in the maternally inherited chromosome is predicted to lead preferentially to skipping of exon 3 and less frequently to partial retention of intron 3.[3,4] If exon 2 of *ATP2A1* were spliced to exon 4 and transcribed, the product would be truncated, consisting of 45 normal amino acids, followed by 5 novel amino acids. If intron 3 were partially retained and transcribed, the product would also be truncated, consisting of 73 normal amino acids, followed by 49 novel amino acids. In Brody patients from both family BD1 and BD2, inheritance of *ATP2A1* mutations are predicted to lead to deletion of essential functional domains and to functional knockouts of the SERCA1 protein.

SERCA1[36] accounts for more than 99% of SERCA isoforms expressed in adult fast-twitch skeletal muscle (type 2) fibers.[37] Despite the predicted absence of SERCA1 in the four Brody patients described here, all were able to relax their fast-twitch skeletal muscles, although at a significantly reduced rate. In cultured muscle cells from F1P3 and F1P4 and controls, the sarcoplasmic Ca^{2+} concentration at rest and the increase in intracellular Ca^{2+} concentration after the addition of acetylcholine were found to be the same.[29] However, the time required to reach resting intracellular Ca^{2+} levels after Ca^{2+} release was increased several-fold in the Brody muscle cells.

Ca^{2+} concentrations in Brody muscle might be lowered through compensatory Ca^{2+} removal[38] by plasma membrane Ca^{2+} ATPases (PMCAs), by Na^+/Ca^{2+} exchangers in the plasma membrane, by mitochondrial Ca^{2+} uptake, or by the proliferation of sarcoplasmic or endoplasmic reticulum containing compensating levels of SERCA2[39] or SERCA3[40] isoforms. Of these possible compensatory processes, only the latter would be predicted to result in Ca^{2+} loading of the sarcoplasmic reticulum, a process necessary for subsequent muscle contraction. As an alternative, refilling of Ca^{2+}-depleted sarcoplasmic reticulum might be possible through some form of capacitative Ca^{2+} entry.[41]

ACKNOWLEDGMENTS

We thank our national and international colleagues who participated in the original studies reviewed in this paper and Vijay K. Khanna, Kazimierz Kurzydlowski, and Stella deLeon for their excellent technical assistance.

REFERENCES

1. MacLennan, D. H., D. M. Clarke, T. W. Loo & I. S. Skerjanc. 1992. Acta Physiol. Scand. **607:** 141–150.
2. Clarke, D. M., T. W. Loo, G. Inesi & D. H. MacLennan. 1989. Nature **339:** 476–478.
3. Vilsen, B., J. P. Andersen, D. M. Clarke & D. H. MacLennan. 1989. J. Biol. Chem. **264:** 21024–21030.
4. Clarke, D. M., T. W. Loo & D. H. MacLennan. 1990. J. Biol. Chem. **265:** 6262–6267.
5. Vilsen, B., J. P. Andersen & D. H. MacLennan. 1991. J. Biol. Chem. **266:** 18839–18845.
6. Andersen, J. P., B. Vilsen & D. H. MacLennan. 1992. J. Biol. Chem. **267:** 2767–2774.
7. Clarke, D. M., T. W. Loo, W. J. Rice, J. P. Andersen, B. Vilsen & D. H. MacLennan. 1993. J. Biol. Chem. **268:** 18359–18364.
8. Andersen, J. P. 1995. J. Biol. Chem. **270:** 908–914.
9. Zhang, Z., C. Sumbilla, D. Lewis, S. Summers, M. G. Klein & G. Inesi. 1995. J. Biol. Chem. **270:** 16283–16290.
10. Chen, L., C. Sumbilla, D. Lewis, L. Zhong, C. Stock, M. E. Kirtley & G. Inesi. 1996. J. Biol. Chem. **271:** 10745–10752.
11. Maruyama, K. & D. H. MacLennan. 1988. Proc. Natl. Acad. Sci. USA **85:** 3314–3318.
12. Maruyama, K., D. M. Clarke, J. Fujii, G. Inesi, T. W. Loo & D. H. MacLennan. 1989. J. Biol. Chem. **264:** 13038–13042.
13. Clarke, D. M., T. W. Loo & D. H. MacLennan. 1990. J. Biol. Chem. **265:** 22223–22227.
14. de Meis, L. & A. L. Vianna. 1979. Annu. Rev. Biochem. **48:** 275–292.
15. MacLennan, D. H., C. J. Brandl, B. Korczak & N. M. Green. 1985. Nature **316:** 696–700.
16. MacLennan, D. H. 1990. Biophys. J. **58:** 1355–1365.
17. Rice, W. R. & D. H. MacLennan. 1996. J. Biol. Chem. **271:** 31412–31419.

18. ANDERSEN, J. P. & B. VILSEN. 1994. J. Biol. Chem. **269:** 15931–15936.
19. ANDERSEN, J. P. & B. VILSEN. 1992. J. Biol. Chem. **267:** 19383–19387.
20. SKERJANC, I. S., T. TOYOFUKU, C. RICHARDSON & D. H. MACLENNAN. 1993. J. Biol. Chem. **268:** 15944–15950.
21. RICE, W. J. & D. H. MACLENNAN. 1996. Biophys. J. **70:** A139 (abstract).
22. TOYOSHIMA, C., H. SASABE & D. L. STOKES. 1993. Nature **362:** 469–471.
23. BRODY, I. A. 1969. N. Engl. J. Med. **281:** 187–192.
24. KARPATI, G., J. CHARUK, S. CARPENTER, C. JABLECKI & P. HOLLAND. 1986. Ann. Neurol. **20:** 38–49.
25. DANON, M. J., G. KARPATI, J. CHARUK & P. P. HOLLAND. 1988. Neurology **38:** 812–815.
26. TAYLOR, D. J., M. J. BROSNAN, D. L. ARNOLD, P. J. BORE, P. STYLES, J. WALTON & G. K. RADDA. 1988. J. Neurol. Neurosurg. Psychiatry **51:** 1425–1433.
27. WEVERS, R. A., P. J. E. POELS, E. M. G. JOOSTEN, G. G. H. STEENBERGEN, A. A. G. M. BENDERS & J. H. VEERKAMP. 1992. J. Inher. Metab. Dis. **15:** 423–425.
28. POELS, P. J. E., R. A. WEVERS, J. P. BRAAKHEKKE, A. A. G. M. BENDERS, J. H. VEERKAMP & E. M. G. JOOSTEN. 1993. J. Neurol. Neurosurg. Psychiatry **56:** 823–826.
29. BENDERS, A. A. G. M., J. H. VEERKAMP, A. OOSTERHOF, P. J. H. JONGEN, R. J. M. BINDELS, L. M. E. SMIT, H. F. M. BUSCH & R. A. WEVERS. 1994. J. Clin. Invest. **94:** 741–748.
30. CALLEN, D. F., S. A. LANE, H. KOZMAN, G. KREMMIDIOTIS, S. A. WHITMORE, M. LOWENSTEIN, N. A. DOGGETT, N. KENMOCHI, D. C. PAGE, D. R. MAGLOTT, W. C. NIERMAN, K. MURAKAWA, R. BERRY, J. M. SIKELA, R. HOULGATTE, C. AUFFRAY & G. R. SUTHERLAND. 1995. Genomics **29:** 503–511.
31. ZHANG, Y., J. FUJII, M. S. PHILLIPS, H.-S. CHEN, G. KARPATI, W.-C. YEE, B. SCHRANK, D. R. CORNBLATH, K. B. BOYLAN & D. H. MACLENNAN. 1995. Genomics **30:** 415–424.
32. ODERMATT, A., P. E. M. TASCHNER, V. K. KHANNA, H. F. M. BUSCH, G. KARPATI, C. K. JABLECKI, M. J. BREUNING & D. H. MACLENNAN. 1996. Nature Genet. **14:** 191–194.
33. MACLENNAN, D. H., D. M. CLARKE, T. W. LOO & I. SKERJANC. 1992. Acta Physiol. Scand. **146:** 141–150.
34. KRAWCZAK, M., J. REISS & D. N. COOPER. 1992. Hum. Genet. **90:** 41–54.
35. EBASHI, S., M. ENDO & I. OHTSUKI. 1969. Q. Rev. Biophys. **2:** 351–384.
36. BRANDL, C. J., N. M. GREEN, B. KORCZAK & D. H. MACLENNAN. 1986. Cell **44:** 597–607.
37. WU, K-D., W-F. LEE, J. WEY, D. BUNGARD & J. LYTTON. 1995. Am. J. Physiol. **269:** C775–C784.
38. CARAFOLI, E. 1987. Ann. Rev. Biochem. **56:** 395–433.
39. LYTTON, J. & D. H. MACLENNAN. 1988. J. Biol. Chem. **263:** 15024–15031.
40. BURK, S. E., J. LYTTON, D. H. MACLENNAN & G. E. SHULL. 1989. J. Biol. Chem. **264:** 18561–18568.
41. BERRIDGE, M. J. 1995. Biochem. J. **312:** 1–11.

Are Pyridoxal and Fluorescein Probes in Lysine Residues of α-Chain in Na^+,K^+-ATPase Sensing ATP Binding?[a]

TAKEO TSUDA, SHUNJI KAYA, TAKESHI YOKOYAMA, AND
KAZUYA TANIGUCHI[b]

*Biological Chemistry
Graduate School of Science
Hokkaido University
Sapporo 060, Japan*

Transport of sodium and potassium ions coupled with hydrolysis of ATP is performed by Na^+,K^+-ATPase,[1–5] which has high- and low-affinity ATP binding sites. To understand the mechanism of energy transduction in the enzyme, ATP-induced conformational changes were studied with an N-(p-(2-benzimidazolyl) phenyl) maleimide (BIPM) probe[6–10] at Cys-964, which is in or near[8] the COOH-terminal transmembrane segment.[11] Pyridoxal probes were used as affinity reagents for ATP binding sites in ATP-requiring enzymes. The fluorescein 5′-isothiocyanate (FITC) probe was also used to inhibit P-type ATPase activities resulting from loss of high-affinity ATP binding.[12–15] Introduction of pyridoxal 5′-diphospho-5′-adenosine (AP_2PL) and FITC probes into Lys-480[10,16] and Lys-501,[13,15] respectively, decreases Na^+,K^+-ATPase activity without influencing K^+-dependent p-nitrophenylphosphatase activity or the phosphorylation capacity from acetyl phosphate (AcP) in the presence of Mg^{2+} and Na^+.[6,14]

Na^+,K^+-ATPase preparations modified with AP_2PL probes at Lys-480 in the α-chain showed dynamic fluorescence changes accompanying formation of E_1P and E_2P by AcP and possibly a Mg-Na-ATP enzyme complex formed by ATP.[10] However, neither the relationship between those dynamic fluorescence changes of pyridoxal probes and the remaining Na^+,K^+-ATPase activity nor conformational changes induced by ATP binding have been studied in detail.

In this paper, the remaining ATPase activity was further reduced to below the 5% level by FITC treatment. The resulting AP_2PL-FITC doubly labeled enzyme preparations contained ~0.5 mol AP_2PL and ~0.9 mol FITC probe/mol α-chain. The preparations showed dynamic fluorescence changes of the AP_2PL probe accompanying ATP binding to high- and low-affinity sites and of the FITC probe accompanying ATP binding to a low-affinity site. The data also suggested the presence of oligomeric interactions between α subunits in Na^+,K^+-ATPase and the simultaneous presence of low-affinity ATP binding sites.

---

[a]This work was supported in part by Grants-in-Aid for Scientific Research (06454648 and 07558220) and for International Scientific Research Program (07044049 and 08044047) from the Ministry of Education, Science and Culture of Japan.

[b]To whom correspondence should be addressed. Tel: 81-11-706-2698; fax: 81-11-736-2074; e-mail: S.K.-kayan@S1.hines.hokudai.ac.jp; K. T.-KTAN@hucc.hokudai.ac.jp

MODIFICATION OF Na$^+$,K$^+$-ATPase BY PLP OR AP$_2$PL

Na$^+$,K$^+$-ATPase preparations from pig kidneys were treated with 0–200 μM PLP or AP$_2$PL in the presence of 5 mM Mg^{2+} and 25 mM Na$^+$, and the resulting schiff-base was reduced with 4 mM NaBH$_4$.[10] Na$^+$,K$^+$-ATPase activity decreased linearly to the 50% level when the amount of the probe was increased from 0.1–0.5 mol pyridoxal probe/mol α-chain; the phosphorylation capacity from acetyl phosphate was little affected. Further increases in the amounts of the probes produced only a slight decrease in activity. The addition of 50 μM AP$_2$PL to enzyme preparations pretreated with 50 μM AP$_2$PL, containing 0.5 mol of AP$_2$PL/mol α-chain, had little influence on the ATPase activity. Lys-480 was shown to be the binding site of PLP or AP$_2$PL probes.

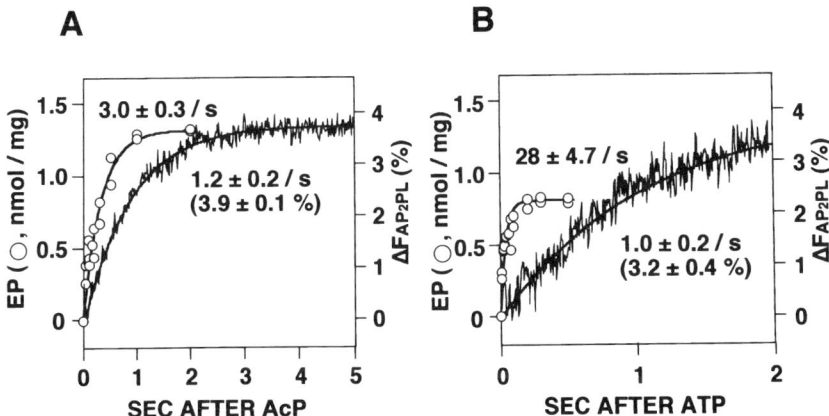

FIGURE 1. Time course of increase in AP$_2$PL fluorescence and phosphorylation. Stopped-flow and rapid-quenching experiments in the presence of 25 mM imidazole-HCl (pH 7.4), 4 mM MgCl$_2$, 16 mM NaCl, and enzyme preparations pretreated with 50 μM AP$_2$PL for 10 minutes (pH 7.8) were performed by adding 1 mM acetyl phosphate (AcP) or 10 μM ATP.[10] Traces of each ratio of accumulated data of AP$_2$PL fluorescence change induced by 1 mM AcP to that by 1 mM Tris-HCl (**A**) or 10 μM ATP to that by 10 μM Tris-HCl (**B**) are shown. Time courses of 1 mM [^{32}P]AcP- (*open circles* in **A**) and 10 μM [^{32}P]ATP-induced phosphorylation (*open circles* in **B**) of the AP$_2$PL-modified enzyme are shown. The *smooth lines* and rate constants in the figures were obtained from a single exponential curve fitting.

FLUORESCENCE CHANGE OF AP$_2$PL PROBE AND PHOSPHORYLATION

To compare the rates of fluorescence changes and phosphorylation, stopped-flow and rapid-quenching experiments were performed in the presence of 4 mM Mg^{2+} and 16 mM Na$^+$. The addition of 1 mM AcP to the AP$_2$PL-modified enzyme induced phosphorylation (3/s) to accumulate 1.3 nmol phosphoenzyme per milligram of protein followed by an increase (1.2/s) in fluorescence intensity (3.9%) (FIG. 1A). Neither increased fluorescence nor phosphorylation occurred in the absence of Mg^{2+}. These data seemed consistent with the idea that AcP-dependent phosphorylation induced the change in fluorescence. The addition of 10 μM ATP induced rapid phosphorylation (28/s) to accumulate 0.8 nmol phosphoenzyme per milligram of protein, a 60% level of phosphorylation from AcP, which was followed by a slow increase (1.0/s) in fluorescence intensity to give a level of 85% of that obtained from AcP (FIG. 1B). The data suggested that part of this change in fluorescence was not directly related to phosphorylation from ATP. As a control experiment, 10 μM ATP was added to the enzyme preparation not treated with AP_2PL. It induced more rapid phosphorylation (60/s) to accumulate 1.5 nmol phosphoenzyme/mg protein.

FITC MODIFICATION OF AP_2PL-MODIFIED ENZYME

To investigate the relation between the rapid ATP-induced phosphorylation (28/s) and the very slow ATP-dependent increase in AP_2PL fluorescence (1.0/s), AP_2PL-modified enzyme preparations were further treated with 5, 10, or 15 μM FITC to inhibit phosphorylation from ATP.[21] The amount of phosphoenzyme from ATP in the presence of Mg^{2+} and Na^+ and Na^+,K^+-ATPase activity was reduced strongly from 50% to 5% with increasing concentrations of FITC. By contrast, the amount of phosphoenzyme from AcP in the presence of Mg^{2+} and Na^+ was not significantly influenced by the FITC treatment. Three different 50 μM AP_2PL-treated enzyme preparations were further treated with 15 μM FITC, which contained 0.53 ± 0.09 mol AP_2PL probe/mol α-chain and 0.88 ± 0.14 mol FITC probe/mol α-chain. The emission spectra of a typical preparation excited at 320 (AP_2PL) and 470 nm (FITC) are shown, the maximum being 390 (FIG. 2) and 520 nm (FIG. 2 and inset) for AP_2PL and FITC fluorescence, respectively.

To investigate the effect of FITC treatment on the AP_2PL fluorescence change, 1 mM AcP or 10 μM ATP was added to AP_2PL-FITC doubly labeled enzyme in the presence of 4 mM Mg^{2+} and 16 mM Na^+. FITC treatment had little influence on the extent and the rate of AP_2PL fluorescence change induced by 1 mM AcP (~1.3/s) or 10 μM ATP (~1.1/s). These data clearly showed that a large part of the increase in AP_2PL fluorescence induced by ATP occurred independent of phosphorylation from ATP.

ATP BINDING TO A HIGH-AFFINITY SITE IN AP_2PL-FITC LABELED ENZYME

It has already been reported that Na^+-bound enzymes accept ATP with high affinity.[1-5] To estimate the apparent affinity for ATP to induce AP_2PL fluorescence change of the AP_2PL-FITC doubly labeled enzyme in the presence of 4 mM Mg^{2+} and 16 mM Na^+, various concentrations (1 ~ 50 μM) of ATP were added. The addition of 1 or 3 μM ATP induced a biphasic fluorescence change, that is, a rapid decrease followed by a slow increase. With increasing concentrations of ATP, the rapid decrease disappeared and seemed to be overcome by a slow increase. The apparent first-order rate constants of the rapid decrease and the slow increase in AP_2PL fluorescence were increased with increasing concentrations of ATP. However, the final extent of the increase in fluorescence was near saturation at around 10 μM ATP. A similar fluorescence increase of the enzyme induced by 10 μM ATP in the presence of 4 mM Mg^{2+} and 16 mM Na^+ was obtained from steady-state measurements. These dynamic

AP$_2$PL fluorescence changes required the simultaneous presence of Mg^{2+}, Na$^+$, and ATP and also occurred in the AP$_2$PL-modified enzyme without FITC treatment. The addition of 10 μM ATP to the AP$_2$PL–FITC-labeled enzyme under the aforementioned conditions did not result in any detectable FITC fluorescence change as in the case of BIPM-FITC doubly labeled enzyme.[14] The data showed that the AP$_2$PL probe, but not the FITC probe, in the enzyme preparations could sense high-affinity ATP binding. The data suggested that ATP binding to a high-affinity site on Na$^+$-bound enzymes induced a sequential change in AP$_2$PL fluorescence, a decrease followed by an increase, or that ATP binding to two high-affinity sites induced parallel changes in fluorescence, that is, a rapid decrease and a slow increase.

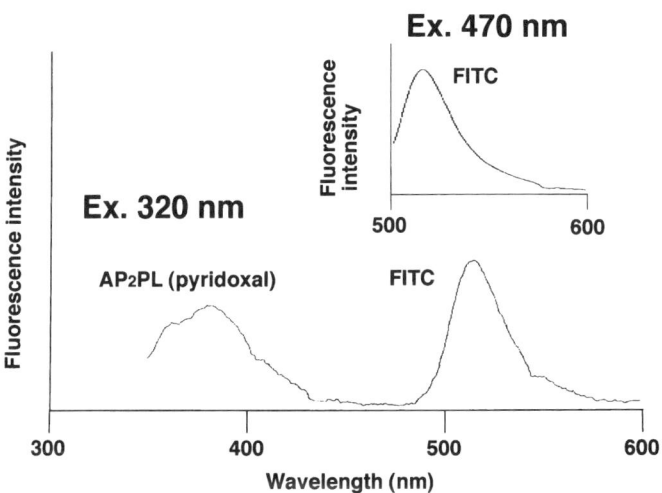

FIGURE 2. Emission spectra of AP$_2$PL-FITC doubly labeled enzyme. The Na$^+$,K$^+$-ATPase preparation pretreated with 50 μM AP$_2$PL as described in FIGURE 1 was further treated with 15 μM FITC (pH 9.2) for 30 minutes. Emission spectra were run at a fixed wave length at 320 or 470 nm (**inset**).

ATP BINDING TO LOW-AFFINITY SITES IN AP$_2$PL-FITC LABELED ENZYME

It has already been reported that K$^+$-bound enzymes accept ATP with low affinity.[2–5] The presence of low-affinity ATP binding to the FITC-modified enzyme[17,18] or vanadate-inhibited enzyme[19] has been reported. To investigate whether the AP$_2$PL-FITC doubly labeled enzyme accepted ATP with low affinity, both AP$_2$PL and FITC fluorescence changes followed the addition of 160 mM Na$^+$ with various concentrations of ATP to the doubly labeled enzyme in the presence of 4 mM Mg^{2+} and 1.6 mM K$^+$. The addition of Na$^+$ without ATP induced a 1.5% monophasic decrease in the AP$_2$PL fluorescence with a rate constant of 3.6/s. The addition of Na$^+$ with ~milli-

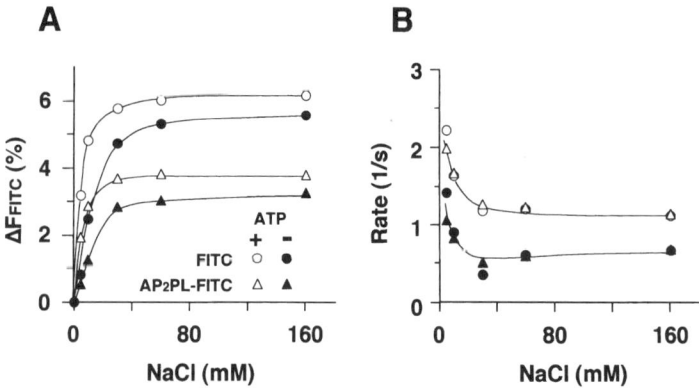

FIGURE 3. Effect of Na^+ concentration on low-affinity ATP binding to K^+-bound AP_2PL-FITC doubly labeled enzyme. Stopped-flow experiments were performed as described in FIGURE 1 except that the solution contained 1.6 mM KCl instead of 16 mM NaCl, and Na^+,K^+-ATPase preparations were AP_2PL-FITC doubly labeled enzymes (*open and closed triangles*) as described in FIGURE 2 or enzymes treated with 15 μM FITC (pH 9.2) for 30 minutes (*open and closed circles*) and that various concentrations of Na^+ were added with 5 mM ATP (*open symbols*) or Tris-HCl (*closed symbols*) to start the reaction. Relative fluorescence intensity (**A**) and the apparent rate constant (**B**) are shown.

molar ATP induced a biphasic change, that is, a rapid decrease followed by a slow increase. The rate constant of each AP_2PL fluorescence change was increased with increasing concentrations of ATP. Data showed that the AP_2PL fluorescence change was accelerated by ATP binding to a low-affinity site in the K^+-bound AP_2PL-FITC doubly labeled enzyme as already shown by Trp fluorescence change in nonmodified Na^+,K^+-ATPase.[20]

The decrease of fluorescence induced by 5 mM ATP was too fast to detect by stopped flow measurement. These data suggested that ATP binding to a low-affinity site of K^+-bound enzymes induced a sequential change in AP_2PL fluorescence, that is, a rapid decreased followed by a slow increase, or that the binding to low-affinity sites induced parallel changes in fluorescence as just described in the case of ATP binding with high affinity to Na^+-bound enzymes.

The addition of Na^+ increased the FITC fluorescence (4.0%) with an apparent rate constant of 1.1/s. The addition of Na^+ with increasing concentrations of ATP slightly increased both the extent and the rate to give maximum values of 4.7% and 1.5/s, respectively, with $K_{0.5}$ values around submillimeters. Similar dynamic fluorescence changes of both probes in the AP_2PL-FITC doubly labeled enzyme also occurred when Mg^{2+} was omitted or 5 mM ATP was replaced with ADP. The addition of ATP without Na^+ to the doubly labeled enzyme induced no change in either the AP_2PL or the FITC fluorescence.

Similar FITC fluorescence changes were observed with FITC-modified enzyme not pretreated with AP_2PL, except that the extent of the fluorescence increased to ~1.7-fold higher in the presence of up to 30 mM Na^+ (FIG. 3A). The rate constants of the FITC fluorescence change in the presence of the same Na^+ were nearly the same independent of AP_2PL modification (FIG. 3B). The apparent affinity for Na^+ also seemed not to be influenced by AP_2PL modification and was ~5 and ~15 mM, respectively, with and without ATP.

OLIGOMERIC NATURE OF THE ENZYME

Biphasic AP_2PL fluorescence changes after the addition of ~micromolar ATP to the Na^+ bound enzyme and of ~millimolar ATP with Na^+ to the K^+-bound enzyme seem to be explained as sequential conformational events to form enzyme-ATP complexes followed by some rearrangements. However, these might be related to the oligomeric nature of the enzyme to be discussed.

The functional unit of the enzyme is reported to be an $\alpha\beta$ protomer[18,21,22] or a diprotomer $(\alpha\beta)_2$ including protomer-diprotomer interaction between them[23–25] or a higher-order oligomer[9,26,28] (see also references in ref. 9), which are presented in FIGURE 4 as several hypothetical models to explain the present data.

There seemed to be at least two possible explanations for the 50% Na^+,K^+-ATPase activity that remained after AP_2PL treatment. One is that 50% of the enzyme molecules were completely unaffected by AP_2PL (model A), but this seems to conflict with the data showing that repeated treatment with AP_2PL had little further influence on the Na^+,K^+-ATPase activity. The other is that half the sites were modified to completely diminish both phosphorylation from ATP and Na^+,K^+-ATPase activity (model B). The modification had little influence on the phosphorylation capacity from ATP and Na^+,K^+-ATPase activity in adjacent unmodified subunits, although the rate of phosphorylation in the presence of 10 μM ATP was reduced by half. These considerations seemed to favor model B.

It was already shown that ~micromolar ATP induced neither BIPM nor FITC fluorescence changes in the BIPM-FITC doubly labeled enzyme,[14] which suggested

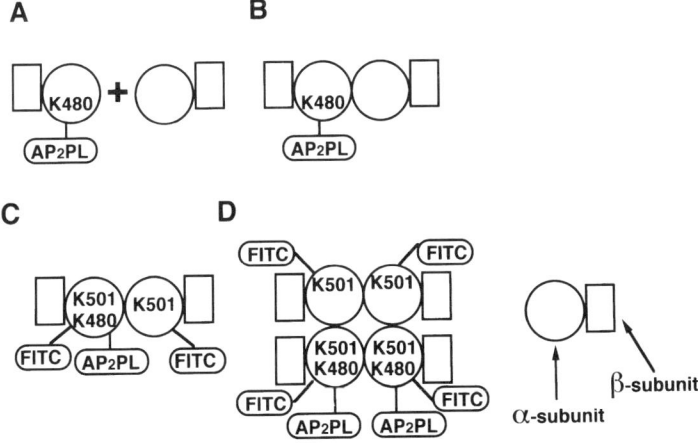

FIGURE 4. Hypothetical models of oligomeric Na^+,K^+-ATPase. Simple hypothetical oligomer models as a protomer (**A**), a diprotomer (**B,C,** and **D**), or a tetramer (**E**) are presented to explain data obtained from the AP_2PL-FITC doubly labeled membrane-bound Na^+,K^+-ATPase preparations.

that high-affinity ATP binding disappeared because of the FITC modification.[12,13] However, FITC-induced inhibition of high-affinity ATP binding was completely protected by AP_2PL probes at Lys-480. Did the modification by AP_2PL and FITC occur in the same α-chain? The addition of ~millimeters ATP with Na^+ to K^+-bound enzymes induced not only a dynamic biphasic change in AP_2PL fluorescence but also an increased FITC fluorescence. These data indicate that each probe sensed each low-affinity ATP binding. It is unlikely that low-affinity ATP binding to a nonmodified half site showed different concentration dependence of ATP on changes in the microenvironments of different probes present near the ATP binding domain (model C). These considerations seemed to favor model D. However, this dimeric model, $(\alpha\beta)_2$, seems insufficient to explain recently reported data[9,27,28] or the amounts of probes/α-chain in the present experiments. These considerations may indicate the possibility of a tetrameric model, $(\alpha\beta)_4$, as shown in the AP_2PL-FITC doubly labeled enzyme having four FITC-modified and two AP_2PL-modified, α-chains (model E). However, it remains unclear whether high- and low-affinity ATP binding sites reside in the same or a different subunit containing the AP_2PL probe at one time or changes alternatively and whether the low-affinity ATP binding site detected by FITC fluorescence resides in the same subunit that contains both probes of AP_2PL and FITC.

SUMMARY

Na^+,K^+-ATPase preparations from pig kidneys were treated with 50 μM pyridoxal 5′-diphospho-5′-adenosine (AP_2PL) in the presence of NaCl. The resulting preparations contained 0.5 mol of the AP_2PL probe at the Lys-480/mol α-chain. This modification reduced both Na^+,K^+-ATPase activity and the amount of Na^+-dependent phosphoenzyme from ATP to around 50% but not that from acetyl phosphate (AcP). The addition of 1 mM AcP to the modified enzyme in the presence of Mg^{2+} and Na^+ induced phosphorylation (3.0/s) followed by an AP_2PL fluorescence increase (1.2/s). The addition of 10 μM ATP instead of AcP induced rapid phosphorylation (28/s) followed by a slow increase in fluorescence (1.0/s). When modified enzyme preparations were treated with fluorescein 5′-isothiocyanate (FITC), the phosphorylation capacity from ATP was reduced to around 5% with little influence on either the AP_2PL fluorescence change by ATP or phosphorylation from AcP. The addition of increasing concentrations of ATP with 160 mM NaCl to the K^+-bound AP_2PL-FITC-labeled enzyme showed different rates for each fluorescence change and different affinities for ATP of the changes. These data and others indicate that the AP_2PL probe at Lys-480 can monitor ATP binding to high- and low-affinity sites and suggest the simultaneous presence of two different low-affinity sites for ATP detected by an AP_2PL probe at Lys-480 and an FITC probe at Lys-501.

ACKNOWLEDGMENT

The authors would like to thank Dr. Y. Hayashi for helpful suggestions in analyzing FITC probe in the α-chain.

REFERENCES

1. POST, R. L., S. KUME, T. TOBIN, B. ORCUTT & A. K. SEN. 1969. J. Gen. Physiol. **54**: 306S–326S.
2. ALBERS, R. W. 1976. *In* The Enzymes of Biological Membranes. A. Martonossi, Ed **3**: 283–301. Plenum Publishing Corp. New York.
3. GLYNN, I. M. 1985. *In* The Enzymes of Biological Membranes. A. Martonossi, Ed. **3**: 35–114. Plenum Publishing Corp. New York.
4. REPKE, K. R. H. & R. SCHON. 1992. Biol. Rev. **67**: 31–78.
5. ROBINSON, J. D. & P. R. PRATAP. 1993. Biochim. Biophys. Acta **1154**: 83–104.
6. TANIGUCHI, K. & S. MÅRDH. 1993. J. Biol. Chem. **268**: 15588–15594.
7. TANIGUCHI, K., D. KAI, S. INOUE, E. SHINOGUCHI, K. SUZUKI, Y. NAKAMURA, Y. ADACHI & S. KAYA. 1994. *In* The Sodium Pump. E. Bamberg & W. Schowner, Eds.: 581–592. Steinkopff Darmstadt & Springer. New York.
8. NAKAMURA, Y., D. KAI, S. KAYA, Y. ADACHI & K. TANIGUCHI. 1994. J. Biochem. (Tokyo) **115**: 454–462.
9. YAMAZAKI, A., S. KAYA, T. TSUDA, Y. ARAKI, Y. HAYASHI & K. TANIGUCHI. 1994. J. Biochem. (Tokyo) **116**: 1360–1369.
10. KAYA, S., T. TSUDA, K. HAGIWARA, T. FUKUI & K. TANIGUCHI. 1994. J. Biol. Chem. **269**: 7419–7422.
11. SCHULL, G. E., A. SCHWARTS & J. B. LINGREL. 1985. Nature **316**: 691–695.
12. KARLISH, S. J. D. 1980. J. Bioenerg. Biomembr. **12**: 111–136.
13. FARLEY, R. A., C. M. TRAN, C. T. CARILLI, D. HAWKE & J. E. SHIVELY. 1984. J. Biol. Chem. **259**: 9532–9535.
14. TANIGUCHI, K., H. TOSA, K. SUZUKI & Y. KAMO. 1988. J. Biol. Chem. **263**: 12943–12947.
15. XU, K. 1989. Biochemistry **28**: 5764–5772.
16. HINZ, H. R. & T. L. KIRLEY. 1990. J. Biol. Chem. **265**: 10260–10265.
17. SCHEINER-BOBIS, G., J. ANTONIPILLAI & R. A. FARLEY. 1993. Biochemistry **32**: 9592–9599.
18. WARD, D. G. & J. D. CAVIERES. 1996. J. Biol. Chem. **271**: 12317–12321.
19. HUANG, W.-H. & A. ASKARI. 1984. J. Biol. Chem. **259**: 13287–13291.
20. KARLISH, S. J. D. & D. W. YATES. 1978. Biochim. Biophys. Acta **527**: 115–130.
21. JØRGENSEN, P. L. & J. P. ANDERSEN. 1986. Biochemistry **25**: 2889–2897.
22. WARD, D. G. & J. D. CAVIERES. 1993. Proc. Natl. Acad. Sci. USA **90**: 5332–5336.
23. BUXBAUM, E. & W. SCHONER. 1991. Eur. J. Biochem. **195**: 407–419.
24. MIMURA, K., H. MATSUI, T. TAKAGI & Y. HAYASHI. 1993. Biochim. Biophys. Acta **1145**: 63–74.
25. GANJEIZADEH, M., N. ZOLOTARJOVA, W.-H. HUANG & A. ASKARI. 1995. J. Biol. Chem. **270**: 15707–15710.
26. HAMER, E. & W. SCHONER. 1993. Eur. J. Biochem. **213**: 743–748.
27. PELUFFO, R. D., P. J. GARRAHAN & A. REGA. 1992. J. Biol. Chem. **267**: 6596–6601.
28. FENDLER, K., S. JARUSCHEWSKI, A. HOBBS, W. ALBERS & J. P. FROEHLICH. 1993. J. Gen. Physiol. **102**: 631–666.

Cation and Cardiac Glycoside Binding Sites of the Na,K-ATPase[a]

JERRY B LINGREL,[b] JOSÉ M. ARGÜELLO,[c]
JAMES VAN HUYSSE, AND THERESA A. KUNTZWEILER

Department of Molecular Genetics, Biochemistry and Microbiology
University of Cincinnati College of Medicine
231 Bethesda Ave, PO Box 670524
Cincinnati, Ohio 45267-0524

The Na,K-ATPase transports sodium and potassium ions across the eukaryotic plasma membrane against their ionic gradients at the expense of ATP. This action is critical for a variety of physiological processes such as the secondary transport of other ions and solutes, the regulation of cell volume and osmotic pressure, and the establishment of membrane potential. In addition to its role in cell physiology, Na,K-ATPase is the receptor for cardiac glycosides. Understanding the mechanisms of enzyme action presents a major challenge and requires the identification of sites that bind Na^+, K^+, ATP, and cardiotonic steroids as well as those involved in the hydrolysis of ATP. To completely understand the enzyme mechanism, the three-dimensional protein structure must be determined for all of the conformational intermediates of the catalytic cycle. However, at this point in time, direct methods for determining these structures are not available. Therefore, mutagenesis studies that probe the interactions of the Na,K-ATPase with its ligands are being employed by our laboratory as well as others to obtain useful information on the relation between binding and catalysis (reviewed in refs. 1 and 2). The resulting structural information will aid in the comprehension of the enzyme mechanism. The general approach has been to replace amino acids in the α subunit, the major catalytic subunit of the Na,K-ATPase, and characterize the altered enzymes by expressing them in mammalian cells. To date, we have identified several residues that are involved in K^+, ouabain, and ATP binding.

METHODS

To identify residues that are important in cation transport, substitutions are introduced into the ouabain-insensitive sheep RD α1 isoform of the Na,K-ATPase, and the recombinant protein is expressed in HeLa cells which carry a ouabain-sensitive endogenous enzyme. The sheep α1 isoform used in these studies (sheep α1 RD) carries two amino acid substitutions in the H1-H2 region which makes the enzyme 1,000-fold less sensitive to ouabain than does the wild-type sheep α1

---

[a]This work was supported by grants HL41496, HL28573, and HL03373 from the National Institutes of Health.

[b]To whom correspondence should be addressed. Tel: 513 558-5324; fax: 513 558-1190; e-mail: Lingrejb@uc.edu

[c]Current address: Worcester Polytechnic Institute, Department of Chemistry and Biochemistry, 100 Institute Rd, Worcester, MA 01609-2280.

subunit.[3] Therefore, HeLa cells expressing this exogenous protein should be resistant to ouabain in the growth medium. An amino acid substitution of interest is made in this protein, and if the recombinant enzyme confers ouabain resistance to the HeLa cells, then this substitution does not prevent enzyme function. Thus, the side chain of the replaced residue is not absolutely required for enzymatic activity. However, through kinetic characterization of the ouabain-resistant ATPase activity, the role of the replaced residues can be determined. In contrast, those substitutions that do not allow the RD sheep α1 isoform to confer resistance must involve a residue that is important for enzyme function or processing. However, these studies do not indicate what feature is altered by the substitution (i.e., protein synthesis or catalytic activity).

Amino acid residues involved in altering ouabain sensitivity are also identified using a variation of this selection scheme. In this case either individual or random base substitutions are introduced into a cDNA which codes for a ouabain-sensitive sheep α1 cDNA and the modified subunit expressed in HeLa cells. If the amino acid substitution prevents the enzyme from being inhibited by cardiac glycosides, the transfected cells will grow in the presence of ouabain. This indicates that the targeted amino acid residue is a key element in ouabain inhibition.

Substitutions that prevent the Na,K-ATPase from providing sufficient ion transport to maintain cell viability must be studied by an approach that does not depend on the activity of the exogenous protein. Base changes encoding these inactivating substitutions are introduced into a cDNA for a ouabain-sensitive enzyme, that is, the wild-type sheep α1 isoform. The mutant cDNA is cotransfected along with a gene encoding resistance to the antibiotic neomycin. Selection with neomycin makes it possible to isolate clonal lines of cells expressing the inactive enzyme, inasmuch as about 25% of the neomycin-resistant cells coexpress the recombinant Na,K-ATPase along with the neomycin resistance protein. The recipient cells for the cotransfections are NIH 3T3 cells which have an endogenous Na,K-ATPase with a low affinity for cardiac glycosides. The exogenous enzyme has approximately 1,000-fold greater sensitivity for ouabain, and thus the binding of [^{3}H]ouabain to the recombinant Na,K-ATPase can be examined without interference from the endogenous pump. The kinetics of ouabain binding depend on the different conformational states that the Na,K-ATPase can assume.[4–7] Because the conformational status is determined by the ligands associated with the protein, ouabain binding can be used as an indirect sensor of such ligand-enzyme interactions.

CATION BINDING SITES

Because the structure of only a few sodium or potassium ion binding sites is known and because no such binding sites in active transport proteins have been fully characterized, it is possible that several types of residues might form the cation binding pocket of the protein. Various functional groups represent potential candidates for forming the binding sites for sodium and potassium ions, such as carboxyl, hydroxyl, backbone peptide carbonyl groups, and possibly aromatic rings. Chemical modification studies and electrophysiology measurements suggest that two carboxyl residues within the transmembrane domains are essential for cation occlusion in the Na, K-ATPase.[8–14] With this in mind, each of the carboxyl-containing residues present in the transmembrane regions of the α subunit have been evaluated in terms of their importance in ion transport and include: Glu327, Glu779, Asp804, Asp808, Asp926,

Glu953, and Glu954 (FIG. 1). Interestingly, substitutions of five of these amino acids, Glu327, Glu779, Asp926, Glu953, and Glu954, give functional enzyme even when the substitution either lacks the carboxyl moiety (i.e., substitution of Glu with Gln or Asp with Asn) or lacks both oxygen moieties (i.e., substitution with Ala or Leu).[15–19] Although the substituted enzymes do not exhibit functional identity to the wild-type protein, only three- to fivefold changes in either K^+ or Na^+ stimulation of Na,K-ATPase activity are observed. Based on these findings it is reasoned that these carboxyl side chains are not absolutely required for enzymatic activity although they do contribute modestly to the overall function of the enzyme. In the case of Glu327, ouabain competition experiments demonstrated that the $E_2P \rightarrow E_2(2K^+)$ conformational change was altered.[20] In electrophysiologic analysis, Glu779 was shown to play an important role in a voltage-dependent step and is likely involved in the ion access channel.[21] Thus, neither Glu779 nor Glu327 appears to be a ligating group within the cation binding site; however, their carboxyl side chains may be essential for distinct partial reactions in the catalytic mechanism. In contrast, all substitutions at Asp804 and Asp808 inactivate the enzyme with the exception of the conservative substitution Asp808Glu.[22] Although the substitution Asp808Glu severely impairs the activity of the Na,K-ATPase, the enzyme is functional, suggesting that the carboxyl group of the glutamic acid may be substituting in the catalytic role for the naturally occurring carboxyl of aspartic acid.

ASP804 AND ASP808 ARE CATION COORDINATING RESIDUES

Because of the essential carboxyl side chains of Asp804 and Asp808, we have characterized several substituted proteins with replacements in these residues using ouabain binding.[23] The aspartic acid residues at positions 804 and 808 of the sheep α1 isoform were replaced by Ala, Asn, and Glu amino acids. Thus, conservative and nonconservative substitutions were examined in addition to the replacement by asparagine which alters these carboxylate groups to amide side chains.

The initial observation that substitutions of Asp804 and Asp808 were unable to support cell growth could be due to phenomena not directly related to specific alterations in enzyme catalytic function. For example, improper folding or lack of targeting to the plasma membrane would result in the inability of the mutant Na,K-ATPase to support cell viability. From studies of the binding of [^{3}H]ouabain to whole cells expressing the heterologous enzymes, it appears that all of the proteins with substitutions in Asp804 or Asp808 are correctly translated and targeted into the plasma membrane.[22] In addition, these substituted enzymes bind ouabain with high affinity which indicates that the structures of the extracellular loops interacting with the drug, namely, H1–H2, H5–H6, and H7–H8, are similar to those of the wild-type protein.

The Asp804 and Asp808 substituted enzymes were also able to bind ATP and undergo phosphorylation by Mg-Pi with only small alterations in the apparent affinities for these ligands. These data suggest not only that the cytoplasmic portions of the enzymes are correctly folded and able to interact with these ligands, but also that the mutant proteins can undergo major conformational transitions associated with the normal enzyme mechanism. Thus, it appears that these substitutions do not produce a major structural disarray of the Na,K-ATPase and that their inability to support cell growth is associated with the specific catalytic roles of Asp804 and Asp808.

EXTRACELLULAR

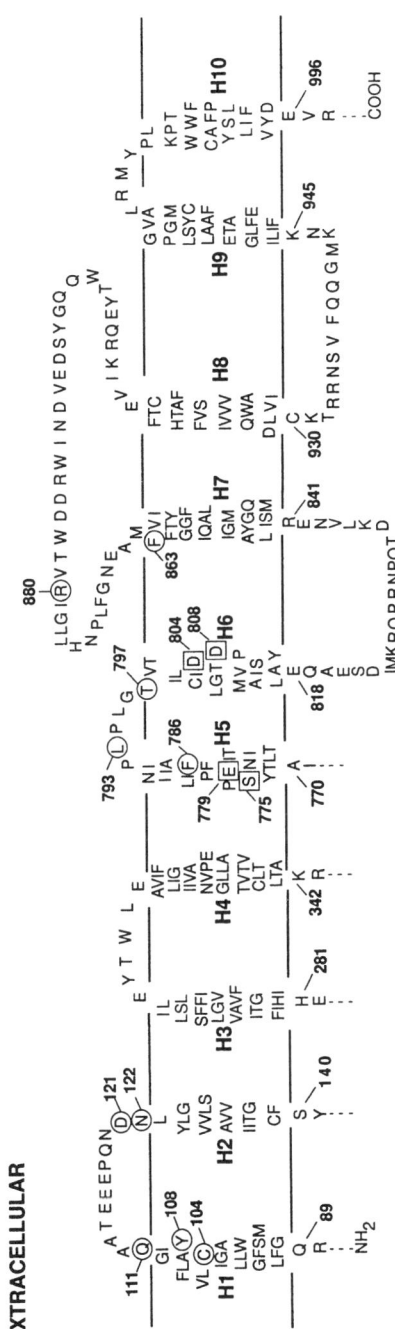

CYTOPLASMIC

FIGURE 1. Transmembrane and extracellular domains of the 10-transmembrane model of the Na,K-ATPase sheep α1 subunit. *Circled residues* have been implicated in ouabain binding. *Boxed residues* have been implicated in cation binding.

Na$^+$ and K$^+$ prevent ouabain binding in the presence of inorganic phosphate in a pseudocompetitive and partially competitive fashion, respectively. We took advantage of this phenomenon to assess the interaction of cations with the Na,K-ATPase containing substitutions at positions 804 or 808 (FIG. 2). Na$^+$ was able to prevent cardiac glycoside binding to both Asp804 and Asp808 substituted enzymes. However, the characteristics of the Na$^+$-interactions with the mutated enzymes were different from those with the wild type, such that alterations in the cation IC$_{50}$ and reductions in cooperativity were observed. The reduction in cooperativity was reflected by a reduced Hill coefficient ($n_{Hill} \cong 1.9$) detected with the nonconservative changes of Asp804 and Asp808 as compared to the wild-type protein ($n_{Hill} \cong 2.8$).[23]

K$^+$ is clearly unable to prevent the binding of the drug to the substituted proteins (FIG. 2). The partially competitive effect of K$^+$ on ouabain binding under our experimental conditions is likely due to displacement of the equilibrium, E$_2$P + 2K $\leftrightarrow$ E$_2$(2K) + Pi, to the right where E$_2$(2K) has a much lower affinity for ouabain than does E$_2$P.[6] The simplest interpretation of our findings is that the substituted proteins, in which Asp804 and Asp808 were replaced, are unable to bind K$^+$, suggesting that these two carboxyl groups compose part of the cation binding site. A more complex explanation involves an interaction between the modified enzymes and K$^+$ which is not translated into a lower affinity for the drug, that is, the enzyme is "locked" in E$_2$P. Another alternative is that the substituted enzyme is dephosphorylated by K$^+$, but a conformation with a low affinity for ouabain is not induced. However, the normal interactions of these mutant proteins with other ligands (i.e., Na$^+$, ATP, Pi, and ouabain) and the specific loss of K$^+$ antagonism supports the involvement of Asp804 and Asp808 in the binding of this cation during transport.

When amino acid substitutions are made in sheep α1 RD Na,K-ATPase, the only substitution at Asp804 or Asp808 with sufficient activity to confer ouabain resistance to HeLa cells is Asp808Glu.[22] Therefore, this protein was initially characterized using ligand-dependent ATPase assays. The apparent cation affinities for the substituted protein Asp808Glu were unaltered compared to those of the wild-type sheep α1 RD protein, as measured in the presence of saturating substrate conditions. Two assays carried out in the presence of saturating ATP and Na$^+$ but in the absence of K$^+$ (Na-ATPase activity and Na-dependent γ-^{32}P-ATP phosphorylation) suggested that high concentrations of Na$^+$ may bind to the protein and induce ATPase turnover and dephosphorylation. For example, Na$^+$-dependent ATP phosphorylation displayed two components, a Na$^+$-activated component at low concentrations (0–10 mM, AC$_{50}$ = 2.33 ± 0.21 mM for Asp808Glu compared to wild-type sheep α1 RD, AC50 = 1.52 ± 0.46 mM) and a Na$^+$ inhibitory component at high concentrations (20–100 mM, IC$_{50}$ = 67 ± 27 mM) (FIG. 3). In addition, this protein demonstrated a 2.7-fold increase in the amount of Na-ATPase activity compared to that of the wild type. Thus, it is tempting to conclude that although the carboxyl group of the glutamic acid replacing Asp808 can substitute in the overall turnover of the ATPase, this replacement alters the K$^+$ binding site or a K$^+$-sensitive intermediate, so that Na$^+$ can bind to the protein (possibly at the K$^+$ site) and mimic the action of K$^+$ (i.e., dephosphorylation). This interpretation is consistent with the conclusion that Asp808 is a K$^+$ coordinating residue as suggested by the ouabain binding analysis just described.

Asp804 and Asp808 are unique carboxyl residues in the transmembrane region of the Na,K-ATPase that may be part of the cation binding sites of the enzyme. Models describing the cation binding sites of the Na,K-ATPase often implicate two negatively

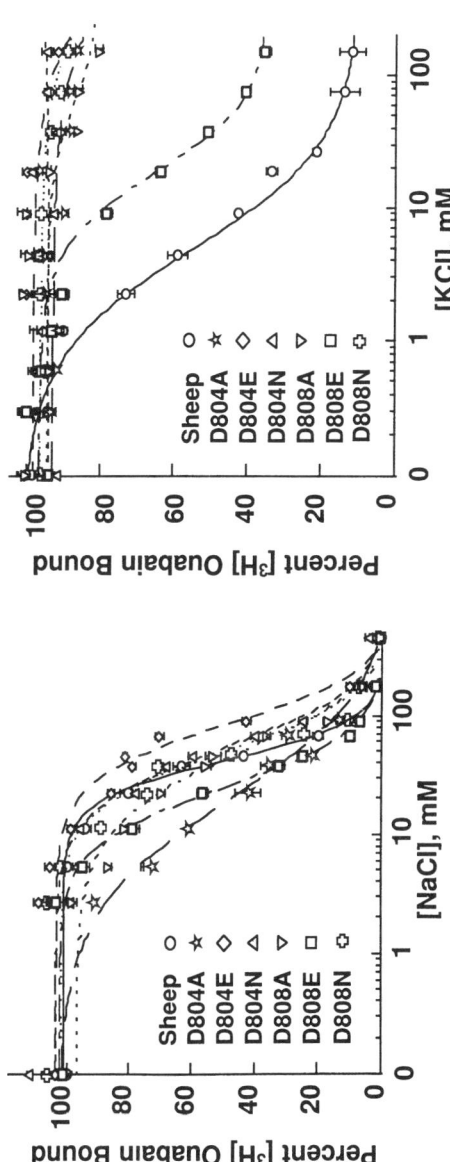

FIGURE 2. Effects of cations on ouabain binding to Na,K-ATPase with substitutions in Asp804 and Asp808. The conditions of the experiments were: 50 mM Tris (pH 7.5), 5 mM Tris-Pi, 5 mM MgCl$_2$, 30 nM [^{3}H]ouabain, 75 µg membrane protein, and various concentrations of cations, incubated at 37°C for 6 hours and filtered. Nonspecific binding in the presence of 30 nM [^{3}H]ouabain and 20 µM unlabeled ouabain was used as a zero, and the data were normalized to 100%.

charged residues within the structure of the binding pocket.[8-14] Thus, it is conceivable that Asp804 and Asp808 are these two negative charges. Although this simple explanation is inviting, recent reports show that both K^+ and Na^+ transport are voltage dependent, challenging the number of negatively charged residues involved in the binding sites.[24,25] Thus, it is possible that a third membrane carboxyl binds cations. In addition, the dielectric constant of the membrane-spanning environment is unknown, which limits our ability to predict whether Asp804 or Asp808 exist in a protonated or charged form. At this time, our current understanding of the enzyme structure does not allow us to speculate about the pKa values for these carboxyls or their possible neutralizing role during cation transport. However, mutagenesis and characterization studies definitively show that the side chains of Asp804 and Asp808 are important for the transport of K^+, regardless of whether they are protonated or charged.

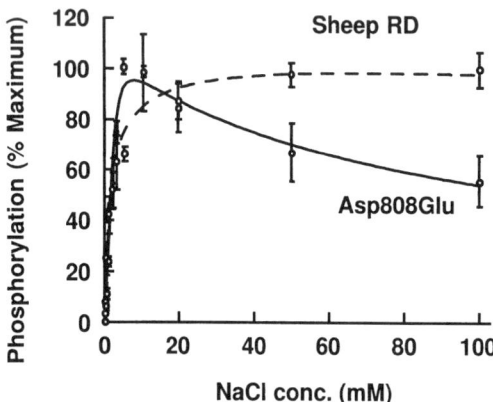

FIGURE 3. Na^+ dependence of phosphorylation by ATP sheep α1 RD and Asp808Glu substituted forms of the Na,K-ATPase. Sodium-activated ATP phosphorylation was measured at different Na^+ concentrations in a medium containing: 0.04 mM EGTA, 75 mM Tris (pH 7.2 at 20°C), 1 mM $MgCl_2$, 0.01 mM [^{32}P] ATP, 0.01 mM ouabain, and 0.2 mg/ml membrane protein, and Choline-Cl was used to maintain the ionic strength at 100 mM.

SER775 IS CRITICAL FOR K^+ INTERACTIONS

We have also examined a number of hydroxyl residues for their roles in binding cations. Many of these substitutions have little or no effect on activity, indicating that they are not part of a cation binding site. Interestingly, when Ser775 was substituted with alanine or cysteine in a ouabain-resistant isoform, the altered protein failed to confer ouabain resistance to HeLa cells.[26] It was reasoned that if Ser775 is involved in binding K^+ and if the substitutions retain some activity, then increasing the extracellular K^+ could stimulate enough enzyme activity to rescue the cells. This is actually the case, and by raising the KCl concentration of the media from 5.4 mM KCl to 10 mM, colonies appear for Ser775Cys. When the external KCl is increased to 20 mM, cells expressing the Ser775Ala enzyme are also rescued. These colonies grow slowly but provide enough cells to isolate the enzyme and characterize its properties. Cells expressing Ser775Cys overproduce the exogenous Na, K-ATPase 4.5-fold, whereas those containing Ser775Ala overexpress the foreign protein 8-fold, corresponding to their reduced turnover numbers (TABLE 1). When Na^+- and K^+-dependent Na,K-ATPase activity is measured, Na^+ dependence is normal, whereas that for K^+ is drastically reduced. For the Ser775Cys substitution, a 13-fold decrease in K^+ affinity is observed, whereas a 34-fold decrease is observed for Ser775Ala. This differential

TABLE 1. Characteristics of SER775 Substituted Proteins[26]

	Expression (μg exogenous/ mg total)	Phosphorylation (nmol/mg exogenous)	Turnover (n)	$K^+ K_{1/2}$ (mM) (at 30 mM Na^+)	$Na^+ K_{1/2}$ (mM)
Sheep α1 RD	34 ± 8	0.580 ± 0.096	12,314 ± 2,822	0.43 ± 0.06	7.06 ± 0.43
Ser775Cys	147 ± 38	0.346 ± 0.085	7,843 ± 1,326	5.82 ± 0.30	5.56 ± 0.53
Ser775Ala	272 ± 71	0.627 ± 0.160	2,669 ± 772	13.32 ± 1.75	5.98 ± 1.01

effect on K^+ versus Na^+ suggests that this amino acid is involved in binding K^+ but not Na^+. The Na^+-dependent phosphorylation is normal, indicating that the defect in the enzyme is past the initial Na^+-dependent phosphorylation step. Thus, the only characteristic of the enzyme that appears to be altered upon replacement of Ser775 is K^+ affinity, indicating a role for this residue in the K^+ binding site.

OUABAIN BINDING SITE

Because ouabain interacts with the extracellular portion of the Na,K-ATPase, initial studies concentrated on substituting amino acids in the extracellular loops of the enzyme (FIG. 1). Species-specific differences between ouabain-sensitive sheep α1 and ouabain-resistant rat α1 subunits were observed in the first extracellular loop of the protein. When the border residues of this loop were substituted in the sheep α1 subunit (Gln111 and Asn122) with those amino acids naturally appearing in the ouabain-resistant rat α1 subunit (Arg and Asp, respectively), a subunit with ouabain-resistant binding properties equal to those of the rat α1 subunit was produced[3] (TABLE 2). Additional studies identified substitutions at Asp121 of the first extracellular loop[27,28] as well as Cys104 and Tyr108 of the first transmembrane region[29,30] as conferring resistance to ouabain. In agreement with these results, when the amino terminal portion of the Ca-ATPase was replaced with the corresponding region of the chicken

TABLE 2. Ouabain-Resistant Mutant Proteins

Amino Acid Substitutions	Fold Resistance
Wild-Type HeLa	1×
-NH$_2$ Terminal Residues	
Q111R	8.3×
D121E, N	25×, 1000×
N122D	13×
C104A, Y	6.3×, 150×
Y108A	9.1×
Q111R and N122D	1150×
-COOH Terminal Residues	
L793P, K	24×, 8.8×
T797A, V	66×, 79×
F786N, I	19×, 11×
F863L	5.7×
R880P, L	7.9×, 3.8×

Na,K-ATPase, the resulting chimeric protein exhibited Ca-ATPase activity which was sensitive to ouabain.[31] However, a chimera carrying the N-terminal half of the ouabain-insensitive rat gastric H,K-ATPase and the C-terminal portion of the rat α1 subunit was also inhibited by ouabain, implicating the C-terminal portion of the Na, K-ATPase in ouabain binding.[32] This finding was supported by identification of Thr797[29,30] and Arg880[35] as required for ouabain inhibition (TABLE 2).

Recently, random mutagenesis coupled with the ouabain selection system has been used as an unbiased approach for determining residues involved in ouabain inhibition.[36,37] For example, we have located three hydrophobic residues in the carboxyl half of the Na,K-ATPase, Leu793, Phe786, and Phe863, that are essential for ouabain inhibition (FIG. 1, TABLE 2).[36] Several substitutions in these residues demonstrate increased I_{50} values for different cardiac glycosides from 3- to 48-fold. Upon identification of these residues we have implicated three new regions of the protein involved in ouabain interaction including H5, H7, and the extracellular loop H5-H6. The amino acids that affect sensitivity are shown in FIGURE 1, and their effects on ouabain inhibition are shown in TABLE 1. Interestingly the substitutions fall into two general regions of the α subunit, the hairpin loop involving the first and second transmembrane regions and H5-H7 transmembrane regions including the extracellular domains H5-H6 and H7-H8.

MODELS OF OUABAIN INHIBITION

Since identification of the species-specific variations in residues that convey ouabain sensitivity within the H1-H2 hairpin loop, this domain of the Na,K-ATPase has been referred to as the "ouabain binding domain." Unfortunately, the mechanism of inhibition associated with this drug-receptor site has remained a mystery. Recently, the H5-H7 region of the Na,K-ATPase was implicated in ouabain inhibition. Interestingly, the largest change in ouabain sensitivity induced by a single amino acid substitution in this region was observed when Thr797 was substituted with a hydrophobic residue.[33,34] Thr797Val demonstrated a 79-fold increase and Thr797Ala displayed a 66-fold increase in the I_{50} value for ouabain (TABLE 2).[34] Thus, both the H1-H2 and the H5-H6 hairpins appear to contribute substantially to the inhibitory action of cardiac glycosides. Unlike the H1-H2 region, mechanisms for ouabain inhibition of the Na,K-ATPase emerge with the identification of ouabain-sensitivity determinants in the carboxyl terminal half of the protein because of the importance of the H5-H6 and H7-H8 hairpin loops in transporting cations.

In recent reports, the H5-H6 hairpin loop was shown to be a crucial domain for cation binding and energy transduction. First, just as described, site-directed mutagencsis of Ser775, Asp804, and Asp808 has shown that these residues may coordinate K^+ ions as they are translocated across the plasma membrane.[23,26] In addition, chemical modification studies have shown that Glu779 is protected from DEAC labeling by Na^+ and K^+.[9,10] Thus, the polar faces of the fifth and sixth transmembrane helices are composed of several residues which coordinate K^+ and form the putative pore through which ions are translocated by the Na,K-ATPase. Second, several studies have suggested that this H5-H6 domain is important in energy transduction as a conformationally flexible region which moves during the catalytic cycle. For example, substitutions of Glu779 resulted in proteins that possess unstable phosphoenzyme intermediates indicative of a conformational role for this region.[21] In addition, modification of

Glu779 by DEAC is greatly increased on phosphorylation of the protein, suggesting that this residue is exposed to different degrees throughout the catalytic cycle.[9,10] Further evidence for a conformational role of the H5-H6 hairpin involves proteolytic digestion studies done in the presence of various substrates.[38,39] These studies demonstrated that two cleavage sites bordering this domain are protected similarly by cations and ouabain, but are exposed on phosphorylation of the protein. Moreover, this H5-H6 hairpin loop remains embedded in the membrane on digestion in the presence of ouabain or cations, but is released into the soluble fraction in the absence of these ligands indicative of the cations or ouabain interactions restricting the free movement of this domain. Thus, it appears that the H5-H6 hairpin loop plays a role in the binding of cations (and ouabain) and moves during the catalytic cycle in a substrate-dependent manner.

Upon assembling all of this structural information, the H5-H6 hairpin emerges as a connecting link for the cardiac glycoside binding site, cation transport sites, and the ATP binding domain (FIG. 4). From this link, one possible mechanism for cardiac glycoside inhibition of Na,K-ATPase can be derived in which drug interactions with the H5-H6 domain paralyze the movement of this domain normally induced by enzyme phosphorylation and required for cation transport. Thus, residues in the H5-H7 domain implicated as ouabain determinants may be important for maintaining flexibility of the protein in these transmembrane regions.[33–37] This mechanism is supported by the antagonistic effects of cations on ouabain binding.[4–6] In addition, linking ouabain binding to the H5-H6 hairpin also suggests a possible explanation for the higher affinity of the phosphorylated intermediate for ouabain.[4–7] Movement of the H5-H6 loop following phosphorylation supports the idea that the residues interacting with ouabain in this domain move and confer the higher affinity for ouabain characteristic of the phosphorylation.

Because of the central role of the H5-H6 domain in ouabain inhibition, one role of the H1-H2 region might be to specifically identify cardiac glycosides as opposed to other steroids. This theory is consistent with species-specific variations in the H1-H2 region which convey the ouabain binding characteristics of the pump and with the conservative nature in several P-type ATPases of the residues in the H5-H6 and H7 domains that are required for cation transport (sequences reviewed in ref. 40). In this sense, chimeric proteins made between the N-terminus of a ouabain-sensitive protein and the C-terminus of different cation pumps appear to be sensitive to ouabain inhibition.[31,32] It is interesting that the H,K-ATPase inhibitors omeprazole and pantoprazole chemically modify Cys813, the residue analogous to Thr797, and Cys892 located four residues away from Arg896 which corresponds to Arg880 in the sheep α1 Na,K-ATPase.[41,42] This suggests that inhibition of P-type ATPases by pharmacological drugs may have a common mechanism.

SUMMARY

From the structural data obtained by systematically altering residues of the Na,K-ATPase, we are beginning to understand portions of how this active cation transporter couples hydrolysis of ATP with the vectorial movement of cations against their ionic gradients. In addition, the inhibitory action of cardiac glycosides and their interaction sites on the protein has focused our attentions on a catalytic core of the protein involving the H5-H6 transmembrane segment. In future investigations, both

A. CONFORMATION WITH HIGH AFFINITY FOR OUABAIN

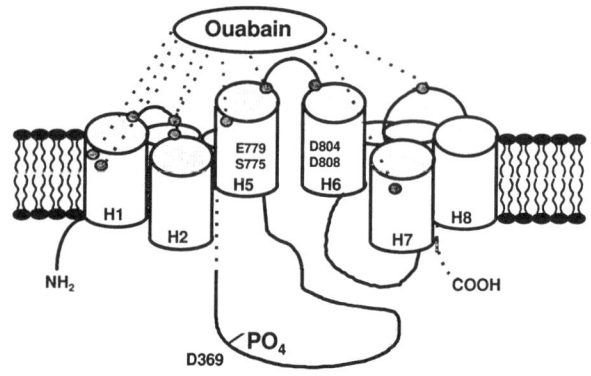

B. CONFORMATION WITH LOW AFFINITY FOR OUABAIN

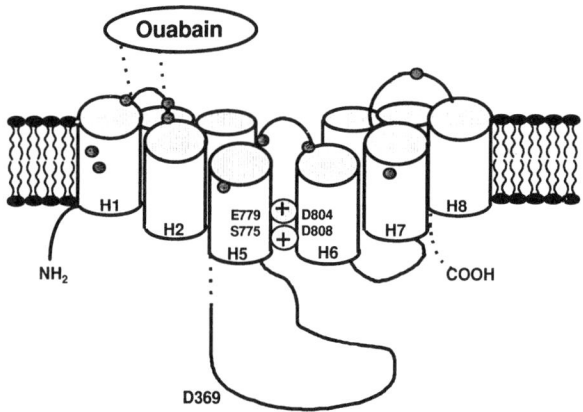

FIGURE 4. Model of cation, ouabain, and phosphate interactions with the Na,K-ATPase. (**A**) This cartoon depicts the high affinity association of ouabain with the extracellular surface of the Na,K-ATPase when the protein is in the E_2P conformation. (**B**) This cartoon describes the low affinity association of ouabain with the Na,K-ATPase ($E_1(2Na)$ or $E_2(2K)$) on cation binding and enzyme dephosphorylation. Asp369 is the site of phosphorylation during the catalytic cycle. Glu779 has been shown to be important in the voltage-dependent steps of cation transport.[21] Whereas Ser775, Asp804, and Asp808 are all proposed to be coordinating groups within the cation binding domain of this protein,[26,23] residues in the H1-H2 domain (Cys104, Tyr108, Gln111, Asp121, and Arg122), the H5-H6 loop (Phe786, Leu793, and Thr797), and the H7-H8 hairpin (Phe863 and Arg880) have all been implicated as residues involved in ouabain binding and are represented as *shaded circles* in this figure.[3,27–30,33–37]

the ATP and the Na$^+$ sites of the Na,K-ATPase must be uncovered to refine the structural picture of this complex transporter.

REFERENCES

1. LINGREL, J. B & T. A. KUNTZWEILER. 1994. J. Biol. Chem. **269:** 19659–19662.
2. PEDERSEN, P. A., J. H. RASMUSSEN & P. L. JORGENSEN. 1996. J. Biol. Chem. **271:** 2514–2522.
3. PRICE, E. M. & J. B LINGREL. 1988. Biochemistry **27:** 8400–8408.
4. HANSEN, O. 1984. Pharmacol. Rev. **36:** 143–163.
5. ERDMANN, E. & W. SCHONER. 1973. Biochim. Biophys. Acta **330:** 302–315.
6. YODA, A. & S. YODA. 1982. Mol. Pharmacol. **22:** 700–705.
7. ASKARI, A., S. S. KAKAR & W. HUANG. 1988. J. Biol. Chem. **263:** 235–242.
8. SHANI-SEKLER, M., R. GOLDSHLEGER, D. M. TAL & S. J. D. KARLISH. 1988. J. Biol. Chem. **263:** 19331–19341.
9. ARGÜELLO, J. M. & J. H. KAPLAN. 1991. J. Biol. Chem. **266:** 14627–14635.
10. ARGÜELLO, J. M. & J. H. KAPLAN. 1994. J. Biol. Chem. **269:** 6892–6899.
11. PEDEMONTE, C. H. & J. H. KAPLAN. 1990. J. Am. Physiol. (Cell Physiol. 27), **258:** C1–C23.
12. GLYNN, I. M. & S. J. D. KARLISH. 1990. Annu. Rev. Biochem. **59:** 171–205.
13. GOLDSHLEDGER, R., S. J. D. KARLISH, A. RAPHAELI & W. D. STEIN. 1987. J. Physiol. **387:** 331–355.
14. DEWEER, P., D. C. GADSBY & R. F. RAKOWSKI. 1988. Annu. Rev. Physiol. **50:** 225–241.
15. VAN HUYSSE, J. W., E. A. JEWELL & J. B LINGREL. 1993. Biochemistry **32:** 819–826.
16. JEWELL-MOTZ, E. A. & J. B LINGREL. 1993. Biochemistry **32:** 13523–13529.
17. VILSEN, B. 1995. Biochemistry **34:** 1455–1463.
18. FENG, J. & J. B LINGREL. 1995. Cell. Mol. Biol. Res. **41:** 29–37.
19. KOSTER, J. C., G. BLANCO, P. B. MILLS & R. W. MERCER. 1996. J. Biol. Chem. **271:** 2413–2421.
20. KUNTZWEILER, T. A., E. T. WALLICK, C. L. JOHNSON & J. B LINGREL. 1995. J. Biol. Chem. **270:** 2993–3000.
21. ARGÜELLO, J. M., R. D. PELUFFO, J. FENG, J. B LINGREL & J. BERLIN. 1996. J. Biol. Chem. **217:** 24610–24616.
22. VAN HUYSSE, J. W., T. A. KUNTZWEILER & J. B LINGREL. 1996. FEBS Lett. **389:** 179–185.
23. KUNTZWEILER, T. A., J. M. ARGUELLO & J. B LINGREL. 1996. J. Biol. Chem. **271:** 29682–29687.
24. SAGAR, A. & R. F. RAKOWSKI. 1994. J. Gen. Physiol. **103:** 869–894.
25. PELUFFO, R. D. & J. R. BERLIN. 1996. Biophy. J. **70:** A18.
26. ARGÜELLO, J. M. & J. B LINGREL. 1995. J. Biol. Chem. **270:** 22764–22771.
27. PRICE, E. M., D. A. RICE & J. B LINGREL. 1989. J. Biol. Chem. **264:** 21902–21906.
28. PRICE, E. M., D. A. RICE & J. B LINGREL. 1990. J. Biol. Chem. **265:** 6638–6641.
29. CANESA, C. M., J. D. HORISBERGER, D. LOUVARD & B. C. ROSSIER. 1992. EMBO J. **11:** 1681–1687.
30. SCHULTHEIS, P. J. & J. B LINGREL. 1993. Biochemistry **32:** 544–550.
31. ISHII, T. & K. TAKEYASU. 1993. Proc. Natl. Acad. Sci. USA **90:** 8881–8885.
32. BLOSTEIN, R., R. ZHANG, C. J. GOTTARDI & M. J. CAPLAN. 1993. J. Biol. Chem. **268:** 10654–10658.
33. BURNS, E. L. & E. M. PRICE. 1993. J. Biol. Chem. **268:** 25632–25635.
34. FENG, J. & J. B LINGREL. 1994. Biochemistry **33:** 4218–4224.
35. SCHULTHEIS, P. J., E. T. WALLICK & J. B LINGREL. 1993. J. Biol. Chem. **268:** 22686–22694.
36. PALASIS, M., T. A. KUNTZWEILER, J. M. ARGUELLO & J. B LINGREL. 1996. J. Biol. Chem. **271:** 14176–14182.
37. BURNS, E. L., NICHOLAS, R. A. & E. M. PRICE. 1996. J. Biol. Chem. **271:** 15879–15883.

38. LUTSENKO, S. & J. H. KAPLAN. 1994. J. Biol. Chem. **269:** 4555–4564.
39. LUTSENKO, S., R. ANDERKO & J. H. KAPLAN. 1995. Proc. Natl. Acad. Sci. USA **92:** 7936–7940.
40. GREEN, N. M. 1994. *In* The Sodium Pump: Structure Mechanism, Hormonal Control, and Its Role in Disease. E. Banberg & W. Schoner, Eds.: 110–119. Springer. New York.
41. BESANCON, M., J. M. SHIN, F. MERCIER, K. MUNSON, M. MILLER, S. HERSEY & G. SACHS. 1993. Biochemistry **32:** 2345–2355.
42. SHIN, J. M., M. BESANCON, A. SIMON & G. SACHS. 1993. Biochim. Biophys. Acta **1148:** 223–233.

ATPase Gene Transfer and Mutational Analysis of the Cation Translocation Mechanism

G. INESI,[a,b] D. LEWIS,[a,b] C. SUMBILLA,[a] A. NANDI,[a]
M. KIRTLEY,[a] AND C. P. ORDAHL[b]

[a]*Department of Biochemistry and Molecular Biology*
University of Maryland
Baltimore, Maryland 21201

[b]*Department of Anatomy and Cardiovascular Research Institute*
University of California
San Francisco, California 94143

The cloning of sarcoplasmic and endoplasmic reticulum Ca^{2+} ATPases (SERCA) has made possible expression of these enzymes by means of wild-type or mutant cDNA transfections.[1] We report here some of our experiences with various transfection procedures, vectors, DNA constructs, and host cells, focusing on mutational analysis of the energy transduction and cation translocation mechanism.

SERCA GENE TRANSFER: TRANSFECTION PROCEDURES AND HOST SYSTEMS

Cos-1 cells constitute a convenient and rapid system for transient expression of mutants, aided by the use of SV40 promoter elements complementing the Cos-1 T-antigen for gene amplification. The efficiency of the current DEAE-dextran transfection procedure, however, is low inasmuch as only 5–10% of the cells in culture are effectively transfected (FIG. 1). Alternatively, stable expression of all cells in culture can be obtained by cotransfection of fibroblasts with SERCA cDNA and a dihydrofolic reductase mutant permitting selective pressure with methotrexate. This system has been useful for *in situ* demonstration of Ca^{2+} transporting organelles formed by the newly expressed ATPase and for characterization of thapsigargin resistance.[3]

High-efficiency gene transfer can be obtained with recombinant viral vectors. Baculovirus is very effective, but the obligatory host insect cells have a limited capacity for correct membrane assembly. Consequently, a significant fraction of the transport ATPase expressed in baculovirus-infected insect cells is not functional. On the other hand, high-efficiency gene transfer can also be obtained by the use of replication-defective adenovirus as a gene transfer vector.[4,5] We recently generated recombinant adenovirus by cotransfection of HEK 293 cells with a replication-defective adenovirus plasmid (pJM17) and a shuttle plasmid (pDeltaE1sp1A) containing promoter and SERCA-1 cDNA.[6] From the standpoint of physiological interest, we found that it is possible to target gene expression to cardiac myocytes (FIG. 2) by the use of the tissue-specific cardiac TnT promoter[7] which is not effective in fibroblasts. This raises the possibility of influencing intracellular Ca^{2+} homeostasis by selective

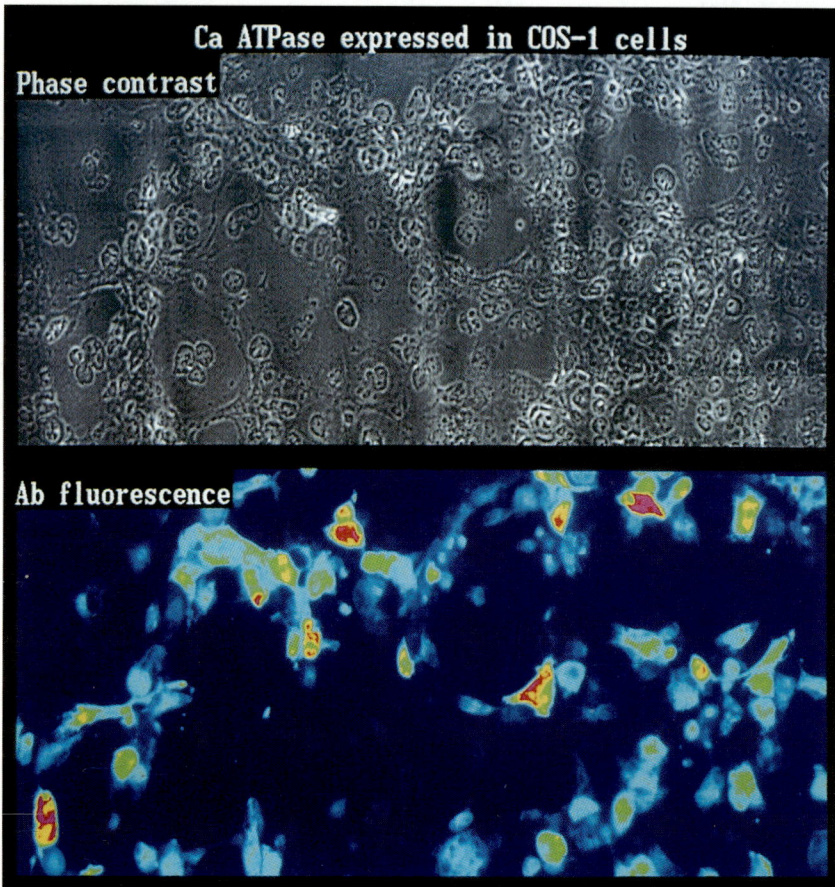

FIGURE 1. ATPase expression in Cos-1 cells. Transfection with SERCA-1 cDNA under control of an SV40 promoter was carried out by the DEAE-dextran method.[2] Three days following transfection the cells were fixed and processed for immunofluorescent staining. (**Top**) Phase contrast image of Cos-1 cells. (**Bottom**) The same field is visualized by fluorescence microscopy, and ATPase expression is rendered by pseudocolor (i.e., red = high expression). Note the relatively low percentage of cells expressing the ATPase.

FIGURE 2. ATPase expression in cardiac myocytes. Cardiac myocytes cultured from chicken embryos were infected with recombinant adenovirus carrying SERCA-1 cDNA under control of the tissue-specific promoter cTnT. Fixing and immunofluorescent staining with a monoclonal antibody specific for SERCA-1 (nonreactive with the endogenous SERCA-2) was carried out 2 days after the infection. Note the high percentage of cells expressing the ATPase (**top**) and the expression targeting on the sarcoplasmic reticulum on the higher magnification image (**bottom**). Similar results were obtained with rat heart myocytes.

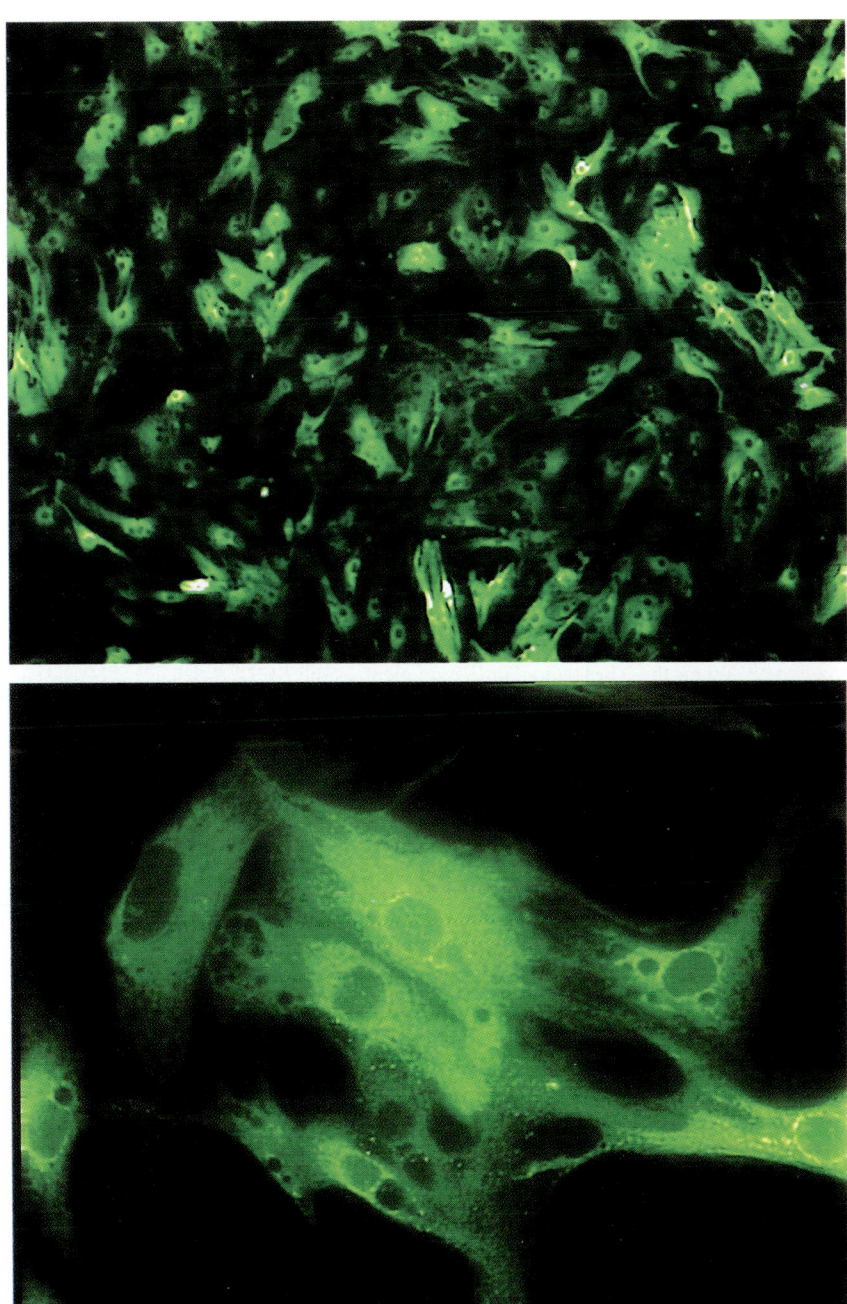

FIGURE 2. *See legend on facing page.*

gene transfer. It is of interest that the chicken cTnT promoter and SERCA-1 cDNA sustain ATPase expression in chicken as well as in rat cardiac myocytes, suggesting that these avian constructs are effective even in mammalian systems.

Although selective pressure and viral vector methods yield efficient gene transfer, the selection and/or recombination work required by these methods limits its use to selected cases. Therefore, transfection of Cos-1 cells is still the most rapid and practical method of expressing large numbers of mutants and of studying their functional characteristics.

MUTATIONAL ANALYSIS OF THE ATPase SEGMENT LINKING PHOSPHORYLATION AND CATION BINDING DOMAINS

Chemical and mutational studies of the Ca^{2+} ATPase show that the catalytic and phosphorylation domain resides within the extramembranous (cytosolic) region of the enzyme,[8,9] whereas the Ca^{2+} binding domain resides within the membrane-bound region.[10,11] As noted in FIGURE 3, the distance between the phosphorylation site (Asp351) and the Ca^{2+} binding domain is approximately 50 A.[12] Therefore, a long-range intramolecular linkage is required for coupling of catalytic and transport functions.[13]

A similar separation of functional domains appears to be present in other cation ATPases,[14,15] and the peptide segment extending from one of the transmembrane helices (M4) to the phosphorylation site retains a high degree of homology in these enzymes.[16] When the Thr316 to Thr357 segment of the SERCA-1 Ca^{2+} ATPase[17] is compared to the corresponding segment of the Na^+,K^+ ATPase,[18] 28 of 42 amino acids are homologous. Nonconservative mutations[19] of *homologous* amino acids in this segment are accompanied by pronounced inhibition of Ca^{2+} transport and ATPase activity (TABLE 1), suggesting a common and important functional role of this segment. In all cases, inhibition of steady-state velocity is related to slow decay of the phosphoenzyme intermediate. Decay of the intermediate is a prominent rate-limiting step in the catalytic and transport cycle and is likely related to E_1-PCa_2 → E_2-PCa_2 transition.

As opposed to mutations of *homologous* residues, mutations of *nonhomologous* residues of the Ca^{2+} ATPase to the corresponding residues of the Na^+,K^+ ATPase do not produce inhibition. This is true even when double or triple mutations are produced. However, if the number of simultaneous mutations is progressively increased to 14, thereby rendering the entire 42 amino acid segment of the Ca^{2+} ATPase identical to that of the Na^+,K^+-ATPase, the catalytic and transport function is reduced to 20–30% of the control (TABLES 2 and 3). Nevertheless, utilization of ATP remains strictly Ca^{2+} dependent, and the steady-state levels of phosphorylated intermediate are similar to those obtained with control Ca^{2+} ATPase.[20] Here again, when overall function is inhibited by the mutations, a proportionally slower phosphoenzyme decay (E_1-PCa_2 → E_2-PCa_2 transition) is observed (FIG. 4). It is of interest that the rate of enzyme inactivation by Ca^{2+} dissociation (likely related to the E_1 → E_2 transition) is also slower in these types of inhibitory mutations (FIG. 4).

The segment between the cation binding domain and the phosphorylation domain apparently can be interchanged between the Ca^{2+} ATPase and the Na^+,K^+-ATPase without a major inhibitory effect, whereas single nonconservative mutations of homologous residues produce major inhibition. These findings suggest that this

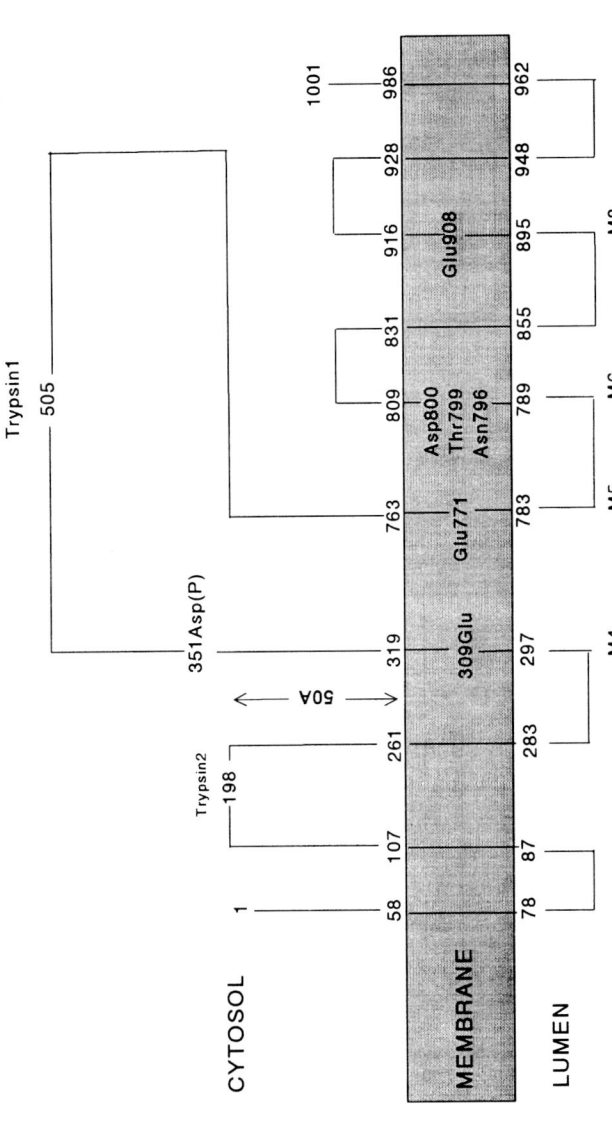

FIGURE 3. Diagram of the Ca^{2+} ATPase sequence and its distribution in membrane-bound and extramembranous domains. The catalytic residue (Asp351) undergoing phosphorylation is shown in the extramembranous domain. The M4, M5, M6, and M8 transmembrane segments are clustered to form a Ca^{2+} binding channel, with participation of Glu309, Glu771, Thr799, Asp800, and Glu908 oxygen functions. Asp796 may also participate, although very weakly. The sequence topology is in conformity with that proposed by MacLennan et al.[17] The ~50 Å distance between the phosphorylation site and the membrane surface is derived from spectroscopic studies.[12]

segment provides a common functional linkage in various cation ATPases, whereby catalytic activation is produced by cation binding and, conversely, vectorial displacement of bound cation is produced by catalytic phosphorylation. The functional linkage occurs by means of protein isomeric transitions that are produced mostly by changes in tertiary rather than secondary structure.[21]

TABLE 1. List of Sarcoplasmic Reticulum (SERCA-1) ATPase Residues Subjected to Mutation, Homologies to Other Cation ATPases, and Levels of Expression and Function, in Percentage Relative to Wild-Type ATPase [19]

Amino Acid Substitution	Original Homology with	Mutated to Homology with	Expression Level	Ca^{2+} Uptake	ATPase Activity
Wild type			1	1	1
Leu321 → Ala	NaK; HK; H(Val)		0.42 ± 0.02	0.32 ± 0.02	0.58 ± 0.16
Thr323 → Ala		NaK; NK; H	0.74 ± 0.25	1.56 ± 0.25	1.30 ± 0.20
Lys329 → Ala	NaK; HK; H		0.48 ± 0.02	0.26 ± 0.08	0.65 ± 0.21
Asn330 → Ala	NaK; HK; H(Gln)		0.82 ± 0.19	0.84 ± 0.39	0.97 ± 0.29
Ala331 → Cys	H	NaK; HK; H	0.60 ± 0.28	1.76 ± 0.18	1.87 ± 0.29
Ala331 → Leu	H		1.03 ± 0.12	0.47 ± 0.04	0.23 ± 0.01
Ala331 → Arg	H		<0.05		
Ala331 → Glu	H		<0.05		
Ile332 → Ala	H; NaK(Leu); HK(Val)		0.41 ± 0.17	0.10 ± 0.01	0.06 ± 0.03
Val333 → Ala	NaK; HK; H		0.77 ± 0.02	0.05 ± 0.02	0.21 ± 0.08
Val333 → Ile	NaK; HK; H		0.91 ± 0.23	0.69 ± 0.36	0.67 ± 0.37
Arg334 → Ala	NaK(Lys); HK(Lys)		0.70 ± 0.18	0.06 ± 0.03	0.22 ± 0.07
Arg334 → Gln	NaK(Lys); HK(Lys)		1.00 ± 0.42	0.17 ± 0.08	0.29 ± 0.09
Ser335 → Ala			0.84 ± 0.11	0.42 ± 0.26	0.59 ± 0.01
Leu336 → Ala	NaK; HK; H		1.01 ± 0.27	0.04 ± 0.03	0.23 ± 0.16
Leu336 → Val	NaK; HK; H		0.86 ± 0.033	0.03 ± 0.01	0.13 ± 0.03
Pro337 → Ala			0.82 ± 0.13	0.30 ± 0.01	0.36 ± 0.06
Ser338 → Ala		NaK; HK; H	0.98 ± 0.32	0.86 ± 0.39	0.84 ± 0.09
Val339 → Ala	NaK; HK; H(Ile)		0.92 ± 0.11	0.35 ± 0.04	0.33 ± 0.04
Glu340 → Ala	NaK; HK; H		1.39 ± 0.14	0.10 ± 0.00	0.11 ± 0.02
Glu340 → Gln	NaK; HK; H		0.79 ± 0.17	0.67 ± 0.26	0.86 ± 0.33
Thr341 → Ala	NaK; HK; H(Ser)		1.05 ± 0.35	0.57 ± 0.49	0.62 ± 0.35
Cys344 → Ala			0.70 ± 0.36	0.79 ± 0.29	0.92 ± 0.09
Cys344 → Ser		NaK; HK; H	0.71 ± 0.30	0.76 ± 0.10	1.08 ± 0.02
Ser346 → Ala	NaK; HK		0.37 ± 0.14	0.27 ± 0.15	0.24 ± 0.04
Ile348 → Ala	NaK; HK; H(Leu)		<0.05		

CATION BINDING AND VECTORIAL TRANSLOCATION

As shown in FIGURE 3, the M4 transmembrane helix (Lys297 to Leu319 in the Ca^{2+} ATPase) precedes and joins the linkage segment (Thr316 to Thr357) just discussed. Most importantly, the M4 helix participates in Ca^{2+} binding by providing the acidic chain of Glu309. Additional Ca^{2+} binding residues are provided by M5 (Glu771), M6 (Thr799, Asp800), and M8 (Glu908). These four helices apparently are clustered to form a transmembrane channel, with the oxygen functions of the Ca^{2+} binding residues facing the lumen of the channel. Evidence for a duplex Ca^{2+} binding site in the membrane-bound region is as follows:

1. Specific site-directed mutations interfere with Ca^{2+} inhibition of enzyme phosphorylation by P_i under equilibrium conditions.[10,15,22]

TABLE 2. Alignment of the Segments Connecting Cation Binding and Phosphorylation Domains in the Ca^{2+} (top) and Na^+,K^+ (bottom) ATPases[a]

	316				320					325					330					335					340					345					350							356
Ca	T	T	C	L	A	L	G	T	R	R	M	A	K	K	N	A	I	V	R	S	L	P	S	V	E	T	L	G	C	T	S	V	I	C	S	D	K	T	G	T	L	T
C2												A																														
C3												A																														
C4												A				C																										
C5								T	A							C																										
C8								T	A				K			C															S											
C10								T	A				K			C	L				K	N			A						S											
C11								T	A				K			C	L				K	N	E		A						S											
C13	V				T			T	A				K			C	L				K	N	E		A						S											
C14	V				T			T	A				K			C	L		K		K	N	E		A						S	T										
Na/K	T	V	C	L	T	L	T	A	K	R	M	A	R	K	N	C	L	V	K	N	L	E	A	V	E	T	L	G	S	T	T	I	C	S	D	K	T	G	T	L	T	

[a]Note that amino acids are homologous. Fourteen nonhomologous amino acids were mutated stepwise to yield total replacement of the Ca^{2+} ATPase segment with the corresponding Na^+,K^+-ATPase segment.[20]

TABLE 3. Functional Characterization of the Mutants Shown in TABLE 2 [20]

Chimeric Mutation	Ca^{2+}-Uptake Rate	ATPase Activity
pCDL (control)	[none]	[none]
Wild type	1.0	1.0
C2	1.04 ± 0.10	1.36 ± 0.20
C3	0.75 ± 0.31	1.08 ± 0.13
C4	0.45 ± 0.09	0.62 ± 0.11
C5	0.38 ± 0.08	0.22 ± 0.03
C8	0.45 ± 0.11	0.44 ± 0.07
C10	0.07 ± 0.01	0.20 ± 0.07
C11	0.19 ± 0.04	0.21 ± 0.02
C13	0.09 ± 0.03	0.28 ± 0.07
C14	0.14 ± 0.10	0.34 ± 0.09

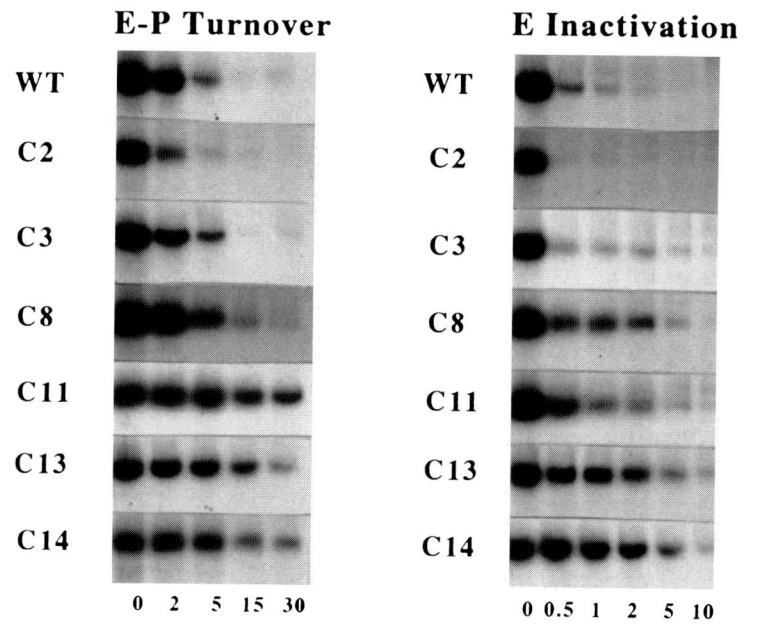

FIGURE 4. Effects of mutations on phosphoenzyme decay (E_1-P → E_2-P transition) and on enzyme inactivation by EGTA (E_1 → E_2 transition). Phosphoenzyme decay was studied by first obtaining steady-state levels of radioactive phosphoenzyme through incubation of SR vesicles with [gamma-32P]ATP in the presence of Ca^{2+} and then adding a nonradioactive ATP chase; samples were then taken at serial time for determination of residual radioactive phosphoenzyme. Enzyme inactivation was obtained by adding EGTA to SR vesicles to dissociate Ca^{2+} and then adding radioactive ATP at serial times to quantitate the residual enzyme ability to form phosphorylated intermediate. In both cases, radioactive phosphoenzyme intermediate was measured by autoradiography of electrophoretic gels. An increasing number of mutations was made in order to render the 41 amino acid segment connecting the cation binding and phosphorylation domains of the Ca^{2+} ATPase identical to the corresponding segment of the Na^+,K^+-ATPase, as explained in TABLE 2.[20]

2. Ca^{2+} binding (measured directly) is inhibited by carbodiimide derivatives producing structural perturbations within the membrane-bound region of the ATPase.[11]
3. The ATPase Ca^{2+} dependence is retained after replacement of most of the extramembranous region with the corresponding Na^+,K^+-ATPase region.[2]

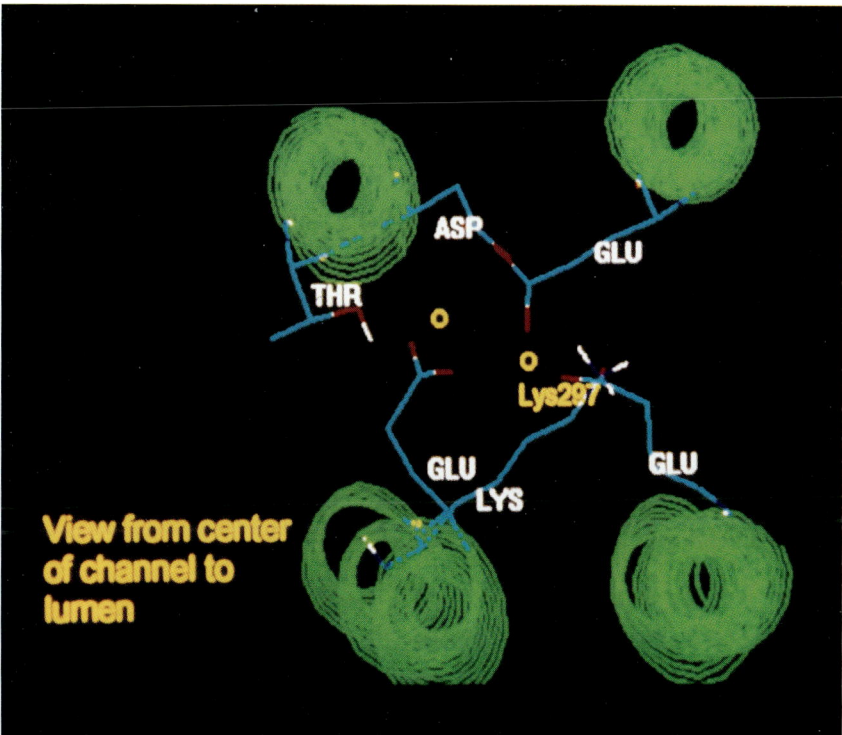

FIGURE 5. Molecular graphics representation of the putative Ca^2 ATPase transmembrane channel, the duplex Ca^{2+} binding site, and the gating lysine side chain at the end of the channel. The channel was constructed by clustering the M4, M5, M6, and M8 transmembrane helices. The side chains of Glu309 (2 oxygens), Glu771 (2 oxygens), Thr799 (1 oxygen), Asp800 (2 oxygens), and Glu908 (1 oxygen) were then positioned by approximation and energy minimization, to construct the duplex Ca^{2+} binding site as explained by Chen et al.[22] Analysis of the coordinates to obtain valence mapping[24] for the entire channel structure yields two discrete areas of valence levels compatible with a duplex Ca^{2+} binding site. The view shown here is from the center of the channel, with the centers of the two bound Ca^{2+} in yellow, and the positively charged Lys297 side chain at the luminal end of the channel. Although the two Ca^{2+} (yellow) appear in this view side by side, they are in fact displaced longitudinally.

4. Molecular modeling shows that it is, in fact, possible to arrange relevant side chains in the lumen of the transmembrane channel, yielding two discrete areas of high valence density as expected of a duplex Ca^{2+} binding site.[22]
5. Extensive mutational analysis of other ATPase regions does not provide evidence of alternate Ca^{2+} sites.

Mutational analysis and molecular modeling reveal that additional residues contribute to the tightness of the channel, even though not involved directly in Ca^{2+} binding.[15,23] Of particular interest is Lys297 whose mutation is accompanied by pronounced inhibition of Ca^{2+} uptake and lesser inhibition of ATPase activity.[22] The Lys297 side chain places a positive charge at the distal end of the lumen (FIG. 5) and thereby seals the channel. This suggests that upon enzyme phosphorylation by ATP, the M4 helix must rotate and possibly move longitudinally to displace the lysine charge from the lumen and render vectorial flow of Ca^{2+} possible. An additional and important consequence of this rotation is destabilization of the helical cluster and of the Ca^{2+} binding domain, with consequent dissociation of bound Ca^{2+}.

pK CHANGES AND $H^+:Ca^{2+}$ EXCHANGE

The pH dependence of Ca^{2+} binding to the ATPase in the absence of ATP and the stoichiometry of $H^+:Ca^+$ countertransport in the presence of ATP[25,26] indicate that two of the acidic functions of the Ca^{2+} binding residues do not bind H^+ throughout the ATPase cycle, whereas two alternatively bind and dissociate H^+ in exchange for Ca^{2+}. In fact, a study[27] of the $H^+:Ca^{2+}$ stoichiometric ratio as a function of luminal and medium (i.e., cytosolic) pH indicates that the pK of these acidic functions is changed,

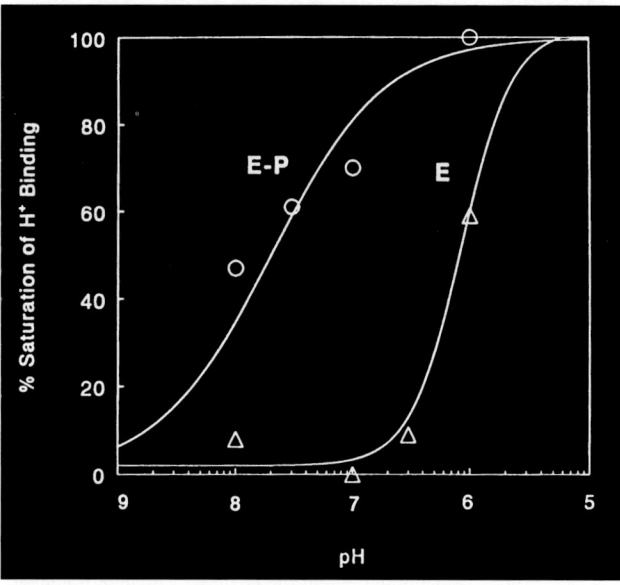

FIGURE 6. H^+ occupancy of two acidic chains participating in $Ca^{2+}:H^+$ exchange, as a function of pH in the ground state and in the phosphorylated state of the enzyme. The experimental points were obtained by determining the stoichiometric ratios of $H^+:Ca^{2+}$ exchange during ATP utilization by ATPase reconstituted in proteoliposomal vesicles.[27] The pH levels of the outer and luminal media were controlled separately. The graph suggests that the pK and possibly even the cooperativity of the two acidic groups are changed as a consequence of ATP utilization and enzyme phosphorylation.

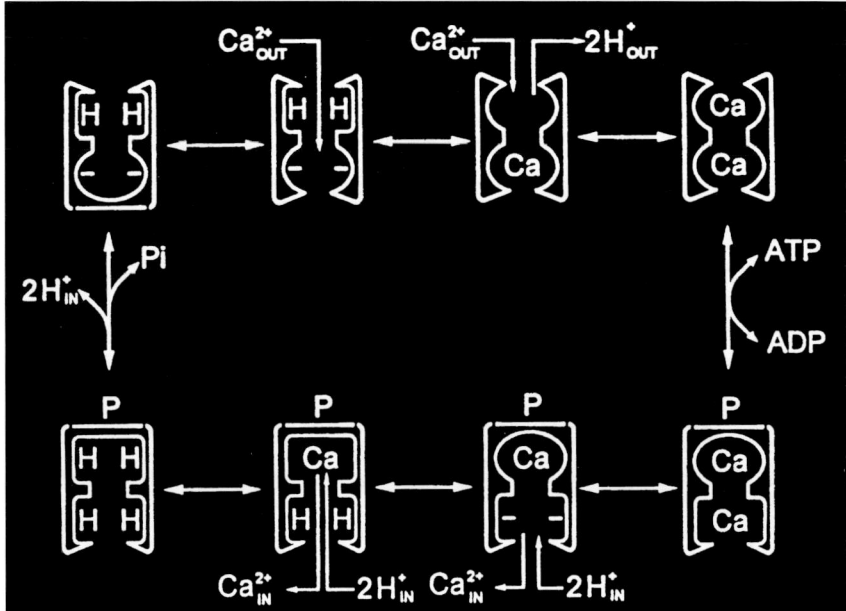

FIGURE 7. Diagram illustrating $H^+:Ca^{2+}$ exchange and vectorial displacement during the ATPase cycle. The transmembrane channel is shown as a cavity open to the cytosolic medium in the ground state of the enzyme. Binding of two Ca^{2+} occurs sequentially[28,29] in exchange for two H^{+}[27] under optimal conditions. This suggests that two of the four acidic chains participating in Ca^{2+} binding are already dissociated. Enzyme phosphorylation by ATP changes the exposure of the channel cavity from the cytosolic to the luminal medium and disrupts the Ca^{2+} binding sites. The affinity for Ca^{2+} is thereby reduced and the affinity for H^+ increased, favoring release of Ca^{2+} into the lumen in exchange for H^+. The cycle is finally completed by hydrolytic cleavage of the phosphoenzyme intermediate.

and their affinity for H^+ is increased, when phosphorylated enzyme intermediate is formed upon ATP utilization (FIG. 6). This pK change is likely produced by displacement of the acidic functions to an environment of different polarity, as the helical cluster is perturbed by enzyme phosphorylation.

Our interpretation of these findings (FIG. 7) is that at the beginning of the ATPase cycle the transmembrane channel is opened to the cytosolic medium and two Ca^{2+} are acquired in exchange for 2 H^+. On utilization of ATP, the channel opens to the luminal medium, the oxygen atoms involved in Ca^{2+} binding are displaced, and Ca^{2+} is released into the luminal medium, whereas 2 H^+ are bound. Finally, hydrolytic cleavage of the phosphoenzyme allows the enzyme to return to the initial state and to undergo another cycle.

CONCLUSIONS

Specific cation binding to transport ATPases occurs within the membrane-bound region of these enzymes, within a channel formed by clustered transmembrane helices. A homologous peptide segment connecting the membrane-bound cation binding domain to the extramembranous (cytosolic) catalytic domain in various ATPases provides the long-range linkage required for two alternating functions comprising the catalytic and transport cycle: (1) catalytic activation by specific cation binding, and (2) vectorial displacement of bound cation upon enzyme phosphorylation by ATP. Segmental protein displacements and pK changes of acidic residues are involved in the linkage and energy transduction mechanism.

SUMMARY

The peptide segment interposed between cation binding and phosphorylation domains retains a high degree of homology in all cation transport ATPases. Mutational analysis and chimeric replacements of Ca^{2+} ATPase components with corresponding Na^+,K^+-ATPase components indicate that this segment is utilized by various cation ATPases as a common structural device for a long-range functional linkage of enzyme phosphorylation and cation transport. Vectorial displacement of bound cation is rendered possible by a transmembrane channel formed by four clustered helices (M4, M5, M6, and M8). Originating from the four helices, the oxygen functions of Glu309, Glu771, Thr799, Asp800, and Glu908 form a duplex Ca^{2+} binding site in the middle of the channel, while Lys297 seals the luminal end of the channel with its positively charged side chain. The perturbation triggered by enzyme phosphorylation is apparently transmitted through the linkage segment to produce rotational displacement of the M4 helix with minimal change of secondary structure. The cation binding site is thereby disrupted and the Lys297 side chain removed, permitting Ca^{2+} to dissociate in exchange for H^+ and to flow through the luminal end of the channel.

REFERENCES

1. MARUYAMA, K. & D. H. MACLENNAN. 1988. Mutation of aspartic acid-35, lysine-352 and lysine-515 alters the Ca^{2+} transport activity of the Ca^{2+}-ATPase expressed in COS-1 cells. Proc. Natl. Acad. Sci. USA **85:** 3314–3318.
2. SUMBILLA, C., L. LU, G. INESI, T. ISHII, K. TAKEYASU, Y. FANG & D. M. FAMBROUGH. 1993. Ca^{2+} dependent and thapsigargin-inhibited phosphorylation of Na^+,K^+-ATPase catalytic domain following chimeric recombination with the Ca^{2+}-ATPase. J. Biol. Chem. **268:** 21185–21192.
3. HUSSAIN, A., C. GARNETT, M. G. KLEIN, J.-J. TSAI-WU, M. F. SCHNEIDER & G. INESI. 1995. Direct involvement of intracellular Ca^{2+}-transport ATPase in the development of thapsigargin resistance by Chinese hamster lung fibroblasts. J. Biol. Chem. **270:** 12140–12146.
4. KASS-EISLER, A., E. FALCK-PEDERSEN, M. ALVIRA, J. RIVERA, P. M. BUTTRICK, B. A. WITTENBERG, L. CIPRIANI & L. A. LEINWAND. 1993. Quantitative determination of adenovirus-mediated gene delivery to rat cardiac myocytes in vitro and in vivo. Proc. Natl. Acad. Sci. USA **90:** 11498–11502.
5. KIRSHENBAUM, L. A., W. R. MACLELLAN, W. MAZUR, B. A. FRENCH & M. D. SCHNEIDER. 1993. Highly efficient gene transfer into adult ventricular myocytes by recombinant adenovirus. J. Clin. Invest. **92:** 381–387.
6. KARIN, N. J., Z. KAPRIELIAN & D. M. FAMBROUGH. 1989. Expression of avian Ca^{2+}-ATPase in cultured mouse myogenic cells. Mol. Cell. Biol. **9:** 1978–1986.

7. MAR, J. H., P. B. ANTIN, T. A. COOPER & C. P. ORDAHL. 1988. Analysis of the upstream regions governing expression of the chicken cardiac troponin T gene in embryonic cardiac and skeletal muscle cells. J. Cell Biol. **107:** 573–585.
8. BASTIDE, F., G. MEISSNER, S. FLEISCHER & R. L. POST. 1973. Similarity of the active site of phosphorylation of the ATPase for transport of sodium and potassium ions in kidney to that for transport of calcium ion in sarcoplasmic reticulum of muscle. J. Biol. Chem. **248:** 8385–8391.
9. DEGANI, C. & P. D. BOYER. 1973. A borohydride reduction method for characterization of the acyl phosphate linkage in proteins and its application to sarcoplasmic reticulum adenosine triphosphatase. J. Biol. Chem. **248:** 8222–8226.
10. CLARKE, D. M., T. W. LOO, G. INESI & D. H. MACLENNAN. 1989. Location of high affinity Ca^{2+}-binding sites within the predicted transmembrane domain of the sarcoplasmic reticulum Ca^{2+}-ATPase. Nature **339:** 476–478.
11. SUMBILLA, C., T. CANTILINA, J. H. COLLINS, H. MALAK, J. R. LAKOWICZ & G. INESI. 1991. Structural perturbation of the transmembrane region interferes with calcium binding by the Ca^{2+} transport ATPase. J. Biol. Chem. **266:** 12682–12689.
12. BIGELOW, D. J. & G. INESI. 1992. Contributions of chemical derivatization and spectroscopic studies to the characterization of the Ca^{2+} transport ATPase of sarcoplasmic reticulum. Biochim. Biophys. Acta Bio-Membr. **1113:** 323–338.
13. INESI, G., D. LEWIS, D. NIKIC & M. E. KIRTLEY. 1992. Long range intramolecular linked functions in the calcium transport ATPase. *In* Advances in Enzymology. A. Meister, Ed.:185–215. John Wiley & Sons. New York.
14. GLYNN, I. M. & S. J. D. KARLISH. 1990. Occluded cations in active transport. Annu. Rev. Biochem. **59:** 171–205.
15. ANDERSEN, J. P. & B. VILSEN. 1995. Structure-function relationships of cation translocation by Ca^{2+}- and Na^+,K^+-ATPases studied by site-directed mutagenesis. FEBS Lett. **359:** 101–106.
16. INESI, G. & M. E. KIRTLEY. 1992. Structural features of cation transport ATPases. J. Bioenerg. Biomembr. **24:** 271–283.
17. MACLENNAN, D. H., C. J. BRANDL, B. KORCZAK & N. M. GREEN. 1985. Amino-acid sequence of a $Ca^{2+} + Mg^{2+}$ dependent ATPase from rabbit muscle sarcoplasmic reticulum, deduced from its complementary DNA sequence. Nature **316:** 696–700.
18. SHULL, G. E., A. SCHWARTZ & J. B LINGREL. 1985. Amino-acid sequence of the catalytic subunit of the (Na^+, K^+)ATPase deduced from a complementary DNA. Nature **316:** 691–695.
19. ZHANG, Z., C. SUMBILLA, D. LEWIS, S. SUMMERS, M. G. KLEIN & G. INESI. 1995. Mutational analysis of the peptide segment linking phosphyorylation and Ca^{2+}-binding domains in the sarcoplasmic reticulum Ca^{2+}-ATPase. J. Biol. Chem. **270:** 1–8.
20. GARNETT, C., C. SUMBILLA, F. FERNANDEZ BELDA, L. CHEN & G. INESI. 1996. Energy transduction and kinetic regulation by the peptide segment connecting phosphorylation and cation binding domains in transport ATPases. Biochemistry **35:** 11019–11025.
21. NAKAMOTO, R. K. & G. INESI. 1986. Retention of ellipticity between enzymatic states of the Ca^{2+}-ATPase of sarcoplasmic reticulum. F.E.B.S. Lett. **194:** 258–262.
22. CHEN, L., C. SUMBILLA, D. LEWIS, L. ZHONG, C. STROCK, M. E. KIRTLEY & G. INESI. 1996. Short and long range functions of amino acids in the transmembrane region of the sarcoplasmic reticulum ATPase. J. Biol. Chem. **271:** 10745–10752.
23. ANDERSEN, J. P. & B. VILSEN. 1994. Amino acids Asn^{796} and Thr^{799} of the Ca^{2+} ATPase of sarcoplasmic reticulum bind Ca^{2+} at different sites. J. Biol. Chem. **269:** 15931–15936.
24. NAYAL, M. & E. DI CERA. 1994. Predicting Ca^{2+}-binding sites in proteins. Proc. Natl. Acad. Sci. USA **91:** 817–821.
25. CHIESI, M. & G. INESI. 1980. Adenosine 5′-triphosphate dependent fluxes of manganese and hydrogen ions in sarcoplasmic reticulum. Biochemistry **19:** 2912–2918.
26. LEVY, D., M. SEIGNEURET, A. BLUZAT & J. L. RIGAUD. 1990. Evidence for proton countertransport by the sarcoplasmic reticulum Ca^{2+}-ATPase during calcium transport

in reconstituted proteoliposomes with low ionic permeability. J. Biol. Chem. **265**(32): 19524–19534.
27. YU, X., L. HAO & G. INESI. 1994. A pK change of acidic residues contributes to cation countertransport in the Ca-ATPase of sarcoplasmic reticulum. Role of H^+ in Ca^{2+}-ATPase countertransport. J. Biol. Chem. **269:** 16656–16661.
28. DUPONT, Y. 1982. Low temperature studies of the SR calcium pump mechanism of calcium binding. Biochim. Biophys. Acta **688:** 75–87.
29. INESI, G. 1987. Sequential mechanism of calcium binding and translocation in sarcoplasmic reticulum adenosine triphosphatase. J. Biol. Chem. **262:** 16338–16342.

Kinetic and Energetic Aspects of Na$^+$/K$^+$-Transport Cycle Steps[a]

HANS-JÜRGEN APELL[b]

Faculty of Biology
University of Konstanz
D-78434 Konstanz, Germany

The Na,K-ATPase is an ion-transporting protein of the cytoplasmic membrane that can couple a scalar chemical reaction, ATP hydrolysis, to vectorial processes, the movement of Na$^+$ and K$^+$ ions. With respect to the molecular mechanism of coupling of enzymatic and transport functions, the Na,K-ATPase is a kind of "black box." A lot of data are available on *what* this black box does, but almost no idea on how it does it.

The question that arises for the physicist at this stage is: How can kinetic experiments be exploited to learn about the inside of the black box? It is possible to measure enzymatic activity and transport functions under a variety of ligand and environmental conditions and to study a number of partial reactions of the pumping cycle.[1-5] Evidence assembled over the years has led to the concept of a pumping cycle that is summarized in the so-called Albers-Post or Post-Albers cycles,[6,7] which have been adjusted continuously to fit new experimental findings. Under physiological conditions the pump transports 3 Na$^+$ ions out of and 2 K$^+$ ions into the cell per molecule of ATP hydrolyzed. The cyclic process is an ordered sequence of three categories of reactions: (1) phosphorylation/dephosphorylation, (2) ion binding/ion release, and (3) switching between the two major conformations, E_1/E_2, which includes translocation of the ion binding sites (or their accessibility) between the two aqueous interfaces of the protein.

In addition to the physiological Na,K-exchange mode, the protein can perform a variety of partial reactions that involve so-called "noncanonical" pumping modes such as Na-only mode, K-K exchange, or uncoupled Na-efflux. A collection of reactions identified experimentally in the Na,K-ATPase from rabbit kidney are shown in comprehensive form in FIGURE 1. This scheme illustrates a unique forward reaction in which phosphorylation is driven by ATP in the presence of 3 Na$^+$ ions, but a series of similar return pathways with differently occupied ion sites. The scheme is characterized completely by three categories of parameters: rate constants, ligand concentrations, and the concentration of protein states. Each reaction step has a forward and a backward rate constant, both of which define the kinetic properties. They depend on temperature and in some cases also on transmembrane electric potential. The dependence of reactions on ligands is noted separately by the corresponding concentrations. Parameters identifying the concentrations of protein states depend

---

[a]This work was supported by the Deutsche Forschungsgemeinschaft (Sonderforschungsbereich 156).

[b]Address for correspondence: Faculty of Biology, University of Konstanz, Postfach 5560 M 635, D-78434 Konstanz, Germany (tel: +49 - 7531 - 882253; fax: +49 - 7531 - 883183; e-mail: h-j.apell@uni-konstanz.de).

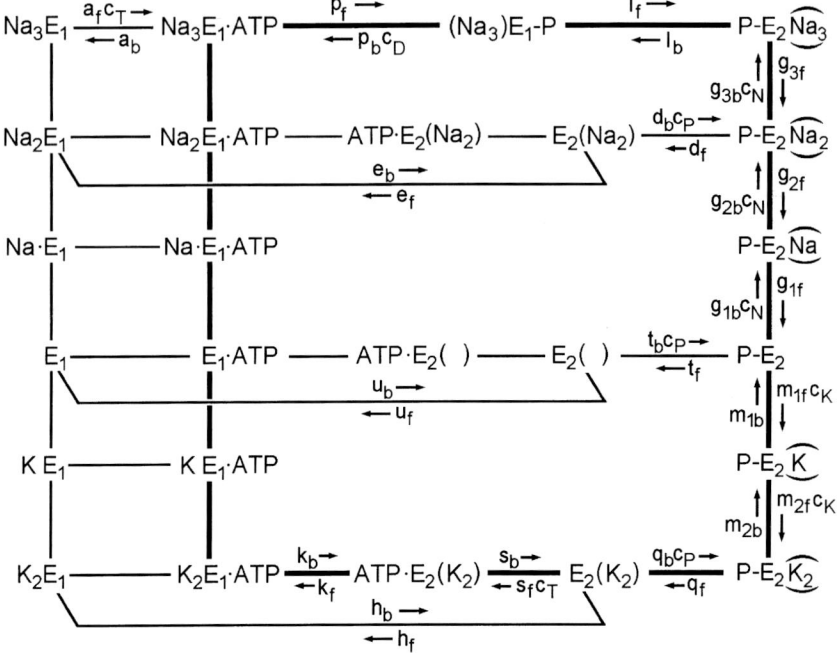

FIGURE 1. Expanded form of the Post-Albers reaction cycle. $a_f, p_f, \ldots$ and $a_b, p_b, \ldots$ are rate constants for transitions in forward and backward directions, respectively. c_T, c_D, c_P, c_N, and c_K are concentrations of ATP, ADP, P_i, Na^+, and K^+. Rate constants a_f and a_b are assumed to be the same for all transitions, $Na_iE_1 \leftrightarrow Na_iE_1 \cdot ATP$ and $K_jE_1 \leftrightarrow K_jE_1 \cdot ATP$ (i = 0, 1, 2, 3; j = 1, 2). For reaction steps, which are not characterized by rate constants, only equilibrium constants, K_x, could be determined so far ($K_x = x_f/x_b$). The *bold line* represents the reaction cycle under physiological conditions.

on the experimental measurements (e.g., the amount of radioactivity bound or specific absorption or fluorescence of the protein or specific labels).

Such measurements were made, for example, in recent investigations in the form of so-called "back-door phosphorylation" in which RH421, an electric field-sensitive fluorescent dye, was used. This dye discriminates between the unphosphorylated E_1 state and the $P-E_2$ state, which shows different specific fluorescent levels.

KINETIC EXPERIMENTS

If the overall phosphorylation reaction of the Na,K-ATPase in the absence of Na^+ and K^+ ions is described by $E_1 + P_i \leftrightarrow P-E_2$ with rate constants k_f and k_b in the forward and backward direction, respectively, then the equation $[E_1] \cdot [P_i] \cdot k_f = [P-E_2] \cdot k_b$ holds for the condition of equilibrium. If phosphate titration experiments are performed, the half-saturating concentration of inorganic phosphate, $[P_i]_{1/2}$, can easily be determined from experimental data. At this concentration the condition $[E_1] = [P-E_2]$ is fulfilled, and, consequently, the following relation is valid: $[P_i]_{1/2} \equiv$

$K_m = k_b/k_f$. Phosphate titration experiments were performed as shown in FIGURE 2.[8] Various concentrations of P_i were added to membrane fragments which contained Na,K-ATPase at a high concentration and which were diluted in a buffer that contained no Na^+ or K^+. The P_i-induced (normalized) fluorescence increase was monitored and the steady-state level after approximately 100 seconds (FIG. 2A) was measured and plotted against the corresponding P_i concentration (FIG. 2B). At pH 7.2 the equilibrium dissociation concentration, K_m, was determined to be 23.2 μM.[8]

Analysis of the time course of the phosphorylation reaction provided additional information. Closer inspection of the fluorescence trace in FIGURE 2A shows a biphasic behavior of the time course, which could be fitted with a sum of two exponentials. The slow process turned out to describe a reaction independent of P_i concentration. The faster process could not be resolved in FIGURE 2A because of the low time resolution of the setup used in this experiment. Therefore, P_i concentration jump experiments were performed in which the P_i concentration was increased from 0 to values between 1 and 60 μM by the release of P_i from "caged phosphate."[8] Caged phosphate[9] is an inactive precursor of inorganic phosphate, which can be released within 1 μs by an UV light flash. A typical experiment is shown in FIGURE 3A. The time course of the RH421 fluorescence can be fitted in the range of 1 second by a single exponential and a linearly increasing component. The latter represents the beginning of the slow phase of the time course shown in FIGURE 2A. The rate of the fast process was determined for different concentrations of P_i released by a light flash

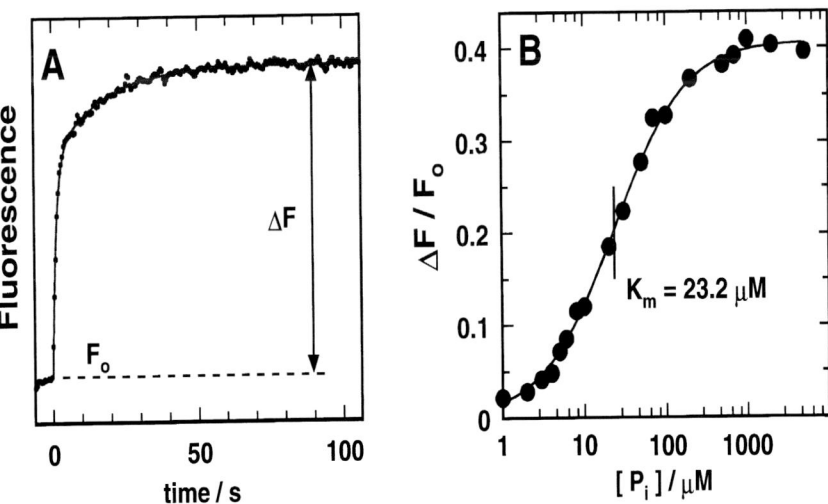

FIGURE 2. (A) Time course of RH421 fluorescence on addition of 0.5 mM P_i at t = 0 to enzyme in a buffer of 230 nM RH421, 25 mM histidine, 10 mM $MgCl_2$, and 0.1 mM EDTA, pH 7.1. The fluorescence signal can be fitted by a sum of two exponentials (*plotted line*). The fast process represents the mixing time of P_i in the cuvette. The slower process with a rate constant of 0.058 s^{-1} could be assigned to the conformational change $E_1 \rightarrow E_2$.[8] The change of the steady-state level of fluorescence after addition of P_i, ΔF, is normalized to the initial level, F_o. (B) Relative fluorescence changes, $\Delta F/F_o$, were determined for various concentrations of P_i. The resulting titration curve of the fluorescence changes was fitted by a Michaelis-Menten function, $\Delta F/F_o = \Delta F_{max}/F_o \cdot ([P_i]/(K_m + [P_i]))$ with a half-saturating concentration of P_i, $K_m = 23.2$ μM.

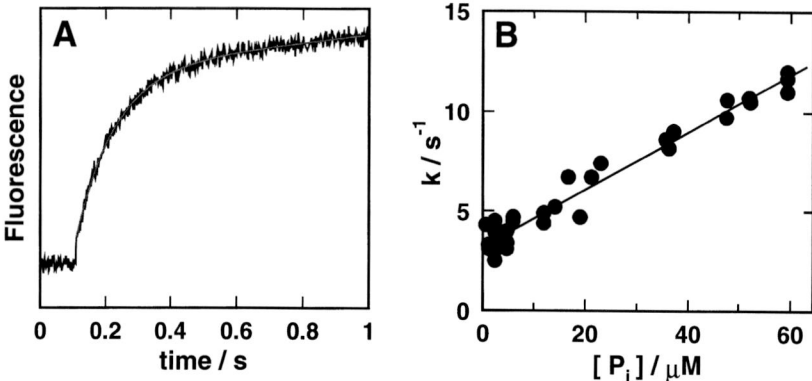

FIGURE 3. (**A**) RH421 fluorescence following a flash photolysis of 129 μM caged P_i that released 17 μM P_i. Details of the experiment are found elsewhere.[8] The best fit to the curve is a single exponential plus a linear component of fluorescence versus time (*solid line* through the data). (**B**) Dependence of the rate of the rapid phase of fluorescence increase on photoreleased P_i concentration. The regression line to the data has a slope of $1.47 \cdot 10^5$ M^{-1} s^{-1} and an ordinate intercept of 3.1 s^{-1}.

and was linearly dependent on the P_i concentration (FIG. 3B). Such dependence is expected for pseudo-first order reaction kinetics according to $k_{fast} = k_f \cdot [P_i] + k_b$, where k_f and k_b are rate constants and $[P_i]$ is the concentration of P_i after the light flash. Analysis of the data as presented in FIGURE 3B showed, with respect to the scheme in FIGURE 1, that the observed reaction is $E_2 \leftrightarrow P\text{-}E_2$ with the rate constants $t_f = 3.1$ s^{-1} and $t_b = 1.47 \cdot 10^5$ $M^{-1}s^{-1}$.

By analysis of steady-state and time-resolved experiments a two-step process was identified, $E_1 \leftrightarrow E_2 \leftrightarrow P\text{-}E_2$, and the corresponding rate constants were evaluated.[8] In recent years, rate constants were determined from numerous experiments, and we tried to obtain as many rate constants as possible, as well as their temperature and voltage dependence, from the same enzyme, Na,K-ATPase from rabbit kidney. The results are shown in TABLE 1. For most of the ion-binding and release steps the available methods so far were not sophisticated enough to determine their (very fast) rate constants by time-resolved measurements. Therefore, only equilibrium constants could be determined. This complete set of parameters allowed us to perform numerical simulations on the basis of the reaction scheme in FIGURE 1. The simulations are in good agreement with results of experiments already performed, and they allowed the predictions of kinetic behavior of the ion pump, which could be tested subsequently by additional experiments.[11,12]

CHANGE OF BASIC FREE ENERGY LEVELS OF A TRANSITION

Hill showed that energy transduction in molecular machines, such as the Na,K-ATPase, is the result not of a single reaction step but of the cycle as a whole.[14,15] It is an interesting question, however, to what extent single reaction steps contribute to storage and consumption of the system's free energy. To gain access to such information, energy levels must be introduced for all states of the pumping cycle. This is possible, because the states (shown in FIG. 1) are long-lived states on the time scale of molecular motions, and are thus in equilibrium with respect to movements of the peptide backbone or amino acid side chains. Accordingly, each state can be treated as a chemical species with a well-defined chemical potential.[14] If we look at the monomolecular transition $ATP \cdot E_2(K_2) \leftrightarrow K_2E_1 \cdot ATP$ (FIG. 1), the corresponding "basic" free energy levels[14] of the two states involved can be defined as $\mu^0(ATP \cdot E_2(K_2))$ and $\mu^0(K_2E_1 \cdot ATP)$. The difference in the basic free energy levels, $\Delta\mu \equiv \mu^0(K_2E_1 \cdot ATP) - \mu^0(ATP \cdot E_2(K_2))$, is directly related to the forward and backward rate constants, k_f and k_b, and to the equilibrium constant of this transition, K_k:

$$K_k = \frac{k_f}{k_b} = \frac{[K_2E_1 \cdot ATP]}{[ATP \cdot E_2(K_2)]} = \exp\left(-\frac{\Delta\mu}{RT}\right) \quad (1)$$

where R is the gas constant and T the absolute temperature. If ligand binding or release is involved in a reaction step, similar treatment is possible, in which the rate constant

TABLE 1. List of Experimentally Determined Rate Constants of Equilibrium Dissociation Constants of Rabbit Kidney Na,K-ATPase[a]

Rate	Forward Rate Constant	Backward Rate Constant	Equilibrium Dissociation Constant	$\Delta\mu$ (kJ/mol)	$\Delta\mu/F$ (mV)	Ref.
n_1			3.0 mM	-1.24	-12.85	10
n_2			5.2 mM	0.10	1.03	10
n_3			6.5 mM	0.64	6.6	10
k_1			2.9 mM^{-1}	14.8	153.4	10
k_2			2.3 mM^{-1}	14.2	147.2	10
a	$1.5 \cdot 10^7$ M^{-1} s^{-1}	1.64 s^{-1}		-26.1	-270.5	11
p	600 s^{-1}	$1.5 \cdot 10^6$ M^{-1} s^{-1}		-6.29	-65.2	[b]
1	25 s^{-1}	2.5 s^{-1}		-5.61	-58.1	12
g_3	1400 s^{-1}	$1.4 \cdot 10^4$ M^{-1} s^{-1}	10 M^{-1}	0.82	8.5	12
g_2	700 s^{-1}	467 M^{-1} s^{-1}	0.67 M^{-1}	-5.77	-59.8	12
g_1	4000 s^{-1}	$4.4 \cdot 10^4$ M^{-1} s^{-1}	11.1 M^{-1}	1.08	11.1	12
m_1			0.2 mM	-7.84	-81.2	13
m_2			0.2 mM	-7.84	-81.2	13
q	$1 \cdot 10^5$ s^{-1}	$5 \cdot 10^6$ M^{-1} s^{-1}		-3.38	-35.0	11
s	$5 \cdot 10^5$ M^{-1} s^{-1}	4 s^{-1}		-15.67	-162.4	11
k	22 s^{-1}	400 s^{-1}		7.06	73.2	11
h	0.1 s^{-1}	100 s^{-1}		16.8	174.3	11
d	10 s^{-1}	$2.5 \cdot 10^4$ M^{-1} s^{-1}		6.15	63.7	11
e	50 s^{-1}	0.1 s^{-1}		-15.13	-156.8	11
u	0.023 s^{-1}	0.059 s^{-1}		-2.29	-23.8	8
t	3.1 s^{-1}	$1.47 \cdot 10^5$ M^{-1} s^{-1}		13.31	138.0	8

[a]From these numbers, basic free energies were calculated on the basis of the following buffer composition: $[Na^+]_{cyt} = 5$ mM, $[K^+]_{cyt} = 150$ mM, $[Na^+]_{ext} = 140$ mM, $[K^+]_{ext} = 5$ mM, $[ATP] = 5$ mM, $[ADP] = 0.1$ mM, $[P_i] = 5$ mM, $U_M = 0$ (Faraday constant F = 96.485 C/mol). Temperature was 20°C. Rate constants in column 1 refer to the reaction scheme in FIGURE 1.
[b]p_f measured by Sokolov et al.[20] p_b determined according to the principle of detailed balance.[11]

has to be replaced by the product of a rate constant and the relevant ligand concentration. It was shown that Equation 1 still holds.[14]

When the rate (or equilibrium) constants and, if necessary, concentrations of ligands are all known, it is possible to calculate the change of the basic free energy from Equation 1. In the case of $ATP \cdot E_2(K_2) \leftrightarrow K_2E_1 \cdot ATP$, it is:

$$\Delta\mu = \mu^0(K_2E_1 \cdot ATP) - \mu^0(ATP \cdot E_2(K_2)) = RT \cdot \ln\left(\frac{k_b}{k_f}\right) = -RT \cdot \ln(K_k) \quad (2)$$

On the basis of the actual numbers determined experimentally for the Na,K-ATPase, as presented in TABLE 1, the change of basic free energy, $\Delta\mu$, can be determined to be 7.06 kJ/mol.

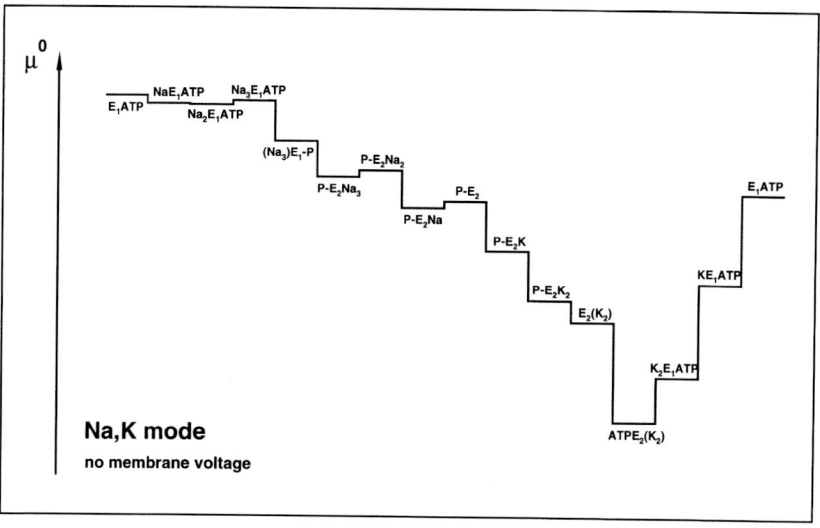

FIGURE 4. Free energy levels of the individual molecular states of the Na,K-ATPase according to the physiological pumping mode of FIGURE 1. The energy levels were calculated on the basis of equation 2 using the parameter values and conditions given in TABLE 1. All levels refer to $E_1 \cdot ATP$ as the reference state.

Analyses of free basic energy levels were performed for ion pumps in general,[16] for the SR Ca-ATPase,[17] and also for the Na,K-ATPase, though on a less elaborate data base than that in TABLE 1.[1,18] A calculation can now be performed for all states of the Post-Albers cycle. Placing state E_1ATP at an arbitrary initial level, the sequence of states around the pumping cycle in the physiological mode led to a lower level after a cycle was completed (FIG. 4). For the sake of simplicity in this figure, the free energy gained by binding ATP, $-\Delta G_{ATP}$, is not included in the change of the basic free energy shown for the ATP-binding step, $E_2(K_2) \rightarrow ATP \cdot E_2(K_2)$; all other conditions are as given in TABLE 1.

The difference between the energy levels of two neighboring states provides a measure of how close to, or how far from, equilibrium the transition resides. No change in free energy means "nothing to lose, nothing to gain", the condition of equilibrium. Several transitions are obviously far from equilibrium under physiological conditions. Of special functional importance is the free-energy decrease upon enzyme phosphorylation and the subsequent conformational change, $Na_3E_1 \cdot ATP \rightarrow (Na_3)E_1\text{-}P \rightarrow P\text{-}E_2(Na_3)$, which effectively drains the states involved in cytoplasmic sodium binding and thus produces dynamically a higher apparent affinity for Na^+ than for K^+ ions, despite the opposite ratio of values determined by steady-state titration experiments (TABLE 1). Other significant "down-hill" reaction steps are extracellular K^+ binding, $P\text{-}E_2 \rightarrow P\text{-}E_2K_2$, and ATP binding, $E_2(K_2) \rightarrow ATP \cdot E_2(K_2)$. In both cases the decrease of free energy is caused by ligand concentrations far above their affinities (which are ~0.1 mM for K^+ and 8 µM for ATP). A reduction of the ligand concentrations would diminish the step size. This is true also in the most prominent "up-hill" reactions, the K^+ release steps, $K_2E_1 \cdot ATP \rightarrow \ldots \rightarrow E_1 \cdot ATP$, in which the ions face a high cytoplasmic concentration of 150 mM from binding sites with an affinity on the order of 0.1 mM. However, K^+ release is not rate limiting because of the high rates of all ion binding and release steps in the reaction sequence $K_2E_1 \cdot ATP \rightarrow \ldots \rightarrow Na_3E_1 \cdot ATP$. In comparison to the preceding and following steps of the cycle, the ion binding and release steps can be treated under turnover conditions as being in equilibrium with each other. It should be mentioned that, in the case of stationary turnover of the pump, the reaction flux, Φ, is the same through each reaction step, regardless of the difference of the basic free energies between two states. With respect to the partial reaction $ATP \cdot E_2(K_2) \leftrightarrow K_2E_1 \cdot ATP$, for example, the reaction flux is defined as $\Phi = k_f \cdot [ATP \cdot E_2(K_2)] - k_b \cdot [K_2E_1 \cdot ATP]$.

Another interesting aspect is the voltage dependence of the basic free energy levels. A few partial reactions in the cycle have turned out to be electrogenic,[11] which means that their rate constants, and as a consequence their basic free energy levels, can be modulated by the membrane potential.[1] This is demonstrated in FIGURE 5A. At the known electrogenic reaction steps the energy levels spread apart when different voltages are applied, because charge is moved with or against the electric potential, a process that produces or consumes additional free energy.

A second important feature to be discussed (FIG. 5B) is, What happens to the energy provided by ATP hydrolysis? Under "physiological" conditions, that is, given concentrations of ATP, ADP, and P_i, the free energy of ATP hydrolysis (ΔG_{ATP}) can be calculated to be -56 kJ/mol, which corresponds to 580 mV ($\Delta G/F$; F is the Faraday constant). ΔG_{ATP} is the energy fed into the protein by phosphorylation, and $\Delta\tilde{\mu}$ is the change in electrochemical potential gradient stored across the membrane (per mol ATP hydrolyzed). The energy dissipated by the pump is the difference between energy made available by ATP hydrolysis and stored energy. At a typical resting potential of -70 mV, the efficiency of the pump is approximately 80% under chosen conditions. And the reversal potential would be around -160 mV. This reversal potential is evidently incorrect when examined in electrophysiological experiments, which evidenced significantly more negative values. Because ΔG and $\Delta\tilde{\mu}$ are energies calculated from basic principles,[1] the origin of the observed discrepancy has to be sought, and it probably indicates that the variables in this calculation, ion and ligand concentrations, are not correctly estimated. A convenient explanation would be that the bulk concentrations determined for ATP and ADP are not applicable in these calculations, because ATP-regenerating systems (such as creatine kinase) are situated

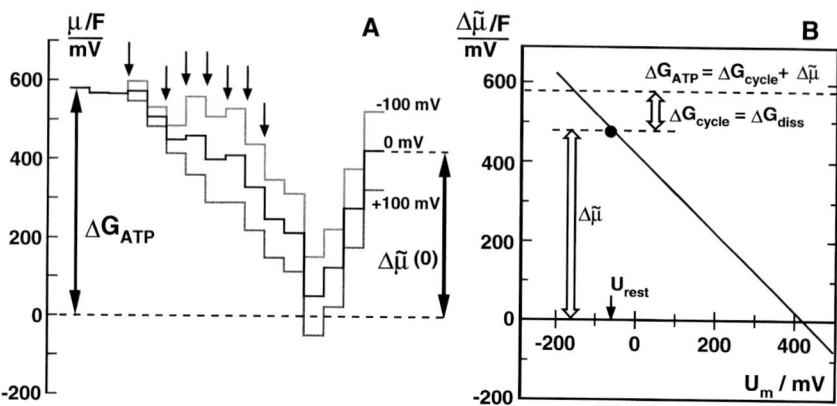

FIGURE 5. Voltage dependence of the basic free energy levels. (**A**) The *black curve* represents the energy levels in the absence of a transmembrane voltage. This curve corresponds to that of FIGURE 4. The free energy of the initial state $E_1 \cdot ATP$ has been set to the free enthalpy of the ATP hydrolysis. The final state after one complete cycle has a free energy equal to the electrochemical potential across the membrane, $\Delta\tilde{\mu}$. This value can also be determined directly from basic physical principles when ion concentrations and membrane voltage are known. Application of ± 100 mV membrane potential leads to diverging traces of the energy levels at the seven electrogenic transitions of the pumping cycle, which are indicated by *arrows*. From *left to right* they are: $Na_2E_1 \cdot ATP \rightarrow Na_3E_1 \cdot ATP$ and $(Na_3)E_1\text{-}P \rightarrow P\text{-}E_2(Na_3) \rightarrow P\text{-}E_2(Na_2) \rightarrow P\text{-}E_2(Na) \rightarrow P\text{-}E_2 \rightarrow P\text{-}E_2(K) \rightarrow P\text{-}E_2(K_2)$.[12] (**B**) Voltage dependence of the electrochemical potential, $\Delta\tilde{\mu}$. The free energy provided by ATP hydrolysis is either stored in $\Delta\tilde{\mu}$ or dissipated in thermal energy, ΔG_{diss}. The yield of the pumping process, $\Delta\tilde{\mu}/\Delta G_{ATP}$, is linearly dependent on the membrane potential, U_m. At the resting potential, U_{rest}, it is ~80%. 100% is reached at $U_m = -156.5$ mV, the reversal potential of the ion pump for the given ion and nucleotide concentrations.

at the membrane surface and locally increase ATP and decrease ADP concentrations, which shifts $-\Delta G_{ATP}$ to larger values.

TEMPERATURE DEPENDENCE AND ENERGETICS

Another way to obtain information from the kinetics of the pump is to investigate the temperature dependence of protein function. According to Arrhenius, the activation energy for a chemical reaction, E_a, can be introduced by the empirical relation

$$k = A \cdot \exp\left(-\frac{E_a}{RT}\right) \quad (3)$$

and determined from the slope of a so-called Arrhenius plot (FIG. 6). Although we do not have a one-step reaction, but a cyclic reaction sequence in the Na,K-ATPase, it is still possible to plot, for example, the hydrolyzing activity in an Arrhenius plot; however, the interpretation is not straightforward. There are good arguments for the claim that the apparent activation energy is mainly that of the rate-limiting process. FIGURE 6 indicates that because of different activation energies below and above temperatures of 25°C, different rate-limiting reaction steps apparently control ATP hydrolysis, with activation energies of 115 kJ/mol (at low temperatures) and 66 kJ/mol (at high temperatures). These data have been obtained under steady-state conditions.

It is also possible to measure the temperature dependence of partial reactions. Experiments similar to those on back-door phosphorylation of the Na,K-ATPase (FIGS. 2 and 3) can be performed under appropriate substrate conditions which single out specific partial reactions. Those experiments were carried out at temperatures between 5 and 35°C. As a typical experiment, the temperature dependence of the reaction ATP · $E_2(K_2)$ ↔ K_2E_1 · ATP is shown in FIGURE 6B. The detected rate of the partial reaction measured can obviously be fitted by a single straight line. This was found for all partial reactions analyzed so far.

Instead of the empirical concept of Arrhenius from the last century, which results in activation energy, representing approximately the height of the potential energy barrier of the rate-limiting reaction step, an alternative concept can be used. The theory of absolute reaction rates describes the rate constant as a function of thermodynamically well-defined energies:

$$k = \frac{k_B T}{h} \cdot \exp\left(-\frac{G^\downarrow}{RT}\right) = \frac{k_B T}{h} \cdot \exp\left(\frac{S^\downarrow}{R}\right) \cdot \exp\left(-\frac{H^\downarrow}{RT}\right) \quad (4)$$

This allows determination of the contributions of entropy $S^\downarrow$ and enthalpy $H^\downarrow$ to a reaction step. From the assumption that the transition state is in equilibrium with the ground state, it is possible to draw the same conclusions from the magnitude of $S^\downarrow$ and

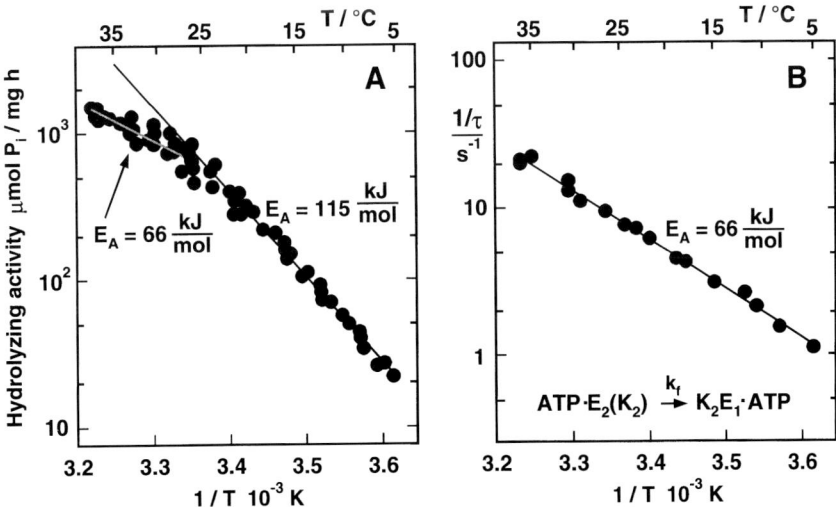

FIGURE 6. (A) Arrhenius plot of ATP hydrolyzing activity of Na,K-ATPase containing membrane fragments in buffer containing 140 mM Na$^+$, 10 mM K$^+$, and 250 μM ATP. The decrease in the slope of the line through the data points above 25°C indicates a change in the rate-limiting reaction step. (B) Arrhenius plot of a selected partial reaction, ATP · $E_2(K_2)$ ↔ K_2E_1 · ATP. Experimental conditions are described elsewhere.[11,19] The *plotted lines* indicate the activation energies noted in the diagrams.

$H^{\downarrow}$ accompanying structural changes during a partial reaction, as can be obtained from thermodynamic parameters of overall reactions.

The reaction steps of the Na,K-ATPase occur in an aqueous environment, and therefore entropy changes are dominated by binding and releasing water molecules or ions. A negative entropy change is caused mainly by immobilization of solvent molecules or ions in the protein. In case of the partial reaction presented (FIG. 6B), for the rate constant k_f the following energetic components were determined: $H^{\downarrow} = 66$ kJ/mol and $S^{\downarrow} = -224$ J/molK (at 25°C).

These results are recent and the project is still being developed. The final information obtained from thermodynamic properties, namely, basic free energies, changes of entropy, and enthalpy, will provide constraints for mechanistic models, which are essential for the development of both microscopic descriptions of protein function as well as coupling of enzymatic activity and transport.

ACKNOWLEDGMENTS

This work is based on experimental collaborations with R. Bühler, S. Heyse, A. Linder, M. Roudna, S. Schulz, W. Stürmer, V. Sokolov, and I. Wuddel in our group.

REFERENCES

1. LÄUGER, P. 1991. Electrogenic Ion Pumps. Sinauer Assoc. :1–313. Sunderland, MA.
2. JØRGENSEN, P. L. 1992. *In* Molecular Aspects of Transport Proteins. J. J. De Pont, Ed. : 1–26. Elsevier Science Publishers. B. V. Amsterdam.
3. GLYNN, I. M. 1993. J. Physiol. **462:** 1–30.
4. HEYSE, S., I. WUDDEL, H.-J. APELL & W. STÜRMER. 1994. J. Gen. Physiol. **104:** 197–240.
5. OR, E., R. GOLDSHLEGER & S. J. D. KARLISH. 1996. J. Biol. Chem. **271:** 2470–2477.
6. ALBERS, R. W. 1967. Ann. Rev. Biochem. **36:** 727–756.
7. POST, R. L., C. HEGYVARY & S. KUME. 1972. J. Biol. Chem. **247:** 6530–6540.
8. APELL, H.-J., M. ROUDNA, J. E. T. CORRIE & D. R. TRENTHAM. 1996. Biochemistry **35:** 10922–10930.
9. CORRIE, J. E. T., G. P. REID, D. R. TRENTHAM, M. B. HURSTHOUSE & M. A. MAZID. 1992. J. Chem. Soc. Perkin Trans. **1:** 1015–1019.
10. SCHULZ, S. & H.-J. APELL. 1995. Eur. Biophys. J. **23:** 413–421.
11. HEYSE, S., I. WUDDEL, H.-J. APELL & W. STÜRMER. 1994. J. Gen. Physiol. **104:** 197–240.
12. WUDDEL, I. & H.-J. APELL. 1995. Biophys. J. **69:** 909–921.
13. BÜHLER, R. & H.-J. APELL. 1995. J. Membr. Biol. **145:** 165–173.
14. HILL, T. L. 1977. Free Energy Transduction in Biology. :1–229. Academic Press. New York.
15. HILL, T. L. 1989. Free Energy Transduction and Biochemical Cycle Kinetics. :1–119. Springer. New York.
16. LÄUGER, P. 1984. Biochim. Biophys. Acta **779:** 307–341.
17. WALZ, D. & S. R. CAPLAN. 1988. Cell Biophys. **12:** 13–28.
18. STEIN, W. D. 1990. J. Theor. Biol. **147:** 145–159.
19. STÜRMER, W., H.-J. APELL, I. WUDDEL & P. LÄUGER. 1989. J. Membr. Biol. **110:** 67–86.
20. SOKOLOV, V., H.-J. APELL, J. E. T. CORRIE & D. R. TRENTHAM. 1997. Biophys. J. **72:** A291.

Charge Translocation by the Na/K Pump[a]

R. F. RAKOWSKI,[b,g] FRANCISCO BEZANILLA,[c]
P. DE WEER,[d] DAVID C. GADSBY,[e] MIGUEL HOLMGREN,[f]
AND JONATHAN WAGG[e]

[b]*Department of Physiology and Biophysics*
Finch University of Health Sciences
The Chicago Medical School
North Chicago, Illinois 60064

[c]*Department of Physiology*
U.C.L.A. Health Sciences Center
Los Angeles, California 90095

[d]*Department of Physiology*
University of Pennsylvania
School of Medicine
Philadelphia, Pennsylvania 19104

[e]*Laboratory of Cardiac/Membrane Physiology*
The Rockefeller University
New York, New York 10021

[f]*Laboratory of Molecular Physiology*
Harvard University
Boston, Massachusetts 02114

ELECTROGENICITY—VOLTAGE DEPENDENCE

The Na/K pump is electrogenic[1] because its $3Na^+:2K^+$ transport stoichiometry (Scheme 1)[2] gives rise to a steady-state current (outward, defined as positive) that contributes directly to the cell membrane potential.

$$
\begin{array}{ccccc}
3Na_i & & 3Na_o & 2K_o & 2K_i \\
\updownarrow & & \updownarrow & \updownarrow & \updownarrow \\
E_1 \leftrightarrow E_1ATP3Na & \leftrightarrow E_1P(Na_3) \leftrightarrow & E_2P & \leftrightarrow E_2(K_2) \leftrightarrow & E_1K_2 \leftrightarrow E_1 \\
\updownarrow & \updownarrow & & \updownarrow & \\
ATP & ADP & & P_i &
\end{array}
\quad \textbf{(Scheme 1)}
$$

If that stoichiometry remains unchanged, thermodynamics requires that net transport cease if the membrane potential were made sufficiently negative to increase

---

[a]This work was supported by National Institutes of Health grants NS22979, GM30376, NS11223, HL36783, and by HHMI.

[g]Address for correspondence: Dr. R. F. Rakowski, Department of Physiology and Biophysics, Finch University of Health Sciences/The Chicago Medical School, 3333 Green Bay Road, North Chicago, IL 60064 (tel: 847-578-3280; fax: 847-578-3265; e-mail: rakowskr@mis.finchcms.edu).

the work needed for active transport until it just balanced the energy available from hydrolysis of ATP.[3,4] Therefore, Na/K pump-mediated current is expected to decline as membrane potential is made more negative, although the form of that voltage dependence cannot be predicted on energetic grounds alone. The unequal cation import and export also means that at least one step in the Na/K transport cycle must involve net charge movement through the membrane's electric field and must therefore occur at a rate that is sensitive to the electric potential across the membrane (e.g., refs 5–8). In general, increasingly negative internal potentials are expected to slow steps involving extrusion of net positive charge. So, on these kinetic grounds, too, outward pump current is expected to diminish at negative potentials, but, again, in a way that cannot easily be anticipated. Fortunately, over roughly the last decade it has proven possible to unequivocally demonstrate the expected voltage sensitivity of Na/K pump function in a variety of tissues, cells, and membrane vesicles (for reviews, see refs 7–11) and even planar lipid bilayers.[12] The consensus is that in the presence of saturating concentrations of substrates (internal ATP and Na and external K) and with physiologically high external [Na], Na/K pump current varies in an approximately sigmoid manner with membrane voltage: the current is negligibly small at large negative potentials and increases steeply over the physiological voltage range towards a plateau near zero potential that is approximately maintained at positive potentials.

But mere demonstrations of voltage sensitivity offer little insight into mechanism, and even fairly extensive Na/K pump current-voltage (I-V) relationships yield disappointingly few quantitative kinetic details,[13] though useful qualitative information may be extracted (e.g., refs 14–17). To identify and characterize charge-moving steps in the reaction cycle of the Na/K pump (summarized in Scheme 1), several techniques have been employed.

Voltage-clamped internally dialyzed giant axons from squid allow simultaneous measurements of pump-mediated unidirectional Na efflux and net current over a range of membrane potentials. With that technique it was shown that during forward Na/K cycling, the voltage dependence of pump-mediated unidirectional ^{22}Na efflux was identical to that of outward pump current, regardless of the presence or absence of extracellular Na ions (Na_o).[18] These data confirmed that the transport stoichiometry is 3Na/2K and demonstrated that it is independent of voltage and of extracellular [Na] ($[Na]_o$). Because, in the absence of external Na, pump current and ^{22}Na efflux both became relatively voltage independent, the steep current decline on hyperpolarization at high $[Na]_o$ could be attributed to an apparent voltage-dependent inhibition by Na_o, slowing the forward cycle rate, rather than an Na_o-dependent stimulation of Na backflux via either Na/Na exchange[19] or pump reversal[20] which would have altered the apparent stoichiometry of transport.

EXTERNAL Na RELEASE/BINDING

It was such measurements of steady-state transport over a range of $[Na]_o$ that initially implicated external Na release/rebinding steps as being voltage sensitive and hence involving charge movement. The apparent voltage-dependent inhibition of forward Na/K exchange by extracellular Na ions (weak inhibition at 0 mV, stronger at increasingly negative potentials; refs 15, 18, and 21) is most straightforwardly explained by external Na ions binding to a site(s) on the pump within the membrane's electric field, thereby reducing the fraction of pumps in the conformation (E_2P;

Scheme 1) capable of binding and translocating external K ions. External Na ion binding within the electric field was more strongly suggested by measurements, using Na/K-ATPase-enriched membrane fragments from kidney cells, of fluorescence changes of the styryl dye RH421,[22,23] which responds to local electric field strength (cf. ref 24).

Additional support came from measurements (FIG. 1) in squid giant axons of the voltage dependence of ^{22}Na efflux[25] mediated by electroneutral, ADP-requiring, Na/Na exchange.[19] The ^{22}Na efflux was stimulated by raising $[Na]_o$ or by making the membrane potential more negative, and the resulting near sigmoid ^{22}Na efflux-voltage curves (approaching saturation at the largest negative potentials) were shifted in parallel fashion along the voltage axis by the same amount (~26 mV) for each

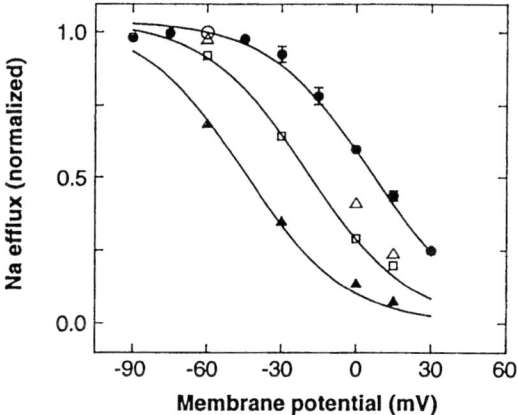

FIGURE 1. Shifts in voltage dependence of Na/Na exchange flux produced by changes in external [Na]. The voltage dependence of dihydrodigitoxigenin-sensitive ^{22}Na efflux was measured under electroneutral Na/Na exchange conditions in internally dialyzed squid giant axons at 400 (*filled circles*), 300 (*open triangles*), 200 (*open squares*), and 100 (*filled triangles*) mM external [Na]. Data were normalized to the value obtained, in each axon, at −60 mV and 400 mM external [Na] (*open circle*). Theoretical curves fitted to the data shifted by 26 mV, without change of shape, for each halving of external [Na]. Best-fit estimate of well depth for Na is 0.69. Figure from Gadsby *et al.*, 1993, with permission.

doubling of $[Na]_o$. The most parsimonious explanation for this sigmoid shape and kinetic equivalence of membrane potential and $[Na]_o$ is that charge translocation is limited to the external Na release step(s). A simple physical interpretation is that external Na ions must pass through an access channel (cf. refs 7, 26, and 27) in the pump molecule before interacting with their binding sites, rendering the probability of external Na ion binding voltage dependent. Because Na ions are assumed to traverse some fraction of the transmembrane electrical field in moving along the access channel between the binding site and the extracellular compartment, the equilibrium binding constant for Na_o becomes voltage dependent and is increased at negative, but decreased at positive, potentials. Fits to the data in FIGURE 1 using the simplest general formalism confined the voltage sensitivity to the pseudo-first-order (lumped) backward rate constant(s) of the release step(s), making Na_o binding, but not ^{22}Na release, appear voltage sensitive. The fits indicated that Na/Na exchange is activated by more than one Na ion (Hill coefficient = 1.8, SD ± 0.13) acting at a relatively low affinity site(s) ($K_{0.5}$ at 0 mV, 335 ± 8 mM) located at a fractional electrical distance of 0.69 ± 0.04. Asymmetry in voltage sensitivity is not required, however, and the conclusions remain unchanged if, instead, one assumes a very slow conformational change (i.e., $E_1P \leftrightarrow E_2P$) immediately preceding rapid (symmetrically) voltage-dependent release/rebinding step(s) (cf. ref 28).

CHARGE MOVEMENTS–Na TRANSLOCATION

Kinetic information can be obtained more directly from analysis of time-resolved relaxations of pump current following jumps of voltage or of substrate concentration,[7,29] particularly if experimental conditions are chosen so as to restrict pump activity to some limited steps of a partial reaction. The first direct measurements of the charge movement associated with Na translocation by the Na/K pump came from experiments in which photo-release of caged ATP caused generation of transient currents by pumps in membrane fragments attached to planar lipid bilayers in the presence of Mg and Na ions, but without K ions.[30] These ATP-induced transient currents seemed to require the $E_1P \leftrightarrow E_2P$ conformational change because they were abolished by mild chymotrypsin pretreatment,[31] known to prevent the $E_1P \leftrightarrow E_2P$ conformational change but not the formation of the occluded state, $E_1P(Na_3)$, or ATP/ADP exchange.[32]

Corresponding transient pump currents were elicited by voltage jumps in the presence of high internal MgATP and Na, and external Na, but no K, in voltage-clamped internally dialyzed cardiac myocytes[33] and also in intact[34] and cut-open[35] *Xenopus* oocytes. The voltage-induced transient pump currents were abolished by oligomycin B[33,35] which also inhibits Na/Na exchange[36] but stimulates ATP/ADP exchange[37] and therefore is believed to prevent the $E_1P \leftrightarrow E_2P$ transition but not the formation of $E_1P(Na_3)$.

These findings established that Na translocation involves substantial charge movement and that the charge-translocating step(s) occur(s) late in the Na transport pathway, that is, after occlusion of the Na ions. The voltage-induced,[33] and ATP-induced,[31,38] transient pump currents decayed with approximately the same rate (~40 s^{-1} at 20°C) as observed for ATP-induced ^{22}Na efflux transients mediated by pumps in right-side-out vesicles exposed to K-free solution[39,40] and for ATP-induced changes in 5-IAF fluorescence thought to reflect $E_1P \leftrightarrow E_2P$ conformational changes.[38,41] In further support of the analogy between transient pump currents elicited by ATP jumps or voltage jumps, under Na/Na exchange conditions, the two were recently compared in the same giant patch of myocyte membrane and shown to be closely similar.[42] The firm conclusion from all of these transient current measurements is that a major charge movement accompanies deocclusion and/or release of the Na ions to the exterior.

EXTERNAL K BINDING

The finding of Goldschleger *et al.*[43,44] that ATP+P_i-activated Rb/Rb (K/K) exchange in Na/K pumps reconstituted in lipid vesicles neither was voltage sensitive nor generated a membrane potential, and the demonstrated lack[45] of transient strophanthidin-sensitive currents in myocytes in the complete absence of Na ions, under K/K exchange conditions, argues that K deocclusion/release to the cell interior, $E_2(K_2) \leftrightarrow E_1K_2 \leftrightarrow E_1 + 2K_i$, must be voltage insensitive. But because the $[K]_o$ ($[Rb]_o$) used in those experiments was supramaximal, the external K binding sites presumably remained saturated and so charge movement associated with the K_o ion binding/ occlusion steps, $E_2P + 2K \leftrightarrow E_2PK_2 \leftrightarrow E_2(K_2) + P_i$, could not be ruled out. Subsequently, a weak negative slope in the Na/K pump I-V curve in *Xenopus* oocytes in the absence of Na_o at nonsaturating K_o[16] demonstrated a voltage dependence of apparent K_o affinity, which suggested a small charge movement during K_o binding/ occlusion. A corresponding shallow negative slope at zero Na_o and subsaturating $[K]_o$ has since been observed in Purkinje myocytes,[46,47] ventricular myocytes,[48,49] and squid axons (FIG. 2). Because, as just described, this voltage sensitivity of apparent

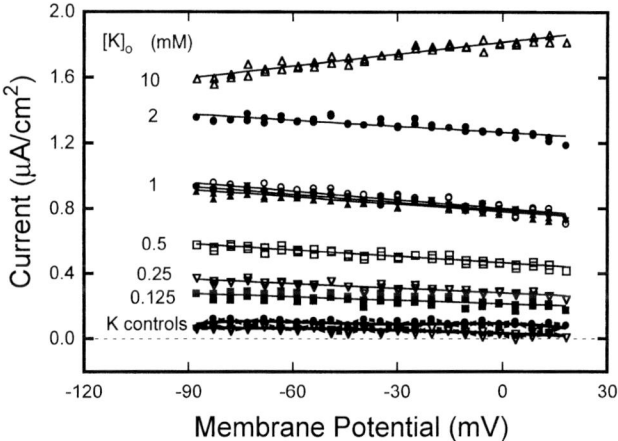

FIGURE 2. Steady-state H_2DTG (100 μM)-sensitive membrane I-Vs in a squid axon in Na_o-free solution at various $[K]_o$ or at zero $[K]_o$ plus 100 μM H_2DTG. A 0.5-s step down-up-down voltage staircase from -30 to -90 mV and $+20$ mV was applied at each $[K]_o$, and each measurement was bracketed by staircases at zero $[K]_o$; I-Vs were then obtained at zero and 10 mM $[K]_o$ in the presence of 100 μM H_2DTG. To obtain H_2DTG-sensitive I-Vs, K_o-activated I-Vs (obtained by forward and backward [in time] subtractions) were averaged, then corrected by subtracting the small H_2DTG-sensitive I-V at zero $[K]_o$. K controls establish virtual absence of K_o-sensitive currents (0–10 mM $[K]_o$) in 100 μM H_2DTG (cf. Rakowski *et al.*, 1989).

affinity for K_o cannot be attributed to voltage dependence of any subsequent step in the K translocation pathway,[43–45] it likely reflects binding of K ions a small distance into the electric field,[17] a conclusion strongly supported by RH-421 measurements in membrane fragments without[22,50] or with[41] parallel measurements of $E_2 \leftrightarrow E_1$ conformational changes using 5-IAF. More recent reexamination of [Rb] dependence of Rb/Rb exchange in vesicles has confirmed that the maximum rate was unaffected by voltage, that apparent affinity for Rb_i was unaffected by a positive potential, and that apparent affinity for Rb_o was barely enhanced by a -70 mV membrane potential.[51]

ENUMERATION/RECAPITULATION OF CHARGE-MOVING STEPS

Thus, the identities and gross characteristics of the charge-moving steps in the Na/K transport cycle have begun to clearly emerge. Beginning with the E_1 conformation, binding of the third Na ion at cytoplasmic sites was found to be weakly voltage sensitive,[22,41,51,52] prompting the suggestion that in the E_1 state two negatively charged binding sites lie outside the membrane's electric field, and one neutral site lies just within the electric field.[53] Present evidence strongly suggests that the major charge movement accompanies the deocclusion/release to the exterior of the first Na ion.[41,53,54] Deocclusion/release of the remaining two Na ions and binding/occlusion of the two external K ions appear to involve relatively small charge movements and so are weakly voltage sensitive.[16,22,41,53,54] FIGURE 2 (top trace) shows that in squid axons (cf. ref 18) under Na_o-free conditions, when the Na/K pump is fully activated by saturating $[Na]_i$ and $[K]_o$, the pump I-V curve displays a weak positive slope, as also observed in other preparations (e.g., ref 15). This very strongly suggests the presence in the Na/K transport cycle of a charge-moving step not associated with ion binding or release, which is, hence, intrinsically voltage dependent and which could be the $E_1P \leftrightarrow E_2P$ conformational change.[53]

MECHANISTIC INSIGHTS BASED ON HIGH-SPEED ELECTROPHYSIOLOGICAL MEASUREMENTS

The widely accepted Post-Albers mechanism (shown simplified in Scheme 1) for active Na/K transport by the pump is based on biochemical analysis of partial reactions of the pump, for example, kinetics of phosphorylation/dephosphorylation. Electrophysiological approaches were used to establish which of the biochemically identified steps of the transport cycle involve charge movement. Recent technical improvements in voltage-clamp techniques (e.g., the giant patch method) and ultrafast analog-to-digital converters have extended the utility of these approaches[54–61] to include direct identification and characterization of novel transport cycle steps (both electrogenic and nonelectrogenic) occurring on a submillisecond time scale and, thus, beyond the temporal resolution of earlier electrophysiological and biochemical techniques.

Using high-speed electrophysiological techniques, we are evaluating and extending a model for translocation and release of Na ions by the pump that incorporates an external access/release channel. An important corollary of confirming such a model will be the implied structural constraints, because structural information related to the cation transport mechanism of the Na/K pump (or, indeed, of any cation pump) though steadily growing is still meager (e.g., refs 62–64). Transient pump-generated currents were measured in voltage-clamped, externally superfused/internally dialyzed squid giant axons. Dialysis and superfusion solutions were designed to restrict the pump to states mediating Na ion translocation and release to the outside and to minimize nonpump-mediated components of charge movement, for example, electrodiffusive ionic currents through Na and K channels. Pump-mediated current was obtained by subtracting currents elicited by voltage steps from -1 mV to potentials in the range of -160 to $+120$ mV in the presence of 100 μM dihydrodigitoxigenin (H_2DTG, a reversible and selective pump inhibitor) from those recorded in its absence. The H_2DTG-sensitive difference currents comprised four distinct components: ultrafast

($\tau_{uf} < 3$ μs), fast (10 μs $\leq \tau_f \leq$ 30 μs), medium (50 μs $\leq \tau_m \leq$ 300 μs), and slow (600 μs $\leq \tau_s \leq$ 5 ms).[59–61] To characterize each component, we applied voltage pulses of variable (0.05–40 ms) duration, and the currents elicited were filtered and sampled at 12.5–200 kHz and 0.02–2 MHz, respectively.

To characterize the slow component, we applied long (20–40 ms) voltage pulses and filtered and sampled current at 12.5 and 20 kHz, respectively.[57] The slow component of pump-generated current was comparable to that previously described in myocytes[15,48,54,65] and oocytes.[34,35] The quantity of charge (Q_S) moved by the slow component on the ON step of the pulses was approximately equal to that moved by this component in the opposite direction on the OFF step. Q_S-V curves saturated at extreme positive and negative voltages and were well described by Boltzmann relations with slopes indicating an equivalent charge of roughly one positive charge. Relaxation rate ($k_S = 1/\tau_s$) versus voltage (k_S-V) curves were also reasonably described by Boltzmann relations but with somewhat shallower slopes (equivalent charge approximately +0.7) and more negative midpoints than the Q_S-V curves. Both the k_S-V and the Q_S-V curves were shifted rightwards about 25 mV per doubling of $[Na]_o$ over the range 50–400 mM, a kinetic equivalence generally expected for any simple model that incorporates an extracellular access channel (see above).

Scheme 2, a subset of Scheme 1, shows the simplest possible model for Na deocclusion and release in the absence of potassium: a single "slow" conformational transition followed by rapid electrogenic release—essentially instantaneous equilibration—of Na ions via an extracellular access channel:

$$E_1P(Na_3) \underset{k_{-1}}{\overset{k_1}{\leftrightarrow}} E_2PNa_3 \leftrightarrow E_2P + 3Na_o \qquad \text{(Scheme 2)}$$

Using such a model, Läuger and Apell[66] obtained a reasonable fit of the k_S-V curve of voltage jump-induced slow transients in cardiac myocytes[33] by assuming a dielectric coefficient (i.e., the equivalent charge in units of elementary charge) of 0.3 for the conformational transition and a relative well depth of 0.5, and a reasonable fit of the Q_S-V curve by assuming a *conformational* dielectric coefficient of 1 (and no ion well). In this interpretation, the saturation of the corresponding Q_S-V curves reflects movements of the enzyme entirely into E_2P forms at extreme positive and entirely into E_1P forms at extreme negative potentials. Our k_S-V curves from squid axons were well fit by assuming an *electroneutral* conformational transition, allowing the interpretation that the observed saturation of Q_S-V curves reflects movements of the enzyme into equilibrium occupancy of the E_1P state (determined by $k_{-1}/[k_1 + k_{-1}]$) at extreme negative potentials.

A testable prediction of access channel models[25,54,55,66] like that in Scheme 2 is that pump-mediated charge movements after a voltage jump should comprise at least two components: first, a spike of current with a time course which follows that of the voltage jump (i.e., with kinetics comparable to those of the decay of the capacity current), reflecting "instantaneous" redistribution of Na ions in the access channel, and perhaps consequent ultrafast binding/unbinding reactions at E_2P binding sites, in response to an altered field, followed by (a) slower current(s), also reflecting Na ion movement along the access channel but now rate-limited by slower conformational change(s). Consistent with this prediction, when pump-mediated current was measured at higher (microsecond or better) temporal resolution, Hilgemann[54] found ultrafast (complete in under 4 μs) charge movements preceding a slow charge movement in cardiac myocytes, and we observed comparable components in the squid axon.[57] More recently, also in squid axon, we observed two rapid components (one ultrafast, $\tau_{uf} < 3$ μs, and one fast, $10\ \mu s \leq \tau_f \leq 30\ \mu s$) which tracked capacity current (the decay of the latter was well described by the sum of two exponential components $\tau_1 < 3$ μs and $10\ \mu s \leq \tau_2 \leq 30\ \mu s$) followed by two slower components, described below, which presumably track the slower conformational transitions.[59–61]

If the observed fast/ultrafast charge movement mainly represents rapid, quasi-equilibrium binding/unbinding of Na ions to E_2P binding sites (and not just redistribution of "free" ions within the channel), then the simple model of Scheme 2 predicts "immobilization" of fast/ultrafast charges at very negative holding potentials for the following reason. A large negative voltage step would cause quasi-instantaneous maximal occupancy of the Na binding sites (and, provided $[Na]_o$ is neither too low nor too high, a maximal ultrafast/fast transient current) followed by slower redistribution of pump molecules to E_1 states, whence they would thus be unavailable to contribute instantaneously to ultrafast/fast charge movements at the OFF step of the pulse. Consistent with this prediction, Hilgemann[54] found Q_{fast} in cardiac myocytes to depend steeply on holding potential, vanishing at extreme negative voltages. In squid axons bathed in 400 mM Na, we found that holding potentials negative enough to saturate the Q_S movement, that is, -100 mV, resulted in almost complete immobilization of ultrafast/fast charge movement (FIG. 3).[60,61]

As already mentioned, $Q_{S(ON)}$ and $Q_{S(OFF)}$ seem approximately matched (e.g., refs 33, 48, 65) over the full range of conditions examined, including pulses at high $[Na]_o$ to potentials sufficiently negative to saturate the amount of slow charge movement. Such behavior is not predicted by the simple model of Scheme 2 unless the entire voltage sensitivity is ascribed to step 1[66] or a second step with appropriate (and finite) rate constants is inserted between deocclusion and actual electrogenic passage through the access channel[25] to prevent extreme negative voltage jumps from causing ultrafast/fast saturation of the binding sites in the E_2PNa_3 state.

A remarkable property of the fast/ultrafast charge component in cardiac myocytes[54] is the shallow slope (dielectric coefficient 0.26) of the Boltzmann function fit to its Q_{fast}-V plot. This finding, which we have confirmed in squid axons,[58–61] requires a shallow (or wide) channel for the fast/ultrafast current transient if indeed it represents movement of Na ions across an electric field. Adapting a model developed by Heyse et al.,[41] Hilgemann[54] proposed representing the deocclusion/release reaction as the succession of a strongly electrogenic release of the first Na ion, rate limited by one conformational transition, followed by the weakly electrogenic release of the other two Na ions, rate limited by another conformational transition. (To preserve the approximate ON/OFF equality of the slow component, certain constraints on the relative values of the rate constants for the two conformational transitions apply.) Scheme 3 is an elaboration of Scheme 2 to show explicitly the release of each of the three Na ions:

$$\begin{array}{ccccccc} & & Na_o & & Na_o & & Na_o \\ (1) & & \updownarrow & (2) & \updownarrow & (3) & \updownarrow \\ E_1P(Na_3) & \leftrightarrow E_2P(Na_2)Na & \leftrightarrow E_2P(Na_2) & \leftrightarrow E_2P(Na)Na & \leftrightarrow E_2P(Na) & \leftrightarrow E_2PNa & \leftrightarrow E_2P \end{array}$$

(Scheme 3)

Resolution of more than one deocclusion step must be expected, in general, to add a third component of charge movement [cf., ref 56]. Consistent with this expectation,

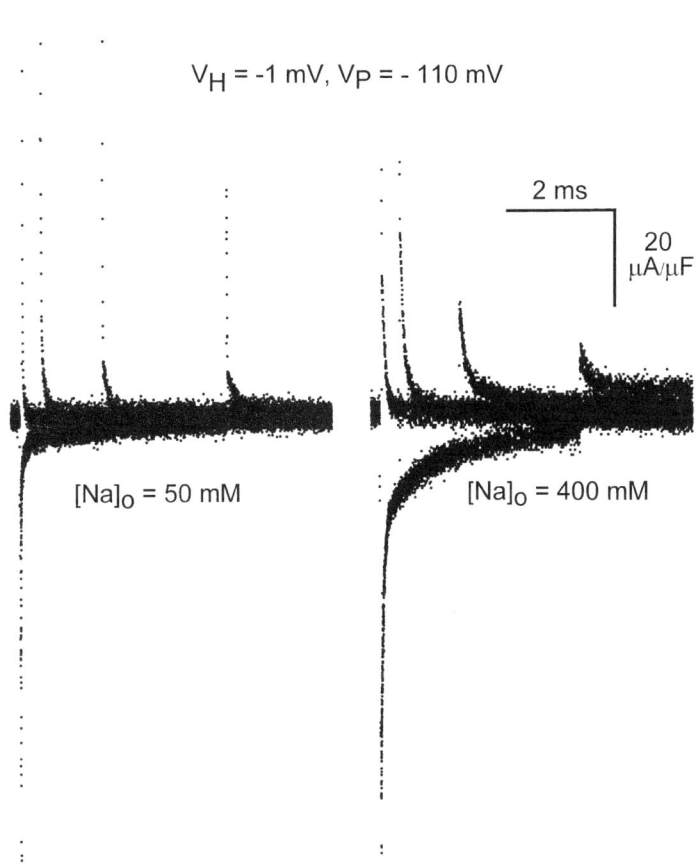

FIGURE 3. Superimposed H_2DTG-sensitive transient currents elicited by voltage jumps (-1 to -110 mV) of 0.05, 0.5, 2, and 5 ms duration recorded in the presence of 50 mM (*left*) and 400 mM (*right*) external Na (filter 100 kHz, sampling 500 kHz). Transients elicited upon return from -110 to -1 mV are dominated by fast and medium components with the amount of charge moved by each component a function of pulse duration. Increasing pulse duration (0.05–0.5 ms) initially enhances but subsequently decreases (0.5 to 2 ms and 2 to 5 ms) the amount of charge moved by the medium component while diminishing that moved by the fast at both 50 mM and 400 mM external Na.

we have found (FIG. 3), in addition to the unresolved fast components (ultrafast and fast components) and the previously described slow component, a third component of charge movement with resolved, medium-speed kinetics.[59–61] The medium and slow components provide direct evidence for the presence of at least two distinct conformational transitions in the Na deocclusion/release limb of the transport cycle. The ultrafast/fast components confirm the presence of at least one ultrafast charge moving event associated with Na release to the exterior, presumably the electrodiffusive migration of Na ions along an extracellular access channel and/or small high-speed conformational changes[55] that accompany Na ion release.

REFERENCES

1. THOMAS, R. C. 1972. Electrogenic sodium pump in nerve and muscle cells. Physiol. Rev. **52:** 563–594.
2. POST, R. L. & P. C. JOLLY. 1957. The linkage of sodium, potassium, and ammonium active transport across the human erythrocyte membrane. Biochim. Biophys. Acta **25:** 118–128.
3. CHAPMAN, J. B. & E. A. JOHNSON. 1978. The reversal potential for an electrogenic sodium pump: A method for determining the free energy of ATP breakdown? J. Gen. Physiol. **72:** 403–408.
4. DE WEER, P. 1984. Electrogenic Pumps: Theoretical and Practical Considerations. In Electrogenic Transport: Fundamental Principles and Physiological Implications. M. P., Blaustein & M. Lieberman, Eds. p. 1–15. Raven Press. New York.
5. LÄUGER, P. & G. STARK. 1970. Kinetics of carrier-mediated ion transport across lipid bilayer membranes. Biochim. Biophys. Acta **211:** 458–466.
6. LÄUGER, P. 1984. Thermodynamic and kinetic properties of electrogenic ion pumps. Biochim. Biophys. Acta **779:** 307–341.
7. LÄUGER, P. 1991. Electrogenic Ion Pumps. Sinauer Associates, Inc., Publishers. Sunderland, Massachusetts.
8. DE WEER, P., D. C. GADSBY & R. F. RAKOWSKI. 1988. Voltage dependence of the Na-K pump. Annu. Rev. Physiol. **50:** 225–241.
9. APELL, H.-J. 1989. Electrogenic properties of the Na,K pump. J. Membr. Biol. **110:** 103–114.
10. GADSBY, D. C. 1990. The Na/K pump of cardiac myocytes. In Cardiac Electrophysiology, from Cell to Bedside. D. P. Zipes & J. Jalife, Eds. p. 35–50. W. B. Saunders Co. New York.
11. HORISBERGER, J.-D. 1994. The Na,K-ATPase: Structure-Function Relationship. R. G. Landes Co. Austin, TX.
12. BAMBERG, E., H. J. BUTT, A. EISENRAUCH & K. FENDLER. 1993. Charge transport of ion pumps on lipid bilayer membranes. Q. Rev. Biophys. **26:** 1–25.
13. HANSEN, U.-P., D. GRADMANN, D. SANDERS & C. L. SLAYMAN. 1981. Interpretation of current-voltage relationships for "active" ion transport systems: I. Steady-state reaction kinetic analysis of Class-I mechanisms. J. Membr. Biol. **63:** 165–190.
14. GADSBY, D. C. & M. NAKAO. 1989. Steady-state current-voltage relationship of the Na/K pump in guinea pig ventricular myocytes. J. Gen. Physiol. **94:** 511–537.
15. NAKAO, M. & D. C. GADSBY. 1989. [Na] and [K] dependence of the Na/K pump current-voltage relationship in guinea pig ventricular myocytes. J. Gen. Physiol. **94:** 539–565.
16. RAKOWSKI, R. F., L. A. VASILETS, J. LATONA & W. SCHWARZ. 1991. A negative slope in the current-voltage relationship of the Na^+/K^+ pump in Xenopus oocytes produced by reduction of external $[K^+]$. J. Membr. Biol. **121:** 177–187.
17. SAGAR, A. & R. F. RAKOWSKI. 1994. Access channel model for the voltage dependence of the forward-running Na^+/K^+ pump. J. Gen. Physiol. **103:** 869–893.
18. RAKOWSKI, R. F., D. C. GADSBY & P. DE WEER. 1989. Stoichiometry and voltage-dependence of the sodium pump in voltage-clamped, internally-dialyzed squid giant axon. J. Gen. Physiol. **93:** 903–941.
19. GARRAHAN, P. J. & I. M. GLYNN. 1967. The behaviour of the sodium pump in red cells in the absence of external potassium. J. Physiol. **192:** 159–174.
20. GARRAHAN, P. J. & I. M. GLYNN. 1967. The incorporation of inorganic phosphate into adenosine triphosphate by reversal of the sodium pump. J. Physiol. **192:** 237–256.

21. BÉHÉ, P. & L. TURIN. 1984. Arrest and reversal of the electrogenic sodium pump under voltage clamp. Proc. 8th International Biophysical Congress, 304 (Abstr).
22. STÜRMER, W., R. BUHLER, H.-J. APELL & P. LÄUGER. 1991. Charge translocation by the Na,K-pump: II. Ion binding and release at the extracellular face. J. Membr. Biol. **121**: 163–176.
23. STÜRMER, W. & H.-J. APELL. 1992. Fluorescence study on cardiac glycoside binding to the Na,K-pump. Ouabain binding is associated with movement of electrical charge. FEBS Lett. **300**: 1–4.
24. CLARKE, R. J., A. ZOUNI & J. F. HOLZWARTH. 1995. Voltage-sensitivity of the fluorescent probe RH421 in a model membrane system. Biophysical J. **68**: 1406–1415.
25. GADSBY, D. C., R. F. RAKOWSKI & P. DE WEER. 1993. Extracellular access to the Na,K pump: Pathway similar to ion channel. Science **260**: 100–103.
26. MITCHELL, P. 1969. Chemiosomotic coupling and energy transduction. Theoret. Exp. Biophys. **2**: 159–216.
27. ANDERSEN, O. S., J. E. N. SILVEIRA & P. R. STEINMETZ. 1985. Intrinsic characteristic of the proton pump in the luminal membrane of a tight urinary epithelium. J. Gen. Physiol. **86**: 215–234.
28. JENNINGS, M. L. & M. A. MILANICK. 1997. Membrane transport in single cells. In Handbook of Physiology: Section 14: Cell Physiology. J. F. Hoffman & J. D. Jamieson, Eds. Oxford University Press.
29. HANSEN, U. P., J. TITTOR & D. GRADMANN. 1983. Interpretation of current-voltage relationships for "active" ion transport systems: II. Non steady-state reaction kinetic analysis of Class-I mechanisms with one slow time-constant. J. Membr. Biol. **75**: 141–169.
30. FENDLER, K., E. GRELL, M. HAUBS & E. BAMBERG. 1985. Pump currents generated by the purified Na^+K^+-ATPase from kidney on black lipid membranes. EMBO J. **4**: 3079–3085.
31. BORLINGHAUS, R., H.-J. APELL & P. LÄUGER. 1987. Fast charge translocations associated with partial reactions of the Na,K-pump. I. Current and voltage transients after photochemical release of ATP. J. Membr. Biol. **97**: 161–178.
32. JØRGENSEN, P. L. & J. H. COLLINS. 1986. Tryptic and chymotryptic cleavage sites in sequence of alpha-subunit of $(Na^+ + K^+)$-ATPase from outer medulla of mammalian kidney. Biochim. Biophys. Acta **860**: 570–576.
33. NAKAO, M. & D. C. GADSBY. 1986. Voltage dependence of Na translocation by the Na/K pump. Nature **323**: 628–630.
34. RAKOWSKI, R. F. 1993. Charge movement by the Na/K pump in *Xenopus* oocytes. J. Gen. Physiol. **101**: 117–144.
35. HOLMGREN, M. & R. F. RAKOWSKI. 1994. Pre-steady-state transient currents mediated by the Na/K pump in internally perfused *Xenopus* oocytes. Biophys. J. **66**: 912–922.
36. GARRAHAN, P. J. & I. M. GLYNN. 1967. The stoichiometry of the sodium pump. J. Physiol. (Lond) **192**: 217–235.
37. FAHN, S., G. J. KOVAL & R. ALBERS. 1966. Sodium potassium-activated adenosine triphosphatase of electrophorus electric organ. J. Biol. Chem. **241**: 1882–1889.
38. STÜRMER, W., H.-J. APELL, I. W. & P. L. 1989. Conformational transitions and charge translocation by the Na,K pump: Comparison of optical and electrical transients elicited by ATP-concentration jumps. J. Membr. Biol. **110**: 67–86.
39. FORBUSH, B., III. 1984. Na+ movement in a single turnover of the Na pump. Proc. Natl. Acad. Sci. USA **81**: 5310–5314.
40. FORBUSH, B., III. 1985. Rapid ion movements in a single turnover of the Na^+ pump. In The Sodium Pump: 4th International Conference on Na,K-ATPase. I. M. Glynn & J. C. Ellory, Eds. p. 599–611. The Company of Biologists. Cambridge, UK.
41. HEYSE, S., I. WUDDEL, H.-J. APELL & W. STÜRMER. 1994. Partial reactions of the Na,K-ATPase: Determination of rate constants. J. Gen. Physiol. **104**: 197–240.

42. FRIEDRICH, T. & G. NAGEL. 1997. Comparison of Na$^+$/K$^+$-ATPase currents activated by ATP concentration or voltage jumps. Biophys. J. **73:** 186–194.
43. GOLDSHLEGER, R., S. J. KARLISH, A. REPHAELI & W. D. STEIN. 1987. The effect of membrane potential on the mammalian sodium-potassium pump reconstituted into phospholipid vesicles. J. Physiol. **387:** 331–355.
44. GOLDSHLEGER, R., Y. SHAHAK & S. J. KARLISH. 1990. Electrogenic and electroneutral transport modes of renal Na/K ATPase reconstituted into proteoliposomes. J. Membr. Biol. **113:** 139–154.
45. BAHINSKI, A., M. NAKAO & D. C. GADSBY. 1988. Potassium translocation by the Na$^+$/K$^+$ pump is voltage insensitive. Proc. Natl. Acad. Sci. USA **85:** 3412–3416.
46. BIELEN, F. V., H. G. GLITSCH & F. VERDONCK. 1991. Dependence of Na$^+$ pump current on external monovalent cations and membrane potential in rabbit cardiac Purkinje cells. J. Physiol. **442:** 169–189.
47. BIELEN, F. V., H. G. GLITSCH & F. VERDONCK. 1993. Na$^+$ pump current-voltage relationships of rabbit cardiac Purkinje cells in Na$^+$-free solution. J. Physiol. **465:** 699–714.
48. GADSBY, D. C., M. NAKAO, A. BAHINSKI, G. NAGEL & M. SUENSON. 1992. Charge movements via the cardiac Na,K-ATPase. Acta Physiol. Scand. Suppl. **607:** 111–123.
49. BERLIN, J. R. & R. D. PELUFFO. 1997. Mechanism of electrogenic reaction steps during K$^+$ transport by the Na,K-ATPase. *In* Na/K-ATPase and Related Transport ATPases: Structure, Mechanism, and Regulation. Ann. N. Y. Acad. Sci., this volume.
50. BÜHLER, R. & H.-J. APELL. 1995. Sequential potassium binding at the extracellular side of the Na,K-pump. J. Membr. Biol. **145:** 165–173.
51. OR, E., R. GOLDSHLEGER & S. J. D. KARLISH. 1996. An effect of voltage on binding of Na$^+$ at the cytoplasmic surface of the Na$^+$-K$^+$ pump. J. Biol. Chem. **271:** 2470–2477.
52. SCHWAPPACH, B., W. STÜRMER, H.-J. APELL & S. J. D. KARLISH. 1994. Binding of sodium ions and cardiotonic steroids to native and selectively trypsinized Na,K pump, detected by charge movements. J. Biol. Chem. **269:** 21620–21626.
53. WUDDELL, I. & H.-J. APELL. 1995. Electrogenicity of the sodium transport pathway in the Na,K-ATPase probed by charge-pulse experiments. Biophys. J. **69:** 909–921.
54. HILGEMANN, D. W. 1994. Channel-like function of the Na,K pump probed at microsecond resolution in giant membrane patches. Science **263:** 1429–1432.
55. HILGEMANN, D. W. 1994. Flexibility and constraint in the interpretation of Na$^+$/K$^+$ pump electrogenicity: What is an access channel? *In* The Sodium Pump: Structure Mechanism, Hormonal Control and its Role in Disease. E. Bamberg & W. Schoner, Eds., 507–528. Steinkopff verlag. Darmstadt.
56. HILGEMANN, D. W. 1997. Recent electrical snapshots of the cardiac Na,K-pump. *In* Na/K-ATPase and Related Transport ATPases: Structure, Mechanism, and Regulation. Ann. N. Y. Acad. Sci., this volume.
57. HOLMGREN, M., F. BEZANILLA, D. C. GADSBY, P. DE WEER & R. F. RAKOWSKI. 1995. Charge translocation mediated by the Na/K pump in internally dialyzed squid giant axon. Biophys. J. **68:** A256.
58. WAGG, J., M. HOLMGREN, D. C. GADSBY, F. BEZANILLA, R. F. RAKOWSKI & P. DE WEER. 1996. Na/K pump-mediated charge movements as an assay of Na$^+$ translocation and occlusion/deocclusion rates. Biophys. J. **70:** A19.
59. WAGG, J., M. HOLMGREN, D. C. GADSBY, F. BEZANILLA, R. F. RAKOWSKI & P. DE WEER. 1996. Na/K pump-mediated charge movements reveal three distinct deocclusion steps accompanying extracellular release of three sodium ions. J. Gen. Physiol. **108:** 30A.
60. WAGG, J., M. HOLMGREN, D. C. GADSBY, F. BEZANILLA, R. F. RAKOWSKI & P. DE WEER. 1997. Na/K pump-mediated charge movements reporting deocclusion of 3 Na$^+$. Biophys. J. **72:** A25.
61. WAGG, J., M. HOLMGREN, D. C. GADSBY, F. BEZANILLA, R. F. RAKOWSKI & P. DE WEER. 1997. Na/K pump-mediated transient charge movements reporting deocclusion and release of three Na$^+$. J. Gen. Physiol. **110:** 43A.
62. KAPLAN, J. H. 1993. Molecular biology of carrier proteins. Cell **72:** 13–18.

63. KAPLAN, J. H., J. M. ARGÜELLO, C. R. G. ELLIS-DAVIES & S. LUTSENKO. 1994. The stabilization of cation binding and its relation to Na^+/K^+-ATPase structure and function. *In* The Sodium Pump: Structure Mechanism, Hormonal Control and its Role in Disease. E. Bamberg & W. Schoner, Eds. Steinkopff verlag. Darmstadt.
64. LUTSENKO, S. & J. H. KAPLAN. 1995. Organization of P-type ATPases: Significance of structural diversity. Biochemistry **34:** 15607–15613.
65. GADSBY, D. C., M. NAKAO & A. BAHINSKI. 1991. Voltage-induced Na/K pump charge movements in dialyzed heart cells. Soc. Gen. Physiol. Ser. **46:** 355–371.
66. LÄUGER, P. & H.-J. APELL. 1988. Transient behaviour of the Na^+/K^+-pump: Microscopic analysis of nonstationary ion-translocation. Biochim. Biophys. Acta **944:** 451–464.

The Na,K-Pump as a Channel

A New Approach to the Study of the Structure-Function Relationship of a P-Type ATPase[a]

JEAN-DANIEL HORISBERGER[b] AND XINYU WANG

Institute of Pharmacology and Toxicology
Bugnon 27
CH-1005 Lausanne, Switzerland

In the absence of three-dimensional structure, the main strategy in getting information about the active site of a protein is to introduce modifications in the domain thought to be involved in its activity and to study by various techniques the effects of these modifications on functional parameters. This approach was extensively used for P-ATPases in order to understand the mechanism of ion translocation across the cell membrane. Structural modifications were obtained mainly by enzymatic treatment of the mature enzyme or by site-directed mutagenesis. Although the same approach has been fairly successful for several cation channels, allowing identification of the part of the protein involved in the structure of the ion pore despite extensive study of various forms of trypsinized enzyme and hundreds of mutants, mostly in Ca-ATPase, Na,K-ATPase, and H-ATPase, the image of their ion-translocating structure is still far from clear.

One reason for this relatively disappointing result may be that the physiological parameters that are usually measured in mutants, such as maximal activity and apparent affinity for substrates or inhibitors, depend not only on a static structure but also on the complex succession of conformational changes, and ligand binding and unbinding, that forms the pump transport cycle.

This problem is illustrated in the detailed functional study that we performed on mutants of the α subunit of the Na,K-pump in which the NH_2-terminus had been truncated by 31 or 40 amino acid residues.[1] The results of this study are briefly summarized. It should be possible to overcome some of this difficulty if it were possible to study properties of the Na,K-pump analogous to those of a channel. Later in this chapter we provide evidence that there may be circumstances in which the Na,K-pump may have "channel-like" properties and propose that investigation of these properties may yield useful information about the cation pathway through the protein.

NH_2-TERMINAL TRUNCATION OF THE α SUBUNIT

Because we were surprised that modification of what is clearly an intracellular domain of the protein produced a large change in the apparent affinity for external K^+

---

[a] This work was supported by grant 31-45867.95 (to J-DH) from the Swiss Fonds National de la Recherche Scientifique.

[b] To whom correspondence should be addressed. Tel: +41 21 692 5362; fax: +41 21 692 5355; e-mail: Jean-Daniel.Horisberger@ipharm.unil.ch

(FIG. 1), we characterized the function of the NH_2-terminal truncation mutants in more detail. All measurements were performed in *Xenopus* oocytes expressing wild-type or mutant *Bufo* Na, K-pumps. (The relative ouabain resistance of the *Bufo* Na,K-pump allows selective inhibition of the endogenous Na,K-pump.) The truncated mutants, obtained by site-directed mutagenesis, had a lower maximal transport rate in the physiological electrogenic (3 Na^+/2 K^+ exchange) transport mode, whereas two indicators of the absolute number of active pumps, the level of expression estimated by the proton-stimulated ouabain-sensitive inward current (ref. 2; also see below) or the maximal translocated charge, demonstrated that the levels of expression were similar. The estimated turnover rate of the N-truncated mutant was $16.3 + 2.7$ s^{-1} ($n = 7$) compared to $59.7 + 9.3$ s^{-1} ($n = 9$) in the wild type.[1]

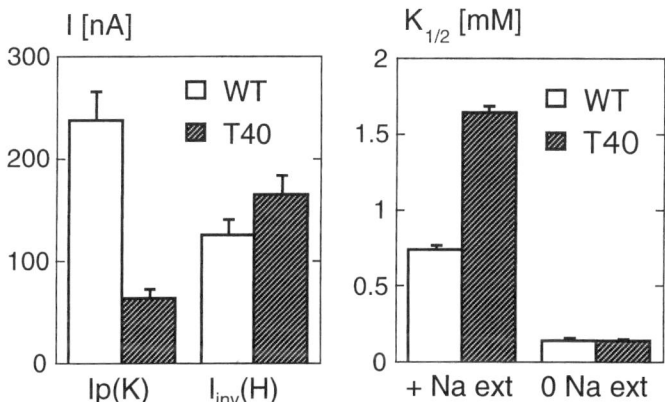

FIGURE 1. Effect of α subunit NH_2-terminal truncation on the Na,K-ATPase activity. (**Left**) Mean values of the K^+-activated pump current, Ip(K), at -50 mV, and the ouabain-sensitive inward current at -100 mV and pH 6.0, $I_{inv}(H)$, in 14 oocytes expressing wild-type (WT) *Bufo* α subunit and 16 oocytes expressing a 40-residue NH_2-terminal truncated α subunit (T40); *Bufo* $\beta1$ subunit was expressed in both groups. (**Right**) Half-activation constant of the Na,K-pump current by external K^+ in the presence (+ Na ext) and the absence (0 Na ext) of extracellular Na^+. (Data from ref. 1.)

We have studied the kinetics of the transient current that follows a fast voltage jump in the absence of external K^+, a situation in which the Na,K-pump functions essentially in the Na^+/Na^+ exchange mode.[3] Data show that the charge-translocating event corresponding to the forward E_1 to E_2 transition and Na^+ release to the extracellular side was about threefold slower in the N-truncated mutant. Analysis of a simplified kinetic model (FIG. 2) following the principle of the Post-Albers scheme indicated that the expected effects of slowing the E_1 to E_2 transition were, first, a reduction of the maximal turnover rate, and, second, a reduction in the apparent affinity for external K^+ when the E_2 to E_1 backward reaction rate had a finite value (reflecting the situation in which the presence of extracellular Na^+ allows this backward reaction to occur) but little effect when the E_2 to E_1 backward reaction rate was set to zero (reflecting the situation in the absence of extracellular Na^+). As all these predicted effects were indeed observed with the N-truncated mutant, we concluded that the reduction in the apparent affinity for external K^+ in this mutant could be entirely explained by the modification of the forward rate constant of a single conformational change, without any modification, direct or indirect, of the K^+ binding site. Our results by no means exclude the possibility that NH_2-terminal truncation may induce other modifications in the pump cycle and therefore do not contradict at all the observations showing modification of the K^+ deocclusion step at low ATP concentration that Daly et al.[4] made with a similarly truncated enzyme.

This example illustrates the difficulty of interpreting physiological data in terms of structure when the functional parameters that we measure depend on a kinetically complex process such as the many-step transport cycle of a P-ATPase.

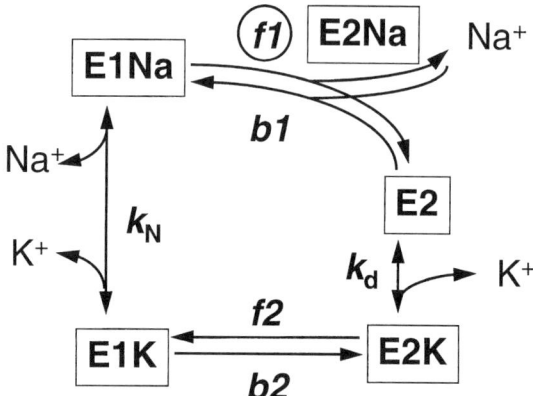

FIGURE 2. Simplified kinetic model of the Na,K-pump transport cycle. This scheme follows the principle of the Albers-Post model. Parameter $f1$ is the rate constant of the forward E1 to E2 conformational change, a step that includes the release of Na^+ to the external side in this simplified scheme. Parameter $b1$ is the rate constant of the reverse step; $b1$ depends on the external Na^+ concentration and has a value of 0 in the absence of external Na^+. K_d is the intrinsic dissociation constant of K^+ for its external binding site and K_N is an equilibrium binding constant that summarizes the binding of Na^+ and K^+ to their internal binding sites. It can be shown[1] that a reduction of $f1$ entails a reduction of the maximal turnover rate of the cycle, an increase of the half-activation constant for external K^+ when $b1$ value is set to 200 s^{-1} (a reasonable value for $b1$ in the presence of extracellular Na^+), but very little effect when $b1$ is 0 (with no extracellular Na^+).

PROTON CURRENT THROUGH THE Na,K-ATPase

A first mode of function of the Na,K-pump that may indicate "channel-like" behavior is provided by the ouabain-sensitive pH-dependent inward current first observed by Efthymiadis et al.[5] and that we characterized in more detail.[2] The fact that this ouabain-sensitive inward current occurs in the absence of any external cation (except for protons) and is associated with an intracellular acidification supports the proposition that it is mediated by an electrogenic inflow of protons. Its relation to other modes of function of the Na,K-pump, such as Na/H exchange[6] or electroneutral "uncoupled" sodium efflux,[7] has still to be defined. However, the large amplitude of the inward ouabain-sensitive current at low pH and at high negative membrane potentials makes it very unlikely that this current is mediated by a fixed stoichiometry cation exchange mechanism. Using the *Xenopus* Na,K-pump, the ouabain-sensitive current at pH 6.0 and at -130 mV amounts to 1.5-fold the absolute value of the current generated by the maximal turnover rate of the "normal" forward Na^+/K^+ exchange mode.[2] We recently observed that this ratio was even higher (>threefold) when we expressed the rat α1 isoform of the Na,K-pump (Horisberger and Wang, unpublished data). Therefore, we conclude that at low pH and large negative membrane potentials, the ouabain-sensitive inward current must be mediated by an "uncoupled" electrogenic inflow of protons, that is, a conductive type of transport, although we cannot rule out the existence of other mode(s) of transport at smaller membrane potentials. We have found no better explanation for this observation than the hypothesis that the E_2 cation-empty state is somewhat leaky for protons. This property of the pump would therefore not depend on the kinetics of the transport cycle. It would require only that the pump can reach and be accumulated in the E_2 cation-empty state.

PALYTOXIN/Na,K-PUMP INTERACTION

The second channel-like function of the Na,K-pump is provided by the effect of palytoxin. This toxin is known to induce cation conductance in membranes containing the Na,K-pump, and because this effect is prevented by ouabain, it has been assumed that this conductance is created by the interaction of PTX with the Na,K-pump.[8,9] Recently, Scheiner-Bobis and colleagues[10,11] provided strong arguments in favor of this hypothesis by showing a specific effect of PTX in yeast expressing a vertebrate Na,K-pump.

We investigated the effect of PTX on *Xenopus* oocytes expressing endogenous or exogenous Na,K-ATPases. As with other cell types,[8,9] we observed that PTX induces a large conductance, which is poorly selective between monovalent cations. Patch clamp single-channel recordings in the presence of PTX in the pipette showed a 7–8 pS channel with a reversal potential compatible with cation conductance and slow gating kinetics. This channel was never observed in the absence of PTX; in particular, there was no channel with this type of conductance and gating kinetics at high negative membrane potential at untreated oocytes. The effects of palytoxin were reversible with a k_{off} of about 0.4 min^{-1}.

We investigated the kinetics of this effect of PTX and its interaction with potassium and ouabain. As observed in other systems,[8] we found that the palytoxin effect was prevented by previous inhibition of the Na,K-pump by ouabain and by the presence of a high extracellular K^+ concentration, as illustrated in FIGURE 3a.

The addition of ouabain after the effect of PTX was established produces a decrease in PTX-induced conductance with a kinetic very similar to that observed when PTX was removed. These results can be explained by the hypothesis that PTX interacts in a reversible manner with the E2 state of the Na,K-pump (FIG. 3b). Binding of PTX and ouabain is mutually exclusive.

Considering the size of the single-channel conductance (7 pS), it is noteworthy that only 1–10 million PTX/Na,K-pump channels are sufficient to account for the macroscopic conductance that we observe after treatment with 5 nM PTX, and this number corresponds to only about 0.1% of the 5 billion Na,K-pumps normally active at the oocyte membrane.

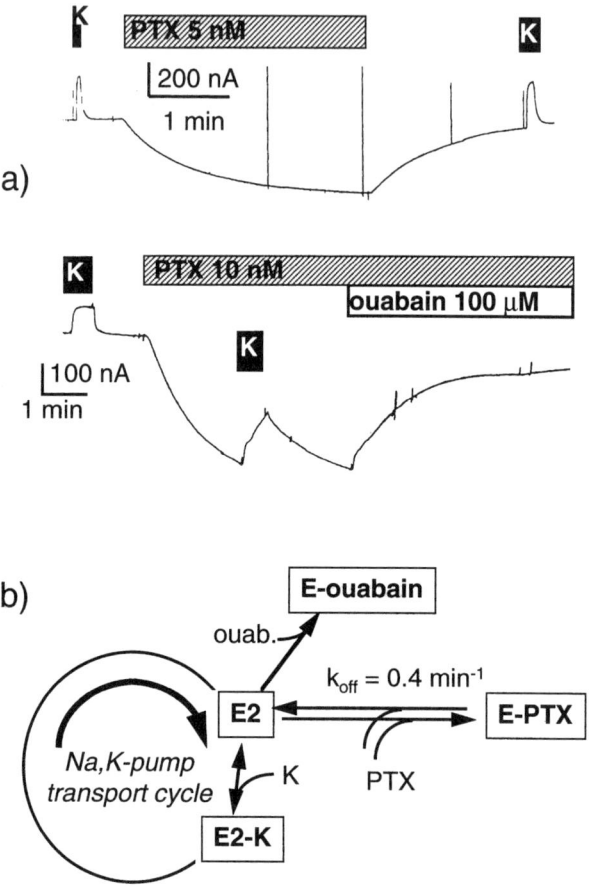

FIGURE 3. Kinetics of PTX binding and interaction with ouabain and potassium. (**a**) **Top trace** shows the effect of 5 nM palytoxin (PTX) on the holding current at -50 mV and on the conductance (current deflection due to a -50 mV to 0 mV voltage step). The outward current activated by 10 mM K^+, the Na,K-pump current, was measured before and after exposure to PTX. The effect of PTX was reversible. **Bottom trace** shows the effect of K^+ (10 mM) and of ouabain (100 μM) on PTX-induced conductance. Both K^+ and ouabain induced a decline in conductance after a time course similar to that observed after the removal of PTX (see *top trace*). (**b**) Kinetic scheme compatible with the observations shown in **a**.

We also studied the effect of PTX on oocytes expressing the *Bufo marinus* Na,K-pump,[12] a form of Na,K-ATPase that is moderately resistant to ouabain with a K_I of about 50 μM in the presence of 10 mM K^+. As shown in FIGURE 4, we compared oocytes expressing either α or β subunits of *Bufo* versus oocytes expressing only the β subunit; these oocytes only have endogenous *Xenopus* ouabain-sensitive α subunits. Using a large dose of ouabain (100 μM) to inhibit close to 100% of the ouabain-sensitive pumps, we observed no effect of PTX on oocytes expressing only ouabain-sensitive pumps. In contrast, PTX induced a large conductance in oocytes expressing the ouabain-resistant *Bufo* Na,K-pump, even though with this concentration of ouabain a significant fraction of pump activity was inhibited.

This finding provides additional proof of the necessity of direct interaction of PTX with the Na,K-pump to induce the cation channel. Using the "alternating access channel" model of ion pump proposed by Läuger,[13] this can be explained by the hypothesis that PTX modifies the pump to allow a temporary *simultaneous* opening of the two gates (internal and external energy barriers). If this were the case, the permeating cation would follow the same pathway in the PTX-induced channel and in the "normal" a$^+$/K$^+$ exchange mode of the Na,K-ATPase as proposed by Redondo *et al.*[11]

EFFECT OF MERCURY ON THE Na,K-PUMP

Several investigators have used site-directed mutagenesis and various cysteine binding molecules to probe the pore of ion channels.[14–16] We plan to use the same strategy to explore the cation pathway through the Na,K-pump using mercury as a probe for cysteine residues exposed in this pathway. We recently showed that when $HgCl_2$ was applied to the external surface of an oocyte, there was an exponential decline in K-activated Na,K-pump current with a rate constant proportional to the Hg^{2+} concentration,[17] indicating a simple first-order kinetic with a K_{on} of about 7.10^3 $M^{-1} \cdot s^{-1}$. This inhibition was practically irreversible, and we could obtain only an upper limit of about 200 nM for the binding constant. This inhibition was prevented by strophanthidin, suggesting that the Hg^{2+} binding site could be covered or hidden by the cardiotonic steroid.[17]

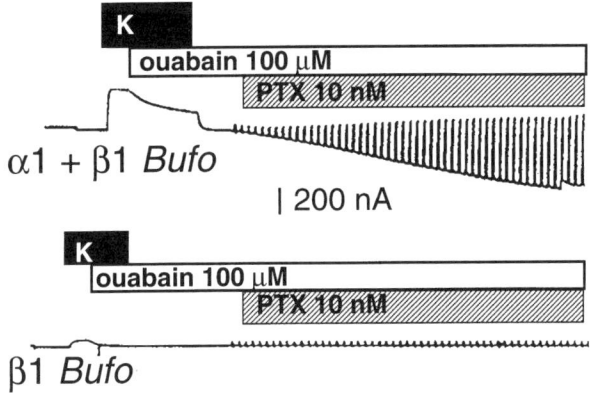

FIGURE 4. Effect of palytoxin (PTX) in oocytes expressing the *Bufo* Na,K-pump. **Top panel** shows a current recording from an oocyte expressing the α1 and the β1 subunit of the *Bufo* Na,K-ATPase, and **bottom trace** shows a recording from an oocyte expressing only the β1 subunit of the *Bufo* Na,K-ATPase. The Na,K-pump current was first measured by the addition of 10 mM K$^+$; then 100 μM ouabain was added, resulting in partial inhibition of oocytes expressing the *Bufo* α subunit but in total inhibition of oocytes expressing only the endogenous *Xenopus* α subunit. The subsequent addition of PTX induced a large increase in conductance in the oocyte expressing the ouabain-resistant *Bufo* Na,K-pump, but it had no effect in oocytes in which all Na,K-pump activity had been inhibited by ouabain.

The first-order kinetics of the inhibition by external application of mercury and the protection by strophanthidin strongly suggested that this inhibition was due to the binding of Hg to an external site. There are no cysteine residues in the extracellular loops according to the current 10 TM segment model. One cysteine is present in the first transmembrane segment (C_{113} in *Xenopus* α1) and is known for its role in ouabain binding,[18] and we have shown that this residue is accessible to the membrane-impermeant cysteine reagent HDMA.[19] We tested the effect of $HgCl_2$ on mutants in which this cysteine had been replaced by either serine or tyrosine. In both cases the inhibition produced by a 1-minute exposure to 10 µM $HgCl_2$ was smaller in the mutant than in the wild-type Na,K-pump.[17] We interpreted these results as indicating that Hg^{2+} can bind to a site accessible from the extracellular side of the membrane, resulting in inhibition of Na,K-pump activity. However, this is probably not the only site of action of Hg^{2+} on the pump. Anner and Moosmayer[20] provided evidence for intracellular sites of action for Hg^{2+}, and interaction of mercury and palytoxin on the Na,K-pump reveals the existence of at least a second Hg^{2+} binding site accessible from the extracellular side of the membrane. These results are presented and discussed by Wang and Horisberger in an extended abstract in this volume.

In summary, we investigated two modes of function of the Na,K-pump, proton-dependent ouabain-sensitive inward current and palytoxin-induced conductance, for both of which there is evidence supporting a channel-like behavior, that is, a conductive mode of ion transport. The possibility of using cysteine-binding reagents such as mercury, may permit exploration of the structure of the cation pathway by the "cysteine scanning" strategy.[14–16]

REFERENCES

1. WANG, X., F. JAISSER & J.-D. HORISBERGER. 1996. J. Physiol. (Lond.). **491:** 579–594.
2. WANG, X. & J.-D. HORISBERGER. 1995. Am. J. Physiol. Cell Physiol. **37:** C590–C595.
3. NAKAO, M. & D. C. GADSBY. 1986. Nature **323:** 628–630.
4. DALY, S. E., L. K. LANE & R. BLOSTEIN. 1994. J. Biol. Chem. **269:** 23944–23948.
5. EFTHYMIADIS, A., J. RETTINGER & W. SCHWARZ. 1993. Cell Biol. Int. **17:** 1107–1116.
6. POLVANI, C. & R. BLOSTEIN. 1988. J. Biol. Chem. **263:** 16757–16763.
7. DISSING, S. & J. F. HOFFMAN. 1990. J. Gen. Physiol. **96:** 167–193.
8. HABERMANN, E. 1989. Toxicon **27:** 1171–1187.
9. TOSTESON, M. T., J. A. HALPERIN, Y. KISHI & D. C. TOSTESON. 1991. J. Gen. Physiol. **98:** 969–985.
10. SCHEINER-BOBIS, G., D. MEYER ZU HERINGDORF, M. CHRIST & E. HABERMANN. 1994. Mol. Pharmacol. **45:** 1132–1136.
11. REDONDO, J., B. FIEDLER & G. SCHEINER-BOBIS. 1996. Mol. Pharmacol. **49:** 49–57.
12. JAISSER, F., C. M. CANESSA, J.-D. HORISBERGER & B. C. ROSSIER. 1992. J. Biol. Chem. **267:** 16895–16903.
13. LÄUGER, P. 1979. Biochim. Biophys. Acta **552:** 143–161.
14. LU, Q. & C. MILLER. 1995. Science **268:** 304–307.
15. KURZ, L. L., R. D. ZUHLKE, H. J. ZHANG & R. H. JOHO. 1995. Biophys. J. **69:** 288.
16. PEREZGARCIA, M. T., N. CHIAMVIMONVAT, E. MARBAN & G. F. TOMASELLI. 1996. Proc. Natl. Acad. Sci. USA **93:** 300–304.
17. WANG, X. & J.-D. HORISBERGER. 1996. Mol. Pharmacol. In press.
18. CANESSA, C. M., J.-D. HORISBERGER, D. LOUVARD & B. C. ROSSIER. 1992. EMBO J. **11:** 1681–1687.
19. ANTOLOVIC, R., W. SCHONER, K. GEERING, C. M. CANESSA, B. C. ROSSIER & J.-D. HORISBERGER. 1995. FEBS Lett. **368:** 169–17.
20. ANNER, B. M. & M. MOOSMAYER. 1992. Am. J. Physiol. **262:** F843–F848.

Mechanism of Electrogenic Reaction Steps during K$^+$ Transport by the Na,K-ATPase

JOSHUA R. BERLIN[a] AND R. DANIEL PELUFFO

Bockus Research Institute
Graduate Hospital
Philadelphia, Pennsylvania 19146

The stoichiometry of ion transport by the Na,K-ATPase, 3 Na$^+$ in exchange for 2 K$^+$ necessitates that some reaction steps move net charge across the membrane electric field.[1] For this reason, the steady-state distribution of enzyme intermediates will depend on the transmembrane potential (V_M). Broadly speaking, electrogenic reactions during ion transport by the Na,K-ATPase are expected to fall into four categories: (1) ion binding, (2) ion translocation, (3) nucleotide binding and dissociation, and (4) phosphorylation and dephosphorylation.[2] These V_M-dependent reaction steps could affect the overall rate of ion transport if they are rate limiting or if they control the amount of enzyme going into rate-limiting steps. In this regard, previous studies showed that ion transport by the Na,K-ATPase is altered by V_M and that one reaction step, extracellular Na$^+$ (Na$_o^+$) binding, is highly electrogenic.[3-6] However, a single electrogenic reaction does not explain the V_M-dependent properties of ion transport by the Na,K-ATPase, particularly the effect of extracellular K$^+$.[7-9]

Using electrophysiological techniques to measure steady-state and transient kinetics of K$^+$ transport, we have been able to show that a charge moving reaction step occurs as part of the extracellular K$^+$ binding/dissociation step. In addition, by comparing the V_M-dependent properties of extracellular Na$^+$- and K$^+$-dependent steps, it appears that under some conditions, these steps should account completely for the V_M dependence of ion transport by the Na,K-ATPase.

V_M AND EXTRACELLULAR ION DEPENDENCE OF STEADY-STATE ION TRANSPORT BY THE Na,K-ATPase

FIGURES 1 and 2 summarize the effects of extracellular Na$^+$ and K$^+$ on the V_M-dependent properties of steady-state ion transport by the Na,K-ATPase. Na,K-pump current was measured in rat cardiac ventricular myocytes that were whole-cell voltage-clamped with low resistance patch electrodes containing a high Na$^+$ intracellular salt solution. Experiments were performed at 35–37°C using a protocol modified from Gadsby et al.[3] so that Na,K-ATPase activity, that is, Na,K-pump current, could be studied in the absence of contaminating ionic currents.[10]

---

[a]Address for correspondence: Joshua R. Berlin, PhD, Department of Physiology, Allegheny University of the Health Sciences, Allegheny University Hospital—Graduate, 415 S. 19th St., Philadelphia, PA 19146 (tel: 215-893-2377; fax: 215-893-4178; e-mail: berlinj@mail.med.upenn.edu).

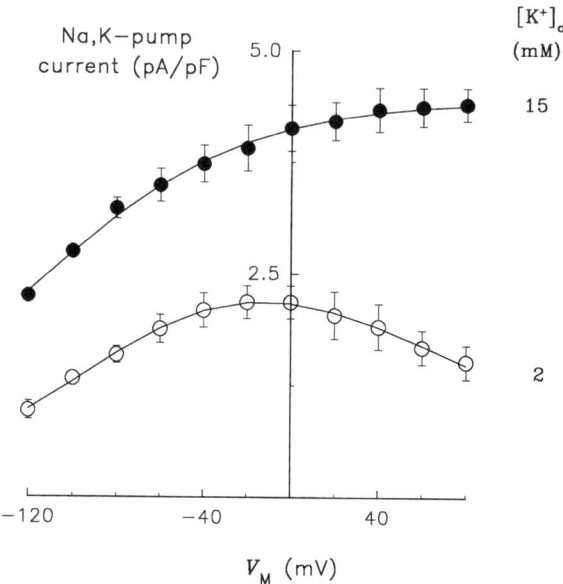

FIGURE 1. Current-voltage relationships for Na,K-pump current in Na$^+$-containing solutions ($n = 5$). Current arising from the Na,K-ATPase is defined as the 1 mM ouabain-sensitive difference current in all figures. *Solid curves* represent the fitting of an equation derived from a pseudo two-state model that assumes Na$^+$ and K$^+$ must pass through an external high-field access channel to reach their binding sites in the pump.[12]

As shown previously,[11,12] the V_M dependence of Na,K-pump current is affected by extracellular Na$^+$. In Na$_o^+$-containing solutions, Na,K-pump current displays a positive slope at negative potentials (FIG. 1) that is abolished in Na$_o^+$-free solutions (FIG. 2A). Detailed studies of Na$^+$ transport by the Na,K-ATPase have also shown that a highly electrogenic reaction step resides in the dissociation and rebinding of Na$^+$ at extracellular facing sites in the enzyme.[5,6]

The V_M dependence of Na,K-pump current is also affected by extracellular K$^+$ (K$_o^+$). In low K$_o^+$, ouabain-sensitive current displays a negative slope with increasingly positive V_M. As K$_o^+$ is increased to concentrations that produce maximal Na,K-pump current, this negative slope disappears (FIGS. 1 and 2A) consistent with previous reports.[7,8] These data show that a charge moving reaction step depends on K$_o^+$. The presence of such an electrogenic step means that the equilibrium distribution of enzyme intermediates will be a function of V_M in a K$_o^+$-dependent fashion. It is then anticipated that the concentration for half-maximal activation of Na,K-pump current by K$_o^+$ ($K_{0.5}$) will depend on V_M (FIG. 2B). In the absence of Na$_o^+$, the fraction (λ) of V_M that drops during this electrogenic reaction step can be estimated by fitting the data with the following function:

$$K_{0.5}^\gamma = K_{0.5}^{0\gamma} * \exp(\gamma\lambda U)$$

where $U = V_M F/RT$. Using this equation, $K_{0.5}^0$ (the $K_{0.5}$ at 0 mV) is 222 ± 6 μM

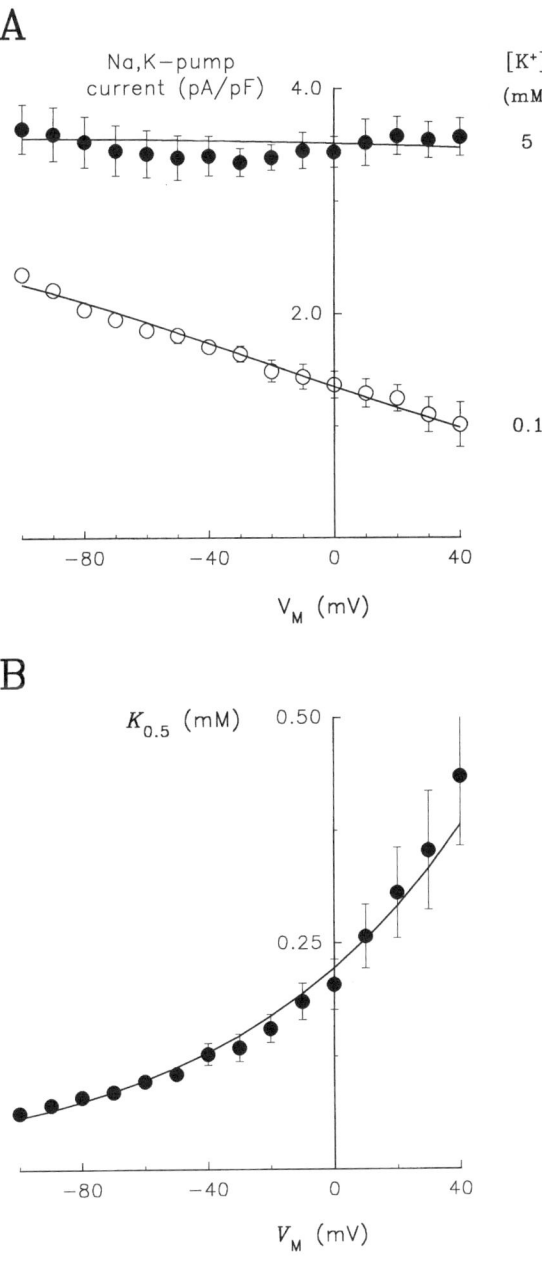

FIGURE 2. Current-voltage relationships for Na,K-pump current in Na^+-free extracellular solutions. (**A**) Steady-state current ($n = 5$). (**B**) V_M dependence of $K_{0.5}$ for extracellular K^+ ($n = 5$).

($n = 5$), γ (a term analogous to a Hill coefficient that accounts for the possible interaction of more than one K^+ during ion transport) is 0.94 ± 0.09 and λ is 0.36 ± 0.04. Thus, the charge-moving step during K^+ transport dissipates approximately one third of the electrical gradient across the membrane. Previous reports also found that $K_{0.5}$ for K_o^+ is V_M dependent.[7,12]

One interesting observation derived from these studies is that K^+ congeners used in biochemical assays for Na,K-ATPase are also electrical congeners for Na,K-pump activation. To illustrate this point, experiments similar to those in FIGURE 2A were repeated with Tl^+, a high affinity K^+ congener (not shown). Using the foregoing equation to fit the resulting data, we found that $K_{0.5}^0$ was 76 ± 2 μM ($n = 5$), threefold smaller than the $K_{0.5}^0$ for K^+. More importantly, γ and λ were 1.12 ± 0.13 and 0.37 ± 0.02, respectively, similar to values calculated for K_o^+. The implication of these data is that the electrogenic step is the same for these congeners. Thus, K^+ congeners can be used to investigate the mechanism of V_M-dependent reaction steps.

PRESTEADY-STATE KINETICS OF K^+-DEPENDENT ELECTROGENIC REACTIONS

Data derived from steady-state current measurements do not distinguish whether the V_M-dependent steps occur during ion binding or translocation processes. To derive this information, transient reaction kinetics were used to study Na^+-dependent charge-moving reaction steps.[4] A similar strategy was employed here to investigate the V_M-dependent kinetics of K^+ transport.

To study the transient reaction kinetics of K^+ transport, presteady-state membrane currents were measured during electroneutral K^+-K^+ exchange by the Na,K-pump. Experiments were conducted with Na^+-free (tetramethylammonium ion-containing) extracellular solutions and a high K^+, Na^+-free intracellular solution containing 1.3 mM magnesium phosphate and 0.7 mM magnesium ATP. These conditions optimally promote K^+-K^+ exchange.[13] The kinetics of ion transport were also slowed by performing experiments at 14°C with Tl^+ included in the superfusion solution.

Under these conditions, Tl^+-dependent transient currents were observed with rapid changes in V_M. FIGURE 3B shows a typical trace of ouabain-sensitive difference current observed with depolarizations from -40 to $+40$ mV in the presence of 100 μM extracellular Tl^+ (Tl_o^+), a concentration near the $K_{0.5}$ for steady-state Na,K-pump current activation. The outward current during depolarization decayed exponentially at a rate that was much slower than charging of linear membrane capacitance. The absence of steady-state current suggests that this process is electroneutral. The quantity of charge moved, calculated as the integral of current during the depolarization ("On" charge), was a saturable function of membrane potential. In addition, the quantity and voltage dependence of "On" charge were similar to ouabain-sensitive charge moved upon returning to the holding potential ("Off" charge). These observations suggest that the electroneutral exchange process contains a reversible electrogenic reaction step arising from a fixed number of charged particles in the membrane. These properties of ouabain-sensitive Tl_o^+-dependent charge movement demonstrate the presence of a V_M-dependent step during electroneutral K^+-K^+ exchange.

The absence of ouabain-sensitive charge movement in Tl_o^+- and Na_o^+-free solutions (FIG. 3A) also demonstrates that intrinsic protein charges do not move through the membrane electric field in the absence of transported ion. Thus, charge moving conformational changes in the protein must be disfavored in the absence of bound ion. The Tl_o^+ dependence of charge movement also suggests that the charge moving step occurs during ion translocation of charges by Tl^+-bound conformations of the Na,K-ATPase (FIG. 4, top) or as Tl_o^+ binds at a site(s) within the membrane electric field, that is, in an ion well (FIG. 4, bottom).

To determine how an electrogenic reaction during ion binding or translocation would affect the V_M dependence of ion transport, K^+-K^+ exchange was modeled as the consecutive sequence of the following reactions:

$$K_i^+ + E_i \underset{\phi_2}{\overset{\phi_1}{\rightleftharpoons}} E_A \underset{\phi_4}{\overset{\phi_3}{\rightleftharpoons}} E_B \underset{\phi_6}{\overset{\phi_5}{\rightleftharpoons}} E_o + K_o^+$$

In this model, intracellular K^+ (K_i^+) binds to an inward-facing enzyme state (E_i). After a series of interconversions among K^+-bound states, E_A and E_B, K_o^+ is released from an outward-facing enzyme state (E_o). In steady-state conditions, the concentration of

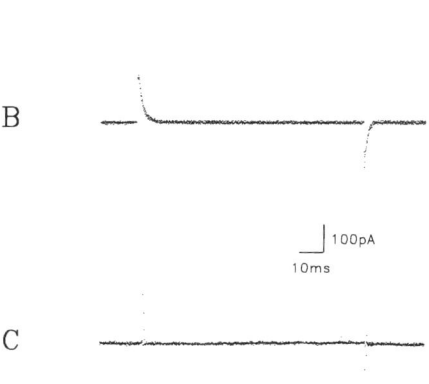

FIGURE 3. Extracellular Tl^+-dependent transient charge movement. Voltage clamp pulses from −40 to +40 mV are displayed. Each trace is a difference current obtained in the presence and absence of ouabain during superfusion with 0 (**A**), 0.1 (**B**), and 2 mM Tl^+-containing solutions (**C**). Reprinted from Peluffo and Berlin,[16] with permission.

E_B (expressed as a fraction of total enzyme) is described as:

$$E_B = \frac{1}{(1 + \phi_2\phi_4/(\phi_1\phi_3[K^+]_i) + \phi_4/\phi_3 + \phi_5/(\phi_6[K^+]_o))}.$$

To express the V_M dependence of the overall reaction, forward rate constants are described as $\phi_i = \phi_i^o \exp((1 - \delta z U)$ for $i = 1, 3, 5$ and reverse rate constants are described as $\phi_i = \phi_i^o \exp(-\delta z U)$ for $i = 2, 4, 6$ where ϕ_i^o is the rate constant at 0 mV.[1] The symmetry factor, δ, apportions the effect of V_M on the reaction ($0 \leq \delta \leq 1$) and z is the effective valence of the charge moved.

If ion translocation steps are electrogenic, that is, ϕ_3 and/or ϕ_4 are V_M-dependent, the steady-state distribution of enzyme intermediates with various K_o^+ is expected to appear as in FIGURE 5A. Increasing K_o^+ does not alter the V_M dependence of steady-state E_B concentration but does control the amount of enzyme intermediate

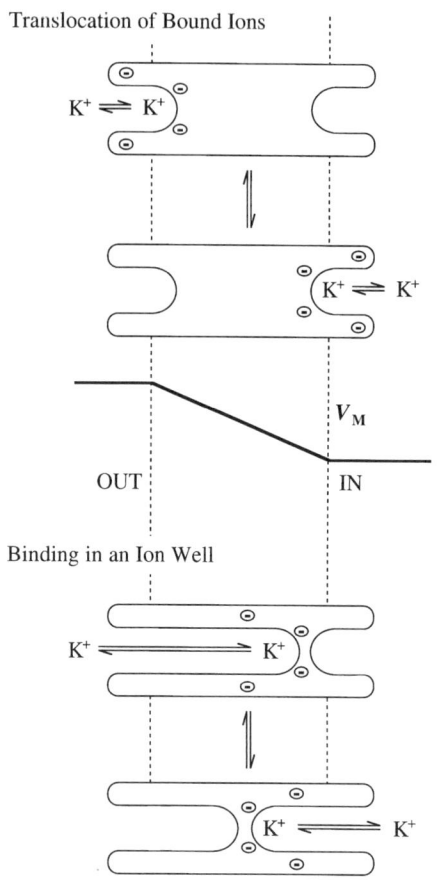

FIGURE 4. Representation of net charge movement during translocation of bound ion (*top*) and during ion binding at a site in the membrane electric field (*bottom*).

available to go through the V_M-dependent step. On the other hand, increasing K_o^+ should cause a rightward displacement in the V_M dependence of steady-state E_B concentration if the extracellular ion binding step is electrogenic, that is, ϕ_5 and/or ϕ_6 are V_M dependent (FIG. 5B). At a high enough K_o^+, a voltage step would be unable to change the steady-state distribution of enzyme intermediates so charge movements should disappear. Such an experiment is shown in FIGURE 3C. Transient charge movements were examined in the presence of 2 mM Tl_o^+, a concentration that activates maximal steady-state Na,K-pump current at all test potentials (-100 to $+40$ mV). With this saturating ion concentration, charge movements were not observed. Thus, charge movements are dependent on V_M and Tl_o^+, consistent with the model that the ion binding step is the electrogenic reaction during K^+ transport.

Although these results demonstrate that the ion binding process is V_M dependent, they do not necessarily mean that extracellular ion binding occurs in a high-field access channel, as was previously proposed.[6] An equally attractive hypothesis, first pointed out by Hilgemann,[14] is that extracellular ion binding occurs through a transitional ion recognition step that is followed by a rapid chelation reaction involving movement of intrinsic protein charges. Such a chelation mechanism can be modeled in the foregoing scheme by forcing the concentration of E_B to remain close to zero. The steady-state distribution of the enzyme intermediate, E_A, then resembles FIGURE 5B for all experimentally realizable V_M and extracellular ion concentrations. Thus, the precise mechanism of this reaction is not clear, even though the data demonstrate the existence of an electrogenic ion binding step.

COMPARISON OF TRANSIENT CHARGE MOVEMENTS DURING Na^+-Na^+ AND K^+-K^+ EXCHANGE

Tl_o^+-dependent charge movement has several properties in common with charge movements observed under conditions of Na^+-Na^+ exchange (for comparison, see refs. 4 and 15). Of particular note is the effect of V_M on the rate constant (k) for decay of the transient currents. Both Na_o^+- (FIG. 6A) and Tl_o^+-dependent charge movements (FIG. 6B) display a highly asymmetric relationship between the rate constant for current decay and V_M. Such an asymmetric relationship is expected when the V_M-dependent properties of the reaction appear to reside entirely with ion association, but not with ion dissociation.[6] This similarity raises the possibility that Na_o^+- and

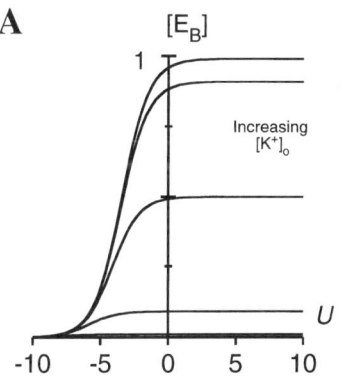

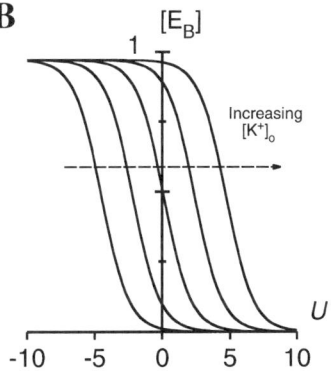

FIGURE 5. Model calculations. The steady-state amount of enzyme intermediate, E_B, was calculated from the four-state consecutive scheme shown in the text. (**A**) Electrogenic interconversions of ion bound enzyme are modeled by V_M dependence in ϕ_4. (**B**) Electrogenic ion binding steps are modeled with V_M dependence in ϕ_6.

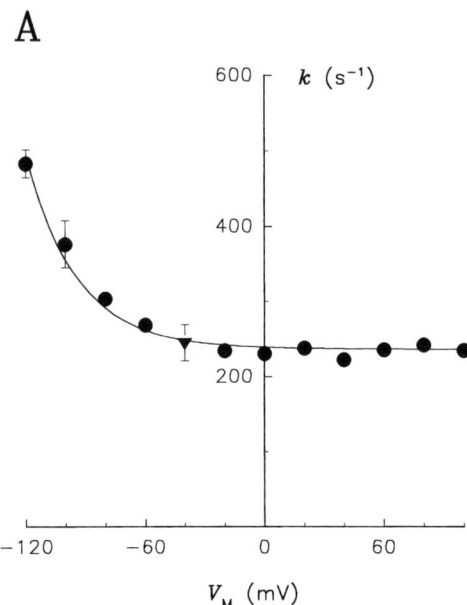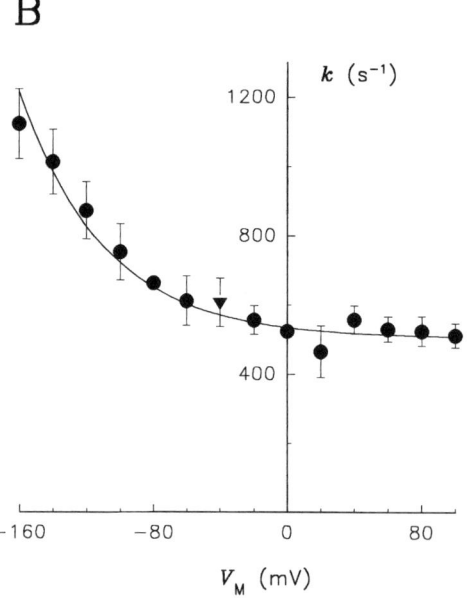

FIGURE 6. Rate constants for extracellular Na^+- (**A**) and Tl^+-dependent charge movements (**B**). The exponential rate constants determined from the transient currents at each V_M are shown. Na^+-dependent charge movements were measured according to Nakao and Gadsby[4] at 25°C ($n = 3$). Tl^+-dependent charge movements were measured at 14°C ($n = 4$). *Solid curves* were calculated with equation 1 in Peluffo & Berlin.[16]

Tl_o^+-dependent charge movements reflect the ion-dependent properties of a single electrogenic reaction step. Two features, however, do distinguish these electrogenic reactions. The effective charge, z, and the activation energy for Na_o^+-dependent transient charge movement are larger than those for Tl_o^+.

The effective valence, z, of charge movement is determined from the V_M dependence of charge movement kinetics shown in FIGURE 6, assuming that the observed rate constant (k) is the sum of pseudo-first order ion association and dissociation rate constants.[16] These calculations demonstrated that z is 0.66 ± 0.09 (n = 8) for Na_o^+-dependent charge movement but only 0.47 ± 0.07 (n = 4) for Tl_o^+-dependent charge movement. Thus, the fraction of the electrical field that is dissipated during these two reactions is different.

The activation energy of charge movement was calculated from the slope of an Arrhenius plot (ln rate vs the inverse of temperature). The values for Na_o^+- and Tl_o^+-dependent charge movements were 87.0 ± 3.8 kJ/mol (n = 7) and 27.5 ± 3.7 kJ/mol (n = 4), respectively, in the range of 11–35°C. Even though Na^+ and Tl^+ (or K^+) have different dehydration energies and/or coordination numbers,[17] the large disparity in activation energies may reflect ion binding to different enzyme conformations or sites. This idea is consistent with the hypothesis that the highly electrogenic step in Na^+ transport represents binding of a single extracellular sodium ion at neutral sites, whereas K^+ binding occurs at charged sites in the protein.[5]

In summary, binding of extracellular K^+ to the Na,K-ATPase is a V_M-dependent process that is, in many respects, similar to V_M-dependent binding of extracellular Na^+. These results suggest that highly V_M-dependent reaction steps during ion transport by the Na,K-ATPase are confined to extracellular ion binding in the presence of high concentrations of intracellular Na^+ and K^+. The molecular mechanism of these V_M-dependent ion binding reactions, in particular the existence of an ion access channel within the enzyme, remains to be determined.

REFERENCES

1. LÄUGER, P. 1991. Electrogenic Ion Pumps. Sinauer Associates. Massachusetts.
2. REYNOLDS, J. A., E. A. JOHNSON & C. TANFORD. 1985. Proc. Natl. Acad. Sci. USA **82:** 6869–6873.
3. GADSBY, D. C., J. KIMURA & A. NOMA. 1985. Nature **315:** 63–65.
4. NAKAO, M. & D. C. GADSBY. 1986. Nature **323:** 628–630.
5. WUDDEL, I. & H.-J. APELL. 1995. Biophys. J. **69:** 909–921.
6. GADSBY, D. C., R. F. RAKOWSKI & P. DE WEER. 1993. Science **260:** 100–103.
7. RAKOWSKI, R. K., L. A. VASILETS, J. LATONA & W. SCHWARZ. 1991. J. Membr. Biol. **121:** 177–187.
8. BIELEN, F. V., H. G. GLITSCH & F. VERDONCK. 1991. J. Physiol. **442:** 169–189.
9. BÜHLER, R., W. STÜRMER, H.-J. APELL & P. LÄUGER. 1991. J. Membr. Biol. **121:** 141–161.
10. ISHIZUKA, N., A. J. FIELDING & J. R. BERLIN. 1996. Jpn. J. Physiol. **46:** 215–223.
11. NAKAO, M. & D. C. GADSBY. 1989. J. Gen. Physiol. **94:** 539–565.
12. SAGAR, A. & R. F. RAKOWSKI. 1994. J. Gen. Physiol. **103:** 869–894.
13. SACHS, J. R. 1988. J. Physiol. **400:** 545–574.
14. HILGEMANN, D. W. 1994. In The Sodium Pump. E. Bamberg & W. Schoner, Eds. :507–516. Darmstadt Steinkopff. New York.
15. FENDLER, K., S. JARUSCHEWSKI, A. HOBBS, W. ALBERS & J. P. FROEHLICH. 1993. J. Gen. Physiol. **102:** 631–666.
16. PELUFFO, R. D. & J. R. BERLIN. 1997. J. Physiol. **501:** 33–40.
17. HILLE, B. 1992. Ionic Channels of Excitable Membranes. Sinauer Associates. Massachusetts.

Recent Electrical Snapshots of the Cardiac Na,K Pump[a]

DONALD W. HILGEMANN[b]

Department of Physiology
University of Texas Southwestern Medical Center at Dallas
5323 Harry Hines Blvd.
Dallas, Texas 75235-9040

Electrical methods allow us to study those partial reactions of the Na,K pump cycle that generate electrical current (i.e., move charges through part or all of the membrane electrical field). However, electrical methods can also be used to study partial reactions that are not electrogenic, namely, reactions that are functionally coupled to the electrogenic ones. The coupling of electrogenic and electroneutral reactions thus provides a great potential for the electrophysiological approach to elucidate pump function. This potential has been exploited by many groups whose results are represented in this volume as well as in previous proceedings of the Na,K pump meeting series.[1,2]

Because multiple reactions of the pump cycle are electrogenic, most prominently extracellular ion binding,[3–6] it can be projected that most, probably all, important partial reactions of the pump can be studied via electrical means under appropriate conditions. To date, Apell and his group[3,7] have exploited this situation most extensively with the kidney Na,K pump, using voltage-sensitive fluorescent dyes to monitor changes in charge distribution in the membrane electrical field. A similarly comprehensive approach was undertaken with the sodium pump in squid axon by Rakowski *et al.* (this volume). The giant membrane patch method[8] now offers important advantages that can be exploited in complementary experiments. It allows fast (MHz) and reliable voltage clamping of a large membrane area, it allows control of solution composition on both membrane sides, and it was recently combined with optical methods to photo-release transport substrates.[9]

Some of our recent results with the cardiac membrane Na,K pump are presented with a perspective on our long-term goal of complete characterization of cardiac pump function by electrical means. First, basic sodium translocation reactions are described, as resolved previously in cardiac giant patches; second, improved results for fast steps in sodium release are presented; third, transient pump behaviors during pump activity are described; and fourth, electrical signals that probably resolve extracellular potassium binding and occlusion steps are described.

METHODS

All results presented were obtained with giant membrane patches from guinea pig cardiac myocytes (25–33 μm in diameter/8–12 pF). Charge transfer rather than

---

[a]This work was supported by a grant-in aid from the American Heart Association.
[b]E-mail: HILGEMAN@UTSW.SWMED.EDU

membrane current was recorded[5]; therefore, membrane current is the first derivative of the charge signals. A rising signal corresponds to an outward current, a falling signal corresponds to an inward current, and a horizontal line indicates zero current. Conditions and solutions were the same as those employed previously. Unless indicated otherwise, N-methylglucamine was the major cation in solutions and was used as a substitute for other cations. The conditions of the experiments were selected to minimize conductances from sources other than the Na,K pump, and a number of experimental tests indicated no significant contamination of records from other mechanisms.

Unless otherwise indicated, charge movements were defined by subtracting signals in the absence of ATP from signals in the presence of ATP. The logic of this subtraction procedure is that the pump takes on an E_1 configuration in the absence of ATP, a state that is evidently electrically silent or nearly so. For all signals presented, otherwise identical experiments were carried out with 0.2 mM ouabain added to the extracellular (pipette) solution; charge signals were absent using the same protocols and subtraction procedures, indicating that the signals presented indeed arise from the Na,K pump.

Sodium Translocation Charge Movements

Charge movements related to sodium release to the extracellular side were obtained in the presence of extracellular sodium and in the absence of extracellular potassium; they are defined by the addition of 0.5 mM of cytoplasmic ATP. The slow components are comparable to those resolved in whole myocytes,[10] but the subtraction procedure is different. FIGURE 1A shows typical charge signals.[5] Subsequent records of fast sodium-dependent signals represent, in essence, an amplification of the portions of these signals that are circled.

In the Na,K pump cycle diagram in FIGURE 1B, the charge movements were interpreted as follows. These signals represent release of sodium to the extracellular side upon depolarization, and they represent sodium binding on hyperpolarization, albeit together with accompanying conformational changes. Although depolarization favors release of sodium from binding sites, this reaction is coupled to the opening of binding sites (deoocclusion); first, one sodium ion is released (#3) and then the binding sites may widen (#4) to release the last two sodium ions (#5). Results can be accounted for assuming that only very fast ion binding reactions are actually electrogenic in the E2P3N and E2P2N states. Other reactions cannot take place under these conditions. The E2P3N state must be assumed to have very low abundance, that is, it must be a high energy transitional state. The fast charge jumps seen upon changing membrane potential are thought to arise largely from the immediate binding and unbinding of sodium from the E2P2N state. Similar signals for the squid axon are described in this volume by Rakowski *et al.*

How can we test this model? Some of the problems have been discussed by others.[11] In fact, a very important signal whose existence is predicted by this model is missing from our records. If significant fractions of the Na,K pump exist in the E2P3N state, the model in FIGURE 1B would predict that a fast charge movement with relatively steep voltage dependence would correspond to rapid sodium ion diffusion across most of the membrane field. The identification of such a "steep" charge movement, even if it were small in magnitude, would add strong support to this model. This was attempted extensively under carefully selected conditions to favor the

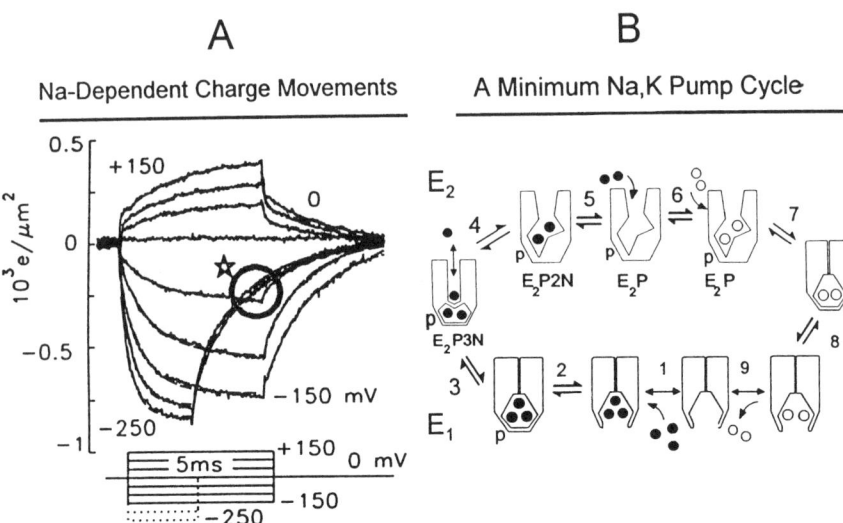

FIGURE 1. (**A**) Typical sodium-dependent charge movements of the cardiac Na,K pump in a giant membrane patch (see ref. 5 for details). Signals are a subtraction of charge records with and without cytoplasmic ATP. (**B**) Schematic diagram of the Na,K pump cycle. The conditions of the experiment in **A** should allow only reactions 3, 4, and 5 to take place. The charge movements in **A** can be simulated from this scheme, assuming that the only charge-moving reactions are instantaneous sodium-binding reactions taking place in the E2P3N state and the binding of two sodium ions to the EDP state, represented by reaction 5. The kinetics of charge movements observed, however, are determined by voltage-independent reactions 3 and 4. According to this scheme, one ion can traverse most of the membrane field in the E2P3N state, which is assumed to be a transitional state. If this is indeed the case, it should be possible to identify a fast charge movement with steep-voltage dependence when experimental conditions favor an accumulation of pumps in the E2P3N state. This would be the case in a midpotential range, where about half the pumps are in an E_1 configuration and half in an E_2 configuration, as marked by a *circle* and *star* in the figure.

presence of the expected fast, strongly voltage-dependent charge movement. Typical results are shown in FIGURE 2, whereby membrane potential was held at −80 mV. This membrane potential should maximize the population of exchangers in the E2P3N state; at more negative Em, most pumps are in the E_1 configuration, and at more positive potentials most pumps are in the E2P2N and EDP configurations. The pipette contains 140 mM sodium. Pulses were applied to potentials from +200 to −200 mV for 50 μs. The signals presented are an average of 10 records with 1-MHz recording; they correspond roughly to the conditions of the circled regions of charge movements in FIGURE 1A.

The charge transfer signals under these conditions show two obvious components: an "ultra fast" component that is not resolved in time, and time-dependent fast components. They are called fast because, if allowed to continue in time, these charge movements would come to a steady state over a much slower 3–5-ms time course, as in FIGURE 1A. The inset in FIGURE 2 shows the voltage dependence of the total charge moved; it is rather shallow. When the entire signals are analyzed as two components, an instantaneous charge and a time-dependent charge, both components are also rather shallow (Boltzmann slopes <0.25). When these signals are analyzed in more detail, however, multiple time-dependent components of the signals can be differentiated. For example, when the voltage dependence of these signals was analyzed using two-exponential fits of the time-dependent components, the steepest voltage-dependent components had Boltzmann slopes of up to 0.6. This is still a smaller slope than the slow components of charge signals in FIGURE 1A, but not by very much. Thus, the faster time-dependent components of these signals might possibly represent the binding of sodium in a deep ion well, as suggested by Gadsby et al.,[4] which may also involve conformational changes of binding sites.[3,5] Much further work is required to be confident about details.

When these signals are analyzed as only two components, the magnitudes of the

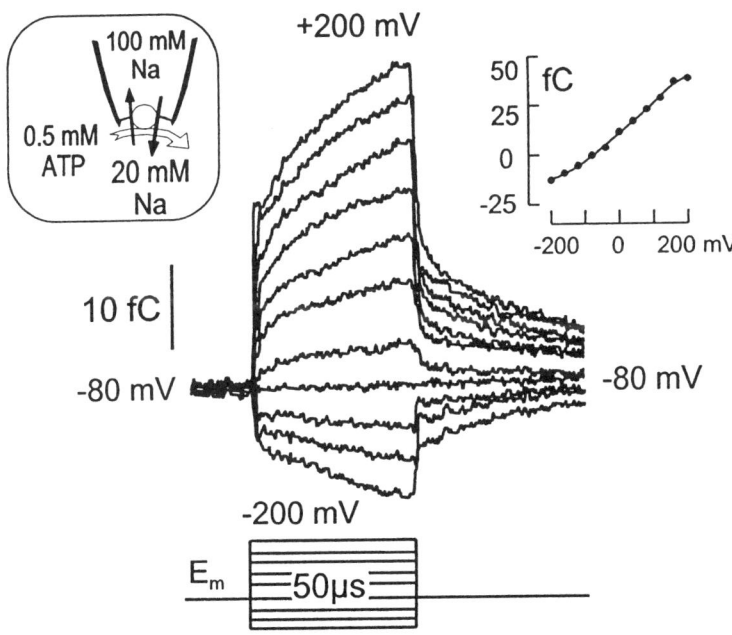

FIGURE 2. Fast sodium-dependent charge movements with a 1-MHz voltage clamp. Same conditions as in FIGURE 1A. In the 50 µs of the voltage step, two components of charge movements are obvious, one that is not resolved in time and one with a rate constant of about 30,000/s at −80 mV (see "off" charge movements). In first approximation, these two components behave as if they are taking place in parallel, not in sequence, because their magnitudes are about the same in the on and off directions. The **inset** shows the voltage dependence of the total charge movement. Note that it is rather shallow (Boltzmann slap). If the signals are divided into instant and time-dependent components only, both components show similarly shallow voltage dependence. However, it is also possible to fit two exponentials to the time-dependent components of these signals; in this case, the faster of the two time-dependent components shows a Boltzmann slope of 0.4–0.6. This steeper slope might represent the major charge-moving step of the pump cycle, namely, release of one sodium ion across most of the membrane electrical field.

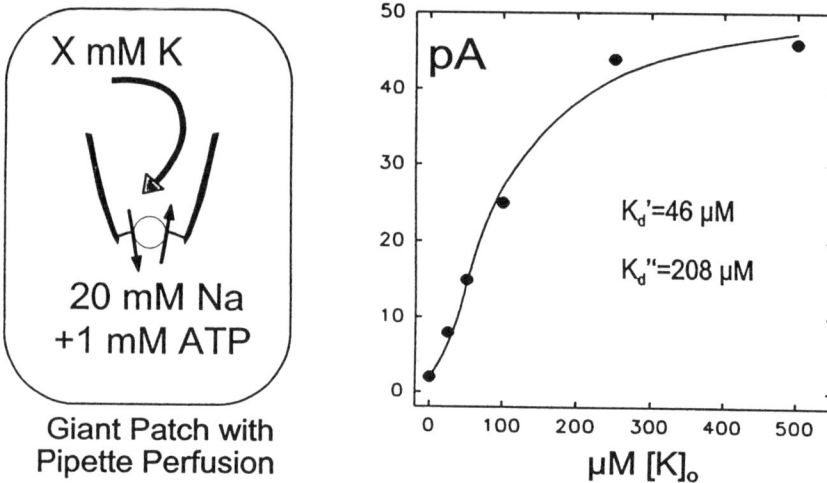

FIGURE 3. Extracellular potassium dependence of Na,K pump current in an excised giant cardiac membrane patch. Results are fitted to the equation:

$$\text{Ipump} = \text{Imax} * K^2 / (K^2 + K * Kd'' + KD' * Kd'')$$

where Ipump is the current activated by cytoplasmic ATP, Imax is the maximum pump current, K is the extracellular pipette (potassium) concentration, Kd' is the dissociation constant for the first potassium ion, and Kd'' is the dissociation for the second potassium ion. See text for further details.

fast (instant) and slow components are nearly the same on application of large voltage steps away from ("on" charge) and on returning to ("off" charge) the holding membrane potential of -80 mV. If valid, this result would indicate that the underlying reactions do not take place in sequential order; for sequential reactions, the kinetics of a slow reaction would be expected to completely dominate the charge movement in one direction (or the other) of the signals. The alternative then is that the underlying reactions take place in parallel, which would present important complications for an understanding of Na,K pump electrogenicity.

One possible interpretation is that two reactions with weak voltage dependence take place independently, but then permit subsequent reactions of ion binding sites to take place in sequence. Thus, the overall process of ion release may appear to have steep voltage dependence, even though no individual step actually has such voltage dependence. Note that this possibility has a parallel in "Hodgkin-Huxley" channel gating, in which multiple independent reactions enable channel opening. The obvious possible mechanisms are the movements of ions in and/or out of binding sites and fast rearrangements of charged residues, possibly including residues that are not immediately involved in ion binding. Some of the results obtained by Rakowski, Gadsby, and Deweer for the Na,K pump in squid axon (personal communication; see article by Rakowski *et al.* in this volume) differ in this respect, and further work in both systems will be required to understand the basis for this difference.

The Cycling Pump

Next, charge movements during steady-state pump activity are presented in dependence on the concentration of extracellular potassium. As a prerequisite to resolving such signals, we examined in detail the extracellular potassium dependence of the Na,K pump current in giant cardiac patches. To do so, we used a pipette perfusion technique to vary extracellular potassium, and we activated the pump with cytoplasmic ATP. All experiments were done in the absence of extracellular sodium, using N-methylglucamine as the only extracellular monovalent cation besides potassium and using 4 mM of magnesium as the only extracellular divalent cation. FIGURE 3 shows the potassium dependence of pump current, defined by the application and removal of cytoplasmic ATP (1 mM) in the presence of cytoplasmic sodium (20 mM). The ATP-induced pump current in the absence of extracellular potassium is <2% of the current with 0.5 mM extracellular potassium, and this result corresponds to the very small effect of ouabain in the absence of extracellular potassium in whole myocytes. In short, the "uncoupled" pump activity in cardiac membrane is very small.

The activation points for extracellular potassium are the average of two or three measurements in one patch, made in random order and differing by <10%. The activation curve shows cooperative behavior. It cannot be described well by a Hill equation because it is not symmetrical, but it can be described by an appropriate equation for sequential binding of two ions. The first dissociation constant (K_d') is 46 µM, and the second (K_d'') is 208 µM. The maximum pump current, about 50 pA,

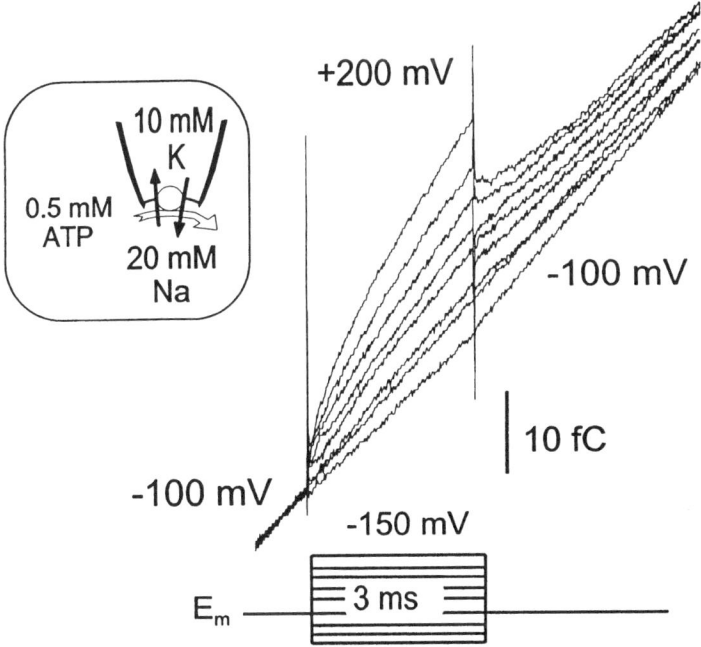

FIGURE 4. Charge records obtained in the presence of a saturating (10-mM) extracellular potassium concentration. See text for details.

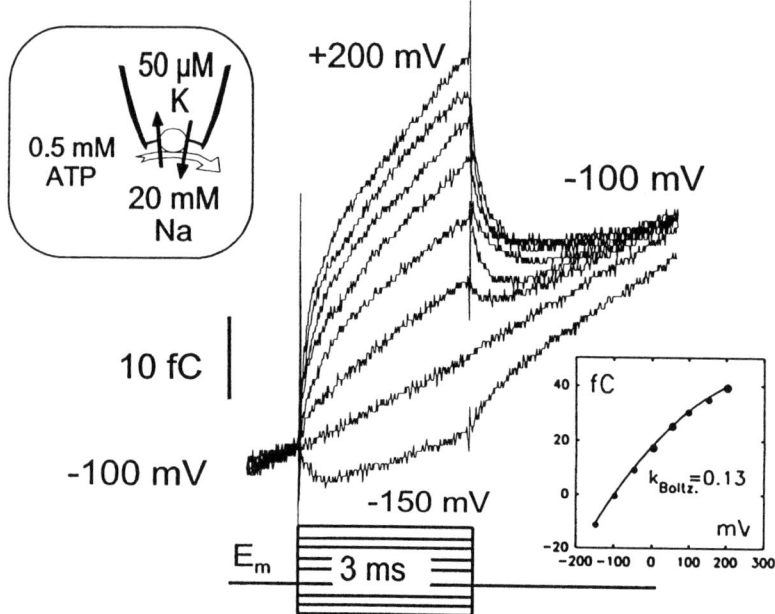

FIGURE 5. Charge records obtained in the presence of a limiting (50-μM) extracellular potassium concentration. See text for details.

corresponds to a pump current density of 5 pA/pF. This is substantially larger than usually obtained in whole cell recording. We speculate that this larger density represents the loss of regulatory factors in the excised patches.

FIGURE 4 shows charge records obtained for the pump when activated with 10 mM extracellular potassium (i.e., maximal potassium), 20 mM cytoplasmic sodium (maximal sodium), and 0.5 mM ATP (nearly maximal ATP). The records are defined by the subtraction of records with ATP from those without ATP. Voltage pulses are from -100 mV to potentials of -150 to $+200$ mV in 50-mV steps. The steady-state pump current, determined as the slope of the charge signals at the end of the pulses, increased weakly with depolarization; the current-voltage relation is plotted in FIGURE 6. The charge records display two types of transient behaviors on changing potential. There is a "slow" increase of charge as the signals approach a new rate on depolarization to positive potentials. The time course is similar to the slow charge signals recorded with extracellular sodium, and this component might reflect the existence of Na,K pumps in the sodium-loaded configurations during steady-state pump operation. The second type of charge signal is a fast "off" component on repolarization to -100 mV from the most positive potentials. The basis of these components is completely unknown. It is important to keep in mind that potassium occlusion and translocation are possible contributors to the kinetics of these signals.

FIGURE 5 shows the charge signals obtained when the same protocols were performed with only 50 μM of extracellular potassium. The steady-state current (i.e., slow of the rising charge signals) is now smaller and the asymptotic charge components (which correspond to transient currents) are now larger than those under high-potassium conditions. The insert in FIGURE 5 shows the voltage dependence of the charge movements, using the off charge to determine their magnitude. Note that the commonly calculated Boltzmann slope is only 0.13, indicating that the underlying reactions have only small voltage dependence.

What do these signals represent? Most pumps will be in the E_2P configuration, according to FIGURE 1B, when extracellular potassium is limiting. Presumably, therefore, the charge movements reflect potassium binding, including fast conformational changes of the pump binding sites and possibly the kinetics of pump dephosphorylation. FIGURE 6 shows the current-voltage relations of the pump current under these two conditions. Notably, we have not observed a negative slope in the current-voltage relation with limiting extracellular potassium in giant cardiac patches; even with longer voltage pulses, only small negative slopes have been detected.

Potassium Translocation (Occlusion) Charge Movements

Finally, FIGURE 7 shows charge movements isolated under conditions that were intended to isolate the potassium binding/occlusion reactions. The records were obtained in the presence of cytoplasmic sodium. Signals presented are a subtraction of records with 2 mM of paranitrophenylphosphate (PNPP) and 10 mM of phosphate from records without these agents. The logic of this subtraction is that PNPP and phosphate can phosphorylate the pump, but the presence of PNPP slows dephosphorylation. In this case, therefore, signals may include larger components which reflect the

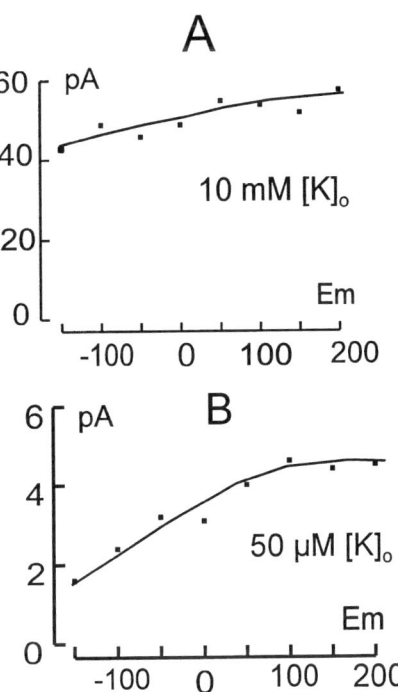

FIGURE 6. (A) Current-voltage relation of the Na,K pump determined from the results of FIGURE 4 with saturating potassium. (B) Current-voltage relation of the Na,K pump determined from the results of FIGURE 5 with limiting potassium. A possible concern is that the current at the end of the pulses in these experiments has not come to a steady state. See text for more details.

binding of potassium per se, because pump dephosphorylation is prevented. In fact, the signals do show large, very fast charge jumps which could not be resolved in time. These fast signals have only very weak voltage dependence, and they are rather similar to signals isolated for sodium binding/unbinding in FIGURE 3.

DISCUSSION

The results presented in this article illustrate some of the range of conditions that can be used to isolate Na,K pump charge movements. It should now be possible to

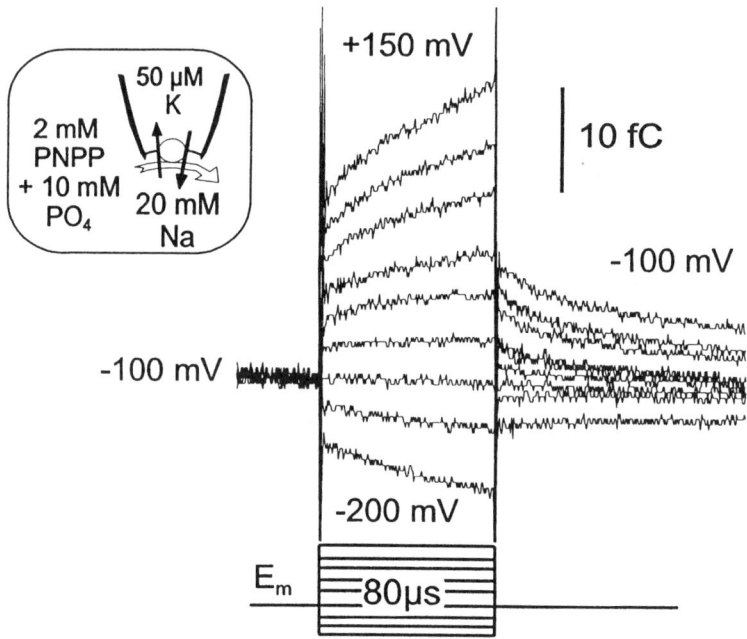

FIGURE 7. Fast charge movements of extracellular potassium binding by the Na,K pump. These signals, while similar to those presented with extracellular sodium, do not contain some of the fast-time dependent components (FIG. 2). See text for details.

step-wise include additional reactions, such as phosphorylation kinetics by including ADP and phosphate on the cytoplasmic side. These results hopefully will provide a basis for improving models of ion translocation. Most importantly, these results show that improved studies of the kinetics and relationships of partial reactions of the native cardiac Na,K pump are now possible.

As outlined in the experimental results, some of the results differ significantly from those of others using other preparations. The reasons for these differences are at present enigmatic; methodological problems are *not* expected to account for the differences. To summarize: (1) Current densities in giant cardiac patches are, with good certainty, larger than can be projected from whole-cell myocyte studies, and the rates of charge movements tend to be larger than those obtained in whole myocytes. (2) ATP activates almost no pump current in the absence of extracellular potassium and sodium, in contrast to some other studies indicating rather large uncoupled pump activities. (3) We detect only very small (or event no) negative slopes of current-voltage relations in the presence of limiting extracellular potassium concentrations. This occurs even though the charge movements recorded with low extracellular potassium are consistent with an electrogenicity of potassium binding reactions. It is expected that important regulational differences are at play in these differences, the basis of which will hopefully be revealed in future experimentation.

Author's note: A number of experimental results suggest that the Na,K pump in cardiac membrane may be regulated by additional membrane proteins and/or lipids which might act as cell signals. As just reported, the cardiacv pump in giant patches can be irreversibly inhibited to about 15% of control by cytoplasmic application of 2 M sucrose for 10 minutes, suggesting loss of associated proteins. Also, the cardiac pump can be irreversibly inhibited to about 15% of control by cytoplasmis application of a phosphatidylinosital-specific phospholipase which is known to cleave phosphatidylinositolglycans and glycosylphosphatidylinositol anchors. Much further work is obviously required to know whether these observations might be a key to identifying the expected regulatory factors at play.

ACKNOWLEDGMENTS

The experiment described in FIGURE 3 was performed by John Cammeron. I am grateful to D. C. Gadsby for discussion, criticism, and sharing unpublished data.

REFERENCES

1. BAMBERG, E. & W. SCHONER, Eds. 1994. The Sodium Pump: Structure, Mechanism, Hormonal Control and its Role in Disease. Steinkopff. Darmstadt.
2. DEWEER, P. & J. KAPLAN, Eds. 1990. The Sodium Pump: Structure, Mechanism and Regulation. Soc. Gen. Physiol. **46**.
3. HEYSE, S., I. WUDDEL, H. J. APELL & W. STURMER. 1994. Partial reactions of the Na,K-ATPase: Determination of rate constants. J. Gen. Physiol. **104**: 197–240.
4. GADSBY, D. C., R. F. RAKOWSKI & P. DE WEER. 1993. Extracellular access to the Na,K pump: Pathway similar to ion channel. Science **260**: 100–103.
5. HILGEMANN, D. W. 1994. Channel-like function of the Na,K pump probed at microsecond resolution in giant membrane patches. Science **263**: 1429–1432.
6. VASILETS, L. A. & W. SCHWARZ. 1993. Structure-function relationships of cation binding in the Na+/K(+)-ATPase. Biochim. Biophys. Acta **1154**: 201–222.
7. APELL, H. J., M. ROUDNA, J. E. CORRIE & D. R. TRENTHAM. 1996. Kinetics of the phosphorylation of Na,K-ATPase by inorganic phosphate detected by a fluorescence method. Biochemistry **35**: 10922–10930.
8. HILGEMANN, D. W. 1995. The Giant Membrane Patch. *In* Single Channel Recording by B. Sakmann and E. Neher. :307–327. Plenum Press. New York.
9. FRIEDRICH, T., E. BAMBERG & G. NAGEL. 1996. Na+,K(+)-ATPase pump currents in giant excised patches activated by an ATP concentration jump. Biophys. J. **71**: 2486–2500.
10. GADSBY, D. C., M. NAKAO & A. BAHINSKI. 1991. Voltage-induced Na/K pump charge movements in dialyzed heart cells. Soc. Gen. Physiol. Series **46**: 355–371.
11. RAKOWSKI, R. F., D. C. GADSBY & P. DE WEER. 1997. Voltage dependence of the Na/K pump. J. Membr. Biol. **155**: 105–112.

Na^+,K^+-ATPase Pump Currents Activated by an ATP Concentration Jump

Comparison of Studies with Purified Membrane Fragments and Giant Excised Patches[a]

GEORG NAGEL,[b] THOMAS FRIEDRICH,[c]
KLAUS FENDLER,[d] AND ERNST BAMBERG[e]

Max-Planck-Institut für Biophysik
Kennedyallee 70
D-60596 Frankfurt am Main, Germany

The Na^+,K^+-ATPase expels 3 Na^+ ions from the cell in exchange for 2 K^+ ions per ATP hydrolyzed, thereby generating an electric current that can directly be monitored by electrophysiological techniques. Several organs, such as kidney, heart, and nervous tissue, contain a very high density of Na^+,K^+-ATPase, and these sources are frequently used for studying the sodium pump's reaction cycle. We compared black lipid membrane (BLM) experiments with photolysis of caged ATP, as first published in 1985,[1–3] with recent patch clamp experiments combined with photolysis of caged ATP.[4] In the patch clamp study we used the giant patch technique[5,6] to record current through the excised plasma membrane patches of isolated ventricular myocytes from rat or guinea pig. This technique enabled us to study the Na^+,K^+-ATPase under the well-controlled conditions of an excised patch.[7]

Photolysis of caged ATP[8] allows fast changes of ATP concentration and has been used in experiments with right-side out vesicles[9] or with Na^+, K^+-ATPase containing membrane fragments attached to a BLM.[1–3,10–12] The BLM experiments, owing to a high signal-to-noise ratio, allowed detailed analysis of partial reactions, yielding rate constants for the electrogenic Na^+ transport step. A disadvantage is the capacitive coupling of the membrane fragments to the BLM which does not allow exact voltage control or direct recording of a steady-state electrical current.

It can be concluded that the rate-limiting step in the Na^+ transport limb, if measured under appropriate conditions, is electrogenic with a rate constant of ~200 s^{-1} at 24°C, confirming earlier work on pig kidney and eel electroplax Na^+,K^+-ATPase adsorbed to a BLM.[3,10] This conclusion is in contrast to a postulated rate of ~25 s^{-1} at 22°C,[2,12] derived from BLM experiments with caged ATP, but not contradictory to an ultrafast electrogenic release of Na^+,[13–15] if preceded by an

---

[a]This work was supported by the Max-Planck-Gesellschaft and the Deutsche Forschungsgemeinschaft (SFB169).
[b]Tel: +49-69-6303-303; e-mail: nagel@biophys.mpg.de
[c]Tel: +49-40-4717-6605; e-mail: tfriedri@uke.uni-hamburg.de
[d]Tel: +49-69-6303-306; e-mail: fendler@biophys.mpg.de
[e]Tel: +49-69-6303-300; e-mail: bamberg@biophys.mpg.de

electrogenic or electroneutral step with a rate of 200 s^{-1}, as this is kinetically equivalent.

MATERIALS AND METHODS

Cell Preparation. Ventricular myocytes from rat or guinea pig were isolated according to the method of Hilgemann.[5,6] For details see ref. 4.

Preparation of Purified Na$^+$,K$^+$-ATPase. Na$^+$,K$^+$-ATPase containing membrane fragments were prepared and purified from pig kidney as described previously.[1] The final solution contained 25 mM imidazole at pH 7.0 and about 2 mg/ml purified Na$^+$,K$^+$-ATPase with a specific activity of 20–30 μmol P$_i$/min/mg protein.

Experimental Solutions for Patch Clamp Experiments. **B-F:** Bath solution for continuous flow fast solution exchange experiments contained (in mM): 40 NaCl, 100 N-methyl-D-glucamine (NMG), 10 EGTA, 10 HEPES (for pH 7.4) or MES (for pH 6.3), 20 tetraethylammonium (TEA) Cl, 2 MgCl$_2$, pH 7.4 or 6.3 adjusted with HCl. **B-P:** Bath solutions for caged ATP photolysis at pH 6.3 or 7.4 contained (in mM): 40 NaCl, 100 NMG, 10 MES (for pH 6.3) or HEPES (for pH 7.4), 10 EGTA, 20 TEA Cl, and 2 MgCl$_2$, pH adjusted with HCl. Glutathione 1 mM or L-ascorbic acid 1 mM were also added. Hexokinase 10 U/ml (Sigma, Munich, Germany) together with glucose 1 mM was added to reduce the concentration of ATP. Caged ATP (P^3-1-[2-nitro]phenyl-ethyladenosine-5'-triphosphate, triethylammonium salt) was used in concentrations of 10–500 μM. It was synthesized as described previously.[1]

Pipette Solutions for Patch Clamp Experiments (in mM): **P-Na-K:** 145 NaCl, 10 HEPES, 5 KCl, 2 BaCl$_2$, 2 MgCl$_2$, 0.5 CdCl$_2$, adjusted to pH 7.4 with NaOH. **P-NMG-K:** 145 NMG, 10 HEPES, 5 KCl, 2 BaCl$_2$, 2 MgCl$_2$, 0.5 CdCl$_2$, adjusted to pH 7.4 with HCl.

Experimental Solutions for Bilayer Experiments (in mM): **Bl-Na-K:** 130 NaCl, 25 imidazole, 10 KCl, 3 MgCl$_2$, 1 DTT, adjusted to pH 6.2 with HCl.

All chemicals were of analytical or higher grade (Merck, Darmstadt, Germany).

Patch Clamp Recording in the Perfusion and Photolysis Chamber. Gigaohm seals were obtained with patch pipettes of ~20 μm tip opening diameter on "blebs" from myocytes, as previously described.[5] After excision of the patch, the pipette was moved into the temperature-controlled perfusion and photolysis chamber (FIG. 1). When superfusion with a Na$^+$- and ATP-containing solution was carried out for the first time after excision of the membrane patch, a slow increase in pump current was frequently observed. Therefore, ATP was always applied at the beginning of an experiment until the pump current became stable. For more details consult ref. 4.

Bilayer Experiments. Electrical currents generated by the Na$^+$,K$^+$-ATPase were measured by adsorbing Na$^+$,K$^+$-ATPase-containing membrane fragments to a lipid bilayer (FIG. 1b). Details of this technique are described elsewhere.[1] Temperature was maintained at 24°C. The lipid bilayer was formed in a cuvette, as shown in FIGURE 1b, from a solution containing 1.5% (w/v) diphytanoylphosphatidylcholine, and 0.025% (w/v) octadecylamine dissolved in n-decane. Na$^+$,K$^+$-ATPase-containing membrane fragments were added to a final concentration of 50 μg/ml of protein. Caged ATP was present at different concentrations as indicated in the figures. The lipid bilayer was connected to the measuring circuit via Ag/AgCl electrodes. The signal was amplified and filtered with a fourth-order, low-pass filter with a cut-off frequency of 500 Hz. Pump currents were initiated by activating the Na$^+$,K$^+$-ATPase with ATP photolytically released from caged ATP.

Photolytical Release of ATP from Caged ATP. To photolyze caged ATP, light pulses of an excimer laser with a duration of 10 ns, a wavelength of 308 nm, and an energy density of 10–200 mJ/cm² were used. The fraction η of caged ATP converted to ATP at different light intensities was calculated according to the relation $\eta = 1 - e^{-\kappa E}$, where E is the energy density of the UV light at the membrane surface (in mJ·cm^{-2}) and $\kappa = 2(\pm 0.2) \cdot 10^{-3}$ cm²·mJ^{-1}. The parameter κ was determined from release measurements using the luciferin luciferase assay (Boehringer, Ingelheim, Germany) as described earlier.[1]

Data Analysis. Reciprocal time constants $1/\tau_i = k_i$ from transient currents were determined by using least-square fitting of a model function I(t) to the data. For simplicity we

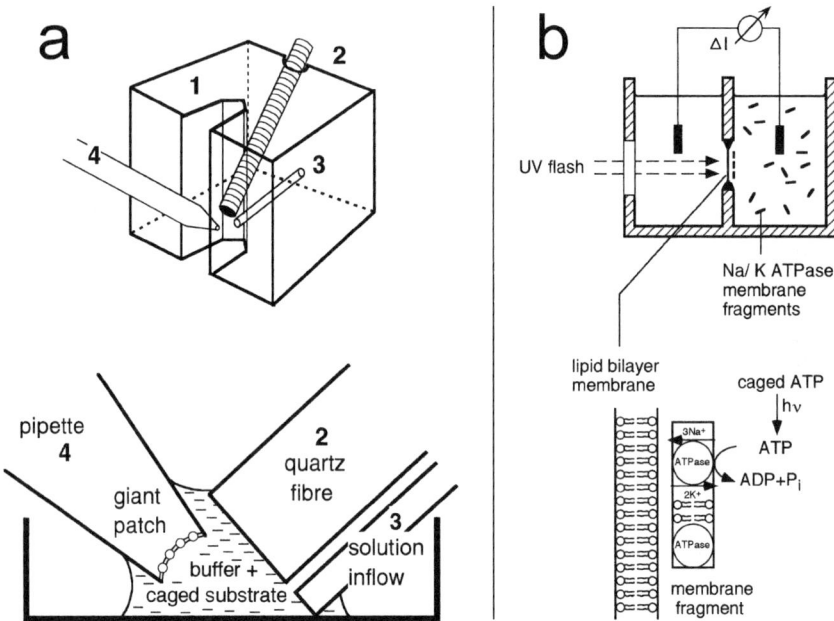

FIGURE 1. (a) (**Upper panel**) Schematic view of the chamber for giant patch experiments: brass cube (1) with quartz-fiber (2) into which the laser flash is sent, and solution inflow line (3) for eight different solutions. An additional duct is drilled into the brass cube to water-jacket the chamber. Also shown is a patch pipette (4) not drawn to scale. (**Lower panel**) An expanded view of the membrane patch with incorporated Na$^+$,K$^+$-ATPase. (**b**) Bilayer setup (**upper panel**) with two-chamber cuvette, electrodes, and measuring circuit. **Lower panel** shows an expanded view of the planar lipid bilayer with an adsorbed membrane fragment containing Na$^+$,K$^+$-ATPase.

refer to k_i as rate constants throughout this paper. For the patch clamp data the model function included a double exponential function and a stationary component:

$$I(t) = A_1 \cdot \exp\left(-(t - t_0) \cdot k_1\right) + A_2 \cdot \exp\left(-(t - t_0) \cdot k_2\right) + I_\infty$$

with the constraints $A_1 + A_2 + I_\infty = 0$ for $t = t_0$

In this formula, I_∞ is the stationary current for $t \to \infty$ and t_0 is a time offset. For the bilayer currents the model function was the sum of 4 exponentials with unconstrained amplitudes:

$$I(t) = \sum A_i \exp(-k_i t).$$

RESULTS

Patch Clamp Experiments. By fast solution switches to different cytoplasmic solutions, an ATP- and Na^+-induced outward current was observed, as shown in FIGURE 2a, which reflects the transport activity of the sodium pump. The ATP dependence of this outward current is shown in FIGURE 2b; it can be well fitted by a simple Michaelis-Menten type saturation function with a K_m value of ~ 150 µM. The nonhydrolyzable (photocleavable) ATP-analog caged ATP[8] binds to the Na^+,K^+-ATPase and therefore acts as a competitive inhibitor.[9–11] To test this inhibition in solution exchange experiments, the pump current was activated by 100 µM ATP or mixtures of 100 µM ATP and increasing concentrations of caged ATP, as shown in FIGURE 3. At higher caged ATP concentrations the pump current is decreased.

Fast Activation of the Pump Current by Photolytic Release of ATP. FIGURE 4a shows the response of the patch current to an ATP concentration jump using photolysis of caged ATP at pH 7.4. FIGURE 4b demonstrates inhibition of this signal by pretreatment of the same patch with ortho-vanadate. Because photolysis of caged ATP is catalyzed by H^+ and has a time constant of ~ 36 ms at pH 7.4 and 2 mM free Mg^{2+},[16] the slow rise of the signal was attributed to this slow ATP release. Further experiments were performed at pH 6.3, where photolytic release of ATP takes place in ~ 2 ms.

FIGURE 5a shows the current response at pH 6.3 together with a fit curve. In contrast to pH 7.4, photolytic release of ATP at pH 6.3 resulted in a fast transient outward current, followed by a stationary outward current, which could be fitted by a sum of two exponential functions and a constant. The laser flash-induced signals were measured at different caged ATP concentrations, but at a fixed conversion ratio (η), and the obtained rate constants k_1 and k_2 were plotted against the concentration of released ATP. FIGURE 6a shows this dependence: ATP-independent rate constant k_1 of $\sim 200\ s^{-1}$ and ATP-saturable rate constant k_2 whereby k_{2max} and K_m depend on η (data not shown).

BLM Experiments. The bilayer experiments were carried out in a cuvette, as described in FIGURE 1b, filled with solution Bl-Na-K and Na^+,K^+-ATPase containing membrane fragments and caged ATP in the rear compartment. An experimental recording is shown in FIGURE 5b. The signals measured on the bilayer contain four exponential components, the two slowest of which cannot be seen in the figure because of the limited time range. Component $k_3 \approx 10\ s^{-1}$ is assigned to the discharging of the capacitances of the planar membrane and membrane fragments.[10] The slowest component, $k_4 \approx 1\ s^{-1}$, is not present in the absence of K^+. It is far too slow to be part of the Na^+,K^+-ATPase reaction cycle and therefore is assigned to the buildup of potential or ion gradients across the membrane fragments.

The rate constants shown in FIGURE 6b were determined at different concentrations of caged ATP corresponding to the ATP concentrations given in the figure. As just detailed, only k_1 and k_2 contain information about partial reactions of the Na^+,K^+-ATPase. They are given as filled symbols in the figure.

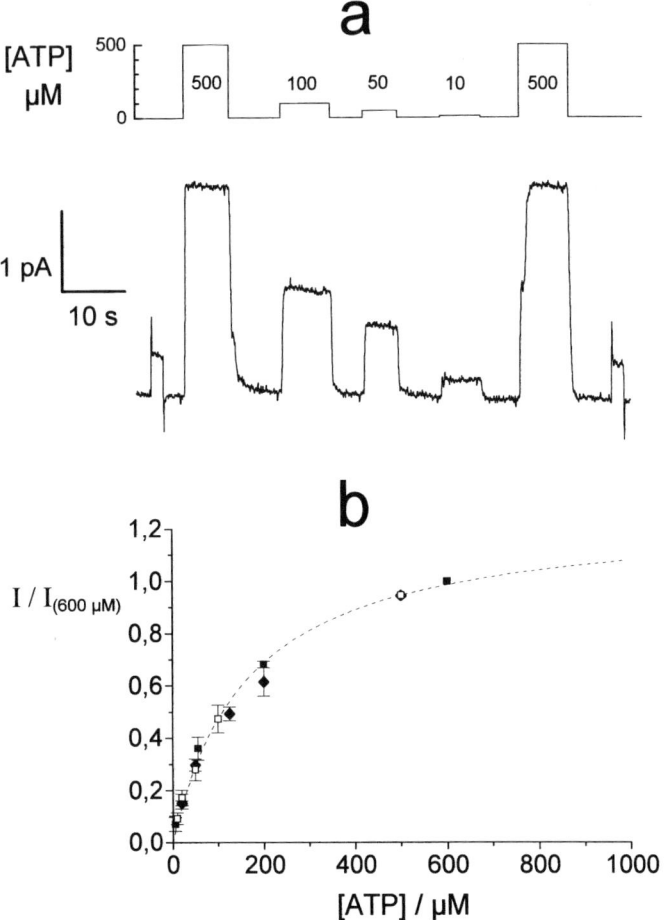

FIGURE 2. (a) Recording of patch current in response to superfusion of the patch with solutions containing different MgATP concentrations, as indicated in the **upper part.** Bath solution: B-F, pipette solution: P-NMG-K. Rat cell, T = 24°C. (b) Dependence of the normalized patch current on ATP concentration; plotted are means ± SD. Included is a least-square fit of a Michaelis-Menten equation to the data (■) (*dashed line*), leading to a V_{max} of 1.26 ± 0.06 (current at 500 μM ATP was set to 1.0), and an apparent K_m of 165 ± 20 μM. T = 24°C. (From ref. 4 with permission.)

DISCUSSION

Caged ATP was used before to measure Na^+ efflux from vesicles in a single turnover of the Na^+,K^+-ATPase[9] or to measure transient currents from membrane fragments containing Na^+,K^+-ATPase which were capacitively coupled to a BLM.[1-3,10-12] We compared the results obtained by flash photolysis of caged ATP at an excised patch to similar experiments with the BLM technique as well as to voltage jump experiments.

ATP Dependence of the Na^+,K^+-ATPase and Inhibition by Caged ATP. The obtained K_m obtained for ATP activation of the pump current is in good agreement with the low affinity ATP site obtained in ATPase and flux measurements. Inhibition by caged ATP at the high affinity site was already postulated[9–11] from photolysis experiments; we now showed it directly by measuring pump current at different caged ATP concentrations. To obtain inhibition constant K_I, more experiments would have to be done; so far, we estimated that K_I has ~13 times the value of K_{m1} (at the high affinity ATP binding site; for details see ref. 4).

Activation of Pump Current by Photolytic Release of ATP at pH 7.4. A fast substrate concentration jump can reveal kinetic constants of the partial reactions of an enzymatic reaction cycle. Some limitations, however, have to be considered: first, the rate of ATP release is pH and Mg^{2+} dependent,[16] that is, magnesium ions slow down the reaction and hydrogen ions speed it up; second, caged ATP binds to the Na^+,K^+-ATPase, therefore acting as a competitive inhibitor.[9,11]

We found a single exponentially increasing pump current on photolysis of caged ATP at pH 7.4 (FIG. 4). The rate of this increase was 70 s^{-1}, but it depends on caged ATP concentration and light intensity, whereas release of ATP at 24°C proceeds at a rate of ~30 s^{-1}. We therefore concluded that photolysis at pH 7.4 cannot provide intrinsic rate constants of the Na^+, K^+-ATPase transport cycle.

Activation of Pump Current by Photolytic Release of ATP at pH 6.3. We chose pH 6.3 because the stationary pump current, that is, enzyme turnover, is still 50% of that at pH 7.4[4] and ATP release at pH 6.3, 2 mM Mg^{2+}, and 24°C proceeds at a rate of ~500 s^{-1}.[16,17] As shown in FIGURE 6, we found that the fast rise in pump current (k_1) is independent of the ATP concentration, whereas the slower decay to a stationary current (k_2) is ATP concentration dependent. Therefore, we conclude that the step that is rate limiting the electrogenic event proceeds with rate k_1 of ~200 s^{-1} at 24°C,

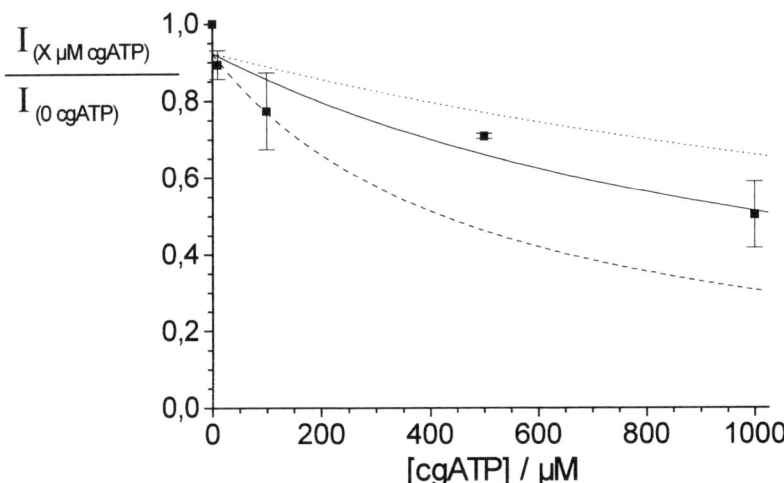

FIGURE 3. Inhibition of pump current by caged ATP. Current was normalized to the value in the presence of 100 μM ATP only; values are means ± SE. Included are three simulations according to equation A5 in Appendix A (ref. 4) in which the mechanistic assumptions about the inhibition are described (*solid line* = best fit: K_I/K_{m1} = 13). Bath solution: B-F plus 100 μM ATP. Pipette solution: P-NMG-K. Guinea pig cells, T = 24°C. (From ref. 4 with permission.)

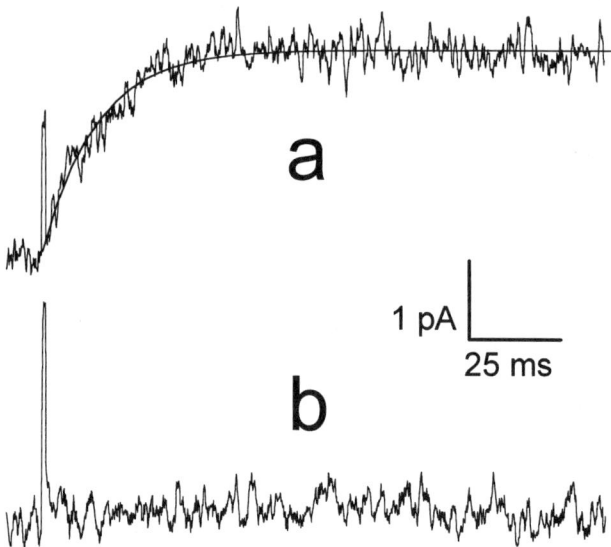

FIGURE 4. (a) Current signal in response to photolysis of 500 μM caged ATP with a release ratio of 30% at pH 7.4. Included is a least-square fit of a monoexponentially rising function to the data, leading to a rise time of 15 ms. (b) Current response to photolysis of 500 μM caged ATP on the same patch as in a, after inhibition of the stationary pump current by 1 mM orthovanadate. Bath solution: B-P. Pipette solution: P-NMG-K. Guinea pig cell, T = 24°C. (From ref. 4 with permission.)

which is in good agreement with voltage jump experiments in oocytes[18] and giant cardiac membrane patches.[14,19]

A rate of 200 s^{-1} was also found for the electrogenic Na^+ translocation in the forward direction in voltage jump experiments on whole cell voltage-clamped myocytes,[20] but at 36°C. This rate was only ~40 s^{-1} at 20°C.[21] That this rate is extremely slow in whole cell experiments on cardiac cells in contrast to excised patch studies[14,19] and to voltage clamp experiments on oocytes from *Xenopus laevis*[18] might be due to some specific cardiac cellular regulation.

Comparison of Patch Clamp and Bilayer Experiments. The current traces in FIGURE 5a and b were recorded under similar conditions with the patch clamp and bilayer technique. They are very similar in the fast time range. At t > 50 ms the patch clamp current has decayed to a stationary current, whereas the bilayer signal decays further and becomes negative at t > 200 ms. The latter behavior is attributable to the capacitive coupling of the membrane fragments to the planar bilayer.[10] Capacitive coupling does not alter the rate constants, but it only adds the time constant of the system and distorts the amplitudes of the signal.[10] Therefore, analysis of the rate constants is meaningful whether or not the ion pump is capacitively coupled to the measuring system.

The rate constants obtained by both methods at different concentrations of released ATP are shown in FIGURE 6a and b. Apparently they depend in the same way on the ATP concentration and have approximately the same values. Both techniques show a rapid rise of the electrical signal with a rate constant of ~200 s^{-1} which is independent of the ATP concentration. On the other hand, the decay of the signal is clearly ATP dependent with an affinity of 10–20 µM. Based on BLM experiments, we previously assigned the rising phase of the signal to a fast electrogenic step with a rate of ~200 s^{-1}.[3,10] The decay of the signal was assigned to ATP binding and exchange with caged ATP.[3,10] These conclusions are now confirmed by the results of our patch clamp study using cardiac myocytes. This rules out tissue-specific behavior as a reason for the discrepancies found in the literature.

Apell and coworkers[12,22] used the results of cardiac whole cell measurements[20] to support their conclusion from caged ATP/BLM experiments of a much slower electrogenic step: 25 s^{-1} at 20°C.[12] However, as just pointed out, the assignment of the

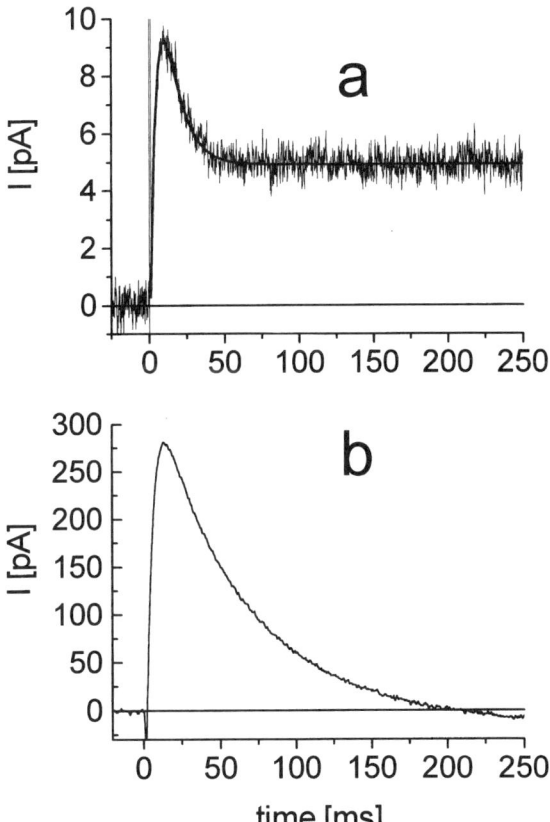

FIGURE 5. Current recorded after rapid release of ATP from caged ATP. (**a**) Patch clamp experiment: Bath solution: B-P at pH 6.3. Pipette solution: P-NMG-K. Rat cell, T = 24°C. Caged ATP concentration 250 µM, released fraction η = 0.25. Parameters of the included fit are: t_0 = 1 ms; k_1 = 220 s^{-1}; and k_2 = 95 s^{-1}. (From ref. 4 with permission.) (**b**) Bilayer experiment: Bath solution: Bl-Na-K at pH 6.2, T = 24°C. Caged ATP concentration 360 µM; released fraction η = 0.22. The fit to the data points yielded the following rate constants: 230, 47, 14, and 1.2 s^{-1}. It is not included in the figure, because it cannot be discriminated from the data points.

observed rates to either ATP release or intrinsic ATPase reaction rates can be misleading if conditions (pH 7.2, 10 mM Mg, 20°C, ref. 22) in which ATP release proceeds with a rate of only ~21 s^{-1} are chosen.[16,17] The substrate (and η) dependence of the rate k_2, in contrast to k_1, makes it a very unlikely candidate for an intrinsic Na$^+$,K$^+$-ATPase rate constant. Therefore, phosphorylation of cardiac Na$^+$,K$^+$-ATPase must be fast, and the electrogenic E$_1$P-E$_2$P conformational change must proceed with a rate constant of ~200 s^{-1} at 24°C. This conclusion is not contradictory to an electroneutral E$_1$P-E$_2$P conformational change (with a rate constant of 200 s^{-1}) followed by a fast electrogenic Na$^+$ release,[14,15] as put forward by access channel

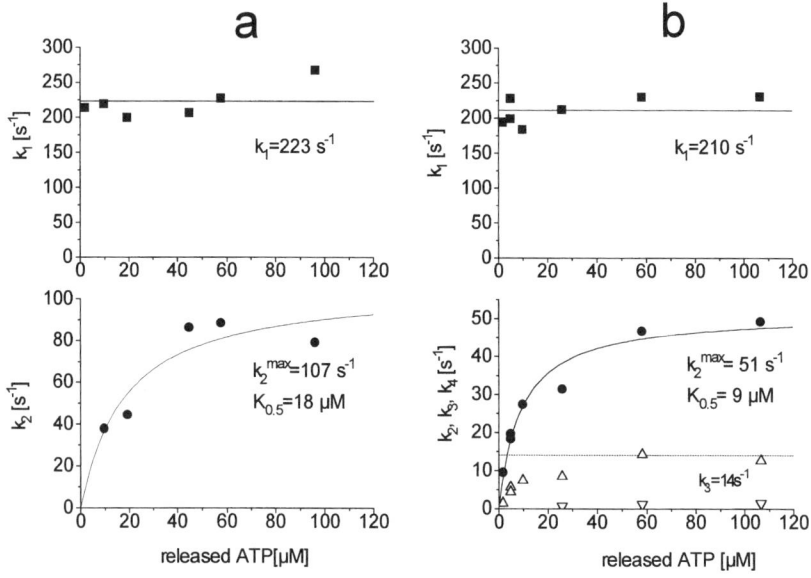

FIGURE 6. Dependence of rate constants k_i of the photolytically induced transient pump currents on the concentration of released ATP at a constant released fraction η. A least-square fit with a Michaelis-Menten equation is included for k_2. Parameters are indicated in the figure. (**a**) Patch clamp experiment: Bath solution: B-P at pH 6.3 with 10–250 μM caged ATP, released fraction η = 0.25. Pipette solution: P-NMG-K. Rat cells, T = 24°C. (**b**) Bilayer experiment: Bath solution: Bl-Na-K at pH 6.2 with 10–660 μM caged ATP, released fraction η = 0.22, T = 24°C.

models for the Na$^+$,K$^+$-ATPase.[13] In our experiments this mechanism would yield the same results because it is kinetically equivalent to an electrogenic E$_1$P-E$_2$P conformational change.

SUMMARY

The lipid bilayer and the giant patch technique were used to study the Na$^+$,K$^+$-ATPase. In excised patches from ventricular myocytes, Na$^+$, K$^+$-pump currents show a saturable ATP dependence with a K_m of ~150 µM at 24°C. Partial reactions in the transport cycle were investigated by generating ATP concentration jumps through photolytic release of ATP from caged ATP at pH 7.4 and 6.3. Transient outward currents were obtained at pH 6.3 with a fast rising phase followed by a slow decay to a stationary current. Experiments with purified pig kidney Na^+,K^+-ATPase attached to a planar lipid bilayer resulted in similar pump current signals and the same outcome. It was concluded that the fast rate constant of ~200 s^{-1} at 24°C reflects a step rate limiting the electrogenic Na^+ release.

ACKNOWLEDGMENTS

We thank E. Grabsch, D. Ollig, A. Schacht, and D. Stiegert for excellent technical assistance.

Note added in proof: A direct comparison of transient Na^+,K^+-ATPase pump currents, elicited by voltage jumps or photorelease of ATP, was recently published: Friedrich, T. & G. Nagel. 1997. Biophys. J. **73**: 186–194.

REFERENCES

1. FENDLER, K., E. GRELL, M. HAUBS & E. BAMBERG. 1985. EMBO J. **12**: 3079–3085.
2. BORLINGHAUS, R., H.-J. APELL & P. LÄUGER. 1987. J. Membr. Biol. **97**: 161–178.
3. FENDLER, K., E. GRELL & E. BAMBERG. 1987. FEBS Lett. **224**: 83–88.
4. FRIEDRICH, T., E. BAMBERG & G. NAGEL. 1996. Biophys. J. **71**: 2486–2500.
5. COLLINS, A., A. V. SOMLYO & D. W. HILGEMANN. 1992. J. Physiol. **454**: 27–57.
6. HILGEMANN, D. W. 1989. Pflügers Arch. **415**: 247–249.
7. HILGEMANN, D. W., G. A. NAGEL & D. C. GADSBY. 1991. *In* The Sodium Pump: Recent Developments. J. H. Kaplan & P. DeWeer, Eds. :543–547. The Rockefeller University Press. New York.
8. KAPLAN, J. H., B. FORBUSH, III & J. F. HOFFMAN. 1978. Biochemistry **17**: 1929–1935.
9. FORBUSH, B. III. 1984. Proc. Natl. Acad. Sci. U.S.A. **81**: 5310–5314.
10. FENDLER, K., S. JARUSCHEWSKI, A. HOBBS, W. ALBERS & J. P. FROEHLICH. 1993. J. Gen. Physiol. **102**: 631–666.
11. NAGEL, G., K. FENDLER, E. GRELL & E. BAMBERG. 1987. Biochim. Biophys. Acta **901**: 239–249.
12. WUDDEL, I. & H.-J. APELL. 1995. Biophys. J. **69**: 909–921.
13. GADSBY, D. C., R. F. RAKOWSKI & P. DE WEER. 1993. Science **260**: 100–103.
14. HILGEMANN, D. W. 1994. Science **263**: 1429–1432.
15. WAGG, J., M. HOLMGREN, D. C. GADSBY, F. BEZANILLA, R. F. RAKOWSKI & P. DE WEER. 1996. Biophys. J. **70**: A18 (abstr.).
16. WALKER, J. F., G. P. REID, J. A. MCCRAY & D. R. TRENTHAM. 1988. J. Am. Chem. Soc. **110**: 7170–7177.
17. BARABÁS, K. & L. KESZTHELYI. 1984. Acta Biochim. Biophys. Acad. Sci. Hung. **19**: 306–309.
18. RAKOWSKI, R. F. 1993. J. Gen. Physiol. **101**: 117–144.
19. FRIEDRICH, T. & G. NAGEL. 1997. Ann. N.Y. Acad. Sci., this volume.
20. NAKAO, M. & D. C. GADSBY. 1986. Nature **323**: 628–630.
21. GADSBY, D. C., M. NAKAO, A. BAHINSKI, G. NAGEL & M. SUENSON. 1992. Acta Physiol. Scand. **146**: 111–123.
22. APELL, H.-J., R. BORLINGHAUS & P. LÄUGER. 1987. J. Membr. Biol. **97**: 179–191.

Complex Kinetic Behavior in the Na,K- and Ca-ATPases

Evidence for Subunit-Subunit Interactions and Energy Conservation during Catalysis

JEFFREY P. FROEHLICH,[a,g] KAZUYA TANIGUCHI,[b]
KLAUS FENDLER,[c] JAMES E. MAHANEY,[d]
DAVID D. THOMAS,[e] AND R. WAYNE ALBERS[f]

[a]*National Institute on Aging*
National Institutes of Health
Baltimore, Maryland 21224

[b]*Department of Chemistry*
Hokkaido University
Sapporo, Japan

[c]*Max-Planck-Institute for Biophysics*
Frankfurt am Main, Germany

[d]*Department of Biochemistry*
University of West Virginia School of Medicine
Morgantown, West Virginia

[e]*Department of Biochemistry*
University of Minnesota School of Medicine
Minneapolis, Minnesota

[f]*National Institute of Communicative Disorders and Stroke*
National Institutes of Health
Bethesda, Maryland 20892

A recurrent problem in analysis of Na,K-ATPase partial reactions is the failure of linear consecutive schemes (such as the Albers-Post mechanism) to account for the presence of complex kinetic behavior in time-resolved measurements. Examples of this behavior include the multiphasic patterns of phosphorylation and dephosphorylation catalyzed by the electric organ[1-4] and mammalian Na,K-ATPases[5] and the inability of rate constants measured in the pre-steady state to account for results in the steady state.[6,7] To resolve these questions, alternatives to the Albers-Post model have been proposed involving parallel independent pathways for ATP hydrolysis[3,4,7] or a unique initiation sequence leading to the steady-state reaction cycle.[6] We summarize here kinetic evidence that requires for explanation elements of both of these models and, in addition, suggests that these complexities arise from homologous interactions between α subunits in an oligomer. In essence, the oligomer is a parallel pathway model with conformational coupling between the catalytic subunits, which prevents

---

[g]Address for correspondence: Jeffrey P. Froehlich, MD, National Institute on Aging, Geronotology Research Center, 4940 Eastern Ave., Baltimore, MD 21224.

their independent operation and controls their kinetic behavior. Because the most stable subunit interactions appear to favor conformational asymmetry (e.g., E_1P/E_2P), the mode of catalytic operation of the subunits is "staggered" so that no two interacting α subunits can carry out the same reaction simultaneously.

Although a variety of methods have uncovered evidence for α chain protein-protein interactions in Na,K-ATPase,[8-10] the functional significance of these interactions and their role in catalysis remain uncertain. One problem in assigning functional significance to these interactions is that they are often measured under conditions in which the enzyme is partially inhibited.[1,10,11] Evidence for oligomeric behavior has also been reported in the Ca-ATPase of sarcoplasmic reticulum (for a review, see ref. 12), which displays multiphasic kinetic patterns similar to those detected in the Na,K-ATPase.[13,14] It will be shown here that homologous protein interactions and staggered catalysis provide a plausible explanation for several of the complex kinetic features found in the ion motive ATPases. The presence of such interactions may have evolved because they improve the efficiency of energy transduction during formation of the cation transport gradient.

METHODS

Membrane Preparation. Eel electric organ Na,K-ATPase was isolated as a broken membrane preparation (native Na,K-ATPase) and solubilized with $C_{12}E_8$ in order to prepare αβ protomers.[15] Pig kidney Na,K-ATPase was purified using sodium deoxycholate followed by NaI and labeled with N-(p-(2-benzimidazolyl)phenyl) maleimide (BIPM) at Cys-964, as previously described.[16] Sarcoplasmic reticulum (SR) Ca-ATPase was isolated from rabbit skeletal muscle and labeled with iodoacetamide spin label (IASL) at Cys-674, as reported by Mahaney et al.[14] Both the BIPM and the IASL labeling procedures employed an N-ethylmaleimide pre-block to enhance the specificity of probe incorporation. Active Ca-ATPase monomers were prepared by solubilizing SR membranes with $C_{12}E_8$ at a detergent-to-protein ratio of 2:1.[17]

Fast Kinetic Measurements. Time-resolved methods, including rapid acid quenching,[1] stopped-flow mixing,[16] the laser flash/electrical bilayer technique,[18] and transient state electron paramagnetic resonance (EPR),[14] were used to investigate the time dependence of the partial reactions catalyzed by the ion motive ATPases. Specific conditions for the experiments are provided in the text and figure legends. In correlative studies involving more than one technique, identical conditions were used with respect to ionic composition, pH, and temperature. Rapid mixing experiments employed ATP as substrate, whereas the planar lipid bilayer and EPR experiments were initiated by the photolytic release of ATP from an inactive analog, caged ATP. Conditions were chosen to minimize kinetic limitations produced by the uncaging reactions.[19]

RESULTS AND DISCUSSION

Kinetics of the E_1P to E_2P Conversion Reaction in the Pre-Steady State. The kinetics of the phosphoenzyme conformational transition were evaluated by rapid mixing, acid quench experiments using the early K^+-activated burst of inorganic phosphate (P_i) production to signal the formation of E_2P. A typical quenched-flow experiment in which electric organ Na,K-ATPase microsomes were mixed with 10 μM ATP in the presence of 130 mM NaCl and 10 mM KCl is shown in FIGURE 1A. The time course of phosphorylation shows a characteristic low-amplitude overshoot which decays during the rapid initial phase of P_i production. All of the phosphate associated with the burst phase originates from the hydrolysis of E_2P, because we were able to completely prevent the P_i burst by blocking the conversion of E_1P to E_2P with oligomycin.[1] The smooth curves through the data points are simulations using the simplified Albers-Post scheme and the rate constants for phosphorylation (k_2) and dephosphorylation (k_4) given in TABLE 1. The values for k_2 and k_4 were measured in separate acid-quench experiments, whereas the rate constants corresponding to ATP binding (k_1) and dissociation (k_{-1}) and the conformational transitions involving the phosphorylated (k_3) and dephosphorylated enzymes (k_5) were determined by the curve-fitting procedure.[20] Prior to running the simulation, a vanadate-resistant, stable phosphoenzyme representing 60–70% of the total phosphoenzyme[21] was subtracted from each data point. This subtraction is crucial to the analysis because we discovered that the rate constant for the phosphoenzyme transition depends on the difference between the level of phosphorylation and the P_i burst amplitude, as shown in FIGURE 1B. At pH 7.4 and 21°C, the best fit of the transport model to the kinetic data was achieved with a transition rate constant of about 3,000 s^{-1}; below this value, the simulated curves overestimated the amount of phosphoenzyme and underestimated P_i production. Large values assigned to k_3 will hasten the formation of E_2P, which is rapidly hydrolyzed to P_i in the presence of K^+. This will increase the burst amplitude while reducing the level of the acid stable phosphoenzyme, thus accounting for the larger difference between the simulated variables at higher values of the transition rate constant. It is important to recognize that the value for k_3 obtained by simulation is not defined by the kinetics of the burst phase, but rather its magnitude. At 10 μM ATP, the kinetics of the burst phase are controlled by the rate of ATP binding, which has a *pseudo*first-order rate constant of 70–80 s^{-1} and is much slower than the rapid phosphoenzyme transition situated downstream in the reaction cycle.

The kinetics of the E_1P to E_2P transition has also been studied at pH 6.2 for comparison with measurements of charge translocation using the laser flash/electrical bilayer technique.[18] These experiments were performed at acidic pH, because the photolytic release of ATP from caged ATP used to synchronously activate the pumps becomes rate-limiting at alkaline pH.[19,22] At pH 6.2 and 24°C, quenched-flow experiments similar to the one shown in FIGURE 1A yielded a rate of conversion of E_1P to E_2P of 350 s^{-1}, whereas the bilayer measurements gave a rate constant of 210 s^{-1} when corrected for an effect resulting from caged ATP photolysis in the catalytic site ("release-in-site"; ref. 23). These rates are about 10 times smaller than the values obtained at pH 7.4, suggesting that the phosphoenzyme conversion reaction has a proton inhibitory site with a linear range of occupation below 1 μM. Although the difference between the biochemically and electrically determined rate constants suggests that the electrogenic step may follow the phosphoenzyme transition, the errors inherent in these measurements are too large to allow this distinction.

For comparison, a rapid mixing, quenched-flow experiment utilizing the BIPM-labeled pig kidney Na,K-ATPase is shown in FIGURE 2. The characteristics of phosphoenzyme formation and P_i release at 10 μM ATP, pH 7.4, and 24°C are very similar to the behavior measured in the electric organ enzyme under similar conditions (FIG. 1A). A separate experiment with unlabeled enzyme (not shown) showed no influence of BIPM labeling on the kinetic behavior. Simulation of these data after an

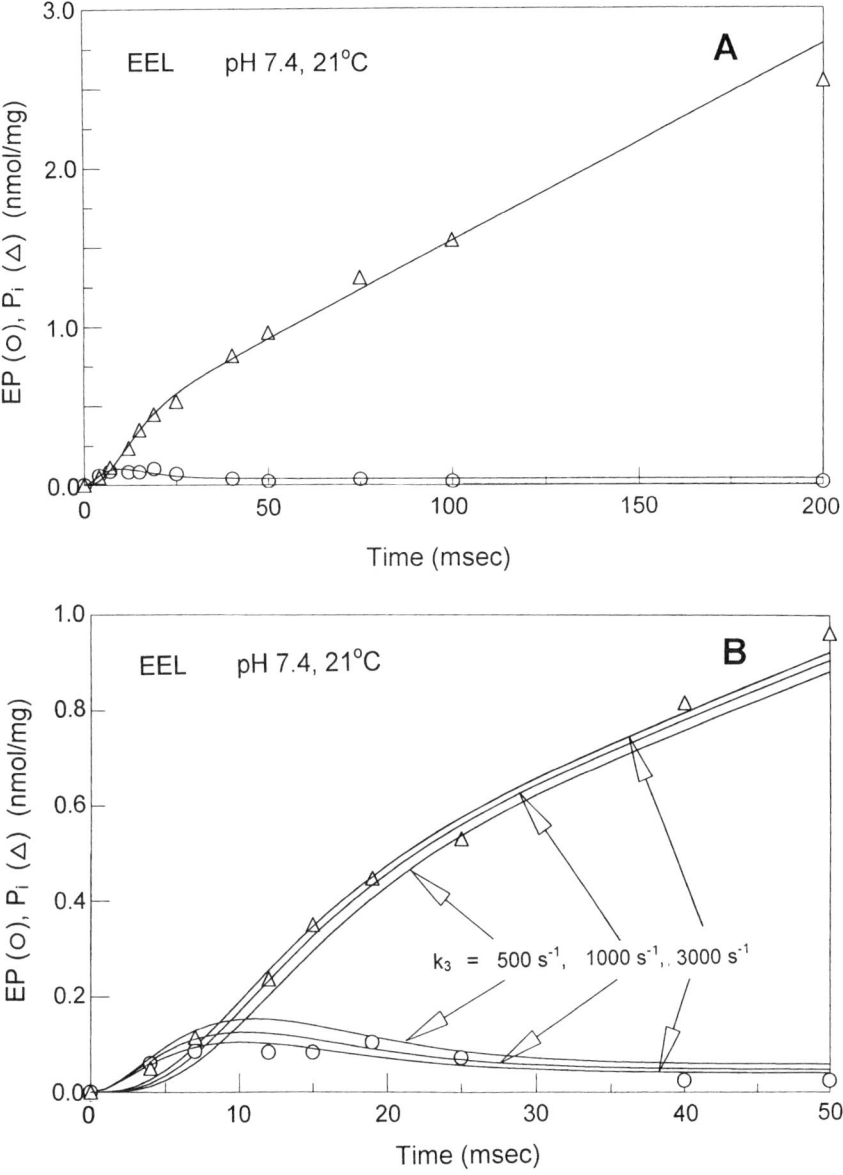

FIGURE 1. Time dependence of phosphoenzyme (EP) formation and inorganic phosphate (P_i) production in native eel electric organ Na,K-ATPase. (**A**) Eel Na,K-ATPase (0.5 mg/ml) suspended in 130 mM NaCl, 10 mM KCl, 3 mM $MgCl_2$, 0.1 mM EDTA, and 25 mM imidazole, pH 7.4, was mixed with an equal volume of an identical solution (without enzyme) containing 20 μM [$\gamma^{32}P$]ATP in a quenched-flow apparatus at 21°C. At the indicated times, 3% perchloric acid + 6 mM K_2HPO_4 was added to quench the reaction, and samples containing [^{32}P]phosphoprotein and [^{32}P]P_i were analyzed as previously described.[1] The curves through the data points represent simulations using the subroutines contained in MLAB[20] and the simplified Albers-Post scheme and rate constants given in TABLE 1 (k_3 = 3,000 s^{-1}). (**B**) Effect of varying the rate of conversion of E_1P to E_2P on simulations of phosphoenzyme formation and P_i release in eel electric organ Na,K-ATPase. Data from (**A**) are plotted on an expanded time scale. The phosphoenzyme transition rate constant, k_3, was varied while the other parameters were held constant (TABLE 1).

TABLE 1. Reaction Rate Constants for the Native and BIPM-Labeled Na,K-ATPases[a]

Reaction Step	Native Eel Na,K-ATPase (k_i/k_{-i})	Pig Kidney Na,K-ATPase (k_i/k_{-i})
1	$1.2 \times 10^7 \text{ M}^{-1} \text{ s}^{-1}/35 \text{ s}^{-1}$	$7.5 \times 10^6 \text{ M}^{-1} \text{ s}^{-1}/35 \text{ s}^{-1}$
2	$196 \text{ s}^{-1}/0$	$197 \text{ s}^{-1}/0$
3	$3,000 \text{ s}^{-1}/0$	$3,000 \text{ s}^{-1}/0$
4	$350 \text{ s}^{-1}/0$	$276 \text{ s}^{-1}/0$
5	$26 \text{ s}^{-1}/0$	$13 \text{ s}^{-1}/0$

[a]Phosphoenzyme formation and inorganic phosphate release were simulated using a simplified Albers-Post model:

$$E1 + ATP \underset{(1)}{\Leftrightarrow} E1ATP \underset{(2)}{\overset{ADP}{\Leftrightarrow}} E1P \underset{(3)}{\Leftrightarrow} E2P \underset{(4)}{\overset{P_i}{\Leftrightarrow}} E2 \underset{(5)}{\Leftrightarrow} E1$$

where k_i and k_{-i} are the forward and reverse rate constants corresponding to the ith step. Initial values for k_1, k_{-1}, k_2, and k_4 were obtained from the literature; values for k_3 and k_5 and the final adjustment in k_1 and k_{-1} were determined by the curve-fitting operation (MLAB; ref. 20). The reaction temperature was maintained at 21°C (eel Na,K-ATPase) or 24°C (pig kidney Na,K-ATPase).

appropriate subtraction of the stable phosphoenzyme (representing two thirds of the total EP) gave a rate of conversion of E_1P to E_2P of $3,000 \text{ s}^{-1}$ at 24°C (TABLE 1). In these simulations, the rate constant for phosphorylation was assigned a value of 197 s^{-1} based on an acid-quench experiment in which the BIPM-labeled pig kidney Na, K-ATPase was mixed with 100 μM ATP in the absence of KCl. When these conditions were used in a stopped-flow experiment to measure the kinetics of the BIPM signal, the initial increase in fluorescence intensity had a rate constant of 180 s^{-1} (D. Kane and R. Clarke, unpublished results), which is very close to the rate constant for the phosphorylation reaction. Because the BIPM transient tracks the formation of E_2P,[16] we conclude that phosphoenzyme conversion occurs immediately after phosphorylation, a finding in agreement with our quenched-flow measurements carried out in the presence of Na$^+$ and K$^+$. Moreover, the similarity between these rate constants implies that the principal rate-determining step in the Na$^+$ translocation sequence at saturating ATP concentrations is the phosphorylation reaction.

Steady-State Kinetics of the E_1P to E_2P Conversion Reaction. In 1993, Rossi and Norby[6] reported a discrepancy between the presteady-state rate of thalium-activated E_2P hydrolysis and the steady-state rate evaluated by comparison of the calculated and measured overall velocities of ATP hydrolysis. The calculated velocity was always 2–4 times larger than the measured velocity, implying that the hydrolytic reaction slows down in the steady state. A similar discrepancy was reported by us[7] in the electric organ Na,K-ATPase which displays higher steady-state levels of E_1P than predicted from the presteady-state rate of conversion of E_1P to E_2P. The conversion rates in this report ($3,000 \text{ s}^{-1}$) exceed our earlier estimates (about 600 s^{-1}) and predict that the ADP-sensitive phosphoenzyme will be a small fraction (<5%) of the total acid-stable phosphoenzyme in the steady state. To further explore this behavior, we conducted measurements of steady-state E_1P formation using an ADP chase to define the level of the ADP-sensitive intermediate. Eel Na, K-ATPase was phosphorylated with 200 μM ATP in the presence of 50 mM NaCl and 10 mM KCl for 116 ms (sufficient time to enter the steady state; see FIG. 1A for comparison) and then dephosphorylated by the addition of a chase containing 10 mM ADP + 10 mM EDTA, as shown in FIGURE 3. The resulting pattern of phosphoenzyme decomposition was triphasic, consisting of fast (1,260 s^{-1}), intermediate (182 s^{-1}), and slow (8 s^{-1}) phases. The fast phase was complete within 4 ms, corresponding to the first time point on the decay curve (excluding the zero time point). Additional evidence for the fast decay phase was obtained by replacing ADP in the chase solution with ATP, which is unable to synthesize ATP from E_1P. As seen in FIGURE 3, dephosphorylation with a chase containing 10 mM ATP + 10 mM EDTA resulted in slower initial decay of the phosphoenzyme (94 s^{-1}) than that observed with ADP. Because this phosphoenzyme can break down only in the direction of P_i and does not exhibit a fast phase (which is detected only when ADP is present in the chase), we conclude that the fast decay phase results from the ADP-stimulated reversal of phosphorylation, namely, ADP + E_1P ⇔ E_1P ADP ⇔ E_1 ATP. The intermediate decay phase following the fast phase resembles the initial ATP-induced decay component kinetically and therefore likely relates to E_2P hydrolysis, whereas the slow phase, which manifests partial ADP sensitivity, turns over at <5% of the intermediate decay rate. These slow kinetics distinguish this species as being in a separate pathway running parallel to the main catalytic pathway.[7]

To establish what fraction of the total phosphoenzyme is ADP sensitive, eel

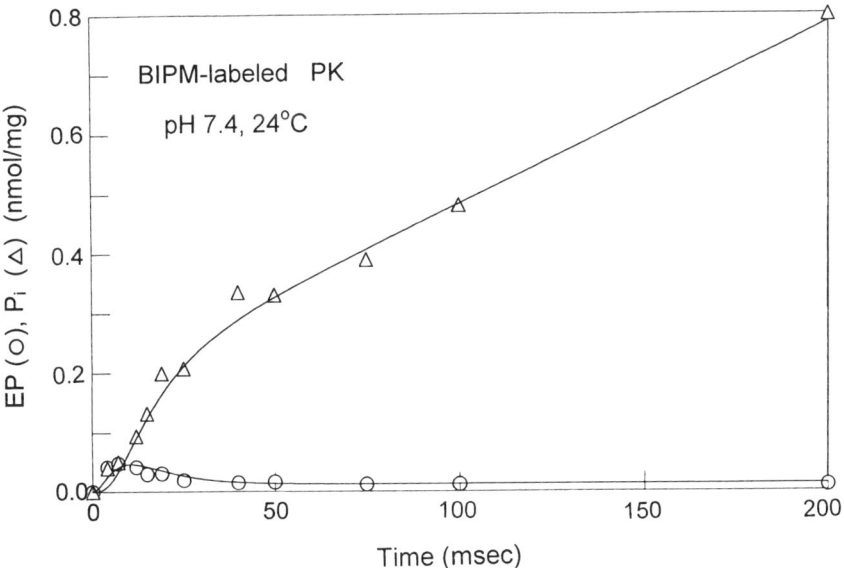

FIGURE 2. Time dependence of phosphoenzyme (EP) formation and phosphate (P_i) production in BIPM-labeled pig kidney Na,K-ATPase. BIPM-labeled pig kidney Na,K-ATPase (100 μg/ml) suspended in 16 mM NaCl, 1 mM MgCl$_2$, 25 mM sucrose, 0.1 mM EDTA, and 25 mM imidazole, pH 7.4, was mixed with an equal volume of an identical solution containing 20 μM [γ^{32}P]ATP in a quenched-flow apparatus at 24°C. The reaction was terminated at the indicated times with 3% perchloric acid + 6 mM K$_2$HPO$_4$, and quenched samples containing [^{32}P]phosphoprotein and [^{32}P]P_i were analyzed as previously described.[1] Curves through the data points are simulations using the Albers-Post model and rate constants given in TABLE 1.

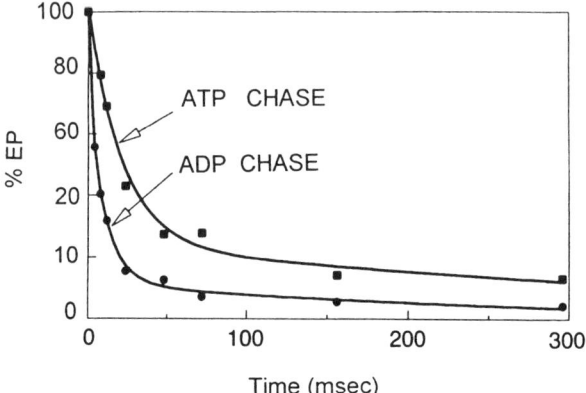

FIGURE 3. Time course of dephosphorylation of eel Na,K-ATPase induced by chasing the phosphoenzyme with ATP or ADP. Eel Na,K-ATPase (1 mg/ml) was phosphorylated with 200 µM [γ^{32}P]ATP in a medium containing 50 mM NaCl, 10 mM KCl, 3 mM MgCl$_2$, 0.1 mM EDTA, and 25 mM TMA-HEPES, pH 7.4, at 21°C. After 116 ms (zero time on the plot), dephosphorylation was initiated by the addition of either 10 mM unlabeled ATP (*squares*) or 10 mM ADP + 10 mM EDTA (*circles*) and the reaction allowed to proceed for the indicated times prior to the addition of acid. Data were normalized to facilitate comparison. The time course of dephosphorylation was fitted using MLAB[20] to a sum of *n* exponentials where *n* = 2 or 3. Fitting the ADP-induced dephosphorylation reaction with three exponentials gave a smaller residual error between the actual and simulated data points (SSQ = 7.05 × 10^{-5}) than a biexponential fit (SSQ = 3.32 × 10^{-4}).

Na,K-ATPase was phosphorylated in the presence of Na$^+$ and K$^+$ at different ATP concentrations (10 and 400 µM) and then dephosphorylated with an ADP + EDTA chase, as shown in FIGURE 4A. At 400 µM ATP, the fast phase corresponding to E$_1$P accounted for about one half the total acid-stable phosphoenzyme after subtraction of the slowly decaying species (TABLE 2). The steady-state rate of conversion of E$_1$P to E$_2$P needed to sustain such high levels of the ADP-sensitive phosphoenzyme can be calculated from the relationship between the steady-state ATPase velocity, v, and the concentrations of E$_1$P and E$_2$P evaluated from the ADP chase experiment:

$$v = k_3[E_1P] - k_{-3}[E_2P]$$

where v = 39 nmol/mg/s, [E$_1$P] = 0.12 nmol/mg and [E$_2$P] = 0.141 nmol/mg. If the back reaction is negligible, that is, k_{-3} = 0, then k_3 = v/[E$_1$P] = 325 s^{-1}. This value is roughly an order of magnitude smaller than the rate of conversion evaluated from the presteady-state data. These results demonstrate that as the system enters the steady state, the rate of the phosphoenzyme transition is dramatically slowed, indicating that something is acting to stabilize the inherently labile E$_1$P state.

To explain the decline in the transition rate constant, we postulated that the conformational stability of E$_1$P is increased by the formation of a strong interaction with a neighboring catalytic subunit in a Na,K-ATPase oligomer. Such a model would require that the adjacent subunit undergo a slow transformation to its product state, constraining E$_1$P turnover to a similar rate. A likely candidate for interaction with E$_1$P is E$_2$P, which is present in similar amounts in the steady state and displays a

K^+-activated rate of hydrolysis (350 s^{-1}) that is very close to the calculated steady-state rate of conversion of E_1P to E_2P (about 300 s^{-1}). To test the oligomer hypothesis, we used the nonionic detergent $C_{12}E_8$ to solubilize the enzyme, reasoning that conditions favorable to active αβ protomer formation would weaken or eliminate any quaternary protein interactions serving to stabilize E_1P. When eel electric organ membranes were solubilized with 3 mg of $C_{12}E_8$/mg of protein (see Methods), phosphorylated with 400 μM ATP in the presence of Na^+ and K^+, and then chased with 10 mM ADP + 10 mM EDTA, as shown in FIGURE 4B, the fraction of E_1P was reduced to less than 15% of the total phosphoenzyme. Neither the sum of the acid-stable species nor the steady-state ATPase velocity was affected by the detergent treatment (TABLE 2); however, [E_1P] turnover (k_{calc}) increased to almost 1,100 s^{-1}.

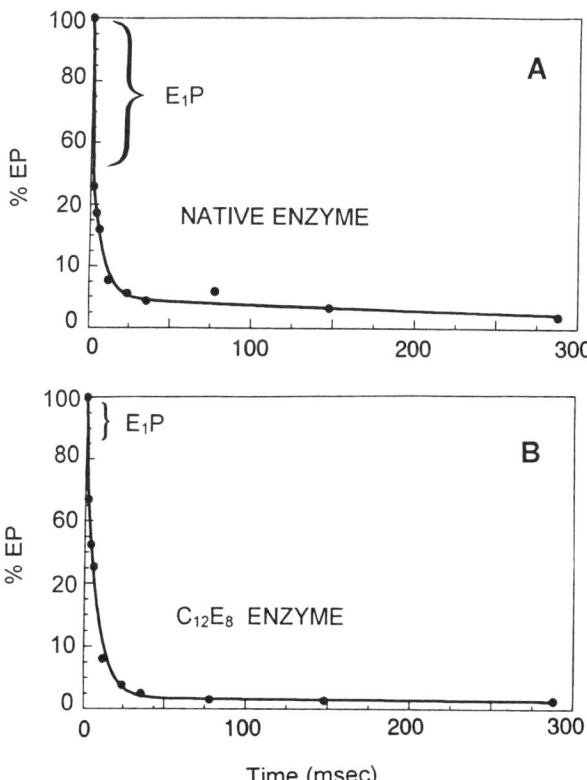

FIGURE 4. Effect of solubilization with $C_{12}E_8$ on ADP-induced phosphoenzyme decay in eel Na,K-ATPase. (**A**) Native eel Na,K-ATPase (0.5 mg/ml) was phosphorylated with 400 μM [γ^{32}P]ATP under the conditions described in the legend to FIGURE 3 and later dephosphorylated by the addition of 10 mM ADP + 10 mM EDTA. The decay curve was fitted using MLAB[20] to a sum of three exponentials to obtain the proportions of E_1P and E_2P, reported in TABLE 2. (**B**) Eel Na,K-ATPase was solubilized with $C_{12}E_8$ in order to prepare αβ protomers,[15] as described in the legend to TABLE 2. The conditions used for phosphorylation and dephosphorylation of the solubilized enzyme and fitting the decay curve are described in **A**.

TABLE 2. Steady-State Formation of E_1P and E_2P in Native and $C_{12}E_8$-Treated Eel Electric Organ Na,K-ATPase[a]

Conditions	[ATP] μM	E_1P nmol/mg	E_2P nmol/mg	v nmol/mg/s	k_{calc} s^{-1}
Native enzyme	10	0.044	0.017	12.5	284
Native enzyme	400	0.120	0.141	39	325
$C_{12}E_8$ (1 mg/mg)	10	0.033	0.020	6.9	209
$C_{12}E_8$ (1 mg/mg)	400	0.132	0.137	37	280
$C_{12}E_8$ (3 mg/mg)	400	0.035	0.218	37	1,057

[a]To prepare αβ protomers,[15] membranous electric organ Na,K-ATPase (1 mg protein/ml) was suspended in 25 mM NaCl, 0.1 mM EDTA and 25 mM TMA-HEPES, pH 7.4, mixed with a solution of recrystallized $C_{12}E_8$ (to achieve a final concentration of 1–3 mg detergent/mg protein) and incubated for 10 minutes at 21°C. Enzyme activity was stabilized by the addition of a sonicated mixture of Folch Fraction III phospholipids containing phosphatidylserine (50–100 μg PS/mg protein) and by the exclusion of $MgCl_2$ from the enzyme solution. k_{calc} (s^{-1}) is the steady-state rate constant for the conversion of E_1P to E_2P calculated from $k_{calc} = v/[E_1P]$ where the back reaction for the conversion reaction is assumed to be negligible.

Although this is three to five times faster than its native counterpart, it does not equal the presteady-state conversion rate, suggesting that the detergent may not have completely prevented these strong binding interactions. Lower detergent concentrations (1 mg detergent/mg protein), which were effective in eliminating complex kinetic behavior from the K^+ dephosphorylation reaction,[4] were unable to reduce the amount of E_1P compared to the native preparation (TABLE 2). These results suggest that the strength of the conformational interactions may vary so that tighter complexes are formed between unlike conformers (e.g., E_1P/E_2P) than between like conformers (e.g., E_2P/E_2P).

Compared to the Na,K-ATPase, the Ca-ATPase of sarcoplasmic reticulum shows higher levels of phosphorylation and a smaller P_i burst, reflecting a slower K^+-activated rate of E_2P hydrolysis (FIG. 5). The steady-state phosphoenzyme consists of ADP-sensitive and ADP-insensitive intermediates,[24] which were evaluated by two-stage rapid mixing experiments similar to those just described. FIGURE 6A shows a chase experiment in which Ca-ATPase vesicles suspended in approximately 10 μM free Ca^{2+} were phosphorylated with 10 μM ATP for 116 ms and then dephosphorylated by the addition of 5 mM ADP. About 40–50% of the phosphoenzyme disappeared immediately upon the addition of the chase and showed no associated P_i production, identifying it as E_1P, whereas the subsequent time-resolved phase was stoichiometrically related to P_i release and thus corresponds to E_2P. If this experiment is carried out in sarcoplasmic reticulum vesicles permeabilized by a Ca^{2+} ionophore, then jumping the [Ca^{2+}] (10 μM to 5 mM) simultaneously with [ADP] caused the slow phase to decay more rapidly and P_i production to decline (FIG. 6A; triangular symbols). This behavior is consistent with Ca^{2+} loading of the low-affinity transport sites on E_2P, which activates the conversion of E_2P to E_1P and increases ATP synthesis (while reducing P_i). We[24] previously showed that reducing the phosphorylation time from 116 to 6 ms left the proportions of the rapid (E_1P) and slow (E_2P) ADP-induced decay components unchanged, suggesting that E_1P is immediately converted to E_2P upon phosphorylation. Paradoxically, E_1P accumulates in the steady state despite the lack of evidence for rapid equilibration of this intermediate with E_2P (which would

produce a monoexponential decay pattern and no P_i upon the addition of ADP). We conclude that E_1P is an inherently labile intermediate as evidenced by the rapid accumulation of E_2P, but that its stability increases dramatically in the steady state. This behavior parallels that observed in the eel electric organ Na,K-ATPase and suggests that oligomeric interactions may be involved in these effects. To test this hypothesis,
Ca-ATPase was solubilized with $C_{12}E_8$ to prepare active monomers[17] and subsequently phosphorylated and dephosphorylated as in the native enzyme. As seen in FIGURE 6B, the phosphoenzyme decay pattern produced by ADP shows no evidence of the fast phase detected in the native enzyme; only a slow phase releasing similar amounts of P_i is observed. These results demonstrate that conditions leading to monomer formation prevent the accumulation of E_1P in the steady state with the implication that quaternary protein interactions are required for the stabilization of this intermediate state.

Biphasic Phosphoenzyme Formation. We[3] previously reported that preincubation of eel Na,K-ATPase with a Na^+-free medium gave a biphasic pattern of phosphorylation when ATP, Mg^{2+}, and Na^+ were subsequently added to initiate phosphorylation. This complex pattern can be eliminated by Na^+ preincubation prior to phosphoryla-

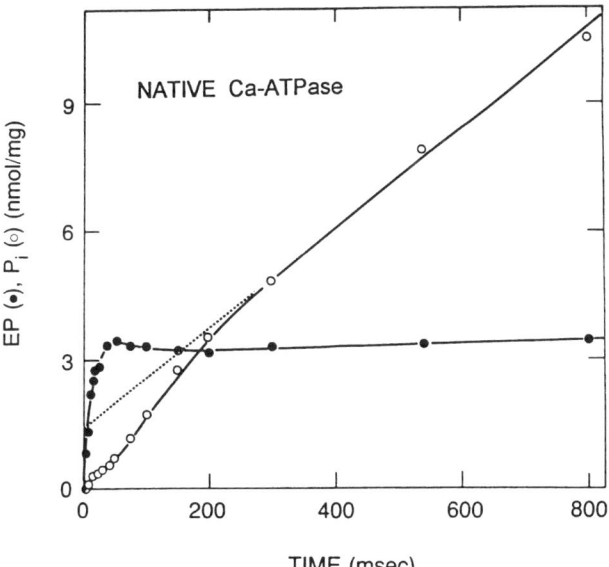

FIGURE 5. Time dependence of phosphorylation (EP) and phosphate (P_i) production in the sarcoplasmic reticulum Ca-ATPase. Ca-ATPase membrane vesicles (0.25 mg/ml) suspended in 100 mM KCl, 1 mM $MgCl_2$, 0.1 mM $CaCl_2$, 0.1 mM EGTA (free Ca^{2+} = 10 μM), and 20 mM Tris-maleate, pH 6.8, were mixed with an equal volume of an identical solution (without enzyme) containing 20 μM [$\gamma^{32}P$]ATP in a quenched-flow apparatus at 21°C. The reaction was terminated at the indicated times with 3% perchloric acid + 6 mM K_2HPO_4, and quenched samples containing [^{32}P]phosphoprotein and [^{32}P]P_i were analyzed, as previously described.[24] The amount of the E_2P_i intermediate that accumulates from the hydrolysis of E_2P (1.5 nmol/mg) was estimated by linear extrapolation of steady-state P_i production (*dashed line*) to the y-axis.

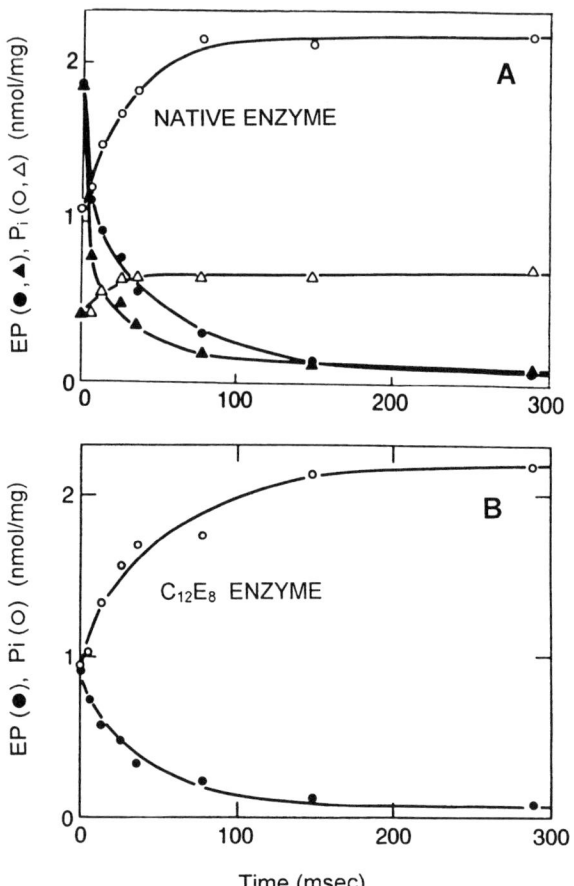

FIGURE 6. Effect of detergent treatment on the time dependence of ADP-induced phosphoenzyme decay in SR Ca-ATPase. (**A**) Native membrane vesicles (0.25 mg/ml) were phosphorylated for 116 ms, as described in the legend to FIGURE 5, and then dephosphorylated by the addition of a chase containing either 5 mM ADP (*circles*) or 5 mM ADP + 5 mM CaCl$_2$ (*triangles*). The charcoal extraction step was repeated in the phosphate assay to remove excess dinucleotide that might interfere with the determination of [^{32}P]P$_i$. (**B**) Ca-ATPase (0.25 mg/ml) suspended in 0.3 M sucrose and 20 mM Tris-maleate, pH 6.8, was mixed with recrystallized C$_{12}$E$_8$ (to achieve a final concentration of 2 mg detergent/mg protein) and incubated for 10 minutes at 21°C. The insoluble residue was removed by centrifugation. The conditions for phosphorylation are described in the legend to FIGURE 5. After 116 ms, dephosphorylation was initiated by the addition of a chase containing 5 mM ADP. Curves through the data points in **A** and **B** are drawn by eye.

tion[3] or by treatment of the enzyme with a low concentration (0.3 mg/mg protein) of the nonionic detergent, dodecylmaltoside. Biphasic ^{32}P labeling has also been observed in the Ca-ATPase at 2°C in the presence of high [KCl] (400 mM), as seen in FIGURE 7C. The slow phase of phosphorylation (5 s^{-1}) parallels a transient conformational change detected by an EPR spin probe (dIASL) covalently attached to Cys 674 of the transport protein.[14] Although not as apparent, biphasic phosphorylation also occurs at low (100 mM) [KCl], and the spin label again tracks the slow phase (22 s^{-1}). Paradoxically, there is no evidence of an EPR signal to match the fast phase of

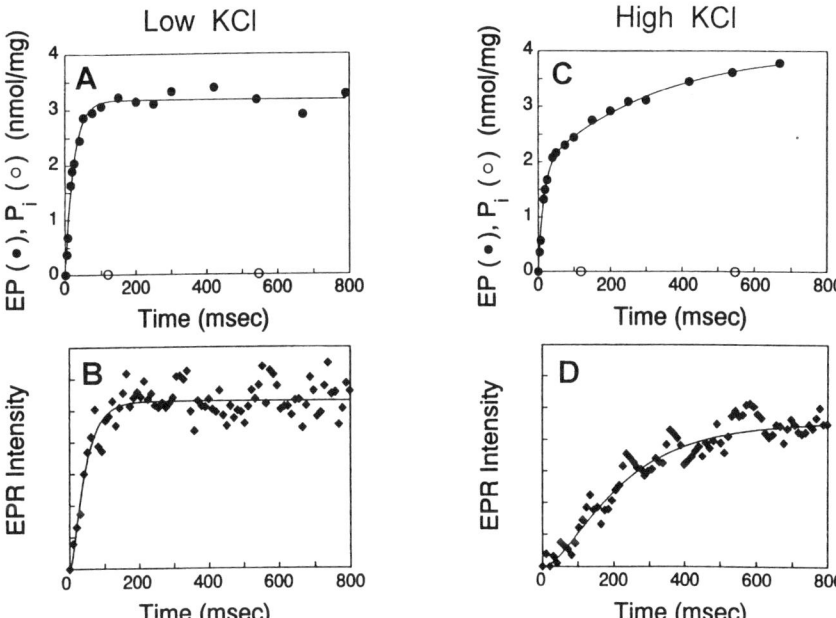

FIGURE 7. Comparison of presteady-state ATP-dependent phosphoenzyme formation with the transient electron paramagnetic resonance (EPR) spectrum of dIASL in sarcoplasmic reticulum (SR) membrane vesicles at 2°C; effect of [KCl]. (**A**) Time dependence of phosphorylation at low (100 mM) KCl. SR Ca-ATPase (0.5 mg/ml) suspended in standard buffer containing 100 mM KCl, 5 mM MgCl$_2$, 0.45 mM CaCl$_2$, 0.5 mM EGTA, and 20 mM MOPS, pH 7.0, was mixed with an equal volume of a standard buffer containing 200 μM [γ^{32}P]ATP in a quenched-flow apparatus at 2°C. The reaction was quenched with acid at the indicated times and the samples assayed for [^{32}P]phosphoenzyme, as previously described.[24] The time course of phosphorylation consists of fast (54.2 s^{-1}) and slow (20.1 s^{-1}) phases, as determined by fitting these data to a sum of n exponentials where $n = 1$–3 (MLAB; ref. 20). (**B**) Transient EPR signal at low (100 mM) KCl. SR vesicles (10 mg/ml) labeled at Cys 674 with dIASL were suspended in standard buffer containing 1 mM caged ATP and placed in a quartz flat cell inside of an EPR spectrophotometer cavity with an optical port. A brief flash from an XeF excimer laser was used to activate ATP release and to generate the appearance of a rotationally restricted EPR spectral component corresponding to E$_1$ATP and E$_1$P.[14] Transient EPR spectra were obtained at a single field position using a 10-ms time constant and 10-ms dwell time. The EPR transient displayed monoexponential behavior with a rate constant of 21 s^{-1}. (**C**) Time dependence of phosphorylation at high (400 mM) KCl. Conditions were identical to those in **A** except that the standard buffer contained 400 mM KCl. The time course of phosphorylation was biexponential with fast (21 s^{-1}) and slow (1.9 s^{-1}) phases. (**D**) EPR transient at high (400 mM) KCl. Conditions were identical to those in **B** except that [KCl] in the standard buffer was 400 mM. The EPR transient displayed monoexponential behavior with a rate constant of 5.3 s^{-1}.

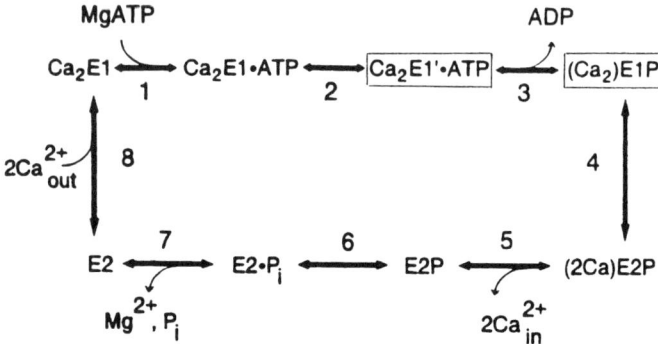

FIGURE 8. E1/E2 linear consecutive model for SR Ca-ATPase. The functional transport unit is a monomer consisting of a single catalytic subunit. Steps 4 and 8 are conformational transitions which transfer Ca^{2+}-loaded sites to the internal membrane surface (E_1P to E_2P) and recycle the unoccupied sites to the external surface (E_2 to E_1). Intermediate states in the boxes produce a rotationally restricted electron paramagnetic resonance transient signal.

phosphorylation at either high or low [salt]. The EPR signal originates from a conformationally restricted fraction of spin probes associated with the E_1ATP and E_1P intermediate states, as shown in FIGURE 8 (boxed states). Because the enzyme cycles through these intermediates during ATP hydrolysis, it is not obvious why the fast phosphorylation reaction fails to show a corresponding signal from the spin probe. A more fundamental issue concerns the origin of the biphasic pattern of phosphorylation. We[14] previously considered linear consecutive schemes that might contribute to this behavior and ruled out a slow conformational transition between E_2 and E_1 as the primary determinant of the slow phase. Because the enzyme medium contains Ca^{2+}, which binds tightly to the transport sites on E_1 and very weakly to those on E_2, essentially all of the enzyme should be trapped in E_1 which will produce only a fast phase of phosphorylation when ATP is added. The biphasic pattern in FIGURES 7A and C cannot be reproduced by a monomeric scheme and the properties of the partial reactions as we know them; neither is there evidence for more than one Ca-ATPase isoform in rabbit skeletal muscle. Thus, it is reasonable to assume that this pattern results from the sequential or "staggered" phosphorylation of two subunits in an oligomer, a hypothesis supported by the 1:1 distribution of fast and slow phases of phosphorylation. A model depicting this behavior is shown in FIGURE 9. The subunits are coupled conformationally so that the subunit on the right carries out the steps of the catalytic cycle ahead of its neighbor on the left. The right-hand subunit produces the fast phase of phosphorylation and is very rapidly converted to E_2P, so that its E_1P state has a very transient existence. Consequently, the EPR signal associated with E_1P is not detected because it is rapidly transformed to E_2P, which has no signal. These events trigger the appearance of the ATP binding site on the left-hand or lagging subunit, which phosphorylates producing the slow phase. Because of a tight interaction between newly formed E_1P on the left-hand subunit and E_2P on the right-hand subunit, the former is stabilized and its EPR signal increases in parallel with the slow phase of phosphorylation. When detergent is present, this tight interaction is lost and all of the phosphoenzyme accumulates in E_2P (FIG. 6B). Because the conversion to E_2P on the right-hand subunit is very fast whereas the accumulation of E_1P on the left is delayed, especially at high [KCl], [E_2P] may exceed [E_1P] at short phosphorylation times. Consistent with this, the addition of ADP to the Ca-ATPase phosphoenzyme during the fast phase revealed that almost 75% of the total intermediate was ADP insensitive after only 10 ms, whereas at longer time intervals the ADP-sensitive phosphoenzyme increased and eventually equalled E_2P (Froehlich, Mahaney, and Thomas, unpublished observations).

A similar type of staggered reaction mechanism can be applied to the electric organ Na,K-ATPase to account for the rapid presteady-state conversion of E_1P to E_2P and stable accumulation of E_1P in the steady state. At high [ATP], about one half the catalytic sites are phosphorylated, and these are about equally divided between E_1P and E_2P (TABLE 2). A similar situation exists in the Ca-ATPase at 2°C in which about 50% (3-4 nmol/mg) of the catalytic sites are phosphorylated in a 1:1 distribution of ADP-sensitive and ADP-insensitive states. We assume that the unphosphorylated sites are active, because at 21°C a P_i burst appears (cf, FIG. 5) which corresponds to either E_2P_i or E_2 and accounts for roughly one fourth the catalytic sites. The remaining sites may exist as an enzyme-substrate complex, namely, E_1ATP. Normally, rapid phosphorylation and ADP release would convert this intermediate to E_1P; however, quaternary protein interactions involving E_2 might allow E_1ATP to accumulate as a stable species. The resulting complex constitutes a tetramer in which the four catalytic subunits are in different states of the catalytic cycle (E_1ATP, E_1P, E_2P, and E_2) at any given point in time during the steady-state cycle. The general features of this description bear a close resemblance to the binding-exchange mechanism of the mitochondrial F_oF_1 ATPase[25] recently supported by the crystal structure of bovine F_1-ATPase.[26]

Although the subject of oligomeric interactions in the ion motive ATPases has aroused much controversy, the question of their possible functional significance has largely been neglected. A clue to this enigma may be found in the dramatic change in kinetic activity occurring at certain reaction steps in the transport cycle as the system proceeds from the presteady state into the steady state of ATP hydrolysis. A possible implication of the decline in the rate of conversion of E_1P to E_2P is that free energy is being transferred from this step and utilized in another portion of the transport cycle to enhance the catalytic activity of an inherently slow, energy-requiring reaction. By coupling a fast reaction to a slow reaction, the former slows down, analogous to the application of a load to a rapidly idling motor (FIG. 10). Energy transfer may occur at multiple points in the cycle, tending to prevent the accumulation of low energy states while enhancing the overall catalytic activity of the transport cycle. Another possible

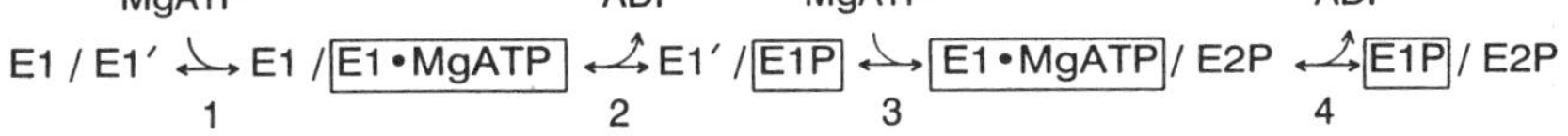

$$\text{E1 / E1}' \xrightarrow{\text{MgATP}} \text{E1 / } \boxed{\text{E1} \cdot \text{MgATP}} \xrightarrow{} \text{E1}' / \boxed{\text{E1P}} \xrightarrow{\text{MgATP}} \boxed{\text{E1} \cdot \text{MgATP}} / \text{E2P} \xrightarrow{\text{ADP}} \boxed{\text{E1P}} / \text{E2P}$$
$$\qquad\qquad 1 \qquad\qquad\qquad 2 \qquad\qquad\qquad 3 \qquad\qquad\qquad 4$$

FIGURE 9. Dimer model for the SR Ca-ATPase. Partial reaction mechanism depicting the first four steps in a dimeric transport model in which the reaction cycles of the conformationally coupled subunits are staggered. Phosphorylation of the right-hand subunit (k_2) produces the fast phase, while the left-hand subunit (k_4) generates the slow phase. Intermediates in the boxes contribute to the motionally restricted electron paramagnetic resonance (EPR) transient. Because $k_3 \gg k_2$, E_2P rapidly accumulates on the right-hand subunit, preventing detection of the EPR signal associated with its precursor. E_1P on the left-hand subunit accumulates at a slower rate defined by k_4, generating the observed EPR transient and a stable intermediate complex (E_1P/E_2P).

$$k' = k_0 \exp(-\Delta F/RT)$$

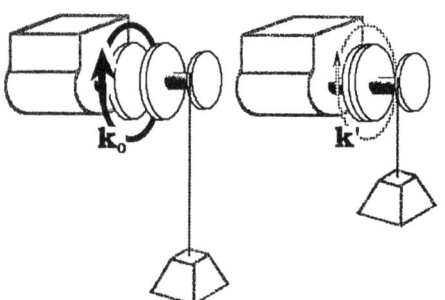

FIGURE 10. Kinetic consequences of coupling a fast reaction to a thermodynamic load. Slowing of the phosphoenzyme conformational transition in the steady state may reflect the utilization of free energy to accomplish work as defined by the above equation where ΔF is the free energy change, k_o is the reaction rate constant in the absence of the work load, and k' is the rate constant after application of the load.

use of this energy is to reduce the probability of back reactions that would allow accumulated cations to leak out of the system through the pump. In this capacity, the subunit interactions might behave like a "rachet and pawl," contributing to the vectorial operation of the system by preventing pump reversal. Specific examples of this might include a reduction in the affinity of cationic discharge sites on E_2P or a slowing of the rate of reversal of the phosphoenzyme conformational transition. It should be noted that the pairing of unequal energy states that allows for intersubunit energy transfer occurs because the catalytic reactions in the subunits are staggered. Thus, intersubunit conformational coupling in the oligomer not only provides the physical basis for energy transfer, but also imposes an order of operation on the protomers that enables states with different energies to accumulate simultaneously and exchange free energy. By conserving chemical energy otherwise dissipated as heat, the efficiency of the system is increased.

REFERENCES

1. HOBBS, A. S., R. W. ALBERS & J. P. FROEHLICH. 1983. Effects of oligomycin on the partial reactions of the sodium plus potassium-stimulated adenosine triphosphatase. J. Biol. Chem. **258:** 8163–8168.
2. HOBBS, A. S., R. W. ALBERS & J. P. FROEHLICH. 1985. Quenched-flow determination of the E1P to E2P transition rate constant in electric organ Na,K-ATPase. *In* The Sodium Pump. I. M. Glynn & C. Ellory, Eds.: 355–361. The Company of Biologists, Ltd. Cambridge, UK.
3. FROEHLICH, J. P. & K. FENDLER. 1991. The partial reactions of the Na^+- and Na^+ + K^+-activated adenosine triphosphatases. *In* The Sodium Pump: Structure, Mechanism and Regulation. J. H. Kaplan & P. De Weer, Eds.: 227–247. The Rockefeller University Press. New York.
4. FROEHLICH, J. P., A. S. HOBBS & R. W. ALBERS. 1994. Parallel pathway models for electric organ Na^+- and Na^+/K^+-ATPases. *In* The Sodium Pump. Structure, Mechanism, Hormonal Control and Its Role in Disease. E. Bamberg & W. Schoner, Eds.: 441–444. Steinkopff Verlag. Darmstadt.
5. NØRBY, J. G., I. KLODOS & N. O. CHRISTIANSEN. 1983. Kinetics of the Na-ATPase activity of the Na,K pump. Interactions of the phosphorylated intermediates with Na^+, $Tris^+$, and K^+. J. Gen. Physiol. **82:** 725–759.

6. ROSSI, R. C. & J. G. NØRBY. 1993. Kinetics of K^+-stimulated dephosphorylation and simultaneous K^+ occlusion by Na,K-ATPase, studied with the congener Tl^+. J. Biol. Chem. **268:** 12579–12590.
7. FROEHLICH, J. P., A. S. HOBBS & R. W. ALBERS. 1983. Evidence for parallel pathways of phosphoenzyme formation in the mechanism of ATP hydrolysis by electrophorus Na, K-ATPase. Curr. Top. Memb. Transp. **19:** 513–535.
8. OTTOLENGHI, P. & J. JENSEN. 1983. The potassium ion-induced apparent heterogeneity of high-affinity nucleotide binding sites in sodium-potassium ATPase can only be due to the oligomeric structure of the enzyme. Biochim. Biophys. Acta **727:** 89–100.
9. NØRBY, J. G. & J. JENSEN. 1991. Functional significance of the oligomeric structure of the Na,K pump from radiation inactivation and ligand binding. *In* The Sodium Pump: Structure, Mechanism and Regulation. J. H. Kaplan & P. De Weer, Eds.: 173–188. The Rockefeller University Press. New York.
10. SCHEINER-BOBIS, G., K. FAHLBUSCH & W. SCHONER. 1987. Demonstration of cooperating a-subunits in working ($Na^+ + K^+$)-ATPase by the use of the MgATP complex analogue cobalt tetraamine ATP. Eur. J. Biochem. **168:** 123–131.
11. SCHONER, W., D. THONGES, E. HAMER, R. ANTOLOVIC, E. BUXBAUM, M. WILLEKE, E. H. SUPERSU & G. SCHEINER-BOBIS. 1994. Is the sodium pump a functional dimer? *In* The Sodium Pump. Structure, Mechanism, Hormonal Control and Its Role in Disease. E. Bamberg & W. Schoner, Eds.: 332–341. Steinkopff Verlag. Darmstadt.
12. ANDERSEN, J. P. 1989. Monomer-oligomer equilibrium of sarcoplasmic reticulum Ca-ATPase and the role of subunit interactions in the Ca^{2+} pump mechanism. Biochim. Biophys. Acta **988:** 47–72.
13. IKEMOTO, N., A. M. GARCIA, Y. KUROBE & T. L. SCOTT. 1981. Nonequivalent subunits in the calcium pump of sarcoplasmic reticulum. J. Biol. Chem. **256:** 8593–8601.
14. MAHANEY, J. E., J. P. FROEHLICH & D. D. THOMAS. 1995. Conformational transitions of the sarcoplasmic reticulum Ca-ATPase studied by time-resolved EPR and quenched-flow kinetics. Biochemistry **34:** 4864–4879.
15. CRAIG, W. S. 1982. Determination of the distribution of sodium and potassium ion-activated adenosinetriphosphatase among the various oligomers formed in solutions of nonionic detergents. Biochemistry **21:** 2667–2674.
16. TANIGUCHI, K., K. SUZUKI, D. KAI, I. MATSUOKA, K. TOMITA & S. IIDA. 1984. Conformational change of sodium and potassium dependent adenosine triphosphatase. Conformational evidence for the Albers-Post mechanism in Na^+,K^+-dependent hydrolysis of ATP. J. Biol. Chem. **259:** 15228–15233.
17. MOLLNER, J. V., K. E. LIND & J. P. ANDERSEN. 1980. Enzyme kinetics and substrate stabilization of detergent-solubilized and membranous ($Ca^{2+} + Mg^{2+}$)-activated ATPase from sarcoplasmic reticulum. Effect of protein-protein interactions. J. Biol. Chem. **255:** 1912–1920.
18. FENDLER, K., E. GRELL & E. BAMBERG. 1987. Kinetics of pump currents generated by the Na^+,K^+-ATPase. FEBS Lett. **224:** 83–88.
19. WALKER, J. W., G. P. REID, J. A. MCCRAY & D. R. TRENTHAM. 1988. Photolabile 1-(2-nitrophenyl)ethyl phosphatase esters of adenine nucleotide analogues: Synthesis and mechanism of photolysis. J. Am. Chem. Soc. **110:** 7170–7177.
20. KNOTT, G. D. 1979. A mathematical modeling laboratory (MLAB). Comput. Prog. Biomed. **10:** 271–280.
21. HOBBS, A. S., J. P. FROEHLICH & R. W. ALBERS. 1980. Inhibition by vanadate of the reactions catalyzed by the ($Na^+ + K^+$)-stimulated ATPase. J. Biol. Chem. **255:** 5724–5727.
22. FENDLER, K., S. JARUSCHEWSKI, A. S. HOBBS, R. W. ALBERS & J. P. FROEHLICH. 1993. Pre-steady state charge translocation in Na,K-ATPase from eel electric organ. J. Gen. Physiol. **102:** 631–666.
23. FENDLER, K., S. JARUSCHEWSKI, J. P. FROEHLICH, R. W. ALBERS & E. BAMBERG. 1994. Electrogenic and electroneutral partial reactions in Na^+/K^+-ATPase from eel electric

organ. *In* The Sodium Pump. Structure, Mechanism, Hormonal Control and Its Role in Disease. E. Bamberg & W. Schoner, Eds.: 495–506. Steinkopff Verlag. Darmstadt.
24. FROEHLICH, J. P. & P. F. HELLER. 1985. Transient-state kinetics of the ADP-sensitive phosphoenzyme in sarcoplasmic reticulum: Implications for transient state calcium translocation. Biochemistry **24:** 126–136.
25. BOYER, P. D. 1993. The binding change mechanism for ATP synthase: Some probabilities and possibilities. Biochim. Biophys. Acta **1140:** 215–250.
26. ABRAHAMS, J. P., A. G. W. LESLIE, R. LUTTER & J. E. WALKER. 1994. Structure at 2.8 angstrom resolution of F1-ATPase from bovine heart mitochondria. Nature **370:** 621–628.

Functional Consequences of Mutations in the Transmembrane Core Region for Cation Translocation and Energy Transduction in the Na$^+$,K$^+$-ATPase and the SR Ca^{2+}-ATPase[a]

BENTE VILSEN,[b] DORTE RAMLOV,
AND JENS PETER ANDERSEN

Department of Physiology
University of Aarhus
DK-8000 Aarhus C, Denmark

Site-directed mutagenesis analysis demonstrates that the putative transmembrane segments M4, M5, and M6 play essential roles in the pump functions of Na$^+$,K$^+$-ATPase and Ca^{2+}-ATPase.[1–4] In addition to providing putative ligands for cation binding, M4 and M5, together with their associated stalk segments S4 and S5, form a direct physical link between the transmembrane region and the large cytoplasmic domain containing the phosphorylated aspartyl residue, suggesting that these are important mediators of signal transmission between the catalytic site and the cation binding sites in the membrane. Extensive studies of the properties of the phosphoenzyme intermediates of SR Ca^{2+}-ATPase mutants transiently expressed in COS-1 cells have indeed confirmed this hypothesis.[1,2] Recently, it proved possible to overexpress Na$^+$,K$^+$-ATPase mutants in COS-1 cells to levels reaching 60 pmol per milligram membrane protein, thus allowing similar kinetic studies of the phosphoenzyme intermediates of mutant Na$^+$,K$^+$-ATPases with alterations to residues in M4 and M5. On this basis, comparative aspects of the two pumps are discussed as are the unique properties of certain Na$^+$,K$^+$-ATPase mutants that display high Na$^+$-ATPase activity in the absence of K$^+$. Focus is directed at the amino acid residues highlighted in the structural model of the Na$^+$, K$^+$-ATPase shown in FIGURE 1.

MUTATIONS AFFECTING PHOSPHOENZYME CONFORMATIONAL CHANGES

That transmembrane segment M4, with its associated stalk segment S4, plays a pivotal role in the transition between ADP-sensitive (E$_1$P) and ADP-insensitive (E$_2$P) phosphoenzyme intermediates and in the associated deocclusion of Ca^{2+} in the

---

[a]This work was supported by the Danish Biotechnology Programme, the Danish Medical Research Council, and the NOVO Nordisk Foundation.

[b]Address for correspondence: Bente Vilsen, Department of Physiology, University of Aarhus, Ole Worms Allé, 160, DK-8000 Aarhus C, Denmark (tel: 45 8942 2832; fax: 45 8612 9065; e-mail: bv@fi.aau.dk).

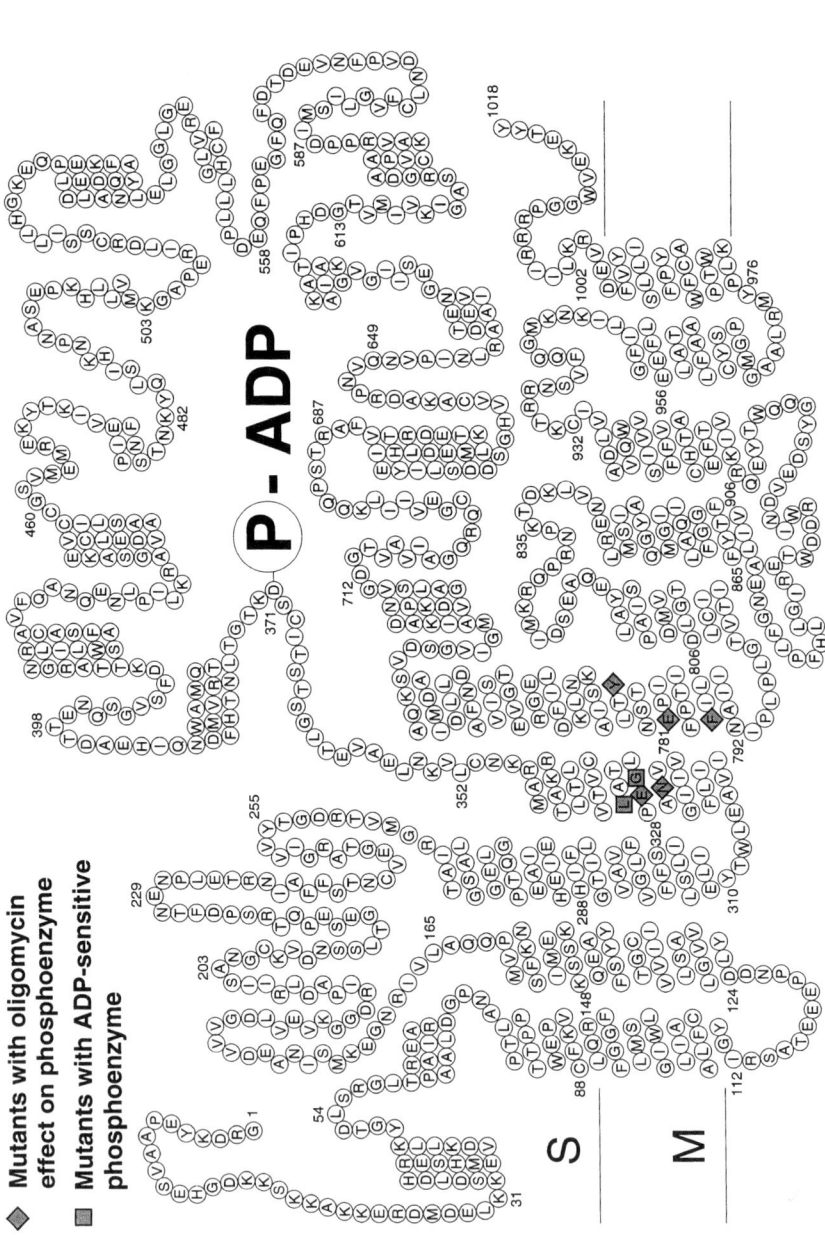

FIGURE 1. 10-Helix structural model of the rat α1-isoform of Na+,K+-ATPase with indication of residues studied in the present mutagenesis work. S and M indicate putative "stalk" and transmembrane segments, respectively (numbering is from the left). *Squares* indicate mutants that accumulate a high level of the ADP-sensitive phosphoenzyme intermediate in steady state. *Diamonds* indicate mutants with a low steady-state level of phosphoenzyme which can be increased by addition of oligomycin.

Ca^{2+}-ATPase was previously deduced from studies showing that replacement of residues at the M4S4 boundary leads to stabilization of E_1P.[5,6] In particular, the proline residue Pro312 was shown to be crucial.[5] In the rat α1 isoform of Na^+,K^+-ATPase, the residue at the equivalent position is Leu332 (FIG. 1). To elucidate the role of this residue in the Na^+,K^+-ATPase catalytic mechanism, it was replaced by alanine as well as by proline. In addition, the nearby residue Gly330 was replaced by alanine. Three mutants, Leu332 → Ala, Leu332 → Pro, and Gly330 → Ala, displayed a significantly reduced maximum Na^+,K^+-ATPase turnover number relative to the wild type, calculated as the ratio between the rate of ATP hydrolysis measured in the presence of saturating Na^+, K^+, and ATP concentrations and the maximum phosphorylation level determined in the phosphorylation assay with oligomycin (TABLE 1). The partition of the phosphoenzyme intermediates between ADP-sensitive/K^+-insensitive and ADP-insensitive/K^+-sensitive species was studied in the experiments shown in FIGURES 2A and 2B. It is well known that for purified Na^+,K^+-ATPases of various species the phosphorylation conditions applied here result in a steady-state distribution of phosphoenzyme intermediates favoring the ADP-insensitive form (E_2P). Consistently, as seen in FIGURE 2A, extrapolation of the slow phase of the ADP-dephosphorylation curve back to ordinate intercept yielded about 70% E_2P for the wild-type rat enzyme expressed in COS-1 cells. A similar steady-state level of E_2P was obtained for mutant Leu332 → Ala. However, the mutant with proline replacing Leu332 accumulated almost exclusively the ADP-sensitive phosphoenzyme species, as indicated by the very fast monoexponential decay observed in the presence of ADP. The Gly330 → Ala mutant displayed intermediate behavior with about 35% E_2P present at steady state.

As expected from the Post-Albers scheme, for each of mutants Leu332 → Pro and Gly330 → Ala, the accumulation of ADP-sensitive E_1P was matched by a loss of K^+

TABLE 1. Na^+,K^+- and Na^+-ATPase Molecular Activities and Oligomycin Effect on Phosphoenzyme Level of Wild-Type and Mutant Rat Na^+,K^+-ATPases Expressed in COS-1 Cells[a]

	ATPase Turnover Number (min^{-1})		
	$Na^+ + K^+$	Na^+	Phosphorylation—OG (%)
Wild type	8500	<800	>80
Leu332 → Pro	5452	<800	102
Leu332 → Ala	4309	<800	97
Gly330 → Ala	7206	<800	99
Glu329 → Gln	2530	705	29
Asn326 → Leu	7203	2304	29
Tyr773 → Leu	7124	1794	41
Glu781 → Ala	6555	2491	31
Phe788 → Leu	2739	2232	32

[a]ATPase activity was measured at 37°C as previously described.[8] Na^+,K^+-ATPase in the presence of 130 mM Na^+, 20 mM K^+, 3 mM MgATP. Na^+-ATPase in the presence of 200 mM Na^+, 3 mM MgATP, absence of K^+. The steady-state phosphorylation level was measured at 0°C in the presence of 150 mM Na^+ with and without oligomycin, as previously described,[8] except that a new filtration procedure was used to separate the acid-quenched phosphorylated protein. The turnover number was calculated as the ratio between the specific ATPase activity and the phosphorylation level measured in the presence of oligomycin. The phosphorylation level measured in the absence of oligomycin (Phosphorylation—OG) is shown as a percentage of that measured with oligomycin.

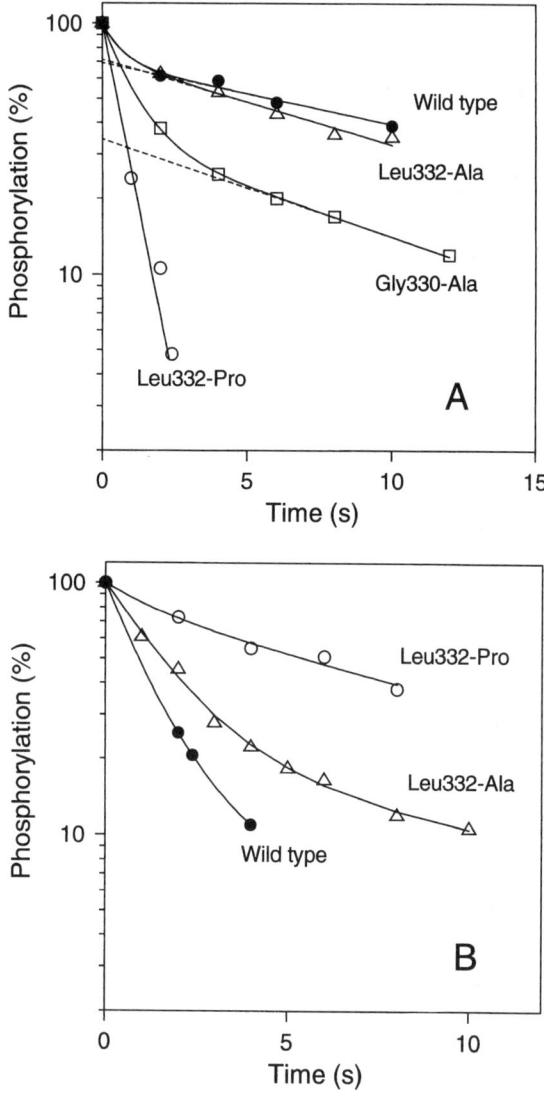

FIGURE 2. Dephosphorylation kinetics in the presence of ADP (**A**) or K^+ (**B**). Phosphorylation was carried out in the presence of 20 mM Na^+, 2 μM [γ-^{32}P]ATP, 3 mM Mg^{2+}, and 20 mM Tris (pH 7.4), and the dephosphorylation was monitored after chase with 1 mM nonradioactive ATP together with 1 mM ADP (**A**) or 20 mM KCl (**B**). Temperature 0°C. Other experimental details were as described previously.[9]

sensitivity of the phosphoenzyme (data shown for mutant Leu332 → Pro in FIG. 2B). For these Na^+,K^+-ATPase mutants the accumulation of E_1P at steady state was due to a reduced rate of E_1P to E_2P conversion, thus explaining the inhibition of ATPase activity. Hence, these mutants behaved very similar to the Ca^{2+}-ATPase mutant Pro312 → Ala.[5]

For the Na^+,K^+-ATPase mutant Leu332 → Ala the situation was more complex, because this mutant did not accumulate E_1P to any further extent than did the wild type in the presence of 20 mM Na^+ (FIG. 2A), but it displayed a reduced rate of K^+-induced dephosphorylation relative to the wild type, at saturating K^+ concentration (FIG. 2B). This suggests either that the rate of K^+-dependent dephosphorylation of E_2P was reduced in this mutant relative to the wild type or that the mutant exhibited a unique response to K^+, driving the interconversion of E_1P and E_2P backwards, thereby depleting the rapidly dephosphorylating E_2P form.

These data pinpoint the Na^+,K^+-ATPase residues at position 330 and 332, and thus the COOH-terminal part of transmembrane segment M4, as central to the control of the conformational changes in the phosphoenzyme involved in ion translocation. This corroborates previous modeling based on Ca^{2+}-ATPase mutagenesis data, implicating the M4S4 sector in the signal transmission between the catalytic site and the transmembrane domain. A rotation or tilting of M4 may provide access for the occluded ions to the extracellular surface and at the same time confer loss of ADP sensitivity of the phosphoenzyme due to long-range effects mediated through the S4 segment linking M4 to the phosphorylated aspartyl residue.[5,6]

MUTANTS WITH OLIGOMYCIN EFFECT ON PHOSPHOENZYME LEVEL

Na^+-ATPase Activity

When the wild-type Na^+,K^+-ATPase is phosphorylated from ATP at low temperature and saturating ATP and Na^+ concentrations in the absence of K^+, the steady-state level of phosphoenzyme built up usually corresponds to around 80–85% of the active site concentration determined by other ligand binding measurements. However, an increase in the phosphorylation level to >95% of the site concentration can generally be achieved by adding oligomycin to stabilize the Na^+-occluded form and thereby reduce the dephosphorylation rate.[7] For the mutants just described, with alterations to Leu332 and Gly330, the addition of oligomycin did not increase the steady-state level of phosphoenzyme significantly (TABLE 1). This is the expected consequence of the reduced rate of phosphoenzyme processing caused by the mutations, mimicking the effect of oligomycin. By contrast, in another series of mutants (indicated by diamonds in FIGURE 1), the relative phosphorylation level detected in the absence of oligomycin was found to be remarkably low, ranging between 29% and 41% of that measured in the presence of oligomycin (TABLE 1). These mutants encompass some with alterations to aromatic residues in transmembrane segment M5 (Tyr773 and Phe788) as well as some with alterations to putative cation binding residues with oxygen-containing side chains in M4 (Asn326 and Glu329) and M5 (Glu781). It was previously shown that mutants Glu781 → Ala[8] and Asn326 → Leu[9] possess an unusual ability to hydrolyze ATP at a high rate in the absence of K^+. As previously shown for mutant Asn326 → Leu[9] and demonstrated here in FIGURE 3 for mutant Glu781 → Ala, this high Na^+-ATPase activity is due to an increased rate of

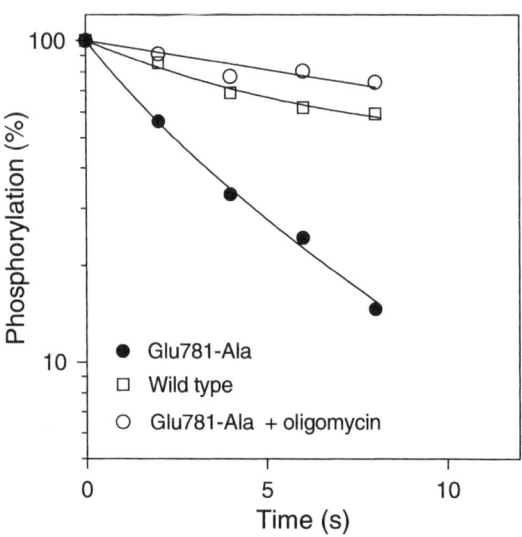

FIGURE 3. Dephosphorylation kinetics in the presence of Na^+ without K^+. Phosphorylation was carried out in the presence of 50 mM Na^+, 2 μM [γ-^{32}P]ATP, 3 mM Mg^{2+}, and 20 mM Tris (pH 7.4), and the dephosphorylation was monitored after chase with 1 mM nonradioactive ATP. As indicated, in one set of experiments, mutant Glu781 → Ala had been preincubated with oligomycin. Temperature 0°C. Other experimental details were as described previously.[9]

K^+-independent dephosphorylation. The addition of oligomycin reduces the dephosphorylation rate of the mutant to a value below that pertaining to the wild type in the absence of oligomycin, thus explaining the large oligomycin effect on the steady-state phosphoenzyme level of the mutant.

As seen in TABLE 1, two mutants with alterations to aromatic residues Tyr773 and Phe788 likewise displayed high Na^+-ATPase activity in addition to the oligomycin effect on the steady-state phosphorylation level. In fact, for mutant Phe788 → Leu, the Na^+-ATPase activity amounted to as much as 82% of the maximum Na^+,K^+-ATPase activity, suggesting that Na^+ is as efficient an activator as K^+ at the sites binding extracellular cations.

The apparent loss of ability to discriminate between extracellular Na^+ and K^+ in the four mutants displaying high Na^+-ATPase activity does not seem to originate from altered Na^+- and/or K^+-affinities at the extracellularly facing sites. (Although mutant Glu781 → Ala displayed some reduction in apparent affinity for K^+ in titrations of Na^+,K^+-ATPase activity, this reduction amounted to only three- to fourfold relative to wild type,[8] and the other three mutants with high Na^+-ATPase activity displayed normal apparent K^+ affinity in the presence of Na^+.) Hence, the K^+-like effect exerted by Na^+, which enhances the dephosphorylation rate and thus the Na^+-ATPase activity as well as the associated Na^+ uptake in cells,[10] may arise primarily from the increased efficacy of Na^+ as activator of dephosphorylation (V_{max} effect on Na^+-ATPase rather than $K_{0.5}$ effect).[8,9] This may be due to disruption of an inhibitory external Na^+ site, or the perturbing influence of the mutations on the conformation of transmembrane segments, and/or on their intersegmental contacts, may in some other way have enhanced the long-range signal transduction to the catalytic domain leading to dephosphorylation.

The Glutamic Acid Residue in M4

Among the mutants with a reduced steady-state phosphorylation level in the absence of oligomycin (described in TABLE 1), there was one, namely, Glu329 → Gln, that did not exhibit a significantly enhanced molecular turnover number for Na^+-ATPase activity relative to the wild type. This raised the question of whether a reduced rate of phosphorylation might contribute to lowering the steady-state phosphorylation level measured for this mutant in the absence of oligomycin. Therefore, kinetic experiments were conducted to compare the rates of phosphorylation of the wild-type Na^+,K^+-ATPase and the Glu329 → Gln mutant. As shown in FIGURE 4, we found a

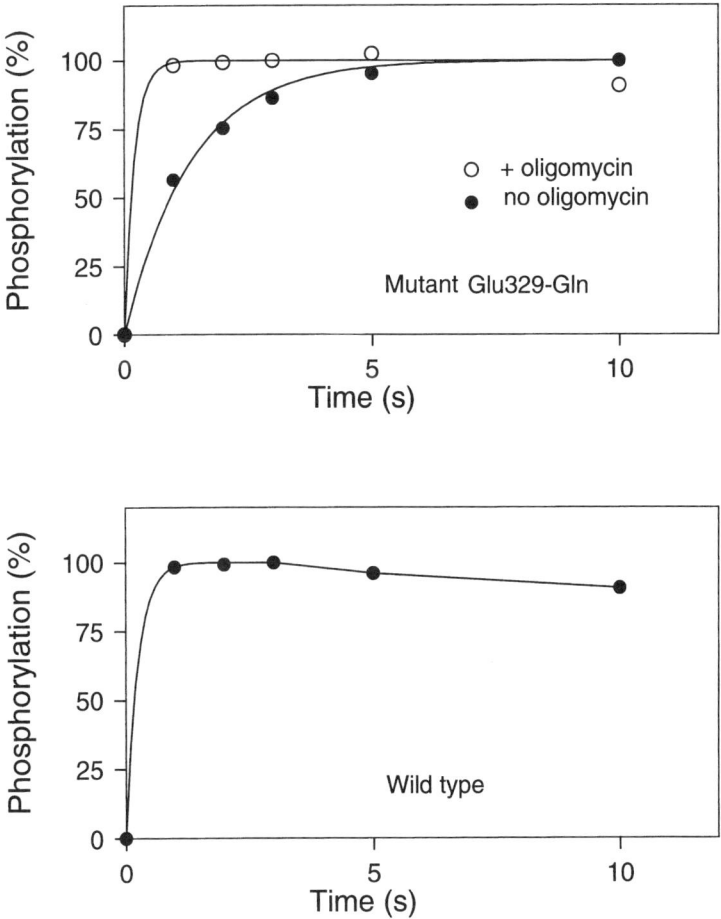

FIGURE 4. Phosphorylation kinetics of wild-type Na^+,K^+-ATPase and mutant Glu329 → Gln measured at 0°C in the presence of 40 mM Na^+, 4 µM [γ-^{32}P]ATP, 3 mM Mg^{2+}, and 20 mM Tris (pH 7.4). As indicated, in one set of experiments, mutant Glu329 → Gln had been preincubated with oligomycin. Other experimental details were as described previously.[9]

significant reduction of the rate constant for the approach to steady-state phosphorylation in the mutant relative to the wild type. It is further seen in FIGURE 4 that the addition of oligomycin increased the rate of phosphorylation in the mutant, thus explaining the promotive effect of oligomycin on the steady-state phosphorylation level. As the measured rate constant for phosphorylation of the mutant was found to be insensitive to a reduction in the Na^+ and ATP concentrations to half those applied in FIGURE 4, it can be excluded that the low phosphorylation rate of the mutant observed in the absence of oligomycin resulted from the lack of saturation of the binding sites for cytoplasmic Na^+ and ATP. This notion seems to be supported by the results of previous studies of this mutant showing that its apparent affinity for Na^+, measured by Na^+ titration of steady-state phosphorylation from ATP, is only three- to fourfold reduced relative to the wild type and that the apparent affinity for ATP is indistinguishable from that of the wild type.[1,2] Hence, it appears that it is the transfer of the phosphoryl group from bound ATP which is slowed down by the mutation and that this effect can be counteracted by oligomycin.

The reduced phosphorylation rate observed for mutant Glu329 → Gln can explain the low maximum Na^+,K^+-ATPase turnover rate of this mutant (approximately 2,500 min^{-1} vs 8,500 min^{-1} for wild type; see TABLE 1). Previously, it was concluded that the maximum turnover number for Na^+, K^+-ATPase activity of the Glu329 → Gln mutant was similar to that of the wild type,[11] but this was before we discovered the oligomycin effect on the phosphorylation level, and the previous calculation of turnover number was based on the site concentration determined by phosphorylation in the absence of oligomycin. Another technical improvement allowing an accurate determination of the absolute value for the turnover number of expressed wild-type and mutant enzymes in the present study has been the use of a filtration method for measurement of the phosphorylation level, instead of the gelelectrophoretic method which gave too high numbers due to loss of part of the phosphoenzyme during centrifugal washing and electrophoresis.

The glutamic acid residue in transmembrane segment M4 is highly conserved among Na^+,K^+-ATPases and Ca^{2+}-ATPases, and studies of mutants of the sarcoplasmic reticulum Ca^{2+}-ATPase have indicated a crucial role of this residue (Glu309) in Ca^{2+} occlusion.[1,12] In mutant Glu309 → Gln of the Ca^{2+}-ATPase, no Ca^{2+}-activated phosphorylation from ATP can be detected even at Ca^{2+} concentrations of several millimolar,[13] thus precluding a study of the phosphorylation rate of this mutant similar to that just presented for the homologous Na^+,K^+-ATPase mutant. However, when Glu309 is replaced by aspartate in the Ca^{2+}-ATPase, the ability to phosphorylate from ATP is retained, although the apparent affinity for activating Ca^{2+} is considerably reduced. Therefore, to examine the significance of Glu309 for the rate of phosphorylation of the Ca^{2+}-ATPase we carried out a kinetic experiment with the Ca^{2+}-ATPase mutant Glu309 → Asp analogous to the experiment just described for the Na^+,K^+-ATPase mutant Glu329 → Gln. Again, care was taken to ensure saturation with substrate and activating cation (Ca^{2+} concentration 5 mM). As seen in FIGURE 5, the result clearly demonstrated a reduced rate of phosphorylation in the Glu309 → Asp mutant relative to the rate for wild-type Ca^{2+}-ATPase, indicating that the glutamic acid residue in M4 plays similar mechanistic roles in Na^+,K^+-ATPase and Ca^{2+}-ATPase.

Whereas it may be questioned whether the three- to fourfold reduction in apparent affinity for Na^+ detected in the steady-state phosphorylation assay with Na^+,K^+-ATPase mutant Glu329 → Gln[1] is sufficient to justify an assignment of this residue as cation ligand analogous to what has been proposed for Glu309 in Ca^{2+}-ATPase, there seems to be no doubt as to its role in ensuring fast phosphorylation. The phosphorylation reaction is closely associated with the binding and occlusion of the cations, and a model explaining this interrelationship has been proposed in which the cation occlusion draws the walls of the binding cavity together, thereby triggering conformational changes in the cytoplasmic domain that bring the γ-phosphoryl group of bound ATP in position for transfer to the aspartyl group.[14] Therefore, it may be surmised that the inefficient phosphorylation of mutants with alteration to the glutamic acid residue in M4 results from a defective cation occlusion, which may not necessarily be associated with a large decrease in apparent affinity. This notion is consistent with the enhancement of the phosphorylation rate of the mutant induced by oligomycin, which is known to promote occlusion of Na^+.[7] Thus, in Na^+,K^+-ATPase a most important role of Glu329 might be to participate in the gating process at the cytoplasmic entrance to the occlusion pocket. In further support of this hypothesis, other functional

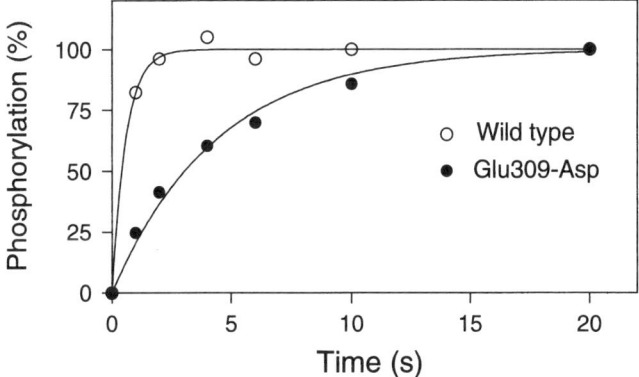

FIGURE 5. Phosphorylation kinetics of wild-type Ca^{2+}-ATPase and mutant Glu309 → Asp measured at 0°C in the presence of 5 mM Ca^{2+}, 4 μM [γ-^{32}P]ATP, 5 mM Mg^{2+}, 0.1 M K^+, and 20 mM MOPS (pH 7.0).

characteristics of the Glu329 → Gln mutant have suggested an increased rate of the E2(K) → E1 transition consistent with a destabilization of the K^+-occluded form, leading to fast release of occluded K^+ at the cytoplasmic surface.[1,11] It is noteworthy in this connection that like the Glu329 → Gln substitution, replacement of Glu329 with leucine led to a Na^+,K^+-pump with partial retention of function, whereas substitution with either alanine or aspartic acid, both of which possess smaller side chains, led to inactive pumps.[4,8] The requirement for a bulky side chain would be explained if Glu329 were part of the occlusion gate. This would be equivalent to the role of the acidic residue at the ninth position of the Ca^{2+}-binding loop in EF-hand Ca^{2+} sites, the substitution of which leads to a considerable change in the rate constant for dissociation of the bound ion (i.e., a change in the height of the energy barrier of the dissociative transition state), while leaving the stability of the ion-site complex (i.e., the affinity) unchanged.[15,16]

Role of Tyr773 in Cation Binding

Replacement in Na^+,K^+-ATPase of the highly conserved tyrosine Tyr773 at the NH_2-terminal end of M5 with leucine resulted not only in increased Na^+-ATPase activity and a low phosphoenzyme level in the absence of oligomycin, as shown in TABLE 1, but, as seen in FIGURE 6, also in a reduction of the apparent Na^+ affinity amounting to as much as 15-fold in the phosphorylation assay, that is, the most conspicuous change in Na^+ affinity observed so far for any Na^+,K^+-ATPase mutant, suggesting the occurrence of a large perturbation of the structure of the Na^+ binding site. It should be noted that oligomycin was included in this assay to stabilize the phosphoenzyme, and therefore the observed change in Na^+ affinity may in fact represent an underestimate of the true effect of the mutation.

In light of the suggestion that interaction of K^+ with π-electrons of aromatic amino acid side chains may confer K^+ selectivity to the pore of K^+-channel proteins,[17] the result of titrating the K^+ dependence of the Na^+,K^+-ATPase mutant Tyr773 → Leu was awaited with interest, but, as seen in FIGURE 6, the apparent affinity for K^+ was unaffected by the mutation. Likewise, we found that replacement of the other aromatic residues in M5, Phe764, Phe785, and Phe788, resulted in mutant Na^+,K^+-ATPases with normal or even increased apparent affinity for K^+. Hence, the side chain π-electrons of these residues do not seem to participate in K^+ coordination in Na^+,K^+-ATPase.

A MODEL FOR THE CATION BINDING POCKET

On the basis of detailed studies of the Ca^{2+} concentration dependencies of phosphorylation from ATP and P_i, in conjunction with measurements of Ca^{2+} occlusion in Ca^{2+}-ATPase mutants, we previously proposed a model for the assignment of five Ca^{2+} liganding residues in transmembrane segments M4, M5, and M6 to two separate binding sites in a single file arrangement in the Ca^{2+} occlusion pocket of the Ca^{2+}-ATPase.[2,18] In this model, Glu309 and Asn796 (equivalent to Na^+,K^+-ATPase residues Glu329 and Asp806, respectively) are associated exclusively with the most superficial binding site facing the cytoplasm in the E_1 form, whereas Glu771 and Thr799 (equivalent to Glu781 and Thr809 in Na^+,K^+-ATPase) are associated exclusively with the deeper site, and Asp800 (equivalent to Asp810 in Na^+,K^+-ATPase) donates ligands to both sites (FIG. 7). A corollary of our model is that the M5-M6 hairpin cannot be α-helical as suggested in most other current structural models (including the one shown in FIG. 1), because it is impossible for the three residues, Asn796, Thr799, and Asp800, all belonging to the M6 segment, to donate ligands to the two sites in the suggested order and at the same time be part of an α-helix orientated with its amino-terminal end at the extracytoplasmic surface. Therefore, we speculated that the actual structure of M5-M6 might be more disorganized, possibly resembling the pore region of cation channel proteins.[18] A mobile loop structure or short "inner" helices in combination with β structure[19] would probably also be much better able to provide the flexibility needed for ion translocation than "stiff" α-helices. In this connection, it is noteworthy that Asn796 together with Glu309 and Glu771, but not Thr799 and Asp800, was found to be essential to the dephosphorylation of the E_2P intermediate,[13,18] suggesting that the former three residues participate not only in Ca^{2+} occlusion but also in mediation of effects triggered by cation binding at extracytoplasmically facing sites in E_2P. This is

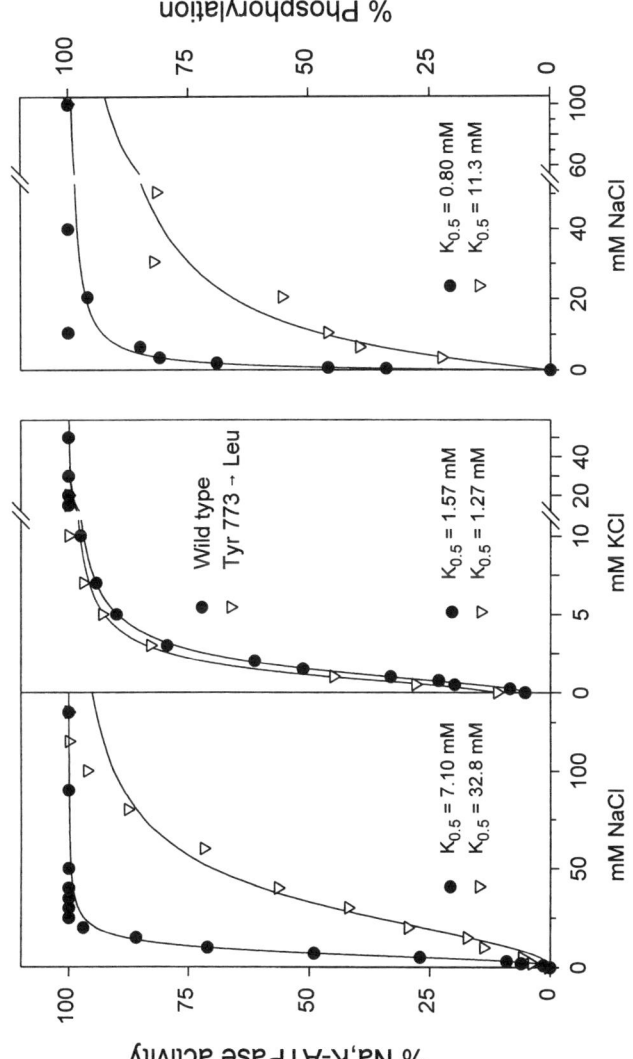

FIGURE 6. Cation dependence of overall Na^+,K^+-ATPase activity (**left and middle panels**) and phosphorylation (**right panel**) of wild-type Na^+,K^+-ATPase and mutant Tyr773 → Leu. Na^+,K^+-ATPase activity was measured at 37°C in the presence of various concentrations of Na^+ and 20 mM K^+ (**left**) or in the presence of various concentrations of K^+ and 100 mM Na^+ (**middle**). Phosphorylation was carried out at 0°C in the presence of various concentrations of Na^+, 2 μM [γ-^{32}P]ATP, 3 mM Mg^{2+}, 20 mM Tris (pH 7.4), and 20 μg oligomycin/ml. Other experimental details were as described previously.[9]

consistent with a model in which the conformational changes associated with cation transport lead to alternating exposure of Glu309 and Asn796 at the cytoplasmic surface in E_1 and at the extracytoplasmic surface in E_2P, as depicted in FIGURE 7. In accordance with our proposal, recent *in vitro* translation-scanning experiments have failed to detect a signal anchor sequence in the M5–M6 region, suggesting that this hairpin might be posttranslationally rather than cotranslationally inserted in the membrane,[20] as expected for a flexible loop structure that is to be positioned in between other transmembrane peptide segments rather than in contact with lipid. Moreover, MacLennan and coworkers have shown by cysteine-scanning mutagenesis that a disulfide bridge forms easily between two cysteines replacing Glu309 and Asn796 in the Ca^{2+}-ATPase, thus confirming the assignment of these two residues to the same (superficial) site in the E_1 form.

Although our model for two Ca^{2+} sites in the Ca^{2+}-ATPase cannot be translated directly into a model assigning liganding groups to three Na^+ and two K^+ sites in the Na^+,K^+-ATPase, striking analogies exist between the functions of the equivalent

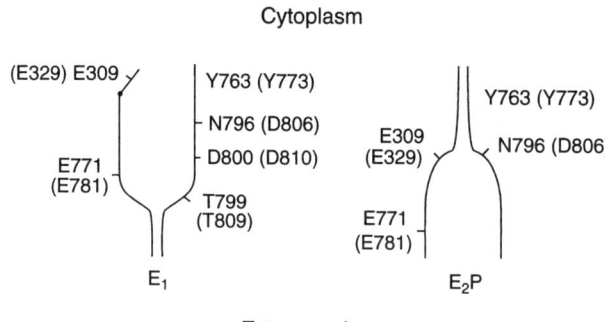

FIGURE 7. Hypothesis for the assignment of conserved residues to various parts of the cation binding pockets of Ca^{2+}-ATPase and Na^+,K^+-ATPase (Na^+,K^+-ATPase residues shown in parentheses) in E_1 **(left)** and E_2P **(right)**.

residues of the two pumps, suggesting that the structures of their cation binding pockets must be highly similar:

1. As just discussed, Glu329 in Na^+,K^+-ATPase is closely associated with the gating mechanism at the cytoplasmic entrance to the cation binding pocket in the E_1 form. In addition, Glu329 has a role in mediation of the effect of binding of extracellular K^{+21} comparable to the role of the Ca^{2+}-ATPase residue Glu309 in dephosphorylation of E_2P.
2. Glu781 is important to Na^+ binding in the E_1 form, because the affinity for Na^+ is reduced at least six- to eightfold in the Glu781 → Ala mutant as determined in the phosphorylation assay in the presence of oligomycin.[8] In addition, Glu781 has a role in connection with the function of the extracellularly facing cation sites, as already discussed. This may be compared to the role of the Ca^{2+}-ATPase residue Glu771 in dephosphorylation of E_2P.

3. Thr809 is important to Na^+ binding in the E_1 form, as indicated by the six- to eightfold reduction of apparent Na^+ affinity in mutant Thr809 → Ala measured in the phosphorylation assay.[8] On the other hand, Thr809 does not seem to play any significant role in K^+ affinity; this is analogous to the lack of a role for the Ca^{2+}-ATPase residue Thr799 in dephosphorylation of E_2P.[2]
4. The aspartic acid residue present in Na^+,K^+-ATPases and H^+,K^+-ATPases at the position equivalent to Asn796 in the Ca^{2+}-ATPase is important for mediation of the effect of K^+ binding. The conspicuous effect of mutation of this residue (Asp806 in FIG. 1) on apparent K^+ affinity, discussed by Lingrel, Jørgensen, and de Pont in this volume, is analogous to the more than 20-fold reduction in the rate of dephosphorylation of the E_2P intermediate observed for the Ca^{2+}-ATPase mutant Asn796 → Ala[18] and confirms our previous proposal (outlined in FIG. 7) that this residue contributes to the extracellularly facing cation binding sites in E_2P.
5. The preferential effect of the Tyr773 → Leu mutation in Na^+, K^+-ATPase on Na^+ affinity places Tyr773 in close association with the cytoplasmically facing (Na^+) sites in the E_1 form and leaves some distance to the extracytoplasmically facing (K^+) sites in the E_2P form. This is similar to the assignment of the homologous Ca^{2+}-ATPase residue Tyr763 to the vicinity of the cytoplasmic inlet of the transmembrane pore in the Ca^{2+}-ATPase which we discuss elsewhere in this volume on the basis of the reduced Ca^{2+} affinity and uncoupling of ATPase activity from Ca^{2+} transport observed for the Ca^{2+}-ATPase mutants Tyr763 → Leu and Tyr763 → Gly, respectively.

REFERENCES

1. VILSEN, B. 1995. Acta Physiol. Scand. **154** (Suppl. 624): 1–146.
2. ANDERSEN, J. P. & B. VILSEN. 1995. FEBS Lett. **359**: 101–106.
3. MACLENNAN, D. H., D. M. CLARKE, T. W. LOO & I. S. SKERJANC. 1992. Acta Physiol. Scand. **146**: 141–150.
4. JEWELL-MOTZ, E. A. & J. B. LINGREL. 1993. Biochemistry **32**: 13523–13530.
5. VILSEN, B., J. P. ANDERSEN, D. M. CLARKE & D. H. MACLENNAN. 1989. J. Biol. Chem. **264**: 21024–21030.
6. VILSEN, B., J. P. ANDERSEN & D. H. MACLENNAN. 1991. J. Biol. Chem. **266**: 18839–18845.
7. SKOU, J. C. 1990. FEBS Lett. **268**: 314–324.
8. VILSEN, B. 1995. Biochemistry **34**: 1455–1463.
9. VILSEN, B. 1995. FEBS Lett. **363**: 179–183.
10. KOSTER, J. C., G. BLANCO, P. B. MILLS & R. W. MERCER. 1996. J. Biol. Chem. **271**: 2413–2421.
11. VILSEN, B. 1993. Biochemistry **32**: 13340–13349.
12. VILSEN, B. & J. P. ANDERSEN. 1992. J. Biol. Chem. **267**: 25739–25743.
13. ANDERSEN, J. P. & B. VILSEN. 1992. J. Biol. Chem. **267**: 19383–19387.
14. TANFORD, C., J. A. REYNOLDS & E. A. JOHNSON. 1987. Proc. Natl. Acad. Sci. USA **84**: 7094–7098.
15. RENNER, M., M. A. DANIELSON & J. J. FALKE. 1993. Proc. Natl. Acad. Sci. USA **90**: 6493–6497.
16. DRAKE, S. K. & J. J. FALKE. 1996. Biochemistry **35**: 1753–1760.
17. DOUGHERTY, D. A. 1996. Science **271**: 163–168.
18. ANDERSEN, J. P. & B. VILSEN. 1994. J. Biol. Chem. **269**: 15931–15936.
19. LODISH, H. F. 1988. Trends Biochem. Sci. **13**: 332–334.
20. BAYLE, D., D. WEEKS & G. SACHS. 1995. J. Biol. Chem. **270**: 25678–25684.
21. KUNTZWEILER, T. A., E. T. WALLICK, C. L. JOHNSON & J. B. LINGREL. 1995. J. Biol. Chem. **270**: 2993–3000.

Eosin as a Probe for Conformational Transitions and Nucleotide Binding in Na,K-ATPase[a]

MIKAEL ESMANN[b] AND NATALYA U. FEDOSOVA

Department of Biophysics
Ole Worms Alle 185
University of Aarhus
DK-8000 Aarhus C, Denmark

Transport of Na^+ and K^+ across the plasma cell membrane by the Na,K-ATPase occurs through a set of kinetically distinguishable conformational states of the enzyme. In the nonphosphorylated enzyme the release of occluded K^+ on the cytoplasmic side is provided by a transition of the enzyme from the E_2 to the E_1 conformation.[1-3] The list of characteristic differences between these two enzyme forms includes the affinity for the nucleotides[4,5]: ADP and ATP bind with very low affinity to the E_2 form (e.g., 20 μM K^+ present) and with high affinity in the presence of, for example, 30 mM Na^+ or certain protonated buffers ("Na^+-like" buffers such as Tris and choline).

The fluorescent dye eosin (tetrabromofluorescein) binds with high affinity to Na,K-ATPase in the presence of Na^+ or Na^+-like buffers (dissociation constant K_{Eo} in the range 0.2–0.5 μM), and binding is associated with an increase in fluorescence yield and a red shift of the emission maximum.[6] In the presence of K^+ the affinity for eosin is low ($K_{Eo} > 5$ μM). Binding of eosin and ADP are mutually exclusive. Eosin can thus be used as a sensitive indicator for conformational transitions between E_2 and E_1 and also as a probe for nucleotide binding.

This paper provides kinetic evidence as well as a structural basis for the idea that eosin and nucleotides bind to the same domain on the Na,K-ATPase and that local charges are essential for high affinity binding of eosin. In addition, we report some results on the temperature dependence of conformational relaxations such as the E_1–E_2 transition and the eosin-binding step. Analysis of the Na^+ effect on the deocclusion of K^+ and eosin binding suggests that the change in affinity for eosin (and nucleotide)—the conformational change in the nucleotide binding site—takes place after the deocclusion of K^+, that is, after the $E_2 \rightarrow E_1$ conformational transition. This means that the E_1 forms of the Na,K-ATPase can exhibit either low affinity (E_{1L}) or high affinity (E_{1H}) for eosin. It is suggested that the nomenclature E_1 and E_2 is used to describe properties of the cation sites (E_2 for occluded [closed] sites and E_1 for open sites with rapid exchange between bound and bulk cations) and that an additional suffix is used to distinguish between low (L) and high affinity states (H) of the nucleotide binding site.

---

[a]This work was supported by Human Frontier Science Program grant RG-511/95M.
[b]Tel: +45 89 42 29 30; fax: +45 86 12 95 99; e-mail: ME@BIOPHYS.AU.DK

MATERIALS AND METHODS

Na,K-ATPase with specific activity of 1,400–1,700 µmol/mg protein per hour[7] is prepared from pig kidney using the SDS method of Jørgensen.[8] Fluorescence of eosin and its derivatives is measured at 20°C in a Spex Fluorolog fluorometer with an excitation wavelength of 530 nm (0.4 nm bandpass) and a 550-nm cut-off filter on the emission side. The time dependence of rapid fluorescence changes is followed in an Applied Photophysics spectrometer (10 nm bandpass for excitation, 550-nm cut-off filter on the emission side) equipped with a rapid mixing device for stopped-flow analysis. Each "shot" is based on mixing of equal volumes (usually 100 µl) of two solutions having the same buffer (0.02 mM KCl, 10 mM histidine, pH 7.0, 1 mM CDTA) and eosin concentration. One syringe also contained Na,- K-ATPase (0.1 mg/ml) and 0–30 mM Na^+. The other syringe contained NaCl, no enzyme, and in some experiments also 200 µM ADP. The dead-time of the instrument under the conditions used was less than 2 ms, which is sufficiently small to allow reactions with rate constants up to 80 s^{-1} to be followed over more than 90% of the total fluorescence change.[9]

Equilibrium binding of eosin is carried out as previously described.[6] Essentially, Na,K-ATPase (concentration 8 µM estimated from the level of high-affinity ADP binding sites) is incubated in the dark at 4°C with eosin in the presence of 150 mM NaCl or KCl with and without 3 mM ADP. Bound eosin is removed by centrifugation (pelleting) of the membranes, and the concentration of free eosin in the supernatant is determined from its fluorescence (after proper dilution). The fluorescence of eosin in buffer is essentially not affected by changes in ionic strength (fluorescence is decreased less than 1% when the fluorescence is compared in 5 and 500 mM of salt), and the fluorescence is identical in NaCl and KCl solutions. Care is taken not to leave the enzyme in light together with eosin for prolonged times in order to minimize photoinactivation of the enzyme (see ref. 10). The photoinactivation also leads to a slow linear decrease in fluorescence, which was clearly seen at long time-scans in the stopped-flow apparatus.

The optimal molecular structures for eosin and ADP, prior to alignment, were calculated using the MM+ molecular mechanics force field method (HyperChem, Hypercube Inc.).

Eosin (Gurr®) was obtained from Hopkin & Williams (Essex, UK) and the derivatives 5-aminoeosin, 5-carboxyeosin, and 6-carboxyeosin from Molecular Probes Inc. (Oregon). The latter two were custom synthesized.

RESULTS AND DISCUSSION

Eosin as a Probe for Nucleotide Binding and Conformational Transitions

Binding and Fluorescence Experiments. Eosin acts as a sensitive fluorescent reporter for nucleotide binding and for cation-induced conformational changes in Na,K-ATPase. This is summarized in FIGURE 1, where the slow fluorescence increase is induced by the addition of NaCl to Na,K-ATPase in a K^+-containing medium, and the rapid fluorescence decrease to initial level is caused by the addition of saturating ADP. The interpretation of the fluorescence changes is that eosin is not bound to the Na,K-ATPase in the presence of KCl and therefore has essentially the same fluores-

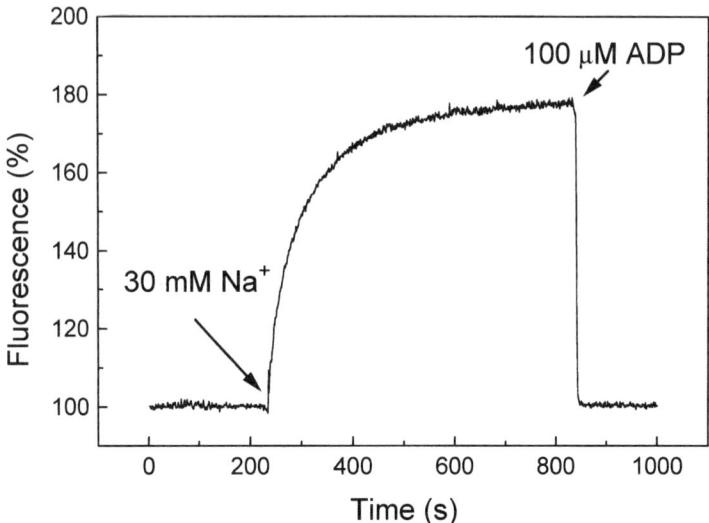

FIGURE 1. Conformational transitions and nucleotide binding monitored by eosin fluorescence. Kidney Na,K-ATPase (0.08 mg/ml) in 10 mM histidine (pH 7.0), 1 mM CDTA, 0.02 mM KCl, and 0.3 μM eosin is transferred to the eosin-binding E_1 form by the addition of 30 mM NaCl, with a rate constant of about 0.05 s^{-1} at 20°C. Eosin is displaced from the high-affinity site by the addition of 100 μM ADP, reducing the fluorescence level to the initial value.

cence as eosin in buffer.[6] The addition of NaCl induces a conformation of the enzyme which binds eosin (and nucleotides) with high affinity, and the fluorescence yield of eosin is increased considerably upon binding.[6] ADP can displace eosin from its binding site; thus, the fluorescence level drops to that of free eosin.

One assumption in this interpretation is that eosin is bound only in the presence of NaCl and not in KCl. This was shown previously for shark Na,K-ATPase (FIG. 6 in ref. 6), and a similar experiment for pig kidney Na,K-ATPase is given in FIGURE 2. Here the binding of eosin is determined at free eosin concentrations up to 7 μM. There is no indication of saturation of a binding site in the presence of 3 mM ADP and 150 mM NaCl or KCl, but rather a linear relationship, indicative of the nonspecific binding or adsorption of eosin to the membranes (○, □). The nonspecific binding was increased at higher ionic strengths (data not shown). In the presence of NaCl alone (■) the data show saturation of a binding site for eosin in addition to the nonspecific binding seen with ADP. A hyperbolic fit of the experiment shown in FIGURE 2 gives a dissociation constant of 1.3 μM for eosin and a binding capacity of 8.6 μM, which with an enzyme concentration of about 8 μM (estimated from ADP binding-site determinations) gives about one eosin bound per ADP site. In the presence of 150 mM KCl (●), the binding is slightly larger than that in the presence of 150 mM KCl + 3 mM ADP. If this additional binding is to the site which in 150 mM NaCl has a dissociation constant of about 1.3 μM, then the data for binding in KCl can be fitted as shown in FIGURE 2, assuming a dissociation constant of about 8.5 μM, which is sixfold higher than that in NaCl. Because fluorescence experiments with eosin are difficult to perform at concentrations higher than about 1 μM due to self-quenching, the interaction of eosin with enzyme in high [KCl] cannot directly be monitored from the fluorescence. It should be noted, however, that low-affinity effects of eosin on Na,K-ATPase activity (similar to those of ATP) were actually reported earlier (FIG. 11 in ref. 6).

The increase in fluorescence on binding to Na,K-ATPase is 4.3-fold, as calculated from the 70% change observed at 0.3 µM eosin (FIG. 1) and a dissociation constant of about 0.25 µM (K_{Eo} in 30 mM NaCl, not shown). The increase in the quantum yield is accompanied by a red-shift of the emission spectrum (λ_{max} shifts from 528 to 532 nm).

Competition between Eosin and ADP. Dissociation constants for eosin and ADP can be determined conveniently from fluorescence titrations such as those shown in FIGURE 3 (note the log-scale for [ADP]). Here the decrease in fluorescence is presented as a function of [ADP]. If a kinetic scheme showing competition between eosin and ADP (Scheme 1) is used for analysis of the data, the displacement curve should be half-saturated at an ADP concentration of $K_{0.5} = K_{ADP} + K_{ADP} \cdot [Eo]/K_{Eo}$. Analysis of the data at 50 mM NaCl for eosin (see FIG. 3 and insert) gives $K_{ADP} = 0.8$ µM and $K_{Eo} = 0.25$ µM. These values for K_{Eo} and K_{ADP} are in (qualitative) agreement with the values obtained from binding experiments if ionic strength effects and temperature differences are taken into account (refs. 11 and 12; see also below).

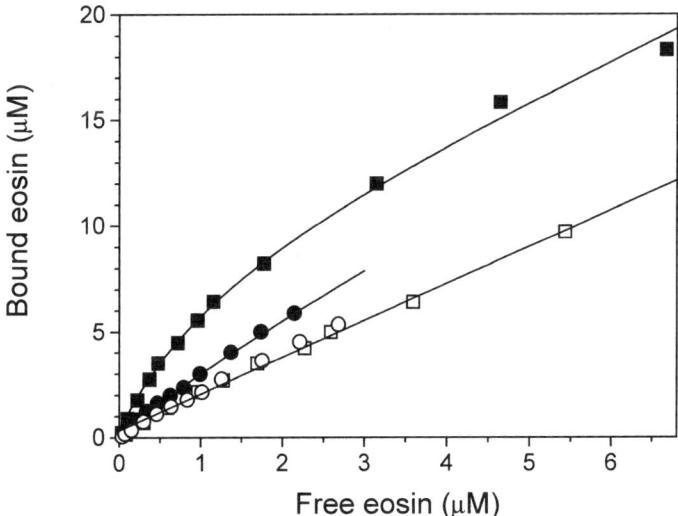

FIGURE 2. Equilibrium binding of eosin. Binding of eosin to kidney Na,K-ATPase at 4°C is determined by centrifugation, and the levels of binding in 150 mM NaCl (■, □) or 150 mM KCl (●, ○) with (○, □) or without (■, ●) 3 mM ADP are compared in a buffer containing 30 mM histidine (pH 7.0) and 1 mM CDTA. For the nonspecific binding of eosin to membranes in the presence of ADP (○ and □, *straight line*), the ratio of 1.1 µM bound eosin per 1 µM free eosin was obtained. The additional high-affinity binding observed in NaCl (■) follows the kinetics of a single-site model with a maximal binding capacity of 1.06 mol bound eosin per mol ADP binding site (see refs. 6 and 10) and $K_{Eo} = 1.3$ µM. The *curved line* through the *filled symbols* thus has the form $[EEo] = [EEo]_{max} \cdot [Eo]/([Eo] + K_{Eo}) + 1.1 \cdot [Eo]$. For KCl a $K_{Eo} = 8.5$ µM was used to fit the data.

$$E \cdot Eo \underset{k_{on, Eo}}{\overset{k_{off, Eo}}{\rightleftarrows}} E \underset{k_{off, ADP}}{\overset{k_{on, ADP}}{\rightleftarrows}} E \cdot ADP$$

SCHEME 1. Kinetic model for the competition between eosin and ADP for high-affinity binding to Na,K-ATPase. Results presented in this paper are interpreted in terms of equilibrium dissociation constants for binding of eosin (Eo) or ADP to Na,K-ATPase (E), that is, K_{Eo} = [E] · [Eo]/[EEo] and K_{ADP} = [E] · [ADP]/[EADP], respectively. The binding and dissociation of the ligands are described by the rate constants $k_{on,ADP}$ ($M^{-1} \cdot s^{-1}$), $k_{off,ADP}$ (s^{-1}), $k_{on,Eo}$ ($M^{-1} \cdot s^{-1}$), and $k_{off,Eo}$ (s^{-1}).

Temperature Dependencies. FIGURE 1 shows that the rate of fluorescence change is much more rapid for the eosin-dissociation step (FIG. 1, right) than for the conformational step involving the binding of Na^+ and eosin (FIG. 1, left). The rate constant for dissociation of eosin by ADP was studied at temperatures ranging from 4 to 37°C, in the form of an Arrhenius plot (log(τ) versus 1/T, FIG. 4) The relaxation time τ_{Eo} is obtained from a single exponential fit of the time course of the fluorescence change. In terms of the simplified model for K^+ occlusion, Na^+ binding, and eosin binding (Scheme 2), the reaction on the addition of saturating ADP (displacing eosin from its site) is characterized by the dissociation rate constant $k_{off,Eo}$ (= $1/\tau_{Eo}$). This increases from about 14 s^{-1} at 4°C to about 41 s^{-1} at 37°C, and the upper line shown in FIGURE 4 corresponds to an activation energy of about 22 kJ/mol. .

The rate constant for deocclusion of K^+ from Na,K-ATPase is very small at low temperatures, such as k_{deocc} ~ 0.01 s^{-1} at 10°C for kidney Na,K-ATPase,[9] increasing to about 0.1 s^{-1} at 20°C.[13] Because the equilibrium $E_2(K_2) \rightleftarrows E_1 \cdot K_2$ is poised heavily towards $E_2(K_2)$ and the apparent dissociation constant for K^+ for E_1 is large,[14,15] it can be assumed that the reciprocal relaxation time describing this transition, $1/\tau_C$, is in the same range as k_{deocc} at low [K^+]. (See refs. 16 and 17 for a kinetic treatment.) It is assumed that the reactions involving binding/dissociation of ions are very fast, and the relaxation times for the occlusion/deocclusion step and eosin binding steps are the only ones experimentally accessible. In such a kinetic scheme a perturbation of the Na^+ and/or K^+ concentrations will lead to a change in the fraction of enzyme molecules in the form that bind eosin with high affinity and therefore to a fluorescence change. The fluorescence change following the addition of 30 mM Na^+ to a solution with 0.02 mM K^+ is characterized by an observed rate constant increasing from 0.006 s^{-1} at 4°C to 0.9 s^{-1} at 37°C, corresponding to an activation energy of about 120 kJ/mol (lower line in FIG. 4).

Evidence for High- and Low-Affinity Nucleotide Sites on the E_1 Form

The complete $E_2 \rightarrow E_1$ transition induced by the addition of 30 mM NaCl to enzyme in 0.02 mM KCl is reported as a uniform slow fluorescence increase from free eosin level (100%) to about 170% (FIG. 1). The addition of the same [Na^+] in two steps, for example, first 2, 4, or 8 mM and then the addition of NaCl up to final concentration of 30 mM, leads to the surprising results shown in FIGURE 5. The first portion of NaCl induces a slow uniform increase in fluorescence, and the amplitude depends on [NaCl]. (For a full titration curve, see, for example, Fig. 3 in ref. 9.) The time dependence is not shown for clarity. The following addition of NaCl gives a marked biphasic fluorescence response, with a very rapid increase (too rapid to be kinetically analyzed in the conventional fluorometer), followed by a slow increase in fluorescence. The proportion of the fluorescence response associated with the rapid change increased with the Na^+ concentration, so that practically all of the fluorescence response was rapid when enzyme in a mixture of 0.02 mM KCl and 8 mM NaCl was exposed to additional Na^+. The ratio between the amplitude of the fast fluorescence change and the total fluorescence increase induced by the second addition of NaCl is shown in FIGURE 5 (insert).

The effect of Na^+ (shown in FIG. 5) and additional experimental evidence led to a suggestion that eosin does not bind with the same affinity to all of the enzyme forms liganded with Na^+ and/or K^+ in the E_1 form. In short, it was found[9] that (a) the time course of the slow fluorescence change was the same as the time course for release of occluded Rb^+ (used as a K^+ substitute), (b) a higher concentration of NaCl was required to induce high affinity eosin binding than to displace occluded Rb^+, (c) binding of more than one Na^+ was required to induce eosin binding, and (d) the observed rate constant for the rapid fluorescence change associated with the second addition of NaCl (determined in a stopped-flow apparatus) is linearly related to the eosin concentration in the same way as the observed rate constant for eosin binding $1/\tau_{Eo} = k_{obs} = k_{off,Eo} + k_{on,Eo} \cdot [Eo]$. The slow fluorescence response thus reflects a step involving K^+ deocclusion, whereas the rapid response reflects an eosin-binding step (FIG. 5).

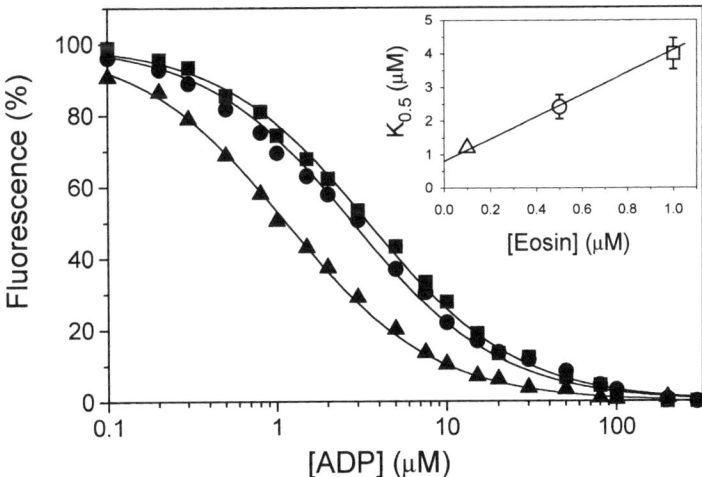

FIGURE 3. Displacement of bound eosin by ADP. The level of eosin fluorescence in the presence of kidney Na,K-ATPase (0.02 mg/ml in 10 mM histidine [pH 7.0], 1 mM CDTA, 30 mM NaCl, t = 20°C) is determined as a function of added ADP at three eosin concentrations (1.0 µM [■], 0.5 µM [●], and 0.1 µM [▲]). The fluorescence change at saturating ADP reflects the amount of bound eosin and was different for different eosin concentrations. The data are referred to that maximal fluorescence drop at each eosin concentration. The lines are drawn according to a binding model (see Scheme 1) with $K_{Eo} = 0.25$ µM and $K_{ADP} = 0.8$ µM. The *insert* shows the found $K_{0.5}$ (± SD, $n = 3$) for the fluorescence decrease as a function of [Eo]. The linear relation has the form $K_{0.5}$ (µM) = $0.8 + 3.2 \cdot [Eo]$ (see text).

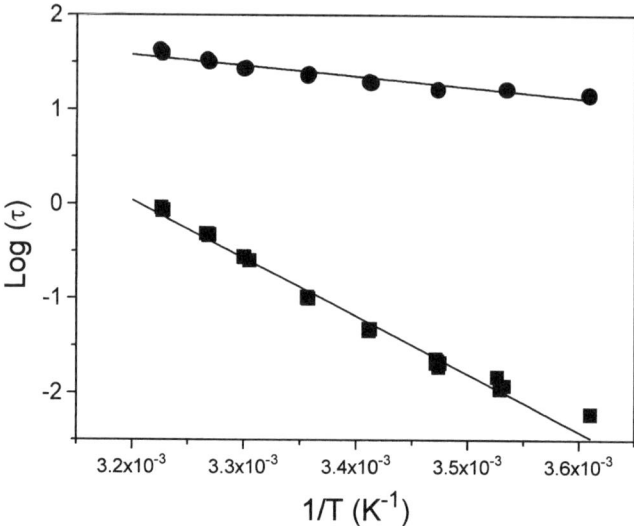

FIGURE 4. Effect of temperature on rates of conformational transitions and eosin dissociation. Results of fluorescence stopped-flow experiments (temperature range 4–37°C) are shown in the form of Arrhenius plot (log($1/\tau$) versus $1/T$) with reciprocal relaxation time constants ($1/\tau$) in s^{-1}. The upper curve (●) shows the relaxation times determined in the experiments where the ADP chase solution was added to the Na,K-ATPase in the presence of 0.3 μM eosin, 30 mM NaCl, 0.02 mM KCl, 10 mM histidine (pH 7.0 at each temperature). The slope of the line corresponds to an activation energy of 22 kJ/mol. The data for the lower curves are obtained in the experiments where Na,K-ATPase in 0.02 mM KCl (■) is mixed with a solution containing 60 mM NaCl and 0.02 mM KCl. The slope of the line drawn through the lower symbols corresponds to 118 kJ/mol.

In a kinetic model which assumes two conformational states with respect to eosin binding, that is, a low-affinity "K-occluded" form ($E_2(K_2)$) and a high-affinity E_1 form (the affinity for eosin being high [and the same] irrespective of the cations bound), the enzyme should have either K^+ occluded or eosin bound. An increase in [Na^+] would displace the equilibrium $E_2(K_2) \rightleftarrows E_1$ towards the E_1-form(s). Assuming that the binding reactions for Na^+ and K^+ are very rapid, the time dependence of eosin binding would reflect the slow relaxation process associated with the $E_2(K_2) \rightarrow E_1$ conformational transition. The biphasic kinetics observed in FIGURE 5 cannot be explained by this model. (See ref. 9 for further discussion.)

Scheme 2 outlines a simplified kinetic model with three basic properties. In the presence of K^+ alone the enzyme is predominantly in the occluded $E_2(K_2)$ state and the observed rate constant for transition to the E_1 form is small. The cation binding sites of the E_1 form can be filled to a different extent with either of the cations. For simplicity, only two cation binding sites are shown. The third property of this model is that high-affinity eosin binding is restricted to E_1 species with certain ligands bound ($E_1 \cdot Na_2$). Binding of eosin to the other species in Scheme 2 is thus assumed to take place with such low affinity that the amount of enzyme residing in these species is practically zero at the used eosin concentrations (indicated by dissociation rate constants much larger than the 20 s^{-1} found for $E_1 \cdot Na_2 \cdot Eo$).

Qualitatively, the data in FIGURE 5 can be interpreted in terms of Scheme 2 as follows: Initially all enzyme molecules are in the $E_2(K_2)$ state (0.02 mM K and 0.3 μM eosin present). An increase in [Na$^+$] to, for example, 2 mM leads to partial deocclusion of K$^+$, that is, transfer of part of the molecules to the E_1 form. Some of them will then bind eosin, giving rise to a fluorescence increase. The observed rate of fluorescence increase will be slow because it is essentially governed by equilibration at the $E_2(K_2) \rightleftarrows E_1 \cdot K_2$ step ($1/\tau_C \sim 0.05$ s^{-1} at 20°C). Equilibration at the cosin binding step is much more rapid ($1/\tau_{Eo} \gg 1/\tau_C$; FIG. 4) and takes place after the conformational transition. A second addition of Na$^+$ (up to 30 mM) will now induce two different processes. One will be further deocclusion of K$^+$, and the second will be displacement of the equilibrium among the E_1 forms ($E_1 \cdot K_2$, $E_1 \cdot K \cdot Na$, and $E_1 \cdot Na_2$) towards $E_1 \cdot Na_2$. The eosin response reflects these two effects: fast phase, a very rapid binding to the $E_1 \cdot Na_2$ molecules formed due to internal reequilibration within the E_1-pool, and slow phase, deocclusion of K$^+$ from the rest of the $E_2(K_2)$ form.

In terms of nomenclature of enzyme conformations, we assigned E_2 to the form with the cation sites in occluded (closed) state and E_1 to the form with the open cation binding site. E_1 is always in rapid equilibrium with the bulk cations. To describe the properties of the nucleotide binding site, we introduced additional characters, H and L, to distinguish between high and low affinity for the nucleotide within the E_1 pool. The change in affinity of the nucleotide binding site is a result of saturation of the cation binding site with Na$^+$ or Na$^+$-like substances (Scheme 2, cartoon). The three

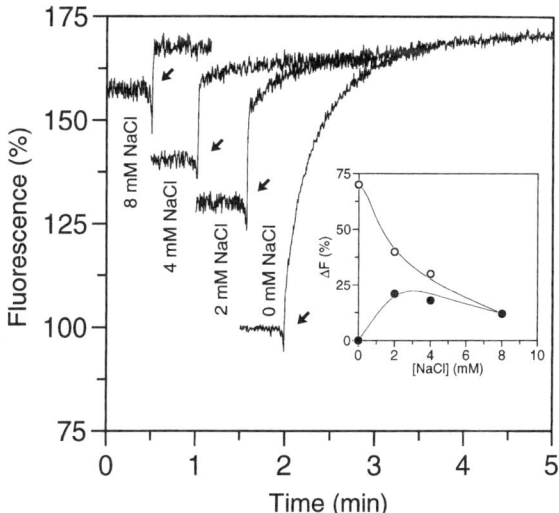

FIGURE 5. Effect of the medium ion composition on Na$^+$-induced eosin binding. Kidney Na,K-ATPase is first incubated with 0.3 μM eosin in 10 mM histidine (pH 7.0), 1 mM CDTA in the presence of 0.02 mM KCl, and 0, 2, 4, or 8 mM NaCl. The fluorescence increase is induced by subsequent addition of NaCl up to 30 mM (marked by *arrows*). The *upper three curves* show the marked biphasic response to NaCl when the preincubation medium contained 2, 4, or 8 mM NaCl. The *insert* shows the total fluorescence change (○) and the amplitude of the rapid phase (●) as a function of [NaCl] in the preincubation medium.

kinetically resolved species discussed here may thus be denoted E_{2L}, E_{1L}, and E_{1H}. This subdivision also implies that the rate of eosin dissociation from the E_1 form depends on the occupancy of the cation binding site. The observed rate constant of a fluorescence drop induced by the addition of a large amount of K^+ may well be faster than the dissociation of eosin from $E_1 \cdot Na_2 \cdot Eo$ (measured by ADP substitution).

Charges on Eosin-Derivatives as Determinants for High-Affinity Binding

Neslund et al.[18] suggested that the observed binding of xanthene dyes (to which eosin belongs) to nucleotide binding sites of a number of proteins stems from the structural similarities of their aromatic and charged parts. FIGURE 6 demonstrates this

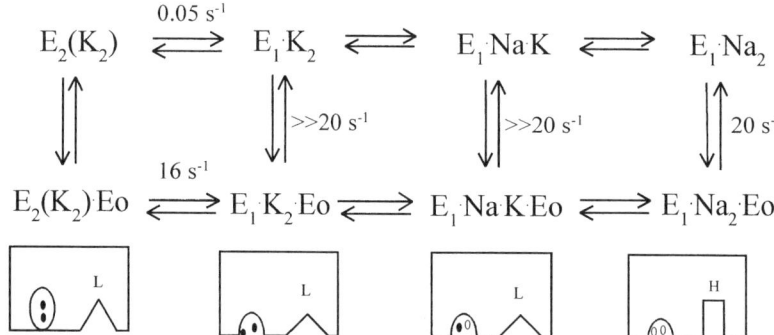

SCHEME 2. A simplified model for interaction of K^+, Na^+, and eosin with Na,K-ATPase. The $E_2(K_2)$ form denotes the enzyme form with 2 K^+ occluded (●). These ions are released slowly (k_{deocc} about 0.05 s^{-1} at 20°C) when free K^+ is reduced to zero or high concentrations of Na^+ (○) are added. The enzyme forms without occluded K^+ are denoted E_1, with either Na^+ or K^+ (or both ions) bound, and possible eosin binding to all forms is indicated together with assumed dissociation rate constants. The rate constant for the $E_2(K_2) \cdot Eo \rightarrow E_1$ transition (about 16 s^{-1}) is assumed to be the same as that observed for the low affinity effect of ATP on deocclusion.[14] Dissociation rate constants for eosin from $E_1 \cdot K \cdot Na$ and $E_1 \cdot K_2$-forms are assumed to be much larger than that of the $E_1 \cdot Na_2 \cdot Eo$ form (about 20 s^{-1}, FIG. 4) due to the lower affinity for eosin in the presence of K^+ (denoted L) than in Na^+ (H).

property for eosin and ADP. The structures of the compounds were optimized using the MM+ method. Overlay of the three fused rings of eosin with the adenine moiety of ADP shows a neat overlap of the negatively charged phosphates of ADP with the negatively charged carboxyl group of eosin.

Recent experiments on ionic strength effects on nucleotide binding to the enzyme suggest that electrostatic interactions play a significant role in the binding process.[12,19] This question was further investigated using eosin derivatives. Introduction of different charged groups in the benzene ring attached to the lower central carbon atom (C1) of the xanthene (see insert in FIG. 7) revealed no correlation between the dissociation constant and the net charge of the molecule (TABLE 1). 6-Carboxyeosin and 5-aminoeosin displayed higher affinity than did both eosin and 5-carboxyeosin.

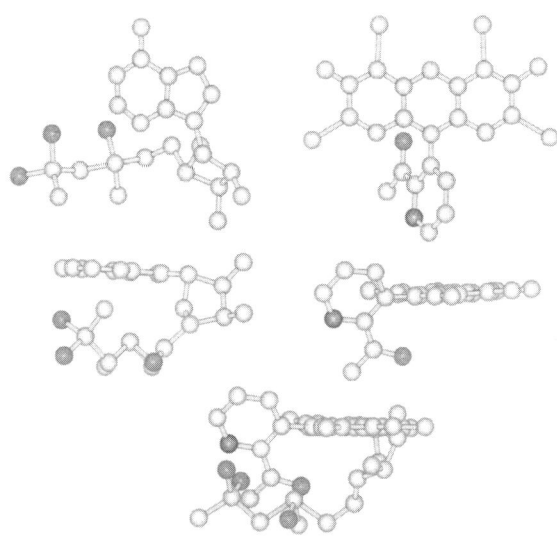

FIGURE 6. Three-dimensional structures of eosin and ADP. The structures of ADP (*left column*) and eosin (*right column*) optimized using the MM+ force-field method (see Methods) are shown in two different projections. Negatively charged atoms are highlighted. Overlay of the structures suggests an overlap of the negatively charged phosphate cluster on ADP with negative charges on eosin.

Analysis of the charge distribution of the four eosins revealed that the substituents on the C5 and C6 positions had a marked influence on the electron distribution in the benzene ring. The partial negative charge on C4 correlated nicely with the observed dissociation constants for eosins (FIG. 7), suggesting that an electrostatic interaction between the binding site and this part of the eosin molecule is a determinant for high-affinity binding.

The kinetic scheme used to describe eosin binding (Scheme 1) implies that the dissociation constant is determined by the ratio of the rate constants for dissociation and binding ($K_{Eo} = k_{off,Eo}/k_{on,Eo}$). If the binding rate constants are diffusion controlled and therefore about the same for all eosin derivatives, the change in (equilibrium) dissociation constant should follow the change in dissociation rate constants. k_{off} for the eosin and its derivatives was determined in stopped-flow fluorescence experiments, as discussed in the legend to FIGURE 4. These data are also shown in FIGURE 7

TABLE 1. Equilibrium Dissociation Constants for Eosin Derivatives

Compound	X^a	Y^a	K_D (μM)[b]
Eosin	H	H	0.25
5-Aminoeosin	NH_2	H	0.12
5-Carboxyeosin	COOH	H	0.29
6-Carboxyeosin	H	COOH	0.11

[a]See FIGURE 7 for definition.
[b]Determined from competition experiments such as those shown in FIGURE 3.

to indicate a qualitative correlation between the partial negative charge on C4 of the eosins and the dissociation rate constant. (Note that the equation $K_{Eo} = k_{off,Eo}/k_{on,Eo}$ is not obeyed using the same k_{on} for the four eosins, suggesting that the detailed structure of the eosin plays a role also for the binding rate constants).

CONCLUSION

Eosin and its derivatives form a group of fluorescent dyes suitable for analysis of the conformational changes in the nucleotide binding site of Na,K-ATPase because of

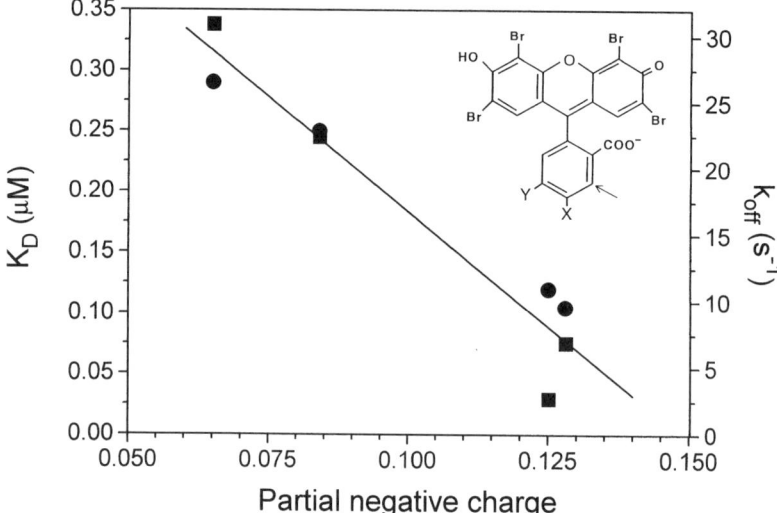

FIGURE 7. Effect of partial negative charge on C4 on the binding constants for eosin derivatives. Equilibrium dissociation constants (●) for eosin, 5- and 6-carboxyeosins, and 5-aminoeosin (TABLE 1) are determined in competition experiments as shown in FIGURE 3 and plotted as a function of the estimated negative charge at carbon atom C4 (marked by an *arrow* in the *insert*). Dissociation rate constants (k_{off} in s^{-1}) for the eosin derivatives from Na,K-ATPase are presented for comparison (■). The line drawn through the data is meant to guide the eye.

both the high affinity in the presence of Na^+ and the marked increase in fluorescence upon binding. Transient kinetic experiments with eosin on the cation-induced conformational changes suggest that the actual change in affinity of the nucleotide binding site occurs after $E_2 \rightarrow E_1$ conformational transition and is induced by the binding of the second Na^+. The effect of structural modification of the eosin molecule on the affinity to Na,K-ATPase could give detailed information about the spatial organization of the nucleotide binding site and the type of interactions (hydrophobic, electrostatic) between the binding site and defined groups on the ligand.

ACKNOWLEDGMENTS

We wish to thank Ann-Dorit Andersen, Angielina Damgaard, and Birthe Bjerring Jensen for excellent technical assistance.

REFERENCES

1. POST, R. L., C. HEGYVARY & S. KUME. 1972. J. Biol. Chem. **247:** 6530–6540.
2. GLYNN, I. M. & S. J. D. KARLISH. 1990. Annu. Rev. Biochem. **59:** 171–205.
3. VASILETS, L. A. & W. SCHWARZ. 1993. Biochim. Biophys. Acta **1154:** 201–222.
4. HEGYVARY, C. & R. L. POST. 1971. J. Biol. Chem. **246:** 5234–5240.
5. NØRBY, J. G. & J. JENSEN. 1971. Biochim. Biophys. Acta **233:** 104–116.
6. SKOU, J. C. & M. ESMANN. 1981. Biochim. Biophys. Acta **647:** 232–240.
7. ESMANN, M. 1988. Methods Enzymol. **156:** 105–115.
8. JØRGENSEN, P. L. 1974. Biochim. Biophys. Acta **356:** 36–52.
9. ESMANN, M. 1994. Biochemistry **33:** 8558–8565.
10. ESMANN, M. 1992. Biochim. Biophys. Acta **1110:** 20–28.
11. MOCZYDLOWSKI, E. G. & P. A. G. FORTES. 1981. J. Biol. Chem. **256:** 2346–2356.
12. NØRBY, J. G. & M. ESMANN. 1997. J. Gen. Physiol. **109:** 555–570.
13. GLYNN, I. M. & D. E. RICHARDS. 1982. J. Physiol. **330:** 17–43.
14. STEINBERG, M. & S. J. D. KARLISH. 1989. J. Biol. Chem. **264:** 2726–2734.
15. KARLISH, S. J. D. 1980. J. Bioenerg. Biomembr. **12:** 111–136.
16. SMIRNOVA, I. N. & L. D. FALLER. 1993. J. Biol. Chem. **268:** 16120–16123.
17. SMIRNOVA, I. N., S.-H. LIN & L. D. FALLER. 1995. Biochemistry **34:** 8657–8667.
18. NESLUND, G. G., J. E. MIARA, J.-J. KANG & A. S. DAHMS. 1984. Cell Regul. **24:** 447–468.
19. PEDERSEN, P. A., J. H. RASMUSSEN & P. L. JØRGENSEN. 1996. J. Biol. Chem. **271:** 2514–2522.

A Two-Site Model of Interacting ATP Sites[a]

DETLEF THOENGES, HOLGER LINNERTZ,
AND WILHELM SCHONER[b]

Institut für Biochemie und Endokrinologie
Justus-Liebig-Universität Giessen
Frankfurter Str. 100
D-35392 Giessen, Germany

The Albers-Post model of the sodium pump explains the mechanism of Na^+/K^+-activated ATP hydrolysis as events of a single ATP binding site that exists consecutively in two different conformations fulfilling two different tasks: The *high affinity ATP binding site* is involved in the Na^+ export that occurs via the formation of a phosphointermediate. On release of ADP and Na^+ it is consequently converted into a low energy phosphointermediate which occludes K^+ on phosphate hydrolysis. The release of K^+ at the inner membrane site needs the binding of ATP at a *low affinity ATP binding site*. An essential of this *single site model* is that the low affinity ATP site (E_2ATP site) is subsequently converted to the high affinity ATP site (E_1ATP site).[1] However, much data obtained mostly by the use of substitution-inert MgATP-complex analogs show that high and low affinity ATP sites coexist.[2] It is not yet definitively proven whether the catalytic α subunit contains a second ATP site,[3] but doubling of phosphorylation sites under certain conditions favors the concept of an $(\alpha\beta)_2$ dimer with cooperating ATP binding sites.[4,5] If high and low affinity ATP sites coexist in time and space, their cooperation during Na^+/K^+-activated ATP hydrolysis and cation transport must follow a two-site competitive model. Therefore, to obtain information on the existence of such kinetics we used substitution-inert MgATP complex analogs as well as fluorescent ATP derivatives.

MATERIALS AND METHODS

The synthesis of MgATP complex analogs[2,6,7] and of the fluorescent ATP analogs TNP-ATP, MANT-ATP, and DANS-N_3-ATP has been described.[8–10] Na^+/K^+-ATPase from pig kidney was inactivated and assayed, as described previously.[2,6,7]

RESULTS AND DISCUSSION

We formerly realized that ATP ($K_d(E_1) = 2.5$ μM, $K_d(E_2) = 360$ μM) and TNP-ATP ($K_d(E_1) = 0.6$ μM, $K_d(E_2) = 175$ μM) show negative cooperativity in their

---

[a]This work was supported by DFG (Bonn-Bad Godesberg) through Graduiertenkolleg "Molekulare Biologie und Pharmakologie Giessen", and the Fonds der Chemischen Industrie, Frankfurt/Main.
[b]Tel: +49-641-99-38170; fax: +49-641-99-38179; e-mail: Schoner@vetmed.uni-giessen.de

binding behavior.[11] Surprisingly, TNP-ATP inhibits Na^+/K^+-activated hydrolysis in a competitive manner at high ATP concentrations and noncompetitively at low concentrations of ATP.[12] MANT-ATP ($K_d(E_1) = 200$ μM, $K_d(E_2) = 35$ μM) is a substrate of Na^+/K^+-ATPase[9] and is hydrolyzed with positive cooperativity.[11] DANS-N_3-ATP ($K_d(E_1) = 500$ μM and $K_d(E_2) = 2.5$ μM) is not hydrolyzed by Na^+/K^+-ATPase. However, incubation of Na^+/K^+-ATPase with DANS-N_3-ATP at 37°C in 50 mM imidazole, pH 7.25, results in a loss of activity in a time- and concentration-dependent way. Evidently, the inactivation by DANS-N_3-ATP proceeds with two different inactivation rate constants which are both decreased in the presence of 1 mM ATP (FIG. 1). A possible way to describe such a phenomenon is to assume that DANS-N_3-ATP binds to two binding sites with differing affinities and that conformational changes may lead to inactive enzyme inhibitor complexes (E*I or IE*I). Surprisingly, when the initial velocity of inactivation (v_i) is plotted against the concentration of DANS-N_3-ATP, a sigmoid concentration dependence is obtained (FIG. 2) which

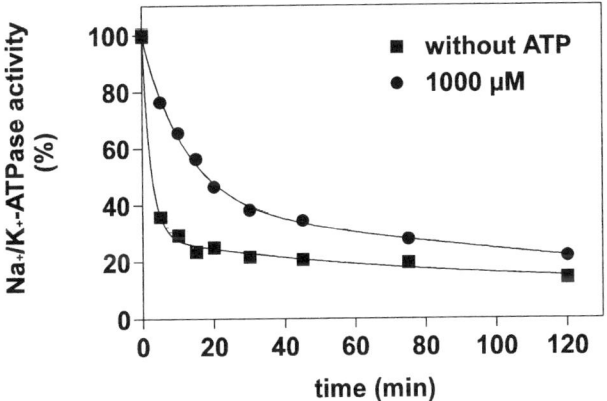

FIGURE 1. Inactivation of Na^+/K^+-ATPase of pig kidney by 75 μM DANS-N_3-ATP in the presence and absence of 1 mM ATP.

indicates a positive cooperativity of the binding sites for DANS-N_3-ATP. Such an interaction can be described by an induced-fit model according to Koshland-Némethy-Filmer (FIG. 3, equation 1). The advantage of such a model is that it describes positive as well as negative cooperativity.

Applying the principles for a cooperative two-site model according to Koshland-Némethy-Filmer for the inactivation process of Na^+/K^+-ATPase, equation 1 can be derived:

$$\frac{v_i}{V_{i,max}} = \frac{\dfrac{[I]}{y \times K_I} + \dfrac{[I]^2}{c \times K_I^2}}{1 + 2 \times \dfrac{[I]}{K_I} + \dfrac{[I]^2}{c \times K_I^2}} \quad (1)$$

In this equation, factor "c" describes the change in the affinity for an inhibitor (I) by

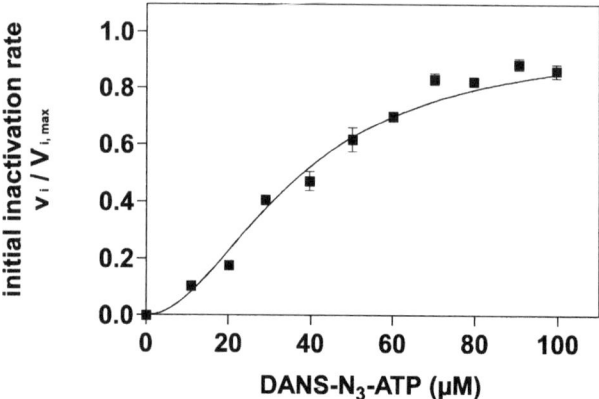

FIGURE 2. Kinetics of inactivation of Na$^+$/K$^+$-ATPase by increasing concentrations of DANS-N$_3$-ATP. The line is the result of a fit according to equation 1, resulting in the parameters K_d = 500 μM, c = 0.005, and y = 100.

interaction of inhibitor binding sites when one inhibitor (EI) is already bound. Factor "y" refers to changes in inactivation rate constant k_i due to inactivation through the IEI complex relative to the EI complex. The $V_{i,max}$ is the maximal rate of inactivation of the enzyme (E) by the double occupied complex and v_i the rate of inactivation at a specific concentration of the inhibitor. When equation 1 is used to generate a diagram of the inactivation rate versus inhibitor concentration, it is evident that the equation describes the sigmoid inactivation process of DANS-N$_3$-ATP (FIG. 2). Because DANS-N$_3$-ATP is an ATP analog, similar considerations like those presented in FIGURE 3 should also apply for the overall hydrolysis of ATP (FIG. 4). A quantitative description of this phenomenon is given in equation 2:

$$\frac{v_p}{V_{p,max}} = \frac{\dfrac{[S]}{z \times K_d} + \dfrac{[S]^2}{a \times K_d^2}}{1 + 2 \times \dfrac{[S]}{K_d} + \dfrac{[S]^2}{a \times K_d^2}} \qquad (2)$$

Factor "a" describes the change in affinity due to the interaction of substrate binding sites when a second substrate is bound. Factor "z" refers to changes in overall rate constant k_p due to the hydrolysis of the SES complex. It is well known that overall ATP hydrolysis shows the phenomenon of negative cooperativity[1] which is consistent with the fact that a high affinity

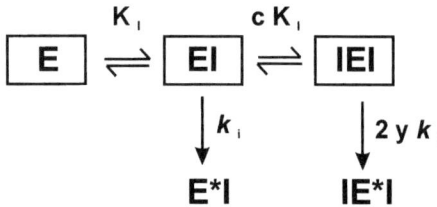

FIGURE 3. Two-site model of inactivation of Na$^+$/K$^+$-ATPase by an ATP analog.

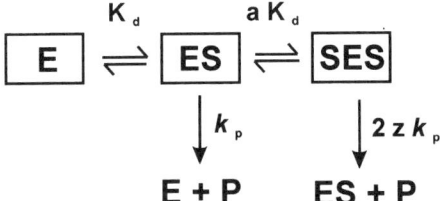

FIGURE 4. Two-site model of ATP hydrolysis by Na^+/K^+-ATPase.

ATP binding site is occupied prior to the low affinity site. It is evident from equation 2 that the two-site model describes the phenomenon of positive or negative cooperativity depending on the parameters chosen (FIG. 5). Therefore, the two-site model (equation 2) should also explain the phenomenon of positive cooperativity of MANT-ATP hydrolysis found previously.[11] Moreover, the observation of a competitive type of inhibition of Na^+/K^+-activated ATP hydrolysis by TNP-ATP at high ATP concentrations but a noncompetitive type of inhibition at low ATP concentrations can probably be described quantitatively by a more detailed version of the model shown in FIGURE 4.

Because the models shown in FIGURES 3 and 4 can describe quantitatively the interaction of ATP sites in Na^+/K^+-ATPase, it was of interest to determine experimentally how phosphorylation of the high affinity ATP binding site may affect the interaction of ATP or phosphate with the second ATP site. We learned that backdoor phosphorylation and binding of ouabain by the phosphate-supported pathway are still possible when the high affinity ATP binding site is blocked by $Cr(H_2O)_4AMP\text{-}PCP$ or FITC[6,7] (FIG. 6). Interestingly, these processes and the remaining K^+-activated phosphatase are blocked by the addition of $Co(NH_3)_4ATP$ which inactivates by tight binding at the low affinity ATP site (E_2ATP site).[6,7] Moreover, blocking of the E_1ATP site by CrAMP-PCP does not alter $^{86}Rb^+$-occlusion which represents a partial reaction involving the low affinity ATP site.[6] All of these findings seem to indicate that the K^+-activated step in the overall reaction refers to the second step in the two-site model, where the ES complex binds a second substrate molecule.

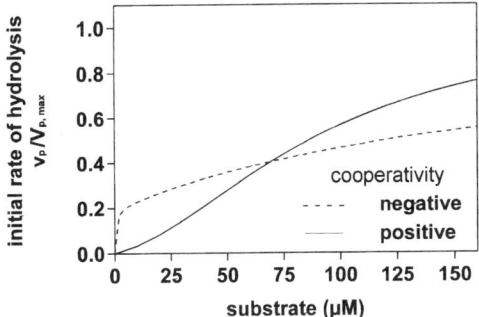

FIGURE 5. The Koshland-Némethy-Filmer model of two interacting ATP sites may show positive or negative cooperativity of substrate hydrolysis. The following parameters in equation 2 show positive cooperativity: $K_d = 1{,}000$ μM, $a = 0.01$, and $z = 0.4$. The following parameters in equation 2 show negative cooperativity: $K_d = 1$ μM, $a = 100$, and $z = 2.5$.

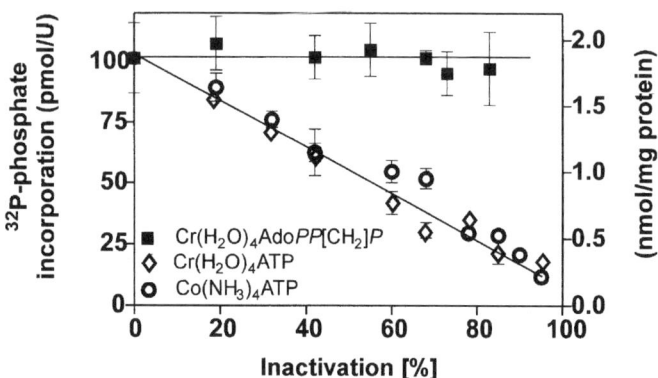

FIGURE 6. Backdoor phosphorylation as a function of the inactivation of Na^+/K^+-ATPase with $Cr(H_2O)_4$AMP-PCP, $Cr(H_2O)_4$ATP, and $Co(NH_3)_4$ATP. (Reprinted with permission from ref. 7.)

It remains to be determined, however, if a more extended version of the two-site model can really explain all of the findings of ATP site interactions. So far, these are the conversion of the E_1ATP site into a Na^+-form when the E_2ATP site is blocked by $Co(NH_3)_4ATP$[13] and the inhibition of backdoor phosphorylation (of the E_2ATP binding site) by phosphorylation of the E_1ATP binding site (FIG. 6).[6,7] Furthermore, we should keep in mind that "ES" in the two-site model (FIG. 4) does not discriminate between the possibilities that ATP is merely bound or that it exists as ADP + phosphointermediate. The two-site model may form a basis for not only quantitative investigation of the hydrolysis of ATP by Na^+/K^+-ATPase but also for quantitative analysis of the interactions between the two sites. The model definitely prompts the study of the location of the two ATP sites on the α subunit and of the distance between them. Fluorescent ATP analogs which have been shown to differ in their cooperativity in ATP hydrolysis[11] may be of help in this respect.

REFERENCES

1. GLYNN, I. M. 1985. *In* The Enzymes of Biological Membranes. 2nd ed. Vol. 3. A. N. Martonosi, Ed.: 35–114. Plenum Press. NY.
2. SCHONER, W. *et al.* 1994. *In* The Sodium Pump. E. Bamberg & W. Schoner, Eds.: 332–341. Springer. New York.
3. WARD, D. G. & J. D. CAVIERES. 1996. J. Biol. Chem. **271:** 12317–12321.
4. PELUFFO, R. D. *et al.* 1992. J. Biol. Chem. **267:** 6596–6601.
5. LIU, G. *et al.* 1996. FEBS Lett. **390:** 323–326.
6. HAMER, E. & W. SCHONER. 1993. Eur. J. Biochem. **213:** 743–748.
7. LINNERTZ, H. *et al.* 1995. Eur. J. Biochem. **232:** 420–424.
8. HIRATSUKA, T. & K. UCHIDA. 1973. Biochim. Biophys. Acta **320:** 635–647.
9. HIRATSUKA, T. 1983. Biochim. Biophys. Acta **742:** 496–508.
10. CHUAN, H. & J. H. WANG. 1988. J. Biol. Chem. **263:** 13003–13006.
11. THÖNGES, D. *et al.* 1994. *In* The Sodium Pump. E. Bamberg & W. Schoner, Eds.: 421–424. Springer. New York.
12. MOCZYDLOWSKI, E. G. & P. A. G. FORTES. 1981. J. Biol. Chem. **256:** 2357–2366.
13. SCHEINER-BOBIS, G. *et al.* 1989. Eur. J. Biochem. **183:** 173–178.

Relationship between Ouabain-Sensitive ATPase Activity and Occluded Rb$^+$ at Micromolar ATP Concentrations[a]

R. C. ROSSI,[b,d] P. J. GARRAHAN,[b] S. B. KAUFMAN,[b]
J. G. NØRBY,[c,e] AND P. J. SCHWARZBAUM[b]

[b]*IQUIFIB*
Facultad de Farmacia y Bioquímica
Universidad de Buenos Aires
Junín 956
(1113) Buenos Aires, Argentina

[c]*Department of Biophysics*
University of Aarhus
DK-8000, Aarhus C, Denmark

This communication describes an attempt to evaluate the obligatory role of the intermediate with occluded potassium, $E_2(K_2)$, during K$^+$ transport by the Na$^+$/K$^+$-ATPase. Using ^{86}Rb to measure the occluded form at steady state and [ATP] from 0.5–10 μM, we found that when [Rb$^+$] ≤ 10 mM, only part of the Na$^+$/K$^+$-ATPase activity takes place through the $E_2(Rb_2)$ and $E_2(Rb_2)$ATP forms, so that at these Rb$^+$ concentrations the preparation has "extra" ouabain-inhibitable ATPase activity.

The scheme in FIGURE 1 represents the currently accepted model for K$^+$ activation and K$^+$ transport by the Na$^+$/K$^+$-ATPase. Equation 4 (FIG. 1, legend) predicts that at saturating K$^+$ concentrations, the ratio v/E_{occl} will increase with the concentration of ATP along a rectangular hyperbola. However, under the conditions of this study, [ATP] ≪ K_{ATP}, so that v/E_{occl} approaches a linear function of [ATP]:

$$v/E_{occl} \cong k_0 + k_\infty [ATP]/K_{ATP} \qquad (5)$$

To assess the competence of the occluded intermediates we measured v and E_{occl} and compared the rate coefficients calculated as v/E_{occl} with deocclusion rate coefficients determined directly under the same experimental conditions.

RESULTS AND DISCUSSION

All experiments were performed with a partially purified Na$^+$/K$^+$-ATPase preparation obtained from pig kidney as previously described.[2] The specific activity of the enzyme was 26 U · mg^{-1} at 37°C. The properties of this preparation were independent of the protein concentration in the range applied.

---

[a]This work was supported by EU-Contract CI1*CT93-0048 and grant FA109, University of Buenos Aires, Argentina.
[d]Tel: +54 1 962 5506; fax: +54 1 962 5457; e-mail: RVROSSI@CRIBA.EDU.AR
[e]Tel: +45 89 42 29 35; fax: +45 86 12 95 99; e-mail: JGN@MIL.AAU.DK

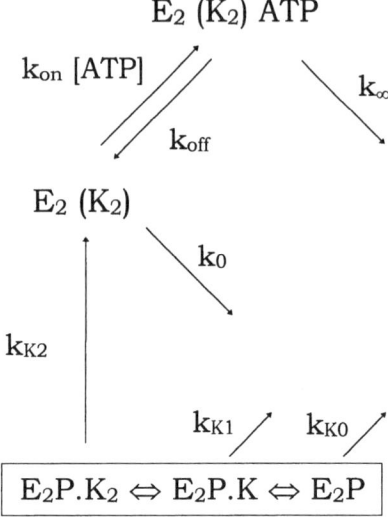

FIGURE 1. The reaction steps involved in the transport of K^+ by Na^+/K^+-ATPase according to the Albers-Post scheme.[1] Dephosphorylation of E_2P is slow (k_{K0}, k_{K1}, about 2 s^{-1}) when either none or only one K^+ is bound, and fast (k_{K2}, about 200 s^{-1}) when 2 K^+ are bound from the external side. Fast dephosphorylation is followed by the formation of the occluded intermediate. Deocclusion is slow (k_0, about 0.1 s^{-1}) from $E_2(K_2)$ and fast (k_∞, 20 to 40 s^{-1}) from $E_2(K_2)$ATP. According to this scheme, in the presence of saturating concentrations of K^+, the steady-state activity (v) will be:

$$v = [E_2(K_2)] \times k_0 + [E_2(K_2)ATP] \times k_\infty \quad (1)$$

and the total level of occluded intermediates, E_{occl}, will be:

$$E_{occl} = [E_2(K_2)] + [E_2(K_2)ATP] \quad (2)$$

As

$$\frac{[E_2(K_2)ATP]}{[E_2(K_2)]} = \frac{[ATP]}{K_{ATP}} \quad (3)$$

then:

$$\frac{v}{E_{occl}} = \frac{k_0 \times K_{ATP} + k_\infty[ATP]}{K_{ATP} + [ATP]} \quad (4)$$

Equation 4 is independent of the rate constants of those elementary steps that do not appear in the scheme.

FIGURE 2 demonstrates that under the conditions used, the steady-state concentration of occluded Rb^+ is virtually independent of [ATP] and [Rb^+]. Controls showed that in the absence of ATP no occlusion was detected. This proves that all occlusion of Rb^+ took place via the physiological route (FIG. 1). In a similar experiment (not shown), the phosphorylation capacity of the enzyme was 2.7 nmol · mg^{-1} and Rb^+ occluded was 5.3 nmol · mg^{-1} ([ATP] = 10 μM, [Rb^+] = 0.5–2.5 mM). The average value for Rb^+ occluded in FIGURE 2 is 5.0 nmol · mg^{-1}, signifying that more than 90% of the enzyme is located in this intermediate and that the deocclusion of Rb^+ is clearly the rate-determining step under all these conditions.

FIGURE 3 shows the ATPase activity determined under the same conditions as those used to measure occluded Rb^+. For all values of $[Rb^+]$, the activity increases with [ATP]. Within the 1–10 μM range, where the high-affinity binding site for ATP is already saturated, the activity increase at each $[Rb^+]$ is linearly related to [ATP] (not shown). This behavior is to be expected if ATP activates a rate-limiting step by binding to a low-affinity site ($K_{diss} > 100$ μM). It allows determination, by extrapolation, of the hypothetical "activity" (and thus v/E_{occl}) at [ATP] = 0, as shown by the open circles in FIGURE 3. It is also obvious from FIGURE 3 that for all values of [ATP], activity decreases towards a plateau with increasing $[Rb^+]$. Before we discuss this we should point out that similar observations on the effect of K^+ (or its congener Rb^+) on the ATPase activity at very low ATP concentrations have been reported by others. In an early study with calf brain microsomes, Neufeld and Levy[5] found that with [ATP] = 2–4 μM and $[Na^+]$ = 100 mM, activity increased with $[K^+]$ from 0–2.5 mM and decreased when $[K^+]$ was further increased to 5 and 15 mM. Also, Post et al.,[6] using enzyme from guinea pig kidney, observed ". . . as did Neufeld and Levy, that increasing concentrations of K^+ first stimulated and then inhibited the activity. . . ."

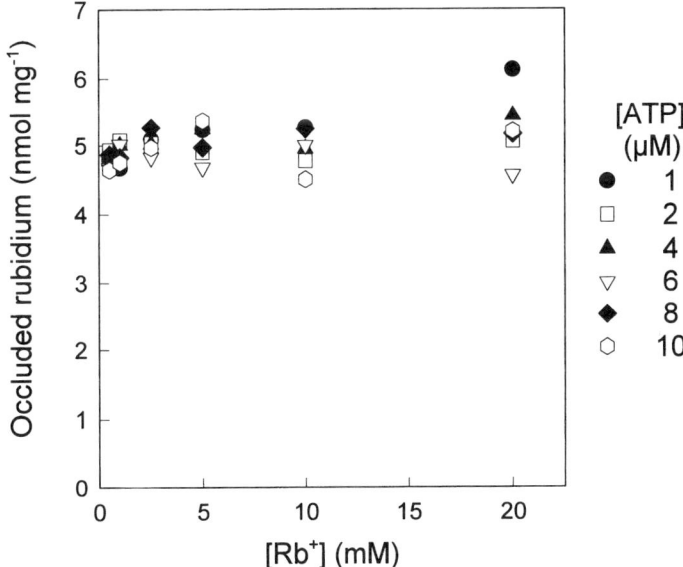

FIGURE 2. The steady-state concentration of occluded Rb^+ as a function of $[Rb^+]$ and [ATP]. Incubation media contained (mM): NaCl 150; imidazole-HCl 25; EDTA 0.2; and $MgCl_2$ 0.7. pH was 7.4 and temperature 25°C. Enzyme concentration was 0.1 mg prot. · ml^{-1}. Occluded rubidium was measured according to Rossi and Nørby,[3] using a rapid-mixing apparatus (BioLogic, France). After 2–3 s of reaction the sample was injected into a quenching-and-washing chamber where the enzyme with occluded ^{86}Rb was retained and washed free of reactants on a filter.

Here the conditions were 0.9, 3, or 13 μM ATP and 16 mM Na^+. Furthermore, Plesner and Plesner[7] found that [K^+] from 1–20 mM inhibited ox brain Na^+/K^+-ATPase when [Mg · ATP] was 0.14–2 μM and [Na^+] < 150 mM. If we analyze these data by the scheme in FIGURE 1, activation by nonsaturating [K^+] is explained by the pathway through the occluded forms being faster than the Na^+-ATPase pathway activity, but only when [ATP] is above a certain value. (Note that k_∞ is much larger than k_0, k_{K0}, and k_{K1}.) Likewise, it is predicted by the scheme that nonsaturating [K^+] would lead to inhibition of the ATPase activity provided that ATP is sufficiently low. (Note that k_0 is

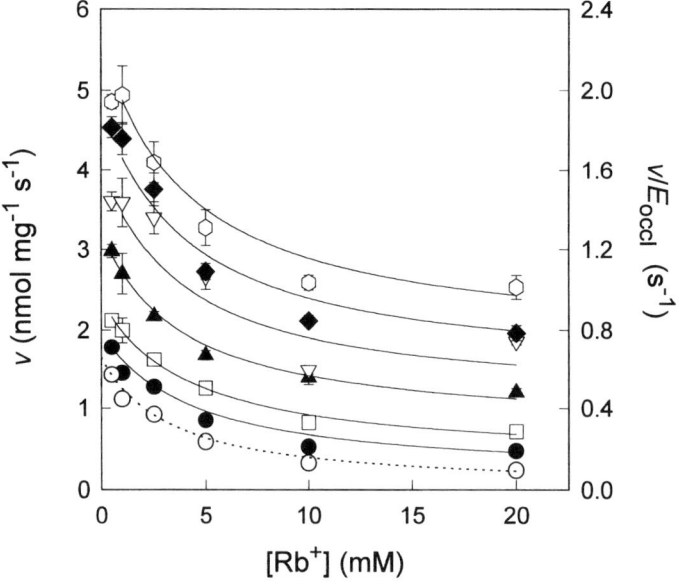

FIGURE 3. The ouabain-inhibitable ATPase activity (v, left axis) as a function of the same [Rb^+] and [ATP] as in FIGURE 2. The values represented by (···O···) are extrapolations to [ATP] = 0, as described in the text. For the meaning of other symbols, see FIGURE 2. The right-hand Y-axis shows the ratio v/E_{occl}. The value for E_{occl} = [Occluded Rb^+]/2 = 2.5 nmol · mg^{-1} is taken by averaging the values in FIGURE 2. The activity was measured according to Schwarzbaum et al.[4] as the amount of ^{32}P released from [$\gamma^{32}P$]-ATP, under the same conditions as those given in FIGURE 2, but with 0.002–0.004 mg prot. · ml^{-1}. Curves were drawn by eye.

smaller than k_{K0} and k_{K1}.) Thus, both of these effects at nonsaturating [K^+] should be accompanied by an increase in E_{occl}.

Inasmuch as [Occluded Rb^+] is already maximum at the lowest [Rb^+] tested (FIG. 2), that is, 1 mM Rb^+ is saturating, the decrease in ATPase activity with increasing [Rb^+] in the present study (FIG. 3) clearly is not related to increased "blockage" of the enzyme in $E_2(Rb_2)$. Because [Occluded Rb^+] is constant, the decrease in v results in a proportional decrease in the calculated deocclusion rate coefficient v/E_{occl}, as shown in

FIGURE 3 (right hand Y-axis). This is not expected from the Albers-Post scheme which predicts values of v/E_{occl} that are independent of [Rb^+] at saturating concentrations.

To further elucidate the possible mechanism for this inhibition by [Rb^+], we directly measured the rate of Rb deocclusion by measuring the time course of the release of occluded ^{86}Rb. The technique used basically the same arrangement as that for the steady-state measurements of occluded Rb (occlusion phase, see legend to FIG. 2), but included an additional 20-fold dilution of the enzyme into media without labeled Rb^+ (dilution phase). The samples were then aged for different times before being injected into the quenching and washing chamber. In some experiments, concentrations of Rb^+ and/or ATP in the dilution-phase medium were different from those in the occlusion-phase medium.

The data (not shown) could all be fitted by single-exponential functions from which deocclusion coefficients (k_{deoccl}) corresponding to the different conditions were derived. In the scheme in FIGURE 1, single-exponential behavior for deocclusion requires rapid-equilibrium dissociation of ATP from $E_2(Rb_2)ATP$, a condition that seems to hold inasmuch as reported values of k_{off} and k_∞ are 833 and 20–40 s^{-1}, respectively (see ref. 4 and FIG. 1).

All occlusion-phase media had 0.5 mM $^{86}Rb^+$, and we observed the same k_{deoccl} whether the dilution-phase medium contained 0.5 or 10 mM Rb^+. At 0.5 mM Rb^+, k_{deoccl} increases with [ATP], being 0.18, 0.72, and 1.3 s^{-1} at [ATP] = 0.5, 5, and 10 µM, respectively.

Extrapolation to [ATP] = 0 of the directly measured k_{deoccl} gives an estimated value for k_0 of 0.13 s^{-1}. Comparison of directly measured deocclusion coefficients with the calculated v/E_{occl} (FIG. 3, right-hand Y-axis) shows a significant discrepancy, because at [Rb^+] = 0.5 mM, the calculated constant v/E_{occl} at [ATP] = 10 µM is two times and at [ATP] = 0 µM it is four to five times higher than those directly measured. On the other hand, for all [ATP] the calculated values for v/E_{occl} at Rb > 10 mM approach the directly measured k_{deoccl}.

This means that only for Rb > 10 mM can the results be explained by the scheme in FIGURE 1, because only under these conditions is the calculated rate coefficient v/E_{occl} equal to that directly measured (as required by equation 5).

These observations together with the fact that the directly measured k_{deoccl} is independent of whether [Rb^+] is 0.5 or 10 mM in the deocclusion medium (see above) exclude the possibility that the inhibition by Rb^+ seen in FIGURE 3 could be due to an inhibitory binding of a third Rb^+ to $E_2(Rb_2)$ and $E_2(Rb_2)ATP$, decreasing the rate of deocclusion.

Rather, the results are most simply explained by extra ouabain-inhibitable ATPase activity when [Rb^+] is less than 10 mM. This activity is increasingly inhibited by [Rb^+] from 0.5–10 mM with an estimated K_i of 3–4 mM, and because v (and v/E_{occl}) is linearly related to [ATP] at all Rb^+ concentrations, it must have a low affinity for ATP (relative to the Na^+-ATPase activity). Currently we can offer no mechanistic explanation for this ATPase activity, but studies along the lines discussed here with higher ATP concentrations are in progress.

REFERENCES

1. GLYNN, I. M. 1988. In The Na^+,K^+-Pump, Part A. J. C. Skou, J. G. Nørby, A. B. Maunsbach & M. Esmann, Eds.: 435–460. Allan Liss. New York.
2. JENSEN, J., J. G. NØRBY & P. OTTOLENGHI. 1984. J. Physiol. (Lond.) **346**: 219–241.

3. Rossi, R. C. & J. G. Nørby. 1993. J. Biol. Chem. **268:** 12579–12590.
4. Schwarzbaum, P. J., S. B. Kaufman, R. C. Rossi & P. J. Garrahan. 1995. Biochim. Biophys. Acta **1233:** 33–40.
5. Neufeld, A. N. & H. M. Levy. 1969. J. Biol. Chem. **244:** 6493–6497.
6. Post, R. L., C. Hegyvary & S. Kume. 1972. J. Biol. Chem. **247:** 6530–6540.
7. Plesner, L. & I. W. Plesner. 1985. Biochim. Biophys. Acta **818:** 222–234.

Site-Directed Mutagenesis Analysis of the Role of the M5S5 Sector of the Sarcoplasmic Reticulum Ca^{2+}-ATPase[a]

JENS PETER ANDERSEN,[b] THOMAS SØRENSEN,
AND BENTE VILSEN

Department of Physiology
University of Aarhus
DK-8000 Aarhus C, Denmark

Structure-function relationships of the sarcoplasmic reticulum Ca^{2+}-ATPase have been studied intensely by site-directed mutagenesis, and it has become feasible to assign specific functional roles to domains and subdomains as well as to individual amino acid residues on the basis of functional characterizations of the mutant enzymes[1–3] (FIG. 1). The largest cytoplasmic domain contains residues crucial to ATP binding and phosphoryl transfer. Transmembrane segments M4, M5, and M6 seem to constitute a central region (pore) for occlusion and translocation of Ca^{2+}. Residues Glu309, Glu771, Asn796, Thr799, and Asp800 found in these transmembrane segments are essential to Ca^{2+} occlusion.[3–5] Glu908 in transmembrane segment M8 is involved in the optimization of Ca^{2+} binding but is less crucial than the aforementioned residues in M4, M5, and M6, because the Glu908 → Ala mutant can occlude Ca^{2+} in the presence of CrATP[5] and displays Ca^{2+} uptake activity corresponding to a rate as high as 47% that of the wild type when assayed at a Ca^{2+} concentration of 200 μM[2]. Stalk segment S4 connecting transmembrane segment M4 with the phosphorylation domain contains several residues (indicated by open squares in FIG. 1) that are important to the conformational rearrangement associated with deocclusion and translocation of Ca^{2+} and transformation of the phosphoenzyme intermediate from an ADP-sensitive to an ADP-insensitive species ($E_1P \rightarrow E_2P$ transition).[3,6,7] Moreover, many residues in M4, M5, and M6, including Ca^{2+}-binding residues Glu309, Glu771, and Asn796,[5] are part of a long-distance signaling pathway that relays information from the ion-binding domain in the membrane back to the phosphorylation site to trigger the dephosphorylation of the E_2P phosphoenzyme intermediate (residues indicated by diamonds in FIG. 1).

Replacement of the highly conserved tyrosine Tyr763 at the M5S5 boundary with glycine uncouples ATPase activity from Ca^{2+} transport in the sense that the microsomes containing this mutant display high Ca^{2+}-activated ATPase activity without any net accumulation of Ca^{2+} in the vesicular space.[8] This could mean that Ca^{2+} is never transported even though ATP is hydrolyzed by the mutant ATPase (true uncoupling) or that following translocation into the microsomal vesicles Ca^{2+} leaks back through an

---

[a]This work was supported by the Danish Medical Research Council and the NOVO Nordisk Foundation.

[b]Address for correspondence: Department of Physiology, University of Aarhus, Ole Worms Allé 160, DK-8000 Aarhus C, Denmark (tel: 45 8942 2814; fax: 45 8612 9065; e-mail: jpa@fi.aau.dk).

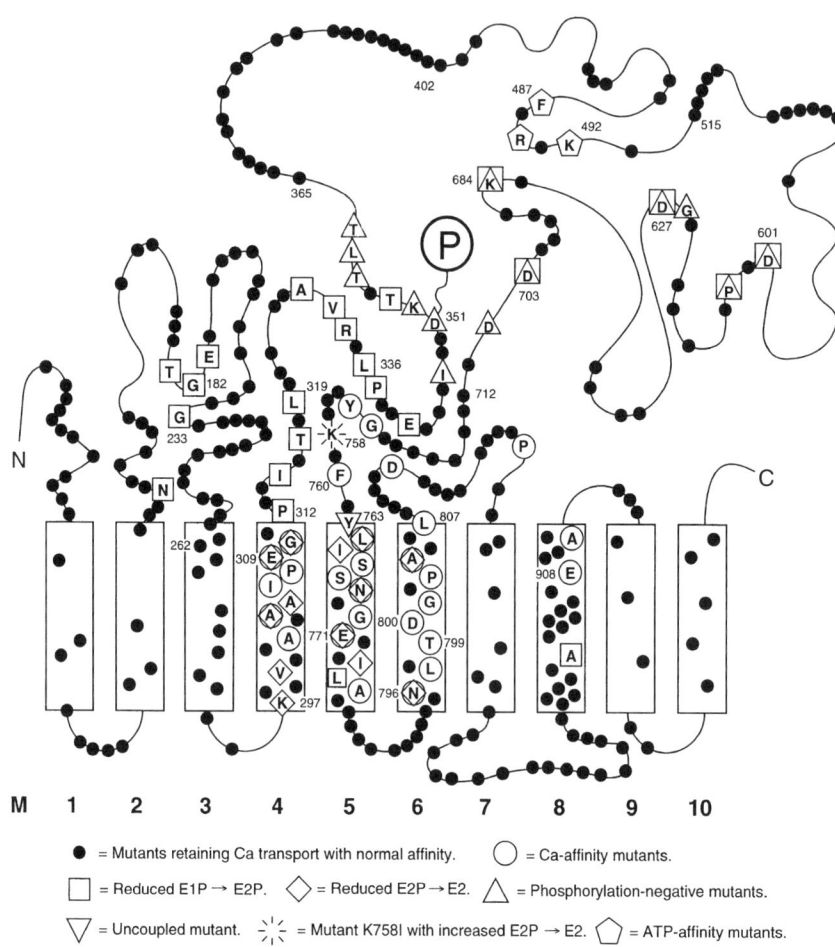

FIGURE 1. Ten-helix model of the topology of the sarcoplasmic reticulum Ca^{2+}-ATPase with classification of residues that have been analyzed by site-directed mutagenesis. *Filled circles* correspond to mutations compatible with normal (or close to normal) function of the expressed SERCA1 enzyme. More deleterious mutations are highlighted by showing the wild-type residue in single-letter code inside a symbol indicating the functional class to which the mutant belongs. Double labeling of residues indicates either that two partial reaction steps are affected by the same mutation (e.g., *open circles with diamonds*) or that two different substituents elicit different effects (e.g., *triangles* pointing upwards with *squares*). This classification is based on published papers reviewed previously[1–3] as well as on unpublished results by the present authors and results personally communicated to us by Dr. David H. MacLennan.

open pore in the mutant protein. Less drastic substitutions of Tyr763 are compatible with Ca^{2+} accumulation in the microsomal vesicles but reduce the apparent affinity for Ca^{2+} significantly, as illustrated in FIGURE 2 for mutant Tyr763 → Leu. Likewise, we found that mutations to Gly750, Tyr754, Lys758, Phe760, or Leu764 result in enzymes with diminished apparent Ca^{2+} affinity. (See data for mutants Gly750 → Ala and Phe760 → Gly in FIG. 2.) These results in conjunction with the observed uncoupling of mutant Tyr763 → Gly suggest that the M5S5 boundary and part of S5

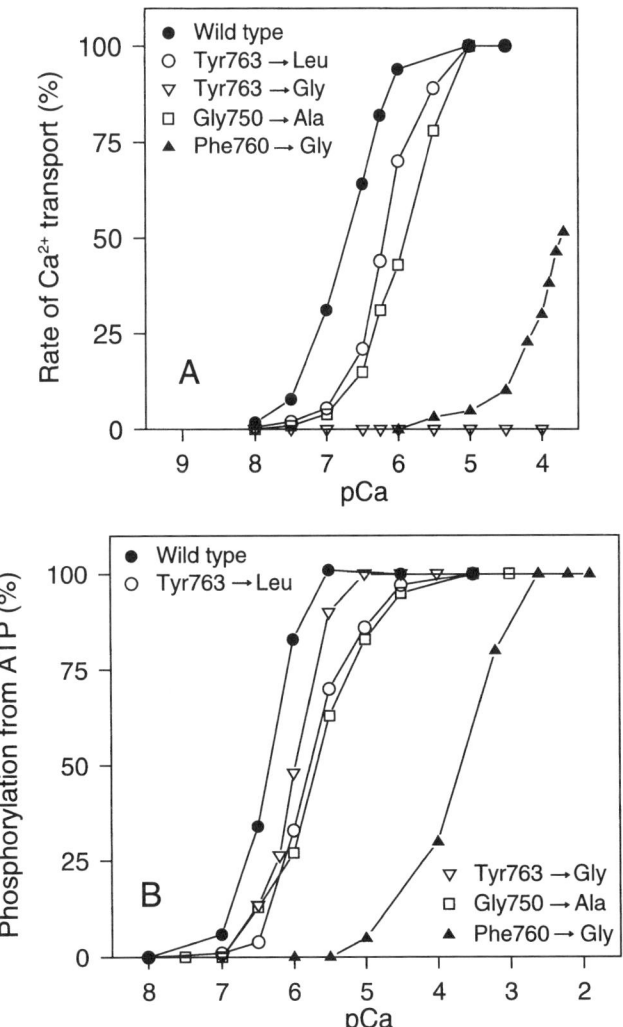

FIGURE 2. Ca^{2+} dependence of Ca^{2+} transport (**A**) and phosphorylation from ATP (**B**) of wild-type Ca^{2+}-ATPase and mutants with alterations to residues at the M5S5 boundary and in S5. Experimental procedures were previously described.[8]

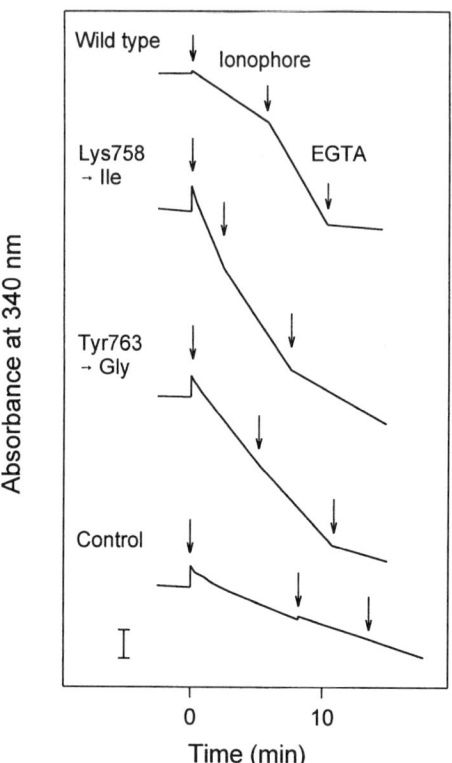

FIGURE 3. ATP hydrolysis monitored spectrophotometrically by an NADH-coupled assay[8] at 37°C in the presence of 20 mM MOPS (pH 7.0), 100 mM KCl, 2 mM $MgCl_2$, 5 mM MgATP, and 100 μM Ca^{2+}. The *bar* in the *lower left corner* indicates 0.1 absorbance unit. At the first *arrow* the microsomes are added to the reaction mixture. At the second *arrow* the calcium ionophore A23187 is added to relieve back-inhibition by accumulated Ca^{2+}. Finally, EGTA is added to show the background ATPase activity not referable to Ca^{2+}-ATPase. "Control" indicates microsomes harvested from cells transfected with vector without insert.

are closely associated with the cytoplasmic inlet of the transmembrane transport pathway.

Recently, we found that replacement of Lys758 by isoleucine results in a mutant with several unique features which in addition to a reduced apparent Ca^{2+} affinity include anomalous pH and ATP dependencies of the overall ATPase reaction, as well as a reduced maximum turnover number at neutral pH and an increased sensitivity to inhibition by vanadate.[9] In the first round of characterization of this mutant, we observed that the ATPase activity measured at a saturating Ca^{2+} concentration of 100 μM is slightly inhibited by the addition of Ca^{2+} ionophore to the sealed right-side-out microsomes (FIG. 3). By contrast, the ATPase activity of the wild-type Ca^{2+}-ATPase is enhanced at least twofold by the addition of Ca^{2+} ionophore due to relief by the ionophore of the "back inhibition" of the rate-limiting $E_1P \rightarrow E_2P$ transition imposed by accumulated Ca^{2+} binding to luminal low-affinity inhibitory sites on E_2P. As seen in FIGURE 3, the anomalous ionophore effect observed for mutant Lys758 → Ile resembles that previously reported for the uncoupled mutant Tyr763 → Gly, where the lack of activation by ionophore is caused by the absence of accumulated inhibitory Ca^{2+} in the vesicular lumen.[8] The similar responses of mutants Lys758 → Ile and Tyr763 → Gly to Ca^{2+} ionophore initially led us to speculate that mutant Lys758 → Ile like mutant Tyr763 → Gly might be uncoupled, but detailed measurements of Ca^{2+} accumulation over a wide range of Ca^{2+} concentrations have clearly revealed that mutant Lys758 → Ile is fully coupled. The failure of this mutant to show ionophore-induced activation of the ATPase activity must therefore be ascribed to insensitivity to inhibition by the Ca^{2+} accumulated in the lumen.

It now appears that the insensitivity of mutant Lys758 → Ile to luminal Ca^{2+} is caused indirectly by two kinetic effects of the mutation. First, this mutant has an unusually slow $E_2 \rightarrow E_1$ transition of the dephosphoenzyme, so that this step rather than the $E_1P \rightarrow E_2P$ transition sensitive to luminal Ca^{2+} becomes rate determining under the standard conditions applied for measurement of ATPase activity. Second, in this mutant the rate of dephosphorylation of the E_2P phosphoenzyme intermediate is enhanced relative to wild type, so that the dephosphorylation competes efficiently with the "back inhibition" by luminal Ca^{2+}. The results illustrated in FIGURE 4 demonstrate the enhanced rate of dephosphorylation of the mutant at pH 8.35 in the absence of K^+. Under these conditions the rate of dephosphorylation is very low for the wild-type enzyme and, as can be seen in the figure, for the uncoupled mutant Tyr763 → Gly as well, so that the E_2P form accumulates at steady state in the presence of ATP and Ca^{2+}. By contrast, the dephosphorylation of mutant Lys758 → Ile is seen to proceed at a more than 10-fold faster rate relative to the wild type.

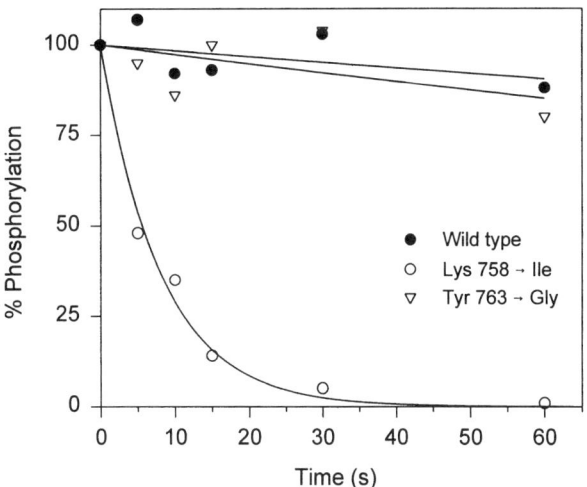

FIGURE 4. Experiment showing the increased dephosphorylation rate of mutant Lys758 → Ile. Phosphorylation in the presence of 2 μM [γ-^{32}P]ATP was carried out for 30 seconds at 0°C in a medium containing 100 mM TES/Tris buffer (pH 8.35), 10 mM $MgCl_2$, and 50 μM $CaCl_2$. Dephosphorylation was monitored after the addition of 1 mM EGTA to chelate Ca^{2+} and thereby terminate phosphorylation. Other experimental details were previously described.[8]

In conclusion, the M5S5 sector seems to play an important role in the coupling of ATP hydrolysis to Ca^{2+} translocation (gate control) and in the optimization of Ca^{2+} binding as well as in conformational changes involved in dephosphorylation of the E_2P phosphoenzyme intermediate. The Lys758 → Ile mutant is the first Ca^{2+}-ATPase mutant for which an increased rate of dephosphorylation of E_2P relative to the wild type has been reported.

REFERENCES

1. ANDERSEN, J. P. & B. VILSEN. 1994. In The Sodium Pump: Structure, Mechanism, Hormonal Control and its Role in Disease. E. Bamberg & W. Schoner, Eds.: 98–109. Steinkopff. Darmstadt.
2. ANDERSEN, J. P. 1995. Biosci. Rep. **15:** 243–261.
3. VILSEN, B. 1995. Acta Physiol. Scand. **154** (Suppl. 624): 1–146.
4. VILSEN, B. & J. P. ANDERSEN. 1992. J. Biol. Chem. **267:** 25739–25743.
5. ANDERSEN, J. P. & B. VILSEN. 1994. J. Biol. Chem. **269:** 15931–15936.
6. VILSEN, B., J. P. ANDERSEN, D. M. CLARKE & D. H. MACLENNAN. 1989. J. Biol. Chem. **264:** 21024–21030.
7. VILSEN, B., J. P. ANDERSEN & D. H. MACLENNAN. 1991. J. Biol. Chem. **266:** 18839–18845.
8. ANDERSEN, J. P. 1995. J. Biol. Chem. **270:** 908–914.
9. SØRENSEN, T., B. VILSEN & J. P. ANDERSEN. Manuscript in preparation.

Changes to Na,K-ATPase α-Subunit E779 Separate the Structural Basis for V_M and Ion Dependence of Na,K-Pump Current

R. D. PELUFFO, J. B LINGREL,[a] J. M. ARGÜELLO,[a] AND J. R. BERLIN[b]

Department of Physiology
Allegheny University of the Health Sciences
Philadelphia, Pennsylvania 19146

[a]*Department of Molecular Genetics, Biochemistry and Microbiology*
University of Cincinnati
Cincinnati, Ohio 45267

Binding of Na^+ and K^+ at extracellular sites appears to be the major electrogenic reaction step during ion transport by the Na,K-ATPase.[1,2] The structural basis for these electrogenic reaction steps and their relation to ion coordination by the enzyme are unknown. To explore these questions, we undertook an electrophysiological study of the ion transport properties of Na,K-ATPase modified by point mutations at charged residues thought to be located in transmembrane regions of the protein. The effects of changing E779, located in the fifth transmembrane-spanning region of the Na,K-ATPase α subunit, on the properties of Na,K-pump current are reported.

HeLa cells were transfected with cDNA coding for E779A, E779D, or E779Q substitutions in a ouabain-resistant sheep α1 subunit (RD) by previously reported methods.[3] After transfection, cell lines expressing the mutant enzymes were selected for propagation in Dulbecco's modified Eagle's medium (DMEM) containing 1 μM ouabain. Cells containing the E779D mutant enzyme were cultured in DMEM that included 20 mM KCl. Activity of the mutant Na,K-pump was measured in patch-clamped cells at membrane potentials (V_M) from −100 to +60 mV as a 10-mM ouabain-sensitive current (Na,K-pump current) upon increasing extracellular K^+ (K_o^+) in the superfusion solution from 0 to 0.2–50 mM (Na^+ plus K^+ = 145 mM) as detailed by Argüello *et al.*[3] K_o^+-sensitive current, termed "Na,K-pump current," was normalized to cell size as estimated from membrane capacitance.

Na,K-pump current in RD cells had a positive slope at negative V_M and a K_o^+-dependent negative slope at positive potentials (FIG. 1A), similar to wild-type Na,K-ATPase.[2] The apparent concentration for half-maximal activation of Na,K-pump current at 0 mV ($K_{0.5}$) was 2.1 ± 0.4 mM with a Hill coefficient, n_H, of 1.00 ± 0.09 (TABLE 1 and FIG. 2). As we previously reported,[3] E779A-substituted enzyme displayed no V_M dependence for Na,K-pump current activated by increasing K_o^+ from

---

[b]Address for correspondence: Joshua R. Berlin, Department of Physiology, Allegheny University of the Health Sciences, Allegheny University Hospital–Graduate, 415 S. 19th St., Philadelphia, PA 19146 (tel: 215-893-2377; fax: 215-893-4178; e-mail: berlinj@mail.med.upenn.edu).

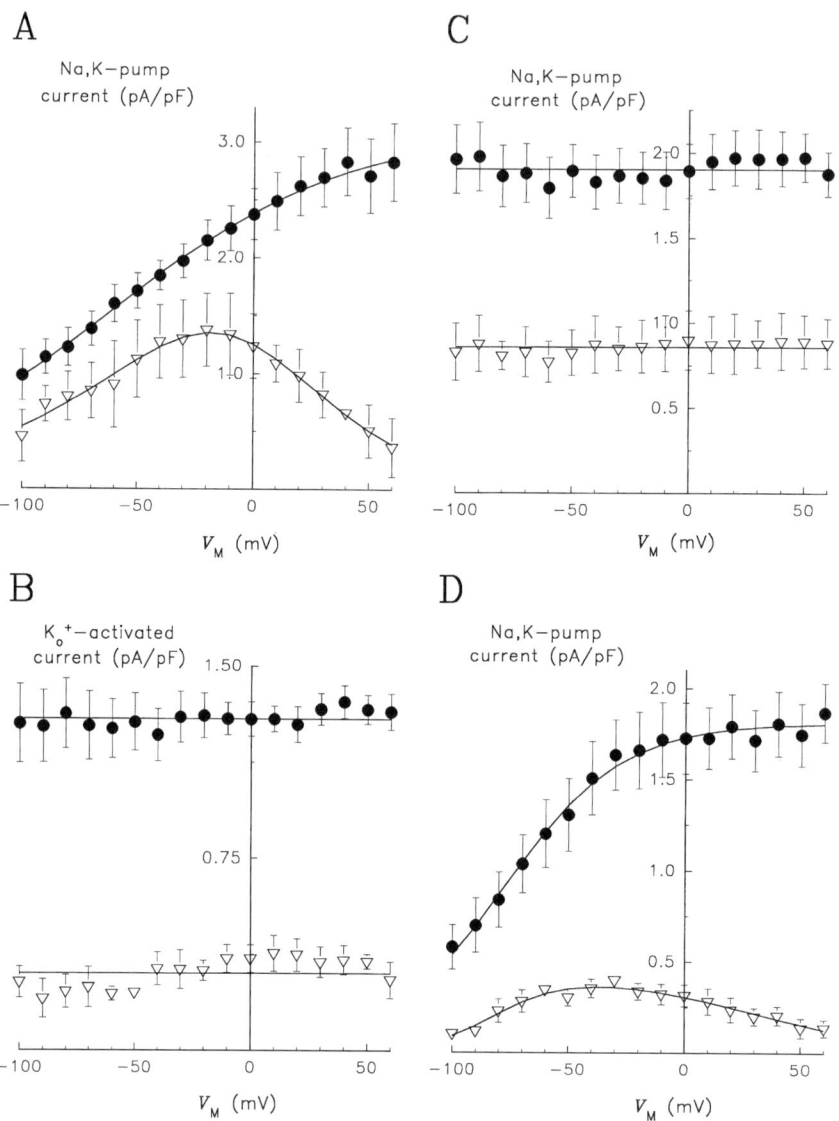

FIGURE 1. V_M and K_o^+ dependence of Na,K-pump current in control (RD) and E779-substituted enzymes. Steady-state current-V_M relationships were obtained by subtracting current measured in K_o^+-free solution from that measured in 2 ($\triangledown$) and 50 mM ($\bullet$) K_o^+-containing solution at each V_M with RD enzyme **(A)** as well as E779A- **(B)**, E779Q- **(C)**, and E779D-substituted enzymes **(D)**. The lines through the data are drawn by eye. TABLE 1 lists the number of replicates for each experiment. The plot for E779A-substituted enzyme **(B)** is labeled "K_o^+-activated current" to emphasize that current arising from electrogenic Na^+-Na^+ exchange is not measured in this experiment.

TABLE 1. Properties of RD and E779-Substituted Enzymes

Substitution	RD	E779A	E779D	E779Q
V_M dependence	Yes	No	Yes	No
$K_{0.5}$ at 0 mV (mM)	2.1 ± 0.4	7.4 ± 1.8	10.4 ± 1.6	2.1 ± 0.3
Na^+-Na^+ exchange	No	Yes	No	No
Number of cells	11	9	12	10

0 to 0.2–50 mM (FIG. 1B). The apparent $K_{0.5}$, however, was 3–4 times greater than that for RD enzyme (TABLE 1) without a significant change in the Hill coefficient (n_H = 1.11 ± 0.03). Furthermore, E779A-substituted enzyme carried out electrogenic Na^+-Na^+ exchange at a rate that was 40 ± 5% of maximal Na^+-K^+ exchange (n = 7). Studies with membrane fractions isolated from RD and E779A cells demonstrated that this marked increase in Na^+-Na^+ exchange is due to acceleration of Na^+-dependent dephosphorylation.[3] Thus, removal of a large polar group at position 779 had dramatic effects on Na,K-ATPase function. Nonetheless, these changes made it difficult to separate the basis of electrogenic reactions from other ion transport properties of this mutant enzyme. For this reason, the effects of conservative substitutions on Na,K-pump current were examined.

Removal of the carboxyl group in the side chain while maintaining the carbonyl moiety in E779Q-substituted enzyme produced no significant changes in $K_{0.5}$ or n_H for K_o^+ (TABLE 1), but the V_M dependence of Na,K-pump current was abolished at all K_o^+ (FIG. 1C). Conversely, when the carboxyl residue was preserved in E779D enzyme, Na,K-pump current had a similar V_M dependence as in RD (FIG. 1D) even though $K_{0.5}$ for K_o^+ was increased approximately fivefold (TABLE 1) and n_H was increased to 1.46 ± 0.18. The effect of these substitutions on K_o^+-dependent activation of the

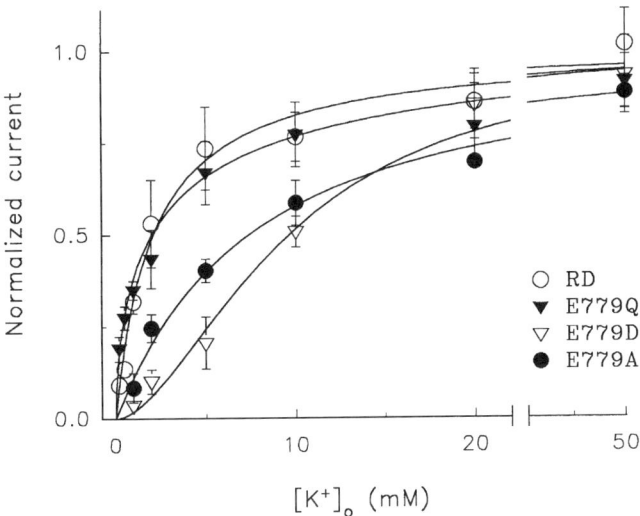

FIGURE 2. K_o^+ activation of Na,K-pump current at 0 mV. Data are normalized to the maximum current calculated by fitting a Hill equation to each data set.

Na,K-pump at 0 mV is shown more completely in FIGURE 2. Neither E779Q nor E779D mutant enzyme displayed the increased Na^+-Na^+ exchange rate observed with E779A.

These data demonstrate that the presence of a carboxyl group on residue 779 is critical for V_M dependence of Na,K-pump current but is unrelated to K_o^+ activation of the Na,K-ATPase. E779 is unlikely to be a K^+ coordination site, because substitutions at this residue had a relatively modest or no effect on the $K_{0.5}$ and n_H for K_o^+ activation. However, substitutions to E779 do alter the structural requirements for extracellular ion-dependent activation of the Na,K-pump, as evidenced by the increased Na^+-Na^+ exchange in E779A enzyme, but these alterations appear to be unrelated to the electrogenic properties of ion transport. Thus, these data show that the structural requirements underlying the electrogenic nature of extracellular ion binding are distinct from the ion binding domain of the Na,K-ATPase.

The glutamate residue at position 779 could affect the electrogenicity of extracellular ion binding in several ways. First, the carboxyl residue could form a portion of the putative extracellular ion access channel. Second, the carboxyl residue could have long- or short-range effects on the structures that determine the V_M dependence of ion binding. Third, the kinetics of the rate-limiting steps controlled by extracellular ion binding could be accelerated. Further experiments are needed to distinguish between these possible mechanisms.

REFERENCES

1. GADSBY, D. C., R. F. RAKOWSKI & P. DE WEER. 1993. Science **260:** 100–103.
2. BERLIN, J. R. & R. D. PELUFFO. 1997. Ann N.Y. Acad. Sci., this volume.
3. ARGÜELLO, J. M., R. D. PELUFFO, J. FENG, J. B LINGREL & J. R. BERLIN. 1996. J. Biol. Chem. **271:** 24610–24616.

Interaction of Palytoxin and Mercury with the Na,K-ATPase on *Xenopus laevis* Oocytes[a]

XINYU WANG AND JEAN-DANIEL HORISBERGER[b]

Institute of Pharmacology and Toxicology
Bugnon 27
CH-1005 Lausanne, Switzerland

Although Na^+ and K^+ ions must follow a pathway across the membrane-associated part of the Na,K-ATPase protein, the structure that constitutes this pathway remains unknown. To define this pathway we are investigating the effect of mercury (Hg^{2+}), a cysteine reagent, on various aspects of the Na,K-pump function. Exposure of the extracellular side of *Xenopus laevis* oocyte membrane to $HgCl_2$ results in rapid and irreversible inhibition of Na,K-pump activity following first-order kinetics (k_{on} 7.10^3 $M^{-1} \cdot s^{-1}$), and we showed recently that a cysteine present in the first transmembrane segment is responsible, at least in part, for this inhibition and therefore probably participates in the structure of the binding site of mercury on the Na,K-pump.[1] Palytoxin (PTX), a nonpeptidic toxin extracted from a marine coelenterate (*Palythoa tuberculosa*), is known to bind to the Na,K-pump and thereby inhibit its ATPase activity and induce a nonselective cation conductance. Indirect evidence indicates that this channel is probably homologous to the cation transport pathway[2] through the Na,K-pump. We hypothesized that we could probe the PTX-induced channel for exposed cysteine residues by studying the effect of Hg^{2+} on PTX-induced conductance.

MATERIALS AND METHODS

Na,K-pump activity was measured using the two-electrode voltage clamp technique as the outward current induced by 10 mM K^+.[3] The effect of PTX on *X. laevis* oocytes was monitored by current changes elicited by voltage jump from -50 to 0 mV. Experiments were performed at room temperature in a solution containing (mM): Na^+ 92.4, Mg^{2+} 0.82, Ba^{2+} 5, Ca^{2+} 0.41, TEA 10, Cl^- 22.4, HCO_3^- 2.4, gluconate 80, HEPES 10, pH 7.4. Palytoxin, purchased from Sigma, was added from a 10 μM in water stock solution immediately before each measurement.

---

[a] This work was supported by grant 31-45867.95 (to J.-D.H.) from the Swiss Fonds National de la Recherche Scientifique.
[b] To whom correspondence should be addressed. Tel: +41 21 692 5362; fax: +41 21 692 5355; e-mail: Jean-Daniel.Horisberger@ipharm.unil.ch

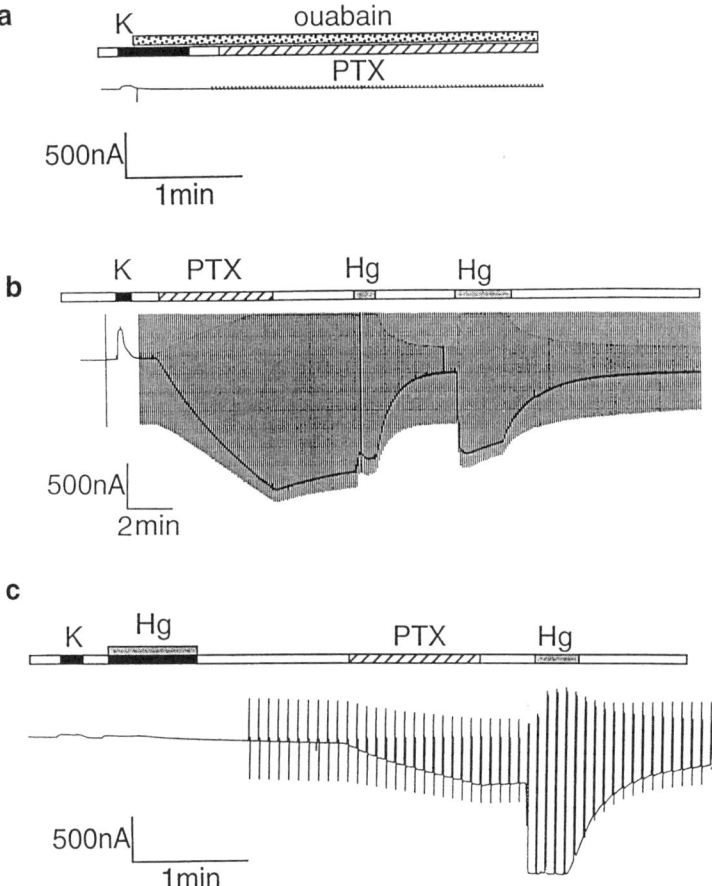

FIGURE 1. PTX-induced ouabain-sensitive conductance in *X. laevis* oocytes. (**a**) After Na,K-pump function was first assessed with 10 mM K^+ (*black bar*) and then inhibited by 100 μM ouabain (*dotted bar*), 2 nM PTX (*hatched bar*) was added in the presence of ouabain. PTX did not induce any change in conductance. (This experiment was performed on a native, noninjected oocyte.) (**b**) Palytoxin 2 nM alone induced a large increase in conductance, an effect that was slowly reversible upon removal of PTX. Subsequent addition of 5 μM Hg^{2+} (*gray bar*) induced a complex series of changes. There was first a small and rapid reduction in conductance followed by partial recovery. Removal of Hg^{2+} resulted in a large reduction in conductance within a 1–2-minute period. Readdition of 5 μM Hg^{2+} instantaneously restored the large conductance observed during the first exposure to Hg^{2+}. This effect of mercury was reversible and reproducible during the duration of the PTX effect. This measurement was performed on an oocyte overexpressing the *Xenopus* WT Na,K-pump. (**c**) After full inhibition of the Na,K-pump by exposure to 5 μM Hg^{2+}, PTX (2 nM) only induced a relatively small conductance, but the subsequent addition of 5 μM Hg^{2+} (*2nd gray bar*) immediately induced a large conductance, an effect that was rapidly reversible (experiment performed on a native, noninjected oocyte).

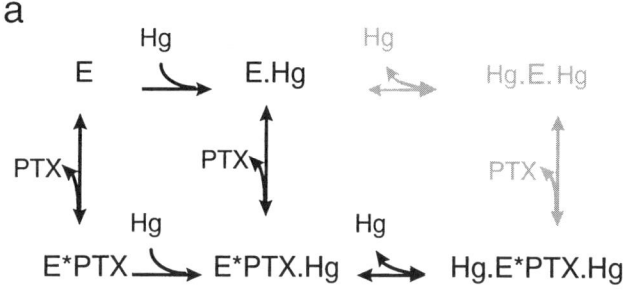

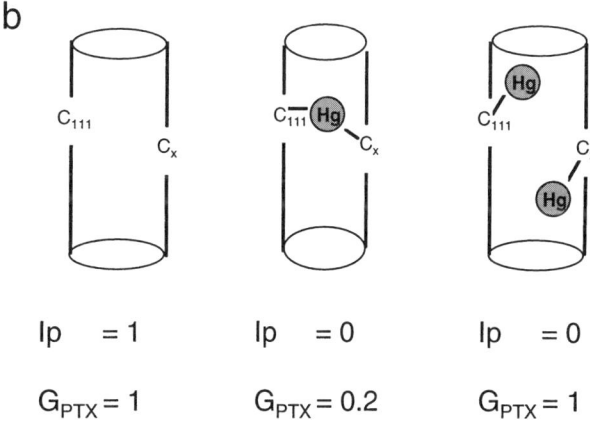

FIGURE 2. Kinetic scheme (a) of the biphasic effect of Hg^{2+} on the palytoxin (PTX)-treated Na,K-pump and a proposed model (b) explaining this interaction. (a) The Na,K-pump (E) can bind to PTX to form a channel-like structure (E*PTX). On the other hand, mercury can bind to the Na,K-pump to form an inactive form of the enzyme (E.Hg). The interaction between PTX and Hg^{2+} reveals that in fact there are at least two binding sites for Hg^{2+}. Binding of Hg^{2+} to the first site is practically irreversible and lead to inhibition of Na,K-pump activity (Na,K-pump current) and to partial inhibition of PTX-induced conductance. Binding of Hg^{2+} to the second site is reversible (k_{off}: 2 min^{-1}) and leads to reactivation of the PTX-induced current. It cannot be decided from current data if this second site is present on the native enzyme (*shaded part* of the scheme) or if it is revealed by the effect of PTX. (b) This scheme proposes a possible structure compatible with the kinetic scheme shown in **a** and in particular with the observation of irreversible inhibition of Na,K-pump activity and partial inhibition of PTX-induced conductance by previous exposure to Hg^{2+} with reactivation of the PTX-induced conductance by a second exposure to Hg^{2+}. Obviously many other, more complicated schemes are possible.

RESULTS

1. Palytoxin (2 nM) induced a large conductance in the *X. laevis* oocyte (FIG. 1b). This conductance was selective for monovalent cations but poorly selective among Na^+, K^+, and Li^+ (data not shown), and this effect was slowly reversible. When PTX was applied after the Na,K-pump had been inhibited by 100 µM ouabain, oocyte conductance did not change (FIG. 1a).

2. Mercury (5 µM) rapidly induced small inhibition of the PTX-induced conductance on the oocyte, but greater inhibition (by about 80%) occurred when Hg^{2+} was removed (FIG. 1b); this inhibition was irreversible within the time course of our experiments. Subsequent addition of the same concentration of Hg^{2+} reactivated the PTX-induced conductance with a fast onset rate ($t_{1/2} < 3$ s), and this effect was reversible with a time course of about 30 seconds. When the Na,K-pump was first inhibited by a 1-minute exposure to 10 µM $HgCl_2$, the effect of PTX was initially reduced, but a large conductance could be activated by a second exposure to the same concentration of mercury (FIG. 1c).

DISCUSSION

Our data show that as in other cell types,[4] PTX binds to the Na,K-pump of *Xenopus* oocytes and induces a large cation conductance. Exposure to Hg^{2+} after PTX treatment indicated two distinct actions of Hg^{2+} on the Na,K-pump. The first effect was irreversible inhibition of Na,K-pump activity accompanied by >80% reduction of PTX-induced conductance. It was observed whether Hg^{2+} exposure was performed before or during PTX exposure. The second effect was reversible (with a half-life of about 30 seconds) and consisted of reactivation of the PTX-induced conductance from the inhibition, resulting from the first effect. In FIGURE 2, we propose a kinetic (a) and a mechanistic (b) model that could explain this biphasic effect of Hg^{2+}. This model assumes that two cysteine residues (one of which is C113 of the first transmembrane segment) are exposed to the external solution. After the proposition of Redondo *et al.*[2] that PTX-induced channel may be homologous with the pathway that the cations follow during the normal transport mode of the Na,K-pump, these two cysteine residues may be part of an aqueous pore forming the external access channel for cation to their binding site inside the protein.

REFERENCES

1. WANG, X. & J.-D. HORISBERGER. 1996. Mol. Pharmacol. **50:** 687–691.
2. REDONDO, J., B. FIEDLER & G. SCHEINER-BOBIS. 1996. Mol. Pharmacol. **49:** 49–57.
3. HORISBERGER, J.-D., P. JAUNIN, P. J. GOOD, B. C. ROSSIER & K. G. GEERING. 1991. Proc. Natl. Acad. Sci. USA **88:** 8397–8400.
4. HABERMANN, E. 1989. Toxicon **27:** 1171–1187.

Voltage-Dependent Inhibition of the Na^+-K^+ Pump by Intracellular Potassium in Rabbit Ventricular Myocytes

PETER S. HANSEN,[a] DAVID F. GRAY,
KERRIE A. BUHAGIAR, AND HELGE H. RASMUSSEN

Department of Cardiology
Royal North Shore Hospital
Sydney, Australia

Voltage dependence of steady-state Na^+-K^+ pump current is thought to arise from several steps in the pump cycle. These include the backward rate constant for the interaction of Na^+ with extracellular pump sites,[1] binding of extracellular K^+ to the pump,[2–5] and binding of Na^+ to intracellular sites.[6] These studies suggest that cations have to travel across a high field access channel to pump sites that are buried within the membrane dielectric before binding can occur. Intracellular potassium (K^+_i) interacts with intracellular pump sites by competing with Na^+ for binding to the E_1 form.[7] Competition with Na^+ is thought to occur at the cytoplasmic surface[8] and is therefore expected to inhibit pump activity in a voltage-independent manner. Intracellular K^+ has also been reported to have an inhibitory effect on Na^+/K^+-ATPase reconstituted into lipid vesicles under some experimental conditions by interacting with the K^+ translocation part of the cycle.[9,10] While competitive inhibition of Na^+ binding by K^+_i near the cytoplasmic surface is not expected to generate voltage dependence, it is conceivable that an alternative site for K^+_i inhibition can contribute to the voltage dependence of the pump. We have examined if K^+_i can generate voltage-dependent pump inhibition.

METHODS

Isolated rabbit ventricular myocytes from male New Zealand White rabbits were voltage clamped at a holding potential of -40 mV using the whole cell patch clamp technique. We used wide-tipped patch pipettes (tip diameter 4–5 µM, tip resistance <1.1 MΩ when filled with pipette-filling solutions). Pipette solutions contained (in mM): 79 Na glutamate, 1 NaH_2PO_4, 0–80 KOH, 5 HEPES, 2 Mg-ATP, 20 TEA.Cl, 31–45 aspartate, and 0–66 TMA.OH titrated to pH 7.2 at 22°C with 1 M HCl. Pipette Na^+ concentration was 80 mM to saturate intracellular Na^+ binding sites. Seven different pipette K^+ concentrations ($[K]_{pip}$) of 0, 2.5, 5, 10, 25, 50, and 80 mM were used.

---

[a] Address for correspondence: Dr P. S. Hansen, Department of Cardiology, Royal North Shore Hospital, Pacific Highway, St Leonards, NSW 2065, Australia (tel: 61-2-99268686; fax: 61-2-99067807).

Osmolality was maintained with TMA.OH. K^+ channels were inhibited by including 20 mM TEA.Cl in pipette-filling solutions and 2 mM Ba^{2+} in superfusate. The whole cell configuration was achieved while myocytes were superfused with modified Na^+- and Ca^{2+}-containing Tyrode's solution. The superfusate was then switched to a solution that was nominally Ca^{2+} free and contained 0.2 mM Cd^{2+} to eliminate Na-Ca exchange. This solution was Na^+ free (NMG.Cl substituted) and contained 15 mM K^+ to minimize the effects of extracellular voltage-dependent steps in the pump cycle. Steady-state membrane currents were recorded at test potentials (V_m) from -100 to $+60$ mV incremented at 20-mV intervals. Each V_m was bracketed by a return to a holding potential of -40 mV for 2 seconds. The pump current (I_p) was identified as the ouabain (100 µM)-sensitive shift in holding current at each V_m. The reference electrode contained 3 M KCl to reduce voltage errors arising from liquid-junction potentials.

RESULTS AND DISCUSSION

After achieving the whole cell configuration, 6–10 minutes were allowed to assure complete washout of Na^+ in the tissue bath and dialysis of the intracellular compart-

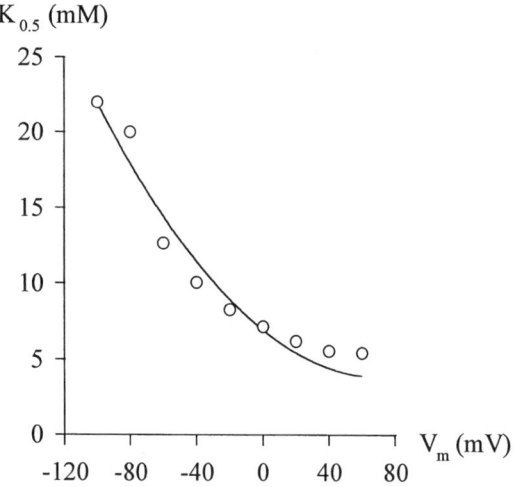

FIGURE 1. V_m-$K_{0.5(Vm)}$ relationship ($\lambda_{Ki} = 0.30$).

ment by the pipette-filling solution. A total of 69 myocytes were used in the study. Mean I_ps were identified at each V_m for each of the 7 $[K]_{pip}$s used. There was a decrease in I_p with an increase in $[K]_{pip}$. The K^+_i inhibitory effect (I_i) for each $[K]_{pip}$ was defined at each V_m by subtracting mean I_ps at each V_m from mean I_ps measured when $[K]_{pip}$ was 0 mM. A $[K]_{pip}$-I_i relationship was determined at each V_m. The Hill equation was fitted to these $[K]_{pip}$-I_i relationships to derive the $[K]_{pip}$ for half-maximal I_i ($K_{0.5(Vm)}$) at each V_m. The V_m-$K_{0.5(Vm)}$ relationship and the fit of a Boltzman distribution is shown in FIGURE 1. The Boltzman distribution used was: $K_{0.5(Vm)} =$

$K_{0.5(Vm = 0mV)} \exp(-\lambda_{Ki}VF/RT)$, where $K_{0.5(Vm=0mV)}$ is the $[K]_{pip}$ for half-maximal I_i when $V_m = 0$ mV and λ_{Ki} the fractional dielectric distance. λ_{Ki} for the best fit was 0.30 ($r^2 = 0.99$). We conclude that intracellular K^+ inhibits the Na^+-K^+ pump in a voltage-dependent manner when the pump operates in its normal forward mode. The formalism of an intracellular access channel to the inhibitory site can describe our findings.

REFERENCES

1. NAKAO, M. & D. GADSBY. 1986. Voltage dependence of Na translocation by the Na/K pump. Nature **323:** 628–630.
2. LAFAIRE, A. V. & W. SCHWARZ. 1986. Voltage dependence of the rheogenic Na^+/K^+ ATPase in the membrane of oocytes of *Xenopus laevis*. J. Membr. Biol. **91:** 43–51.
3. GADSBY, D. C., M. NAKAO, A. BAHINSKI, G. NAGEL & M. SUENSON. 1992. Charge movements via the cardiac Na,K-ATPase. Acta Physiol. Scand. **146:** 111–123.
4. SAGAR, A. & R. F. RAKOWSKI. 1994. Access channel model for the voltage dependence of the forward-running Na^+/K^+ pump. J. Gen. Physiol. **103:** 869–894.
5. PELUFFO, R. D. & J. BERLIN. 1996. Transient charge movement during K^+-translocating steps by the Na pump. Biophys. J. **70:** A18.
6. HEYSE, S., I. WUDDEL, H.-J. APELL & W. STÜRMER. 1994. Partial reactions of the Na,K-ATPase: Determination of rate constants. J. Gen. Physiol. **104:** 197–240.
7. WUDDEL, I. & H.-J. APELL. 1995. Electrogenicity of the sodium transport pathway in the Na,K-ATPase probed by charge-pulse experiments. Biophys. J. **69:** 909–921.
8. OR, E., P. DAVID, A. SHAINSKAYA, D. M. TAL & S. J. D. KARLISH. 1993. Effects of competitive sodium-like antagonists on Na,K-ATPase suggest that cation occlusion from the cytoplasmic surface occurs in two steps. J. Biol. Chem. **268:** 16929–16937.
9. VAN DER HIJDEN, H. T. W. M. & JAN J. H. H. M. DE PONT. 1989. Cation sidedness in the phosphorylation step of Na^+/K^+-ATPase. Biochim. Biophys. Acta **983:** 142–152.
10. CORNELIUS, F. & J. C. SKOU. 1991. The effect of cytoplasmic K^+ on the activity of the Na^+/K^+-ATPase. Biochim. Biophys. Acta **1067:** 227–234.

Effect of pH on Charge Movement by the Na/K Pump in *Xenopus* Oocytes[a]

KEVIN A. KHATER[b] AND R. F. RAKOWSKI[b]

Department of Physiology and Biophysics
Finch University of Health Sciences
The Chicago Medical School
North Chicago, Illinois 60064

Ouabain-sensitive presteady-state transient currents were first described by Nakao and Gadsby[1] in cardiac myocytes under conditions for electroneutral Na/Na exchange. The transient currents were also measured in *Xenopus* oocytes using the two-microelectrode[2] and cut-open oocyte vaseline seal technique.[3] Here we describe the extracellular pH dependence of the transient currents using the cut-open technique. Little difference was noted in the slow component of the relaxation rate constants or charge versus voltage (Q/V) relationship in the pH range 6.6–9.6. However, at more alkaline pH, the poorly resolved intermediate component appeared to peak later and was smaller in amplitude as pH was increased to 8.6 and 9.6. When the external [Na$^+$] was decreased from 100 to 50 mM, the midpoint voltage (V_q) of the intermediate component shifted from -80 ± 8 mV to -163 ± 15 mV, indicating that the dielectric distance of this Na$^+$ binding/release site is 0.21 ± 0.08. Consistent with this determination is the observation that the apparent valence of the Q/V relationship (z_q) of this component at 100 mM Na$^+$ was 0.29 ± 0.05 (in close agreement with Hilgemann's[4] value of 0.26), and the relaxation rate constants appear to be relatively voltage independent.

MATERIALS AND METHODS

Isolated ovarian tissue obtained from *Xenopus* oocytes was treated with Type I (Sigma, St. Louis, Missouri) collagenase to facilitate manual defolliculation. Oocytes were mounted and voltage clamped using an open-oocyte, guarded-seal technique.[5] This technique allows steps in membrane potential to reach the new steady state within 100 µs. After mounting the oocyte and allowing 30 minutes for equilibration of the internal solution, the membrane current response to step changes in membrane potential was recorded. Data were sampled at 100 kHz for the first 1.8 ms and at 16.7 kHz for the remaining 48.2 ms. Data were low-pass filtered at 50 kHz. Each trace represents the average of 20 successive trials.

The external solution contained (in mM): 100 Na sulfamate, 20 tetraethylammonium sulfamate, 3 Mg sulfamate, 2 Ni(NO$_3$)$_2$, 5 Ba(NO$_3$)$_2$, and 10 TRIS/HEPES. Solutions were titrated to the appropriate pH using 2.8 M TEAOH. The internal

---

[a]This work was supported by National Institutes of Health grant NS22979.
[b]Address for correspondence: Department of Physiology and Biophysics, Finch University of Health Sciences/The Chicago Medical School, 3333 Green Bay Road, North Chicago, IL 60064 (tel: 847-578-3280; fax: 847-578-3265; e-mail: rakowskr@mis.finchcms.edu).

solution contained (in mM): 50 sodium sulfamate, 20 TEA sulfamate, 10 Mg(SO$_4$), 5 BAPTA, 5 TrisADP, and 5 MgATP (pH 7.3). The solution components were chosen to promote electroneutral Na/Na exchange.[6–8]

Transient currents were calculated by subtraction of current records obtained after the addition of 20 µM ouabain from those obtained immediately before the addition of ouabain.

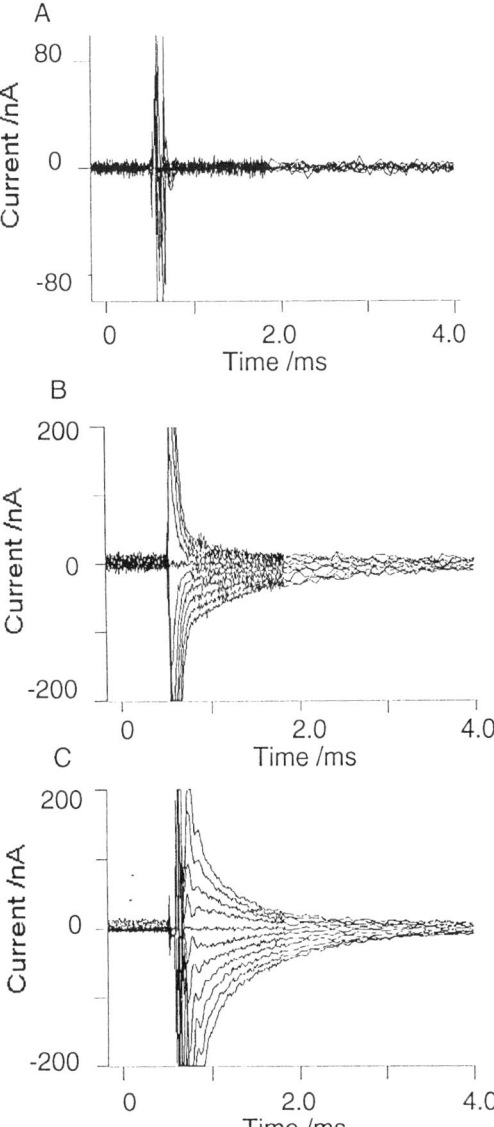

FIGURE 1. Time only control (**A**) and ouabain-sensitive transient current records obtained at pH 7.6 (**B**) and 8.6 (**C**). Current records were obtained every 5 minutes and the last two sets of current records were subtracted from one another until the time only control (**A**) became nearly flat, indicating that most of the system had come to a steady state. The Pool compartment was then perfused for 4 minutes with external solution that contained 20 µM ouabain. The current response to command voltage steps was sampled, and the ouabain-sensitive transient currents were measured (**B,C**).

RESULTS

The slow component of the presteady-state transient current remained unaffected over the pH range of 6.6–9.6 (data not shown). FIGURE 1A shows a central record

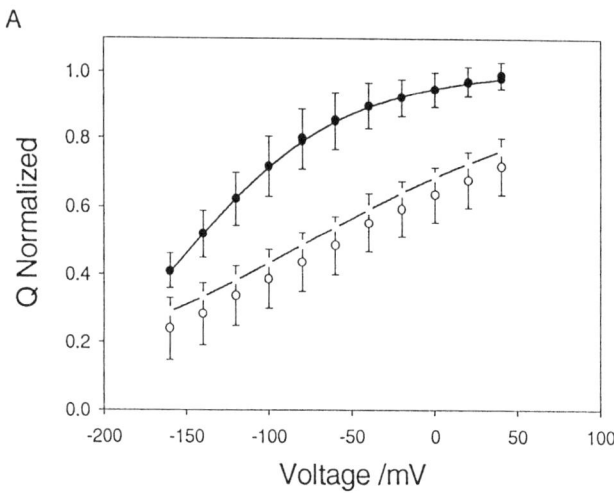

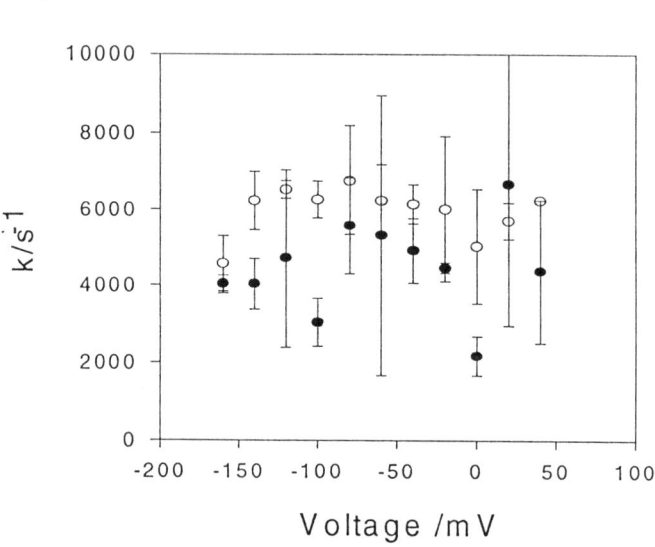

FIGURE 2. Normalized charge movement with least-squares fit curve (**A**) and relaxation rate constants (**B**) of the on-pulse intermediate component at pH 8.6 with 100 (○) and 50 mM external (●) [Na^+]. The best fit curve parameters for FIGURE 2A are: at 100 mM Na^+, z_q = 0.29 ± 0.05, V_q = −80 ± 8 mV, and at 50 mM Na^+, z_q = 0.42 ± 0.04, V_q = −163 ± 15 mV.

demonstrating the stability of the current measurement prior to the addition of ouabain. FIGURE 1B at pH 7.6 shows that preceding the slow component is a fast transient current that occurs with the time course of the change in voltage and is therefore not temporally resolved. At pH 8.6 the intermediate component decreases in maximum amplitude and has a slower relaxation rate (FIG. 1C).

FIGURE 2A shows the normalized Q/V relationship of the intermediate component at pH 8.6 at 100 and 50 mM external [Na^+]. Normalization was performed by doing a least-squares fit of the calculated charge movement to the following equation, $Q = (Q_{max} - Q_{min})/(1 + \exp(Fz_q (V_q - V_m)/RT))$, where Q = charge moved at a given membrane potential, Q_{min} is the minimum charge moved, Q_{max} represents the maximum amount of charge moved, V_q is the membrane potential at which one half the total charge is moved, and V_m is the membrane potential. Q was then normalized with respect to $Q_{max} - Q_{min}$. The parameters obtained by the least-squares fit are given in the legend. FIGURE 2B shows the relaxation rate constants of the intermediate component obtained by performing a double exponential fit to the data.

CONCLUSIONS

If the three sodium ions have different binding sites, then presteady-state transient currents would be expected to have at least three exponential components. At pH 7.6, the transient currents have a fast unresolved component and a slower exponential component. At pH 8.6 and 9.6, the slow component is unaffected, but an intermediate component is slowed and is resolvable. The voltage dependence of the intermediate component shifts as external [Na^+] is decreased. The large shift in midpoint voltage at different Na^+ concentrations, the relatively small value for z_q, and the shallow voltage dependence of the rate constants all suggest that the binding/release site of the Na^+ ion responsible for the intermediate component has a dielectric coefficient of about 0.25. This is consistent with charge-pulse studies that have reached a similar conclusion.[9] These results also suggest that the binding site of the Na^+ ion has a dissociable protonation site with a pK_a in the range of 7.6 to 8.6.

REFERENCES

1. NAKAO, M. & D. C. GADSBY. 1986. Voltage dependence of Na translocation by the Na/K pump. Nature **323:** 628–630.
2. RAKOWSKI, R. F. 1993. Charge movement by the Na/K pump in *Xenopus* oocytes. J. Gen. Physiol. **101:** 117–144.
3. HOLMGREN, M. & R. F. RAKOWSKI. 1994. Pre-steady-state transient currents mediated by the Na/K pump in internally perfused *Xenopus* oocytes. Biophys. J. **66:** 912–922.
4. HILGEMANN, D. W. 1994. Channel-like function of the Na,K pump probed at microsecond resolution in giant membrane patches. Science **263:** 1429–1432.
5. TAGLIALATELA, M., L. TORO & E. STEFANI. 1992. Novel voltage clamp to record small, fast currents from ion channels expressed in *Xenopus* oocytes. Biophys. J. **61:** 78–82.
6. DE WEER, P. 1970. Effects of intracellular adenosine-5'-diphosphate and orthophosphate on the sensitivity of sodium efflux from squid axon to external sodium and potassium. J. Gen. Physiol. **56:** 583–620.
7. GLYNN, I. M. & J. F. HOFFMAN. 1971. Nucleotide requirements for sodium-sodium exchange catalysed by the sodium pump in human red cells. J. Physiol. **218:** 239–256.
8. CAVIERES, J. D. & I. M. GLYNN. 1979. Sodium-sodium exchange through the sodium pump: The roles of ATP and ADP. J. Physiol. **297:** 637–645.
9. WUDDEL, I. & H.-J. APELL. 1995. Electrogenicity of the sodium transport pathway in the Na,K-ATPase probed by charge-pulse experiments. Biophys. J. **69:** 909–921.

Sodium Pump of Cultured Guinea Pig Atrial Myocytes

JENS KOCKSKÄMPER[a] AND
HELFRIED GÜNTHER GLITSCH[a]

*Arbeitsgruppe Muskelphysiologie der Ruhr-Universität
D-44780 Bochum, Germany*

Cardiac atrial cells are frequently used in studies on membrane currents, the most prominent probably being $I_{K(ACh)}$, but little is known about the Na/K pump of these cells. In order to fill this gap we studied the characteristics of I_p, the current produced by the atrial Na/K pump, by means of whole-cell recording.[1,2]

Experiments were performed at 30°C on cultured, isolated guinea pig atrial myocytes ("cardioballs"[3]) in Na-containing media. Their spherical shape and small size make cardioballs ideally suited for whole-cell recording. Their geometric properties favor the control of the ionic composition of the cell interior, which is of particular importance for measurements of the dependence of I_p on $[Na]_i$. Low-resistance (1.5–3.5 MΩ) patch pipettes contained (in mM): 110 CsCl; 40 NaCl; 10 NaOH; 3 MgCl$_2$; 6 ethylene glycol-bis(β-aminoethyl ether) N,N,N',N'-tetraacetic acid (EGTA); 16 N-2-hydroxyethylpiperazine-N'-2-ethanesulfonic acid (HEPES); 10 adenosine 5'-triphosphate (ATP), pH 7.4 (CsOH). The extracellular solution included (in mM): 144 NaCl; 0–10.8 KCl; 5 NiCl$_2$; 2 BaCl$_2$; 1.8 CaCl$_2$; 0.5 MgCl$_2$; 10 HEPES; pH 7.4 (NaOH). CsCl, BaCl$_2$, and NiCl$_2$ suppressed K and Ca conductances and/or the sarcolemmal Na/Ca exchange. Rapid changes of extracellular solution (<1 s) were performed with a multibarreled pipette positioned close to the cell under investigation. Measurements were carried out at 0 mV except for studies on the voltage dependence of I_p.

I_p was identified as current blocked by $2 \cdot 10^{-4}$ M dihydro-ouabain (DHO) or K-free solution. Under the experimental conditions chosen, either I_p estimation yielded almost identical results. I_p density declined with time in culture. On day 7 it amounted to $\approx$20% of that measured on day 1 (0.66 pA$\cdot$pF^{-1} at 50 mM Na$_{pip}$; 5.4 mM K$_o$; $n = 8$). Half-maximal I_p activation occurred at 1.2 mM K$_o$ ($n = 5$–14), in line with earlier observations on multicellular atrial preparations.[4,5] Internal Na half-maximally activated I_p at 14 mM Na$_{pip}$ ($n = 4$–9). Similar $K_{0.5}$ values have been reported for single cardiac ventricular[6] and Purkinje cells.[7] FIGURE 1 illustrates the voltage dependence of I_p in atrial cardioballs. FIGURE 1A shows blockade and reactivation of I_p during and after, respectively, a brief pulse of K-free solution (not indicated) at various membrane potentials. I_p increases with depolarization at negative membrane potentials and shows no unequivocal variation at positive voltages (50 mM Na$_{pip}$). FIGURE 1B displays normalized mean I_p-V curves at different [Na]$_{pip}$ and reveals an increasing slope of the I_p-V relationships with decreasing [Na]$_{pip}$, confirm-

---

[a]Authors' addresses: Arbeitsgruppe Muskelphysiologie, Fakultät für Biologie der Ruhr-Universität, D-44780 Bochum, Germany (tel: +234/7003983 or 7005838; fax: +234/7094-129).

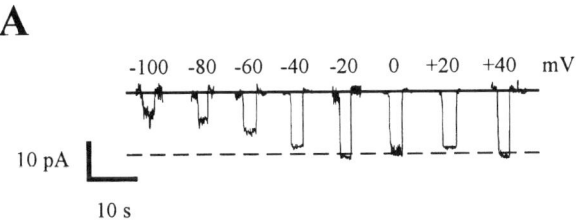

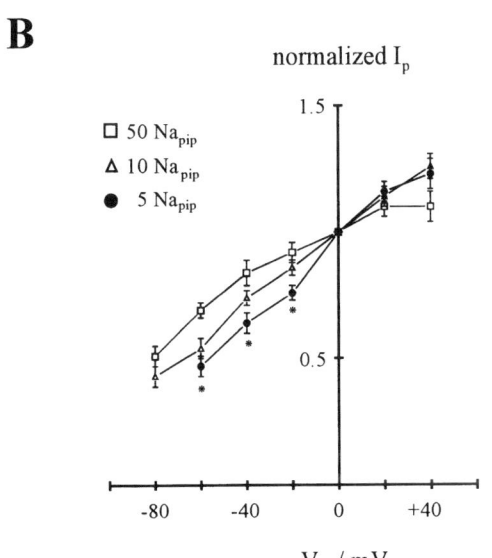

FIGURE 1. Voltage dependence of I_p in guinea pig atrial cardioballs in Na-containing solution. (**A**) Original record of membrane current at various membrane potentials (indicated above the current traces). The cell was internally perfused with 50 mM Na. I_p identified as current activated by 5.4 mM K_o following a brief pulse of K-free medium (not indicated). (**B**) Normalized mean I_p-V curves of cardioballs internally perfused with pipette media containing various Na concentrations ($[Na]_{pip}$). I_p normalized to the respective I_p density (mean ± SEM) at 0 mV (50 mM Na_{pip}: 0.66 ± 0.05 pA · pF^{-1} ($n = 8$); 10 mM Na_{pip}: 0.39 ± 0.05 pA · pF^{-1} ($n = 9$); 5 mM Na_{pip}: 0.23 ± 0.05 pA · pF^{-1} ($n = 6$)). *Asterisks* mark significant differences ($p < 0.05$) between data points at 5 and 50 mM Na_{pip}.

ing results from other isolated cardiac cells.[6,7] The effect of $[Na]_{pip}$ on the slope of the I_p-V curve persisted in Na-free solution (Na replaced by choline; 10 or 50 mM Na_{pip}). In these experiments, I_p was estimated as current activated by 1 mM K_o. The observation that the slope of the normalized I_p-V curves in Na-containing and Na-free solution increases with decreasing $[Na]_{pip}$ suggests that Na binding at intracellular sites of the Na/K pump is voltage dependent.

REFERENCES

1. HAMILL, O. P., A. MARTY, E. NEHER, B. SAKMANN & F. J. SIGWORTH. 1981. Pflügers Arch. **391:** 85–100.
2. GADSBY, D. C. & M. NAKAO. 1989. J. Gen. Physiol. **94:** 511–537.
3. BECHEM, M., L. POTT & H. RENNEBAUM. 1983. Eur. J. Cell. Biol. **31:** 366–369.
4. RASMUSSEN, H. H., D. H. SINGER & R. TEN EICK. 1986. Am. J. Physiol. **251:** H331–H339.
5. GLITSCH, H. G., W. GRABOWSKI & J. THIELEN. 1978. J. Physiol. **276:** 515–524.
6. NAKAO, M. & D. C. GADSBY. 1989. J. Gen. Physiol. **94:** 539–565.
7. GLITSCH, H. G., T. KRAHN & H. PUSCH. 1989. Pflügers Arch. **414:** 52–58.

Dipole Potential Drop due to RH-Dye Adsorption on the Lipid Bilayer and Its Influence on Na$^+$/K$^+$-ATPase Activity[a]

D. YU. MALKOV, K. V. PAVLOV, AND V. S. SOKOLOV[b]

A. N. Frumkin Institute of Electrochemistry
Russian Academy of Sciences
Moscow, 117071 Russia

Fluorescent voltage-sensitive styryl dyes of the RH series (RH-421, RH-237, and RH-160) are successful in detecting local electric field changes related to the Na$^+$/K$^+$-ATPase activity.[1,2] However, because of the amphiphilic structure and considerable dipole moment of the dye molecules (FIG. 1A), these dyes themselves can alter the electric potential drop at the membrane/water boundary. In attempting to establish whether such an effect exists, we found out that these probes created dipole potential drops, positive in the hydrophobic part of the membrane[4] (FIG. 1B). Dye adsorption led to a considerable increase in the rate constant of the transport of hydrophobic anions (dipicrylamine or tetraphenylborate) through the bilayer lipid membrane (BLM) and did not affect the partition coefficient between membrane and water. This indicates that the potential drop induced by the dye is located inside the membrane deeper than the adsorption plane of the hydrophobic ions (FIG. 1C). When the dye was added to the solution on one side of the membrane (intramembrane field compensation method[3]), the values of the boundary potential difference measured in asymmetric conditions were lower than those obtained under the symmetric condition (current relaxation method[3]). These results suggest that RH dye molecules penetrate the lipid bilayers.

As these dyes are employed as probes of local electric fields inside the membrane, it is necessary to assess the extent to which the electric field induced by oriented dye molecules can disturb the object under investigation. We estimated the effects of the electric field of the dye dipole layer on a dipole located in the same layer and on ion transport through a membrane protein like Na$^+$/K$^+$-ATPase.[4] The local electric field of each dye dipole decayed so rapidly in the lateral direction that a neighboring dye molecule did not feel it. We concluded that RH dyes had a minor effect on the electrogenic transport performed by the sodium pump (FIG. 1C) in the examined range of dye concentrations.

We checked our estimates of the dye's influence on ion transport by Na$^+$/K$^+$-ATPase in direct experiments in the model system consisting of a bilayer lipid membrane with adsorbed planar fragments of cellular membranes containing the purified enzyme.[5,6]

---

[a]This work was supported by ISSEP grants a796-x and a96-947.

[b]Address for correspondence: Dr. Valerij S. Sokolov, A. N. Frumkin Institute of Electrochemistry, Russian Academy of Sciences, Leninsky prospect 31, Moscow, 117071, Russia (tel: (095)-952-55-82; fax: (095)-952-55-82; e-mail: sokolov@bioel.glas.apc.org; dima@neurs.siobc.ras.ru).

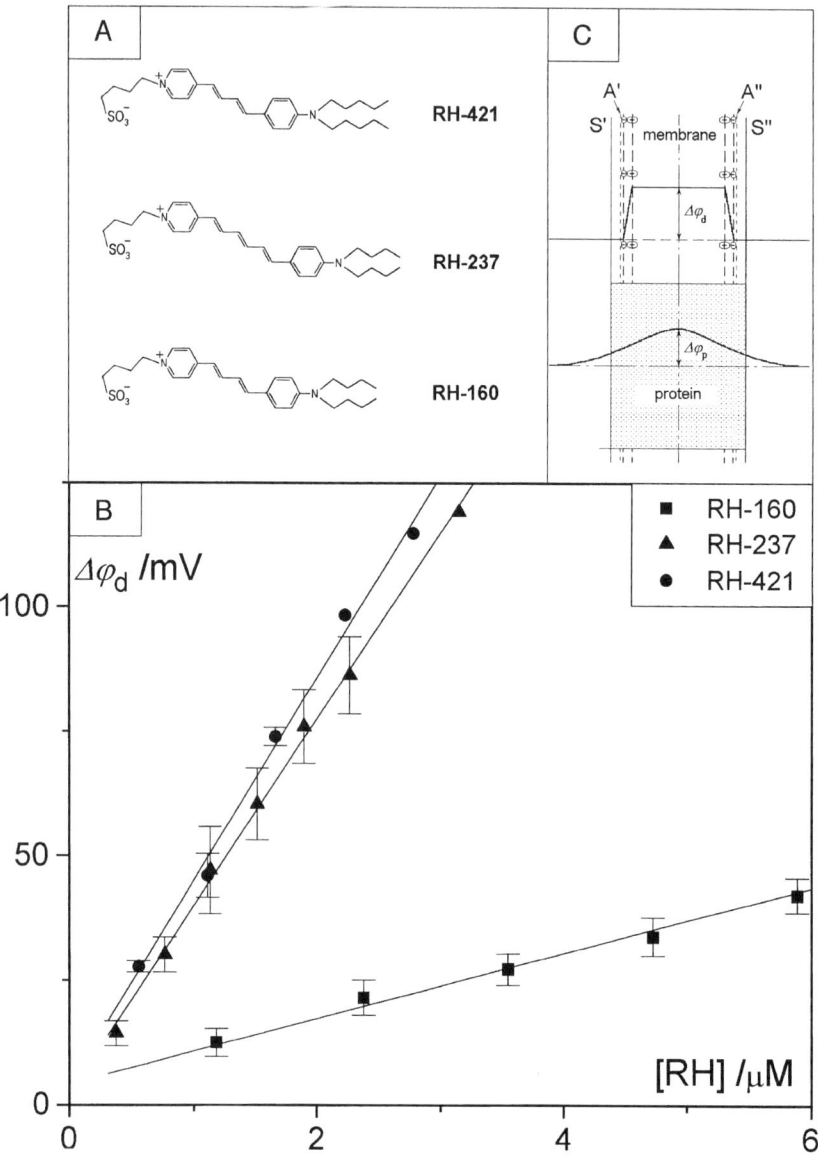

FIGURE 1. (**A**) Structures of potential-sensitive RH styryl dyes. (**B**) Dipole potential changes, $\Delta\varphi_d$, induced by adsorption of RH-160, RH-237, and RH-421 as a function of dye concentration. $\Delta\varphi_d$ was calculated from measurements of current relaxation in the presence of 1.25 µM tetraphenylborate under the addition of RH dyes on both sides of the membrane. The solution containing 100 mM KCl, 5 mM K_3PO_4, 5 mM Na citrate, 5 mM Tris + HCl, pH 7.7, was used. (**C**) Electrostatic model of electric field distribution induced by dipole dye molecules adsorbed on both sides of the membrane with a transmembrane protein molecule. $\Delta\varphi_d$ and $\Delta\varphi_p$ are the electric potentials; A′,A″ adsorption planes of hydrophobic ions; S′,S″ membrane surfaces. $\Delta\varphi_d/\Delta\varphi_p \approx 6$ in the case of Na^+/K^+-ATPase.[4]

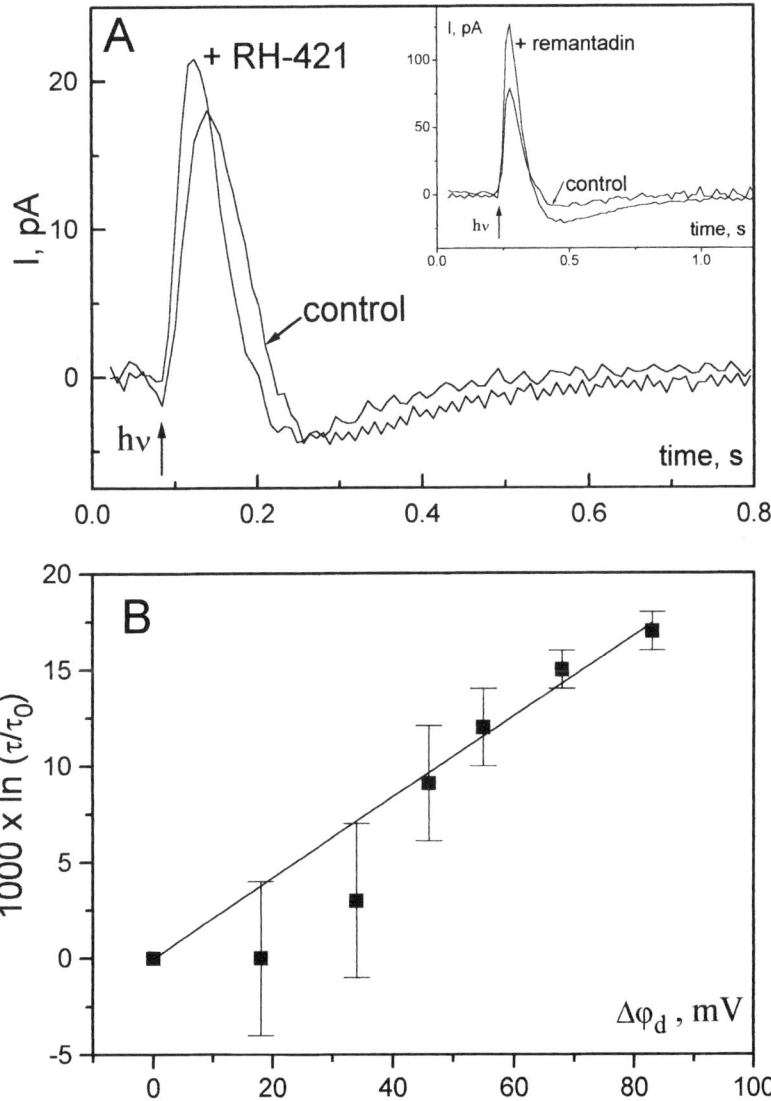

FIGURE 2. Influence of adsorption of RH-421 on charge transfer by Na^+/K^+-ATPase in the model system consisting of a bilayer lipid membrane with adsorbed planar fragments of cellular membranes containing the purified enzyme. **(A)** Current transients after fast photorelease of ATP from cage-ATP at the time indicated by the *arrow*. The aqueous solution contained (in mM): 130 NaCl, 10 $MgCl_2$, 30 imidazole, 1 EDTA, 1 dithiotritol, 100 µM caged ATP, pH 6.2. Concentrations of RH-421 and remantadin were 10 µM and 0.3 mM, respectively. **(B)** The relative rate constant of charge transfer by Na^+/K^+-ATPase as a function of dipole potential changes, $\Delta\varphi_d$, induced by adsorption of RH-421 (see FIG. 1B).

The addition of RH-421 in high concentration increased the amplitude of the short circuit current occurring due to the fast release of ATP. The kinetics of the increased amplitude correlated with the growth of potential during adsorption of RH-421.[4] The net transferred charge was not affected, but the time required to reach peak current notably decreased (FIG. 2A). Hence, the created dipole potential accelerated charge transfer in the sodium limb of the ATPase cycle. The maximal increment of the charge transfer rate constant was about 2–3% of the initial value and was achieved at RH-421 concentrations corresponding to the potential drop in the BLM of the order of 100 mV (FIG. 2B). Such weak dependence of the charge transfer rate constant on the potential confirmed the existence of considerable shielding of the potential within the protein. By contrast to dipole molecules of RH dyes, we used remantadin. The electrostatic potential induced by adsorption of this positively charged molecule is expected to be shielded much less. We observed that it also caused the amplitude of the current to increase (FIG. 2A, insert). However, in contrast to the dipole potential, it increased the transferred charge without having a measurable influence on the kinetics. Shielding factor was smaller than in the case of RH-421 and amounted to more than 0.5.

This difference in the effects of dipoles and charged molecules on ion transport was discovered earlier in investigations of ion channels. Our results therefore support the hypothesis that Na^+/K^+-ATPase has a channel-like structure.

REFERENCES

1. BUHLER, R. *et al.* 1991. J. Membr. Biol. **121:** 141.
2. FEDOSOVA, N. U. & I. KLODOS. 1994. *In* The Sodium Pump: Structure Mechanism. E. Bamberg & W. Schoner, Eds.: 561. Springer. New York.
3. SOKOLOV, V. S. *et al.* 1990. Bioelectrochem. Bioenerg. **23:** 27.
4. MALKOV, D. Y. & V. S. SOKOLOV. 1996. Biochim. Biophys. Acta **1278:** 197.
5. FENDLER, K. *et al.* 1985. EMBO J. **4:** 3079.
6. BORLINGHAUS, R. *et al.* 1987. J. Membr. Biol. **97:** 161.

Electrogenic Reactions of Na^+/K^+-ATPase Investigated on Solid Supported Membranes

J. PINTSCHOVIUS,[a] K. SEIFERT,[b] AND K. FENDLER[c]

Max-Planck-Institut für Biophysik
Kennedyallee 70
D-60596 Frankfurt/M, Germany

Charge translocation by the Na^+/K^+-ATPase has been measured using a novel technique that can generate concentration jumps by rapid solution exchange on solid supported membranes (SSM), exploiting their high mechanical stability. This method is particularly useful for substrates that are not available in "caged" form.[1] Concentration jumps of ATP and Na^+ were applied to obtain capacitive pump currents of the Na^+/K^+-ATPase with a time resolution of 10 ms.

MATERIALS AND METHODS

Adsorption of membrane fragments containing Na^+/K^+-ATPase to lipid membranes allows the detection of electrogenic events during the Na^+/K^+-ATPase reaction cycle with high sensitivity and time resolution.[2] Very high stability preparations can be obtained using SSM as a carrier for the membrane fragments. Preparation of the SSM and its layer structure (FIG. 1) was described previously.[3] This bilayer acts as a high capacitance carrier membrane for adsorbed membrane fragments containing pig kidney Na^+/K^+-ATPase.

A schematic illustration of the cuvettes is shown in FIGURE 1. An o-ring is sandwiched between the cuvette and the SSM glass plate to obtain a closed reaction volume (17 μl). The solution in the volume can rapidly be exchanged using an electrically switched valve at the inlet. In the outbound flow is an Ag/AgCl counter electrode separated from the cuvette solution by a gel salt bridge. Capacitive currents between the SSM electrode and the Ag/AgCl electrode are amplified (10^9 V/A), filtered (rise time <10 ms), and recorded on a digital oscilloscope.

RESULTS AND DISCUSSION

ATP concentration jumps could be performed by switching from an ATP-free solution to a solution containing ATP (data not shown). Also, Na^+ concentration jumps in the presence of ATP (no K^+ present) were used to initiate transient pump currents of the Na^+/K^+-ATPase (FIG. 2). These signals were similar to the ATP-

---

[a]Tel: 49-69-6303302; e-mail: pinti@biophys.mpg.de
[b]Tel: 49-69-6303306
[c]Tel: 49-69-6303306; e-mail: fendler@biophys.mpg.de

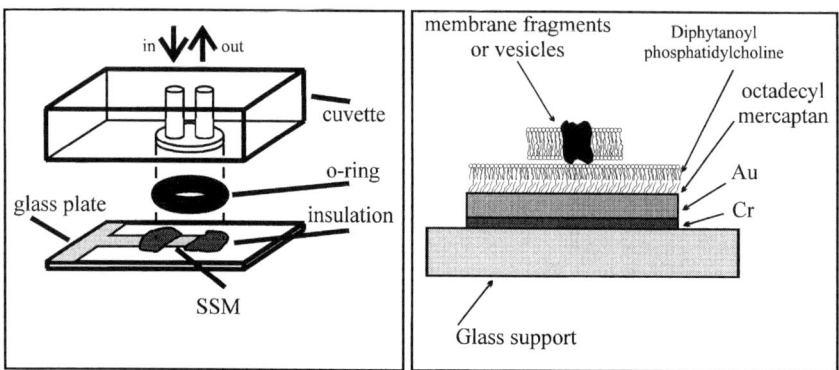

FIGURE 1. Schematic illustration of the cuvette and the layer structure of the solid supported membrane (SSM).

induced signals. Surprisingly, when no ATP was present, the rapid addition of Na^+ generated an electrogenic event (FIG. 2). Washing out the solution containing sodium produced an electric current in the reverse direction. Integration of on-peak and off-peak yields the same absolute values for the translocated charge. In these experiments, ionic strength was kept constant at 210 mM by adding an appropriate

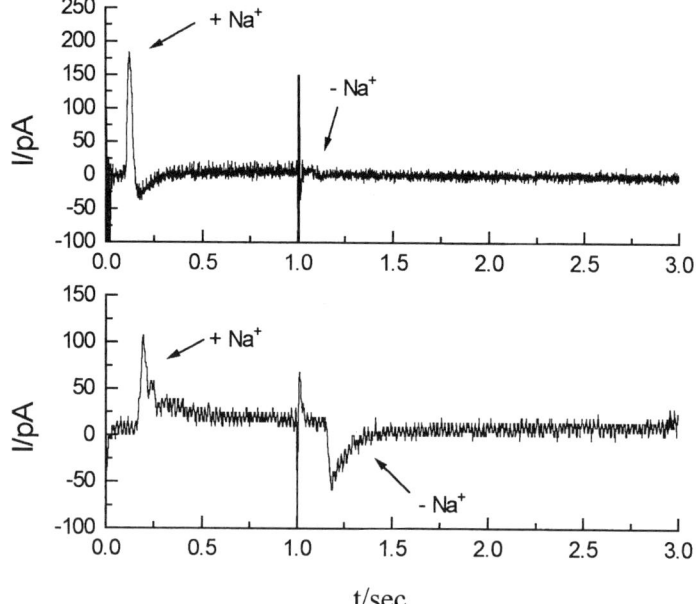

FIGURE 2. Na^+ concentration jump in the presence and absence of ATP. The peak at $t = 1$ s represents the switching of the valve.

amount of choline chloride to the solutions. The electrical signals are inhibited in the presence of 1 mM orthovanadate. The presence of 1 mM ouabain did not affect the signal, whereas membrane fragments preincubated with ouabain did not show a response to Na^+ concentration jumps. This demonstrates that the membrane fragments contributing to the charge translocation are all oriented with the extracellular surface facing the lipid layer.

Signals were obtained only with Na^+, not with K^+, Rb^+, or Cs^+. The half-saturation concentration of the translocated charge after an Na^+ concentration jump was 15 mM. The results are explained by electrogenic binding of Na^+ at the cytoplasmic surface of the Na^+/K^+-ATPase. Comparison of the charge measured in the presence and absence of ATP shows that the charge translocated during Na^+ binding is about 30% the charge translocated during Na^+-transport (assuming that they are transported over the same distance in both cases). These data represent direct evidence for electrogenic Na^+ binding at the intracellular binding sites of the Na^+/K^+-ATPase.

REFERENCES

1. FRIEDRICH, T., G. NAGEL, K. FENDLER & E. BAMBERG. 1997. Ann. N.Y. Acad. Sci., this volume.
2. BAMBERG, E., H.-J. BUTT, A. EISENRAUCH & K. FENDLER. 1993. Quart. Rev. Biophys. **26:** 1–25.
3. SEIFERT, K., K. FENDLER & E. BAMBERG. 1993. Biophys. J. **64:** 384–391.

Study of Electrogenic Transport of Sodium Ions Inside the Na,K-ATPase by Means of Membrane Capacitance Measurements

V. S. SOKOLOV,[a] S. M. STUKOLOV,[a] N. M. GEVONDYAN,[b] AND H.-J. APELL[c]

[a]*A. N. Frumkin Institute of Electrochemistry RAS*
117071 Moscow, Russia

[b]*M. M. Shemyakin and Yu. A. Ovchinnikov Institute*
of Bioorganic Chemistry RAS
117871 Moscow, Russia

[c]*Department of Biology*
University of Konstanz
D-78434 Konstanz, Germany

The aim of this work was to study electrogenic transport in a model system consisting of bilayer lipid membrane with adsorbed membrane fragments containing Na,K-ATPase in the Na-only mode by application of alternating voltage. A fast concentration step of ATP after its release from caged ATP gives rise to a transient electric

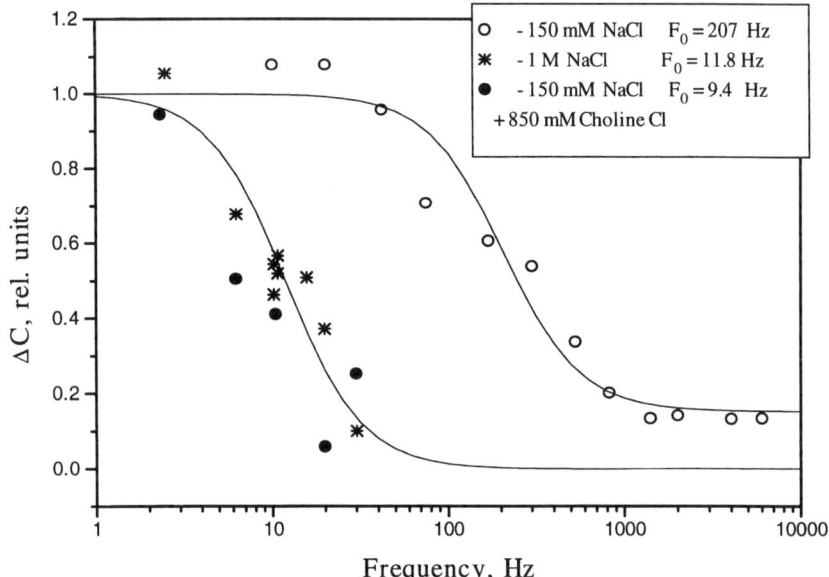

FIGURE 1. *See legend on facing page.*

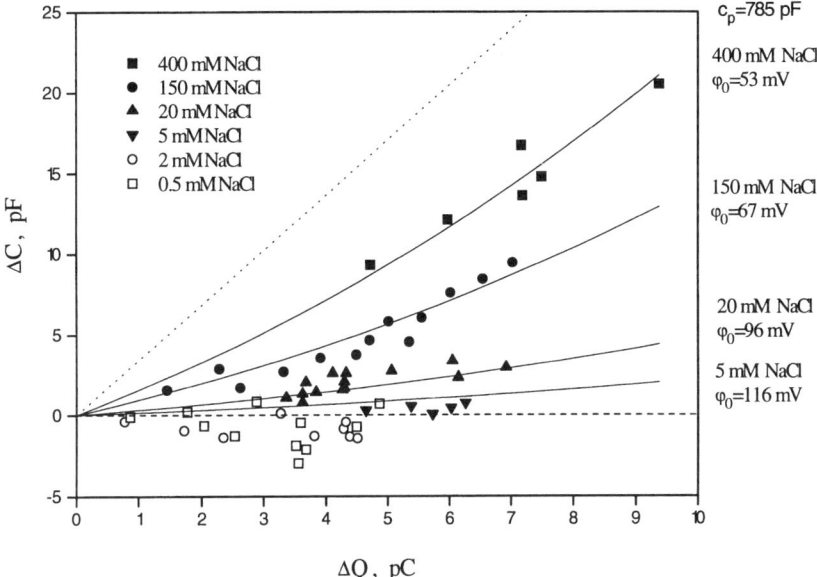

FIGURE 2. Dependence of capacitance increments, ΔC, on charge transferred across the membrane upon release of different ATP concentrations from 100 µM caged ATP induced by a UV light flash of varied intensity. Capacitance was determined at 15 Hz, at which frequency the change of capacitance still corresponds to the low-frequency maximum of FIGURE 1. The charge was determined by integration of the current measured in the absence of an applied voltage. Except for differences in NaCl concentration, all other conditions are the same as in FIGURE 1. Theoretical curves according to the model of ref. 4 were plotted with two free parameters: the (same) total capacitance of adsorbed membrane fragments C_p of 785 pF and the potential φ_0 at which the maximum of capacitance change is reached and which depends on the NaCl concentration (values shown in the figure).

current in this system.[1,2] We showed in previous studies that this process is coupled with a change in capacitance of the compound membrane system.[3] This effect can be explained by the process of sodium ion redistribution by the alternating electric field,[4] similar to that discovered by voltage-jump experiments.[5]

The capacitance increment decreased with higher frequencies of applied voltage. The frequency dependence could be described by a Lorenzian function with corner frequency F_0. This finding corresponds to a single charge-relaxation process with the rate constant $k = F_0$.[6,7] F_0 depends on the ionic strength of the solution. In standard

FIGURE 1. Dependence of capacitance increment, ΔC, after photorelease of ATP from 100 µM caged ATP in a model system consisting of a planar lipid bilayer with adsorbed membrane fragments containing Na,K-ATPase (from rabbit kidney) on the frequency of applied transmembrane voltage in buffers (30 mM imidazole, 10 mM $MgCl_2$, and 1 mM EDTA, pH 7.0) with different ion concentrations: 150 mM NaCl (*open circles*), 1 M NaCl (*filled circles*), and 150 mM NaCl + 850 mM choline Cl (*stars*). The curves are Lorenzian functions with different values of characteristic frequency F_0 shown in the right upper corner of the figure.

solution with 150 mM NaCl, F_0 was about 200 Hz; it decreased to ~10 Hz in solutions with 1 M ionic strength, achieved by the addition of either NaCl or choline chloride (FIG. 1). The decelerated charge relaxation could be due to the influence of high ionic strength on the rate-limiting conformational transition.

The value of the capacitance increment at the low frequency limit reflects the charge distribution close to equilibrium. Its dependence on applied voltage is equivalent to the dependence of charge transferred due to the applied voltage. It follows that the potential at which capacitance change reaches a maximum is equivalent to the midpoint potential φ_0 of the equilibrium charge distribution between two states.[5,8] To evaluate φ_0, the dependence of ΔC on the total charge, ΔQ, transported across the membrane after a light flash was measured at various flash energies (to alter the concentration of released ATP) and at a series of sodium concentrations (FIG. 2). Nonlinear dependencies of ΔC as a function of ΔQ can be explained by a voltage bias across the membrane fragments due to pumping.[4] The fitting of experimental data by theoretical curves allowed the determination of (1) the fraction of the bilayer area covered by membrane fragments with active Na,K-ATPase, which is in the range of 20% of the total membrane area, and (2) the values of φ_0, which depended on Na^+ ion concentration. The shift of φ_0 with Na ion concentration was explained by voltage-dependent binding of these ions due to their distribution in the narrow access channel at the extracellular side of the protein.[8] The dependence of φ_0 on sodium concentration in our work is weaker than that predicted by Rakowski[8] because of the limitation of our experimental system, in which the extracellular side of the protein is facing a small buffer volume in the gap between the lipid bilayer and membrane fragments. At a low concentration of Na^+ the capacitance increment became negative. This observation cannot be explained by a simple model[4] and forces us to suppose that in highly diluted solutions there is another charge relaxation process suppressed by ATP.

REFERENCES

1. FENDLER, K., E. GRELL, M. HAUBS & E. BAMBERG. 1985. EMBO J. **4:** 3079–3085.
2. BORLINGHAUS, R., H.-J. APELL & P. LAUGER. 1987. J. Membr. Biol. **97:** 161–178.
3. SOKOLOV, V. S., K. V. PAVLOV, K. N. DZHANDZHUGAZYAN & E. BAMBERG. 1992. Biol. Membr. **6:** 1263–1272.
4. SOKOLOV, V. S., S. M. STUKOLOV, A. S. DARMOSTUK & H.-J. APELL. 1997. Biol. Membr. **14:** 529–548.
5. NAKAO, M. & D. C. GADSBY. 1986. Nature **323:** 628–630.
6. SOKOLOV, V. S., K. V. PAVLOV & K. N. DZHANDZHUGAZYAN. 1994. *In* The Sodium Pump. E. Bamberg & W. Schoner, Eds.: 529–532. Springer. New York.
7. LU, C.-C., A. KABAKOV, V. S. MARKIN, S. MAGER, G. A. FRAZIER & D. W. HILGEMANN. 1995. Proc. Natl. Acad. Sci. USA **92:** 11220–11224.
8. RAKOWSKI, R. F. 1993. J. Gen. Physiol. **101:** 117–144.

Transposing Results from an Artificial Minicell to a Real Cell

Experimental Evidence for a Working Hypothesis Linking Na,K-ATPase Permeability States to Specific Alterations of Cell Life[a]

BEATRICE M. ANNER[b]

Department of Pharmacology
Geneva University Medical School, CMU
CH-1211 Geneva 4, Switzerland

The Na,K-ATPase is an ancestral and ubiquitous membrane transport system; its presence is indissociable from cell life because it establishes and maintains a steep reciprocal Na/K gradient across the cell membrane, a fundamental biological mechanism for biophysical, osmotic, and signaling purposes. In view of its importance, specific alterations in the functioning and structural integrity of the Na,K-ATPase will predictably be echoed by distinct changes of cell life.

Three different classes of compounds produce well-defined alterations in the Na,K-ATPase turnover: (1) the cardioactive steroids (e.g., ouabain) block ATPase activity with negatively cooperative dose-effect curves spanning several orders of magnitude,[1] (2) the highly toxic nonpeptidic and hydrophilic palytoxin from coral inhibits the Na,K-ATPase reversibly by a mechanism antagonized by cardioactive steroids,[2,3] and (3) mercury and silver inactivate it with narrow dose-effect curves.[4,5]

To assess the effects of these three different patterns of Na,K-ATPase inhibition on its permeability, the system must be inserted into tight artificial phospholipid vesicles (liposomes) where transmembrane gradients can be built up or dissipated. In a newly developed liposome system, artificial minicells, in which the Na,K-ATPase functions as in cells, the ouabain-blocked Na,K-ATPase expresses a conformation tight to Rb ions,[6] whereas the forms carrying palytoxin, mercury, or silver form Na,K-ATPase-mediated leakage pathways across the otherwise tight phospholipid bilayer.[3,5,7] The same inhibitors were then applied to isolated peripheral human lymphocytes. The present work constitutes a synthesis of the data obtained with isolated, reconstituted, or cellular Na,K-ATPase.

RESULTS

Ouabain Keeps the Pump Tight and Preserves Lymphocyte Life

Artificial minicells consist of predominantly single-walled liposomes containing purified Na,K-ATPase molecules inserted randomly in their membranes (FIG. 1A);

---

[a]This study was supported by Swiss National Science Foundation grants 31-30317.90 and 31-37552.93

[b]Tel: 41 22 702 54 61; fax: 41 22 70254 52; e-mail: beatrice.anner@medecine.unige.ch

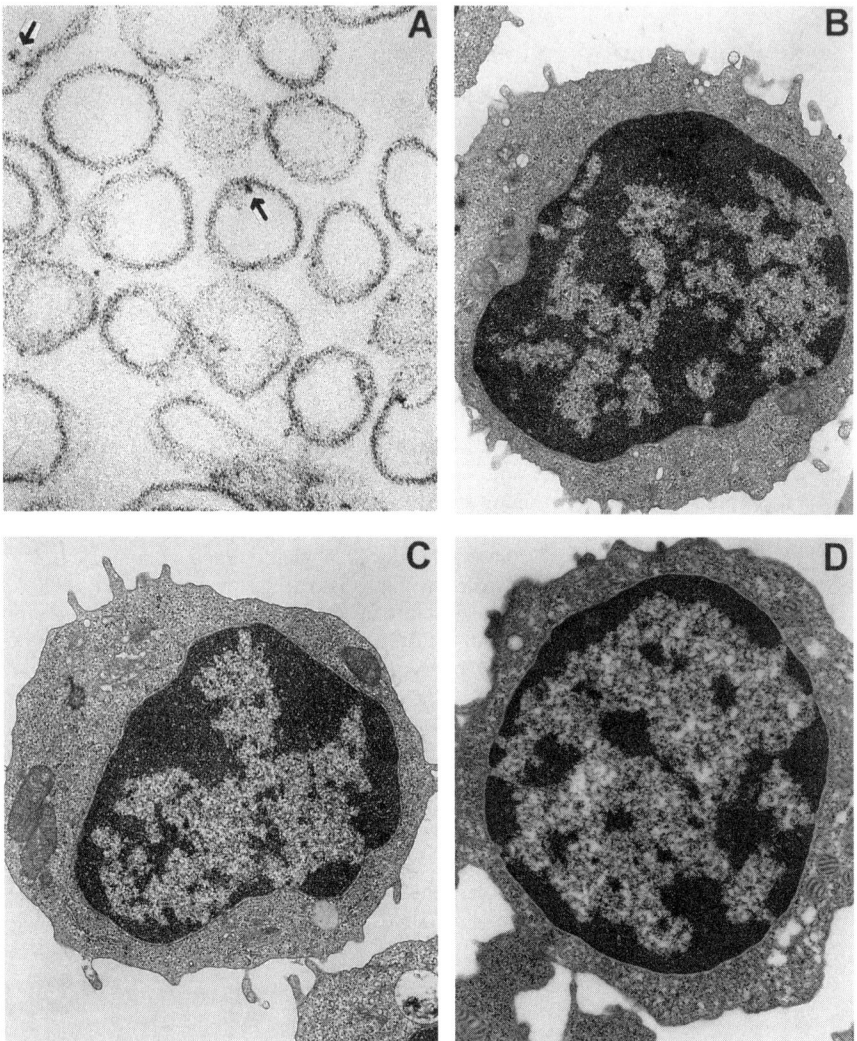

FIGURE 1. Na,K-ATPase reconstituted with egg-phosphatidylcholine by the cholate-dialysis procedure forms a homogeneous population of predominantly single-walled vesicles containing noninteracting randomly oriented[14] Na,K-ATPase molecules (**A**, *arrows*). By including a reservoir of ATP molecules and Na ions during dialysis, artificial minicells are formed in which the Na,K-ATPase functions as in cells. Peripheral human lymphocytes were isolated and incubated without (**B**) or with 1 mM ouabain for 3 hours (**C**) or with 3 nM palytoxin for 15 minutes (**D**) and processed for electron microscopy as described.[11,12,16] A cell representative for each incubation condition is shown.

they contain a reservoir of ATP and Na for activating the right-side-oriented Na,K-ATPase upon external Rb or K addition.[8] When ouabain is added to artificial minicells which are in the process of Rb accumulation, the uptake is interrupted[1] and the Rb ions remain entrapped throughout a 15-minute gel-filtration step despite an inside-out Rb gradient,[8] indicating that no Rb ions leak across the E_2-ouabain form. The gel filtration procedure used for removal of external isotope is a sensitive indicator of pump tightness.[9] Ouabain added before Rb fully prevents Rb accumulation by the artificial minicells[8] with dose-effect curves corresponding to the unreconstituted form.[1,4,10]

In analogy to artificial minicells, ouabain inhibits entirely the [^{86}Rb]K uptake by isolated human lymphocytes in nanomolar concentrations (TABLE 1). Yet, despite the potent inhibition of active cellular K uptake, resulting in increased intracellular Na and decreased K concentrations, survival of the cells is not affected as assessed by exclusion of ethidium homodimer or trypan blue[11] or by uptake and intracellular hydrolysis of calcein.[12] In conformity to the viability measurements, the morphology of the ouabain-treated cells is unaltered[13] (FIG. 1C) compared to that of control cells (FIG. 1B).

Palytoxin Induces Pump-Leak Conversion and Lymphocyte Death

In sharp contrast to ouabain, palytoxin opens a leakage pathway in the Na,K-ATPase molecule of artificial minicells membranes, whereas Na,K-ATPase-free phospholipid bilayers remain tight in the presence of palytoxin.[3] Ouabain antagonizes the palytoxin-induced leak.[3] Conversely, ouabain is poorly effective when the artificial minicells contain the ouabain-resistant renal α1 Na,K-ATPase, supporting the concept that specific ouabain binding and not some unspecified palytoxin-ouabain interaction underlies the ouabain-antagonized palytoxin action.[3]

Palytoxin applied to isolated human lymphocytes causes rapid cell swelling and bursting.[11] The fine structure of the cell gave away to a progressively amorphous aspect (FIG. 1D). In peripheral lymphocytes, ouabain antagonizes the pore-forming effect of palytoxin[11] as in artificial minicells[3] and in other cell systems.[2]

TABLE 1. Na,K-ATPase in Artificial Minicells and Human Lymphocytes

	Artificial Minicells	Human Lymphocytes
Average diameter	100 nm	10 μm
Surface	3.14×10^{-14} m^2	3.14×10^{-10} m^2
Na,K-ATPase molecules, average number[14,15]	10	30,000
Na,K-ATPase molecules, n/μm^2	320	95
Ouabain-sensitive ^{86}Rb-uptake	Yes[1,8]	Yes[11,12]
Ouabain sensitivity (IC$_{50}$)	50 nM–100 μM[1,4,10]	25 nM[16]
Ouabain-palytoxin antagonism	Yes[3]	Yes[11,12]
Na,K-ATPase-ouabain form	Tight[6]	Cell survival[11,12]
Na,K-ATPase-palytoxin form	Leaky[3]	Cell death[11]
Na,K-ATPase-mercury form	Leaky[7]	Cell death[13]
Na,K-ATPase-silver form	Leaky[5]	Cell death[13]

SUMMARY AND CONCLUSION

The present work sets in parallel data obtained with the Na,K-ATPase in artificial vesicular membranes (artificial minicells) and in peripheral human lymphocytes *ex vivo*. The Na,K-ATPase was purified and reconstituted into single-walled, tight liposomes filled with a reservoir of ATP and Na in which the Na,K-ATPase functions as in cells, that is, the receptor is accessible on the liposome surface (artificial minicells). In this system, an E_2-ouabain state impermeable to Rb ions and an Na,K-ATPase–palytoxin state leaky for Rb ions were characterized. The tight E_2-ouabain form preserves the viability of isolated lymphocytes, whereas the leaky Na,K-ATPase–palytoxin induces rapid cell bursting and death by a ouabain-sensitive mechanism. Thus, the effect of Na,K-ATPase inhibitors on lymphocyte survival can be predicted from permeability measurements in artificial minicells as verified also with the leak-inducing metals mercury[7,13] and silver.[5,13]

REFERENCES

1. ANNER, B. M., H. G. REY, M. MOOSMAYER, I. MESZOELY & G. T. HAUPERT, JR. 1990. Hypothalamic Na,K-ATPase inhibitor characterized in two-sided liposomes containing pure renal Na,K-ATPase. Am. J. Physiol. **258:** F144–F153.
2. HABERMANN, E. 1989. Palytoxin acts through Na^+,K^+-ATPase. Toxicon **27:** 1171–1187.
3. ANNER, B. M. & M. MOOSMAYER. 1994. Na,K-ATPase characterised in artificial membranes. 2. Successive measurement of ATP-driven Rb-accumulation, ouabain-blocked Rb-flux and palytoxin-induced Rb-efflux. Mol. Membr. Biol. **11:** 247–254.
4. ANNER, B. M., M. MOOSMAYER & E. IMESCH. 1992. Mercury blocks Na,K-ATPase by a ligand-dependent and reversible mechanism. Am. J. Physiol. **262:** F830–F836.
5. HUSSAIN, S., E. MENEGHINI, M. MOOSMAYER, D. LACOTTE & B. M. ANNER. 1994. Potent and reversible interaction of silver with pure Na,K-ATPase and Na,K-ATPase-liposomes. Biochim. Biophys. Acta **1190:** 402–408.
6. ANNER, B. M., M. MOOSMAYER & E. IMESCH. 1994. Na,K-ATPase characterised in artifical membranes. 1. Predominant conformations and ion fluxes associated with active and inhibited states. Mol. Membr. Biol. **11:** 237–245.
7. IMESCH, E. & B. M. ANNER. 1992. Mercury weakens the membrane anchoring of Na,K-ATPase. Am. J. Physiol. **262:** F830–F836.
8. REY, H. G., M. MOOSMAYER & B. M. ANNER. 1987. Characterization of Na,K-ATPase-liposomes. III. Controlled activation and inhibition of symmetric pumps by timed asymmetric ATP, RbCl and cardiac glycoside addition. Biochim. Biophys. Acta **900:** 27–37.
9. ANNER, B. M. 1983. Hypothesis: A bar model for the pump and channel function of the reconstituted Na,K-ATPase. 1983. FEBS Lett. **158:** 7–11.
10. ANNER, B. M., E. IMESCH & M. MOOSMAYER. 1991. Defective Na transport of murine ouabain-resistant alpha-1 Na,K-ATPase. Soc. Gen. Physiol. **46/2:** 479–482.
11. FALCIOLA, J., B. VOLET, R. M. ANNER, M. MOOSMAYER, D. LACOTTE & B. M. ANNER. 1994. Role of cell membrane Na,K-ATPase for survival of human lymphocytes in vitro. Biosci. Rep. **14:** 189–204.
12. FALCIOLA, J. C. 1995. Survie de Lymphocytes Isolés du Sang Humain: Rôle de la Na,K-ATPase. M.D. thesis, Geneva University, Geneva, Switzerland.
13. HUSSAIN, S., B. VOLET, C. BURRUS, R. ANNER, D. LACOTTE, M. MOOSMAYER & B. M. ANNER. 1995. Uptake of silver by isolated human lymphocytes in presence of L-cysteine or *N*-acetyl-cysteine. In Vitro Toxicol. **8:** 377–388.
14. ANNER, B. M., H. P. TING-BEALL & J. D. ROBERTSON. 1984. Characterization of (Na + K)-ATPase liposomes. I. Effect of enzyme concentration and modification on liposome

size, intramembrane particle formation and Na,K-transport. Biochim. Biophys. Acta **773:** 253–261.
15. BERNTOP, E., K. BERNTOP, D. NELSON & N. HENNINGSEN. 1985. [^{3}H]Ouabain binding and sodium content in lymphocyte sub-populations and the demonstration of increased binding in type 2 diabetes mellitus. Scand. J. Clin. Invest. **45:** 27–36.
16. ANNER, B. M., D. LACOTTE, R. M. ANNER & M. MOOSMAYER. 1994. Interaction of hypothalamic Na,K-ATPase inhibitor with isolated human peripheral blood mononuclear cells. Biosci. Rep. **14:** 231–242.

Transport Activity of a Chimeric Na$^+$,K$^+$-ATPase with Ca^{2+}/Calmodulin Binding Domain from Ca^{2+}-ATPase in *Xenopus* Oocytes[a]

JIANXING ZHAO,[b] LARISA A. VASILETS,[c] QUANBAO GU,[b]
TOSHIAKI ISHII,[d] KUNIO TAKEYASU,[d]
AND WOLFGANG SCHWARZ[e]

Max-Planck-Institut für Biophysik
60596 Frankfurt/M, Germany

Chimeric ATPases formed from the α1 subunit of chicken Na$^+$/K$^+$-ATPase with COOH-terminal 165 amino acids of plasma membrane Ca^{2+}-ATPase-II of rat brain (NNN-CBS) were expressed in *Xenopus* oocytes together with the β subunit of the Na$^+$ pump of *Torpedo* electroplax; the COOH-terminal region includes the calmodulin binding site. It was demonstrated previously[1] that this chimera shows Na$^+$,K$^+$-dependent ATPase activity but only in the presence of Ca^{2+}/calmodulin, and this activity is inhibited by ouabain. This was taken as additional evidence that the COOH terminus with the calmodulin binding site interacts with regions that are conserved in the Ca^{2+}- and Na$^+$,K$^+$-ATPases; the interaction blocks ATPase activity and is released by Ca^{2+}/calmodulin. To further characterize function of the chimera we measured (^{3}H)ouabain binding to determine the number of pump molecules in the plasma membrane and measured pump-mediated ^{86}Rb uptake, ^{22}Na efflux, and current. All determinations of transport activity were performed under conditions of maximized turnover. (For details see ref. 5.)

FIGURE 1 summarizes measurements of (^{3}H)ouabain binding and of ^{86}Rb uptake on the same batches of oocytes. The increased binding of ouabain demonstrates that the chimeric ATPase is incorporated into the membrane of the oocyte to about the same extent as is the wild-type *Torpedo* pump. These additional ATPases are functioning as pumps by transporting Rb$^+$ as K$^+$ congener into the cell. For a Na$^+$,K$^+$-transporting ATPase, in addition to the uptake of Rb$^+$, the transport of Na$^+$ out of the cell has to be demonstrated. This can be monitored directly if oocytes are injected with ^{22}Na$^+$ and

---

[a]This work was supported by Deutsche Forschungsgemeinschaft (SFB 169 to W.S.) and National Institute of Health (GM44373 to K.T.).
[b]Present address: Shanghai Institute of Cell Biology, Chinese Academy of Science, 200031 Shanghai, China.
[c]Present address: Institute of Chemical Physics, Russian Academy of Science, 142 432 Chernogolovka, Russia.
[d]Present address: Ohio State Biotechnology Center, Ohio State University, Columbus, OH 43210-1002.
[e]Address for correspondence: W. Schwarz, Max-Planck-Institut für Biophysik, Kennedyallee 70, D-60596 Frankfurt/Main, Germany (tel: (+49) 69-6303 339; fax: (+49) 69-6303 340; e-mail: 100711.750@compuserve.com).

the content of radioactivity in the oocyte is continuously recorded during superfusion with radioactivity-free solution (FIG. 2). The pen recording shows that application of 5 mM K^+ to the superfusion medium activates the release of Na^+ and outward current; both signals can be inhibited by the addition of ouabain. In oocytes not injected with RNA a current signal of similar magnitude can be induced by K^+, but the corresponding loss of $^{22}Na^+$ is much smaller and can hardly be detected within the employed time intervals. (See also ref. 3.) Therefore, the detectable outward transport of $^{22}Na^+$ seems to be predominantly mediated by the newly expressed chimeric pumps. This finding is

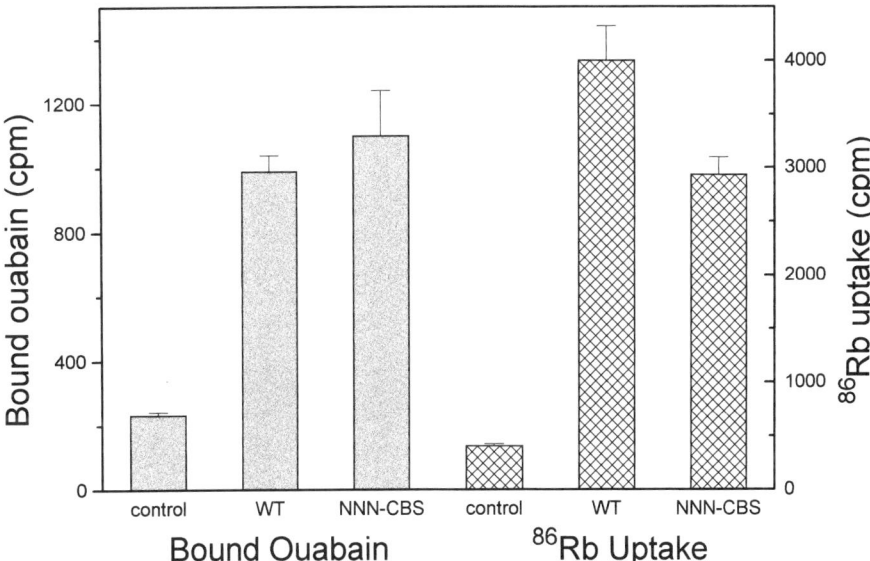

FIGURE 1. Ouabain-binding and ^{86}Rb-uptake experiments (for details see ref. 2) for noninjected control oocytes, for oocytes injected with cRNA for wild-type α subunit and β subunit of the pump of *Torpedo* (WT), and for oocytes injected with cRNA for the chimeric NNN-CBS α subunit and the β subunit of *Torpedo*. Experiments were performed on the same batches of oocytes. Before measurements, oocytes were loaded with Na^+ by incubating in K^+-free solution. For Rb uptake, oocytes were incubated for 12 minutes in Na^+-free solution in the presence of 5 mM Rb^+ (^{86}Rb as tracer), and pump-specific contribution was determined as ouabain-sensitive uptake. For ouabain binding; oocytes were incubated for 20 minutes in K^+-free solution in the presence of 5 μM ouabain (2.5 μM ^3H-ouabain).

supported by the fact that microinjection of calmodulin leads to further stimulation of Na^+ efflux, suggesting that the oocytes contain endogenous calmodulin at a nonsaturating concentration. Interestingly, whereas injection of calmodulin increases ^{22}Na efflux by a factor of two, pump current increases only by a factor of about 1.2. Because pump current reflects the extra charges transported out of the cell compared to the ions transported across the cell membrane, comparison of Na^+ efflux and current allows an estimate of the Na^+/K^+ stoichiometry. Whereas for the *Torpedo* pump the ratio of the number of Na^+ ions coming out of the cell to the net charges flowing outward is close

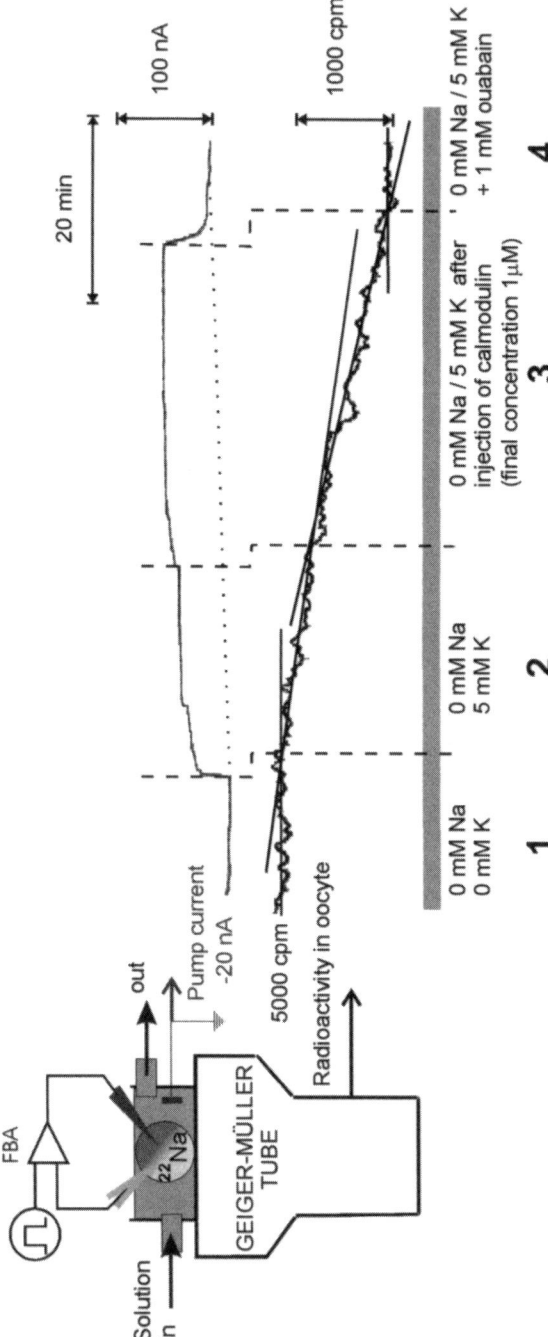

FIGURE 2. Simultaneous measurements of ^{22}Na efflux and pump current. An oocyte injected with ^{22}Na$^+$ is placed into a chamber, the bottom of which is formed by the mica window of a Geiger-Müller tube (*right part*). The decline of radioactivity in the oocyte was recorded on a pen recorder (*left part, lower trace*) and was fitted piecewise by an exponential: $y = y_0 \cdot e^{-k \cdot t}$. The fitted k values are proportional to the efflux; the respective rates are (in 10^{-6} s^{-1}) $k_1 \approx k_4 \approx 0$, $k_2 = 60$, and $k_3 = 92$. Simultanously the respective membrane current (*upper trace*) was recorded at a fixed holding potential ($V_H = -60$ mV).

to $n_e = 3$,[3,5] for the chimeric pump the electrogenicity is much reduced. From an average rate of efflux of $(106 \pm 25)\ 10^{-6}\ s^{-1}$ and with an intracellular concentration of free Na^+ of 80 mM,[2] a total efflux equivalent to 407 ± 95 nA can be calculated and has to be compared with an electrogenic pump current of 51 ± 4 nA determined in the same experiments. Because this current is, to a large extent, mediated by nonchimeric pumps, the ratio of efflux to current yields a lower limit for the ion-to-charge ratio of $n_e \geq 8$.

In conclusion, the chimeric NNN-CBS ATPase is incorporated into the oocyte membrane as an ouabain- and calmodulin-sensitive ATPase transporting Na^+ out of and K^+ (Rb^+) into the cell. Although none of the amino acids of the α1 subunit of the Na^+/K^+-ATPase was altered, the addition of the calmodulin binding site to the COOH-terminus seems to have modified Na^+,K^+ stoichiometry, leading to nearly electrically silent transport.

ACKNOWLEDGMENTS

We would like to thank Heike Biehl and Heike Fotis for their technical assistance.

REFERENCES

1. ISHII, T. & K. TAKEYASU. 1995. EMBO J. **14:** 58–67.
2. SCHMALZING, G., H. OMAY, S. KRÖNER, S. GLOOR, H. APPELHANS & W. SCHWARZ. 1991. Biochem. J. **279:** 329–336.
3. SCHWARZ, W. & Q.-B. GU. 1988. Biochim. Biophys. Acta **945:** 167–174.
4. VASILETS, L. A. & W. SCHWARZ. 1992. J. Membr. Biol. **125:** 119–132.

Solute Effects on the Sodium Pump

An Evaluation of the Osmotic Dehydration Hypothesis

PETER R. ARULANANTHAM, ZACHARY V. EDMONDS, AND
R. WAYNE ALBERS

Laboratory of Neurochemistry
NINDS
National Institutes of Health
Bethesda, Maryland 20892

Many small, neutral, polar solutes can act on P-type cation pumps to stabilize catalytic-site aspartyl phosphorylation by P_1.[1] In the case of the Na^+ pump, this effect is accompanied by ATPase inhibition and enhancement of alkaline K^+-*p*-nitrophenylphosphatase activity. We investigated the effects of pH and solute size on these three parameters as manifest by the electrophorus sodium pump. In the series of mono- to penta-ethylene glycols, solute effectiveness in all three assays increases with size up to tetraethylene glycol. FIGURE 1 shows ATPase and NPPase data.

K^+-NPPase, measured in dilute aqueous buffers, consists of a neutral component and a minor alkaline component. Neutral polar solute activation of K^+-NPPase consists of marked enhancement of the alkaline component activity with little effect on the neutral component. The solute-induced alkaline component can be selectively eliminated by low concentrations of dodecylmaltoside (FIG. 2).

These data are consistent with the following: (1) neutral polar solutes that are

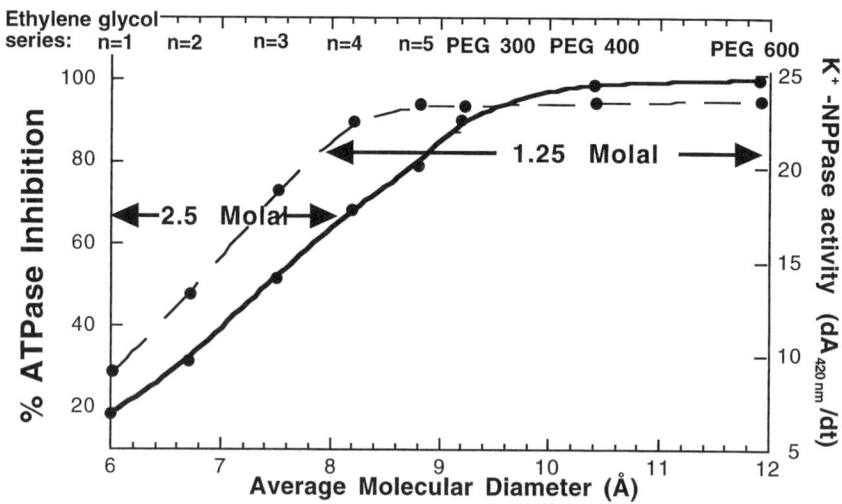

FIGURE 1. ATPase and NPPase data.

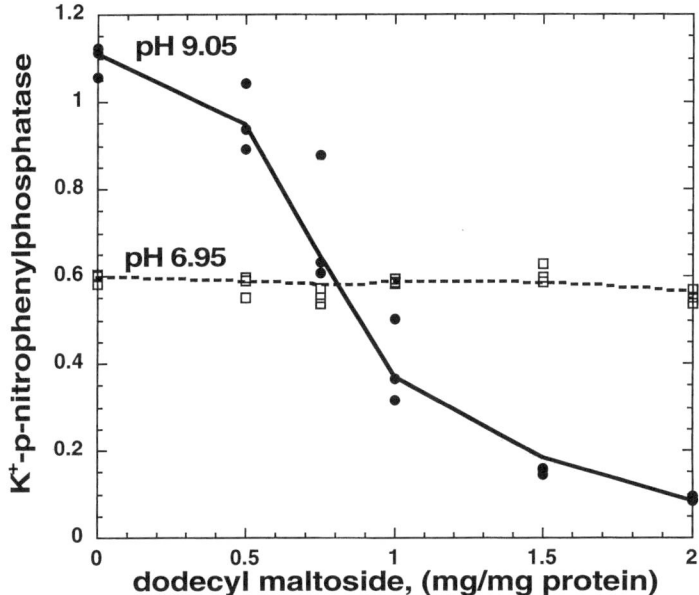

FIGURE 2. Elimination selectively of solute-induced alkaline component by low concentrations of dodecylmaltoside.

excluded from critical intramolecular pump spaces will osmotically dehydrate these spaces[2]; (2) dehydration stabilizes a P_i-reactive E_2 conformational state; and (3) the induced alkaline NPPase is a property of this state. Concomitant ATPase inhibition follows from the Albers-Post model. Suppression of solute-induced K^+-NPPase by neutral detergents, which also dissociate diprotomers, suggests that interfacial dehydration of promoters may be involved in these solute effects.

Stabilization of the E_1 conformation by chaotropic anions[3] appears to be a related phenomenon because we have shown that: (1) solute activation of alkaline K^+-NPPase is antagonized by chaotropic anions and (2) stabilization of E_1P by, for example, nitrate, is antagonized by DMSO.

REFERENCES

1. DE MEIS *et al.* 1980. Biochemistry **19**: 42–52.
2. LEIKIN *et al.* 1993. Annu. Rev. Phys. Chem.
3. KLODOS *et al.* 1994. J. Biol. Chem. **269**: 1734.

ADP Dephosphorylation of the E_1P Form of the Na^+,K^+-ATPase[a]

MARTA CAMPOS AND LUIS BEAUGÉ

Instituto M. & M. Ferreyra
C.C. 389
5000 Córdoba, Argentina

The reverse reaction of the Na^+,K^+-ATPase leads to ATP synthesis. Although Mg^{2+} ions are essential for ATP-ADP exchange, concentrations higher than 25 µM become inhibitory.[1] The mechanism of that inhibition is not known, but one possibility is that the true substrate is free ADP. This is consistent with the facts that E_1P has Mg^{2+} bound to it and that this Mg^{2+} is not released until dephosphorylation takes place.[2] This and other aspects of the ADP-stimulated dephosphorylation of E_1P, like substrate specificity, are described in this work. To simplify analysis of the data we performed the experiments in enzyme partially digested with chymotrypsin. Under this condition the whole reaction cycle is reduced to the following scheme:

$$E_1 \leftarrow (f_1 \text{ ATP}, b_1) \rightarrow E_1 \text{ ATP} \leftarrow (f_2, b_2) \rightarrow E_1P \leftarrow (f_3) \rightarrow E_1$$

where all unidirectional rate constants can be independently estimated.[3]

METHODS

Pig kidney Na^+,K^+-ATPase was partially purified as previously described.[4] Chymotryptic digestion was carried out.[5] The experiments were performed with an Intermekron DSM 3-SF Rapid Mixing apparatus[3] at 20–22°C. NaCl (100 mM) and imidazole-HCl (70 mM; pH 7.4 at 20°C) were always present. ATP-ADP exchange was performed as in ref. 6. Concentrations of free and Mg-complexed nucleotides were calculated from the K_{DS} for MgATP (86 µM) and MgADP (621 µM) estimated with Arsenazo III.[3] Curve fitting and simulations were performed with the Scop program (National Biomedical Simulation Resource, Berrien Spring, Michigan).

RESULTS AND DISCUSSION

Unless f_2 is zero or b_1 very large, cases in which a single exponential is seen, the scheme just given predicts a biphasic ADP-dependent dephosphorylation in a semilogarithmic plot. The initial fast phase is governed by b_2, whereas the slow component is related to b_1. This biphasic behavior was experimentally verified. Because there is no way to estimate the "on" and "off" rates for ADP binding to E_1P, we took $b_2 = b_{2mx}$.

---

[a]This work was supported by CONICET PID-BID 1053, CONICOR, and Volkswagen Stiftung I/68 788.

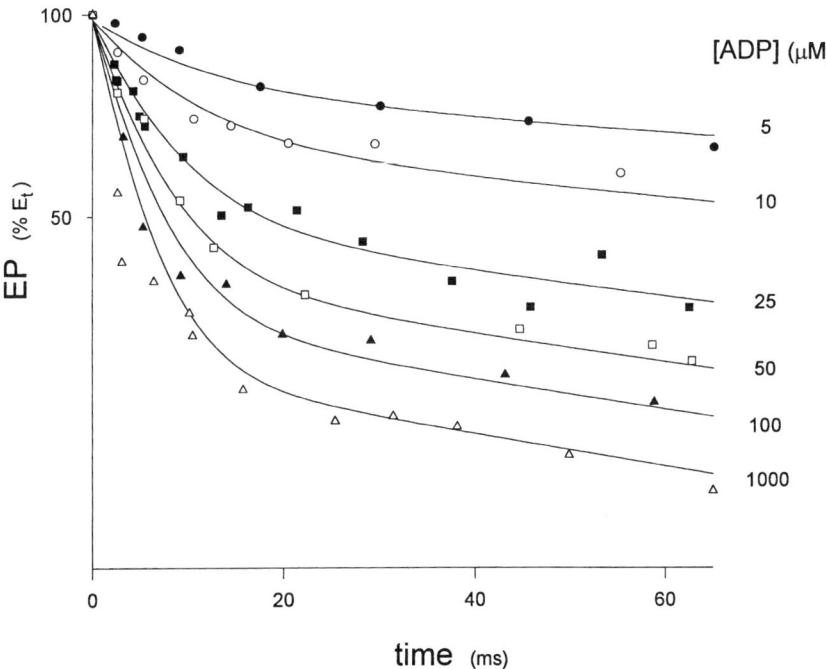

FIGURE 1. Dephosphorylation of E_1P in chymotrypsin-digested Na^+,K^+-ATPase at several [ADP] in the absence of free Mg^{2+}. Lines correspond to a simultaneous best fit of all data points to the equation resulting from the model depicted in the text. The fitting parameters were b_{2max} and K_{mADP}; the other unidirectional rate constants were independently determined and are also listed in TABLE 1. Note the biphasic response to ADP.

$[ADP]/(K_{mADP} + [ADP])$. The unidirectional rate constants other than b_{2max} were determined[3] at 0, 2 mM, and 5 mM free $[Mg^{2+}]$. The b_{2max} and K_{mADP} values for the different $[Mg^{2+}]$ were determined from the simultaneous fit of data such as those in FIGURE 1. The effects of ADP analogs were studied only in the absence of free Mg^{2+}. All kinetics parameters for ADP are displayed in TABLE 1.

TABLE 1. Kinetic Constants for Phosphorylation and Dephosphorylation of Chymotrypsin-Treated Na^+,K^+-ATPase

	Concentration of Ionized Magnesium		
Constant and Units	0 Mg^{2+}	2 mM Mg^{2+}	5 mM Mg^{2+}
f_1 ms^{-1} μM^{-1}	0.0248 ± 0.0010	0.0264 ± 0.0012	0.0151 ± 0.0012
b_1 ms^{-1}	0.0075 ± 0.0018	0.0069 ± 0.0019	0.0088 ± 0.0029
f_2 ms^{-1}	0.0593 ± 0.0022	0.0620 ± 0.0017	0.0684 ± 0.0009
f_3 ms^{-1}	0.0006 ± 0.0003	0.0013 ± 0.0004	0.0007 ± 0.0001
b_2 max ADP ms^{-1}	0.1660 ± 0.0080	0.1910 ± 0.0050	0.2170 ± 0.0040
K_m ADP_{free} μM	32.0 ± 0.80	44.0 ± 2.00	76.0 ± 3.00
$K_m ADP_{total}$ μM	32.0 ± 0.80	237 ± 2.0	780 ± 5.0

ADP stimulated along a hyperbolic function. The K_m for total ADP increased 20-fold, whereas that for free ADP only doubled when [Mg^{2+}] was raised from 0–5 mM. The b_{2max} values increased 33%. Thus, Mg^{2+} inhibiton seems to occur mainly by a reduction in free [ADP]. However, we cannot rule out the presence of a Mg^{2+} inhibitory site in the native enzyme which was removed by chymotrypsin. The K_m for free-ADP stimulation of ATP-ADP exchange was 104 µM; this is in the range of, although somehow higher than, that for dephosphorylation of E_1P. Therefore, from the foregoing, we can conclude that the true substrate for ATP synthesis in the Na^+,K^+-ATPase cycle is free ADP.

These experiments also show that the nucleotide structure is crucial for dephosphorylation of E_1P. All nucleotides tested are much less effective, although their K_m and b_{2max} values do not always follow the same sequence as do those of ADP. For K_m the sequence is ADP ≪ CDP < UDP < GDP, whereas for b_{2max} it is ADP ≫ UDP > GDP > CDP. Finally, the dephosphorylation rates obtained in the presence of ADP plus other nucleotides diphosphate concur with those anticipated on the basis of simple competition for the E_1P site.

REFERENCES

1. BEAUGÉ, L. & I. M. GLYNN. 1979. J. Physiol. **289:** 17–31.
2. CAMPOS, M. & L. BEAUGÉ. 1988. Biochim. Biophys. Acta **944:** 242–248.
3. CAMPOS, M. & L. BEAUGÉ. 1992. Biochim. Biophys. Acta **1105:** 51–60.
4. JORGENSEN, P. 1974. Biochim. Biophys. Acta **256:** 36–52.
5. JORGENSEN, P. & J. PETERSEN. 1985. Biochim. Biophys. Acta **821:** 319–333.
6. BEAUGÉ, L. & M. CAMPOS. 1986. J. Physiol. **375:** 1–25.

K⁺ Induces an Acid-Labile Phosphoenzyme (*or* an Occluded P_i Form) in Na,K-ATPase[a]

J. D. CAVIERES, E. BUXBAUM, D. G. WARD, AND T. J. H. WALTON

Transport ATPase Laboratory
Department of Cell Physiology & Pharmacology
Leicester University
P.O. Box 138
Leicester LE1 9HN, UK

Both steady-state and transient kinetic studies of the ATPase activity of the sodium pump have pointed to quantitative difficulties with the role of the acid-stable phosphoenzyme as a reaction intermediate.[1,2] There is agreement that E_1P seems "kinetically competent" as an intermediary during ATP hydrolysis in K^+-free conditions. In the presence of K ions, however, the result of multiplying steady-state E_1P levels and the rate constant for E_1P decay work out to be too low by a factor of between 2 and 10 when compared to steady-state measurements of Na,K-ATPase activity.[2] Evidently, if E_1P is not turning over fast enough, it cannot be an intermediary for the hydrolysis reaction in these conditions.

We speculated whether either factor in this product could have been systematically underestimated. In the first analysis, it seemed unlikely that the values for the dephosphorylation rate constants could be too far out. On the other hand, the steady-state total E-P level was already a small number in the presence of K ions, before considering what fraction E_1P represented, and there was room here for an error large enough to account for the paradox. One possibility was that in the presence of K^+ the acid used to stop the reaction might be instantly breaking down some, or even most, of the phosphoenzyme. To find out if this was the case, a method was needed to measure phosphoenzyme levels in a nondestructive manner.

As our pig kidney Na,K-ATPase is purified in membrane-bound form,[3] we decided to immobilize the fragments on syringe-tip filters and follow with a wash with sodium medium.[4] The reaction occurs when $[\gamma^{32}P,^3H]$-ATP is injected through the filter with a 5-ml Hamilton syringe. The effluent jet is collected in a circular train of 100 disposable cuvettes containing half-frozen nucleotide/P_i carrier and that rotates on a calibrated turntable. The product [³H]-ADP and [³²P]-P_i in each time interval are then measured: nucleotides and P_i are separated by HPLC on a SAX column and counted simultaneously for ³H and ³²P. This "flow kinetics" device was built as a modification to a quenched-flow apparatus[5] and takes advantage of its motorized ram to drive the 5-ml syringe at a constant linear speed. Filter and syringe are kept at 5°C.

---

[a]This work was supported by The Wellcome Trust, the Medical Research Council, and the Biotechnology and Biological Science Research Council.

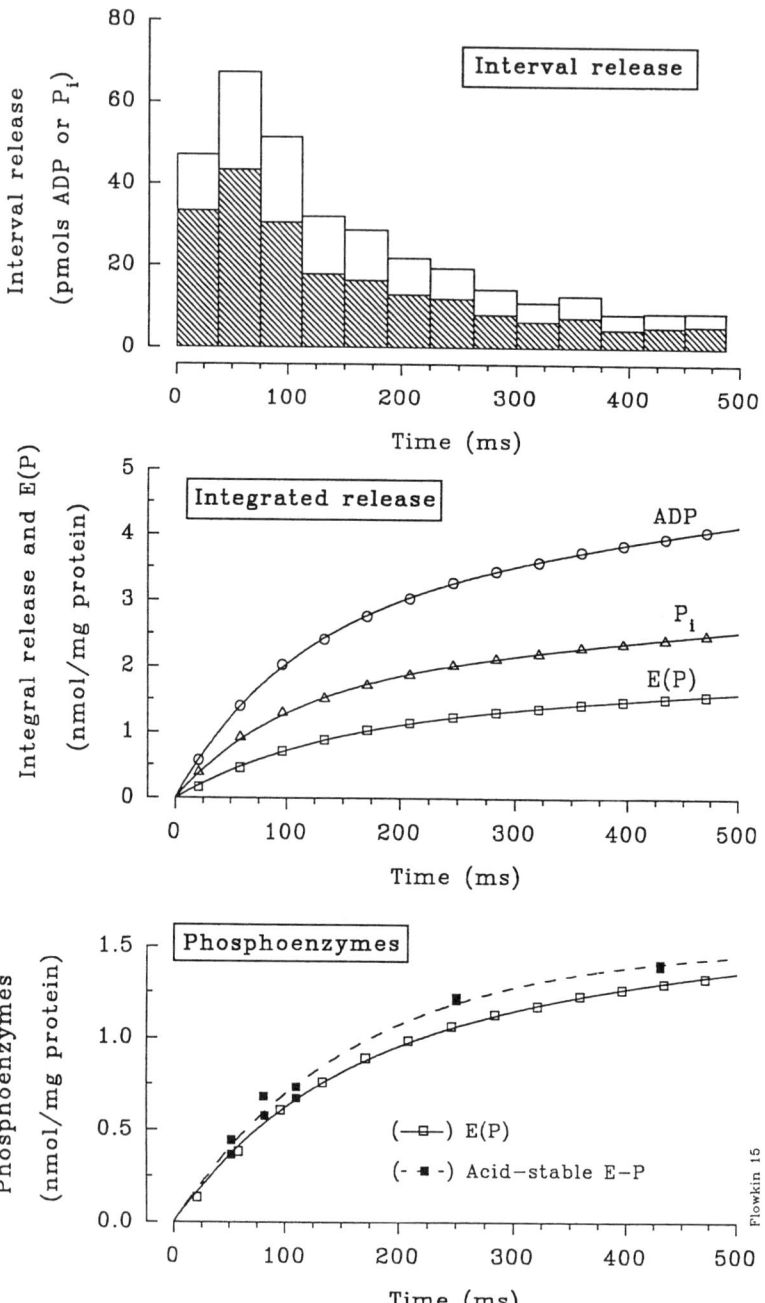

FIGURE 1. *See legend on facing page.*

Several experiments like those in FIGURES 1 and 2, conducted at 5 μM [γ^{32}P,^{3}H]-ATP, have shown that both [^{3}H]-ADP and [^{32}P]-P$_i$ are released with bursts. In the top panels we show the ADP and P$_i$ contents in each time fraction. The summation against time, now divided by the mass of enzyme in the filter, is plotted in the middle panels. Control experiments show that the deficit in [^{32}P]-P$_i$ release during the transient represents γ^{32}P transferred to the enzyme. This difference, named E(P), is also plotted in the middle panels and, on an expanded axis, again in the lower panels (open symbols). On the other hand, the results for acid-stable phosphorylation with 5 μM [γ^{32}P]-ATP, measured with the quenched-flow configuration of the apparatus,[5] show characteristic biphasic or exponential time-courses[6] with or without K$^+$, respectively (lower panels, filled symbols).

What is distinctive about the separate experiments in FIGURES 1 and 2 is that the quenched-flow measurements were started immediately after the flow-kinetics run, using the same enzyme suspension and the same master solutions. The results show that in K$^+$-free Na$^+$ medium (FIG. 1) there is remarkably good agreement between both phosphorylation estimates. However, when K$^+$ is added (FIG. 2), the congruence is limited to the initial 100–150 ms. Thereafter, the two techniques return their usual profiles, and this leads to a substantial difference. At around 410 ms, for instance, the nondestructive method shows an E(P) level 2.5 times higher than the acid-stable phosphorylation.

These results are certainly in the right direction to explaining the kinetic discrepancy just mentioned, and future experiments shall explore times into steady-state phosphorylation. However, the chemical nature of this acid-labile component of E(P), formed only in the presence of K$^+$, is unknown. It may represent an acid-labile phosphoenzyme[7] appearing late in the cycle, but the possibility of a tightly bound, or occluded, P$_i$ should also be considered.[8] Analysis of the product release curves will be presented elsewhere.

FIGURE 1. Pre-steady-state ADP and P$_i$ release and phosphoenzyme formation, by Na,K-ATPase in K$^+$-free medium at 5°C. *Flow kinetics.* Washed purified enzyme (83 μg protein from a suspension in Na$^+$ medium*) was trapped in a cellulose-nitrate filter (0.2 μm, μStar, Costar); 5 μM [γ^{32}P, ^{3}H]-ATP in Na$^+$ medium was injected at a constant flow rate of about 5 ml · s^{-1}, and the jet was collected in a 100-cuvette train rotating at 16 rpm. **Top panel:** [^{32}P]-P$_i$ (*hatched bars*) and [^{3}H]-ADP contents (*hatched plus clear bars*) in the earliest cuvettes. **Middle panel:** Time summation of the data in the *top panel,* referred to enzyme mass. E(P) represents the difference between ADP and P$_i$ release curves. *Continuous lines* denote fitting of the data to exponential-plus-linear functions. **Lower panel,** *clear symbols:* E(P) on an expanded vertical axis. *Quenched flow* (**lower panel,** *filled symbols*): Syringe I contained 44 μl (5.5 μg) of the same enzyme suspension above; Syringe II had 44 μl of 10 μM [γ^{32}P]-ATP in Na$^+$ medium. The effluent from the aging tube was quenched with chilled perchloric acid (4%) containing 1 mM Na$^+$ pyrophosphate. The phosphoenzyme was filtered through RAWP Millipore filters and washed with a chilled solution containing 1% trichloroacetic acid, 10 mM phosphate, 10 mM pyrophosphate, and 1 mM ATP and counted. All E-P levels (duplicates for each quench time) were individually corrected for percent delivery (85–90%) from the aging tube; the average of duplicate blanks (in high-K$^+$, Na$^+$-free medium) for each quench time was then deducted (correction ≈ 5%). *Na$^+$ medium:* 145 mM NaCl, 0.1 mM MgCl$_2$, 0.1 mM EGTA, and 25 mM histidine (pH 7.4, measured at 5°C).

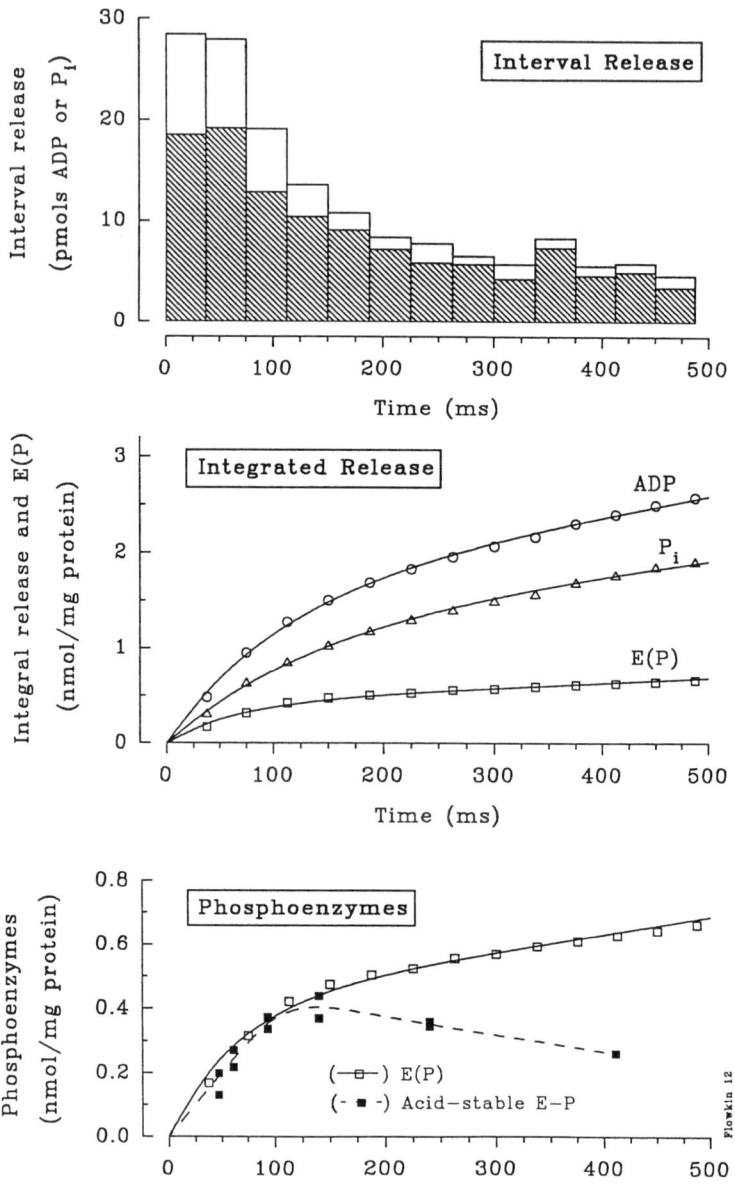

FIGURE 2. Another experiment, conducted as that in FIGURE 1, except that K^+ was also added with the radioactive ATP. *Flow kinetics:* Enzyme in Na^+ medium (60 μg protein immobilized); $[\gamma^{32}P,^3H]$-ATP in Na^+ medium also containing 15 mM KCl. *Quenched flow:* 9.2 μg enzyme in Na^+ medium were mixed with $[\gamma^{32}P]$-ATP also containing 30 mM KCl.

ACKNOWLEDGMENTS

We thank C. Bagshaw and A. Maxwell (Biochemistry Department, Leicester) for the loan of their quenched-flow machine.

REFERENCES

1. PLESNER, I. W., L. PLESNER, J. G. NØRBY & I. KLODOS. 1981. Biochim. Biophys. Acta **643:** 483–494.
2. NØRBY, J. G., I. KLODOS & N. O. CHRISTIANSEN. 1983. J. Gen. Physiol. **82:** 725–759.
3. JØRGENSEN, P. L. 1974. Biochim. Biophys. Acta **694:** 27–68.
4. WALTON, T. J. H. & J. D. CAVIERES. 1991. In The Sodium Pump: Recent Developments. J. H. Kaplan & P. De Weer, Eds.: 401–404. Rockefeller University Press. New York.
5. ECCLESTON, J. F., D. B. DIX & R. C. THOMPSON. 1985. J. Biol. Chem. **260:** 16237–16241.
6. HOBBS, A. S., R. W. ALBERS & J. P. FROEHLICH. 1985. In The Sodium Pump. I. Glynn & C. Ellory, Eds.: 355–361. The Company of Biologists. Cambridge.
7. FORBUSH, B. 1988. J. Biol. Chem. **263:** 7961–7969.
8. FALLER, L. D. & R. A. DIAZ. 1989. Biochemistry **28:** 6908–6914.

Diversity of the E_2P Phosphoforms of Na,K-ATPase[a]

NATALYA U. FEDOSOVA,[b] FLEMMING CORNELIUS,
BLISS FORBUSH III, AND IRENA KLODOS

Department of Biophysics
University of Aarhus
DK-8000 Aarhus C, Denmark
Department of Cellular and Molecular Physiology
Yale University School of Medicine
New Haven, Connecticut 06510

Phosphorylation of the Na,K-ATPase either through the physiological route (from MgATP in the presence of low Na^+) or from P_i in the presence of Mg^{2+} but in the absence of alkali cations leads to the formation of phosphorylated intermediates. Traditionally, both phosphoforms are referred to as E_2P although they differ in their sensitivity to K^+: dephosphorylation of E_2P formed from ATP is greatly accelerated by K^+, whereas E_2P formed from P_i is essentially insensitive to K^+.[1] In both cases the enzyme-ligand bond has been characterized as an anhydride bond between Asp369 and phosphate with an identical rate of spontaneous hydrolysis of ~0.07 s^{-1} at 0°C. A characteristic feature of the acyl-phosphate bond is its sensitivity to hydroxylamine.[2] Hydroxylaminolysis should lead to the formation of a stable hydroxamic acid with the release of inorganic phosphate. In the following experiments, N-methyl hydroxylamine was used to characterize and to compare the properties of the two phosphoforms.

RESULTS AND DISCUSSION

Interaction of N-Methyl Hydroxylamine with the Dephosphorylation Reactions. As seen in FIGURE 1A, N-methyl hydroxylamine greatly accelerates the rate of dephosphorylation of E_2P formed from ATP: the rate constant increases from 0.062 ± 0.003 s^{-1} to 36.4 ± 4.5 s^{-1}, that is, more than by a factor of 500. By contrast, dephosphorylation of the phosphoenzyme formed from P_i is slightly inhibited by N-methyl hydroxylamine (FIG. 1B). This stabilizing effect of N-methyl hydroxylamine is even more obvious in experiments in which the rate of dephosphorylation was slowed down by ouabain (FIG. 1C). The observed effects are not due to an increase in ionic strength (not shown).

---

[a]This work was supported in part by the Danish Research Academy (fellowship to N.U.F.), the Danish Medical Research Council, and the Danish Biomembrane Research Centre, University of Aarhus, Aarhus, Denmark.
[b]Tel: +45 89422933; fax: +45 86129599; e-mail: nf@biophys.aau.dk

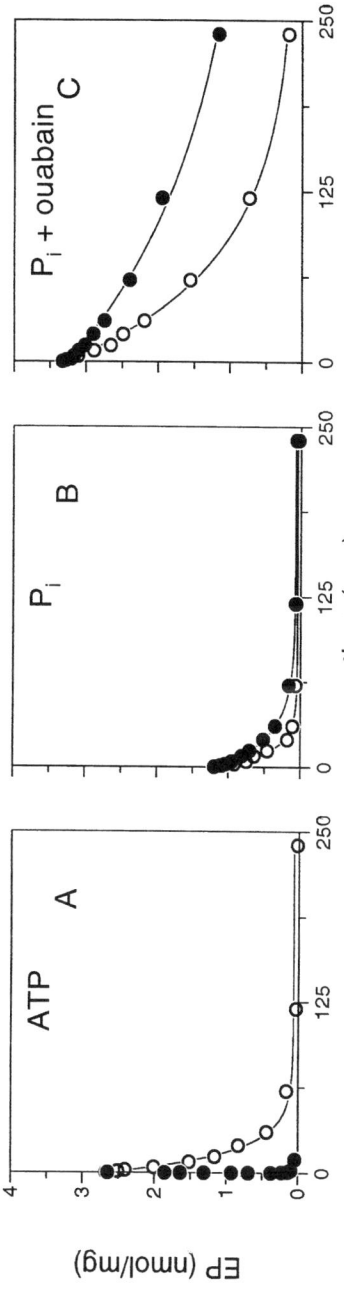

FIGURE 1. Interaction of N-methyl hydroxylamine with the dephosphorylation reactions. Phosphorylation of Na,K-ATPase from $Mg[\gamma\text{-}^{32}P]ATP + Na^+$ (**A**), $^{32}P_i + Mg$ (**B**), and $^{32}P_i + Mg + $ ouabain (**C**) was stopped by dilution of the labeled substrate, and dephosphorylation was followed in the absence (*empty circles*) or the presence (*filled circles*) of 100 mM N-methyl hydroxylamine by the acid precipitation method.

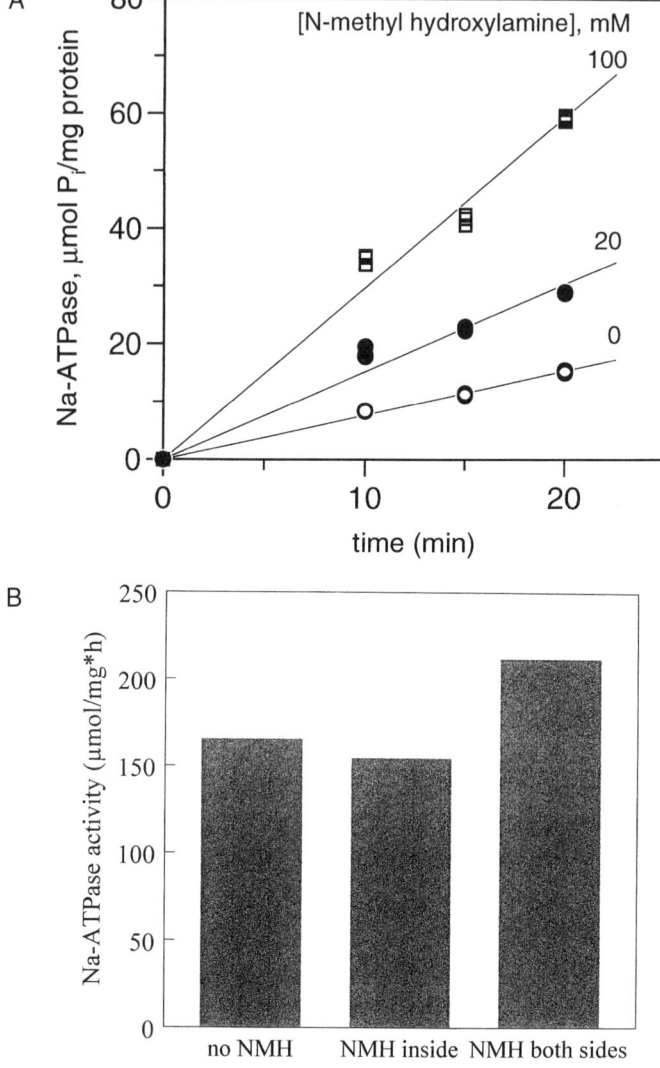

FIGURE 2. Activation of Na-ATPase reaction by N-methyl hydroxylamine. (**A**) Time course of P_i release measured in 10 mM HEPES, 10 mM MES, pH 7.4, 4 mM $MgCl_2$, and 16 mM NaCl in the presence of different concentrations of N-methyl hydroxylamine. (**B**) Activation of Na-ATPase is not due to interaction of N-methyl hydroxylamine with cation-binding sites. Proteoliposomes produced in the presence of N-methyl hydroxylamine were passed through a Penefsky column to remove external N-methyl hydroxylamine. Preincubation with MgP_i and ouabain ensured that the only active enzyme is oriented inside out. Proteoliposomes were tested for Na-ATPase activity with or without N-methyl hydroxylamine in the test solution. Therefore, in one case N-methyl hydroxylamine had access only to the extracellular high-affinity K^+ sites of the inside-out enzyme, and in the other case N-methyl hydroxylamine had access to both the extracellular and the cytoplasmic side of the enzyme.

Activation of the Na-ATPase by N-Methyl Hydroxylamine. To determine the site of the nucleophilic attack on the phosphoenzyme formed from ATP, we investigated the effects of *N*-methyl hydroxylamine on the Na-ATPase activity in which the dephosphorylation step is considered to be rate determining (FIG. 2A). It should be pointed out that release of NH_4^+, a very potent congener of K^+, is prevented by the use of *N*-methyl hydroxylamine. Nucleophilic attack on the carboxylic carbon will result in the formation of a stable hydroxamic acid and irreversible inhibition of the enzyme activity. By contrast, activation of the Na-ATPase reaction indicates that the nucleophilic attack is on the phosphorus atom. As seen from FIGURE 2A, Na-ATPase activity was increased with increasing concentrations of *N*-methyl hydroxylamine. Any direct K^+-like effect of *N*-methyl hydroxylamine, which is likely because it was previously shown that various amines can interact with K^+ binding, was excluded as demonstrated by using a sided enzyme preparation. FIGURE 2B clearly demonstrates that *N*-methyl hydroxylamine increases Na-ATPase activity by direct interaction with the phosphorylation site, but it had no effect mediated by the extracellular K^+ binding sites.

The results show that phosphorylation from P_i leads to formation of an E_2P subconformation (E'_2P) with different accessibility of the acyl-phosphate bond and structure of the substrate site than the phosphoenzyme formed from MgATP. In the latter the site is accessible to the bulk phase, whereas in the E'_2P the site is shielded and the acyl-phosphate bond is not accessible for *N*-methyl hydroxylamine. Apparently, *N*-methyl hydroxylamine also has a secondary effect on the phosphoenzyme as revealed by the slowing down of its dephosphorylation reaction.

REFERENCES

1. POST, R. L., G. TODA & F. N. ROGERS. 1975. J. Biol. Chem. **250:** 691–701.
2. LIPMANN, F. & L. C. TUTTLE. 1945. J. Biol. Chem. **149:** 21–28.

Interaction between Substrate Site and Cation Binding Sites in P_i Phosphorylation of Na,K-ATPase[a]

FLEMMING CORNELIUS,[b] NATALYA U. FEDOSOVA, AND
IRENA KLODOS

Department of Biophysics
University of Aarhus
DK-8000 Aarhus C, Denmark

The phosphoforms formed in the reaction of Na,K-ATPase with ATP or P_i are regarded as chemically identical. However, essential differences between these phosphoenzymes were found in their kinetics and their susceptibility to *N*-methyl hydroxylamine, leading to the proposal of an E'_2P subconformation formed in the reaction with P_i (Fedosova *et al.*, this volume). Here we describe how the phosphorylation-dephosphorylation kinetics of EP formed from P_i are modified by interaction between the substrate site and the cation binding sites as measured both chemically and with the styryl dye RH421.

RESULTS AND DISCUSSION

Chemical Identification. Experiments were performed with Na,K-ATPase from pig kidney and shark rectal glands. Both phosphorylation and dephosphorylation were

FIGURE 1. **A** and **B** depict P_i phosphorylations of shark Na^+,K^+-ATPase at 0°C in the presence of 5 mM alkali cations as indicated on the curves compared to controls in the absence of alkali cations. The best fits as evaluated by F-tests to the data are indicated by the curves. All except Na^+ were biexponential. Fitted constants for the curves in **A** were: In the nominal absence of alkali cations (control) the fitted rate constants were: 0.241 ± 0.049 s^{-1} (0.68 nmol/mg) and 0.023 ± 0.007 s^{-1} (0.51 nmol/mg). With Li^+: 0.91 ± 0.25 s^{-1} (0.38 nmol/mg) and 0.037 ± 0.004 s^{-1} (0.95 nmol/mg). With Na^+: 0.087 ± 0.004 s^{-1} (0.74 nmol/mg). **B** shows phosphorylations using rapid mixing. With K^+: 116.4 ± 9.3 s^{-1} (0.47 nmol/mg) and 3.4 ± 1.1 s^{-1} (0.11 nmol/mg). With Cs^+: 64.7 ± 5.7 s^{-1} (0.51 nmol/mg) and 4.2 ± 1.0 s^{-1} (0.22 nmol/mg). **C** and **D** show dephosphorylation after P_i phosphorylation of shark Na^+,K^+-ATPase at 0°C in the presence of different alkali cations both during initial phosphorylation (5 mM) and in subsequent dephosphorylation (all 5 mM). The dephosphorylations could all be adequately fitted to monoexponential functions. Fitted rate constants were (i) control: 0.091 ± 0.006 s^{-1}; (ii) Na^+: 0.064 ± 0.004 s^{-1}; (iii) Li^+: 0.096 ± 0.005 s^{-1}; (iv) Cs^+: 13.9 ± 1.0 s^{-1}; (v) K^+: 20.3 ± 1.8 s^{-1}.

[a]This work was supported in part by the Danish Research Academy (fellowship to N. U. F.), the Danish Medical Research Council, and the Danish Biomembrane Research Centre, University of Aarhus, Aarhus, Denmark.
[b]Tel: +45 8942 2926; fax: +45 8612 9599; e-mail: fc@biophys.aau.dk

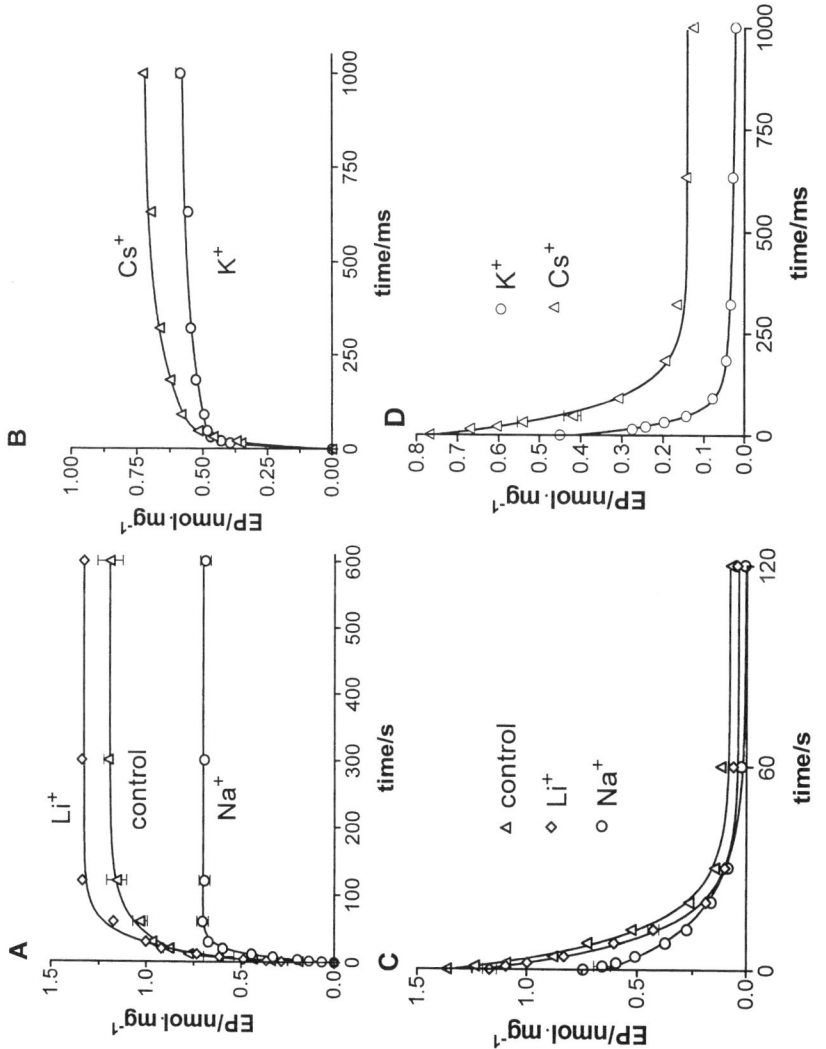

FIGURE 1. *See legend on facing page.*

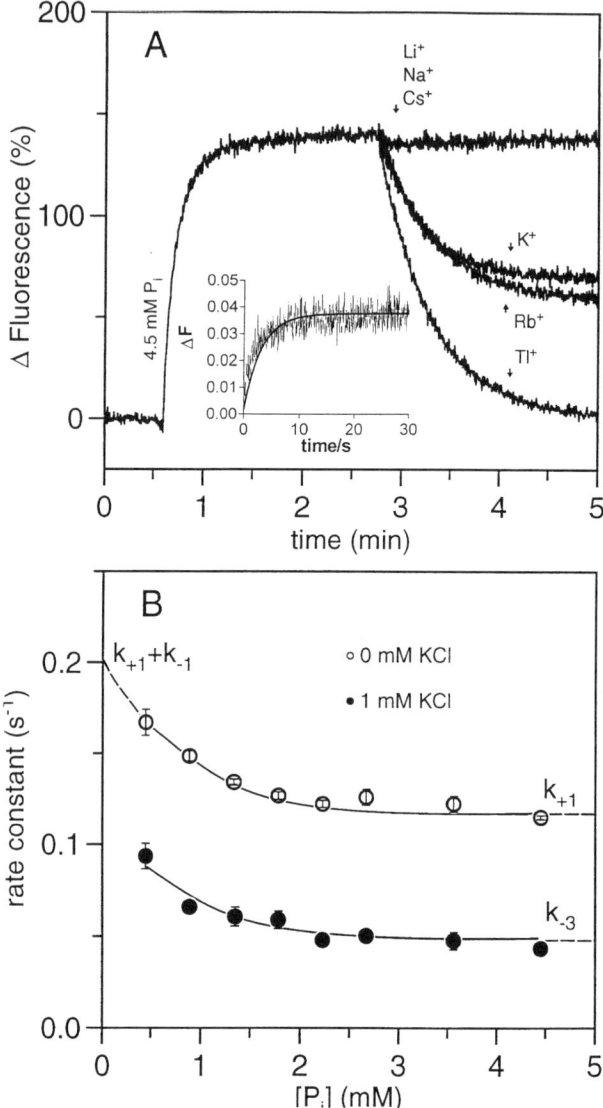

FIGURE 2. (**A**) Fluorescence response of RH421 to phosphorylation from P_i (4.5 mM) of shark Na^+,K^+-ATPase at 22°C. After attaining a steady-state level, 1 mM of K^+ or its congeners Tl^+, Rb^+, Cs^+, or Li^+ and Na^+ are added. K^+, Tl^+, and Rb^+ cause a decrease in fluorescence, whereas Cs^+, Li^+, and Na^+ only cause an insignificant decrease. (**Insets**) The initial rise in fluorescence and the monoexponential fit to the data. The fit parameters (rate constant, λ, and steady-state level, F_{max}) were $\lambda = 0.138 \pm 0.007$ s^{-1} and $F_{max} = 0.0109 \pm 0.0003$. (**B**) Rate constants for the monoexponential increase in fluorescence of RH421 of shark enzyme phosphorylated by increasing concentration of inorganic phosphate in the presence and absence of 1 mM K^+.

modified by cations. K^+ and its congeners Rb^+, Tl^+, and Cs^+ strongly increased the phosphorylation rate, whereas Li^+ and Na^+ did not (FIG. 1A and 1B). The formation of $E_2(K_2)$-P from $E_2(K_2)$ is 400 times faster than the formation of E'_2P in the absence of K^+. The order of affinities for the effect on phosphorylation rate is: $Na^+ <$ control $\leq Li^+ \ll Cs^+ < K^+ \cong Rb^+ \cong Tl^+$.

The dephosphorylation of E'_2P is independent of added K^+. As indicated by fluorescence measurements to be described, E'_2P binds K^+, but the rate of K^+ binding is slower than spontaneous dephosphorylation. The presence, and thus binding, of alkali cations during phosphorylation affected the rate of subsequent dephosphorylation with the same order of affinity as that for phosphorylation (FIG. 1C and 1D). With K^+ and its congeners, subsequent dephosphorylation was more than 400 times faster than that of E'_2P ($k \sim 30$ s^{-1} for $E_2(K_2)$-P and 0.07 s^{-1} for E'_2P). Dephosphorylation was not affected by Na^+ or Li^+. Accordingly, for shark enzyme the steady-state phosphoenzyme level was decreased with all cations except Li^+.

Fluorescence. The RH421 response to the formation of E'_2P in shark enzyme is shown in FIGURE 2A. In the absence of alkali ions, a slow increase in ΔF ($k \sim 0.14$ s^{-1}) is observed. The addition of alkali cations induces a graded decrease in ΔF according to ion species in the order: $Na^+ \cong Li^+ \cong Cs^+ \ll K^+ \cong Rb^+ \cong Tl^+$. The decrease in ΔF is caused by cation binding to dephosphoenzyme, resulting in a decrease in the steady-state E'_2P level and formation of low/none fluorescent $E_2(K_2)$-P: $E'_2P \leftrightarrow E_2 \leftrightarrow E_1 \leftrightarrow E_2(K_2) \leftrightarrow E_2(K_2)$-P. This K^+ binding is low affinity ($K_D \approx 1$ mM). At saturating K^+ the rate-constant measured ($k \sim 0.04$ s^{-1}) approaches the rate constant for the rate-limiting step $E_2 \rightarrow E_1$. Note the absence of effect of Cs^+ here as opposed to that in FIGURE 1D and to its accelerating effect on ATP hydrolysis.

In both the absence and the presence of K^+ the observed rate constant for P_i phosphorylation decreased with P_i (FIG. 2B). In the absence of K^+, phosphorylation is preceded by the $E_1 \leftrightarrow E_2$ transition, and the value reached at saturating P_i corresponds to k_1 (≈ 0.13 s^{-1}) and the initial value to $k_1 + k_{-1}$ (≈ 0.20 s^{-1}), giving a $k_{-1} \approx 0.07$ s^{-1}. In the presence of K^+, the increase in fluorescence is due to phosphorylation of E_2 preceded by $E_2(K_2) \leftrightarrow E_1 \cdot K_2 \leftrightarrow E_1 \leftrightarrow E_2$. At saturating P_i, k_{obs} will approach the rate-limiting deocclusion constant ($k_{-3} \sim 0.05$ s^{-1}). Thus, the rate constants of the reactions $E_2 \rightarrow E_1$ and $E_2(K_2) \rightarrow E_1 \cdot K_2$ are about equal.

CONCLUSIONS

We suggest that E_2 represents conformation with an occluded empty site, $E_2(\ )$, and that opening of the occlusion site is independent of the presence of cation in the occlusion pocket. The difference observed in the reactivity in P_i phosphorylation/dephosphorylation towards a series of alkali cations indicates that binding of the cations has a different ability to increase the water disorder, thereby destabilizing the acyl bond in the substrate site. With E'_2P, the most effective ions, K^+, Tl^+, and Rb^+, all have comparable ion radii (1.33 Å, 1.40 Å, and 1.48 Å, respectively), whereas the ineffective Li^+ and Na^+ are smaller (0.60 Å and 0.95 Å) and Cs^+ much larger (1.68 Å). The Cs^+ effect indicates different properties of the extracellular (E_2P) and cytoplasmic (E'_2P and E_1) access pathways.

Fluorescent Styryl Dyes as Probes for Na,K-ATPase Reaction

Enzyme Source and Fluorescence Response[a]

IRENA KLODOS,[b] NATALYA U. FEDOSOVA,
AND FLEMMING CORNELIUS

Department of Biophysics
University of Aarhus
DK-8000 Aarhus C, Denmark

Fluorescent styryl dyes (RH-dyes), when bound to Na,K-ATPase-enriched broken membranes, change their fluorescence in response to some molecular events in the Na,K-ATPase reaction cycle. Although the mechanism of dye response is still controversial, they have recently found increasing application in studies of the reaction mechanism of P-type transport ATPases. It has been suggested[1,2] that the dyes' responses to changes in the intramembranous electric field are modified by interactions between these dyes and the Na,K-ATPase protein. It was previously shown[1] that the excitation spectra of RH160 depend on the concentration of protein reconstituted into liposomes, and the same is found for absorbance (not shown), an indication of a dye-protein interaction.

To evaluate the significance of these interactions we studied RH421 responses to (1) Na^+ binding, (2) phosphorylation from either ATP or P_i, or (3) vanadate binding to Na,K-ATPase (FIG. 1). Experiments were performed with membrane-bound Na,K-ATPase, solubilized enzyme, and enzyme reconstituted into liposomes. Furthermore, we used enzymes from different sources, shark rectal gland and pig kidney, and varied their lipid surroundings in reconstituted preparations. For both enzymes we measured the specific activities at 23°C and 37°C, the phosphorylation level in the presence of 4 mM $MgCl_2$, 16 mM NaCl, and 30 µM [γ-^{32}P]-ATP in 30 mM imidazole (pH 7.5) at 0°C and in the presence of inorganic ^{32}P in the same medium, but without NaCl. Both enzyme preparations and all applied methods were described by Cornelius[3] and Fedosova *et al.*[1] Shark enzyme was solubilized in $C_{12}E_8$. The activity of kidney enzyme, solubilized in CHAPS, deteriorated quickly. Therefore, measurements of its catalytic activity and fluorescence measurements were processed in parallel.

The results are shown in TABLE 1. The experimental observations can be summarized as follows:

1. RH421 responses depend on the enzyme source. Formation of several intermediates is reported differently with the two enzymes, for example, formation of the enzyme-vanadate complex is "seen" with the shark enzyme but not with the kidney enzyme (TABLE 1). RH421 bound to proteoliposomes containing kidney enzyme

---

[a]These studies were supported by the Danish Medical Research Council, the Novo Foundation, the Danish Research Academy, and the Danish Biomembrane Research Centre, University of Aarhus, Aarhus, Denmark.
[b]Tel: +4589422937; fax: +4586129599; e-mail: ik@biophys.aau.dk

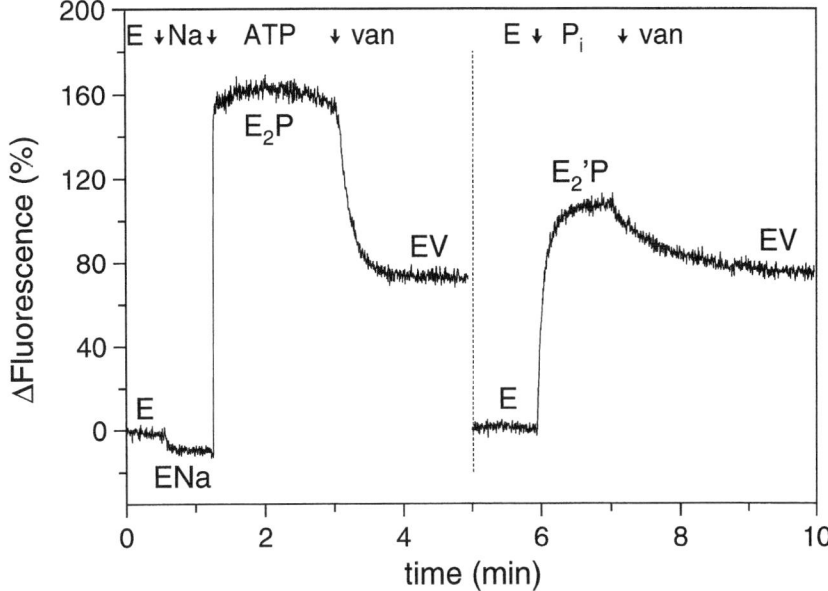

FIGURE 1. Fluorescence response of RH421 to binding of Na$^+$ to Na,K-ATPase (ENa), to formation of E$_2$P from ATP and the following binding of vanadate (EV), or to phosphorylation from P$_i$ (E'$_2$P) followed by binding of vanadate (EV). Fluorescence was measured with a Spex Fluorolog 2 fluorometer at 20°C under continuous stirring. The excitation wavelength was 580 nm, and a 630 nm cut-off filter was used on the emission site. The experiment was performed with the membrane-bound shark enzyme. At the start of the experiment the medium contained 50 µg of enzyme preparation, 10 mM HEPES/10 mM MES (pH 7.4 at 20°C), and 4 mM MgCl$_2$. Final concentrations of added ligands: 16 mM NaCl, 30 µM ATP, 100 µM vanadate, or 4 mM P$_i$, 100 µM vanadate.

reports formation of E'$_2$P by a slight decrease in fluorescence, but with the shark enzyme vesicles, the same event is reported by an increase in fluorescence.

2. No correlation exists between steady-state concentrations of several intermediates and the magnitude of fluorescence responses even for enzyme from the same

TABLE 1. Characterization of Na,K-ATPase from Shark Rectal Glands and Pig Kidney Outer Medulla[a]

Enzyme Preparation	Specific Activity at 23°C U/mg	ENa ΔF (%)	E$_2$P from ATP nmol/mg	E$_2$P from ATP ΔF (%)	E$_2$'P from P$_i$ nmol/mg	E$_2$'P from P$_i$ ΔF (%)	EP from P$_i$ + Ouabain nmol/mg	EV ΔF (%)
Shark								
Membranes	10.5 ± 0.5	−5.5 ± 0.6	2.5 ± 0.3	177 ± 2	1.25 ± 0.05	107	3.1 ± 0.2	74.3
Reconstituted	9.4 ± 0.4	−7.3	6.9 ± 0.6	75.3	6.1 ± 0.2	18.7		8.5
Solubilized	10.5 ± 0.4	−3.0 ± 0.4	4.2 ± 0.3	77 ± 1.6		25.7 ± 1.1		24 ± 1.1
Kidney								
Membranes	~8	−7.7 ± 1.2	2.35 ± 0.02	35 ± 3	2.5 ± 0.1	15.2	2.82 ± 0.05	1.1
Reconstituted	6.5	−4.3		8.0		−1.2		−3.8
Solubilized	1.5	−5.2 ± 0.6		6.4 ± 0.6		0.3 ± 0.6		−1.7 ± 0.3

[a]Data are given as means ± SEM (3 ≤ n ≤ 6).

source (e.g., compare the level of phosphorylated intermediates and the corresponding ΔF in TABLE 1).

3. The lipid environment of the enzyme affects RH421 responses, as shown by the fact that both the magnitude and the ratio between responses to several molecular events are modified on solubilization and reconstitution of Na,K-ATPase.

4. Fluorescence responses observed with the proteoliposomes depend on the Na,K-ATPase concentration in the vesicles (not shown).

5. We currently have no simple explanation for these observations; however, the complexity of the results indicates that an interpretation of RH421 signals in terms of sensitivity exclusively to the electric field seems to be an oversimplification.

REFERENCES

1. FEDOSOVA, N. U., F. CORNELIUS & I. KLODOS. 1995. Biochemistry **34:** 16806–16814.
2. VISSER, N. V., A. VAN HOEK, A. J. W. G. VISSER, J. FRANK, H.-J. APELL & R. J. CLARKE. 1995. Biochemistry **34:** 11777–11784.
3. CORNELIUS, F. 1995. Biochim. Biophys. Acta **1235:** 197–204.

Kinetic Properties of the Na,K-ATPase of Goldfish Kidney[a]

MARIELA P. GARCÍA,[b] PABLO J. SCHWARZBAUM,
ROLANDO C. ROSSI, AND SERGIO B. KAUFMAN

IQUIFIB (UBA-CONICET)
Junín 956
(1113) Buenos Aires, Argentina

Studies on the reaction cycle of the Na,K-ATPase postulate that the hydrolysis of ATP includes the formation of phosphorylated intermediates (EP) followed by hydrolysis of EP. In media without K^+, ATP acts only at a high-affinity site as shown in the following scheme:

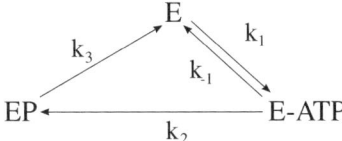

It can be demonstrated that in steady state, EP is a hyperbolic function of [ATP] as follows:

$$EP = \frac{EP_m \times [ATP]}{K_m^{EP} + [ATP]} \quad (1)$$

with

$$K_m^{EP} = \frac{k_3 \times (k_{-1} + k_2)}{k_1 \times (k_3 + k_2)} \quad (2)$$

and

$$EP_m = ET \times \frac{k_2}{k_2 \times k_3} \quad (3)$$

The aim of the present work was to study, in microsomal fractions from goldfish kidney, the kinetics of the Na,K-ATPase in the absence of K^+ (Na-ATPase mode) through the quantification of EP. Although several studies on goldfish kidney ATPase have been published,[1] none of them was able to assess the kinetic properties of the enzyme at submicromolar ATP concentrations where the Na,K-ATPase of goldfish might function under physiological conditions.

---

[a]This work was supported by grant FA109 from the University of Buenos Aires, Buenos Aires, Argentina.
[b]To whom correspondence should be addressed.

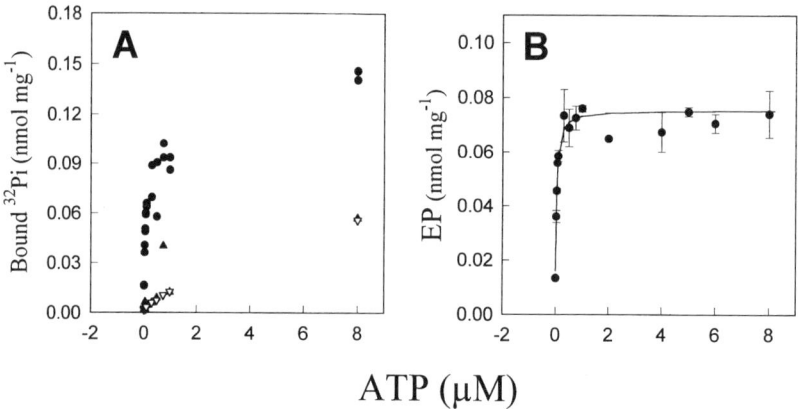

FIGURE 1. EP vs [ATP] in steady state. (**A**) The level of bound ^{32}Pi was measured at 0°C in the presence of Mg^{2+} and Na^{+} (●) or omitting either Mg^{2+} (▲) or Na^{+} (▽). (**B**) EP as a function of [ATP].

MATERIALS AND METHODS

EP was measured at 0°C as the acid-stable amount of ^{32}P incorporated into the enzyme from [γ^{32}P]ATP as previously described.[2] Assay medium contained: NaCl 150 mM, EDTA 0.2 mM, free Mg^{2+} 0 (blanks) or 0.5 mM, ATP 0.01–10 μM, 25 mM imidazole-HCl (pH 7.4), and 100 μg ml^{-1} of protein.

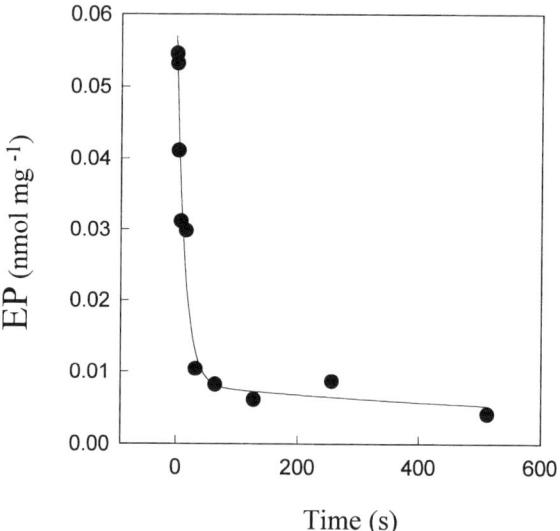

FIGURE 2. Dephosphorylation kinetics. The enzyme suspension was phosphorylated in 1 μM [γ^{32}P]ATP and 0.5 mM Mg^{2+} at 0°C. After 10 seconds, 20 volumes of 1 μM ATP, 150 mM NaCl, 0.2 mM EDTA, 0 (blanks) or 0.5 mM free Mg^{2+}, and 25 mM imidazole-HCl (pH = 7.4) were added, and the decay of EP was measured at different times thereafter.

RESULTS

Substrate Curves of EP in Steady State. FIGURE 1A shows the steady-state level of bound $^{32}P_i$ in media with (●) and without (▲) Mg^{2+} (blanks). In Mg^{2+}-containing media in which Na^+ was replaced by K^+ (▽), the level of bound $^{32}P_i$ was similar to that of the blanks. FIGURE 1B shows that EP followed a rectangular hyperbola for [ATP] from 0.01–8 μM, with $EP_m = 0.075 \pm 0.002$ nmol mg^{-1} and $K_m^{EP} = 0.031 \pm 0.067$ μM ($n = 42$).

Dephosphorylation Kinetics. The enzyme suspension was phosphorylated in 1 μM [γ^{32}P]ATP and 0.5 mM free Mg^2 for 10 seconds. FIGURE 2 shows the dephosphorylation kinetics of EP after the addition of unlabeled ATP (t = 0). The curve fitted to the points represents:

$$EP = 0.057[0.95(e^{-(0.072s^{-1})t} 0.9 + e^{-(0.001s^{-1})t} 0.1) + 0.05] \text{nmol mg}^{-1}$$

CONCLUSIONS

(1) EP was identified as an acid-stable intermediate of the Na,K-ATPase reaction cycle which requires Na^+ as well as Mg^{2+} for phosphorylation (FIG. 1A); (2) As predicted by Equation 1, EP was a hyperbolic function of [ATP] with $K_m^{EP} = 0.03$ μM; and (3) dephosphorylation kinetics was biphasic, with apparent dephosphorylation rate constants of 0.072 s^{-1} and 0.001 s^{-1}. Future experiments on ATP binding to the enzyme and on phosphorylation kinetics will allow us to complete the set of elementary constants necessary to compare the K_m^{EP} obtained from substrate curves with that predicted by the model (Equation 2).

REFERENCES

1. PAXON, R. & B. L. UMMINGER. 1983. Comp. Biochem. Physiol. **74B:** 503–506.
2. SCHWARZBAUM, P. J., S. B. KAUFMAN, R. C. ROSSI & P. J. GARRAHAN. 1995. Biochim. Biophys. Acta **1233:** 33–40.

Mn as Cosubstrate for the Phosphorylation of the Sarcoplasmic Reticulum Ca-ATPase by P_i

D. A. GONZÁLEZ,[a,c] G. L. ALONSO,[a,c] AND
J.-J. LACAPÈRE[b,d]

[a]Cátedra de Biofísica
Facultad de Odontología
Universidad de Buenos Aires
M. T. de Alvear 2142
1122 Buenos Aires, Argentina

[b]Section de Recherche
UMR-168 CNRS
Institut Curie
11 rue P. et M. Curie
75231 Paris cedex 05, France

Phosphorylation of the sarcoplasmic reticulum Ca-ATPase with ATP occurs when the Ca transport sites of the enzyme are saturated, whereas phosphorylation with P_i occurs with the protonated form. Both reactions require Mg as cosubstrate, but it can be replaced by other divalent cations. The binding of P_i or ATP and Mg to the enzyme occurs randomly to form a noncovalent complex before the formation of the covalent phosphoenzyme. The stoichiometry of cations bound to the catalytic site remains an open question. In this study we used Mn as an Mg analog and combined fluorescence and radioactive techniques to study the phosphorylation of the Ca-ATPase by P_i.

At pH 5.5, EDTA, P_i, or Mn alone has little or no effect on the fluorescence signal, whereas the addition of Ca alone or the simultaneous addition of P_i and Mn markedly increases it. These experiments indicate that the ATPase is Ca free at pH 5.5, which is a good condition in which to study the phosphorylation of the Ca-ATPase with P_i using Mn as cosubstrate in the absence of a chelating agent. The increase in the fluorescence signal is a sigmoidal function of $\log[P_i]$, or $\log[Mn]$, for several constants [Mn] and [P_i], respectively (data not shown). From these functions, the apparent affinities for P_i and Mn at various [Mn] and [P_i] were obtained (FIG. 1A and 1B, respectively). ^{54}Mn binding was measured in both the presence and the absence of P_i (FIG. 1C). The curves were drawn up to 1 mol of bound Mn per mol of ATPase (5–6 nmol/mg). The Scatchard plot supports the assumption that the first points indicate the binding of Mn to a first site with affinity ≥ 40 µM in the absence of P_i; the presence of P_i decreases the apparent affinity of the enzyme for this Mn.

---

[c]Tel: 54 1 961 0350; fax: 54 1 962 0176; e-mail: debora@biofis.odon.uba.ar or alonso@biofis.odon.uba.ar
[d]Tel: 33 1 42 34 67 81; fax: 33 1 40 51 06 36; e-mail: lacapere@curie.fr

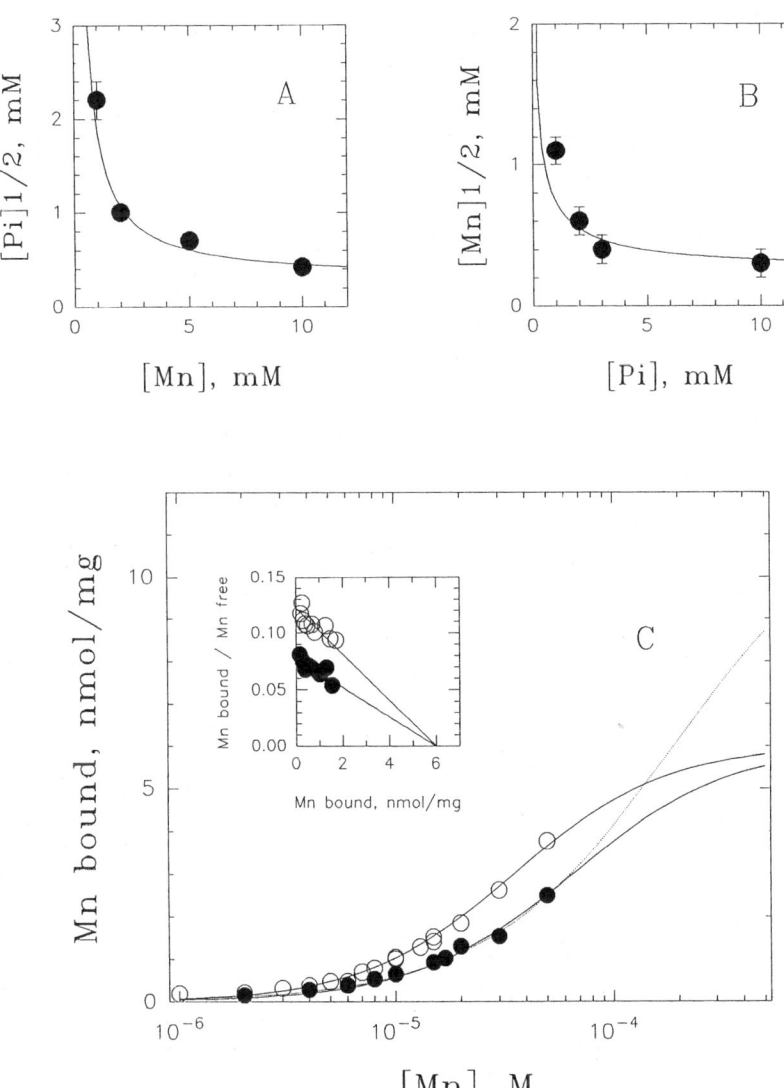

FIGURE 1. (**A** and **B**) The amplitude of the fluorescence change is a sigmoidal function of log[P_i], or log[Mn], at several constants [Mn] and [P_i] (not shown). From these curves, the apparent affinities for P_i (**A**) and Mn (**B**) are deduced and represented as functions of [Mn] and [P_i], respectively. Curves were obtained with the FIGURE 2 reaction model and equilibrium constants, as previously described.[2] (**C**) [Mn] dependence of Mn binding with (●) or without (○) P_i; Mn binding was measured with ^{54}Mn. The curves are sigmoidal functions drawn up to 1 (———) or 2 (------) mole of bound Mn per mole of ATPase. (**Inset**) Scatchard plot of the data. Sarcoplasmic reticulum membranes were isolated from rabbit skeletal muscles.[2] Intrinsic fluorescence changes and binding of ^{54}Mn to the Ca-ATPase were measured as described in ref. 2. Reprinted with permission from González et al.[2]

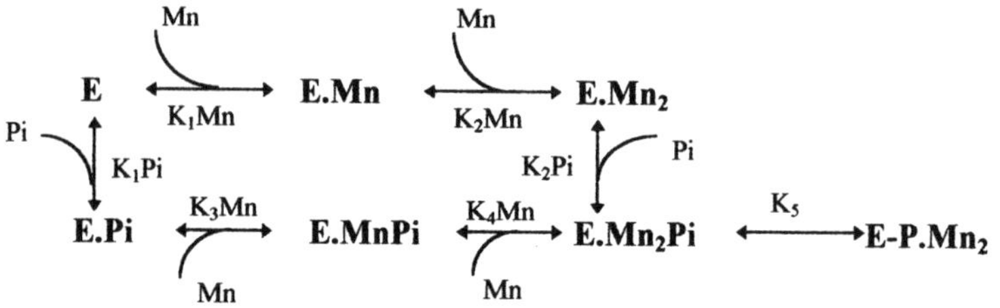

```
                    Mn              Mn
          E    ⇌    E.Mn    ⇌    E.Mn₂
       Pi ↑  K₁Mn        K₂Mn      ↑
          ↓ K₁Pi              K₂Pi │  Pi
                K₃Mn         K₄Mn          K₅
        E.Pi  ⇌    E.MnPi   ⇌    E.Mn₂Pi  ⇌    E-P.Mn₂
              Mn              Mn
```

FIGURE 2. Reaction model of the Ca-ATPase phosphorylation with P_i, with Mn as cosubstrate. Mn and P_i bind randomly before covalent phosphoenzyme formation. The following first-order equilibrium constants derived from the experimental data fit the reaction model (mM): $K_1Mn = 0.04–0.10$; $K_2Mn = 4–6$; $K_1P_i = 6–8$; $K_3Mn = 0.15–0.30$; $K_4Mn = 8–12$; $K_2P_i = 12–16$; $K_5 = k_{-5}/k_5 = 0.025$. Reprinted with permission from González et al.[2]

The fluorescence levels are an index of the covalent phosphoenzyme, and the radioisotopic measurements indicate the amount of bound Mn. Although the fluorescence experiments give an apparent affinity for Mn higher than mM in the presence of the lowest [P_i] (FIG. 1B), the radioisotopic ones yielded a µM affinity for Mn in the absence of P_i (FIG. 1C). These different affinities for Mn support the requirement of 2 Mn ions to form the covalent phosphoenzyme (FIG. 2). The high affinity bound ^{54}Mn is not chased by 1 mM Ca, which discards Mn binding to the Ca transport sites. FIGURE 2 indicates the random binding of 1 P_i and 2 Mn per mole of ATPase to form the noncovalent complex. To obtain a set of equilibrium constants, some kinetic data are required ($K_5 = k_{-5}/k_5$). The dephosphorylation rate constant was obtained by the addition of an excess EDTA to a phosphoenzyme preformed with Mn as cosubstrate and measuring the fluorescence decay. The phosphorylation rate constant was obtained from determinations of the observed rate of fluorescence increase in the presence of several Mn and P_i concentrations, as previously described[1] for Mg. The equilibrium of reaction 5 strongly favors the covalent phosphoenzyme. The entire set of equilibrium constants given in FIGURE 2 was obtained as previously described.[2] The continuous traces shown in FIGURE 1A and B are theoretical functions[2] obtained with this set of equilibrium constants. Similar results are obtained by applying numerical calculus to the FIGURE 2 reaction scheme. Assuming this model, the radioactive data obtained in the presence of high [P_i] were fitted with a sigmoidal curve drawn up to 2 mol of Mn per mole of ATPase (FIG. 1C, dotted line). The apparent affinity for the 2 Mn in our model, with nonlimiting P_i, is the same (0.2 mM) when obtained from either the ^{54}Mn binding curve or the fluorescence determinations.

The stoichiometry of 2 Mn and 1 P_i proposed in our model agrees with previous observations of the simultaneous binding of 2 Mg and 1 fluoride to the catalytic site.[3] It also correlates with nuclear magnetic resonance experiments performed on the Na/K-ATPase which show two divalent cation sites at the active site.[4] We determined (data not shown) that only 1 Mn is required for the phosphorylation of the Ca-ATPase with ATP, as previously described.[5] Therefore, the binding of the second Mn during the forward catalytic cycle might be involved in a regulation process.

REFERENCES

1. LACAPÈRE, J.-J., M. P. GINGOLD, P. CHAMPEIL & F. GUILLAIN. 1981. J. Biol. Chem. **256:** 2302–2306.
2. GONZÁLEZ, D. A., G. L. ALONSO & J. J. LACAPÈRE. 1996. Biochim. Biophys. Acta **1276:** 188–194.
3. KUBOTA, T., T. DAIHO & T. KANAZAWA. 1993. Biochim. Biophys. Acta **1163:** 131–143.
4. STEWART, M. D. J., P. L. JORGENSEN & C. M. GRISHAM. 1989. Biochemistry **28:** 4695–4701.
5. OGURUSU, T., S. WAKABAYASHI & M. SHIGEKAWA. 1991. J. Biochem. **109:** 472–476.

Comparative Aspects of Ligand Stoichiometry in Na^+,K^+-ATPase from Kidney and in Ca^{2+}-ATPase from Rabbit Sarcoplasmic Reticulum

OTTO HANSEN[a] AND JØRGEN JENSEN

Department of Physiology
Aarhus University
DK-8000 Aarhus C, Denmark

The two P-type ATPases, the Ca^{2+}-ATPase from sarcoplasmic reticulum (SR) and the α_1 peptide of Na^+, K^+-ATPase from mammalian kidney, have a considerable degree of homology. For some specific domains, for example, the phosphorylation domain and the putative nucleotide binding region, the homology is nearly 100%. More or less identical behavior of the corresponding ligands could therefore be expected. Pure Na^+, K^+-ATPase preparations consisting of 147 kD ($\alpha + \beta$)-protomers are supposed to bind ~6.8 nmol ligand per milligram protein and pure Ca^{2+}-ATPase (110 kD) is supposed to bind ~9 nmol of the relevant ligands per milligram protein. For highly purified Na^+, K^+-ATPase from kidney, about half the protein is not accounted for in ligand binding, and the ligand stoichiometry of ADP/ATP:ouabain:vanadate: phosphorylation is 1:1:1:1.[1] In search of an explanation of the inert protein of purified Na^+, K^+-ATPase, comparative ligand binding studies on Ca^{2+}-ATPase from SR were carried out. For Ca^{2+}-ATPase purified to a similar degree as Na^+, K^+-ATPase, the ratio between the relevant ligands does not seem that simple, however.[2]

Ca^{2+}-ATPase was prepared from rabbit fast-twitch skeletal muscle according to de Meis and Hasselbach[3] and gently extracted with a low concentration of deoxycholate.[2] The final product is partly characterized by the parameters shown in TABLE 1. Only the specific 110-kD peptide was visible in SDS-PAGE. [^{14}C]-ADP and [^{14}C]-ATP binding to Ca^{2+}-ATPase took place at 0–2°C in the absence of Mg^{2+} and EGTA. After separation by centrifugation, free nucleotide was determined on the supernatant and bound nucleotide was calculated as total minus free ligand. In binding experiments with thapsigargin, enzyme was preincubated for 10 minutes with an equimolar concentration of thapsigargin before the addition of [^{14}C]-ADP. ^{32}P-phosphorylation was carried out according to traditional methods.[2] $^{85}Sr^{2+}$ binding was determined by a procedure similar to that used for nucleotide binding,[2] and Ca^{2+} binding was carried out by combining [^{45}Ca]Ca^{2+} and determination of Ca^{2+} by atomic absorption spectrophotometry.[4]

The conditions traditionally used for nucleotide binding to Ca^{2+}-ATPase include the presence of Mg^{2+} and EGTA, that is, Mg-nucleotide and the absence of Ca^{2+}. In our experience, hardly any [^{14}C]ADP binding was seen under such conditions. ADP

---

[a]Address for correspondence: Otto Hansen, Department of Physiology, Aarhus University, Ole Worms Allé 160, DK-8000 Aarhus C, Denmark (tel: +45 89 42 28 06; fax: +45 86 12 90 65; e-mail: oh@fi.aau.dk).

TABLE 1. Steady-State Phosphorylation and Nucleotide and Ca^{2+} (Sr^{2+}) Binding Capacity of Purified Ca^{2+}-ATPase from Sarcoplasmic Reticulum

Source of Material	Specific Activity (μmol $P_i \cdot$ mg$^{-1} \cdot$ min^{-1})	Ligand Binding Capacity (nmol $\cdot$ mg^{-1})			^{32}P-ATP Phosphorylation (nmol $\cdot$ mg^{-1})	Ref.
		ADP	ATP	Ca^{2+} (Sr^{2+})		
Fast-twitch rabbit skeletal muscle	30	8.49 ± 0.43	8.4	10.5	6.5 ± 0.2	2
—	—			10.5		4

binding with much higher affinity was obtained in the absence of Mg^{2+}. Binding data plotted according to Scatchard were curvilinear, however, which might indicate heterogeneity of the receptor, but probably not that half of the binding took place to an alternative, such as a regulatory nucleotide receptor, because this binding would imply phosphorylation of the enzyme. The specificity of ADP binding to Ca^{2+}-ATPase under such conditions was documented in an experiment in which enzyme was preincubated with an equimolar concentration of thapsigargin, because this specific inhibitor competitively reduced the affinity for ADP to high- as well as low-affinity receptors.

Binding capacities of Ca^{2+}-ATPase for a number of ligands are summarized in TABLE 1. It is seen that (1) nearly 100% of a 110-kD protein is apparently disposed of in nucleotide binding, (2) the nucleotide binding capacity does not differ much from the Ca^{2+}-binding capacity, and (3) the phosphorylation capacity neither matches the nucleotide nor half the Ca^{2+}-binding capacity. TABLE 2 compares binding capacities for nucleotide and ouabain to Na^+,K^+-ATPase isolated from kidney and purified to a similar degree as SR-Ca^{2+}-ATPase.[5,6] Also with Na^+,K^+-ATPase, high-affinity nucleotide binding took place in the absence of Mg^{2+}. The binding capacities for nucleotide and ouabain are equal (and equal to vanadate binding and phosphorylation) and only roughly half the $\alpha\beta$-protomers are disposed of.

In conclusion, Na^+,K^+-ATPase as well as Ca^{2+}-ATPase has no Mg^{2+} requirement for high-affinity ADP or ATP binding. The nucleotide binding capacity of Ca^{2+}-ATPase corresponds to the total number of 110-kD peptides, whereas the same parameter in Na^+,K^+-ATPase corresponds to roughly half the 147-kD $\alpha\beta$-protomers. Acid-stable ^{32}P-phosphorylation of Ca^{2+}-ATPase neither matches the nucleotide binding capacity nor half the Ca^{2+}-binding capacity and may be a dubious parameter for this enzyme. In the absence of cooperativity in Ca^{2+} or Sr^{2+} binding to

TABLE 2. Nucleotide and Ouabain Binding Capacity of Purified Kidney Na,K-ATPase

Source of Material	Specific Activity (μmol $P_i \cdot$ mg$^{-1} \cdot$ min^{-1})	Ligand Binding Capacity (nmol $\cdot$ mg^{-1})		Ref.
		ADP	Ouabain	
Pig kidney	33.7 ± 0.6 ($n = 7$)	3.61 ± 0.01 ($n = 7$)		⎤ 5
Pig kidney	33.5		3.4	
Pig kidney	40.1		3.6	⎦
Mink kidney	36.2		4.0	⎤ 6
Mink kidney	35.5		3.6	⎦

Ca^{2+}-ATPase and with a Ca^{2+} binding capacity similar to the nucleotide binding capacity, a monomeric structure and a ratio of 1 Ca^{2+} per 110 kD peptide of Ca^{2+}-ATPase is proposed.

REFERENCES

1. HANSEN, O., J. JENSEN, J. G. NØRBY & O. OTTOLENGHI. 1979. Nature **280:** 410–412.
2. HANSEN, O. & J. JENSEN. 1995. Cell Calcium **18:** 557–568.
3. DE MEIS L. & W. HASSELBACH. 1971. J. Biol. Chem. **246:** 4759–4763.
4. HANSEN, O. & J. JENSEN. 1994. The Sodium Pump. Structure Mechanism, Hormonal Control and its Role in Disease. E. Bamberg & W. Schoner, Eds.: 139–142. Steinkopff/Springer.
5. JENSEN, J. 1992. Biochim. Biophys. Acta **1110:** 81–87.
6. HANSEN, O. 1992. Acta Physiol. Scand. **146:** 229–234.

On the Mechanism of Inhibition of the PMCa^{2+}-ATPase by Lanthanum

CLAUDIO J. HERSCHER[a] AND ALCIDES F. REGA

Instituto de Química y Fisicoquímica Biológicas
Facultad de Farmacia y Bioquímica
Junín 956
1113 Buenos Aires, Argentina

Lanthanum ion (La^{3+}) inhibits the plasma membrane Ca^{2+}-ATPase (PMCa^{2+}-ATPase) with high affinity.[1,2] Presteady-state effects of La^{3+} on the partial reactions of ATP hydrolysis by the PMCa^{2+}-ATPase were investigated to determine the mecha-

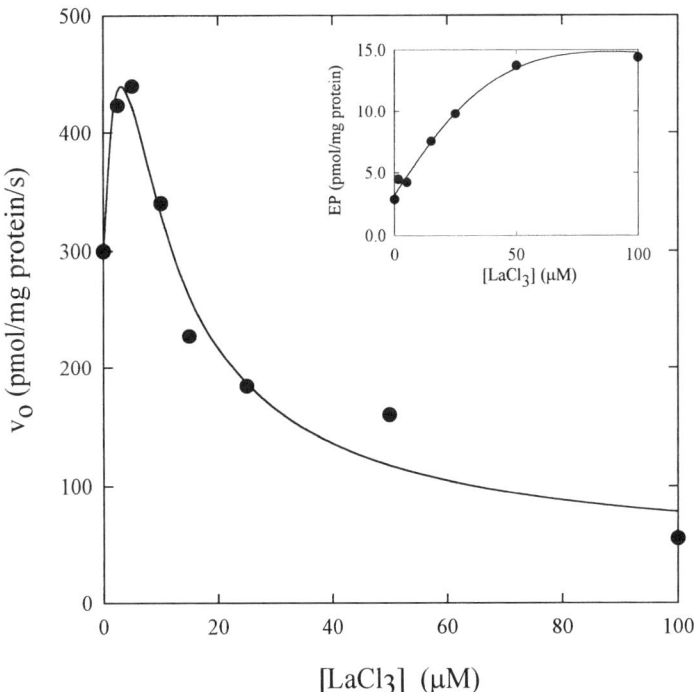

FIGURE 1. Effect of LaCl$_3$ added during phosphorylation. Membranes were preincubated in buffer with 0.2 mM CaCl$_2$ and 0.5 mM MgCl$_2$ and phosphorylated 2.5 ms in the same medium with 10 μM (γ-^{32}P)ATP plus different concentrations of LaCl$_3$. v_o was the ratio between the amount of EP and 2.5 ms. (**Inset**) Effect of LaCl$_3$ on steady-state level of EP. Conditions were identical except that phosphorylation time was 5 seconds.

[a]Tel: (541) 9615810; fax: (541) 9625457.

nism of such inhibition. Experiments were performed at 37°C on fragmented intact membranes from pig red cells as previously described.[3-5]

RESULTS AND DISCUSSION

Effects of La^{3+} before Phosphorylation. Since among conformers E_1 and E_2, the former is the only catalyzing phosphorylation, the initial rate of phosphorylation (v_o) should be directly related to the concentration of E_1, and its rate of increase should be the rate of transformation of E_2 into E_1. As a function of preincubation time, v_o raised along exponential curves with k_{app} of 14 and 23 s^{-1} in the absence and presence of $LaCl_3$, respectively. The effect of La^{3+} during preincubation was the same whether or not preincubation media contained added $CaCl_2$ and $MgCl_2$. Results showed that La^{3+} before phosphorylation raised the k_{app} of the E_2 to E_1 transition, increasing the concentration of E_1.

Effects of La^{3+} Added during Phosphorylation. FIGURE 1 shows that as a function of the concentration of $LaCl_3$, v_o raised, passed through a maximum at about 8 µM $LaCl_3$, and then lowered to about 20% of the value without $LaCl_3$. The decrease in v_o

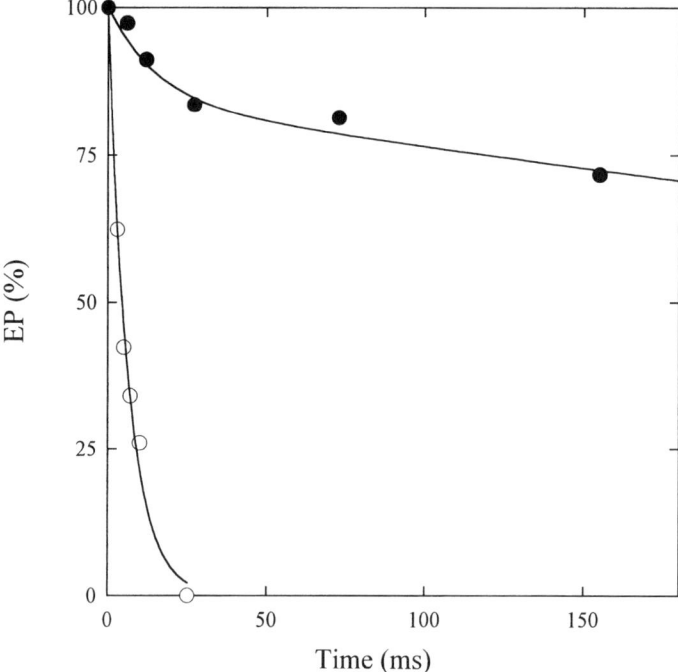

FIGURE 2. Effect of $LaCl_3$ on the time course of dephosphorylation of EP. Membranes were phosphorylated in 0.2 mM $CaCl_2$, 0.5 mM $MgCl_2$ plus 75 (●) or 0 µM $LaCl_3$ (○). Dephosphorylation was started by the addition of a concentrated solution of ATP to obtain a final concentration of 0.3 mM ATP.

was half-maximum at about 20 μM $LaCl_3$ and had to be attributed to slower phosphorylation because a 5-second long phosphorylation made $LaCl_3$ raise the concentration of EP up to fivefold the level without $LaCl_3$ (inset to FIG. 1). $CaCl_2$ lowered the relative drop in v_o, but increasing concentrations of $MgCl_2$ protected the enzyme from this effect of La^{3+}. The values of $K_{0.5}$ for Mg^{2+} increased linearly with the concentration of $LaCl_3$ up to 54-fold, as expected if La^{3+} displaced Mg^{2+}.

Dephosphorylation. FIGURE 2 shows that $LaCl_3$ slowed the decomposition of EP with $k_{app} = 0.8$ s^{-1}, a value characteristic of phosphoenzyme without magnesium.[5] The concentration of $LaCl_3$ for half-maximum inhibition of dephosphorylation increased linearly with the concentration of $MgCl_2$, suggesting that La^{3+} inhibited dephosphorylation by displacing Mg^{2+} from the site from which Mg^{2+} accelerates this reaction.[5] Since La^{3+} accelerated the transition of E_2 to E_1 and slowed phosphorylation to a lesser extent than did dephosphorylation, it seems possible that La^{3+} inhibited steady-state $PMCa^{2+}$-ATPase activity by displacing Mg^{2+} from the site at which it combines to accelerate dephosphorylation.

REFERENCES

1. SCHATZMANN, H. J. & H. BÜRGIN. 1978. Ann. N.Y. Acad. Sci. **307**: 125–147.
2. LUTERBACHER, S. & H. J. SCHATZMANN. 1983. Experientia **39**: 311–312.
3. ADAMO, H. P., A. F. REGA & P. J. GARRAHAN. 1988. J. Biol. Chem. **263**: 17548–17554.
4. ADAMO, H. P., A. F. REGA & P. J. GARRAHAN. 1990. J. Biol. Chem. **265**: 3789–3792.
5. HERSCHER, C. J., A. F. REGA & P. J. GARRAHAN. 1994. J. Biol. Chem. **269**: 10400–10406.

Nucleotide Binding to Na,K-ATPase

Effect of Ionic Strength and Charge[a]

JENS G. NØRBY[b] AND MIKAEL ESMANN

Department of Biophysics
University of Aarhus
DK-8000 Aarhus C, Denmark

The physiological ligands for the substrate site of Na,K-ATPase are ions (Mg^{2+}, ATP^{4-}, ADP^{3-}, and $HPO^{2-}_4/H_2PO^-_4$) and a molecule (H_2O) with a prominent dipole moment. Electrostatic effects are therefore likely to play an important role in their interaction with the enzyme. Such effects depend on the charge at the enzyme binding site, Z_E, the charge of the ligand, Z_A, as well as the ionic strength, I, of the medium. In the present communication we quantify this phenomenon by using the minimal law of Debye and Hückel for the relation between activity coefficient, γ, charges Z_E and Z_A, and I (1): $\log \gamma_i = -0.5 \cdot Z_i^2 \cdot \sqrt{I}$, and we compare the effect of I on ADP-binding to the enzyme with literature data on the effect of charge on both ADP and ATP binding.

RESULTS AND DISCUSSION

Binding of ADP to pig kidney Na,K-ATPase[2] was performed by a centrifugation method as previously described.[3] The basal ionic strength, I, of the medium was 0.075 M, and it was increased by NaCl. Binding of ADP and ATP to Na,K-ATPase, E, follows a one-site model[3] and the binding capacity is independent of I. The dissociation constant of EADP increases with I and is 25 times higher at $I = 0.5$ M than at $I = 0.075$ M.[4] The dependence of $K_{diss} = [E][ADP]/[EADP] = K_0 \cdot \gamma_{EADP}/(\gamma_E \cdot \gamma_{ADP})$ on I is described by the minimal law of Debye and Hückel. K_0 is independent of I, and if it is assumed that $Z_{EADP} = Z_E + Z_{ADP}$, insertion of the minimal law into the equation for K_{diss} results in: $\log(K_{diss}) = \log(K_0) - Z_E \cdot Z_{ADP} \cdot \sqrt{I}$. We found that $\log(K_{diss})$ was a linear function of $\sqrt{I}$ with a slope $-Z_E \cdot Z_{ADP} \cong +3$ (FIG. 1). Because $Z_{ADP} = -3$, we could estimate $Z_E \cong +1$. This is presumed to be the net charge at the substrate binding site, and $K_0 = 39$ nM for ADP. The observed higher affinity for ATP at 0.075 M ionic strength of 60–109 nM[5,6] suggests that $K_0 \approx 6.9$ nM for ATP, assuming $Z_{ATP} = -4$ (FIG. 1).

Recently, Pedersen *et al.*[6] obtained the enzymatically inactive pig kidney Na,K-ATPase mutant Asp369 → Asn369, which according to the foregoing model should have a net charge of $+2$ instead of $+1$ at the substrate site. They measured K_{diss} for EATP and EADP for the native enzyme and for the mutant at $I = 0.075$ M and reported that removal of one negative charge resulted in a decrease in K_{diss} for EATP from 109 to about 5.9 nM, whereas K_{diss} for EADP was unchanged by the mutation.[6] In analogy with the analysis just given, this leads to $K_0 \approx 0.038$ nM for ATP-binding to the mutant (with $Z_E = +2$; FIG. 1), that is, an increase in affinity for ATP by a factor

---

[a]This work was supported by EU contract CI1*CT93-0048 and HFSP grant RG-511/95M.
[b]Tel: +45 8942 2935; fax: +45 8612 9599; e-mail: jgn@mil.au.dk

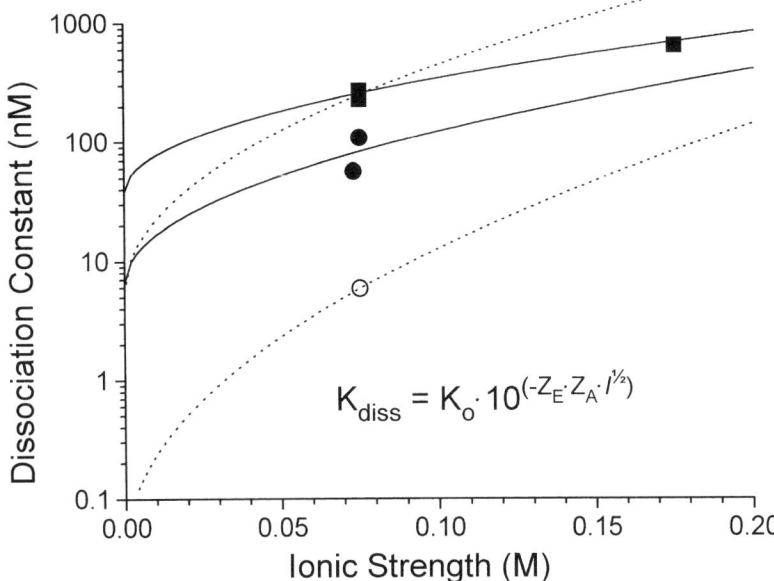

FIGURE 1. Dependence of dissociation constants for EADP and EATP on ionic strength. Data for K_{diss} for ADP (*filled blocks*) are taken from Jensen et al.[2] and Nørby and Esmann,[4] and data for ATP binding from Nørby and Jensen[5] and Pedersen et al.[6] with the *filled circles* giving K_{diss} for native Na,K-ATPase and the *open circle* the value for K_{diss} for the Asp369 → Asn369 mutant. *Full lines* are drawn according to the Debye-Hückel relationship $\log(K_{diss}) = \log(K_0) - Z_E \cdot Z_A \cdot \sqrt{I}$, with $Z_E = +1$ for native Na,K-ATPase, $Z_A = -3$ and $K_0 = 39$ nM for ADP, and $Z_A = -4$ and $K_0 = 6.9$ nM for ATP. For the mutated enzyme (*dashed lines*), $Z_E = +2$ and $K_0 dl$ is 5.9 for ADP and 0.038 nM for ATP.

180 (at $I = 0$). It is possible that the lack of effect of mutation on K_{diss} for ADP indicates a specific electrostatic interaction between Asp369 and the γ-phosphate of ATP (and no such interaction with the negative charges of ADP).

In light of our analysis of the effect of ionic strength on K_{diss} for EADP an alternative explanation is possible, namely, that K_0 for ADP for the mutant is decreased to about 6 nM. With $Z_E = +2$ this would give the same K_{diss} at $I = 0.075$ M as that of the native Na,K-ATPase, about 250 nM (FIG. 1). The increase in affinity for ADP upon mutation would thus be about sevenfold, considerably less than the 180-fold increase in affinity for ATP.

REFERENCES

1. ROBINSON, R. A. & R. H. STOKES. 1970. Electrolyte Solutions, 2nd Ed. Butterworth. London.
2. JENSEN, J., J. G. NØRBY & P. OTTOLENGHI. 1984. J. Physiol. (Lond.) **346:** 219–241.
3. NØRBY, J. G. & J. JENSEN. 1988. Methods Enzymol. **156:** 191–201.
4. NØRBY, J. G. & M. ESMANN. 1997. J. Gen. Physiol. **109:** 555–570.
5. NØRBY, J. G. & J. JENSEN. 1974. Ann. N.Y. Acad. Sci. **242:** 158–167.
6. PEDERSEN, P. A., J. H. RASMUSSEN & P. L. JØRGENSEN. 1996. J. Biol. Chem. **271:** 2514–2522.

Two Unexplained Kinetic Features of Na,K-ATPase May Be Understood as Indicating K$^+$-Induced Cooperativity between Subunits in a Dimeric Enzyme

IGOR W. PLESNER[a]

Department of Chemistry
University of Aarhus
Langelandsgade 140, Bldg. 510
DK-8000 Aarhus C, Denmark

In 1994 Klodos *et al.*[1] published a detailed study of the transient dephosphorylation kinetics of Na,K-ATPase as a function of the NaCl concentration in the dephosphorylating medium (chase solution). Experiments were performed both by initiating the dephosphorylation by the addition of ADP and by adding K$^+$.

Most experiments were performed at 0°C. On the basis of their experiments the authors concluded that new features in a model, such as "heterogeneous kinetics," that is, processes involving several unmixed lipid phases, or, alternatively, sudden temporary changes in rate constants on changing the salt concentration and subsequent relaxation to the initial values, were necessary to interpret the results.

In 1995 Schwarzbaum *et al.*[2] observed that although the steady-state velocity of the enzyme divided by the total phosphoenzyme concentration in the absence of K$^+$ was independent of the ATP concentration, a property characteristic of the Post-Albers model, that quantity was a hyperbolic function of ATP in the presence of K$^+$. These two sets of properties are interpreted in this presentation.

PROCEDURES

The objective is not to present a quantitative fitting of a model. Rather, it is shown that with reasonable values of the rate constants, the models to be presented possess the properties exhibited in the experiments and that, therefore, with sufficient data a quantitative fit could be obtained.

One key feature used here is based on the experimentally determined rate constant for spontaneous dephosphorylation. Data at 20°C are given in ref. 1 and at 37°C in ref. 3. In both cases the rate constant decreases greatly as the NaCl concentration is increased. On the basis of these data the behavior of the same rate constant at 0°C can be calculated using straightforward thermodynamics, and consistency in the model then requires that also the rate constant governing the addition of ADP to the enzyme must depend on the NaCl concentration, leading to a strong *increase* as the salt concentration increases. Based on these considerations, rate constants are assigned to the models. The simulations were carried out by simply solving numerically the

[a]Tel: +45 8942 3858; fax: +45 8619 6199; e-mail: IWP@kemi.aau.dk

differential equations of the model: the simple Post-Albers scheme (E1-E1ATP-E1P-E2P-E1) in the experiments in which dephosphorylation was initiated by a chase of ADP in the absence of K^+, and a simple prototype dimer model previously analyzed[4] in the K^+-chase experiments. In simulations of experiments during which the concentration of NaCl was suddenly changed, the NaCl-dependent rate constants, only two in both cases, were replaced by new values at appropriate time points corresponding to those of the experiments, whereas all other rate constants were

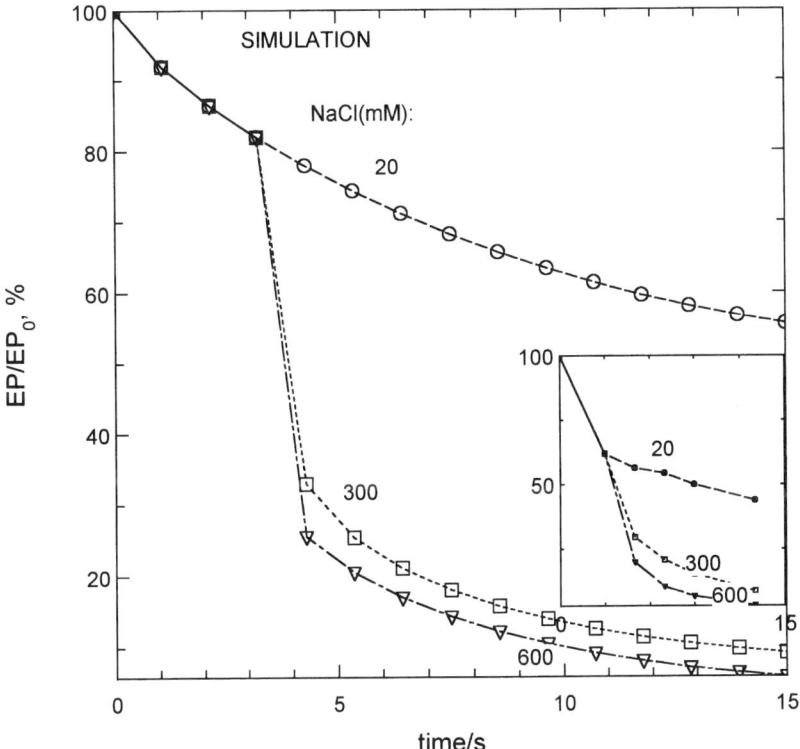

FIGURE 1. Simulations of "ADP-chase experiments" using the simple Post-Albers scheme. The enzyme was phosphorylated with 20 mM NaCl and 20 µM labeled ATP. At t = 0 a chase solution with 2.5 mM ADP and 1 mM unlabeled ATP was added. Subsequently, at t = 3 seconds the NaCl concentration was suddenly changed to (a) 20 mM; (b) 300 mM; and (c) 600 mM. Experiments are shown in the **inset**. Data were taken from Fig. 12 of ref. 1.

unchanged. Also, to correspond with the experiments, the proper steady-state distribution of phosphoenzyme between E_1P and E_2P at time = 0 must be used. This distribution, in turn, depends on some of the rate constants. In the K^+-chase experiments in which a dimer model is necessary to ensure the desired qualities in the resulting simulations, it is assumed that all the monomer units (αβ-protomers) are phosphorylated initially. Also, in this case it is necessary to work with a model in an expanded form[11] to account properly for the fate of the label during the kinetic runs.

The simulated curves resemble the experimental curves in all cases, indicating that the simple models used here are qualitatively satisfactory and that no new principles need be invoked to explain the data. Two examples from the simulations are shown in FIGURES 1 and 2, and a comparison is made with the relevant experimental data. The property of the steady-state rate, v, divided by the total phosphoenzyme, EP, will in these models be in accord with the experimental findings: independent of ATP in the case of the Na^+-ATPase and a parabolic dependence in the presence of K^+. The results demonstrate that (1) the dephosphorylation kinetics in the absence of K^+ can be well simulated using a simple Post-Albers model provided account is taken of the

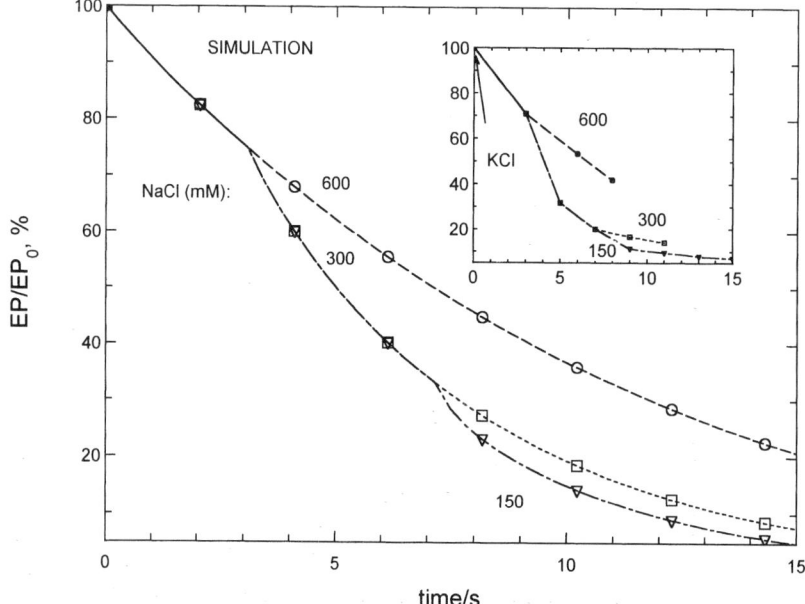

FIGURE 2. Simulations of "K^+-chase experiments" using a simple dimer model. Data (**inset**) were taken from Fig. 9 of ref. 1. Enzyme was phosphorylated with 600 mM NaCl and 20 μM labeled ATP, and 20 mM KCl was added, together with 1 mM unlabeled ATP, at t = 0. Subsequently, at t = 3 and 7 seconds, respectively, the KCl concentration was decreased by a factor of 2.

experimentally determined variation with [NaCl] of certain rate constants, and the ratio v/EP is independent of ATP in this case; and (2) the observed kinetics in the presence of K^+ cannot be accounted for with a Post-Albers scheme, but a primitive model for a dimeric enzyme possesses the correct qualitative features, including the variation with ATP of the quantity v/EP.

It is therefore suggested that (1) the enzyme is monomeric under K^+-free conditions and the Post-Albers scheme is satisfactory, and (2) the presence of K^+ induces cooperative interactions between the monomers, resulting in a dimeric function of the enzyme.

Evidence is mounting to support the notion of a dimeric Na,K-ATPase.[5–10] Even a simple dimer model leads to the existence of a presteady-state cycle for the enzyme, in which the first phosphorylation (of one subunit) takes place. The existence of such a cycle has been suggested by Rossi and Nørby.[12]

REFERENCES

1. KLODOS, I., R. L. POST & B. FORBUSH, III. 1994. J. Biol. Chem. **269:** 1734–1743.
2. SCHWARZBAUM, P. J., S. B. KAUFMAN, R. C. ROSSI & P. J. GARRAHAN. 1995. Biochim. Biophys. Acta **1233:** 33–40.
3. PLESNER, I. W. 1991. *In* The Sodium Pump: Recent Developments. J. H. Kaplan & P. de Weer, Eds.: 339–343. The Rockefeller University Press. New York.
4. PLESNER, I. W. 1987. Biophys. J. **51:** 69–78.
5. FROEHLICH, J. P. & R. W. ALBERS. 1995. Biophys. J. **68:** A316.
6. GANJEIZADEH, M., N. ZOLOTARJOVA, W.-H. HUANG & A. ASKARI. 1995. J. Biol. Chem. **270:** 15707–15710.
7. BUXBAUM, E. & W. SCHONER. 1991. Eur. J. Biochem. **195:** 407–419.
8. LINNERTZ, H., D. THÖNGES & W. SCHONER. 1995. Eur. J. Biochem. **232:** 420–424.
9. BOLDYREV, A. & P. J. QUINN. 1994. Int. J. Biochem. **26:** 1323–1331.
10. LIU, G., Z. XIE, N. N. MODYANOV & A. ASKARI. 1996. FEBS Lett. **390:** 323–326.
11. PLESNER, I. W. 1992. Biochem. J. **286:** 295–303.
12. ROSSI, R. C. & J. G. NØRBY. 1993. J. Biol. Chem. **268:** 12579–12590.

Models for ATPase-Substrate Activation Kinetics

GRETEL ROBERTS AND LUIS BEAUGÉ

Instituto M. & M. Ferreyra
CC 389
5000 Córdoba, Argentina

Depending on the enzyme structure and chemical environment, H^+-ATPase of higher plants may present a complete variety of behaviors towards ATP, including positive or negative cooperativity and michaelian kinetics.[1] These activation patterns require two ATP effects on the enzyme: catalytic and regulatory. However, they may result from the existence of two simultaneous sites or a single site that consecutively changes its properties. We describe three models for two substrate sites in the plant H^+-ATPase: (1) Two simultaneous and initially identical catalytic sites, the second of which becomes regulatory after occupancy of the first; (2) two simultaneous and initially different sites, one always catalytic and the other always regulatory. This offers two alternatives: (a) one with potential cooperative binding (system does not discriminate between binding to the catalytic or to the regulatory site) and (b) one with the affinities of the catalytic and regulatory sites intrinsically different (system discriminates between binding to the sites); (3) a single site that consecutively becomes catalytic and regulatory as the cycle proceeds. The results show that the three mechanisms give equivalent rate equations.

METHODS

Curve fittings and simulations were done with the Scop Program (Simulation Resources, Inc., Berrien Springs, Michigan). Models 1 and 2 were analyzed by rapid equilibrium, whereas model 3 has a steady-state solution.[2] Equations were:

Model 1: $v = V_{max}[(aS/Ks) + (S^2/\alpha Ks^2)]/[1 + (2S/Ks) + (S^2/(\alpha Ks^2)]$

Model 2a: $v = V_{max}[(aS/Ks) + (S^2/\alpha Ks^2)]/[1 + (2S/Ks) + (S^2/(\alpha Ks^2)]$

Model 2b: $v = V_{max}[(aS/Kc) + (S^2/\alpha KcKr)]/[1 + (S/Kc) + (S/Kr) + (S^2/(\alpha Ks^2)]$

Model 3: $v = V_{max}(A[S] + B[S]^2)/(1 + C[S] + D[S]^2)$

where A, B, C, and D are the quotients of sums and products of unidirectional rate constants. In the fitting of the data they were chosen arbitrarily.

We used the v/s versus v Eadie-Scatchard plot due to its sensitivity to deviations from michaelian behavior: (1) concavity facing upward = negative cooperativity, (2) concavity facing downward = positive cooperativity, and (3) straight line = michaelian kinetics.[2] The results correspond to ATP activation of the AHA2 isoform of H^+-ATPase obtained by heterologous expression in the endoplasmic reticulum (ER) of yeast strains MP-142 and MP-194.[3] The ER-enriched membrane fraction was prepared according to Villalba et al.[4] ATP hydrolysis was estimated as in ref. 5.

RESULTS AND DISCUSSION

The three patterns of enzyme-substrate interactions are shown in FIGURE 1. The values of the fitting parameters for Models 1 and 2 are in TABLE 1. For Model 3 we describe only ratios of rate constants which are given in the text. Positive cooperativity is demonstrated in FIGURE 1A (native enzyme), michaelian kinetics in FIGURE 1B (detergent-solubilized enzyme), and negative cooperativity in FIGURE 1C (COOH-terminal truncated enzyme).[1] All models can produce excellent fittings. Another important conclusion is that the form of the curve might resemble but does not necessarily mean a certain cooperativity type. For instance, a curve with concavity

TABLE 1. Fitting Parameters Obtained with Two Simultaneous ATP Site Models to Fit ATP Activation Kinetics of Plant H^+-ATPase[a]

Model	Parameter	ER AHA2	ER aha2Δ92	ER AHA2 with Detergent
(1)	K_s (μM)	74	15	60
	α	1.08	12	1.3
	a	0.17	0.8	0.98
	V_{max} (μmol P_i/min × mg)	2.4	0.47	4.3
(2)	K_c (μM)	65	9	42
	K_r (μM)	250	70	140
	α	0.5	7	1.2
	a	0.15	0.4	0.7
	V_{max} (μmol P_i/min × mg)	2.6	0.53	4.3

[a](1) Two initially catalytic sites, the second becoming regulatory after occupancy of the first. (2) Two simultaneous ATP sites, one catalytic and one regulatory. The patterns of enzyme-substrate interactions are shown in FIGURE 1A, B, and C as Eadie-Scatchard plots (v/s versus v). Note: (i) "a" is the ratio of v1/v2 (nonregulated over ATP-regulated pathway); and (ii) "α" is the cooperativity factor (>1 for negative and <1 for positive cooperativity).

facing downwards can result from a v1 velocity pathway much slower than the v2 path, even with the cooperativity factor $\alpha > 1$. Two simultaneous ATP sites (Models 1 and 2) are possible provided the substrate also has regulatory properties in the sense that ES (or SE) leads to a slower velocity path than does SES (v1 < v2). With the two consecutive ATP sites (Model 3), the resemblance to a given cooperativity or a michaelian kinetics will depend on the relative affinities of the catalytic and regulatory loci and the relative velocity of the path without and with ATP bound to the regulatory site. We obtain an excellent fit to positive cooperativity resemblance (FIG. 1A) when the ATP affinity is about 100 times higher for the regulatory than for the catalytic site and the "regulated" path is about 10 times slower than the "nonregulated" path. Reverse relationships will account for negative cooperativity. Taken together, these results show that kinetic analyses by themselves are unlikely to provide unambiguous evidence on the mechanisms underlying enzyme-ATP interactions in any transport ATPase.

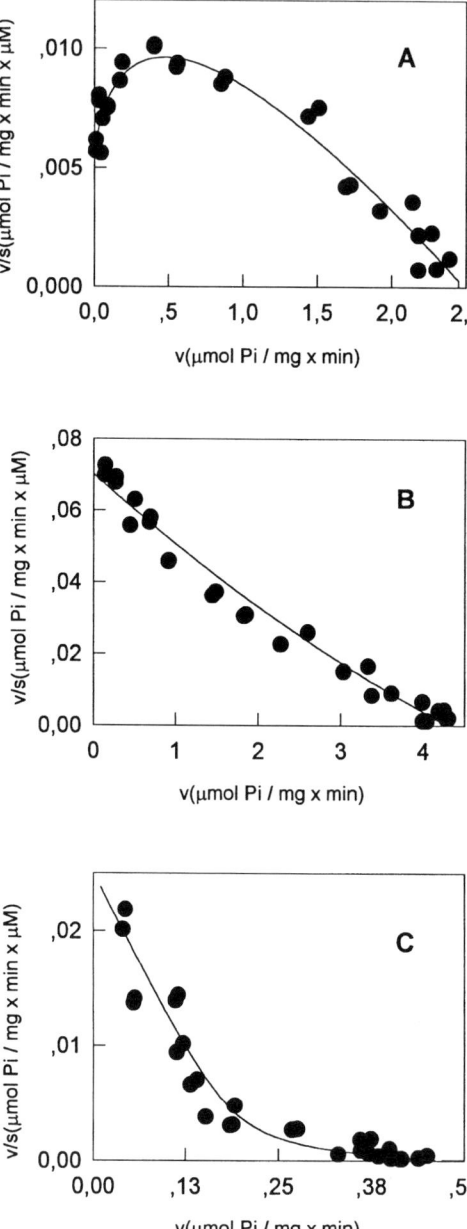

FIGURE 1. Enzyme-substrate interactions of AHA2 H^+-ATPase plotted in an Eadie-Scatchard graph (v/s versus v). Positive cooperativity is represented in **A** (native enzyme), michaelian kinetics in **B** (detergent-solubilized enzyme), and negative cooperativity in **C** (COOH-terminal truncated enzyme). Note that the three models illustrated in the text, two simultaneous sites (initially both catalytic or one catalytic and one regulatory), and two consecutive (catalytic and regulatory) ATP sites can account for all of these results.

REFERENCES

1. ROBERTS, G., G. BERBERIAN & L. BEAUGÉ. 1995. Biochem. Biophys. Acta **108**: 813–819.
2. SEGEL, Y. 1975. Enzyme Kinetics. Wiley Interscience. New York, NY.
3. PALMGREN, M. & G. CHRISTENSEN. 1993. FEBS. Lett. **317**: 216–222.
4. VILLALBA, J., M. PALMGREN, G. BERBERIAN, C. FERGUSON & R. SERRANO. 1992. J. Biol. Chem. **267**: 12341–12349.
5. ROBERTS, G., G. BERBERIAN & L. BEAUGÉ. 1991. Biochem. Biophys. Acta **1064**: 131–138.

Properties of the Cytoplasmic Ion Binding Sites[a]

WOLFGANG DOMASZEWICZ, ANNE SCHNEEBERGER, AND HANS-JÜRGEN APELL[b]

Faculty of Biology
University of Konstanz
D-78434 Konstanz, Germany

A so far unresolved question concerning the Na,K-ATPase is the molecular structure of the ion binding sites, their physicochemical properties, and the detailed mechanism of their action. Two approaches to contribute to this field are presented in this paper: (1) a newly developed experimental device using "total internal reflection fluorescence" and (2) a comparison of results from experiments with two fluorescence labels of the Na,K-ATPase, RH421 and FITC.

(1) To obtain information on the dielectric properties, we developed an experimental setup to measure optical and electrical signals simultaneously. The optical component of the setup detects the fluorescence of styryl dye RH421 which reports changes in the local electrical field in the Na,K-ATPase.[1] Analysis of electrical signals is similar to the method in which membrane fragments are coupled capacitively to lipid bilayers.[2]

In our setup (FIG. 1), Na,K-ATPase containing membrane fragments stained with RH421 are added to the buffer solution in the cuvette. They adsorb to the quartz interface and are oriented supposedly with the extracellular side towards the quartz. Adsorption is completed after 30–60 minutes. Adsorption of the fragments is stable for several hours, allowing us to exchange buffers and to perform repeated experiments with the same proteins by returning to the initial conditions. The RH421 fluorescence in the adsorbed fragments is excited by the evanescent wave of a laser beam (594 nm) that is totally reflected at the quartz interface. Therefore, the depth of excitation is confined to the order of the wavelength of excitation. The fluorescence of RH421 is detected at 660 nm by a photomultiplier. Electrical potentials can be applied to the adsorbed fragments by a pulse generator connected to the ITO electrode on the quartz and to a platinized platinum electrode immersed into the buffer (FIG. 1). The ITO layer has a high conductance and is transparent at 594 nm. The insulating quartz layer on top of the ITO layer prevents a short circuit between both electrodes. Adsorption of the membrane fragments on the quartz layer leads to a capacitive coupling to the applied voltage and to the detection system.

Fluorescence changes of RH421 observed upon the addition of substrates (Na^+, ATP, and K^+) to adsorbed membranes were in good agreement with similar measurements of fragment suspensions in a standard fluorimeter. Titrations of cytoplasmic ion

---

[a]This work was supported by the Deutsche Forschungsgemeinschaft (Sonderforschungsbereich 156).

[b]Address for correspondence: Faculty of Biology, University of Konstanz, Postfach 5560 M 635, D-78434 Konstanz, Germany (tel: +49-7531-882253; fax: +49-7531-883183; e-mail: h-j.apell@uni-konstanz.de).

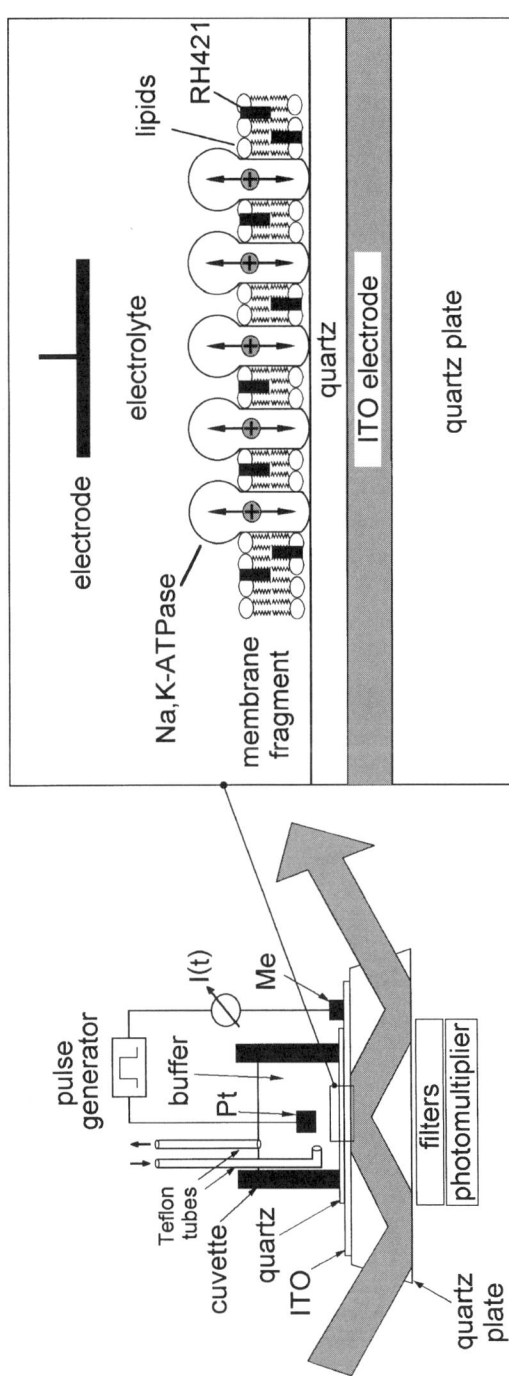

FIGURE 1. Schematic representation of TIRF setup (total internal reflection fluorescence). Laser light of 594 nm is led through a quartz trapezoid. In the area of the second total reflection a cuvette is built by a Teflon cylinder pressed against the quartz surface (**left panel**). Two Teflon tubes are used to exchange the buffer in the cuvette using a peristaltic pump. To perform electrical measurements a platinized platinum electrode (Pt) is immersed into the buffer and a metal pin (Me) is pressed against the transparent Indium Tin Oxide electrode (ITO) deposited on the trapezoid. The 100-nm ITO layer is covered by a 50-nm quartz layer to prevent a short circuit between both electrodes. The quartz layer forms the bottom of the cuvette to which the membrane fragments adsorb, which contain the Na,K-ATPase and the fluorescent dye RH421 (**right panel**). The emitted light is collected by a photomultiplier underneath the quartz trapezoid. By cut-off and interference filters the fluorescence at 660 nm is selected which is sensitive to the electrogenic effects in the ion pump. Therefore, (voltage-induced) charge movements can be detected by RH421 fluorescence.

binding sites with Na$^+$ were performed. The resulting half-saturating concentration $K_{1/2}$ of 6.9 mM in the presence of 5 mM Mg^{2+} is in good agreement with previously published data. Voltage jumps of varying amplitude have been applied to the adsorbed fragments in the presence of various Na$^+$ concentrations to compare optical and electrical effects for differently occupied cytoplasmic sites. These experiments are in progress. Cytoplasmic Na$^+$ binding is of particular importance, because in contrast to the results with rabbit kidney enzyme, Lu et al.[3] concluded from electrical measurements with guinea pig cardiac cells that cytoplasmic Na$^+$ binding is not electrogenic.

(2) In its E_1 conformation the Na,K-ATPase presents three binding sites to the cytoplasm. The enzyme can bind two cations (Na$^+$ or K$^+$) to two negatively charged

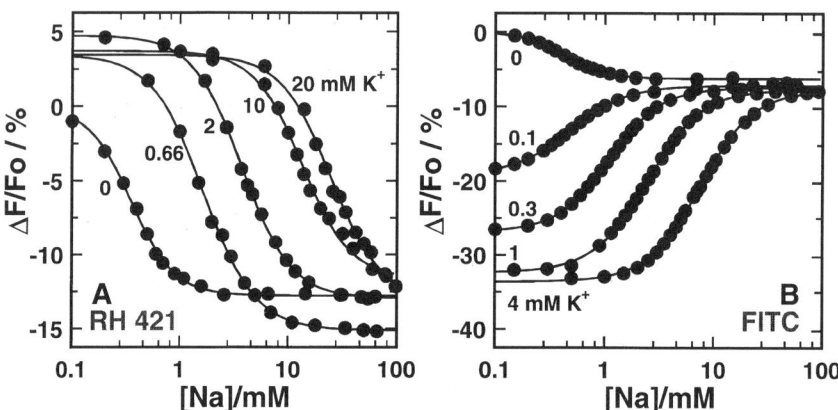

FIGURE 2. Titration of the cytoplasmic ion binding sites with Na$^+$ ions detected by RH421, a fluorescence dye sensitive to charge movements within the ion pump (**left panel**), and by FITC, a covalently bound fluorescence dye sensitive to conformational changes of the protein. Membrane fragments were suspended in buffer containing 25 mM histidine, 0.5 mM EDTA, 300 mM choline chloride, pH 7.3, and the indicated concentration of KCl (T = 17°C). Fluorescence was normalized to the level before the addition of KCl. (**A**) RH421 fluorescence as a function of Na$^+$ concentration in the presence 0, 0.66, 2, 10, and 20 mM KCl. The detected reaction is assumed to be the electrogenic transition Na$_2$E$_1$ → Na$_3$E$_1$. (**B**) FITC fluorescence as a function of the Na$^+$ concentration in the presence 0, 0.1, 0.3, 1, and 4 mM KCl. In the presence of K$^+$ the detected reaction is assumed to be the conformational transition E$_2$(K$_2$) → K$_2$E$_1$. In the absence of K$^+$ it might be Na$_2$E$_1$ → Na$_3$E$_1$.

binding sites which are placed in the (dielectric) surface of the protein.[4] The third site is selective for Na$^+$ and is uncharged. Binding of Na$^+$ to this site is electrogenic and is detected by the electrochromic dye RH421.[1] Na$^+$ binding can also be studied by the use of FITC (fluoresceine 5′-isothiocyanate), a fluorophore bound covalently to the pump, which allows discrimination between the two conformations, E_1 (high fluorescence intensity) and E_2 (low intensity).[5] When applying this dye, ion binding is indirectly measured monitoring actually the preceding conformational change.

With FITC-labeled Na,K-ATPase confined to state E_1 by high ionic strength, the addition of K$^+$ led to a fluorescence drop. Subsequent additions of Na$^+$ reversed the K$^+$-induced quench (FIG. 2B). By titration experiments, apparent affinities $K_{1/2}$ for

Na$^+$ as a function of the K$^+$ concentration were determined. Corresponding measurements using the styryl dye RH421 (FIG. 2A) yielded similar K$_{1/2}$ values. This finding indicates that both methods led to comparable results, although the functional mechanisms of the two dyes are quite different.

Interestingly, with our FITC-labeled enzyme we observed a distinct fluorescence decrease of approximately 7% upon the addition of Na$^+$ when the pump was already confined to state E$_1$ (FIG. 2B). Correspondingly, reversal of the K$^+$-induced fluorescence drop by Na$^+$ ended at the same fluorescence level as was reached upon the sole addition of Na$^+$. Correction for dilution effects and control experiments at even higher ionic strengths than 300 mM choline chloride allowed us to rule out FITC-induced artifacts. Moreover, titrating this Na$^+$-induced fluorescence decrease and fitting the Hill equation to the data led to an apparent affinity for Na$^+$ of 0.4 mM (at 0 Mg^{2+}) which is in good agreement with the 0.36 mM derived by the RH421 method. Inasmuch as RH421 is thought to detect the transition to state Na$_3$E$_1$, comparable results with both labels suggest that the additional fluorescence level observed with FITC enzyme might also represent state Na$_3$E$_1$. Additional evidence was obtained by performing Na$^+$ titration experiments in the presence of various Mg^{2+} concentrations. The apparent affinities derived by RH421 and FITC experiments, respectively, decreased in the same way with increasing Mg^{2+} concentrations, and the K$_{1/2}$ values obtained by the two methods were almost identical (not shown).

Thus, we conclude that the additional fluorescence level could be attributed to state Na$_3$E$_1$. If we assume that FITC, which is bound proximately to the ATP binding site,[5] detects conformational changes that also affect the nucleotide binding site, then the binding of the third Na$^+$ resulting in a fluorescence drop of 7% is caused by a small conformational change close to the ATP site. A possible explanation for this change is that it represents the transition into a state in which the enzyme is competent for phosphorylation by ATP after all three of the ion binding sites are occupied by Na$^+$, the crucial constraint for this partial reaction.

REFERENCES

1. BÜHLER, R., W. STÜRMER, H.-J. APELL & P. LÄUGER. 1991. J. Membr. Biol. **121:** 141–161.
2. BORLINGHAUS, R., H.-J. APELL & P. LÄUGER. 1987. J. Membr. Biol. **97:** 161–178.
3. LU, C.-C., A. KABAKOV, V. MARKIN, S. MAGER, G. A. FRAZIER & D. W. HILGEMANN. 1995. Proc. Natl. Acad. Sci. USA **92:** 11220–11224.
4. WUDDEL, I. & H.-J. APELL. 1995. Biophys. J. **69:** 909–921.
5. KARLISH, S. D. J. 1980. J. Bioenerg. Biomembr. **12:** 111–136.

Interactions of Palytoxin with the Na,K-ATPase

Where Are Those Sites?

M. T. TOSTESON,[a] D. R. L. SCRIVEN, A. BHARADWAJ,
J. ARNADOTTIR, AND D. C. TOSTESON

Laboratory for Membrane Transport
Department of Cell Biology
Harvard Medical School
240 Longwood Avenue
Boston, Massachusetts 02115

Palytoxin (PTX), which binds specifically to the Na,K-ATPase, is one of the most poisonous (LD_{50}): 10–250 ng/kg), nonproteinaceous toxins.[1] To determine the site of action of PTX, we have continued our studies of the interactions of PTX with intact human red blood cell and with isolated enzyme preparations.

Binding of PTX to human red blood cells produces an increase in sodium and potassium electrodiffusion dependent on the concentration of PTX.[2,3] Our data could be fitted to a model in which PTX has two binding sites: N_1 = 300/cell and N_2 = 4,000/cell with apparent affinities: K_{m1} = 25 pM and K_{m2} = 1,500 pM. The increase in the permeability of the cells to cations was further shown to be reversed by ouabain and prevented by the addition of vanadate to the inside of human red cells.[2] These results suggest that a consequence of the binding of PTX to its receptor sites on the Na,K-ATPase molecule might be to alter states of the enzyme that are directly or indirectly involved with the transport of ions across the cell membrane.

We report here the effects of PTX on the rates of occlusion and deocclusion of Rb^+ (as a surrogate for K^+) and of Na^+ by the purified Na,K-ATPase and on the activity of the K^+-activated *p*NPPase.

We, in collaboration with Liqin Liu and Amir Askari from the Medical College of Ohio, Toledo, Ohio, found that PTX accelerates the deocclusion of K^+ in the absence and, more strikingly, in the presence of 10 mM Rb^+, suggesting that the PTX-enzyme complex releases K^+ from the "fast" but not the "slow" K^+ occlusion sites,[4] but has no effect on the occlusion or deocclusion of Na^+ by the Na,K-ATPase. These results raise the possibility that the conformation of the PTX–Na,K-ATPase complex allows access to the K^+ occlusion sites by both Na^+ and K^+ on both sides of the membrane. Moreover, the data suggest that sites for cation occlusion in the native enzyme might be different.

In the presence of PTX, both Na^+ and K^+ increase the activity of the *p*NPPase. These results indicate that the cationic selectivity of the *p*NPPase is greatly diminished by PTX. Our findings are consistent with the notion that within the PTX–Na,K-ATPase complex, the K^+ activation site can also accept Na^+ and becomes accessible to both Na^+ and K^+ on both sides of the plasma membrane.

---

[a]Tel: 617 432-1264; fax: 617 432-0933; e-mail: mtt@warren.med.harvard.edu

Clearly more work is needed to assess the effects of PTX on the various conformational states of the Na,K-ATPase.

REFERENCES

1. SCHEUER, P. J. 1964. Fortschr. Chem. org. Nat Stoffe **22:** 265–283.
2. HABERMANN, E. 1989. Toxicon **27:** 1171–1187.
3. TOSTESON, M. T., J. A. HALPERIN, K. KISHI & D. C. TOSTESON. 1991. J. Gen. Physiol. **98:** 969–985.
4. FORBUSH, B., III. 1988. In The Na^+,K^+-pump, Part A: Molecular Aspects. J. C. Skou, J. G. Nørby, A. B. Maunscbach & M. Esmann, Eds.: 229–248. AR Liss, Inc. New York.

Ordered Interaction of Ions with Na/K-Pump May Confound Interpretation of Unidirectional Fluxes[a]

J. WAGG AND D. C. GADSBY

The Rockefeller University
1230 York Avenue
New York, New York 10021

Normally, in each complete Na/K transport cycle, three Na^+ ions are exported and then two K^+ ions imported in a consecutive manner. The transport pathways for each ion type consist of steps in which ions are bound at one surface of the membrane, occluded within the pump protein, and then deoccluded and released to the other surface. The reversibility of these steps underlies the electroneutral Na_{in}^+-Na_{out}^+ exchange[1,2] and K_{out}^+-K_{in}^+ exchange[3] partial reactions. Although electroneutral, Na_{in}^+-Na_{out}^+ exchange is sensitive to membrane potential (V), and the V dependence of ^{22}Na efflux mediated by Na_{in}^+-Na_{out}^+ exchange in squid axons has been characterized.[4] Over the accessible voltage range (≈ -90 to $+30$ mV), increasingly negative potentials increased Na^+ efflux along a saturating sigmoid curve. A simple interpretation is that Na^+ release to the exterior is accompanied by charge movement, as would arise if, upon release, these ions passed through a high-field access channel.[4]

In the present work, three simplified mechanisms for electroneutral Na_{in}^+-Na_{out}^+ exchange, based on either random (mechanism 1, denoted M1) or strictly ordered Na^+ binding/release (mechanisms 2 and 3, denoted M2 and M3), were used to simulate Na^+ efflux-V relationships. All three mechanisms incorporated an extracellular access channel, and Na^+ ion movements along this channel were assumed responsible for all pump-mediated transmembrane charge movement. For random binding/release (M1), the efflux-V curve was predicted to be sigmoidal with a protracted plateau at negative voltages. By contrast, for ordered binding/release (M2 and M3), curves showed a monophasic rise, as the membrane voltage was made less positive, to a pronounced peak at negative potentials with a subsequent fall to a lower, maintained, level at extreme negative V. These curves differed depending on the ordering relationship assumed, with the peak being sharper, and the negative basal level lower, for M2 (assuming 1st-Na_{in}^+-on = 1st-Na_{out}^+-off) than for M3 (assuming 1st-Na_{in}^+-on = 3rd-Na_{out}^+-off). These results provide important constraints on mechanistic interpretations of the observed voltage dependence of pump-mediated electroneutral Na_{in}^+-Na_{out}^+ exchange.[4]

METHODS

Simulations were based on three simplified mechanisms for electroneutral Na_{in}^+-Na_{out}^+ exchange, M1, M2, and M3 (FIG. 1). Each incorporates two distinct pump

[a]This work was supported by HHMI and National Institutes of Health grant HL-36783.

M1: Random

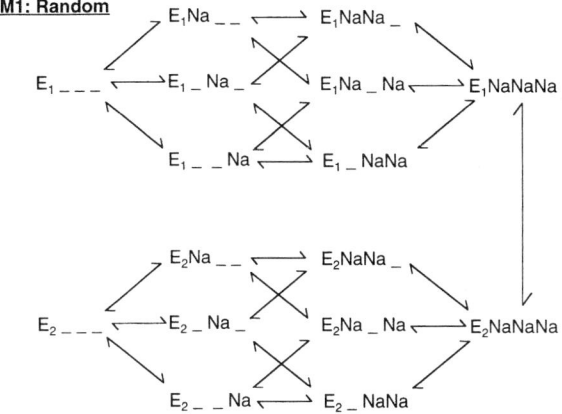

M2: 1st-on-1st-off

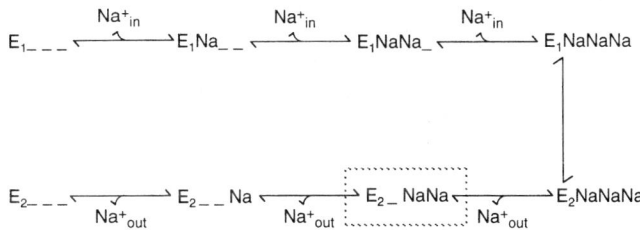

M3: 1st-on-3rd-off

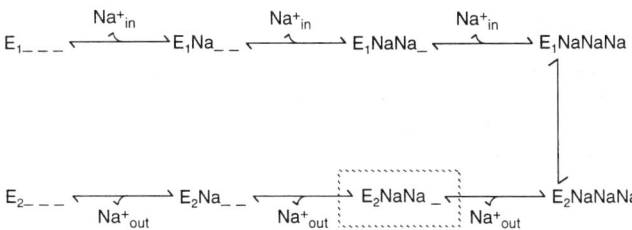

FIGURE 1. Reaction schemes for electroneutral Na_{in}^+-Na_{out}^+ exchange used to simulate Na^+ efflux-V relationships. Each mechanism is based on a different set of assumptions as to the nature of Na^+ binding/dissociation. Mechanism 1 (M1): random; mechanism 2 (M2): strictly ordered so that the first, second, and third ions to bind at one side correspond to the first, second, and third ions to dissociate from the other side; mechanism 3 (M3): strictly ordered so that first, second, and third ions to bind at one side correspond to third, second, and first ions to dissociate from the other side. *Dashed boxes* in mechanisms 2 and 3 indicate sites of inclusion of an additional conformational transition between dissociation of the first and second Na^+. For simplicity, Na_{in}^+ and Na_{out}^+ are omitted from the state diagram for mechanism 1.

conformations, E_1 and E_2, with three intracellularly and extracellularly accessible binding sites, respectively, and consists of (reversible) steps by which: (1) 3 Na_{in}^+ bind to E_1; (2) bound ions are translocated during an $E_1 \rightarrow E_2$ transition (for all mechanisms, it was assumed that only the fully loaded E_1NaNaNa state undergoes transition to the E_2NaNaNa state and vice versa); and (3) translocated ions dissociate from E_2 to the exterior. Na_{in}^+-Na_{out}^+ exchange is a result of "shuttling" back and forth across these steps. Each ion binding site is denoted by "_" (unoccupied) or "Na" (occupied by a Na^+ ion). The three sites, from left to right, will be referred to as sites 1, 2, and 3, respectively; for example, E_1Na_ _ denotes an E_1 state with a Na^+ ion bound to site 1 and the remaining two intracellularly accessible sites unoccupied by Na^+. Differences between the three mechanisms reflect differences in assumptions about the nature of Na^+ ion interaction with these sites.

For M1 (FIG. 1, top), Na^+ ions are assumed to bind/dissociate randomly to/from three independent binding sites. One Na^+ ion binds to either site 1, 2, or 3 of E_1, a second ion then binds to either of the two unoccupied sites, and a third ion binds to the remaining site. Following the E_1NaNaNa $\rightarrow$ E_2NaNaNa transition, a Na^+ ion dissociates from any one of the three sites (site 1, 2, or 3), another ion then dissociates from either of the two remaining occupied sites, and then the final ion dissociates. For this random binding/dissociation, no unique mapping between the order in which ions bind at one side of the membrane and the order in which they subsequently dissociate from the other side (i.e., an ordering relationship) can be defined; thus, the first ion to dissociate from E_2 may correspond to the first, second, or third ion that bound to E_1.

For M2 (FIG. 1, middle), ion binding/dissociation to/from the three binding sites was assumed to be strictly ordered such that the first Na^+ binds to E_1 at site 1, the second at site 2, and the third at site 3; and, following the E_1NaNaNa $\rightarrow$ E_2NaNaNa transition, the first Na^+ is released from site 1, the second from site 2, and the third from site 3. This ordering relationship will be denoted as: 1st-Na_{in}^+-on = 1st-Na_{out}^+-off; 2nd-Na_{in}^+-on = 2nd-Na_{out}^+-off; 3rd-Na_{in}^+-on = 3rd-Na_{out}^+-off. For M3 (FIG. 1, bottom), Na^+ interaction was also assumed to be strictly ordered but with the following ordering relationship: 1st-Na_{in}^+-on = 3rd-Na_{out}^+-off; 2nd-Na_{in}^+-on = 2nd-Na_{out}^+-off; 3rd-Na_{in}^+-on = 1st-Na_{out}^+-off.

In all three mechanisms, upon dissociation from E_2, each Na^+ ion was assumed to traverse 0.33 of the membrane's electrical field via an extracellular access channel (see ref. 4). All other events were assumed electroneutral. E_1 binding sites were assigned equilibrium dissociation constants of 3 mM and association rate constants of 10^7 M^{-1} s^{-1}. Corresponding parameters for E_2 sites were voltage-dependent products of equilibrium dissociation constants, and association rate constants, assigned, respectively, values at 0 mV of 100 mM and 10^7 M^{-1} s^{-1}, and exponential functions of V incorporating an effective valence of 0.33; the association rate constant also incorporated a symmetry factor of 0.5.[5] Ion dissociation rate constants were calculated from assigned association rate constants and equilibrium dissociation constants. Forward and reverse rate constants assigned for the Na translocation ($E_1 \leftrightarrow E_2$) step were both 100 s^{-1}, and internal and external Na^+ concentrations (denoted $[Na]_{in}$ and $[Na]_{out}$) were set at 100 mM. The steady-state distributions of pump enzyme across each step were calculated from steady-state equations as V was varied between -800 and $+200$ mV in 1-mV increments. For each steady-state distribution, forward and backward reaction rates across each step were evaluated and then used to calculate, via a general (applied Markov process/matrix theory) method,[6] the rate of unidirectional Na^+ efflux, as might be measured using the radiotracer ^{22}Na.

These mechanisms are not proposed as realistic descriptions of pump-mediated electroneutral Na_{in}^+-Na_{out}^+ exchange, but they do provide a basis for specific simulations of Na^+ efflux-V relationships which can help identify some general kinetic principles that may govern the form of measured Na^+ efflux-V relationships.

RESULTS AND DISCUSSION

FIGURE 2A presents Na^+ efflux-V relationships simulated, using the parameters and methods just outlined, for each of the three mechanisms (M1, M2, and M3). For random binding (M1) the efflux-V curve was sigmoidal with a protracted plateau at

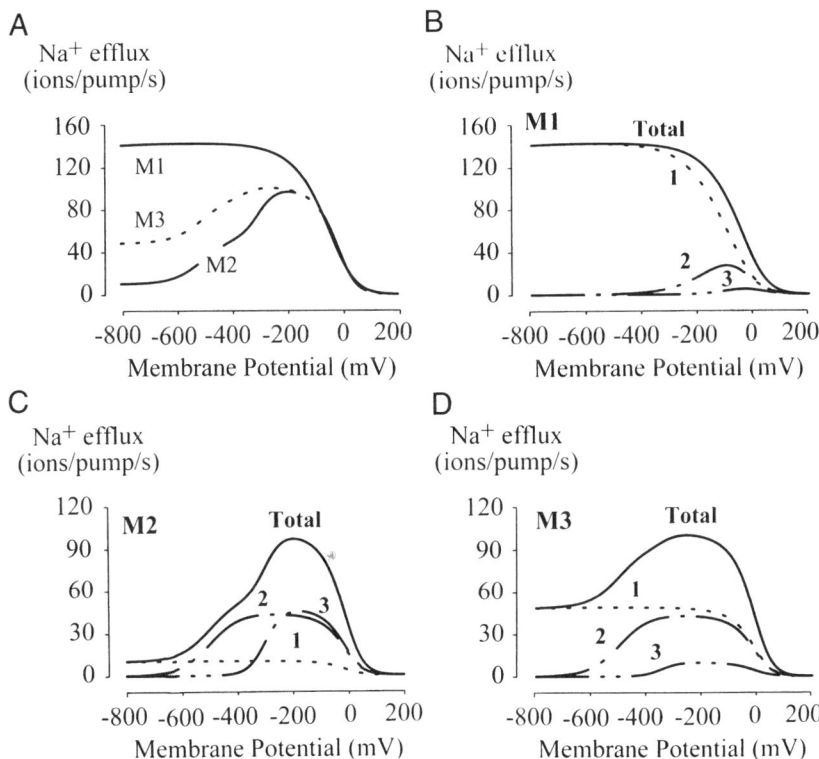

FIGURE 2. Simulated Na^+ efflux-V relationships. All kinetic parameters were held constant as outlined in Methods. **(A)** Efflux-V relationships for differing assumptions as to the nature of Na^+ binding (Mechanisms 1 to 3). Relationships showed marked differences in terms of both magnitude and qualitative behavior. Random binding (Mechanism 1: —) yields a protracted plateau in the efflux relationship, ordered 1st-on = 1st-off; 2nd-on = 2nd-off; 3rd-on = 3rd-off binding/release (Mechanism 2: -----) somewhat diminishes the plateau, and ordered 1st-on = 3rd-off binding/release (Mechanism 3: — —) greatly reduces the plateau. **(B-D)** Breakdown of simulated Na^+ efflux (—) into components mediated by the first (----), second (—·—), and third (—··—) external Na^+ release steps for mechanisms M1 **(B)**, M2 **(C)**, and M3 **(D)**.

negative voltages. This differed from the efflux-V curves simulated on the basis of ordered binding (M2 and M3) which both showed a monophasic rise, as the membrane voltage was made less positive, to a pronounced peak at negative potentials with a subsequent fall to a lower level at extreme negative V. The curves for the two ordered mechanisms differed depending on the ordering relationship assumed, with the peak being sharper and the lower level at negative potentials being smaller for M2 than for M3 (compare FIG. 2C with FIG. 2D).

These differences reflect the presence of a kinetic effect comparable to the long-pore effect described for ion channels.[7] For M2, the second and third Na^+ ions that bind to E_1 can be released to the exterior only after the first, and first and second, respectively, Na^+ ions have dissociated from E_2. Likewise, for M3, the second and first Na^+ ions to bind to E_1 can be released from E_2 only after the first and first and second Na^+ ions have dissociated from E_2. Consequently, intracellular Na^+ ions entering M2 or M3 and proceeding towards the second and third extracellular Na^+ dissociation steps risk having their progress impeded by Na^+ ions rapidly rebinding via reversal of the first and first and second, respectively, extracellular Na^+ dissociation steps. Because all three released ions are assumed to pass through the external access channel, rebinding of ions via reversal of the first and second dissociation steps becomes more likely as V is made more negative, thereby decreasing the likelihood of Na^+ release via the second and third dissociation steps.

To illustrate this effect, for each mechanism, efflux was partitioned into the three components carried by the Na^+ ions released to the exterior via the first ($E_2NaNaNa \leftrightarrow E_2NaNa + Na_{out}^+$), second ($E_2NaNa \leftrightarrow E_2Na + Na_{out}^+$), and third ($E_2Na \leftrightarrow E_2 + Na_{out}^+$) dissociation steps (labels 1, 2, and 3 in FIGS. 2B-D). Note that the effect does not influence release via the first dissociation step which always shows a protracted plateau in the efflux-V relationship. This plateau occurs because Na^+ release to the exterior is assumed much faster than the $E_1NaNaNa \leftrightarrow E_2NaNaNa$ transitions, however, at sufficiently negative potentials (beyond the range shown), even this rate of Na^+ release via the first dissociation step must decline until it begins to rate limit this component of efflux, which then decreases with further hyperpolarization. By contrast, Na^+ release to the exterior via the second and third release steps is more strongly reduced at increasingly negative membrane potentials, because rebinding of ions via reversal of the first and second dissociation steps becomes more likely (see FIGS. 2B-D, curves 2 and 3). The decrease is more pronounced for release via the third dissociation step because this is impeded by reversal of both the first and second dissociation steps (compare curves 2 and 3 in FIGS. 2B-D).

For all mechanisms, the simulated efflux-V relationship is a composite of the efflux-V relationships of each of the three components, whose relative magnitudes are mechanism dependent. For M1, the component representing the first dissociation step predominated, whereas for M2 that component was smallest: in the latter mechanism, the very high $[Na]_{in}$ (relative to the small dissociation constant) permits only a fraction of the sites to which internal Na^+ ions first bind to become freshly occupied by radiolabeled Na^+, and those sites are the first to be unloaded at the external surface. For similar reasons, for M3, components 1 and 2 are relatively large.

The sigmoidal simulated efflux-V relationship that assumes random Na^+ binding/dissociation (M1) seems more comparable to that observed experimentally[4] than do those predicted on the basis of ordered binding (M2 and M3), and yet ordered Na^+ binding/dissociation seems likely. Thus, there is evidence for ordered binding/dissociation of extracellular $K^{+[8-10]}$ so that if Na^+ and K^+ share a common cation-binding domain, it would not seem unreasonable that external Na^+ binding/dissociation is also ordered. Also, a recent model[11] based on charge-pulse experiments proposed that the last Na^+ ion to bind from the interior is the first released to the exterior. What simple modification(s) of M2 and/or M3 would yield a more sigmoidal efflux-V relationship? In M3, but not in M2, inserting a conformational transition between the first and second external release steps enhances the component of efflux mediated by the first Na^+ ion released at the expense of the other components, resulting in a more sigmoidal efflux-V curve. A relative increase in the effective valence of the first release step, with respect to that of the subsequent step(s), enhances this effect. The implication is that the near sigmoidal ^{22}Na efflux-V relationship observed in squid axons[4] under Na^+-Na^+ exchange conditions reflects an ordered Na^+ binding/release mechanism in which the last intracellular Na^+ ion to bind is the first released to the exterior (likely in a strongly voltage-dependent manner), and that a conformational transition is interposed between release of the first and second Na^+ ions.

The long-pore effects presented here are also evident in simulations based on more complex mechanisms for Na_{in}^+-Na_{out}^+ exchange incorporating additional steps such as ATP binding, enzyme phosphorylation, and ADP dissocation, and multiple Na^+ deocclusion steps and in simulations of other types of flux experiments, such as activation of Na^+ efflux by increases in extracellular Na^+ concentration. The existence of these effects might confound interpretations of measured unidirectional flux data from a variety of systems.

ACKNOWLEDGMENT

The authors thank Mari Kuwabara for assistance in the preparation of the figures.

REFERENCES

1. GARRAHAN, P. J. & I. M. GLYNN. 1967. J. Physiol. **192:** 159–174.
2. DE WEER, P. 1970. J. Gen. Physiol. **56:** 583–620.
3. GLYNN, I. M., V. L. LEW & U. LUTHI. 1970. J. Physiol. **207:** 371–391.
4. GADSBY, D. C., R. F. RAKOWSKI & P. DE WEER. 1993. Science **260:** 100–103.
5. LAUGER, P. 1991. Soc. Gen. Physiol. Ser. **46:** 303–315.
6. WAGG, J. 1987. J. Theor. Biol. **128:** 375–385.
7. HODGKIN, A. L. & R. D. KEYNES. 1955. J. Physiol. **128:** 61–88.
8. GLYNN, I. M., J. L. HOWLAND & D. E. RICHARDS. 1985. J. Physiol. **368:** 453–469.
9. FORBUSH, B., III. 1987. J. Biol. Chem. **262:** 11104–11115.
10. FORBUSH, B., III. 1987. J. Biol. Chem. **262:** 11116–11127.
11. WUDDEL, I., & H.-J. APELL. 1995. Biophys. J. **69:** 909–921.

Nucleotides Trigger the Release of Co(NH$_3$)$_4$ATP Tightly Bound to Inactivated Na,K-ATPase[a]

D. G. WARD,[b] W. SCHONER,[c] AND J. D. CAVIERES[b]

[b]Transport ATPase Laboratory
Department of Cell Physiology and Pharmacology
Leicester University
P.O. Box 138
Leicester LE1 9HN, UK

[c]Institut für Biochemie und Endokrinologie
Justus-Liebig-Universität
Giessen, Germany

The high- and low-affinity effects of ATP on the sodium pump[1] are inherent in the behavior of the αβ protomeric unit.[2] One approach to study these effects in isolation is to make use of substitution-inert ATP analogs. Co(NH$_3$)$_4$ATP, for instance, inactivates the overall Na,K-ATPase cycle by blocking E$_2$-type reactions, and ATP protects the enzyme by competing at an appropriately low-affinity site.[3–5] During inactivation, the analog becomes incorporated in the α chain, as revealed by SDS-PAGE of the labeled enzyme.[3]

We now find that the radioactive cobalt ATP-analog can be released by incubating the labeled enzyme with ATP or other nucleotides. Two methods were used (with similar results) to measure the loss of tightly bound radioactive analog from the purified, membrane-bound enzyme at 20°C: (1) filtration through 0.22-μm filters (cellulose-nitrate, Sartorius), followed by wash with a chilled solution and measurement of trapped radioactivity, or (2) suction through 0.22-μm syringe-end filters (cellulose-nitrate, Costar) and counting of the filtrate. FIGURE 1 shows an experiment conducted using the second method with enzyme labeled with Co(NH$_3$)$_4$[2,8-^{3}H]-ATP and suspended in a solution containing 30 mM imidazole (pH 7.2) and 1 mM EDTA. Panel **A** shows that the addition of ATP at time zero causes release of the labeled analog. So far, all time courses of nucleotide-triggered release can be represented by double-exponential functions. This apparent "slowing down" of release is not due to unexpected ATP hydrolysis, as ADP as well as the nonhydrolyzable analog βγ-methylene ATP (albeit less efficiently) can also cause cobalt ATP-analog release (panel **B**). Experiments to measure the release of [γ^{33}P]-labeled Co(NH$_3$)$_4$ATP led to curves that could be superimposed on those obtained with the Co(NH$_3$)$_4$[^{3}H]-ATP-labeled enzyme. Neither Mg^{2+} nor Na$^+$ promotes release nor do they influence the pattern of ATP-induced release.

It is interesting that release of the analog does not affect the extent of inactivation of the sodium pump. This was evaluated from its Na,K-ATPase activity, which was

---

[a]This work was supported by grants from The Wellcome Trust and the Medical Research Council (to J.D.C.) and the Volkswagen Foundation, Hannover, Germany (to W.S.).

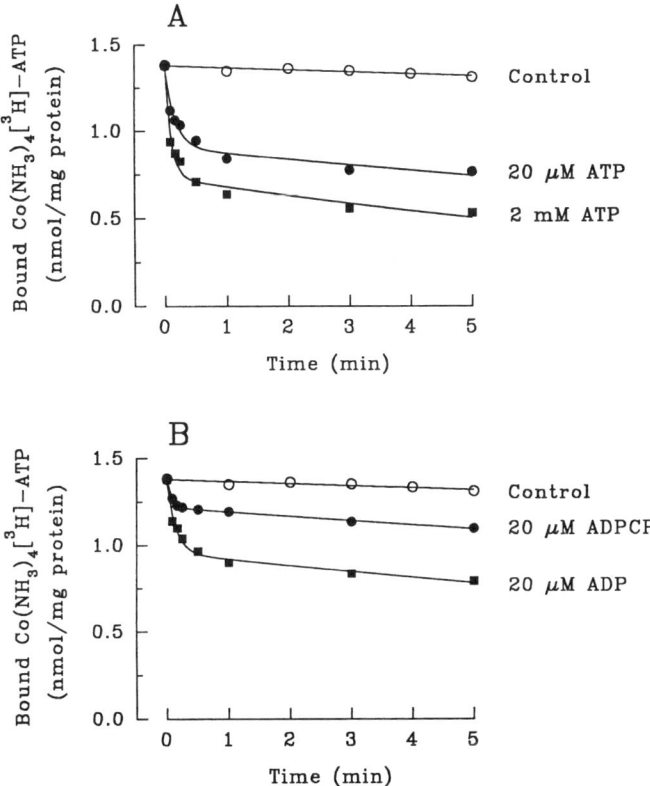

FIGURE 1. Nucleotide-induced release of radioactivity at 20°C from Na,K-ATPase inactivated with Co(NH$_3$)$_4$[2,8-^{3}H]-ATP. (For method, see text.) **(A)** Effect of 20 µM and 2 mM ATP. **(B)** Effects of 20 µM ADP and 20 µM βγ-methylene ATP (ADPCP). Nucleotides added at time zero. *Control:* No nucleotide added.

measured at 1 mM [γ^{32}P]-ATP, 3 mM MgCl$_2$, 1 mM EDTA, 130 mM Na$^+$, 20 mM K$^+$, and 30 mM imidazole (pH 7.2 at 20°C). The cobalt ATP-analog was released with 100 µM ATP, and the enzyme was spun down and washed in a Beckman TL-100 ultracentrifuge. The Co(NH$_3$)$_4$[^{3}H]-ATP bound to the inactivated enzyme was 1.99 ± 0.07 nmol/mg before and 0.42 ± 0.01 nmol/mg after (mean ± SEM, n = 3) incubation with 100 µM ATP. The specific Na,K-ATPase activity of the inactivated enzyme was 8.0 ± 0.3% before and 8.3 ± 0.6% after treatment with 100 µM ATP (100% was the specific activity of native, control enzyme processed in parallel).

As illustrated in FIGURE 1A, increasing the ATP concentration mainly leads to an increase in amplitude of the first exponential. When this quantity was plotted against a range of ATP concentrations, apparently hyperbolic curves were obtained, with $K_{0.5(ATP)}$ values between 2 and 5 µM. We interpret this behavior as arising from an equilibrium binding of the cobalt ATP-analog to the enzyme. In this context, the apparently irreversible binding (which survives two or three washes in the ultracentrifuge) would rather be the result of a very low dissociation constant. The effect of ATP and the other nucleotides would then be to increase that dissociation constant, leading to fast, transient cobalt ATP-analog release until a new equilibrium is established. Considering the foregoing $K_{0.5(ATP)}$ values and assuming $K_{diss(ATP)} = 0.2$ µM, a $K_{diss(CoATP)}$ (i.e., in the absence of ATP) around 10–20 nM can be calculated. This is far lower than $K_{0.5}$ for $Co(NH_3)_4ATP$ binding at the inactivation site ($=360$ µM, ref. 3) or even than its $K_{0.5}$ for the high-affinity ATP site (250 nM), where $Co(NH_3)_4ATP$ binds at equilibrium and without causing inactivation.[5] We may speculate that both $Co(NH_3)_4ATP$ and the enzyme experience reciprocal chemical changes during the inactivation step, changes that lead to the large increase in binding affinity.

We have shown that $Co(NH_3)_4ATP$ can randomly inactivate all protomers in the enzyme[6,7] and have offered evidence for more than one class of ATP sites in the soluble protomeric enzyme.[8] A plausible guess is, therefore, that whereas the Co-analog is released from the low-affinity ATP site, the triggering effects of ATP, ADP, or ADPCP take place at the high-affinity site. In that case, the mutual effects of Co-analog and added nucleotide in decreasing each other's affinity may result from ATP-site interaction rather than straight competition.

ACKNOWLEDGMENT

We thank Tim Walton for help with the preparation of the enzyme.

REFERENCES

1. KANAZAWA, T., M. SAITO & Y. TONOMURA. 1967. J. Biochem. **61:** 555–566.
2. WARD, D. G. & J. D. CAVIERES. 1993. Proc. Natl. Acad. Sci. USA **90:** 5332–5336.
3. SCHEINER-BOBIS, G., K. FAHLBUSCH & W. SCHONER. 1987. Eur. J. Biochem. **168:** 123–131.
4. SCHEINER-BOBIS, G., M. ESMANN & W. SCHONER. 1989. Eur. J. Biochem. **183:** 173–178.
5. SERPERSU, E. H., S. BUNK & W. SCHONER. 1990. Eur. J. Biochem. **191:** 397–404.
6. WARD, D. G., W. SCHONER & J. D. CAVIERES. 1994. J. Physiol. (Lond.) **480.P:** 84P.
7. LINNERTZ, H., D. THÖNGES & W. SCHONER. 1995. Eur. J. Biochem. **232:** 420–424.
8. WARD, D. G. & J. D. CAVIERES. 1996. J. Biol. Chem. **271:** 12317–12321.

Transient Currents of Na^+/K^+-ATPase in Giant Patches from Guinea Pig Cardiomyocytes Induced by ATP Concentration Jumps or Voltage Pulses

THOMAS FRIEDRICH[a] AND GEORG NAGEL[b]

Max-Planck-Institut für Biophysik
Kennedyallee 70
D-60596 Frankfurt/M, Germany

In recent years two relaxation techniques were developed to measure electric currents generated by the Na^+/K^+-ATPase in the millisecond range. The results led to different conclusions about the rate constant of the step, which is rate limiting the electrogenic event.

One technique uses an ATP concentration jump, which is generated by the release of ATP from a nonhydrolyzable ATP analog (caged ATP) by short irradiation with intense ultraviolet light.[1] Application of this method to purified Na^+/K^+-ATPase containing membrane fragments adsorbed to an artificial lipid bilayer was pioneered by Fendler et al.[2]

Another technique uses protocols of fast voltage steps. The sodium pump current is obtained by subtracting control currents with inhibited Na^+/K^+-ATPase (by extracellular ouabain or in the absence of cytoplasmic ATP and/or Na^+) from currents measured under conditions in which the Na^+/K^+-ATPase can perform Na^+/Na^+ exchange. This method reveals Na^+- and ATP-dependent difference currents, which can be attributed to electrogenic events in the pump cycle.[3,4]

Fendler et al.[5] measured a rate of ≥ 200 s^{-1} at 24°C and pH 6.2 for the electrogenic conformational change, whereas Apell et al.[6] concluded the rate to be slower with about 20 s^{-1} in similar experiments (pH 7.0, 22°C). Nakao and Gadsby[3] arrived at 200 s^{-1} at 35°C (50 s^{-1} at 24°C, pH 7.4) for the sum of the forward and backward electrogenic rates in voltage jump experiments on whole cell patch clamped heart cells.[3] Hilgemann,[7] however, measured ~600 s^{-1} at 37°C (pH 7.4) in voltage jump experiments on cardiac giant excised patches.[4]

METHOD AND RESULTS

We combined the giant patch technique[7] with a method to photolyse caged ATP at an excised patch (as described in more detail in ref. 8). Using this technique we were able to apply the two fast relaxation methods just described on the same membrane patch in order to compare the rate constants of the involved reaction steps directly ($[Na_o^+] = 145$ mM, $[K_{o,i}^+] = 0$).

[a]Present address: ZMNH, D-20246 Hamburg, Germany. Tel: +49-40-4717 6605; e-mail: tfriedri@uke.uni-hamburg.de
[b]Tel: +49-69-6303-303; e-mail: nagel@biophys.mpg.de

Fast ATP concentration jumps were generated by photolysis of caged ATP with laser flashes of 308-nm wavelength and 10-ns duration at holding potential $U = 0$ mV. Under our experimental conditions (pH 6.3, T = 24°C, 2 mM free Mg^{2+}), ATP release proceeds with a rate of ~400 s^{-1}. Transient outward currents with a fast rising phase, followed by a slower decay, were obtained (FIG. 1). Analysis of the current traces was performed by fitting a model function to the data consisting of the sum of two exponentially decaying functions and a constant. Whereas the slow component, showing distinct substrate dependence, can be attributed to caged ATP/ATP binding/dissociation, the substrate-independent fast rate constant of about 200 s^{-1} (24°C) reflects a step that is rate limiting the electrogenic event.

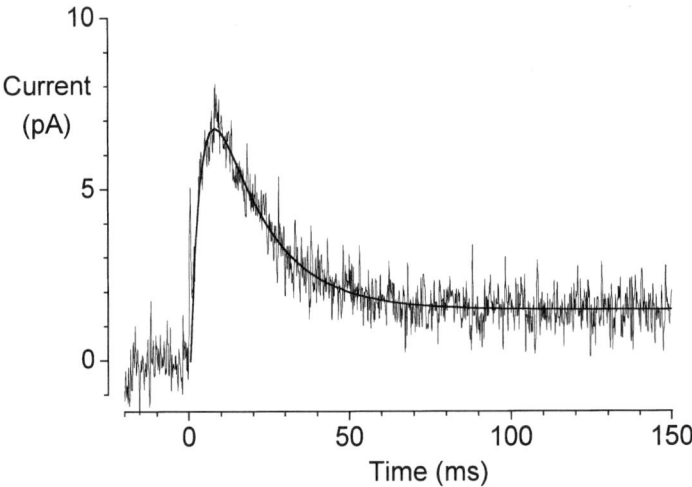

FIGURE 1. Current signal after photolysis of 50-μM caged ATP with a release ratio of 0.28 (pH 6.3, 24°C, $[K_{o,i}^+] = 0$, $[Na_o^+] = 145$ mM, $[Na_i^+] = 40$ mM). After a short light-induced artifact the signal shows a fast rising phase followed by a slower decay to a (small) stationary current. Fit of equation $I_{(t)} = A_1 \cdot e^{-(t-t_0)/\tau_1} + A_2 \cdot e^{-(t-t_0)/\tau_2} + I_\infty$ to the data yields: $A_1 = -12.2 \pm 0.6$ pA, $\tau_1 = 3.6 \pm 0.3$ ms, $A_2 = 10.8 \pm 0.7$ pA, $\tau_2 = 16.4 \pm 0.8$ ms, $I_\infty = 1.46 \pm 0.03$ pA, leading to a charge integral of $Q = 133$ fC.

Voltage pulses were applied on the same membrane patch either with or without ATP at the cytoplasmic side. Subtraction of the voltage jump induced currents in the absence of ATP from those taken in the presence of ATP yielded monoexponential current signals, which depended on external Na^+ but did not differ between pH 6.3 or pH 7.4 (data not shown).

Rate constants obtained from fits of a monoexponential decay function to the data showed a characteristic voltage dependence, saturating at positive potentials (175 s^{-1}, 24°C) and exponentially rising towards increasing negative potentials (FIG. 2a). The fitted rate constants for $U \geq 0$ mV agreed well with values for the fast rate constants obtained from current signals induced by an ATP concentration jump. The dependence of the amount of charge moved during such voltage-induced transients on the membrane potential U can be described by a Boltzmann distribution (FIG. 2b). The

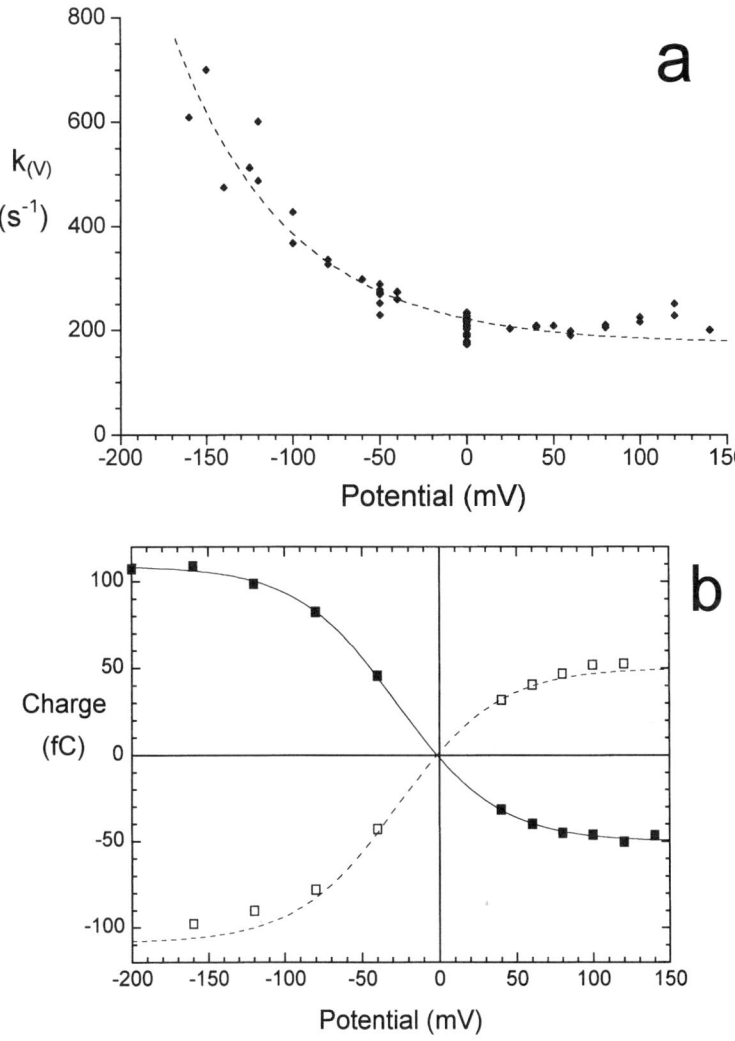

FIGURE 2. (a) Dependence of rate constants from voltage-jump induced transient currents on membrane potential. Values were derived from fits of a monoexponential decay function to the data traces (experimental conditions as in FIG. 1, same experiment, pH 7.4, 500 μM ATP). Fit function $k_{(V)} = k_1 + k_{-1(V)} = k_1 \cdot (1 + e^{z_k}(V_k - V) \cdot F/RT)$ yields $k_1 = 175 \pm 13$ s^{-1}, $z_k = 0.38 \pm 0.04$, $V_k = -87.6 \pm 12.9$ mV. (b) Boltzman distribution $(Q_{(V)} = Q_{min} + \Delta Q_{max} \cdot (1 + e^{z_q}(V_q - V) \cdot F/RT)^{-1})$ of charge moved during voltage-induced transients (full symbols: *off*-phases; open symbols: *on*-phases of voltage pulse). *Fit constants* (*solid line*): $\Delta Q_{max} = 159 \pm 4$ fC, $V_q = -26.9 \pm 1.6$ mV, $z_q = 0.78 \pm 0.03$. The *dashed line* is the fit curve multiplied by (-1) to indicate symmetry of *on*- and *off*-charge movements.

difference in the corresponding saturation values at $U \rightarrow \pm\infty$ agrees well with the amount of charge calculated as the integral of the current response after an [ATP]-jump. Both voltage-induced rate constants (at $U \geq 0$ mV) and the fast rate constant from [ATP]-jump experiments exhibit an activation energy of about 80 kJ/mol ($17°C < T < 30°C$).

CONCLUSIONS

Voltage- and [ATP]-jumps on the same membrane patch revealed the same rate constant for the step rate limiting the electrogenic Na^+ translocation of about 120 s^{-1} at 21°C and 0 mV.

The data presented here corroborate the findings of Fendler *et al.*[2,5] for pig kidney and eel electric organ Na^+/K^+-ATPase and now also for rat and guinea pig Na^+/K^+-ATPase in heart cells: the major electrogenic event (or the step, which is rate limiting the electrogenic event) is at least 200 s^{-1} fast at 24°C (120 s^{-1} at 21°C) and 0 mV (pH 6.3 and pH 7.4).

Note added in proof: A more detailed study on the comparison of ATP concentration jumps and voltage jumps was recently published: Friedrich, T. & G. Nagel. 1997. Biophys. J. **73**: 186–194.

REFERENCES

1. KAPLAN, J. H., B. FORBUSH III. & J. F. HOFFMANN. 1978. Biochemistry **17**: 1929–1935.
2. FENDLER, K., E. GRELL, M. HAUBS & E. BAMBERG. 1985. EMBO J. **12**: 3079–3085.
3. NAKAO, M. & D. C. GADSBY. 1986. Nature **323**: 628–630.
4. HILGEMANN, D. W. 1994. Science **263**: 1429–1432.
5. FENDLER, K., S. JARUSCHEWSKI, A. HOBBS, W. ALBERS & J. P. FROEHLICH. 1993. J. Gen. Physiol. **102**: 631–666.
6. APELL, H. J., R. BORLINGHAUS & P. LÄUGER. 1987. J. Membr. Biol. **97**: 179–191.
7. HILGEMANN, D. W. 1989. Pflügers Arch. **415**: 247–249.
8. FRIEDRICH, T., E. BAMBERG & G. NAGEL. 1996. Biophys. J. **71**: 2486–2500.

Fluorescence Quenching of IAF-Na$^+$/K$^+$-ATPase via Energy Transfer to TNP-Labeled Nucleotide[a]

EDWARD H. HELLEN[b] AND PROMOD R. PRATAP

Department of Physics and Astronomy
University of North Carolina at Greensboro
Greensboro, North Carolina 27412

Fluorescence quenching of 5-iodoacetamidofluorescein (IAF)-labeled Na$^+$/K$^+$-ATPase by trinitrophenyl-adenosine diphosphate (TNP-ADP) was used to investigate models of nucleotide binding. Sets of fluorescence quenching curves obtained by titrating the enzyme with TNP-ADP in the presence of various concentrations of ADP could not be adequately fit using a simple model with a single nucleotide binding site. Therefore, we examined two-site models and a model with a single site plus nonspecific binding (SSPNB) where nonspecific binding is defined as nonsaturable binding that is not blocked by competitive ligand. These models explicitly include: (1) competition between nucleotide and TNP-nucleotide for specific nucleotide binding sites; and (2) an "inactive" fraction of enzyme whose specific binding site or sites cannot bind nucleotide. The SSPNB model also includes: (1) quenching due to nonspecific binding of TNP-nucleotide to the enzyme; and (2) separate energy transfer parameters for specifically bound and nonspecifically bound TNP-nucleotide. On the basis of the goodness of fit, on the number of parameters in each model, and on how the best fit parameters vary under different experimental conditions, we find that the interaction of TNP-ADP with IAF-labeled Na$^+$/K$^+$-ATPase is best described by the SSPNB model. These results are relevant to the question of whether observed high- and low-affinity binding of nucleotide to the enzyme is due to two distinct sites or a single site that changes affinity during the enzyme's cycle.[1,2]

MATERIALS AND METHODS

TNP-ADP and IAF were from Molecular Probes (Eugene, Oregon); other chemicals were from Sigma (St. Louis, Missouri). Frozen dog kidneys were from Pel-Freez (Rogers, Arkansas). IAF-labeled Na$^+$/K$^+$-ATPase was obtained as described earlier.[3–5] Enzyme concentrations used for fluorescence measurements were about 10 nM. The buffer contained 25 mM imidizole, pH 7.0, 140 mM choline chloride, 20 mM NaCl, and 1 mM EDTA.

Fluorescence intensities were measured on a QM-1 fluorescence spectrophotometer (PTI, Brunswick, New Jersey), with excitation at 490 nm and emission at 520 nm

---

[a]This work was supported by a grant from the National Institutes of Health (GM-47550) to P. R. P.

[b]Tel: (910) 334-5844 office; (910) 334-4279 lab; fax: (910) 334-5865; e-mail: ehhellen@dirac.uncg.edu

at 25°C. Intensities were corrected for the fluorescence of TNP, dilution caused by the titrations, and the inner filter effect for excitation and emission using measured absorbances at 490 and 520 nm, respectively.

The SSPNB model for the quenched IAF-enzyme fluorescence is[6]:

$$F = \frac{1 - \dfrac{x_a T K_a}{K_a K_t + A K_t + T K_a}}{1 + q T^a} + \frac{\dfrac{x_a T K_a}{K_a K_t + A K_t + T K_a}}{1 + q T^a + k_s}$$

where $K_a(K_t)$ is the dissociation equilibrium constant for ADP (TNP-ADP) interacting with a single nucleotide binding site on the enzyme, $A(T)$ is the free concentration of ADP (TNP-ADP), x_a is the fraction of enzyme that can specifically bind nucleotide, k_s is the normalized (with respect to the inverse of the unquenched IAF fluorescence lifetime) rate of decay due to energy transfer to TNP-ADP bound to the specific site, and q and a parameterize energy transfer to nonspecifically bound TNP-ADP. The data were also analyzed using two-site models in which each site is characterized by affinity constants for ADP and TNP-ADP, and normalized energy transfer decay rates

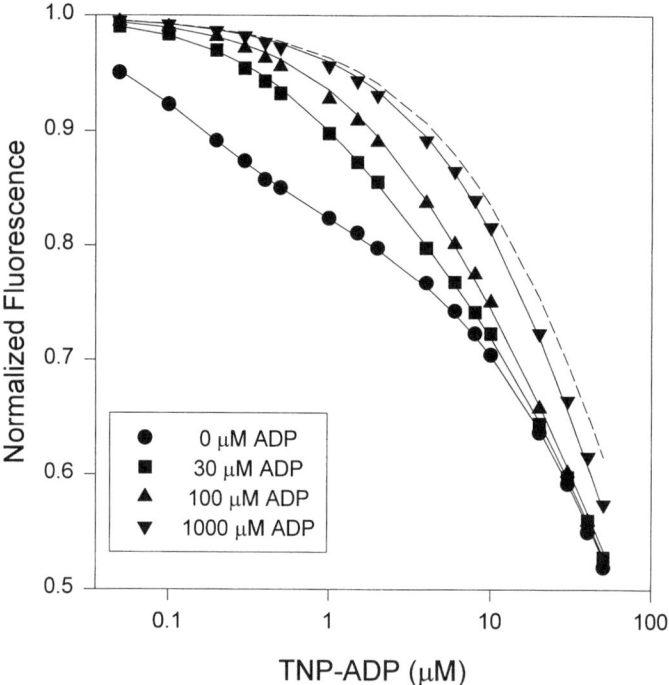

FIGURE 1. Normalized fluorescence quenching data for IAF-labeled Na$^+$/K$^+$-ATPase being titrated by TNP-ADP in the presence of 0 (●), 30 (■), 100 (▲), and 1,000 (▼) μM ADP and best fit using the single-site plus nonspecific binding model. *Dashed line* is the quenching due to nonspecifically bound TNP-ADP predicted by the single-site plus nonspecific binding model using the best fit parameters. Note that TNP-ADP is plotted on a logarithmic scale because its concentration spans three decades.

TABLE 1. Best Fit Parameters for Dissociation Equilibrium Constants for Nucleotide K_a and TNP-Labeled Nucleotide K_t Using the Single-Site Plus Nonspecific Binding Model[a]

	K_a (µM)	K_t (µM)
TNP-ADP vs ADP	2.5	0.13
TNP-ATP vs ATP	0.5	1.5

[a]Buffer contained 25 mM imidizole, pH 7.0, 140 mM choline chloride, 20 mM NaCl, 1 mM EDTA, and approximately 15 nM IAF-labeled Na$^+$/K$^+$-ATPase.

for TNP-ADP bound at the site. For one of the two-site models the "inactive" fraction of enzyme could not bind nucleotide at either site, whereas for the other two-site model only the high affinity site of the inactive enzyme could not bind nucleotide. Curve fitting was done by simultaneously minimizing the sum of the square of the errors for complete sets of fluorescence quenching data curves including all the concentrations of ADP.

RESULTS AND DISCUSSION

FIGURE 1 shows fluorescence quenching data and best fit using the SSPNB model for IAF-labeled Na$^+$/K$^+$-ATPase as a function of TNP-ADP concentration in the presence of 0, 30, 100, and 1,000 µM ADP. The two-site models have one more free parameter than the SSPNB model, yet their best fits had larger sums of the square of the errors.[6] Therefore, the SSPNB model best fits the data. TABLE 1 shows the best fit values for the equilibrium constants in addition to the results for analogous experiments using TNP-ATP and ATP.

Our results show that the binding of TNP-ADP to the Na$^+$/K$^+$-ATPase is comprised of two parts: saturable binding to a single specific site, and nonspecific binding which is not saturable out to 50 µM TNP-ADP. At 1 µM TNP-ADP the quenching due to nonspecific binding is about 20% the total quenching, whereas it is 50% at 10 µM and 85% at 50 µM. Reports of nonspecific binding of TNP-nucleotide to the Na$^+$/K$^+$-ATPase,[7] the cytoplasmic loop of Ca^{2+}-ATPase,[8] and the ATP binding domain of H$^+$-ATPase[9] suggest that its explicit inclusion in a binding model is of general use.

REFERENCES

1. WARD, D. G. & J. D. CAVIERES. 1996. J. Biol. Chem. **271:** 12317–12321.
2. SCHEINER-BOBIS, G., J. ANTONIPILLAI & R. A. FARLEY. 1993. Biochemistry **32:** 9592–9599.
3. JORGENSEN, P. L. 1974. Biochim. Biophys. Acta **356:** 36–52.
4. KAPAKOS, J. G. & M. STEINBERG. 1982. Biochim. Biophys. Acta **693:** 493–496.
5. STEINBERG, M. & S. J. D. KARLISH. 1989. J. Biol. Chem. **264:** 2726–2734.
6. HELLEN, E. H. & P. R. PRATAP. 1997. Biophys. Chem. In press.
7. MOCZYDLOWSKI, E. G. & P. A. G. FORTES. 1981. J. Biol. Chem. **256:** 2346–2356.
8. MOUTIN, M., M. CUILLEL, C. RAPIN, R. MIRAS, M. ANGER, A. LOMPRE & Y. DUPONT. 1994. J. Biol. Chem. **269:** 11147–11154.
9. CAPIEAUX, E., C. RAPIN, D. THINES, Y. DUPONT & A. GOFFEAU. 1993. J. Biol. Chem. **268:** 21895–21900.

Eosin, Energy Transfer, and RH421 Report the Same Conformational Change in Sodium Pump as Fluorescein[a]

SHWU-HWA LIN, IRINA N. SMIRNOVA,
VLADIMIR N. KASHO, AND LARRY D. FALLER[b]

*CURE: Digestive Diseases Research Center
Department of Medicine
University of California at Los Angeles and
Wadsworth Division
Department of Veterans Affairs Medical Center
West Los Angeles, California 90073*

The rate of the conformational change in sodium pump reported by fluorescein covalently attached to lysine-501 with fluorescein 5′-isothiocyanate (FITC) depends sigmoidally on [K^+], and the ratio of the macroscopic K^+ dissociation constants is 4:1. To explain these observations, we proposed that K^+ causes a concerted change from the E_1 to the E_2 conformation of the unphosphorylated enzyme by binding competitively with Na^+ to identical and independent sites.[1] Analytical equations were derived for the rate and amplitude of the fluorescence change and used to predict their dependence on [Na^+]. The mechanism was tested by experiments that confirmed the predictions for FITC-labeled enzyme.

Because our proposal conflicts with published mechanisms based on studies with other fluorescent probes, we also studied the reactions reported by eosin, fluorescence resonance energy transfer (FRET), and the voltage-sensitive styryl dye RH421. Each fluorescent probe reports the conformational change by a different mechanism.

Eosin also reports the conformational change because of different fluorescence quantum yields in the E_1 and E_2 conformations, but unlike FITC binds reversibly.[2] This was proven by showing that (1) K^+ quenches fluorescence faster than eosin can dissociate and (2) eosin binds with approximately the same affinity to E_1 and E_2. The relative amplitudes of the fast (eosin binding) and slow (conformational change) fluorescence changes observed with this probe can be predicted with equations for the concerted mechanism by assuming that eosin binds to all E_1 forms independent of the number of Na^+ and/or K^+ ions bound.

Energy transfer reports the conformational change because the distance between donor and acceptor increases when E_1 changes to E_2.[3] The measured fluorescence increase caused by K^+ was shown to result entirely from FRET by confirming that the quantum yield of the donor, fluorescein attached to cysteine-457 of hog enzyme with 5′-(iodoacetamido)fluorescein (IAF), does not change. Therefore, the change in

---

[a]This work was supported by the American Heart Association, National Science Foundation and VA Merit Review.
[b]Tel: 310-268-3896; fax: 310-268-4963.

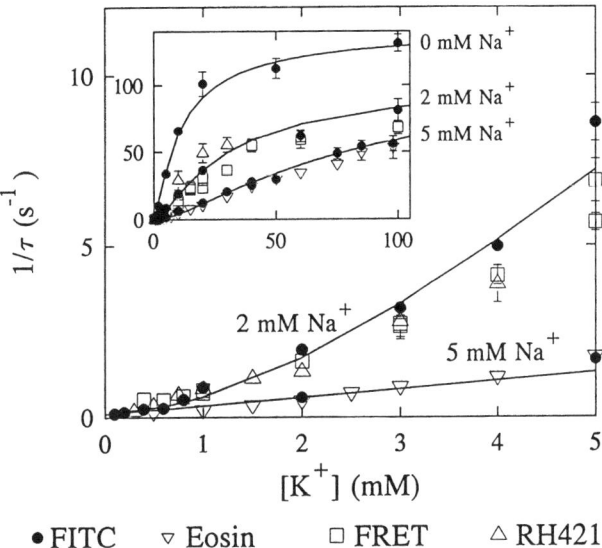

FIGURE 1. Data for the rate of the $E_1 \rightarrow E_2$ transition obtained by all four methods of following the reaction can be superimposed on theoretical curves calculated with the analytical equations derived for the concerted mechanism.

distance (2.9 Å) between IAF and the acceptor, a substrate analog 2′ (or 3′)-O-(trinitrophenyl)fluorescein (TNP-ATP) bound at the active site, could be estimated with sufficient precision (±0.6 Å) to obtain the first indication of the size of the conformational change.

In contrast to the other probes, RH421, which is nonspecifically absorbed by membranes, reports the conformational change indirectly because the $E_1 \rightleftharpoons E_2$ equilibrium is coupled to Na$^+$ binding to E_1.[4] In the $E_2 \rightarrow E_1$ direction (+Na$^+$), the protein rearrangement is rate limiting. In the opposite direction, there are two fluorescence enhancements. The faster, unresolved enhancement reports Na$^+$ reequilibration with E_1 after mixing with K$^+$, but before the protein rearranges. The slower enhancement reports the conformational change because additional reequilibration with Na$^+$ is rate limited by redistribution of the enzyme between E_1 and E_2.

FIGURE 1 shows that data for the rate of the $E_1 \rightarrow E_2$ transition obtained by all four

TABLE 1. Estimated Kinetic Parameters

Parameter	Fluorescent Reagent or Method			
	FITC	Eosin	FRET	RH421
k_f (s^{-1})	150 ± 19	118–150	118 ± 4	134 ± 12
k_r (s^{-1})	0.13 ± 0.03	0.10 ± 0.13	0.16 ± 0.04	0.07 ± 0.03
K_K (mM)	4.8 ± 1.2	3.6 ± 0.2	4.3 ± 0.3	1.8–3.0
K_{Na} (mM)	0.38 ± 0.12	0.41 ± 0.01	0.18–0.44	0.24–0.55

Abbreviations: FITC = fluorescein 5′-isothiocyanate; FRET = fluorescence resonance energy transfer.

methods of following the reaction can be superimposed on theoretical curves calculated with the analytical equations derived for the concerted mechanism. TABLE 1 shows that the kinetic parameters estimated from data obtained by following the conformational change different ways are experimentally indistinguishable.

We conclude that the equivalent-site mechanism provides a simple, quantitative, and general description of the conformational change in the second half of the reaction cycle that can explain (1) rate limitation of the pump, (2) specificity of the enzyme for Na^+ and K^+, and (3) stoichiometry of ion countertransport.

REFERENCES

1. SMIRNOVA, I. N., S.-H. LIN & L. D. FALLER. 1995. Biochemistry **34:** 8657–8667.
2. SMIRNOVA, I. N. & L. D. FALLER. 1995. Biochemistry **34:** 13159–13169.
3. LIN, S.-H. & L. D. FALLER. 1996. Biochemistry **35:** 8419–8428.
4. LIN, S.-H. & L. D. FALLER. 1996. FASEB J. **10:** A561.

Binding of TNP-ATP to IAF-Labeled Na$^+$/K$^+$-ATPase as Examined by Fluorescence Quenching[a]

P. R. PRATAP,[b] E. H. HELLEN, AND A. PALIT

Department of Physics and Astronomy
University of North Carolina at Greensboro
Greensboro, North Carolina 27412

We examined stopped-flow kinetics of substrate binding to Na$^+$/K$^+$-ATPase enzyme labeled with 5-iodoacetamidofluorescein (IAF enzyme) using the fluorescent analog of ATP, trinitrophenyl-ATP (TNP-ATP). IAF enzyme has been used to examine steady-state and transient kinetics of conformational changes.[1–4] TNP-ATP has been used with unlabeled enzyme to examine substrate binding.[5,6] IAF fluorescence quenching by TNP-ATP has been used to estimate the distance between the substrate binding site and the IAF binding site.[7]

Here, we use this quenching as a signal of binding. Under steady-state conditions, quenching data were consistent with TNP-ATP binding to two classes of sites: one, termed specific, is saturable and can bind ATP; the other, termed nonspecific, is nonsaturable and does not bind ATP. We report that the transient kinetics of TNP-ATP is nonexponential with time and is consistent with the existence of binding sites with a distribution of transition energies.

MATERIALS AND METHODS

IAF enzyme was isolated from dog kidneys using a modification of method C of Jørgensen[8]; specific activity ranged from 4–7 µmol $P_i \cdot$ (mg protein)$^{-1} \cdot$ min^{-1} at 24°C. Steady-state enzyme activity was measured with a fluorescence assay.[9] IAF labeling[1] and stopped-flow experiments[1,3] were performed as previously described. Under each condition, 7–9 fluorescence traces were recorded. Averages and standard errors of the mean (SEM) were then calculated and fitted to a stretched exponential:

$$F(t) = F_\infty + \Delta F \exp[-Bt^\alpha] \qquad (1)$$

using a Gauss-Newton algorithm. Here, $0 < \alpha \leq 1$; F_∞ is the fluorescence at infinite time; ΔF is the change in fluorescence; and B is a characteristic parameter of the stretched exponential. If $\alpha = 1$, then F(t) would decay exponentially and B would be the rate constant of the exponential. We refer to α as the stretch parameter.

---

[a]This work was supported by a grant from the National Institutes of Health (to P.R.P.).
[b]Tel: (910) 334-5844; fax: (910) 334-5865; e-mail: pratapp@dirac.uncg.edu

RESULTS

Effect of Increasing [TNP-ATP]

The inset in FIGURE 1 shows a typical fluorescence trace when IAF enzyme was rapidly mixed with TNP-ATP. The solid line in the trace is the fit of the data by equation 1. The magnitude of the fluorescence change increased and α (FIG. 1, ■) decreased with increasing [TNP-ATP]; α decreased by approximately the same amount when enzyme + 1 mM ATP was mixed with varying amounts of [TNP-ATP] (FIG. 1, ●). When enzyme in the presence of 1 μM TNP-ATP was mixed with varying amounts of ATP, α did not change significantly with ATP (FIG. 1, ▲).

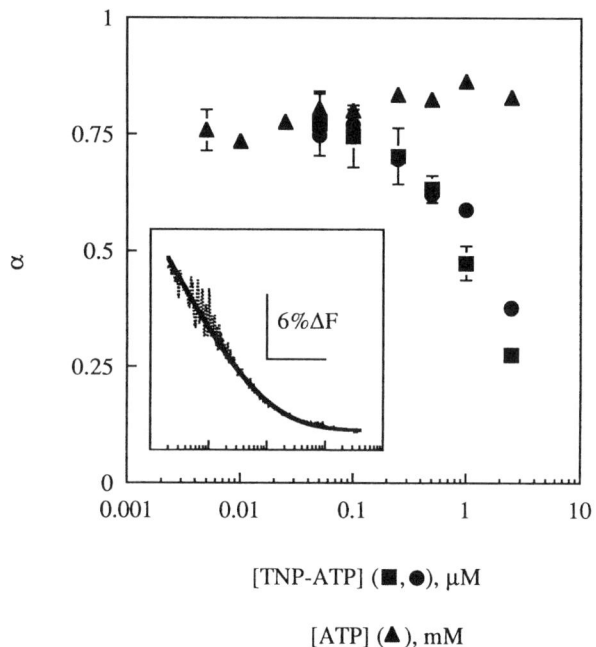

FIGURE 1. α as a function of [TNP-ATP] in the absence (■) and presence (●) of 0.1 mM ATP. Experimental conditions as described for the inset. Also shown (▲): α as a function of [ATP] in the presence of 1 μM TNP-ATP. Enzyme + 1 μM TNP-ATP in the same buffer as in the inset was rapidly mixed with varying amounts of ATP, and the fluorescence recovery was measured. *Inset:* Fluorescence change when IAF enzyme (100 μg/ml in 25 mM imidazole-HCl, pH 7, 1 mM EDTA, 155 mM choline chloride) was rapidly mixed with 10 μM TNP-ATP in the same buffer (final concentration 5 μM). *Solid line* is a fit to equation 1, with parameters: %ΔF (100 · ΔF/(F_∞ + ΔF)) = 33 ± 1; B = 5.09 ± 0.02 $s^{-\alpha}$; and α = 0.269 ± 0.005. The horizontal (time) axis is logarithmic to show the rapid early decrease in fluorescence followed by the slower decrease at a later time, a characteristic of stretched exponentials; the axis runs from 1 ms to 10 seconds. The horizontal scale bar is one decade, whereas the vertical scale bar is a 6% change in fluorescence.

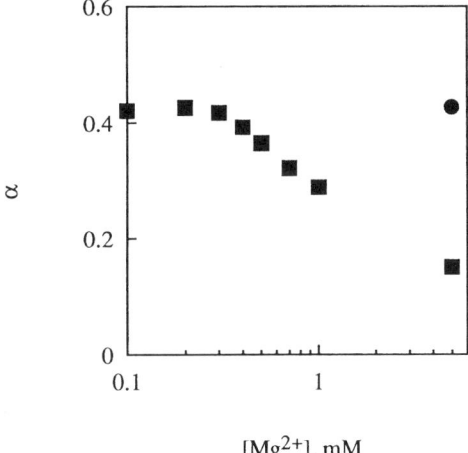

FIGURE 2. 100 µg/ml enzyme in buffer (25 mM imidazole-HCl, pH 7, 0.1 mM EDTA, 155 mM choline chloride, varying [Mg^{2+}]) was rapidly mixed with 1 µM TNP-ATP in the same buffer (■). The fluorescence change was fitted with equation 1 and α plotted against total [Mg^{2+}]. At 5 mM Mg^{2+}, the experiment was repeated in the presence of 1 mM ATP (●).

Effect of [Mg^{2+}]

IAF enzyme was rapidly mixed with 1 µM TNP-ATP in the presence of varying amounts of Mg^{2+} (FIG. 2). As the concentration of Mg^{2+} increased, α decreased. This effect was equivalent to increasing the concentration of TNP-ATP in the absence of Mg^{2+} (FIG. 1).

DISCUSSION

In a previous report,[10] we showed that TNP-ADP binds the IAF enzyme at two distinct classes of sites: one, termed specific, is saturable and can bind ADP also, while the other, termed nonspecific, is not saturable (over the range of concentrations examined) and does not bind ADP. These results hold for TNP-ATP and ATP.

In this work, we examined the transient kinetics for TNP-ATP binding to the enzyme. We found that this binding has a nonexponential time dependence and that the fluorescence traces could be fitted with a stretched exponential (equation 1). The stretched exponential reflects a reaction in which the rate constant varies with time in the following manner:

$$k(t) = k_0 t^{\alpha-1} \tag{2}$$

The functional form of equation 1 can be derived a first-order reversible reaction is assumed in which both reaction rates are described by equation 2 with identical α.[11]

Under steady-state conditions, we found that TNP-ATP binds to the specific site with well-defined affinity.[10] In our experiments, the specific site was characterized by an α of about 0.75. These results are consistent with: (1) the forward and reverse rate constants for TNP-ATP binding to this site varying with time according to equation 2, and (2) the time dependence of both rate constants being identical (because their ratio, the equilibrium constant, is time independent).

The stretched exponential can be expressed as a superposition of exponentials:

$$\exp[-\beta t^\alpha] = \int_0^\infty \Phi(k) \exp[-kt] dk \qquad (3)$$

Each exponential represents a reaction with a slightly different observed rate constant (k). $\Phi(k)$ is the distribution of these rate constants and therefore of the energy of the transition state. α represents the width of this distribution. Our experiments indicate that both the specific and the nonspecific sites exhibit some distribution of rate constants.

One interpretation of equation 3 is as follows: The binding sites (both classes) can exist in many conformations. In the initial (i.e., free) and final (i.e., bound) states, the energy differences between these conformations are small, much less than the thermal energy. These assumptions imply a well-defined energy difference between the bound and the free state (and hence a well-defined equilibrium constant). On the other hand, the time dependence of the forward and reverse rate constants implies that the transition states corresponding to these conformations have widely differing energies.

In experiments with Mg^{2+}, α decreased with increasing Mg^{2+} at a constant TNP-ATP concentration (FIG. 2). This observation implies that Mg^{2+} reduces the energy of the transition state, thus stabilizing it. This effect of Mg^{2+} is lost in the presence of 1 mM ATP, when the specific sites are blocked. This observation implies that Mg^{2+} affects TNP-ATP binding to the specific site.

ACKNOWLEDGMENT

Our thanks to Sufi Zafar of Motorola Corp. for illuminating discussions on stretched exponentials.

REFERENCES

1. PRATAP, P. R., J. D. ROBINSON & M. I. STEINBERG. 1991. Biochim. Biophys. Acta **1069:** 288–298.
2. BÜHLER, R., W. STÜRMER, H.-J. APELL & P. LÄUGER. 1991. J. Memb. Biol. **121:** 141–161.
3. PRATAP, P. R. & J. D. ROBINSON. 1993. Biochim. Biophys. Acta **1151:** 89–98.
4. PRATAP, P. R. & J. D. ROBINSON. 1996. Biochim. Biophys. Acta, in press.
5. MOCZYDLOWSKI, E. G. & P. A. G. FORTES. 1981. J. Biol. Chem. **256:** 2346–2356.
6. MOCZYDLOWSKI, E. G. & P. A. G. FORTES. 1981. J. Biol. Chem. **256:** 2357–2366.
7. FORTES, P. A. G. & R. AGUILLAR. 1988. Prog. Clin. Biol. Res. **268A:** 197–204.
8. JØRGENSEN, P. L. 1974. Biochim. Biophys. Acta **356:** 36–52.
9. BANIK, U. & S. ROY. 1990. Biochem. J. **266:** 611–614.
10. HELLEN, E. H. & P. R. PRATAP. 1997. Ann. N.Y. Acad. Sci., this volume.
11. PLONKA, A. 1986. Lecture Notes in Chemistry, vol. 40. Springer-Verlag. New York.

A Mutant of the Plasma Membrane Ca^{2+} Pump Highly Sensitive to Inhibition by Mg^{2+}[a]

HUGO P. ADAMO,[b,d] ALCIDES F. REGA,[b]
AND JOHN T. PENNISTON[c]

[b]*IQUIFIB-Facultad de Farmacia y Bioquímica (UBA)*
Junín 956, 1113 Cap. Fed.
Buenos Aires, Argentina

[c]*Department of Biochemistry and Molecular Biology*
Mayo Clinic/Foundation
200 First Street South West
Rochester, Minnesota 55905

Val^{674} in the "nucleotide binding domain" of the plasma membrane Ca^{2+} pump (PMCA) is replaced by Pro in most of the other P-type ATPases. We previously found that restitution of the consensus Pro at position 674 of the PMCA causes a substantial reduction in the enzyme's activity, and this effect seemed unrelated to ATP binding.[1]

In the present study we constructed and expressed in COS-1 cells the mutant V^{674}P (ct120) after knowing that removal of the COOH-terminal 120 amino acid residues of hPMCA4b[e] results in a mutant enzyme called hPMCA4b (ct120) which has higher activity independently of calmodulin.[3] The V^{674}P and V^{674}P (ct120) Ca^{2+} pumps could readily be detected in membranes from transfected COS-1 cell membranes by immunoblotting with monoclonal antibody 5F10,[4] as single bands with the expected migration according to their molecular mass. The intensity of the bands was comparable to that of the hPMCA4b and hPMCA4b (ct120), respectively, indicating that the substitution V^{674}P did not affect the level of expression of the Ca^{2+} pump.

The capacity of the V^{674}P and V^{674}P (ct120) enzymes to pump Ca^{2+} was compared to that of the hPMCA4b and hPMCA4b (ct120), respectively, by measuring the Ca^{2+} uptake rate by microsomal vesicles derived from transfected COS-1 cells. The results in TABLE 1 show that substitution of Val^{674} by Pro reduced the activity of the hPMCA4b and that of the hPMCA4b (ct120) to a similar extent.

Despite its lower activity, the apparent affinity for Ca^{2+} of the V^{674}P (ct120) enzyme was at least as high as that of hPMCA4b (ct120), indicating that the lower activity of the V^{674}P (ct120) enzyme could not be accounted for by a reduction in the affinity for Ca^{2+} at the transport site.

Study of the Mg^{2+} dependency of the Ca^{2+} transport of the V^{674}P (ct120) mutant

---

[a]This work was supported in part by the Consejo Nacional de Investigaciones Científicas y Tecnológicas, Fundación Antorchas, National Institutes of Health grant GM28835, and National Science Foundation grant INT 93-02981.
[d]To whom correspondence should be sent. Fax: 54-1-962-5457; e-mail: rvadamo@criba.edu.ar
[e]Abbreviations: hPMCA4b, human plasma membrane Ca^{2+} pump, isoform 4b; which has also been called hPMCA4CI.[2]

TABLE 1. Ca^{2+} Transport by Various Types of Plasma Membrane Ca^{2+} Pump[a]

Ca^{2+} Pump	Ca^{2+} Transport	
	− Calmodulin	+ Calmodulin
	(nmol/mg protein/min)	
hPMCA4b	1.46 ± 0.16	4.20 ± 0.16
$V^{674}P$	0.55 ± 0.12	1.87 ± 0.21
hPMCA4b (ct120)	4.21 ± 0.31	4.24 ± 0.17
$V^{674}P$ (ct120)	1.84 ± 0.16	2.07 ± 0.28

[a] Ca^{2+} uptake by microsomal vesicles was measured in a medium containing 1 μM free Ca^{2+} and 0.8 mM free Mg^{2+} with or without 240 nM calmodulin. Ca^{2+} uptake by the endogenous Ca^{2+} pump from COS-1 cells transfected with the empty plasmid PMM_2 was 0.05 ± 0.14 nmol/mg protein/min (− calmodulin) and 0.32 ± 0.04 nmol/mg protein/min (+ calmodulin). Values were obtained by subtracting the Ca^{2+} uptake by endogenous enzyme from the total. The activities shown are the average (± standard deviation) of three experiments conducted in duplicate using membranes from different transfections.

showed that it reached maximum activation at 100 μM Mg^{2+} in contrast with 500 μM in the hPMCA4b (ct120) (FIG. 1). Furthermore, whereas at 2 mM Mg^{2+} the hPMCA4b (ct120) showed no signs of inhibition, activity of the mutant decreased to less than 50% the maximum activity observed at 100 μM Mg^{2+}. A higher sensitivity to inhibition by Mg^{2+} accounts for the lower activity of the mutant shown in TABLE 1 in media with 0.8 mM Mg^{2+}, a concentration that is optimal for the wild type. These results indicate that the decrease in activity observed upon substitution of Val^{674} by Pro was due to the higher sensitivity to Mg^{2+} as inhibitor.

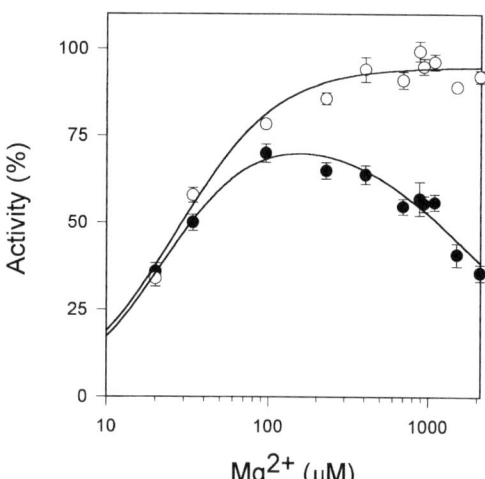

FIGURE 1. Ca^{2+} transport by the hPMCA4b (ct120) and $V^{674}P$ (ct120) Ca^{2+} pumps as a function of Mg^{2+} concentrations. Values shown were normalized to the activity of hPMCA4b (ct120) at 0.8 mM Mg^{2+} (3 nmol/mg/min), which was taken as 100%. ○ = hPMCA4b (ct120); ● = $V^{674}P$ (ct120).

REFERENCES

1. ADAMO, H. P., A. G. FILOTEO, A. ENYEDI & J. T. PENNISTON. 1995. J. Biol. Chem. **270:** 30111–30114.
2. CARAFOLI, E. 1994. FASEB J. **8:** 993–1002.
3. ENYEDI, A., A. K. VERMA, A. G. FILOTEO & J. T. PENNISTON. 1993. J. Biol. Chem. **268:** 10621–10626.
4. ADAMO, H. P., A. J. CARIDE & J. T. PENNISTON. 1992. J. Biol. Chem. **267:** 14244–14249.

Amino Acid Residues 18–75 Are Essential for Expression of an Active Plasma Membrane Ca^{2+} Pump[a]

MIRTA E. GRIMALDI,[b] HUGO P. ADAMO,[b,d]
ALCIDES F. REGA,[b] AND JOHN T. PENNISTON[c]

[b]*IQUIFIB-Facultad de Farmacia y Bioquímica (UBA)
Junin 956, 1113 Cap. Fed.
Buenos Aires, Argentina*

[c]*Department of Biochemistry and Molecular Biology
Mayo Clinic/Foundation
200 First Street South West
Rochester, Minnesota 55905*

The NH_2- and COOH-terminal segments of the P-type ATPases are the most variable regions of these enzymes. Recently a mutant of the plasma membrane Ca^{2+} pump (PMCA) called hPMCA4b (ct120) with a deletion of the COOH-terminal 120 amino acid residues was expressed in COS-1 cells and found to be fully active.[1] We have now investigated the functional relevance of the NH_2-terminal region of the PMCA which extends from the initial methionine to about the beginning of the first transmembrane segment. With this aim, we constructed a mutant called hPMCA4b(d18–75) (ct120) containing a deletion of the NH_2-terminal 18–75 and the COOH-terminal 120 amino acid residues.

Immunoblots of COS-1 cell membranes transfected with hPMCA4b(d18–75) (ct120) or hPMCA4b (ct120) DNA[2,3] showed that both proteins were expressed in similar amounts, indicating that the NH_2-terminal deletion did not significantly affect the level of expression of the PMCA. The expressed hPMCA4b(d18–75) (ct120) was found in one major band of the expected molecular mass (114 kD), indicating that it was as stable as the wild-type enzyme to degradation by intracellular proteases.

Because the removal of several amino acid residues may lead to proteins that are not correctly folded or inserted in the membrane, we tested the topology of the mutant by partial proteolysis. Tryptic digestion of the hPMCA4b(d18–75) (ct120) mutant resulted in the appearance of the same fragments obtained by proteolysis of the hPMCA4b (ct120) enzyme, indicating that both proteins exposed the same limited number of sites to the protease and suggesting that deletion of residues 18–75 neither impeded the insertion in the membrane nor extensively affected the folding of the mutant protein.

Results in TABLE 1 show that Ca^{2+} transport activity of microsomes from cells expressing the hPMCA4b(d18–75) (ct120) protein was not significantly different

---

[a]This work was supported in part by the Consejo Nacional de Investigaciones Científicas y Tecnológicas, Fundación Antorchas, National Institutes of Health grant GM28835, and National Science Foundation grant INT 93-02981.

[d]To whom correspondence should be sent. Fax: 54-1-962-5457; e-mail: rvadamo@criba.edu.ar

TABLE 1. Ca^{2+} Transport and Ca^{2+} ATPase Activities of Microsomes Isolated from Cells Transfected with PMM$_2$, hPMCA4b (ct120) or hPMCA4b(d18–75) (ct120)[a]

DNA Transfected	Ca^{2+} Transport (nmol Ca^{2+}/mg membrane protein/min)	Ca^{2+}-ATPase (nmol P_i/mg membrane protein/min)
PMM$_2$	0.14 ± 0.12	0.45 ± 0.25
hPMCA4b (ct120)	1.80 ± 0.34	6.08 ± 1.08
hPMCA4b(d18–75) (ct120)	0.25 ± 0.12	0.33 ± 0.43

[a]For the Ca^{2+} transport assays, the reaction was initiated by the addition of 6 mM ATP to 5 µg of vesicles preincubated at 37°C for 5 minutes in a reaction mixture containing 100 mM KCl, 50 mM Tris-HCl (pH 7.3 at 37°C), 5 mM NaN$_3$, 400 nM thapsigargin, 20 mM sodium phosphate, 95 µM EGTA, 800 µM free Mg^{2+}, and 1 µM free Ca^{2+} and terminated after 5 minutes by filtration. The ^{45}Ca taken up by the vesicles was determined in a liquid scintillation counter. Ca^{2+}-dependent ATP hydrolysis was measured by monitoring the (^{32}P)P_i liberated from (γ-^{32}P)ATP for 30 minutes in a reaction mixture containing 50 mM Tris-HCl (pH 7.4 at 37°C), 100 mM KCl, 0.1 mM MgCl$_2$, 0.1 mM EGTA, 0.1 mM CaCl$_2$, 5 mM NaN$_3$, 0.5 mM ouabain, 4 µg/ml oligomycin, 400 nM thapsigargin, and 10 µg of membranes. Average (± standard deviation) of duplicate measurements from three to five experiments with different membrane preparations is shown.

from that of the control microsomes from cells transfected with the empty vector, indicating that the mutant enzyme was not able to transport Ca^{2+}. TABLE 1 also shows that the Ca^{2+} ATPase activity of microsomes from cells transfected with the hPMCA4b(d18–75) (ct120) DNA was similar to that of the control, indicating that the hPMCA4b(d18–75) (ct120) mutant was also incapable of hydrolyzing ATP in a Ca^{2+}-dependent manner.

As previously reported,[4] a mutant plasma membrane Ca^{2+} ATPase lacking the 105 NH$_2$-terminal residues (PMCA105) expressed in COS-1 cells is inactive, probably because of the absence of the portion of the pump containing the first two transmembrane domains and the so-called transducing domain. Our results show that a smaller deletion in the portion of the molecule between the NH$_2$-terminus and the first transmembrane domain exposed to the cytosol, a region previously assumed irrelevant for the function of the enzyme, suffices for inactivation; indicating that residues 18–75 are essential for the expression of a functional Ca^{2+} pump.

REFERENCES

1. ENYEDI, A., A. K. VERMA, A. G. FILOTEO & J. T. PENNISTON. 1993. J. Biol. Chem. **268:** 10621–10626.
2. ADAMO, H. P., A. J. CARIDE & J. T. PENNISTON. 1992. J. Biol. Chem. **267:** 14244–14249.
3. ADAMO, H. P., A. G. FILOTEO, A. ENYEDI & J. T. PENNISTON. 1995. J. Biol. Chem. **270:** 30111–30114.
4. HEIM, R., T. IWATA, E. ZVARITCH, H. P. ADAMO, B. RUTISHAUSER, E. E. STREHLER, D. GUERINI & E. CARAFOLI. 1992. J. Biol. Chem. **267:** 24476–24484.

Increase in Affinity for ATP and Change in E_1-E_2 Conformational Equilibrium after Mutations to the Phosphorylation Site (Asp^{369}) of the α Subunit of Na,K-ATPase

PER AMSTRUP PEDERSEN,[a] JAKOB H. RASMUSSEN, AND PETER LETH JØRGENSEN

Biomembrane Research Center
August Krogh Institute
University of Copenhagen
Universitetsparken 13
DK-2100 Copenhagen OE, Denmark

The side chain of phosphorylated residue D369 is essential for ATPase activity of all P-type ATPases and mutations to residue D369 have been described to prevent assembly of αβ units at the cell surface of mammalian cells[1] and *Xenopus* oocytes.[2]

We analyzed whether the recently developed yeast expression system[3] was able to target the enzymatically inactive mutations α(D369N)β and α(D369A)β to the plasma membrane. We furthermore investigated the effects of these mutations on equilibrium [^{3}H]ATP binding, ADP binding, and [^{3}H]ouabain binding and on the conformational equilibrium between the E_1 and E_2 forms.

Equilibrium [^{3}H]ouabain binding to intact yeast cells expressing α(wt)β, α(D369N)β, and α(D369A)β demonstrated the expression of all these enzymes at the cell surface at comparable densities (FIG. 1). The discrepancy between these results and data obtained in mammalian cells[1] and oocytes[2] might be due to competition between the endogenous α subunit and the α(D369N) or α(D369A) subunits for a limiting number of β subunits in those cells. The absence of endogenous α subunits in yeast cells circumvents this problem.

For the interpretation of site-directed mutagenesis experiments it is essential to have a homogeneous population of heterologously expressed protein. The yeast expression system was shown to fulfill this criterion as the density of expressed α-subunit protein in yeast membranes, as quantified by Western blotting, equaled the density of [^{3}H]ouabain sites and [^{3}H]ATP sites. Additionally, size exclusion chromatography on a TSK 3000 SW column demonstrated that the hydrodynamic properties of recombinant α(wt)β, α(D369N)β and α(D369A)β enzymes were identical to those of purified pig kidney enzyme.

Removal of the negative charge on D369 and introduction of a more hydrophobic amino acid had a dramatic effect on [^{3}H]ATP binding at equilibrium (TABLE 1). The

---

[a]Tel: +4535321667/1678/1670; fax: +4535321567; e-mail addresses: PAPedersen@aki.ku.dk; JHRasmusse@aki.ku.dk; PLJorgense@aki.ku.dk

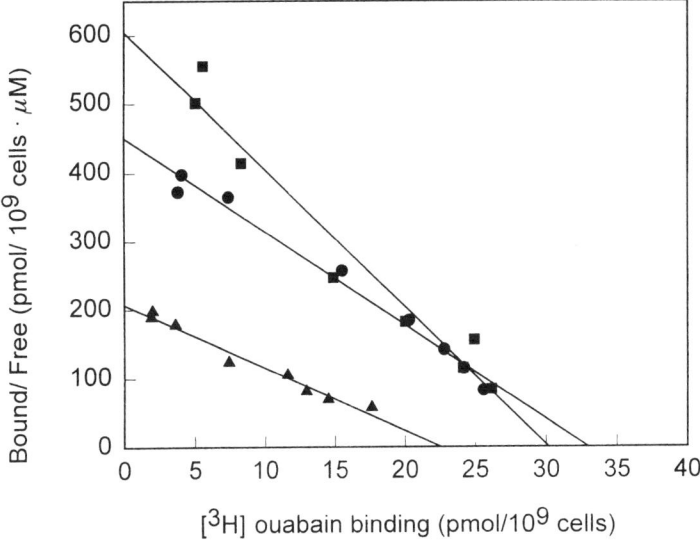

FIGURE 1. [^{3}H]ouabain binding assay of the expression of wild-type α1β1 (●) and the mutations α1(D369N)β1 (▲) and α1(D369A)β1 (■) in cell membranes of intact yeast cells. Transformed yeast cells were produced in the fermentor, and aliquots containing 10 mg cells were incubated for [^{3}H]ouabain binding. The data were fitted by nonlinear least-squares regression to the lines with K_D 72 ± 4 nM for wild type, 106 ± 7 nM for α1(D369N)β1, and 48 ± 4 nM for α1(D369A)β1. Binding capacities (pmol/10^9 cells) are shown in TABLE 1.

affinity for ATP increased 18-fold for α(D369N)β compared to α(wt)β and 39-fold for α(D369A)β compared to the wild type. The affinity for ADP was unaffected by mutation.

Equilibrium binding of [^{3}H]ATP in the presence of varying concentrations of potassium to α(D369N)β, α(D369A)β, and α(wt)β revealed a two- to fourfold increase in the $E_2(K)$ concentration for the mutations compared to the wild type. The shift in conformational equilibrium was confirmed by data obtained from equilibrium binding of [^{3}H]ouabain in the presence of varying concentrations of potassium.

Magnesium is required for [^{3}H]ouabain binding to wild-type Na,K-ATPase. However, the recombinant α(D369A)β enzyme was demonstrated to bind [^{3}H]oua-

TABLE 1. Density of [^{3}H]Ouabain Sites at the Surface of Yeast Cells Expressing Wild-Type and Mutated Na,K-ATPase and the Dissociation Constants for Equilibrium Binding of ATP and ADP

	[^{3}H]Ouabain Sites (pmol/10^9 cells)	K_{ATP} (nM)	K_{ADP} (nM)
α(wt)β	33	109 ± 11	152 ± 10
α(D369N)β	23	5.9 ± 0.4	196 ± 33
α(D369A)β	30	2.8 ± 0.5	122 ± 30

bain with relatively high affinity in the absence of Mg^{2+} ions. This agrees with data obtained from ATP-K antagonism and ouabain-K antagonism, as high affinity ouabain binding is a recognized feature of the E2 conformation.

CONCLUSIONS

1. The yeast expression system allowed targeting of mutations to D369 to the plasma membrane at densities equal to those found for wild-type enzyme.
2. Residue 369 is a major determinant of the affinity for ATP and determines the affinity for ATP relative to ADP.
3. Residue 369 is strongly involved in determining the conformational equilibrium between the E_2 and the E_1 forms.

REFERENCES

1. KUNTZWEILER, T. A. *et al.* 1995. J. Biol. Chem. **270:** 2993–3000.
2. OHTSUBO, K. *et al.* 1990. Biochim. Biophys. Acta **1021:** 157–160.
3. PEDERSEN, P. A. *et al.* 1996. J. Biol. Chem. **271:** 2514–2522.

α1T Can Support Na$^+$,K$^+$-ATPase: Na$^+$ Pump Functions in Expression Systems

JULIUS C. ALLEN,[a] XUN ZHAO, TIMOTHY ODEBUNMI,
SANDRA JEMELKA, RUSSELL M. MEDFORD,
THOMAS A. PRESSLEY, AND ROBERT W. MERCER

Baylor College of Medicine
One Baylor Plaza
Houston, Texas 77030

In four previous papers[1–4] we identified and studied a truncated isoform of the α1 subunit of the Na pump, termed α1T, generated by alternative gene splicing. The resulting protein (MW 66 kD) appears to be identical to the full-length α1 up to aa 554. There follows a unique 27 amino acid tail at the carboxy terminus that is encoded by a retained intron 12. Based on circumstantial evidence we suggested that in canine vascular smooth muscle, this protein appeared to have functional capabilities despite the fact that it lacked 30% of the carboxy end of the protein. The current work demonstrates that the truncated α1T can support pump function when expressed with an appropriate β subunit in heterologous expression systems.

We first used insect *Sf*-9 cells and asked four fundamental questions: (1) Can α1T be expressed in *Sf*-9 cells? (2) Can α1T complex with a β subunit? (3) Is α1T localized to the cell membrane? (4) Does α1T demonstrate functions that can be related to the Na$^+$ pump and therefore related to Na$^+$ transport?

Immunoblotting was used to demonstrate the expression of exogenous α1T by the baculovirus-infected *Sf*-9 cells. Lysates of these cells after infection with either full-length α1 cDNA or α1T cDNA were probed with two antibodies, one that was made to a sequence on the amino side of the splice site and thus would recognize either full-length α1 or α1T and the other that was constructed to 13 amino acids from the retained intron sequence just referred to and would therefore be specific for α1T. Expression of both proteins was demonstrated, that is, α1 at 100 kD and α1T at 66 kD. In addition, immunoprecipitation studies showed complex formation between the α1T and β2 subunit, and confocal microscopy demonstrated membrane localization of the complex as well.

While these data showed that the truncated protein could attain appropriate cellular localization and complex with the β subunit, they said nothing about function. Additional studies showed that the α1T-β complex could be phosphorylated by ATP^{32} (Na$^+$ dependent) and dephosphorylated (K$^+$ dependent) in the ion-specific manner indicative of "classical" Na$^+$,K$^+$-ATPase, with specific autoradiographic localization at 66 kD. Furthermore, enzymatic assays for ouabain-sensitive ATPase activity showed an increase over background control: 0.26 μmol P$_i$/mg protein/hr vs 123 μmol P$_i$/mg protein/hr. Because the α1T was from rat, a species insensitive to ouabain, successful infection also rendered the preparation less sensitive to the inhibitory effects of the cardiac glycoside: K$_i$ (control) = 0.45 μM; K$_i$ (α1T,β) = 1.2 μM. Thus,

---

[a]Tel: 713–798–4977; fax: 713–790–0681; e-mail: juliusa@bcm.tmc.edu

these data strongly suggest that in this expression system, α1T can support a variety of functions indicative of an active pump.

It was still important, however, to try to express the truncated protein in a mammalian system. We began these important expression studies in Hela cells. Because we were able to make an antibody specific for the truncated protein and also had available a specific cDNA, we could monitor message formation and protein expression. We assayed the formation of functional pumps by measuring ouabain-sensitive ^{86}Rb uptake. We used the specific method of Lingrel and co-workers[5] which utilized the differences in ouabain sensitivity between rat and Hela (human) α1 isoforms. Our assumption was that the rat α1T would have the same ouabain resistance as the full-length rat α1. The cells were transfected with α1T and β and then selected in 0.1 μM ouabain for 2 weeks, using vector only and full-length rat α1 as controls.

As expected, the mock transfected cells died when exposed to ouabain, and the cells transfected with the full-length rat α1 survived. However, the cells transfected with the α1T cDNA survived as well, with the concomitant appearance of the α1T protein as determined by Western blots. Ouabain-sensitive ^{86}Rb uptake of the surviving transfected cells was then assayed. Control cells had a K_i of 0.5×10^{-7} M, whereas both the α1 and the α1T transfected cells had a K_i much greater than 10^{-4} M. Indeed, complete inhibition was not achieved with this high ouabain concentration in either of these transfected cells.

These data strongly suggest that the α1T, despite the lack of almost 30% of the carboxy terminus, can complex with an appropriate β subunit to form functional pump sites in both Sf-9 cells and Hela cells. While we are fully aware of the extensive data indicating the importance of many sites in the absent segment of the α1T, we can only suggest that in the absence of these domains the remaining portion of the molecule can effectively complex with a β subunit. At present, we cannot speculate on the specific mechanism whereby this occurs, but a variety of hypotheses are under investigation.

REFERENCES

1. ALLEN, J. C., R. M. MEDFORD, X. ZHAO & T. A. PRESSLEY. 1991. Disproportionate alpha and beta mRNA content in canine vascular smooth muscle. Circ. Res. **69:** 39–44.
2. MEDFORD, R. M., M. AHMAD, R. HYMAN, J. C. ALLEN, T. A. PRESSLEY, P. D. ALLEN & B. NADAL-GINARD. 1991. Vascular smooth muscle expresses a truncated Na$^+$,K$^+$-ATPase α subunit isoform. J. Biol. Chem. **266:** 18308–18312.
3. ALLEN, J. C., T. A. PRESSLEY, T. ODEBUNMI & R. M. MEDFORD. 1994. Tissue specific membrane association of α1T in smooth muscle cells. FEBS Lett. **337:** 285–288.
4. MEDFORD, R. M., T. A. PRESSLEY, R. M. MERCER & J. C. ALLEN. 1994. Expression of a truncated α subunit of the Na$^+$,K$^+$-ATPase (α1T) in smooth muscle cells. *In* The Sodium Pump. E. Bamberg & W. Schoner, Eds. :192–200. Springer. New York.
5. VAN HUYSSE, J. W., E. A. JEWELL & J. B. LINGREL. 1993. Site directed mutagenesis of a predicted cation binding site of the Na$^+$,K$^+$-ATPase. Biochemistry **32:** 819–826.

Involvement of Different Sites for Nucleotide Analogs in the Phosphatase Activity of the Red Cell Calcium Pump

CLAUDIA DONNET,[a,c] ARIEL J. CARIDE,[b]
SILVINA A. TALGHAM,[a] AND JUAN P. F. C. ROSSI[a]

[a]IQUIFIB, Instituto de Química y Fisicoquímica Biológicas
(UBA-CONICET)
Facultad de Farmacia y Bioquímica
Junín 956
1113 Buenos Aires, Argentina

[b]Department of Biochemistry and Molecular Biology
Mayo Clinic Foundation
Rochester, Minnesota 55905

The calcium pump of plasma membranes catalyzes the hydrolysis of ATP along a curve that is fitted by the sum of two hyperbolic components: a high affinity one (so-called catalytic site) and a low affinity one. Also, in the presence of ATP and/or calmodulin and Ca^{2+}, the pump catalyzes the hydrolysis of phosphoric esters such as p-nitrophenyl phosphate (pNPP). We have studied the effect of known nucleotide analogs and/or chemical modification of nucleotide binding sites on Ca^{2+}-pNPPase activity.

MATERIALS AND METHODS

Membranes stripped of calmodulin were prepared by the method of Gietzen et al.[1] Chemical modification was carried out by incubating for 30 minutes (unless otherwise indicated) membranes (2 mg protein/ml) in bicine buffer, pH 8.4, in the presence of 10 μM fluorescein isothiocyanate (FITC) or 2.5 mM N-hydroxysuccinimidyl acetate (SA). Membranes were then washed three times with Tris-ClH (pH 7.4 at 37°C). Ca^{2+}-ATPase and Ca^{2+}-pNPPase activities were measured as previously described.[2]

RESULTS AND DISCUSSION

Treatment of erythrocyte membranes with FITC abolished both Ca^{2+}-ATPase and ATP-dependent Ca^{2+}-pNPPase. However, the FITC-labeled enzyme retained 50% of its calmodulin-dependent Ca^{2+}-pNPPase activity (FIG. 1). The calmodulin-dependent Ca^{2+}-pNPPase activity was also measured as a function of pNPP concentration (FIG. 1), showing no change in its apparent affinity for the phosphoric ester.

[c]Tel: 54 1 964 8289; fax: 54 1 962 5457; e-mail: cdonnet@mail.retina.ar; caride@rcf.mayo.edu; rtjpaul@criba.edu.ar

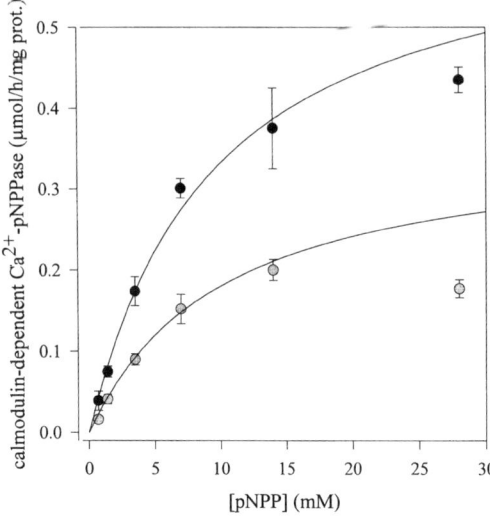

FIGURE 1. Effect of FITC labeling on Ca^{2+}-pNPPase. Erythrocyte membranes were labeled with 0 (●) or 10 (●) µM FITC, and calmodulin-dependent Ca^{2+}-pNPPase activity was measured in the presence of different concentrations of pNPP.

We previously demonstrated that acetylation of lysine residues with N-hydroxysuccinimidyl acetate (SA) inactivates Ca^{2+}-ATPase by: (1) modifying the catalytic site and (2) impairing stimulation by modulators by reaction with other residue(s) located outside the catalytic site.[2] Incubation of the enzyme with SA fully inactivated the ATP-dependent Ca^{2+}-pNPPase, following biphasic kinetics, with a fast and a slow component (FIG. 2). The presence of ATP during SA treatment canceled the fast component. SA inhibited only partially the calmodulin-dependent Ca^{2+}-pNPPase following monophasic kinetics, and ATP during treatment did not have any effect on this inactivation (FIG. 2). The presence of pNPP during incubation with SA prevented

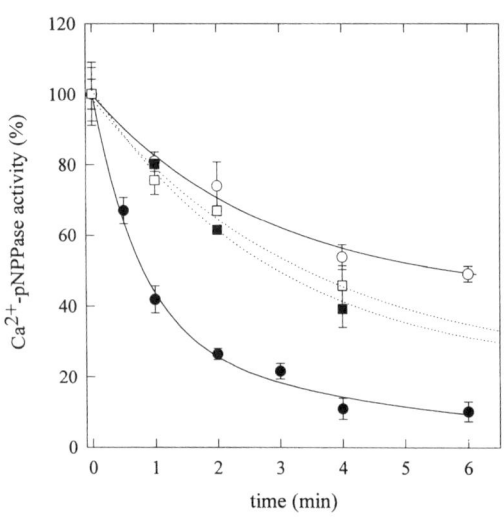

FIGURE 2. Effect of acetylation with N-hydroxysuccinimidyl acetate (SA) on Ca^{2+}-pNPPase. Erythrocyte membranes were pretreated with SA in the presence of 0 (●, ■) or 3 (○, □) mM ATP. Ca^{2+}-pNPPase activity was measured in the presence of 0.5 mM ATP (*circles*) or 120 nM calmodulin (*squares*). Control activities were 0.55 and 0.35 µmol/h/mg protein, respectively.

Ca^{2+}-ATPase from inhibition by SA, while having no effect on inactivation of the calmodulin-dependent Ca^{2+}-pNPPase by SA. Inhibition of the ATP-dependent Ca^{2+}-pNPPase was partially protected by pNPP. These results suggest that binding of pNPP to the catalytic site protects both the Ca^{2+}-ATPase and the ATP-dependent pNPPase; this finding agrees with previous work which showed competition between ATP and pNPP for the catalytic site.[3]

These results lead to the following conclusions: (1) ATP-dependent pNPPase depends on binding of ATP to its catalytic site; (2) although FITC decreases the calmodulin-dependent pNPPase by 50%, protection of the catalytic site with either ATP or pNPP during acetylation by a less bulky reagent such as SA does not prevent inactivation of this activity. Therefore, it can be said that the catalytic site plays no role in calmodulin-dependent pNPPase.

REFERENCES

1. GIETZEN, K., M. TEJCKA & H. U. WOLF. 1980. Biochem. J. **189:** 81–88.
2. DONNET, C., A. J. CARIDE, H. N. FERNÁNDEZ & J. P. F. C. ROSSI. 1994. Biochem. J. **302:** 133–140.
3. CARIDE, A. J., A. F. REGA & P. J. GARRAHAN. 1982. Biochim. Biophys. Acta **689:** 421–428.

Nonaqueous Nature of the Environment Surrounding the Acyl-phosphate of the Na^+/K^+-ATPase Deduced from Nucleophilic Attack by Hydroxylamine on the Phosphate Group

YOSHIHIRO FUKUSHIMA,[a,c] YASUO SHINOHARA,[b,d]
AND MAKOTO USHIMARU[a,c]

[a]*Department of Chemistry*
Kyorin University School of Medicine
Mitaka, Tokyo 181, Japan

[b]*Laboratory of Medicinal Biochemistry*
Faculty of Pharmaceutical Sciences
University of Tokushima
Tokushima 770, Japan

The phosphate group is widely distributed in biosystems as a constituent moiety of ATP and plays a key role in processes responsible for decreases in entropy, such as active transport of ions mediated by Na^+/K^+-ATPase. Investigation of the characteristics of the phosphate group incorporated into this enzyme is therefore essential for clarifying the molecular mechanism of Na^+/K^+-ATPase as the functional unit of active ion transport.

The present work represents a return to research on the nucleophilic substitution of the acyl-phosphate group of the phosphoenzyme (EP) of Na^+/K^+-ATPase by hydroxylamine.[1] We compared the reaction mechanism of the EP in the active state and that of acetylphosphate, which is a classic model compound for the phosphate bond in the EP of so-called P-type ATPases. The reaction mechanism of hydroxylamine with the phosphate group of the EP would be expected to reflect only the static chemical state of the acyl-phosphate bond in the active-site cleft of the enzyme, because hydroxylamine, unlike water, is not a physiological ligand.

RESULTS AND DISCUSSION

Methylhydroxylamine or hydroxylamine accelerated the cleavage of the acyl-phosphate linkage of both active E_1P and E_2P of dog kidney Na^+/K^+-ATPase (FIG. 1A). On the other hand, these amines did not cleave the phosphate bond of EPs of Ca^{2+}-ATPase from the sarcoplasmic reticulum of rabbit skeletal muscle (FIG. 1B).

[c]Tel: +81-422-44-1981; fax: +81-422-44-1981.
[d]Tel: +81-886-33-7278; fax: +81-886-33-5196.

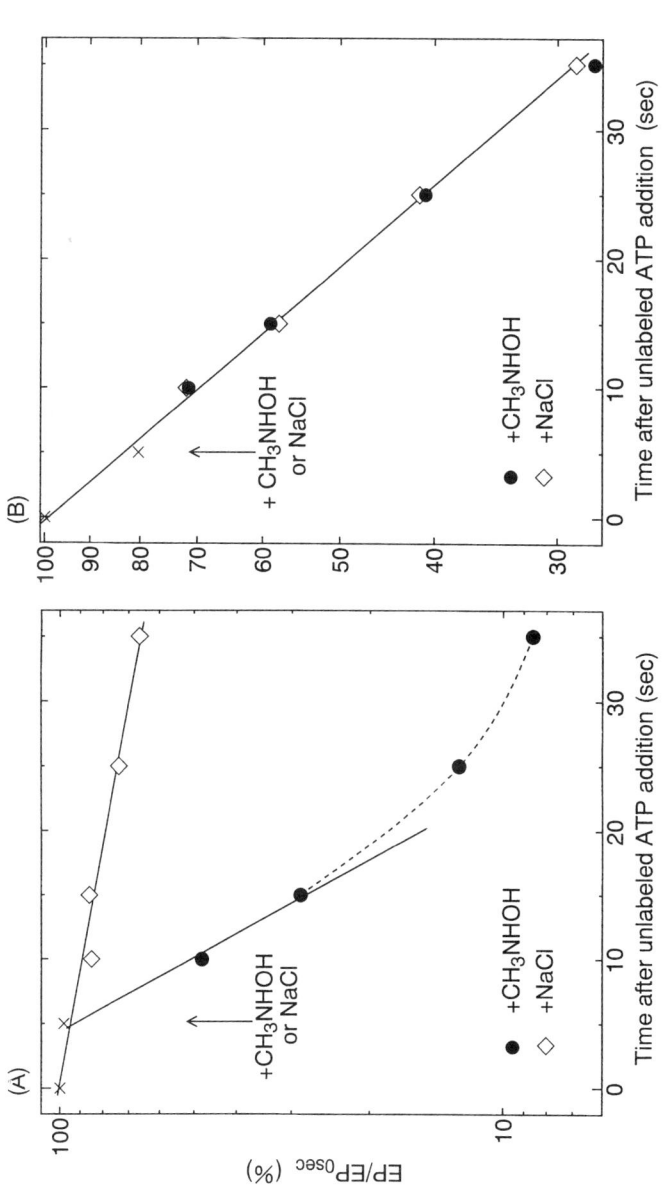

FIGURE 1. Comparison of the reactivity with methylhydroxylamine of E_1Ps of Na^+/K^+-ATPase and Ca^{2+}-ATPase. **(A)** Na^+/K^+-ATPase (10 μg/ml) and 20 μM [γ-^{32}P]ATP were incubated at 0°C and pH 7.3 in the presence of 2 M NaCl, 0.2 mM $MgCl_2$, 0.5 mM EGTA, and 20 mM histidine. At 60 seconds, about 80% of the total EP was E_1P, which was sensitive to ADP. Unlabeled ATP (1 mM) was added to the EP at 60 seconds to stop further labeling (X). Then 0.1 M methylhydroxylamine was added (●) 5 seconds after the unlabeled ATP. As a control, NaCl was added (◇), because the solution of $CH_3NHOH \cdot HCl$ was neutralized with NaOH. **(B)** E_1P of Ca^{2+}-ATPase was prepared in the presence of 1.5 mM $CaCl_2$ and 100 mM KCl, but otherwise the conditions were the same as those in A. Chasing and testing of reactivity with methylhydroxylamine were also performed in the same way as described for A. Similar results were obtained when the labeling was prevented with CDTA instead of unlabeled ATP and for the E_2P of both Na^+/K^+-ATPase and Ca^{2+}-ATPase.

(A) Net Atomic Charge of Acetylphosphate

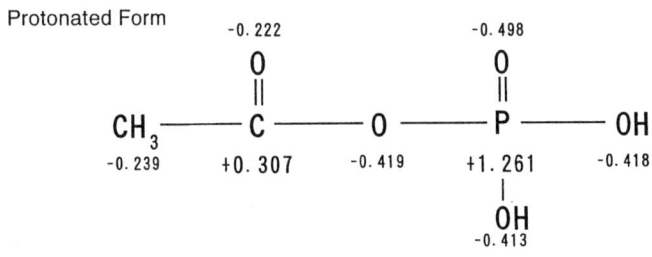

(B) Overlap Population of Acetylphosphate

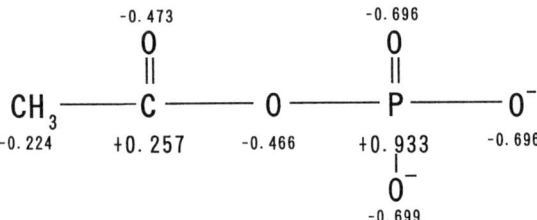

FIGURE 2. (A) Net atomic charge of acetylphosphate; (B) overlap population of acetylphosphate.

To determine if hydroxylamine mediates a nucleophilic attack on the carbonyl C atom of the acyl-phosphate linkage of the essential aspartate residue in the EPs of Na^+/K^+-ATPase, the enzyme was pretreated with hydroxylamine under conditions in which E_1P or E_2P was the dominant form of the phosphoenzyme, respectively. Under these circumstances, ATPase activity should have been inhibited by derivatization of the residue to the corresponding hydroxamic acid; however, there was no inhibition. The nucleophilic center of the native EPs was probably not the carbonyl C atom but the P atom,[2] so that an N-P compound was released from the EP by attack with hydroxylamine, thus producing active enzyme.

It was confirmed that the classic model compound for the phosphate bond of EP, acetylphosphate, whose respective pK_1 and pK_2 values are 1.2 and 4.8, formed hydroxamic acid at pH values ranging from 0.3–7.0.[3] These results were not modified in the presence of Mg^{2+} up to a concentration of 2 M, where more than 90% of the acetylphosphate would have been coordinated with Mg, according to the value of the association constant, 4.4 M^{-1}, for acetylphosphate and Mg. These two results indicated that the discrepancy of the nucleophilic centers between EPs and acetylphosphate was not due to differences in the acid dissociation state of their respective phosphate groups or to the presence of Mg coordinated to the phosphate groups.

To clarify the electronic state of the acyl-phosphate bond, we calculated the molecular orbital of acetylphosphate by the nonempirical (*ab initio*) method. The net atomic charge of the P atom was more positive than that of the carbonyl C atom regardless of the acid dissociation state of the phosphate group (FIG. 2A). The value of the overlap population of the C-O bond was larger than that of the O-P bond regardless of the acid dissociation state of the phosphate bond (FIG. 2B). These results suggested that the P atom is more likely to be the nucleophilic center than the C atom and that the O-P bond is cleaved more easily than is the C-O bond. This is consistent with the results obtained for the active EPs, but it is inconsistent with these obtained for acetylphosphate in solution.

As no water molecule hydrated to acetylphosphate was considered in the molecular orbital calculations, the suggested electronic state of the acyl-phosphate group is similar to that in a nonaqueous environment. Hence the hydroxylamine attack on the P atom of the EP means that the phosphate group of the Na^+/K^+-ATPase EP is located in a cleft in an almost nonaqueous environment.[4]

Although hydroxylamine access to the phosphate group from the exterior of the enzyme is still allowed in the case of Na^+/K^+-ATPase, the active site cleft in Ca^{2+}-ATPase appears to be completely segregated from the exterior water phase (FIG. 1B).

ACKNOWLEDGMENT

We are grateful to Dr. Yutaro Hayashi for kindly supplying purified Na^+/K^+-ATPase from dog kidney outer medulla.

REFERENCES

1. HOKIN, L. E. & J. L. DAHL. 1972. *In* Metabolic Pathways, 3rd Ed. Vol 6. Metabolic Transport. L. E. Hokin, Ed.: 269–315. Academic Press. New York.
2. POST, R. L. & S. KUME. 1973. J. Biol. Chem. **248**: 6993–7000.
3. SABATO, G. D. & W. P. JENCKS. 1961. J. Am. Chem. Soc. **83**: 4393–4399.
4. DE MEIS, L. 1981. *In* The Sarcoplasmic Reticulum.: 134–143. Wiley & Sons. New York.

Study of Structure-Functional Organization Features in the Pig Kidney Na,K-ATPase at Different Conformational States

NATALIA M. GEVONDYAN,[a] ASYA V. GRINBERG,[b] AND VLADIMIR S. GEVONDYAN[a]

Shemyakin and Ovchinnikov Institute of Bioorganic Chemistry
Russian Academy of Sciences
ul. Miklukho Maklaya 16/10
117871 Moscow, Russia

The Na,K-ATPase reaction cycle is a chain of consecutive correlated conformational transitions in the enzyme molecule induced by Na-dependent phosphorylation and K-dependent dephosphorylation conjugated with occlusion and translocation of sodium and potassium ions through the plasma membrane. Every stage in this process is characterized by a unique, thermodynamically stable structure of polypeptide chain organization; the description of this organizational peculiarity may be key to interpreting enzyme function. Despite extensive studies of enzyme function at various stages, little is known about the enzyme's thermodynamic potential or the molecular mechanisms underlying its cyclic conformational transitions. This work is dedicated to the study of those aspects of Na,K-ATPase function. For this purpose we chose two approaches: measurement of thermodynamic characteristics of the various enzyme structural domains and analysis of reactivity of SH groups affected by intramolecular ligand-dependent conformational transitions.

The thermal unfolding and domain structure of the Na,K-ATPase from pig kidney were studied by high-sensitivity differential scanning calorimetry. We demonstrated that thermal denaturation of Na,K-ATPase was displayed in the unfolding of three independent cooperative units (domains 1, 2, and 3) with midpoint transition temperatures (T_d) of 320.6, 327.5, and 331.5 K, respectively. Domains 1 and 3 were identified as corresponding to the α subunit, domain 2 to the β subunit.

The total enthalpy of ion-free native Na,K-ATPase is 1,753 kJ/mol or 2.8 cal/g, including α subunit enthalpy 1,259 kJ/mol (340 kJ/mol for domain 1 and 919 kJ/mol for domain 3) and β subunit enthalpy 494 kJ/mol.

To clarify whether conformations of the domains are affected by natural ligands, we examined the thermal unfolding of membrane-bound Na,K-ATPase with monovalent cations Na^+ and K^+ and with ATP. Na^+ ions induce a conformational transition, stabilizing the enzyme structure, with total enthalpy increase of 155 kJ/mol. The main increase (105 kJ/mol) corresponds to domain 1 of α subunit. Domains 2 and 3 are

[a]Tel: 095 336 7844; fax: 095 330 6456; e-mail: gevond@ibch.siobc.ras.ru
[b]Present address: Max-Delbrück-Centre for Molecular Medicine, Robert Rössle Str, 10, 13122 Berlin, Germany (tel: 49 681 302 2045; fax: 49 681 302 4739; e-mail: agrin@orion.rz.mdc-berlin.de).

subjected to structural changes of lesser extent (38 and 12 kJ/mol, respectively). K^+ ions induce sharp structural changes in the enzyme, whereas the total enthalpy increase is 388 kJ/mol and is proportional for all three domains. The enzyme-substrate complex formation in the E1ATP state is accompanied by an enzyme enthalpy decrease to 1,670 kJ/mol mainly because of changes in the domain 3 structure, whereas after transition to the potassium form, enthalpy increases substantially. Thus, changes in the stability of all three Na,K-ATPase domains suggest the conformation of both subunits to be dependent on physiological ligands. K^+ binding produces a sharp unfolding enthalpy increase in all three domains. In addition, a similar increase is detected in domain 1 unfolding enthalpy upon Na^+ binding. As for domain 2, the Na^+ ion-induced conformation is much less stable than that induced by K^+ ions. Domain 3 enthalpy changes are especially substantial in the presence of ATP.

The obtained enthalpy change data, when compared with the states of the Albers-Post scheme, reveal an appropriate cyclic nature, so that these states could be characterized quantitatively. The magnitude of their enthalpy is shown to correlate with previously registered cyclic changes in enzyme hydrophily and hydrophobicity.[1]

We previously showed the presence of 5 S-S bonds and 20 Cys residues differing in reactivity in Na,K-ATPase.[2-5] It would be of interest to estimate their contribution to the enzyme structural organization. Thus, we applied the complex approach and simultaneously measured the enzyme activity under reducing reagents and studied the conformational changes of the enzyme.

We studied the kinetics of the inhibition of enzyme activity by dithiothreitol and β-mercaptoethanol at 40°C, excluding the thermal denaturation of the protein. The data obtained suggest that there are two types of bonds sufficiently different in their accessibility to thiol reagents. The calorimetric experiments at the end of the first stage of the inhibition process in which 25% of the activity is lost revealed a total enzyme enthalpy decrease of 570 kJ/mol, mainly because of the disappearance of the conformation transition peak of domain 1 of the α subunit. A slight enthalpy decrease of β subunit (26%) and of domain 3 of α subunit (11%) was also observed. The enzyme's high sensitivity to the action of reducing agents under mild conditions, resulting in disturbance of the enzyme's structured organization, may be caused both by S-S bond reduction and by destruction of hydrophobic bonds involving Cys residues.

Our results on conformational state-dependent SH group reactivity appear to confirm this hypothesis. The study shows cyclic recurrence in reactivity changes in the SH group. The greatest number of rapidly reactive SH groups (12 Cys residues) is detected in the E1 state at the highest molecular hydrophilicity. Investigation of the E_1ATP SH group reactive capacity indicates that interaction of Na,K-ATPase and ATP causes a decrease in the number of rapidly reactive SH groups and an increase in the number of rapidly masked ones. ATP binding is accompanied by the shielding of about five SH groups that therefore can take part (directly or indirectly) in the ATP binding process. The same masking tendency was observed in the $E_1ATP \rightarrow E_2ATP$ transition (all rapidly reactive SH groups become masked ones). As a result of the $E_2ATP \rightarrow E_2$ process, six SH groups become rapidly reactive.

Therefore, 12 Cys residues participate in some way in the molecular conformational transitions. Comparison of enthalpy values (hydrophobicity) and the number of masked SH groups reveals a direct proportional correlation.

We thus managed for the first time to quantitatively estimate Na,K-ATPase subunit structural organization characteristics and to reveal SH-dependent mechanisms underlying the cyclic conformational transitions of the molecule.

REFERENCES

1. MODYANOV, N., S. LUTSENKO, E. CHERTOVA, R. EFREMOV & D. GULYAEV. 1992. Acta Physiol. Scand. **146:** 49–58.
2. GEVONDYAN, N. M., V. S. GEVONDYAN, E. E. GAVRILYEVA & N. N. MODYANOV. 1989. FEBS Lett. **255:** 265–268.
3. GEVONDYAN, N. M., V. S. GEVONDYAN & N. N. MODYANOV. 1993. Biochem. Mol. Biol. Int. **29:** 327–337.
4. GEVONDYAN, N. M., V. S. GEVONDYAN & N. N. MODYANOV. 1993. Biochem. Mol. Biol. Int. **30:** 337–346.
5. GEVONDYAN, N. M., V. S. GEVONDYAN & N. N. MODYANOV. 1993. Biochem. Mol. Biol. Int. **30:** 347–355.

High Yield Fermentation of Pig α1β1 Na,K-ATPase in *Saccharomyces cerevisiae*

JAKOB H. RASMUSSEN,[a] PER AMSTRUP PEDERSEN,
AND PETER L. JORGENSEN

*Biomembrane Research Center
August Krogh Institute
Universitetsparken 13
2100 Copenhagen OE, Denmark*

We systematically analyzed the capacity of *Saccharomyces cerevisiae* as host for large scale production of recombinant Na,K-ATPase by fermentation.[1] We characterized the expression level with respect to its dependence on gene copy number of the expression plasmid, promoter strength, and growth medium composition.

We constructed a 15.7-kb yeast expression plasmid in which transcription of α1 pig cDNA and β1 pig cDNA are controlled by identical galactose-regulated promoters to achieve stoichiometric expression of subunits. Inducible expression was preferred to prevent potential problems with toxic expression of the membrane-spanning Na,K-ATPase. The expression plasmid carries two selective markers, URA3 and *leu2-d*, a poorly expressed allele of the LEU2 gene. Selection for leucine autotrophy increases the plasmid copy number 5–10 times relative to selection for uracil autotrophy.[2] An expression plasmid with a regulated copy number was favored to obtain a low plasmid copy number in the yeast growth phase to avoid selection against the presence of the expression plasmid and to obtain a high plasmid copy number in the expression phase to optimize conditions for high level expression.

Yeast galactose-inducible promoters are controlled at the transcriptional level by the GAL4 and GAL80 proteins.[3] The GAL4 protein is a transcription factor present at a copy number of 2–4 per cell, whereas the GAL80 protein has a copy number exceeding 100. Expression from high copy number plasmids carrying transcriptional fusions to a galactose-regulated promoter will consequently be limited by the availability of GAL4 protein.[3] We therefore constructed an integrative yeast plasmid carrying a transcriptional fusion between the GAL4 gene and a galactose-regulated promoter. This plasmid was targeted to the chromosome of protease-deficient yeast strain BJ5457.[4] Induction of galactose-regulated promoters in this strain transformed with the expression plasmid should initiate a cascade reaction with respect to production of GAL4 protein and Na,K-ATPase α1 and β1 subunit protein.

It is well known that growth medium composition and fermentation conditions can dramatically influence the expression level and yield of recombinant proteins. We therefore analyzed how growth medium composition and fermentation conditions influenced the expression level of recombinant Na,K-ATPase. TABLE 1 shows the influence of the aforementioned parameters on the expression level. It can be seen that

---

[a]Tel: +4535321678; fax: +4535321567; e-mail addresses: JHRasmusse@aki.ku.dk; PAPedersen@aki.ku.dk; PLJorgense@aki.ku.dk.

TABLE 1. Expression Levels of Fully Active Recombinant Na,K-ATPase Expressed in Yeast

	−GAL4	+GAL4	Low Copy Number	High Copy Number	−Amino Acids	+Amino Acids
BJ5457 (pmol/mg)	2.2	22	11	22	22	43

[a] ± GAL4, a strain with or without a GAL4 fusion and a high plasmid copy number; low/high copy number, low or high copy number of the expression plasmid in a GAL4 strain; ± amino acids; growth of a GAL4 strain in medium supplemented with or lacking amino acids.

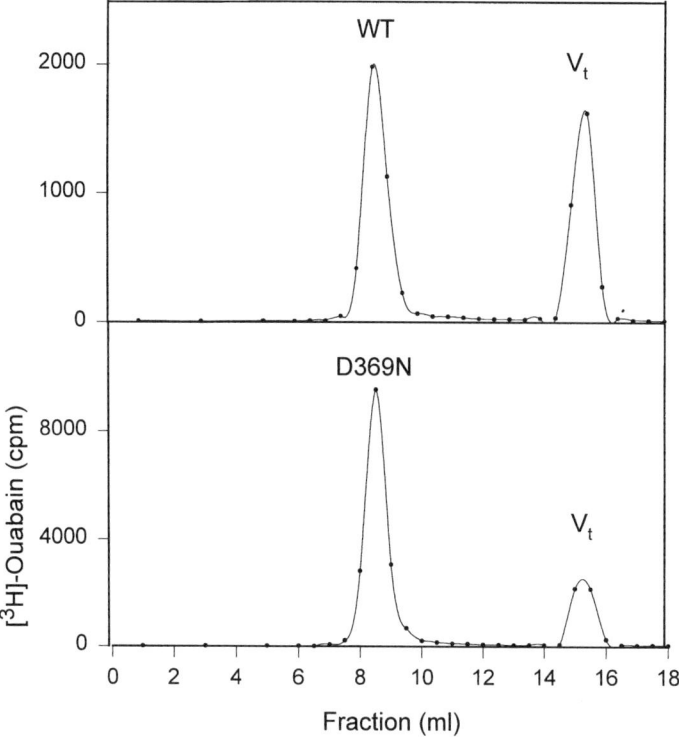

FIGURE 1. Size-exclusion chromatography of recombinant Na,K-ATPase from yeast. Solubilized wild-type and D369N Na,K-ATPase from yeast were separated using a 7.5 × 300 mm plus 7.5 × 75 mm TSK-gel G 3000 SW column equilibrated with $C_{12}E_8$ 5 mg/ml, 150 mM potassium acetate, pH 6.0, and operated at 4°C. Both mutant D369N and wild-type enzyme were eluted at equal volumes as pig kidney enzyme.[1] This provides evidence that the Na,K-ATPase expressed in *Saccharomyces cerevisiae* is organized in αβ—units, which in the solubilized state have hydrodynamic properties similar to those of pig kidney Na,K-ATPase. The peak at V_t is comprised of free [^{3}H]ouabain which was trapped in the pellet prior to solubilization.

introduction of a transcriptional fusion between a galactose-inducible promoter and the GAL4 gene increased the expression level tenfold, whereas increasing plasmid copy number caused a twofold increase in expression level. Supplementing the growth medium with amino acids further increased the expression level twofold.

Combined efforts to increase the density of recombinant Na,K-ATPase in yeast resulted in an expression level of 40 pmol/mg protein determined as [^{3}H]ouabain sites. All the expressed Na,K-ATPase protein was active as determined from quantitative Western blotting, [^{3}H]ATP binding, phosphorylation, and Na,K-ATPase activity.

[^{3}H]Ouabain binding to intact cells showed that the wild-type enzyme and inactive mutants D369N and D808N were expressed in the plasma membrane at identical densities. Wild-type enzyme, α(D369N)β, and α(D808N)β were shown to have identical hydrodynamic properties after solubilization in $C_{12}E_8$, indicating that mutations to D369 or D808 did not interfere with protein folding (FIG. 1).

REFERENCES

1. PEDERSEN, P. A., J. H. RASMUSSEN & P. L. JORGENSEN. 1996. J. Biol. Chem. **271:** 2514–2522.
2. EBERHART, E. & C. P. HOLLENBERG. 1983. J. Bacteriol. **156:** 625–635.
3. SCHULTZ, L. D. *et al.* 1987. Gene **54:** 113–123.
4. JONES, E. W. 1991. Methods Enzymol. **194:** 182–187.

Involvement of Glutamic Acid 820 in K^+ and SCH 28080 Binding to Gastric H^+,K^+-ATPase[a]

HERMAN P. G. SWARTS, CORNÉ H. W. KLAASSEN, AND JAN JOEP H. H. M. DE PONT[b]

Department of Biochemistry
Institute of Cellular Signalling
University of Nijmegen
P.O. Box 9101
6500 HB Nijmegen, The Netherlands

Negatively charged residues present in transmembrane segments of P-type ATPases may be involved in cation binding and transport. There are several glutamic and aspartic acid residues in or around the fifth and sixth transmembrane domain of the catalytic subunit of gastric H^+,K^+-ATPase that are conserved in other P-type ATPases.[1] The role of six of these amino acid residues was investigated by site-directed mutagenesis, resulting in conversion of these acid groups into their corresponding acid amides. Sf9 cells were used as an expression system using a single baculovirus with coding sequences for both the α and β subunits of H^+,K^+-ATPase.[2,3] Both subunits of all mutants, like the wild-type enzyme, were expressed in intracellular membranes of Sf9 cells as indicated by western blotting experiments, an enzyme-linked immunosorbent assay, and confocal laser scan microscopy studies.

The amount of the specific H,K-ATPase phosphorylated intermediate was measured as the difference between the ATP-phosphorylation level in the absence of SCH 28080 and in its presence. In mutants D824N, E834Q, E837Q, and D839N the amount of specific phosphorylated intermediate was not significantly different from zero. In all preparations a considerable amount of mutated H^+,K^+-ATPase protein (0.8–4.1% of total) was produced, indicating that the lack of activity of these mutants is not due to a lack of biosynthesis. These findings suggest that each of the D824, E834, E837, and D839 residues is essential for the enzyme to become phosphorylated. It might be that these residues are involved in H^+ binding which is essential for ATP phosphorylation.

The two mutants E795Q and E820Q, which show an SCH 28080 sensitive phosphorylated intermediate, were studied in more detail. The effect of preincubation with either K^+, or SCH 28080, or ouabain or vanadate on the $AT^{32}P$-phosphorylation level is shown in the autoradiograms of the SDS-PAGE gels (FIG. 1). In the Sf9 membranes both a 100- and a 140-kD phosphorylated protein are found. The 140-kD protein is also present in membranes of uninfected cells and does not originate from

---

[a]This work was supported by the Netherlands Foundation for Scientific Research, Division of Medical Sciences (NWO-GMW), under grant 902-22-086 (to C.H.W.K.).

[b]To whom correspondence should be addressed. Tel: +31243614260; fax: +31243540525; e-mail: J.dePont@bioch.kun.nl

FIGURE 1. Autoradiogram of SDS-PAGE of the AT^{32}P phosphorylated Sf9 membranes infected with wild-type and mutated baculoviruses. Membranes isolated from Sf9 cells infected with either wild-type virus or mutant viruses E795Q and E820Q were phosphorylated at 0°C with 0.1 μM [γ-^{32}P]ATP in the presence of 1 mM MgCl$_2$ and 20 mM Tris-acetic acid (pH 6.0) after preincubation for 60 minutes at 0°C with the following compounds: 100 μM SCH 28080, 10 mM KCl, 1 mM ouabain, and 1 mM vanadate. The acid-quenched samples were solubilized and subjected to SDS-PAGE at pH 6.5. Purified pig gastric H$^+$,K$^+$-ATPase was used as a control.

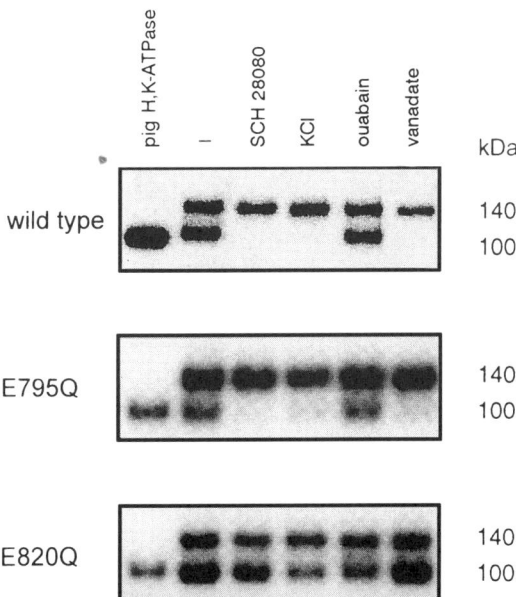

H$^+$,K$^+$-ATPase. Formation of the 100-kD phosphorylated intermediate of mutant E820Q is hardly affected by each of the used ligands in contrast to both the wild-type enzyme and mutant E795Q, where these ligands except ouabain prevent formation of the 100-kD band. The K$^+$ sensitivity of mutant E795Q (I_{50} = 0.45 ± 0.10 mM; n = 3) is similar to that of the wild-type enzyme (I_{50} = 0.38 ± 0.04 mM; n = 5). The I_{50} value of mutant E820Q, however, is 10 times higher (I_{50} = 4.5 ± 1.2 mM; n = 4), and complete inhibition is not reached at 30 mM. This mutant has additionally a 100 times lower sensitivity towards SCH 28080. An I_{50} value of 1.7 ± 0.6 μM (n = 4) was found, whereas the wild-type enzyme has an I_{50} value of 14 ± 3 nM (n = 4) and mutant E795Q an I_{50} value of 8 ± 3 nM (n = 4).

Dephosphorylation studies show that the phosphorylated intermediate of the wild-type enzyme is K$^+$ sensitive and ADP insensitive. The E795Q mutant demonstrated similar behavior. The phosphointermediate of the E820Q mutant, however, was insensitive towards added K$^+$ up to 100 mM. The E820Q mutant also showed no sensitivity for ADP, suggesting that this mutation did not lead to blockade of the E_1-P → E_2-P conversion, which would have resulted in an ADP-sensitive phosphorylated intermediate. The phosphorylated intermediate of the wild-type enzyme and of mutants E795Q and E820Q showed similar hydroxylamine sensitivity, indicating that the K$^+$-insensitive mutant E820Q had also formed an acyl-phosphate.

The overall ATPase activity, determined at 10 μM ATP, of both the wild-type enzyme and mutant E795Q could be activated by K$^+$ ($K_{0.5}$ = 0.2 mM). In the membranes of uninfected Sf9 cells and mutant E820Q, no activation by K$^+$ of the ATPase activity was observed.

The aforementioned findings suggest that Glu795 is not involved in K$^+$ binding, but binding of extracellular K$^+$ to Glu820 is essential in the long range of conformational changes that enhance the hydrolysis rate of the phosphorylated intermediate at Asp386.

REFERENCES

1. LUTSENKO, S. & J. H. KAPLAN. 1995. Biochemistry **34:** 15607–15613.
2. KLAASSEN, C. W. H., T. J. F. VAN UEM, M. P. DE MOEL, G. L. J. DE CALUWE, H. G. P. SWARTS & J. J. H. H. M. DE PONT. 1993. FEBS Lett. **329:** 277–282.
3. KLAASSEN, C. W. H., H. G. P. SWARTS & J. J. H. H. M. DE PONT. 1995. Biochem. Biophys. Res. Commun. **210:** 907–913.

Copper-Stimulated Adenosine Triphosphatase from Rat Liver

Isolation and Kinetic Characterization

JULNAR USTA, HANA BARAKEH, HASHEM MAHFOUZ,
AND NADIM CORTAS

Departments of Biochemistry, Pharmacology, and Internal Medicine
American University of Beirut
Faculty of Medicine
Beirut, Lebanon

Copper is an essential heavy metal, being a cofactor of many critical enzymes such as cytochrome oxidase, tyrosinase, and superoxide dismutase.[1] Copper is also a highly toxic element capable of generating free radicals that may damage nucleic acids and oxidize proteins and lipids.[2] A fine homeostatic mechanism should therefore exist to ensure an optimal copper supply to cellular sites. Defects in cellular copper export underly both the copper deficiency of Menkes' disease and the excess copper toxicity of Wilson's disease. Both were found recently[3-5] to result from mutations in two closely related genes that encode proteins belonging to the P-ATPase family. No candidate protein product for either of the genes has been identified. We report the isolation of a copper-activated Mg^{2+} requiring ATPase (Cu^{2+} ATPase) from rat liver that may be the candidate gene product of Wilson's disease.

MATERIALS AND METHODS

Rat livers were homogenized in 1 mM $NaHCO_3$ pH 7.4[6] fractionated by centrifugation (1,500 g × 20 minutes). The resulting upper brown pellet overlying a tight red pellet was carefully separated and subfractionated (24,000 g × 30 minutes) on discontinuous sucrose density gradient 56%:44%:36% (5:20:10 ml, respectively). The resulting band at the center of the 44% sucrose was collected, treated for 60 minutes with penicillamine (3.27 µmol/mg protein) and mercaptoethanol (1.46 µg/mg protein), washed, suspended in 250 mM sucrose, 10 mM Hepes, 50 mM Tris (SHT) buffer, pH 7.2, labeled as MF, and assayed for Cu^{2+}-stimulated ATPase activity.

Cu^{2+} ATPase ASSAY

The assay that was optimized for SDS, pH, Mg, ATP, protein, and time contained in a final volume of 0.3 ml: Tris-maleate (7.5 mM); MF (10 µg); ouabain (2 mM); oligomycin (1.4 µg); $MgCl_2$ (3 mM); and $CuCl_2$ (0–300 µM). The reaction was started by the addition of ATP to a final concentration of 3 mM. Incubation was carried at 37°C and terminated after 10 minutes by the addition of trichloroacetic acid. The amount of P_i released was measured colorimetrically.[7] The activity of Cu^{2+} ATPase was expressed as μmoles P_i/mg protein/hour above the basal Mg^{2+} ATPase activity. The specificity of Cu^{2+} ATPase was determined under optimal assay conditions by substituting Cu^{2+} with Co^{2+}, Zn^{2+}, Mn^{2+}, or Cd^{2+} and for anions by substituting chloride by sulfate or nitrate.

RESULTS AND DISCUSSION

The effect of SDS on Cu^{2+} ATPase and Mg^{2+} ATPase activities was determined to optimize signal to noise ratio. At the optimal SDS/MF protein ratio of 0.4, the maximal apparent specific activity was 58.75 ± 3.64 μmol/mg protein/hour. A copper activation curve was then obtained by varying $CuCl_2$ from 0–300 μM in the assay (FIG. 1). The curve best fits a Hill plot with an apparent maximal velocity of 63.8 ±

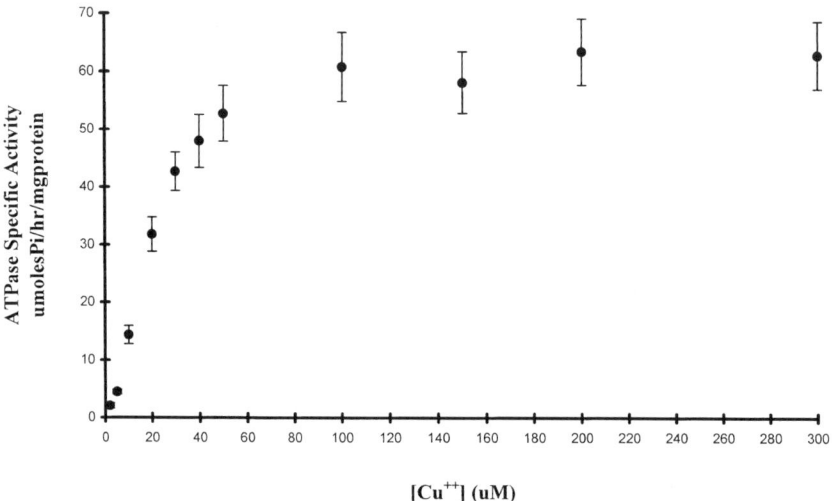

FIGURE 1. SDS-treated membranes were assayed for Cu^{2+}-ATPase activity under optimal assay conditions, varying $CuCl_2$ concentration from 0–300 μM. Each point represents the mean ± SEM of 16 determinations from 8 membrane preparations.

5.70 μmol/mg/hour (>180% activation) and an apparent half-maximal activation of 20 μM [Cu^{2+}]. A Hill coefficient n of 1.76 suggests that this Cu^{2+} ATPase has more than one binding site interacting positively. The specificity of the Cu^{2+} ATPase was determined with respect to cations and anions. Results show no significant change in activity when the chloride salt of Cu^{2+} was substituted with nitrate or sulfate (FIG. 2). Substituting copper by Co^{2+}, Zn^{2+}, or Mn^{2+} did not result in any stimulation above the basal Mg^{2+} ATPase activity. Cadmium, however, resulted in 44% activation relative to that produced by Cu^{2+}.

In conclusion we report the isolation of Cu^{2+}-activated Mg^{2+} requiring ATPase from rat liver. This ATPase may be implicated in hepatocyte copper transport, ensuring optimal [Cu^{2+}] at critical hepatic cellular sites. It is a candidate for Wilson's

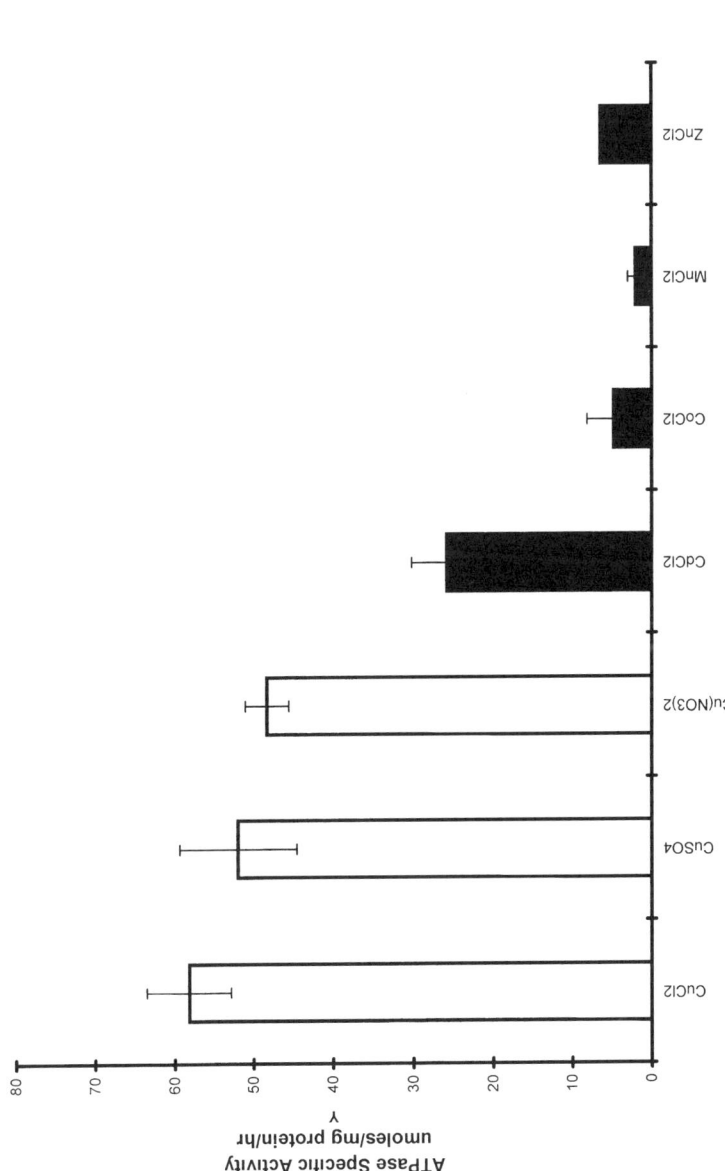

FIGURE 2. SDS-treated membranes were assayed for Cu^{2+}-ATPase activity under optimal assay conditions in the presence of 150 μM of one of the following salts: $CuCl_2$, $Cu(NO_3)_2$, $CuSO_4$, $MnCl_2$, $CoCl_2$, $ZnCl_2$, or $CdCl_2$. Each bar represents the mean ± SEM of 6 determinations from 3 membrane preparations.

disease gene product. To confirm this further, we are purifying the enzyme to obtain NH_2-terminal sequencing and will assay for enzyme activity in liver biopsies from patients with the disease or from Long-Evans Cinnamon rats known to lack the Wilson's gene.

REFERENCES

1. DANKS, D. M. 1993. Disorders of Copper Transport. *In* The Molecular and Metabolic Bases of Inherited Diseases, 7th Ed. A. Beaudet, W. Sliy & D. Valle, Eds.: 1411–1431. McGraw Hill. New York.
2. HALLIWELL, B. 1994. Nutr. Rev. **52:** 253–265.
3. BULL, P. C., G. R. THOMAS, J. M. ROMMENS, J. R. FORBES & D. W. COX. 1993. Nature Genet. **5:** 327–337.
4. TANZI, R. E., K. PETRUKHIN, I. CHERNOV, J. L. PELLEQUER, W. WALSO, B. ROSS, D. M. ROMANO, E. PARANO, L. PARCONE, L. M. BRZUSTOWICZ, M. DEMOTO, J. PEPPERCORN, A. I. BUSH, I. STERNLEIB, M. PIRASTU, J. F. GUSELLA, O. EVGRAFOV, G. K. PENCHAS ZADEH, B. HONIG, I. S. EDELMAN, M. B. SOARES, I. H. SCHEINBERG & T. C. GILLIAM. 1993. Nature Genet. **5:** 344–350.
5. MERCER, J. F., J. LIVINGSTON, B. HALL, J. A. PAYNTER, B. CATHERINE, S. CHANDRASEKHARAPPA, P. LOCKHART, A. GRIMES, M. BHARE, D. SIEMIENIAK & T. W. GLOVER. 1993. Nature Genet. **3:** 20–25.
6. MEIER, P. J., E. S. SZTUL, A. REUBEN & J. C. BOYER. 1984. J. Cell. Biol. **98:** 991–1000.
7. CORTAS, N. & M. WALSER. 1971. Biochim. Biophys. Acta **249:** 181–187.

Phosphorylation of Na,K-ATPase by Protein Kinases

Sites, Susceptibility, and Consequences[a]

MARINA S. FESCHENKO, RANDALL K. WETZEL,
AND KATHLEEN J. SWEADNER[b]

Laboratory of Membrane Biology
Neuroscience Center
Massachusetts General Hospital
149 13th St.
Charlestown, Massachusetts 02129

It was previously shown that both protein kinase C (PKC) and cAMP-dependent protein kinase (PKA) can phosphorylate the Na,K-ATPase α subunit.[1-4] The site(s) of phosphorylation and the functional consequences of phosphorylation were not known, however. We set out to determine the sites of phosphorylation by direct biochemical analysis of purified Na,K-ATPase treated with purified protein kinases. Unique sites were identified for each kinase, and it was observed that Na,K-ATPase conformation affected the susceptibility to phosphorylation by PKA and PKC in different ways. PKC was also observed to be modified by autophosphorylation as a result of its interaction with the Na,K-ATPase. It was determined that activation of endogenous PKC causes phosphorylation in intact cells. Using conditions that stabilize phosphorylation against phosphatases, we observed no detectable effect of PKC on Na,K-ATPase hydrolytic activity or affinity for Na^+.

SITES OF KINASE-MEDIATED PHOSPHORYLATION

Purified rat kidney Na,K-ATPase was phosphorylated by PKA to a measured stoichiometry of close to 1 mol/mol of α subunit, and digested with trypsin to identify the smallest possible peptide fragment containing the labeled phosphate.[5] This fragment was separated by gel electrophoresis, blotted to PVDF, and submitted directly to sequence analysis. The phosphorylated serine was identified by the disappearance of serine from the expected position in the sequence in the second cycle of Edman degradation. The phosphorylation was located at RRNSVFQQ, which is 81 amino acids from the COOH-terminus; the tryptic fragment began with the asparagine. The site is a classic consensus sequence for phosphorylation by PKA (RR/KXS) and is predicted to lie in the cytoplasm between the eighth and ninth transmembrane spans.[6] The phosphorylated serine is Ser 938 in rat α1 if numbering begins at the mature NH_2-terminus, Ser 943 if numbering begins with the initiation methionine.

[a]This work was supported by National Institutes of Health grant NS 27653.

[b]Address for correspondence: Dr. Kathleen Sweadner, 149-6118, Massachusetts General Hospital, 149 13th St., Charlestown, MA 02129 (tel: 617-726-8579; fax: 617-726-7526; e-mail: sweadner@helix.mgh.harvard.edu).

For PKC, we phosphorylated purified rat kidney enzyme again to a measured stoichiometry of close to 1 mol/mol of α subunit. Limited digestion with trypsin at the T1 and T2 sites, as well as proteolytic fingerprinting with trypsin during gel electrophoresis, indicated that the phosphorylation site was close to the NH_2-terminus.[7] In rat α1 there are two serine residues that could be targets of PKC in this stretch of amino acids (FIG. 1). To discriminate between them, we needed quantitative data on the amount of phosphate incorporated into either site. Normally it is impractical to directly sequence ^{32}P-labeled peptides on an automatic sequencer because the phosphorylated amino acids are too hydrophilic and fail to elute from the column. Our collaborators in the HHMI microchemistry facility developed a modified buffer elution system and altered cycle times to make direct sequencing possible, however.[8] Phosphate was found at both candidate serines (FIG. 1), but more than 80% of the phosphate was at the second serine (Ser 18 if numbering is from the first amino acid of mature rat α1; Ser 23 if numbering is from the initiation methionine).

We have never seen a stoichiometry greater than 0.15–0.2 for the first serine, Ser 11 (Ser 16 if numbered from the methionine) in rat α1 or in α1 from other species. This is generally consistent with the literature for phosphorylation stoichiometries observed in species that have a homolog of Ser 11 but not of Ser 18. The surrounding sequence is not typical for a site recognized by PKC because of the lack of positively charged amino acids. Only the histidine two residues after Ser 11 might assist with PKC recognition, as supported by the work of Dr. K. Geering (this volume). If stoichiometric phosphorylation cannot be obtained at this residue, it will be difficult to quantitatively evaluate any evidence for the functional consequences of phosphorylation, negative or positive.

PKA was observed to phosphorylate rat, pig, and dog α1 to the same extent.

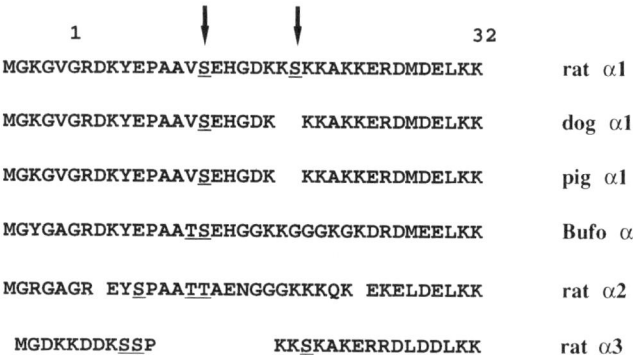

FIGURE 1. *Serine and threonine residues near the NH_2-termini of Na,K-ATPase.* The NH_2-terminal sequences of rat, dog, pig, and *Bufo* α1, as well as those of rat α2 and α3, are aligned. The number 1 above the sequences indicates the first amino acid in the mature protein for rat, dog, and pig α1 and rat α2, because the methionine and next four amino acids are known to be removed during biosynthesis. Such processing is not known for *Bufo* α1 or rat α3, hence some differences in numbering convention. The *two arrows* indicate the serines phosphorylated by PKC in rat α1; Ser 18 is utilized much better than Ser 11.[7] It can be seen that Ser 18 is not present in any of the other sequences. Other serines and threonines that could be phosphorylated are underlined. The threonine-serine pair in *Bufo* α1 has been implicated by the work of Geering and collaborators (this volume).

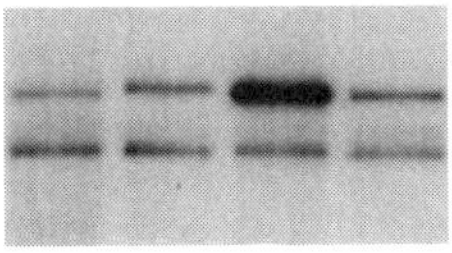

PK RAx RK DK

α
PKC

FIGURE 2. *PKC phosphorylation of kidney α1 compared to rat axolemma α2 and α3.* Equivalent amounts of α subunit were phosphorylated by PKC *in vitro*, separated by gel electrophoresis, and detected by autoradiography. The buffers, enzyme concentrations, and activators of PKC (Ca^{2+}, phospholipid, and PMA) were as described previously,[5,7] and the medium additionally contained 10 mM P_i to promote a conformation favorable for phosphorylation by this kinase. PK, RK, and DK represent pig, rat, and dog kidney preparations, respectively, all containing α1. RAx represents rat axolemma, which contains α2 and α3. Both of these isoforms migrate more slowly than α1 on SDS gels, which can be seen by the higher apparent molecular weight of the phosphorylated band in the axolemma sample. It is typical that the rat α1 sample incorporated more labeled phosphate than did any of the others. It is of interest that autophosphorylation of PKC was comparable in all of the samples (see below).

Unlike that for PKA, the level of PKC phosphorylation of rat, pig, and dog α1 subunits differed greatly (FIG. 2). This is readily explained by the absence of the KKSKK phosphorylation site (Ser 18) in pig and dog α1 (FIG. 1). It was routinely possible to obtain phosphorylation stoichiometries of 0.15 in these species in optimal ligand conditions, presumably at Ser 11. FIGURE 2 also shows that the level of phosphorylation of rat axolemma Na,K-ATPase, which contains predominantly α2 and α3, was comparable to that of the α subunits that have only Ser 11. As seen in FIGURE 1, α2 has one serine and two threonine residues close to the NH_2-terminus, but lacks basic amino acids nearby; it does not even have the histidine seen after Ser 11 in α1. On the other hand, α3 has serines in what should be a good consensus sequence for PKC: KSSPKKSKAK.

EFFECTS OF Na,K-ATPase CONFORMATION

When PKA was the kinase used, the presence of Triton X-100 was required to see phosphorylation at all (as observed by others), something that is still not understood. The presence of Na,K-ATPase ligands also affected the level of phosphorylation. Mg^{2+} (3 mM) and ATP (20 μM) were required for kinase activity and were present in all samples. The level of phosphorylation was highest in the absence of additional ligands or in the presence of 3 mM P_i (Tris salt). There was much more phosphorylation in the presence of Na^+ than of K^+ or (Na^+ plus K^+). PKA phosphorylation was also reduced by ouabain. For rat α1, ouabain caused a small reduction in the level seen in Na^+ or K^+, but it had a more dramatic effect in P_i or with no added ligands. With pig α1, which has a higher ouabain affinity, ouabain was able to largely prevent phosphorylation. The effect of ouabain must be on conformation rather than on ATP hydrolysis, because ATPase activity is inhibited by Triton X-100 in these conditions. A quantitative analysis of the ligand sensitivity of the phosphorylation was published previously.[5]

As with PKA, PKC phosphorylation of rat α1 was affected markedly by the presence and absence of Na,K-ATPase ligands. Ligand effects on PKC phosphorylation were more or less the opposite of those on PKA, however. ATP (20 µM) and Mg^{2+} (3 mM) was present in all conditions as required by the kinase. Ouabain, instead of reducing phosphorylation, enhanced it when no other ligands were present or when Na^+ or P_i was added. The presence of Na^+ gave the lowest level of phosphorylation for PKC, whereas K^+ gave the lowest level for PKA. The absence of ligands apart from Mg^{2+} and low ATP gave low levels of phosphorylation with PKC and high levels with PKA. Triton X-100, which was required for PKA, was inhibitory for PKC.

Since ATP is a substrate of Na,K-ATPase, the Na,K-ATPase and kinase compete for the same ATP during the phosphorylation reaction. We naturally wondered whether the presence of different ligands of Na,K-ATPase affected the rate of hydrolysis of ATP, thereby having an artifactual effect on the level of ^{32}P incorporated into the protein. Since Triton X-100 inhibited Na,K-ATPase, ATP hydrolysis was not a factor for PKA. For PKC, whether or not Na^+ and K^+ were present, the rate of hydrolysis of 20 µM ATP by the Na,K-ATPase was slow at 30°C, and direct measurement of hydrolysis after the reaction demonstrated that ATP was never more than 60% consumed.[5] Slow hydrolysis at low ATP concentrations, as well as paradoxical inhibitory effects of ion ligands, is well known. The addition of Na^+ in basal conditions would be expected to support active site phosphorylation, however. FIGURE 3 demonstrates the contrasting effects of ligands and of the addition of extra ATP to compensate for ATP consumption during the phosphorylation reaction. Twelve identical samples of Na,K-ATPase and kinase were incubated with NaCl, without ions, with P_i, and with KCl as shown. In the middle lane, ouabain (3 mM) was added. This had the most dramatic effect with NaCl and no ions, only a modest effect with P_i, and little effect at all in KCl, as reported previously.[5] When the concentration of ATP was doubled, the ATP remaining at the end of the experiment was never less than 16 µM. The addition of ouabain made the largest difference in phosphorylation with NaCl, and in this specific instance, the residual ATP was the same in 20 µM ATP with

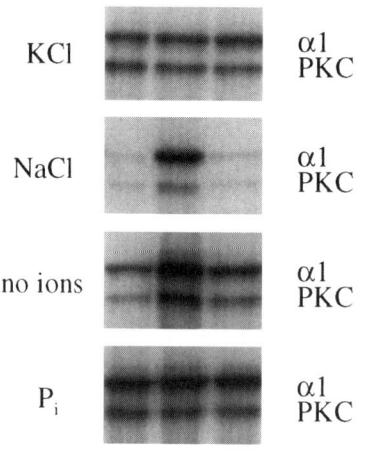

FIGURE 3. *ATP hydrolysis is not the reason for ligand effects on phosphorylation.* This is an autoradiogram showing phosphorylation of rat kidney Na,K-ATPase and autophosphorylation of PKC in different ligands. All samples contained Mg^{2+} (3 mM), buffer, and the reagents needed for activation of PKC. The additional ligands present in all three lanes are shown on the *left*. The addition of ouabain and the concentration of ^{32}P-ATP are indicated underneath. Ouabain had effects on phosphorylation levels that were separable from any effects on the consumption of ATP.

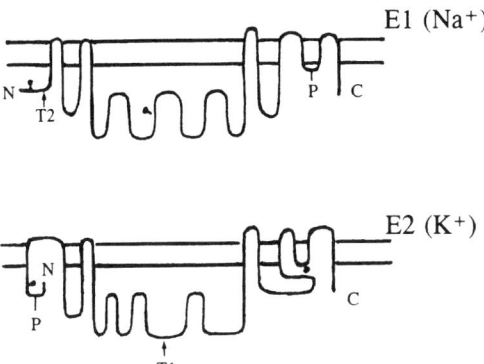

FIGURE 4. *Conformation affects kinase-mediated phosphorylation.* This cartoon depicts a model for the reciprocal exposure of PKA and PKC sites and of tryptic cleavage sites as a function of Na,K-ATPase conformation. In the E1, or Na$^+$ conformation, the PKC site is relatively inaccessible, the PKA site is exposed, and the T2 tryptic cleavage site is digested most rapidly. In the E2, or K$^+$ conformation, the PKC site is exposed, the PKA site is relatively inaccessible, and the T1 tryptic cleavage site is digested most rapidly. These conformation differences are only approximated by the experimental conditions, and the presence of Mg^{2+} and P$_i$, without Na$^+$, K$^+$, or ouabain, permits good phosphorylation with both kinases.[5]

ouabain as in 40 μM ATP without ouabain. This makes it unambiguous that enzyme conformation, rather than an effect of ouabain on ATP hydrolysis, is what controls the level of phosphorylation.

FIGURE 4 is a diagram illustrating that exposure of kinase phosphorylation sites is affected by E1/E2 conformation. Although pure E1 and E2 cannot be obtained in phosphorylation conditions, the data are consistent with the hypothesis that E1 favors phosphorylation by PKA and E2 favors phosphorylation by PKC. As shown in the diagram, the T1 and T2 tryptic cleavage sites also differ in their exposure in the two conformations. The observations suggest that prior phosphorylation by one kinase may influence susceptibility to phosphorylation by the other. This may explain some of the controversies in the literature.

Na,K-ATPase INFLUENCES PKC AUTOPHOSPHORYLATION

FIGURE 5 illustrates the intriguing observation that interaction with the Na,K-ATPase has a lasting impact on PKC. PKC phosphorylation of rat, pig, and dog α1 samples is shown. As above, pig and dog α1 were phosphorylated to a much lower extent, even in the presence of ouabain. PKC autophosphorylation is known to be influenced by interaction with PKC substrates, and in this case autophosphorylation was markedly increased in parallel with the increase in Na,K-ATPase phosphorylation caused by the addition of ouabain. The effect of ouabain is apparently mediated by its ability to alter Na,K-ATPase conformation, putting the enzyme into a conformation (E2-P) with optimal exposure of the phosphorylation site. Exactly how this impacts the PKC is not clear: PKC is thought to be recruited to the membrane when activated, and autophosphorylation could be influenced by augmented recruitment or by a conformational response to having interacted with a susceptible substrate.

What is most intriguing is that the effect on PKC autophosphorylation is as large in the pig and dog α1 samples as in the rat α1 sample, despite the lower net level of phosphorylation. All three species' α1 subunits are inferred to be phosphorylated on Ser 11 (visible in dog and pig samples after longer exposures). The effect on PKC autophosphorylation appears to be mediated by phosphorylation of Ser 11 rather than of Ser 18, which is absent in pig and dog α1. The molar ratio of Na,K-ATPase to kinase in these experiments was approximately 60:1, and so even if only a fraction of the Na,K-ATPase units is phosphorylated on Ser 11 at any given time, more than enough phosphorylated protein is present to interact with the PKC.

As just discussed, an artifactual effect of ouabain to preserve ATP levels during the reaction and thus support PKC autophosphorylation better was unlikely because final ATP concentrations were 8 μM in the absence of ouabain and 16 μM in its presence.[5] In other experiments, a direct relationship was observed between the level of Na,K-ATPase phosphorylation and the level of PKC autophosphorylation, without

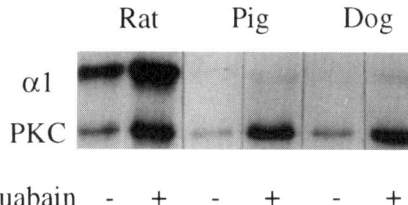

FIGURE 5. *Autophosphorylation of PKC is influenced by phosphorylation of Na,K-ATPase.* Autoradiography was used to detect phosphate incorporated into both α1 and PKC as indicated. Phosphorylation was enhanced by the addition of ouabain. In the conditions used (Mg^{2+}, 20 μM ATP, Ca^{2+}, lipid, and PMA; no P_i) phosphorylation of the dog and pig kidney α1s was faint but readily detected upon longer exposure (not shown). This exposure of the film, however, best illustrates the parallel increase in PKC autophosphorylation seen when Na,K-ATPase phosphorylation was enhanced. The effect has the same magnitude in dog and pig preparations despite their lower incorporation of phosphate into α1. The three Na,K-ATPase preparations are hypothesized to all incorporate similar amounts of phosphate into Ser 11, and it may be this event that most significantly affects PKC autophosphorylation. (Reproduced with permission from Feschenko and Sweadner.[5])

any relationship to the amount of ATP remaining.[5] In FIGURE 3, PKC autophosphorylation paralleled Na,K-ATPase phosphorylation except in KCl, where there was a relatively higher ratio of autophosphorylation to Na,K-ATPase phosphorylation than in the other conditions. We also observed that KCl caused a higher rate of PKC phosphorylation of H1 histone, suggesting that this effect of KCl is nonspecific. A control for nonspecific effects of ouabain is shown in FIGURE 6. Parallel enhancement of phosphorylation of Na,K-ATPase and autophosphorylation of PKC was seen upon the addition of ouabain, but the addition of ouabain had no effect on PKC autophosphorylation when PKC was activated by Ca^{2+}, phorbol ester, and the addition of lipid as usual, but the substrate Na,K-ATPase was omitted. This rules out any unexpected direct effect of ouabain on the kinase.

The Na,K-ATPase effect on PKC autophosphorylation is evidence for mutual interaction between the kinase and its substrate. Autophosphorylation has been observed to correlate with increased kinase activity and higher affinity for Ca^{2+}.[9] We

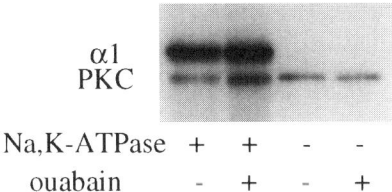

FIGURE 6. *Ouabain has no direct effect on PKC autophosphorylation.* As shown here, when PKC was activated by Ca^{2+}, lipid, and PMA, it did undergo some autophosphorylation even in the absence of an added protein substrate, but this autophosphorylation was unaffected by the addition of ouabain. In the control with added rat kidney Na,K-ATPase, ouabain enhanced phosphorylation of α1 and autophosphorylation as before.

can speculate that activated PKC can interact with other cellular components after its interaction with Na,K-ATPase, complicating the physiological response. This is a possible factor to be borne in mind for investigating controversies in the literature that arise from the responses of intact cells.

NONRADIOACTIVE MEASUREMENT OF PHOSPHORYLATION OF SER 18

We developed a new method for assessing the extent of phosphorylation of Ser 18 based on the use of an antibody.[13] The antibody McK1 was previously determined to bind to the epitope DKKSKK in rat α1.[10,11] The serine in this epitope is the same one phosphorylated by PKC and that accounts for 80% of the phosphate incorporated. We were able to demonstrate that phosphorylation completely blocked the binding of the antibody (FIG. 7). The figure shows samples of purified rat kidney Na,K-ATPase with and without phosphorylation by PKC *in vitro*. On the top, an autoradiogram demonstrates the phosphorylation of α and the autophosphorylation of PKC, as usual. This

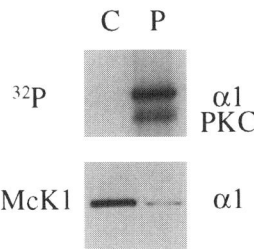

FIGURE 7. *Antibody-based detection of phosphorylation of Ser 18.* The *two panels* show the autoradiography (*top*) and immunostain (*bottom*) of the same blot, which contains samples of Na,K-ATPase alone (C, control) and Na,K-ATPase plus PKC (P, phosphorylated). The bands labeled with ^{32}P were the α1 subunit and autophosphorylated PKC, as indicated. Phosphorylation of Na,K-ATPase under conditions that routinely give close to stoichiometric levels of phosphate incorporation (80% in Ser 18, the rest in Ser 11) also nearly abolished the binding of the McK1 antibody. This forms the basis for a sensitive, nonradioactive assay for phosphorylation at this specific site.

exposure required overnight. The same blot was later stained with McK1 antibody and the image developed with HRP-conjugated goat-anti-mouse antibodies with luminol reagent (Pierce Chemical Co.); this exposure required only minutes. Although identical amounts of Na,K-ATPase were loaded in the lanes, it can be seen that McK1 binding was practically abolished. To make this a semiquantitative assay, blots were later restained with an antibody (6F) that binds about 40 amino acids away[11] and in a manner that is not affected by phosphorylation. When care is taken to obtain exposures that give fluorographic signals in the linear range, the ratio of stain by McK1 to that by 6F gives a reproducible estimate of the fraction of Ser 18 that remains unphosphorylated. In some ways this assay is more satisfactory than is using an antibody that recognizes phosphorylated enzyme but not unphosphorylated enzyme, because in that case it is harder to determine when an increase in signal means that phosphorylation is close to stoichiometric. With our assay, any signal at all with McK1 indicates the presence of some unphosphorylated enzyme. We have observed that endogenous phosphatase activity is high in our preparations, making it unlikely that any basal level of phosphorylation is preserved after manipulation of tissue or cell samples when no phosphatase inhibitors are used.[13]

PHOSPHORYLATION OF SER 18 IN INTACT CELLS

With this antibody-based assay, we used intact cultured cells to ask if phosphorylation by PKC at the DKKSKK site in rat α1 would occur after stimulation of endogenous PKC.[13] Four kinds of cultured cells were used: three cell lines, C6 (rat glioma), NRK (rat kidney epithelial cells), L6 (rat myoblasts), and primary cultures of rat cerebellar granule neurons. All of these cells express α1.

Endogenous PKC was stimulated by the addition of phorbol ester (PMA). PMA alone showed a slight decrease in McK1 staining. Calyculin A was used to inhibit endogenous phosphatase activity; it alone had almost no effect on the level of McK1 stain. Combined, PMA and calyculin A caused a substantial disappearance of McK1 binding (data not shown). This demonstrates close to stoichiometric phosphorylation at this site in intact cells under conditions in which protein phosphatase activity is inhibited. Similar phosphorylation by PKC was seen in C6, NRK, L6, and primary cultures of cerebellar granule neurons. The immunological assay is simple enough to be used in every experiment to quantitatively evaluate the level of phosphorylation achieved. It is also specific for Ser 18, eliminating the concern that any ^{32}P label incorporated in the protein in intact cells might be at a different site, introduced by a different kinase as part of a complex cascade.

We also investigated the use of a more intact tissue preparation for studies of PKC activation. The distribution of Na,K-ATPase α isoforms has been determined for the rat retina,[12] and it is known that the plexiform layers, where most of the synaptic activity occurs, are particularly rich in α1. We have also localized the β isoforms in the retina and find that β1 predominates in the same plexiform layers, whereas β2 is mainly in the photoreceptors. The isoform combination, α1β1, is thus the same as in the rat kidney. Using pieces of retina incubated *in vitro* with PMA plus calyculin A, a reduction in McK1 staining was seen consistent with phosphorylation in the range of 10–20%.

FUNCTIONAL CONSEQUENCES OF PHOSPHORYLATION OF SER 18

In our initial studies utilizing phosphorylation of purified rat kidney Na,K-ATPase by purified PKC,[5] we were unable to detect a significant effect on the V_{max} for ATP hydrolysis as measured in a coupled assay. One could postulate that phosphorylation was labile under those conditions, however, due to the presence of phosphatase activity. To further study the functional consequences of phosphorylation, we wanted to demonstrate the retention of stoichiometric levels of phosphorylation. Initially, stoichiometry was measured by quantifying the phosphate incorporated with ^{32}P and by quantifying the protein present in α by collecting the protein band on PVDF membrane and subjecting it to amino acid analysis.[5,7] This method is very accurate but cumbersome and not amenable to use on a daily basis. For Ser 18, a correspondence was demonstrated between close-to-stoichiometric phosphorylation and close-to-complete reduction in McK1 binding, however, and this was applied to the evaluation of preparations made by stimulating endogenous PKC in intact cells.

We observed that when phosphorylation was induced by stimulation of endogenous PKC in intact cells, isolated membranes that were subsequently incubated in ATPase assay conditions rapidly lost the phosphate from Ser 18. This problem was solved by adding the phosphatase inhibitor calyculin A to all buffers for the rest of the experiment, with the result that full phosphorylation was preserved (data not shown). Under these conditions, ATPase activity showed no effect from stoichiometric phosphorylation. Activity was assayed by hydrolysis of ATP in membrane preparations in both test tube and coupled assays, and in permeabilized cells, with the same result.[13]

ATPase activity was also measured at different concentrations of Na^+ to determine whether phosphorylation of Ser 18 resulted in a change in Na^+ affinity. Again, there was no effect (data not shown).

CONCLUSIONS

We recognize that regulation of Na,K-ATPase activity caused by activation of PKC has been demonstrated by a number of investigators. We are struck by the fact that either decreases or increases in activity have been seen in complex cellular systems, and that even with purified enzyme, results have not been consistent from one laboratory to another. We have attempted to define the locus of phosphorylation and measure the stoichiometry, to answer the most stringent formulation of the question: Does phosphorylation of this site result in direct inhibition of ATPase activity or reduction of Na^+ affinity? The answer was no. We conclude that regulation of activity must entail interaction of the Na,K-ATPase with other proteins. Phosphorylation of the Na,K-ATPase could in principle facilitate the binding of other proteins, just as phosphorylation of receptors by tyrosine kinases creates docking sites for other regulatory components.

Assessment of phosphorylation of Ser 18 with McK1 binding is blind to the level of phosphorylation of Ser 11, but all evidence of which we are aware suggests that this site is not normally fully phosphorylated. We assume that the conditions we use to stimulate phosphorylation in intact cells also stimulates phosphorylation of Ser 11 to a proportional extent, but this has not been experimentally verified.

On the other hand, we have demonstrated that interaction of PKC with Na,K-ATPase results in a lasting modification of PKC itself, which could have other functional consequences that ultimately impact the activity of the Na,K-ATPase. This effect of the Na,K-ATPase on PKC evidently is sensitive to Na,K-ATPase conformation. If it in fact requires Ser 11, as implied by its presence in pig and dog α1 preparations, it could be a general mechanism applicable to many animal species.

REFERENCES

1. MARDH, S. 1979. Phosphorylation by the catalytic subunit of protein kinase of a preparation of kidney Na,K-ATPase. *In* Na,K-ATPase Structure and Kinetics. J. C. Skou & J. G. Norby, Eds.: 359–370. Academic Press. London.
2. LOWNDES, J. M., M. HOKIN NEAVERSON & P. J. BERTICS. 1990. Kinetics of phosphorylation of Na+/K(+)-ATPase by protein kinase C. Biochim. Biophys. Acta **1052:** 143–151.
3. BERTORELLO, A. M., A. APERIA, S. I. WALAAS, A. C. NAIRN & P. GREENGARD. 1991. Phosphorylation of the catalytic subunit of Na+,K(+)-ATPase inhibits the activity of the enzyme. Proc. Natl. Acad. Sci. USA **88:** 11359–11362.
4. CHIBALIN, A. V., L. A. VASILETS, H. HENNEKES, D. PRALONG & K. GEERING. 1992. Phosphorylation of Na,K-ATPase α subunits in microsomes and in homogenates of *Xenopus* oocytes resulting from the stimulation of protein kinase A and protein kinase C. J. Biol. Chem. **267:** 22378–22384.
5. FESCHENKO, M. S. & K. J. SWEADNER. 1994. Conformation-dependent phosphorylation of Na,K-ATPase by protein kinase A and protein kinase C. J. Biol. Chem. **269:** 30,436–30,444.
6. ARYSTARKHOVA, E., D. L. GIBBONS & K. J. SWEADNER. 1995. Topology of the Na,K-ATPase: Evidence for externalization of a labile transmembrane structure during heating. J. Biol. Chem. **270:** 8785–8796.
7. FESCHENKO, M. S. & K. J. SWEADNER. 1995. Structural basis for species-specific differences in the phosphorylation of Na,K-ATPase by protein kinase C. J. Biol. Chem. **270:** 14072–14077.
8. KHATRI, A., T. R. OSTREA, M. S. FESCHENKO & K. J. SWEADNER. 1995. Sequence analysis and radiometric quantitation of phosphorylation sites in the alpha subunit of Na,K-ATPase on PVDF blot: A non-covalent approach. Protein Sci. 4(Suppl. 2): 153.
9. SANDO, J. J., M. C. MAURER, E. J. BOLEN & C. M. GRISHAM. 1992. Role of cofactors in protein kinase C activation. Cell. Signalling **4:** 595–609.
10. FELSENFELD, D. P. & K. J. SWEADNER. 1988. Fine specificity mapping and topography of an isozyme-specific epitope of the Na,K-ATPase catalytic subunit. J. Biol. Chem. **263:** 10932–10942.
11. ARYSTARKHOVA, E. A. & K. J. SWEADNER. 1996. Isoform-specific monoclonal antibodies to Na-K-ATPase α subunits: Evidence for a tissue-specific post-translational modification of the α subunit. J. Biol. Chem. **271:** 23,407–23,417.
12. MCGRAIL, K. M. & K. J. SWEADNER. 1989. Complex expression patterns for Na,K-ATPase isoforms in retina and optic nerve. Eur. J. Neurosci. **2:** 170–176.
13. FESCHENKO, M. S. & K. J. SWEADNER. 1997. Phosphorylation of Na,K-ATPase by protein kinase C at Ser 18 occurs in intact cells but does not result in direct inhibition of ATP hydrolysis. J. Biol. Chem. **272:** 17726–17733.

Cytoplasmic Regions of the Alpha Subunit of the Sodium Pump Involved in Modulating the Na,K-ATPase Reaction[a]

STEWART E. DALY,[b] LOIS K. LANE,[c]
AND RHODA BLOSTEIN[b,d]

[b]Department of Medicine
McGill University
Montreal, Canada

[c]Department of Pharmacology and Cell Biophysics
University of Cincinnati College of Medicine
Cincinnati, Ohio 45267-0575

The topography and secondary structure of the Na,K-ATPase have been predicted from analyses of hydropathy, products of proteolytic digestion and chemical modification, and epitope mapping. (For review, see ref 1.) To date, experiments using these approaches as well as site-specific mutagenesis have localized determinants of cation binding to residues in transmembrane helices, and sites of ATP binding and phosphorylation, to sites within the cytoplasmic loop between helices H4 and H5. Extracellular sites of cardiac glycoside binding have been localized to a number of extracellular regions (ref. 2 and citations therein).

This paper summarizes our recent studies of the functional consequences of alterations in cytoplasmic regions of the α1 subunit of the Na,K-ATPase distinct from the H4-H5 loop. One region is the highly charged amino terminus, a region of marked diversity among the otherwise highly homologous isoforms of the α subunit. The other is the first cytoplasmic loop between helices 2 and 3 (H2-H3 loop) in which we characterized a spontaneous point mutation (E233K) in the putative β-strand region. The distinct behavior of this mutant provides evidence for a role of this region in conformational coupling.

ALTERATIONS IN THE AMINO TERMINUS

We reported earlier that removal of ≈ 30 amino acids from the NH_2-terminus by tryptic cleavage of the enzyme in the E_1 conformation according to the procedure of Jorgensen[3] does not alter overall Na,K-ATPase activity.[4] Deletion of this terminus by mutagenesis of the α1 subunit results in a functional enzyme (α1M32; ref 5) with similar apparent affinities for intracellular Na^+ and extracellular K^+ under physiologi-

---

[a]This work was supported by grants from the Medical Research Council of Canada, the Quebec Heart and Stroke Foundation (to RB), and the National Institutes of Health (to LKL), and a postdoctoral fellowship from the Canadian Heart and Stroke Foundation (to SED).

[d]Address for correspondence: Dr. R. Blostein, Montreal General Hospital, 1650 Cedar Avenue, Montreal, Que. Canada, H3G 1A4.

cal conditions of transport measurements carried out with intact HeLa cells transfected with the individual cDNAs for α1 and α1M32.

We also showed that the truncated enzyme has distinctive kinetics apparent at micromolar ATP concentration[4,5] under which condition ouabain-sensitive Na^+-dependent ATPase activity is inhibited by K^+.[6] This inhibition by K^+ reflects its well-documented effect on the dephosphorylation pathway of the reaction. Accordingly, K^+ stimulates the dephosphorylation step ($E_2P + K^+ \rightarrow E_2(K) + P_i$) and becomes occluded within the pump protein. Its rate of deocclusion is extremely slow and is increased by ATP binding with low affinity according to the reaction $E_2(K) + ATP \rightarrow ATP.E_1 + K^+$.[6] In fact, as shown earlier[4,5] and described below, at micromolar ATP concentration, the response of Na-ATPase to K^+ is a sensitive means of characterizing isoform- and mutant-specific differences in the K^+ deocclusion pathway of the reaction.

As shown previously,[5] Na-ATPase activity of the α1M32 truncated enzyme is stimulated by low concentrations of K^+, whereas the wild-type α1 enzyme is inhibited. Interestingly, the α2 isoform, heretofore referred to as α2*,[a] resembles the α1M32 mutant in this respect. These observations suggested that the NH_2-terminus of the α subunit is involved in regulating K^+ deocclusion and that isoform-specific differences in primary structure in this region might account for the difference between α2* and α1M32, on the one hand, and α1, on the other.

Mechanistic Basis for the Kinetic Differences between α1, α2*, and α1M32

K^+ deocclusion and release from $E_2(K)$ can be described by a branched pathway whereby ATP binds with either (1) low affinity ('ATP_L') to the K^+-occluded form of the enzyme, $E_2(K)$, at a step preceding the release of potassium and the formation of $ATP.E_1$ (pathway **a**), or (2) high affinity ('ATP_H') at a step following the slow release of potassium from $E_2(K)$ (pathway *b*) as follows:

pathway **a** (low affinity): $E_2(K) \xrightarrow{ATP_l} ATP.E_2(K) \rightarrow ATP.E_1.K \underset{K^+}{\rightarrow} ATP.E_1$

pathway **b** (high affinity): $E_2(K) \rightarrow E_1K \underset{K^+}{\rightarrow} E_1 \xrightarrow{ATP_H} ATP.E_1$ (*Scheme 1*)

Accordingly, K^+ activation or inhibition at low ATP concentration is a function of the apparent affinity of the enzyme for ATP at its low affinity site and/or the rate of release of K^+ from $E_2(K)$.

The results of kinetic analysis of the effect of ATP concentration on Na,K-ATPase are summarized in TABLE 1. For this analysis, the data points were fitted to a simple two-component reciprocal Michaelis-Menten relationship of which one component was linear in the range of 1–10 μM and the other, linear in the range of 25–500 μM ATP. Kinetic parameters (K'_{ATP} or V_{max}) for pathways involving high and low apparent affinities for ATP are denoted by subscripts H and L, respectively. As shown, K'_H values are similar for the three forms. However, K'_L values for α2* and α1M32 are 2.5-fold lower than that for α1. A notable difference is the ratio V_H/V_L, which indicates that hydrolysis through the high affinity pathway **b** is considerably greater for α2* and α1M32 than for α1. As discussed elsewhere,[8] the distinct ratios

---

[a] α2* is the ouabain-resistant mutant form of α2 developed by Jewell and Lingrel.[7]

TABLE 1. Comparison of Kinetic Parameters for High and Low ATP Affinity Pathways of Na,K-ATPase Catalyzed by α1, α1M32, and α2* [a]

α Subunit	V_H/V_L	V_H/V_L (normalized)[b]	K'_L (μM)	K'_H (μM)
α1	0.042 ± 0.009 (6)	1.0	331 ± 44 (6)	5.44 ± 1.9 (6)
α1M32	0.099 ± 0.03 (5)	4.4	130 ± 32 (5)	4.57 ± 2.3 (5)
α2*	0.086 ± 0.01 (4)	2.4	133 ± 19 (4)	4.77 ± 0.85 (4)

[a] Values for apparent kinetic constants were determined from the y-intercepts and slopes of the low (25–500 μM ATP) and high (1–10 μM ATP) affinity components of biphasic reciprocal plots. Values shown are the means ± SD of the number of experiments in parentheses. Reproduced with permission from ref. 8.

[b] V_H/V_L were normalized to account for differences in Na-ATPase/Na,K-ATPase activity ratios which were: 0.032 ± 0.003 (α1), 0.018 ± 0.002 (α1M32), and 0.029 ± 0.005 (α2*). Na-ATPase was measured at 100 mM NaCl.

(normalized to account for differences in Na-ATPase/Na,K-ATPase activity ratios) of the three enzyme forms account for the characteristic differences in their K^+-activation/inhibition profiles (FIG. 1 in ref. 5).

Further support for the conclusion that the rates of K^+ deocclusion through pathway **b** are faster for α2* and α1M32 than for α1 was obtained from experiments designed to measure (1) K^+ dependence of K^+ occlusion via the reaction $E_1 + K^+ \leftrightarrow E_2(K)$ and (2) the rate of formation of E_1P from the K^+-occluded enzyme $E_2(K)$. The low specific activity of the Na,K-ATPase expressed in HeLa cell membranes compared to pump-rich tissue sources prevents the direct measurement of occluded K^+ ions and, therefore, K^+ deocclusion. Therefore, an indirect method was used as described by Daly et al.[8] Briefly, formation of $E_2(K)$ is reflected by the decrease in phosphoenzyme ($E^{32}P$) formed following equilibration of the enzyme at room temperature (1) without and (2) with varying concentrations of K^+. The reduction in $E^{32}P$ resulting from preincubation with K^+ ($\Delta E^{32}P$) is a measure of the amount of $E_2(K)$. The results summarized in TABLE 2 indicate that the $K_{0.5}$ for K^+ occlusion ($K_{0.5(Kocc)}$) is 8-fold higher for α2* and 10-fold higher for α1M32 than for α1.

The shift in the $E_1 + K \leftrightarrow E_2(K)$ equilibrium towards E_1 observed with α1M32 and α2* relative to α1 must be due to either slower occlusion and/or faster deocclusion. Evidence in support of faster deocclusion was obtained by measuring the rate of E_1 formation from $E_2(K)$. $E_2(K)$ was first formed by preincubating the enzyme with optimal K^+ concentration. The difference ([$E^{32}P$ formed following preincubation in the absence of K^+] minus [$E^{32}P$ formed following preincubation with optimal, 8 mM, K^+]), $\Delta E^{32}P$, was taken to represent 100% $E_2(K)$. Deocclusion was then measured by following the rate of increase in $\Delta E^{32}P$ at 10°C, a temperature at which deocclusion is sufficiently slow to permit manual assays. Oligomycin was present

TABLE 2. Comparison of α1, α2*, and α1M32 with Respect to $K_{0.5}$ for K^+ Occlusion, Rate of K^+ Deocclusion, and Phosphoenzyme Turnover

α Subunit	$K_{0.5(Kocc)}$ (mM)	$k_{K^+Deoccl}$ (sec^{-1})	EP Turnover (min^{-1})
α1	0.12	0.02	7,598
α2*	0.94	0.07	4,652
α1M32	1.20	0.08	3,646

during the phosphorylation reaction to block the $E_1P \to E_2P$ transition. Under these conditions, it can be assumed that phosphorylation of E_1 is rapid and not rate-limiting. As shown in TABLE 2, the rate constants for E_1 formation from $E_2(K)$ with the $\alpha 2^*$ and $\alpha 1M32$ enzymes are 0.07 and 0.08 sec^{-1}, respectively, which are 3.5- and 4-fold faster than that for $\alpha 1$ (0.02 sec^{-1}). Similar results were obtained when deocclusion was allowed to proceed first in ATP-free Na^+ medium after which E_1 formed was measured by rapid dilution and conversion to E_1P at 0°C.

The foregoing kinetic differences between $\alpha 1$ and either $\alpha 2^*$ or $\alpha 1M32$ indicate that they may be regarded as E_1/E_2 conformational isoforms. The higher affinity of $\alpha 2^*$ and $\alpha 1M32$ for ATP at its low affinity site and the shift in $E_1 + K \leftrightarrow E_2(K)$ towards E_1 are consistent with a shift in the conformational equilibrium towards E_1 as described earlier. Our results concur with and extend the earlier studies of Jorgensen[9] and co-workers showing that tryptic removal of the NH_2-terminus alters the E_1-E_2 conformational equilibrium.

Evidence also suggests a higher ratio of E_1P to E_2P with $\alpha 2^*$ and $\alpha 1M32$ than with $\alpha 1$, which may indicate a slower step in the reaction sequence involved in formation of E_2P from E_1. The evidence is that oligomycin increased the level of $\alpha 1$ more than twofold, but had little effect on EP of $\alpha 2^*$ and $\alpha 1M32$ (experiments not shown). Furthermore, $\alpha 2^*$ and $\alpha 1M32$ had lower turnovers than did $\alpha 1$. Turnovers estimated from the ratio of V_{max} to the maximal level of EP observed in the presence of oligomycin were 3,646 ± 262 and 4,652 ± 424 min^{-1} for $\alpha 2^*$ and $\alpha 1M32$, respectively, compared to 7,598 ± 1,331 min^{-1} for $\alpha 1$.

Structural Basis for the Kinetic Differences between $\alpha 1$ and $\alpha 2^*$

This issue was addressed by comparing the kinetic behavior of $\alpha 1/\alpha 2$ chimeric enzymes. In one chimera, $\alpha 1(1\text{-}32\alpha 2)$, the NH_2-terminal 27 residues of $\alpha 1$ were replaced by residues 1–25 of $\alpha 2$. In the other chimera, $\alpha 2^*(1\text{-}32\alpha 1)$, the NH_2-terminal 25 residues of $\alpha 2^*$ were replaced by residues 1–27 of $\alpha 1$. The names of the two chimeras (i.e., 1–32) reflect the fact that residues 28–34 in $\alpha 1$ are identical to residues 26–32 in $\alpha 2^*$.

As shown in FIGURE 1a, at 1 µM ATP, the K^+ activation/inhibition profile of $\alpha 1(1\text{-}32\ \alpha 2)$ is the same as that of $\alpha 1$, whereas that of $\alpha 2^*(1\text{-}32\ \alpha 1)$ resembles that of $\alpha 2^*$. Since deletion of residues 1–32 from $\alpha 1$ yields an enzyme with $\alpha 2^*$-like kinetics[8] whereas substituting the equivalent region of the $\alpha 2^*$ sequence into $\alpha 1$ has no effect, it is concluded that it is not a difference in the NH_2-terminal sequence, per se, that is responsible for the observed kinetic differences between $\alpha 1$ and $\alpha 2^*$. Rather, it is the interaction of other isoform-specific domain(s) of $\alpha 1$ with (certain) residues of the NH_2-terminal that is responsible for the distinct $\alpha 1$ kinetic profile.

To identify the $\alpha 1$ NH_2-terminal residues involved in $\alpha 1$-specific behavior, a series of NH_2-terminal deletion mutants were characterized with respect to K^+ inhibition/activation of Na^+-ATPase. These are shown in FIGURE 1b. Although the lysine cluster within the highly charged, hydrophilic, and flexible amino terminus was initially postulated to function as a cation gate,[10] deletion of residues up to and including the lysine cluster, namely, deletion mutants $\alpha 1M16$ and $\alpha 1M23$, does not alter the kinetic behavior as shown in FIGURE 1b. Since removal of the next nine residues (residues 23–32: MERDMDELK) does alter the behavior, it is concluded that this segment has a role in modulating the kinetic behavior.

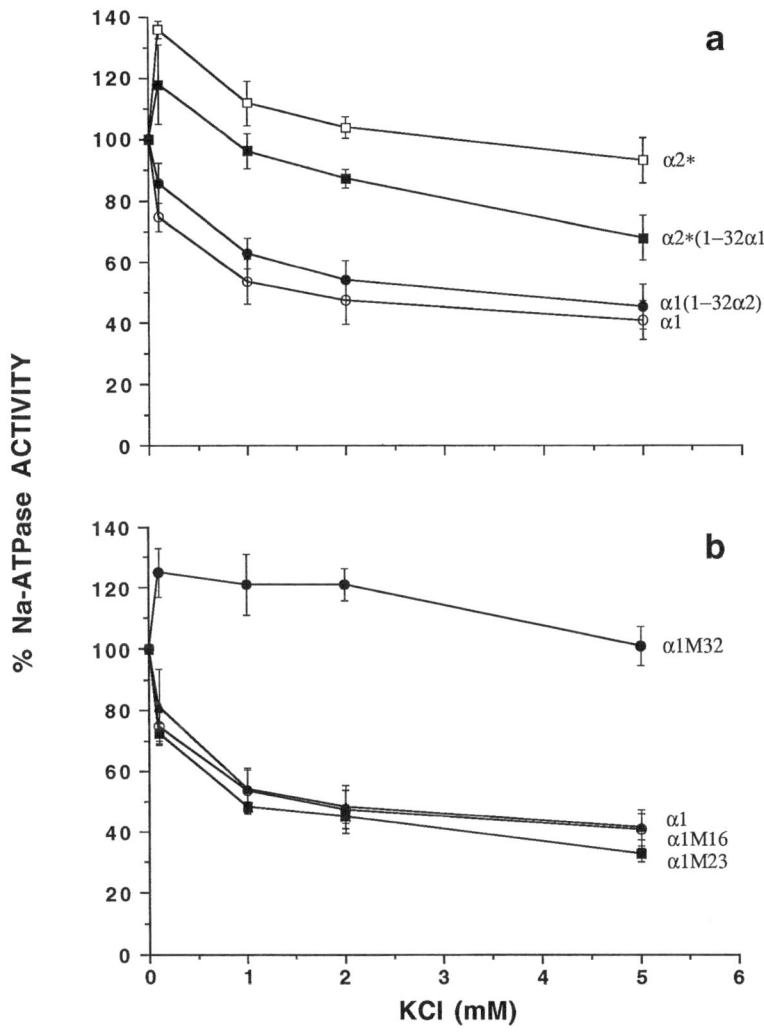

FIGURE 1. K⁺ sensitivity of the Na-ATPase. ATP hydrolysis was assayed in the presence of 1 μM ATP, 20 mM NaCl, and various concentrations of KCl as described in ref. 5. Data are presented as a percentage of Na-ATPase activity (control) measured in the absence of added KCl. (**a**) α1/α2* chimeric enzymes. Control activities in the presence of 20 mM NaCl, 1 mM MgSO$_4$, and 20 mM choline chloride [nmol/(mg × h)] are 118 ± 41, 139 ± 32, 116 ± 9.2, and 98.2 ± 36 for α1, α1(1-32α2), α2*, and α2*(1-32α1), respectively. Symbols: ○ = α1; ● = α1(1-32α2); □ = α2*; ■ = α2*(1-27α1). (**b**) α1 deletion mutants. Control activities in the presence of 20 mM NaCl, 1 mM MgSO$_4$, and 20 mM choline chloride [nmol/(mg × h)] are 118 ± 41, 150 ± 58, 184 ± 29, and 207 ± 43, for α1, α1M16, α1M23, and α1M32, respectively. Symbols: ○ = α1; ▲ = α1M16; ■ α1M23; ● = α1M32. (Reproduced with permission from ref. 8.)

MUTATION OF GLU[233] → LYS LOCATED IN THE H2-H3 CYTOPLASMIC LOOP

During the course of studies with rat α1 mutant cDNAs transfected into HeLa cells, we observed that one out of several replicate clones had kinetic properties different from those of rat α1. This observation suggested that a spontaneous mutation had occurred. The alteration has been identified as the substitution of lysine for glutamate at position 233 (E233K) in the cytoplasmic region between transmembrane helices H2 and H3.

The functional difference between E233K and α1 was readily apparent when the response of Na-ATPase to K^+ was tested. In contrast to that of α1, the Na-ATPase activity of E233K was stimulated markedly by the addition of K^+. The response profile was similar to that of α1M32 and α2*, except that stimulation by K^+ was even greater: 200% increase with the addition of 1 mM K^+. As discussed below, kinetic difference(s) between E233K and α1 are not primarily the result of differences in cation interactions; as shown in TABLE 3, under conditions of transport assays with

TABLE 3. Kinetic Parameters for E233K

Parameter Measured		Difference from α1
$K_{0.5(Kext)}$ (mM)	0.38	None
$K_{0.5(Nacyt)}$ (mM)	20.9	None
K'_L (mM)	56.3	6-fold decrease
K'_H (mM)	Not apparent	—
$K_{0.5}$ (K^+ occlusion) (mM)	1.0	8-fold increase
k (K^+ deocclusion) (sec^{-1})	0.09	4.5-fold increase
EP turnover (min^{-1})	4481	1.7-fold decrease

intact cells, they behave similarly with respect to apparent affinities for cytoplasmic Na^+ and extracellular K^+.

The functional distinction between E233K and α1 was analyzed further as described previously for α2* and the α1M32 deletion mutant. The results are shown in TABLE 3. Like α1M32 and α2*, the affinity of E233K for ATP at its low affinity binding site was increased, although to a greater extent (sixfold; $K'_L = 56.3 \pm 14.0$). Moreover, in contrast to the other forms, the reciprocal plot of the Na,K-ATPase activity of E233K mutant as a function of ATP concentration is a straight line within the entire range of 1–500 μM. Analysis of the K^+ occlusion/deocclusion characteristics of this mutant was carried out as just described for α1, α2* and the α1M32 deletion mutant. The results summarized in TABLE 3 indicate that E233K resembles α1M32 and α2*: compared to α1 it has a lower $K_{0.5}$ for occlusion as well as a faster rate of K^+ deocclusion.

The results shown in TABLE 3 also indicate a 1.7-fold lower turnover of E233K than of α1. As observed for α2* and α1M32, this may reflect a slower step leading to E_2P formation.

DISCUSSION

The changes effected by the E233K mutation in the H2-H3 cytoplasmic loop and by deleting residues 1–32 from the cytoplasmic amino terminus of α1 are consistent with the conclusion that these structural changes alter the equilibrium between the major conformational states of the enzyme. Thus, when deocclusion of K^+ from the K^+-occluded state, $E_2 (K)$, is analyzed as a branched pathway reaction (Scheme I; c.f. ref. 11), it is evident that the relative rate of the reaction via the high ATP affinity pathway **b** is increased in these mutations, consistent with their faster rate of formation of E_1 from $E_2(K)$.

The failure to detect a high affinity ATP component in the analysis of the E233K mutant is not surprising. According to the following simple Michaelis-Menten relationship

$$v = \{V_L[S]/([S] + K'_L)\} + \{V_H[S]/([S] + K'_H)\}$$

the increase in affinity of E233K for ATP at the low affinity site, which is attributed to pathway **a**, is greatest for the E233K mutant. In fact, the activity apparent through pathway **b** is largely masked even when the ATP is reduced to 1 μM. Nevertheless, an increase in the rate of deocclusion via release of K^+ through pathway **b** $[E_2(K) \rightarrow \rightarrow E_1 + K^+]$ is also observed with E233K under conditions in which this reaction is measured, that is, with $E_2(K)$ first formed by equilibrating the enzyme with a saturating concentration of K^+.

Glu233 is in a region previously identified as having a role in the $E_1P \rightarrow E_2P$ conformational change. (For review, see ref. 12.) When Na,K-ATPase in the E_1 conformation is exposed to cleavage at either Leu266 or Arg262 in the H2-H3 cytoplasmic loop as described by Jorgensen and coworkers (reviewed in ref. 9), the $E_1P \rightarrow E_2P$ conformational transition is blocked. Similarly, tryptic cleavage at Arg198 of the SR Ca-ATPase as well as site-specific mutation of residues in the predicted β-strand domain of the H2-H3 loop of this enzyme also blocked the $E_1P \rightarrow E_2P$ transition (reviewed in ref. 13).

The concept that conformational coupling in P-type ATPases occurs via interaction of the β-strand in the H2-H3 loop with the catalytic domain, as envisioned by Green and Stokes[14] and based largely on studies of the Ca-ATPase,[15] is supported by earlier studies of the Ca-ATPase.[16] In studies of Na,K-ATPase, involvement of the H2-H3 loop in structural rearrangements associated with ligand binding and phosphorylation was also evident in distinctive conformational changes revealed by proteolytic cleavage patterns.[17] Similarly, perturbations of residues in the β-strand region of the H2-H3 loop domain of the yeast plasma membrane H^+-ATPase not only alters the distribution of conformational intermediates during steady-state catalysis, but also decreases the sensitivity of this P-type pump to vanadate inhibition, consistent with the conclusion that this region interacts with the catalytic phosphorylation domain.[18,19]

These cytoplasmic mutants contrast with those in which substitutions in transmembrane helices also result in active, functionally altered enzyme, most of which are characterized by changes in affinities for Na^+ and/or K^+. These mutations are located in transmembrane helices H4, H5, H6, H8, and H9.[20–25]

In all of these mutants, substitutions causing a decrease in affinity for K^+ are associated with an increase in apparent affinity for ATP. The basis for this behavior is readily evident from an analysis of the mutual interactions of ATP and K^+ as described by Eisner and Richards.[26] In their analysis, pump-mediated K^+ influx was empirically described by a relationship which showed that decreasing K_{ext} concentration (presumably equivalent to increasing K_{Kext}) increased the apparent affinity for ATP. Similarly, decreasing ATP concentration (presumably equivalent to increasing K'_{ATP}) increased the apparent affinity for external K^+ (and also increased V_{max}). This being the case, it is also intuitively obvious that changes in K'_{ATP} are complicated functions of changes in rate constants of steps leading to the sequential binding, occlusion, and deocclusion of two K^+ ions as in the following expanded sequence of partial reactions via pathway **a** (Scheme 1):

$$E_2P \xrightarrow{K_{ext}} E_2P.K \underset{P_i}{\rightleftarrows} E_2(K) \xrightarrow{ATP_L} ATP.E_2(K) \rightarrow ATP.E_1.K \underset{K_{cyt}}{\rightleftarrows} ATP.E_1$$

Accordingly, it is not surprising that the magnitude of reciprocal changes in K'_{ATP} and K'_K are different in the different mutants, depending on the nature of the affected reaction step.

Although the cytoplasmic mutations described in the present study caused increases in the apparent affinity for ATP, changes in K'_K were not observed as long as the ATP concentration was well above saturation. Thus, neither α1M32 nor E233K has notably altered affinities for K_{ext} or Na_{cyt} under physiological conditions in which the ATP concentration is saturating. This behavior is consistent with the conclusion that the primary functional alteration effected by mutating glutamate 233 to lysine is probably the increase in the rate of E_1 formation from $E_2(K)$ and the decrease in the rate of E_1P to E_2P conversion. We suggest that this nonconservative substitution in the cytoplasmic H2-H3 loop alters its interaction(s) with other regions of the α subunit, resulting in a change in the conformational equilibria such that the apparent affinity for ATP is increased. Changes in apparent affinity for K^+ observed at suboptimal ATP concentration are probably secondary to changes in K'_{ATP} as well as to steps involved in the conversion of conformational forms.

ACKNOWLEDGMENTS

We thank Dr. Jerry B Lingrel for generously providing us with α2*-transfected HeLa cells and Rosemarie Scanzano and Ania Wilczynski for technical assistance.

REFERENCES

1. LINGREL, J. B. & T. KUNTZWEILER. 1994. J. Biol. Chem. **269:** 19652–19662.
2. PALASIS, M., T. A. KUNTZWEILER, J. N. ARGUELLO & J. B. LINGREL. 1996. J. Biol. Chem. **271:** 14176–14182.
3. JORGENSEN, P. L. 1977. Biochim. Biophys. Acta **466:** 97–108.
4. WIERZBICKI, W. & R. BLOSTEIN. 1993. Proc. Natl. Acad. Sci. USA **90:** 70–74.
5. DALY, S. E., L. K. LANE & R. BLOSTEIN. 1994. J. Biol. Chem. **269:** 23944–23948.
6. POST, R. L., C. HEGYVARY & S. KUME. 1972. J. Biol. Chem. **247:** 6530–6540.
7. JEWELL, E. A. & J. B. LINGREL. 1991. J. Biol. Chem. **266:** 16926–16930.
8. DALY, S. E., L. K. LANE & R. BLOSTEIN. 1996. J. Biol. Chem. **271:** 23683–23689.
9. JORGENSEN, P. L. 1994. *In* The Sodium Pump: Structure, Mechanism, Hormonal Control, and Its Role in Disease. E. Bamberg & W. Schoner, Eds.: 297–308. Springer. New York.

10. SHULL, G. E., G. GREEB & J. B. LINGREL. 1986. Biochemistry **25:** 8125–8132.
11. SACHS, J. R. 1994. Biochim. Biophys. Acta **1193:** 199–211.
12. ANDERSEN, J. P. & B. VILSEN. 1995. FEBS. Lett. **359:** 101–106.
13. MACLENNAN, D. H., D. M. CLARKE, T. W. LOO & S. SKERJANC. 1992. Acta Physiol. Scand. **146:** 141–150.
14. GREEN, N. M. & D. L. STOKES. 1992. Acta Physiol. Scand. **146:** 59–68.
15. MACLENNAN, D. H. 1990. Biophys. J. **58:** 1355–1365.
16. DUX, L. & A. MARTONOSI. 1983. J. Biol. Chem. **258:** 10111–10115.
17. LUTSENKO, S. & J. H. KAPLAN. 1994. J. Biol. Chem. **269:** 4555–4564.
18. HARRIS, S. L., D. S. PERLIN, D. SETO-YOUNG & J. E. HABER. 1991. J. Biol. Chem. **266:** 24439–24445.
19. BANDELL, M., M. J. HALL, G. WANG, D. SETO-YOUNG & D. S. PERLIN. 1996. Biochim. Biophys. Acta **1280:** 81–90.
20. VILSEN, B. 1993. Biochemistry **32:** 13340–13349.
21. FENG, J. & J. B. LINGREL. 1995. Cell. Mol. Biol. Res. **41:** 29–37.
22. JEWELL-MOTZ, E. A. & J. B. LINGREL. 1993. Biochemistry **32:** 13523–13530.
23. VILSEN, B. 1995. Biochemistry **34:** 1455–1463.
24. KOSTER, J. C., G. BLANCO, P. B. MILLS & R. W. MERCER. 1996. J. Biol. Chem. **271:** 2413–2421.
25. ARGUELLO, J. M. & J. B. LINGREL. 1995. J. Biol. Chem. **270:** 22764–22771.
26. EISNER, D. A. & D. E. RICHARDS. 1981. J. Physiol. **319:** 403–418.

Subunit Interactions in the Sodium Pump[a]

T. COLONNA, M. KOSTICH, M. HAMRICK, B. HWANG,
J. D. RAWN,[b] AND D. M. FAMBROUGH

Department of Biology
The Johns Hopkins University
Baltimore, Maryland 21218

[b]*Department of Chemistry*
Towson State University
Baltimore, Maryland 21204

Our laboratory has been studying the process of biosynthesis and assembly of sodium pump subunits and the targeting of assembled, processed sodium pump molecules to the plasma membrane. This line of inquiry began with pulse-chase studies of sodium pumps in tissue-cultured chick skeletal muscle and sensory nerve.[1-3] Among the conclusions from these studies were: (1) assembly of the α and β subunits occurs very rapidly, either during or immediately after biosynthesis of the α- and β-subunit polypeptide chains; (2) assembly occurs in the endoplasmic reticulum (ER); (3) N-glycosylation of the β subunit is not necessary for assembly; and (4) N-glycosylation is not necessary for transport of assembled sodium pumps through the secretory pathway to the plasma membrane and has little effect on the kinetics of the transport process.

ANALYSIS OF SODIUM PUMP SUBUNIT ASSEMBLY THROUGH IMMUNE PRECIPITATION OF α–β SUBUNIT COMPLEXES AFTER EXPRESSION IN MAMMALIAN CELLS

With cloning of encoding cDNAs for the chicken α and β subunits,[4-7] it became possible to study the assembly process in greater detail by expressing these cDNAs in mammalian cells. In a typical experiment, cDNA encoding one or both chick sodium pump subunits would be used to transfect HeLa cells or mouse L-cells. Once biosynthesis of the chick sodium pump subunits began in the mammalian cells, the cells could be pulse-labeled with [^{35}S]methionine, the membrane proteins solubilized in solution containing nonionic detergent, and the chick sodium pump β subunit isolated by immune precipitation with a monoclonal antibody specific for avian β subunits (as diagrammed in FIGURE 1). The immune precipitate would be analyzed by SDS-PAGE and fluorography. Recovery of [^{35}S]methoinine-labeled chick Na,K-ATPase β subunit was evidenced by the presence of a set of bands in the fluorograph

---

[a]This research was supported by grant NS-23241 from the National Institute of Neurological Diseases and Stroke, National Institutes of Health, Bethesda, Maryland.

[b]Address for correspondence: Dr. Douglas M. Fambrough, Department of Biology, The Johns Hopkins University, 3400 N. Charles Street, Baltimore, MD 21218 (tel: 410 516-5174; fax: 410 516-6157; e-mail: fambro@jhu.edu).

with apparent molecular weights in the 40–50-kD region. These were shown to represent N-glycosylated forms of the β subunit, including high-mannose forms of lower apparent molecular weight and complex oligosaccharide forms of higher apparent molecular weight. Assembly of the chick β subunit with α subunits was assessed by the presence of a [^{35}S]methoinine-labeled band of approximately 100 kD. This experimental protocol served as the basis for exploring what elements within each subunit of the sodium pump are necessary and sufficient for α-β subunit assembly.[8–12]

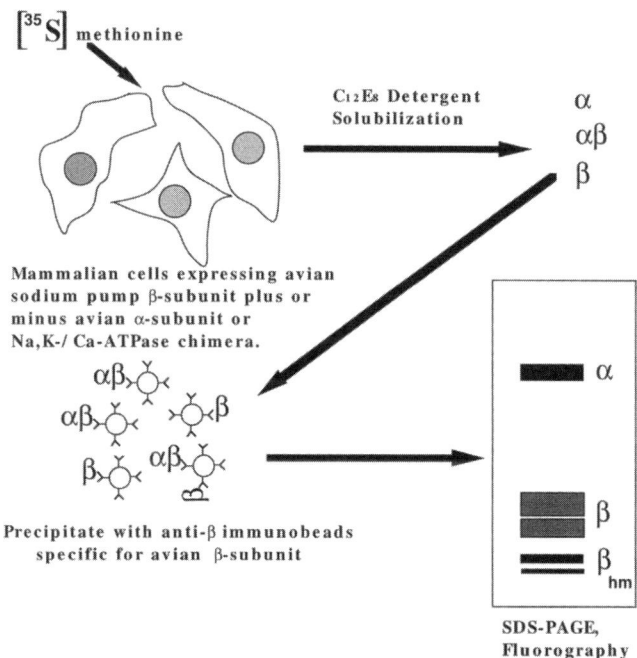

FIGURE 1. Diagram of Na,K-ATPase subunit assembly assay by coimmunoprecipitation of α and β subunits. After expression of chick β-subunit DNA in mouse L-cells, for example, assembly of the avian β subunit with mouse α subunits is assayed by immunoprecipitation of ^{35}S-labeled, detergent-solubilized avian β subunits with a monoclonal antibody specific for the avian β subunit. Assembly with mouse α subunits is detected after SDS-PAGE and fluorography in which the α-subunit polypeptide chain is identified as a ~100-kD band.

To examine the issue of targeting of assembled sodium pumps to the plasma membrane, the same expression systems were used. The steady state distribution of the chicken sodium pump subunits was assayed by immunofluorescence microscopy. As FIGURE 2 shows, a marked difference in distribution of exogenously expressed sodium pumps is noted when most of the subunits are retained in the ER as when the subunits are transported through the membrane biogenesis pathway to the plasma membrane. More quantitative analysis was easily done with the use of [^{125}I]-labeled monoclonal antibodies to the ectodomain of the avian β subunit. The fraction of

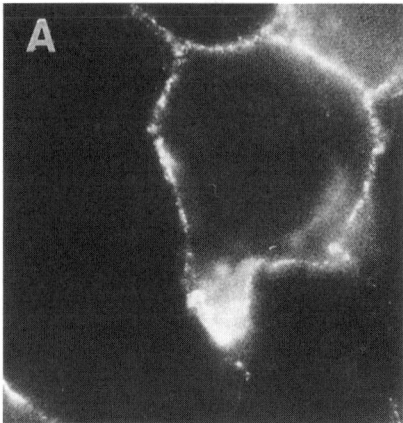 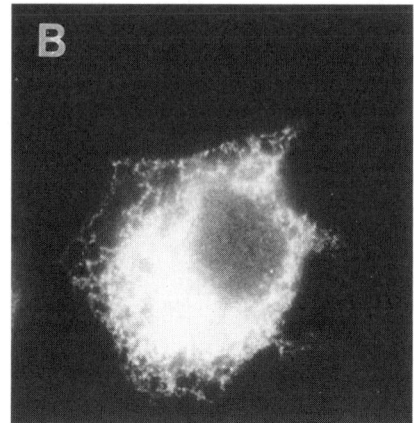

FIGURE 2. Immunofluorescence microscopy of mouse L-cells expressing (**A**) avian β subunits or (**B**) avian β subunits truncated by 92 residues at the COOH-terminal end. A monoclonal antibody to the avian β subunit was used to detect its expression. Both full-length and truncated β subunits assemble with mouse α subunits (FIG. 3). However, the truncated form is retained in the ER, here shown as a reticular pattern of fluorescence. Surface expression of the full-length avian β subunit is shown as sharp edge fluorescence when the focal plane passes through a somewhat rounded cell. Adapted from reference 12.

expressed β subunits that reach the plasma membrane can be quantified by comparing the specific binding of the [^{125}I]-labeled monoclonal antibodies to the surfaces of intact cells with specific binding to cells made permeable to antibody molecules by treatment with saponin.

With this expression/immune precipitation assay, the minimal structural requirements in the α subunit for assembly with the β subunit were explored. For these studies, a large series of plasmids encoding chimeric catalytic subunits were constructed and expressed together with full-length chick β subunits in HeLa cells, generally in acute experiments with expression driven by T7-RNA polymerase expressed in the HeLa cells through infection with recombinant Vaccinia virus.[13] Assembly of the chick β1 subunit with the endogenous HeLa α subunit gave a very weak, sometimes undetectable signal, whereas assembly with the chimeric catalytic subunit was typically robust, provided that the Na,K-ATPase α-subunit region necessary for α-β subunit assembly were present in the chimera.[10] For these experiments the chimeras were formed between the avian Na,K-ATPase α1 subunit and the avian SERCA1 Ca-ATPase.[14] As expected, the Ca-ATPase did not assemble with the Na,K-ATPase β1 subunit. However, when 26 amino acids of the Na,K-ATPase were substituted for the corresponding region within the extracellular loop between the seventh and eighth membrane spans (henceforth H7 and H8) of the Ca-ATPase, the chimera assembled robustly with the β subunit.[9] Thus, these 26 amino acids of the ectodomain of the α subunit are sufficient for α-β subunit assembly. In more recent work, the 26 amino acid region was also shown to contain the amino acyl residues not only sufficient but also necessary for α-β subunit assembly. This was determined by constructing the reverse chimera, in which the 26 amino acid region of the Na,K-ATPase was replaced by the corresponding region of the Ca-ATPase and expressed together with chick β1 subunits in HeLa cells. No subunit assembly could be measured, although expression of the chimera was strong. Thus, the H7H8 loop of the Na,K-ATPase α subunit contains a region of 26 or fewer amino acyl residues that are necessary and sufficient for α-β subunit assembly.

Assembly of the chick β1 subunit with endogenous mouse α1 subunits in mouse L-cells was a rather efficient process, simplifying the protocol for assessing what region(s) of the β subunit are necessary and sufficient for assembly with the α subunit. Altered chick β-subunit forms with NH_2-terminal or COOH-terminal truncations or internal deletions were expressed in mouse L-cells, and assembly with the endogenous mouse Na,K-ATPase α subunit was assayed by immune precipitation and SDS-PAGE/ fluorography after pulse labeling the transfected cells with [^{35}S]methionine. Neither the NH_2-terminal cytosolic domain nor any region of the membrane-spanning domain of the β subunit was found to be essential for α-β subunit interactions.[11] However, despite robust assembly with α subunits, β subunits containing deletions within the transmembrane region remained in the ER. These results showed that although α-β subunit assembly seems to be required for sodium pumps to leave the ER, assembly can occur between subunits that are, nevertheless, retained in the ER, presumably because there are quality control mechanisms in the ER that recognize imperfection in the α-β complex and prevent exit from the ER and transport through the membrane biogenesis pathway to the plasma membrane. Similarly, complexes between the avian Na,K-ATPase β subunit and the Ca-ATPase, an ER resident protein that was modified to contain the critical assembly domain in the H7H8 loop, remained in the ER.

COOH-terminal truncations of the β subunit that remove one or two of the three cysteine pairs involved in disulfide bridges remain competent for assembly with the α subunit,[12] although these complexes, too, remained in the ER. FIGURE 3 shows SDS-PAGE fluorographs that document α-β assembly involving these COOH-terminally truncated β subunits. This figure also illustrates a limitation of the coimmune precipitation strategy for detecting α-β subunit assembly. As the size of the truncated β subunit decreased, the recovery of α-β complexes also diminished. This

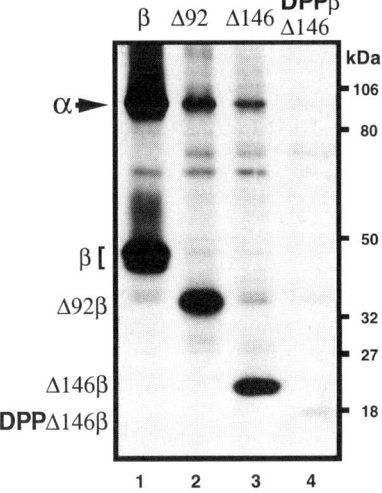

FIGURE 3. Fluorograph of SDS-PAGE analysis of immune precipitates of avian β subunits from mouse cells expressing full-length avian β1 subunits (β; lane 1) or β1 subunits truncated to remove one (Δ92; lane 2) or two (Δ146; lane 3) of the disulfide bridging cysteine pairs. Note that the smaller, truncated β subunits assemble with mouse α subunits but that the yield of α subunits diminishes with increasing truncation. No α subunit is evident in the case of the β-subunit truncation that removed 146 amino acids and, in addition, substituted the cytosolic and transmembrane domains with those of dipeptidylpeptidase IV (DPPΔ146β; lane 4). Adapted from reference 12.

might be due to a decline in efficiency of subunit assembly in the ER as the β subunit is more and more severely modified. Alternatively, it might be due to increasingly weak α-β interactions such that, while subunit assembly remained approximately as efficient as that between native α and β subunits, the α-β complexes underwent significant dissociation in detergent solution, resulting in lower recovery of α subunit in the immune precipitates of β subunits. This latter explanation seems to be at least partially correct, based on the observation that by changing the detergent from Triton X-100 to $C_{12}E_8$ and by shortening the time between detergent extraction and completion of the immune precipitation, greater yields of α subunit were achieved.[10] It appeared that a more sensitive assay for subunit assembly might be needed to define the regions of α-β association more exactly.

The same conclusion, that a more sensitive assay system was needed, came from studies of subunit assembly between chimeric catalytic subunits and modified β subunits and from studies of α subunits assembling with the ectodomain of the β subunit. In the latter experiments, mouse α subunits were found to assemble well with β subunits in which the NH_2-terminal cytosolic domain and the transmembrane span were both replaced with those of the unrelated protein dipeptidyl-peptidase IV (DPP). This DPP-β chimera, in the immune precipitation assay for assembly, appeared to assemble with the α subunit at about 30% the efficiency of the normal chick β1 subunit, whereas the reverse chimera showed no ability to assemble. It was concluded that although the NH_2-terminus and/or transmembrane regions may be involved in interactions with the α subunit, these regions were not sufficient for subunit assembly measured by coimmunoprecipitation.[12] Furthermore, a cleavage site for the ER enzyme signal peptidase was generated by making a point mutation in the DPP transmembrane span of the DPP-β chimera. This resulted in conversion of the chimera to a secretory protein consisting entirely of the ectodomain of the sodium pump β subunit. This secretory β subunit was able to assemble with α subunits[15] and even progress through the membrane biogenesis pathway to the plasma membrane. But, as FIGURE 4 shows, compared with DPP-β, secretory β subunit appeared to assemble with mouse α subunits much less efficiently or less strongly. To find the minimal region of the β subunit sufficient for assembly with the α subunit, a more sensitive assay was needed. We attempted to employ the yeast two-hybrid system[16] to explore sodium pump α-β subunit interactions further.

BRIEF DISCUSSION OF THE YEAST TWO-HYBRID SYSTEM

The two-hybrid system is based on the modular nature of many eukaryotic transcriptional activators, such as the yeast Gal4 transcription factor. Earlier experiments by Ptashne and colleagues[17] established that the Gal4 transcription factor contains separable domains for DNA binding and transcriptional activation. The DNA binding domain localizes the transcription factor to specific DNA sequences present in the upstream region of genes that are regulated by this factor. The activation domain contacts other components of the transcription machinery in order to initiate transcription. Brent and Ptashne[18] demonstrated that a hybrid transcriptional activator could be generated. In their experiment, a fusion gene was constructed encoding a hybrid of the *Escherichia coli* repressor LexA with the yeast Gal4. This hybrid protein could activate transcription in yeast of a gene containing a LexA operator, indicating that the

LexA portion conferred the site-specific DNA binding and the Gal4 portion contributed the transcriptional activation function. This demonstrated that for these proteins, the structure was sufficiently modular that totally different proteins could be cut and pasted together through recombinant DNA technology to generate functional hybrids.

The feasibility of the two-hybrid approach was shown by Fields and Song[16] in studies in which the interaction of two yeast proteins, SNF1 and SNF4, was used to reconstitute Gal4 activity. SNF1, a protein kinase, was fused to the DNA-binding domain of Gal4, and SNF4, a protein associated with this kinase, was fused to the activation domain of Gal4. Although neither hybrid protein alone could activate transcription of a gene containing Gal4 sites, the presence of both hybrids in the same cell allowed the SNF1-SNF4 interaction to bring the two Gal4 domains into sufficient proximity for transcriptional activation. This experiment showed that not only

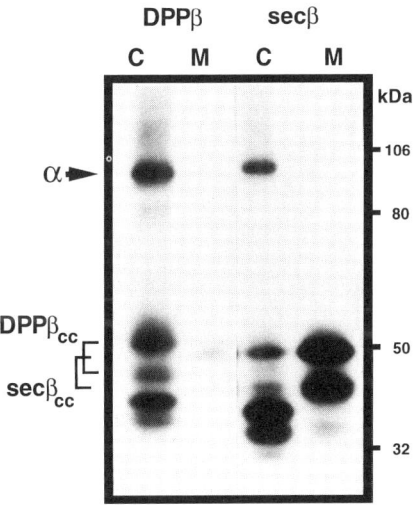

FIGURE 4. Assembly of avian membrane-bound and secretory β-subunit ectodomains with mouse α subunits after expression in mouse L-cells. Mouse L-cells expressing DPP-β and secretory β subunits were positive for α-β subunit assembly, assayed by the coimmunoprecipitation method, illustrated in FIGURE 1. The figure shows fluorograph of SDS-PAGE analysis of ^{35}S-labeled proteins after immunoprecipitation with anti-avian β-subunit monoclonal antibody. C above lanes refers to detergent extracts of cells, whereas M refers to medium conditioned by the cells. Because signal peptidase clips the secretory β subunits at the COOH-terminal end of the membrane span, each glycosylated form of the secretory β subunit is smaller than the corresponding form of the DPP-β subunit. Note secretion (appearance in the medium) of the complex-carbohydrate (cc) forms of secretory β subunits (sec$β_{cc}$). Adapted from reference 15.

transcriptional activators, but also many, if not most, proteins could be made into hybrid proteins. These hybrids would each possess two functions, one involved in transcription (i.e., DNA binding or activation) and the other involved in protein-protein interactions.

A diagram of how the two-hybrid system works is shown in FIGURE 5. Plasmids are constructed that encode two hybrid proteins: one consists of the DNA-binding domain of Gal4 fused to one test protein X and the other consists of the Gal4 activation domain fused to another test protein Y. These plasmids are transformed into a *Saccharomyces cerevisiae* strain that contains a *lacZ* reporter gene whose regulatory regions contain Gal4 binding sites. Either hybrid protein alone must be unable to activate transcription of the reporter genes. The DNA-binding domain hybrid should not activate transcription because it does not provide activation function, and the activation domain hybrid also should not activate transcription because it cannot localize to the Gal4 binding sites. Interaction of the two test proteins reconstitutes the function of the Gal4 and results in expression of the reporter genes, which is detected by assays for the reporter gene products.

The two-hybrid system has many advantages for the analysis of protein-protein interactions. (1) The assay is performed *in vivo,* presumably under conditions similar to those in which protein interactions normally occur. Some examples of proteins successfully used in the two-hybrid assays include those normally found in the nucleus, cytoplasm, or mitochondrion, and membrane-associated and extracellular proteins.[19] It is not known if the hybrid proteins first interact within the nucleus or interact in the cytoplasm and are transported into the nucleus.[19] (2) Purified target protein or antibody against this protein is not required to detect the interaction. (3) The most important advantage of the two-hybrid system is that it appears to be more sensitive than coimmunoprecipitation and therefore may be able to detect low-affinity

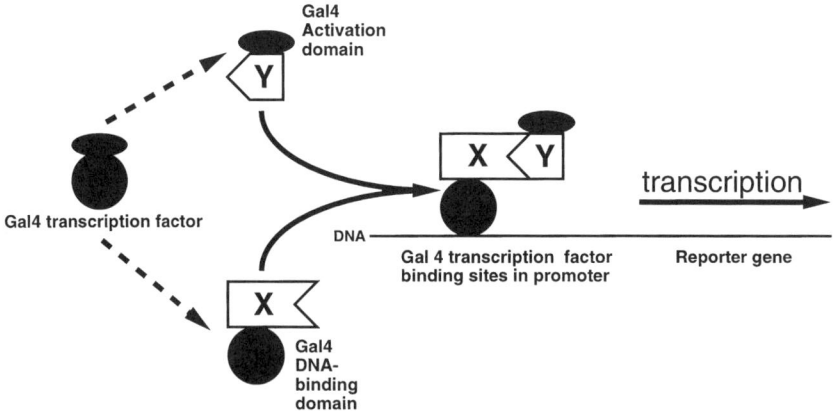

FIGURE 5. The yeast two-hybrid system. Diagram illustrates that the Gal4 transcription factor, shown in *black,* can function even when it is expressed as separate DNA binding and activation domains, provided that the domains are fused to proteins that are capable of interaction to bring the Gal4 domains into a common complex. Proteins X and Y, in our experiments, were parts of the α and/or β subunits of the Na,K-ATPase. Reconstitution of Gal4 function results in activation of transcription of reported genes that have been incorporated into the yeast strain.

interactions. Phizicky and Fields[20] estimate that the minimal binding constant required to detect an interaction in their version of the system is on the order of 1 μM. The sensitivity of this system probably arises from the fact that transient interactions are sufficient to trigger a small amount of transcription, but then amplification occurs based on the potential for repeated rounds of translation and the stable nature of the reporter proteins. (4) It is possible to detect extremely small amounts of these reporter proteins. The minimal binding affinity of two proteins necessary to obtain a two-hybrid signal is not known, and it is likely that this value will vary significantly depending on the protein combination being assayed.[21] This variability is due to the many factors that contribute to the signal, such as folding of the hybrid proteins, stability of each hybrid, entry into the nucleus, accessibility of the interacting domains to each other, and accessibility of the activation domain to the transcription apparatus.[21] Therefore, the levels of transcription from two different protein combinations cannot be compared to yield information about relative affinities. However, within a given combination, mutations that reflect increased or decreased affinity can be assayed. (5) The use of genetically based organisms such as yeast cells as the host for studying interactions allows both a direct selection for interacting proteins and the screening of many variants to detect those that might interact either more or less strongly. This benefit enables this system to be used to map protein domains that are responsible for protein interactions and to identify mutations that affect these interactions.[21]

The two-hybrid system is limited by several factors. First, since the proteins must be localized to the nucleus, this may prevent its use with certain extracellular or membrane proteins. Second, the proteins must be able to fold into the correct conformation and be metabolically stable as fusion proteins in yeast cells. Third, the use of fusion proteins raises the possibility that the site of interaction may be occluded by one of the transcription factor domains. Fourth, it may not be possible to reproduce interactions involving proteins that are glycosylated and/or contain disulfide bonds.[19] Similarly, interactions that are mediated by posttranslational modification, such as phosphorylation, may not be detected.[19] Additionally, many proteins not normally involved in transcription will activate transcription when fused to a DNA-binding domain, and this activation precludes the screening of a library. Fifth, in the analysis of mutations in which semiquantitative information is needed, it must be assumed that the amount of reporter gene transcription is proportional to the affinity of the two test proteins for each other. Experiments by Li and Fields[22] have shown that a rough correlation exists between the affinity of two proteins for each other and the amount of reporter gene transcription in the two-hybrid assay.

Results of two-hybrid assays performed with various combinations of two-hybrid vectors can significantly vary in sensitivity. Legrain et al.[23] showed that for five different pairs of interactions, assay with the pGAD424/pGBT9 set of vectors supplied with the Clonetech Matchmaker kit was roughly 50–100 times less sensitive than that with the original Fields vectors pGAD2f/pMA424 and the Elledge vectors pAS2/pACT2. Therefore, an interaction may sometimes escape detection when assayed with the pGAD424/pGBT9 set. Their results suggest that certain combinations of vectors may introduce a bias because weak interactions may be overlooked. In our experiments, we found that the pGAC424/pGBT9 vectors did not give sufficient signal to allow reproducible detection of α-β subunit interactions.

In assays of defined protein combinations, one orientation of the hybrids (i.e., X fused to the DNA-binding domain, Y to the activation domain) often activates transcription much more efficiently than the reverse orientation.[19] This may reflect differences between the stability of hybrids containing X and those containing Y. Transcription is optimal when the activation domain hybrid is in excess over the DNA-binding domain hybrid.[19] When the reverse is true, DNA-binding domain hybrids bound to the reporter gene promoters are less likely to be engaged in the X-Y protein-protein interaction. In some cases, as deletions of one hybrid are constructed and assayed to define a minimal domain for interaction, transcription increases significantly.[19] Although in certain instances this may be because residues that are not available in the intact protein are exposed in a shorter domain, it may also be that smaller proteins or domains simply work better in this system.

Finally, the two-hybrid system can be used to assay interactions in which components are small peptides. When the Rb protein is fused to the Gal4 DNA-binding domain, a 13-residue peptide that is derived from SV40 large T antigen and contains the Leu-X-Cys-X-Glu motif found in several Rb-binding proteins is sufficient to activate transcription when fused to the Gal4 activation domain.[24]

ANALYSIS OF SODIUM PUMP α-β SUBUNIT INTERACTIONS WITH THE YEAST TWO-HYBRID SYSTEM ASSAY

The first test of the two-hybrid system as a tool for studying sodium pump α- and β-subunit assembly was to determine if in two-hybrid assays one could obtain positive signals for interaction between the regions of the α and β subunit that had already been shown to interact by coimmunoprecipitation assays. Thus, Gal4 fusions were made with the H7H8 extracellular loop of the α subunit and the ectodomain of the β subunit. Coexpression of these in yeast gave positive two-hybrid assays. Similarly, two-hybrid assays were positive for interactions between the H7H8 α loop and COOH-terminal truncations of the β-subunit ectodomain that removed one or two of the three pairs of cysteines involved in disulfide bridging. All of these assays were positive regardless of which subunit domain was fused to the DNA binding domain of Gal4. These results confirmed the utility of the yeast two-hybrid system for exploring α-β interactions.

The two-hybrid system was next used in a search for the residues of the α-subunit H7H8 loop that are critical for α-β subunit interaction. A set of alanine-scanning mutations was made in the region of the α loop that is most conserved in α-subunit evolution. This region lies within the 26 amino acid stretch of the H7H8 loop previously shown to be necessary and sufficient for assembly with the β subunit, as measured in the coimmunoprecipitation assay. FIGURE 6 diagrams the H7H8 loop regions that were scanned. For each construct, four consecutive alanyl residues replaced four amino acyl residues of the loop. Each of these altered forms was tested for interactions with the entire β-subunit ectodomain and with the two COOH-terminally truncated forms. Only mutation of the sequence SYGQ to AAAA abolished α-loop interactions with the β-subunit constructs, and this was true for both orientations of the two-hybrid assay. The SYGQ sequence is virtually invariant among α subunits of Na,K and H,K-ATPases. As we know from previous studies, the α-subunit H7H8 loop can interact not only with the β1- and β2-subunit isoforms of the Na,K-ATPase but also with the β subunit of the H,K-ATPase, as determined in the coimmunoprecipitation assay.[10] We conclude that the sequence SYGQ includes residues that are probably critical for α-β subunit interactions in both the Na,K- and H,K-ATPases.

Next we sought to define a maximum COOH-terminal truncation of the β-subunit ectodomain that retained affinity for the α-subunit H7H8 loop. To do this, we adopted a more conventional use of the two-hybrid system, that is, to screen libraries for DNAs encoding proteins that can interact with a protein of interest. In our case, we prepared a library of β-subunit COOH-terminal truncations by performing a set of timed exonuclease III digestions of the plasmid containing the Gal4/β-subunit chimera DNA. Exonuclease digestion began at the 3' end of the β-subunit coding DNA and chewed back towards the Gal4 activation domain DNA. After digestion, the free ends of the surviving plasmid DNA were blunt-ended and relegated to generate a library of 3' truncations of the Gal4/β-subunit hybrid. These plasmids were transfected into

COLONNA et al.: SUBUNIT INTERACTIONS IN SODIUM PUMP

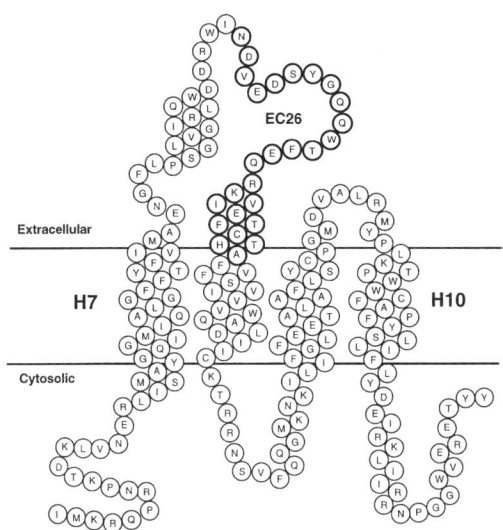

FIGURE 6. Scanning alanine mutagenesis of the H7H8 region of the avian Na,K-ATPase α1 subunit. The chicken α1-subunit H7H8 loop is predicted to be a pair of α-helical regions separated by a region lacking predictable structure (unpublished algorithm of J.D. Rawn). The critical region of the loop for subunit assembly, as shown by coprecipitation assays, is indicated by the heavier borders of the *circles*. Below are the aligned amino acid sequences of these critical 26 residues in the three avian Na,K-ATPase α-subunit isoforms, the rat gastric H,K-ATPase α-subunit, and the avian SERCA1 Ca-ATPase. Sequences of the four alanine-scanning mutants tested in this study are also shown. Replacement of SYGQ by AAAA prevented interaction between the α loop and the ectodomain of the β subunit, as measured in the two-hybrid system.

yeast carrying the Gal4/α-subunit loop hybrid. As diagrammed in somewhat more detail in FIGURE 7, the transfected cells were selected for growth on media lacking histidine and subsequently for expression of β-galactosidase, tests that involve α-loop interaction with the β subunit to reconstitute Gal4 transcription factor function. Finally, DNA was recovered from the positive yeast colonies, and polymerase chain reactions (PCR) were used to determine how much of the β-subunit encoding DNA remained in each case. A set of 3'-end PCR primers that were six bases apart in the β-subunit DNA sequence were used. By determining the primer closest to the 5' end of the β-subunit DNA that yielded a PCR fragment in combination with a primer within the Gal4 region, it was possible to estimate, within a couple of amino acids, the remaining length of the β-subunit ectodomain. By this method it was determined that the β-subunit ectodomain truncated to contain no more that 61 amino acids remained positive in the two-hybrid assay for α-β interactions. FIGURE 8 shows this region of the β subunit highlighted in the structure of the total β subunit. As can be seen in this figure, the maximal COOH-terminal truncation removed all of the cysteines from the ectodomain as well as all of the N-glycosylation sites. The minimal β-subunit domain

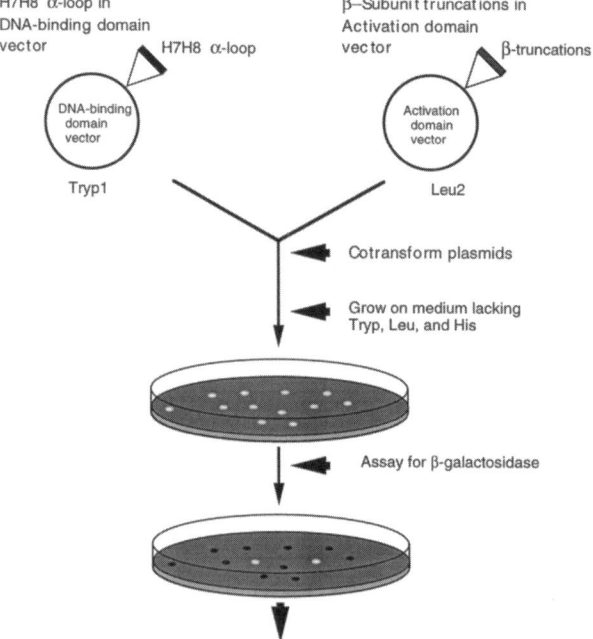

FIGURE 7. Diagram of the screen for the maximal β-subunit COOH-terminal truncation that remains positive for interaction with the H7H8 α loop in the two-hybrid assay. After construction of a library of β-subunit COOH-terminal truncations, these were coexpressed with the α loop/Gal4 construct in yeast. Selection for the presence of both plasmids and for α-β interaction was first made by growth on agar plates with medium lacking tryptophan, leucine, and histidine. Then a second reported gene, *β-gal,* was assayed. Blue colonies, positive in two tests for α-β interaction, were subsequently analyzed by rounds of polymerase chain reaction to define the length of the remaining β-subunit ectodomain.

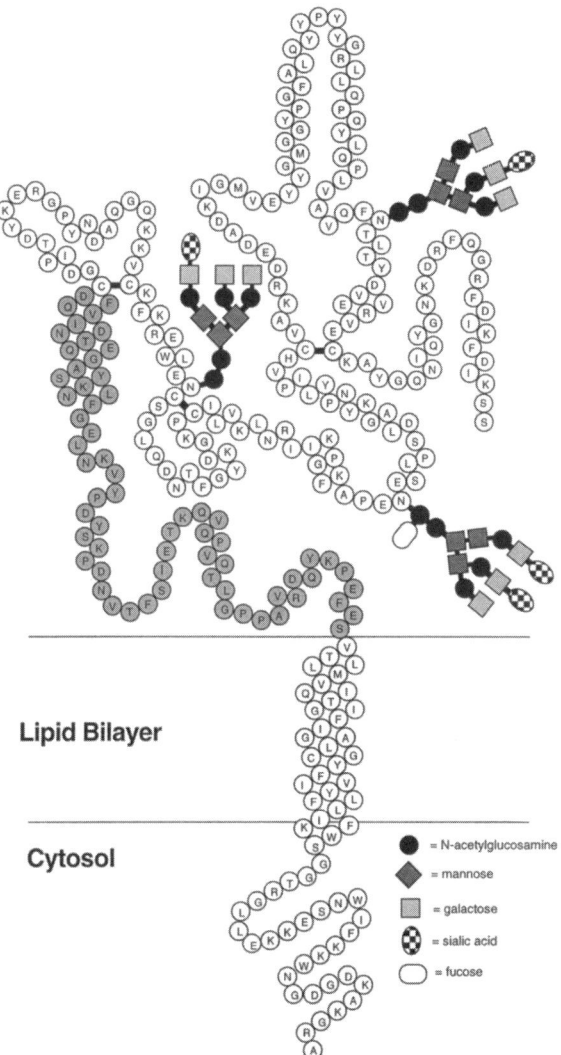

FIGURE 8. Minimal β-subunit region so far found to interact with the Na,K-ATPase α subunit is highlighted in the context of the structure of the entire β subunit. Secondary structure predictions for the ectodomain of the β subunit are indicated in the cartoon (unpublished algorithm of J. D. Rawn). Each *circle* represents an amino acid. The *N*-linked oligosaccharides in the cartoon are merely examples of such structures, not ones known to occur in the avian β subunit. Disulfide bridging cysteines are shown linked.

involved in α-β subunit assembly is therefore proposed to involve only the proximal stem of the β-subunit ectodomain.

In retrospect, the success of the two-hybrid assay for examining α-β interactions can now be understood better. In the two-hybrid system the components must fold into forms capable of interaction while in the cytosol or nucleoplasm. In these environments, there are probably reducing conditions so that no disulfide bridges form, and there is no possibility of conventional NH_2-glycosylation (which occurs in the lumen of the ER). If α-β interactions had required disulfide bridges or glycosylation, the two-hybrid assays would have been negative.

A SEARCH FOR HOMOTYPIC α- AND β-SUBUNIT INTERACTIONS

There is substantial evidence for the existence of $\alpha_2\beta_2$ dimers (composed of two α-β heterodimers) of the Na,K-ATPase in cell membranes. During the VIIIth International Sodium Pump Conference, additional evidence for the biological significance of the $\alpha_2\beta_2$ form of the Na,K-ATPase was offered by Hayashi, Mercer, Taniguchi, and Froehlich, and the reader is urged to read their contributions in this volume. In particular, Mercer and colleagues defined a region within the α-subunit large cytosolic loop that seems to be involved in α-α interactions when Na,K-ATPase subunits are expressed in Sf9 insect cells. We employed the yeast two-hybrid system to seek elements of the α subunit that might be involved in α-α interactions; similarly we sought evidence for β-β interactions. Some possible types of α-α and β-β interactions are depicted in FIGURE 9.

For examining α-α subunit interactions, Gal4 fusions were constructed with each of the two large cytosolic loops of the α subunit, as highlighted in the α-subunit cartoon in FIGURE 10. In two-hybrid assays, neither of these loops appeared to self-associate. However, in tests for interaction between loop-1 and loop-2, positive results were obtained for one orientation of the loops *vis à vis* the Gal4 domains, but not for the other orientation. As we discussed earlier in this paper, there is good reason why interacting domains might yield a positive result in only one orientation in the yeast two-hybrid assays. Thus, our results suggest an interaction between the two large α loops, but the suggestion is not so strong as would have been the case if positive results had been obtained for both orientations. Whether such an interaction bespeaks an inter-subunit or an intra-subunit association cannot be determined at this

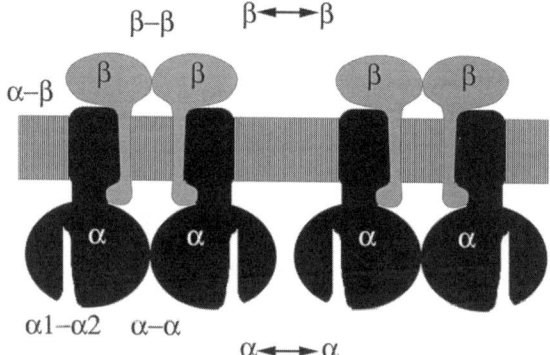

FIGURE 9. Cartoon showing possible α-α interactions and β-β subunit interactions. A variety of inter- and intramolecular interactions are possible. Some of these were analyzed by two-hybrid system assays in this study.

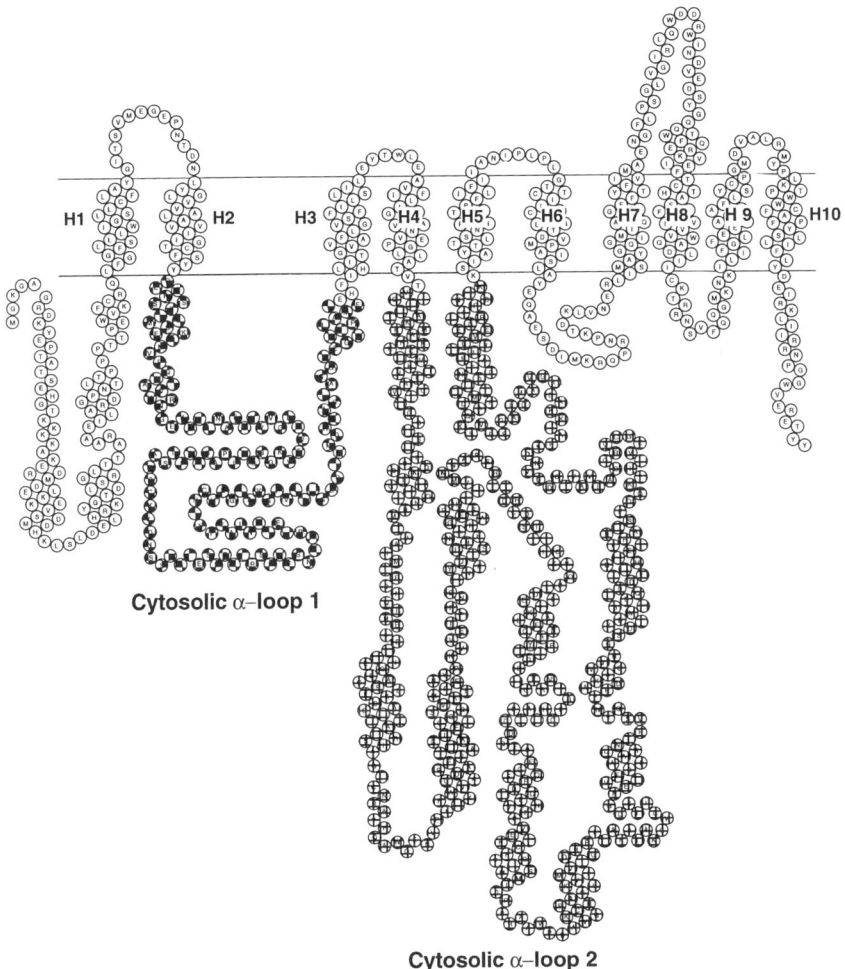

FIGURE 10. Cartoon of the avian Na,K-ATPase α1 subunit, showing presumed topology in the membrane. Each *circle* represents an amino acid. The first and second large cytoplasmic loops of the α subunit are highlighted. These highlighted regions were tested in the two-hybrid system in this study.

time, but an intra-subunit interaction would be expected, given evidence from point-mutation studies that changes in the first cytosolic loop can have consequences on functions mapped to the second cytosolic loop.

The β-subunit ectodomains truncated to 96 amino acids gave positive results for homotypic interactions, but the longer β-subunit ectodomain pieces were negative. These results are inconclusive as to whether β-β-subunit interactions occur, but the lack of positive results with the larger ectodomain suggests that β-β interactions are not biologically meaningful. The full-length β-subunit ectodomain could not be tested for β-β interactions, because the β-subunit ectodomain fused to the Gal4 DNA binding domain was positive when expressed in yeast by itself.

CONCLUSION

We are greatly encouraged by the two-hybrid system results. The congruence of these results with data obtained from subunit assembly in mammalian cells supports the validity of the approach. New data from the two-hybrid system include identification of a region with the α-subunit H7H8 loop that may be critical for α-β interactions. This region, SYGQ, is the most evolutionarily conserved region of the loop, another reason to propose that this region may, indeed, be a key one for α-β interactions. The results for the β-subunit ectodomain likewise seem reasonable. The region of the β-subunit ectodomain found to interact with the H7H8 α loop includes the first hydrophilic residues past the membrane-spanning domain and therefore must be close to the lipid bilayer. It is reasonable to suppose that this region occurs in proximity to the α-subunit external face in the native α-β heterodimer and therefore is positioned for interaction with the α-subunit H7H8 loop. It is also reasonable to find interaction between the two largest cytosolic loops of the α subunit. The two-hybrid system may be useful in defining much more precisely the nature of this interaction.

Finally, it will be of great interest to determine if other elements of the α subunit and β subunit are involved in inter- or intrasubunit interactions. The NH_2-terminus of both the α subunit and the β subunit should be tested. In addition, each of the major structural elements of each subunit could conceivably interact specifically with other molecules in the cell. The α subunit is known to interact with ankyrin. The β subunit may interact with a "receptor" on other cells.[25] The yeast two-hybrid system is the most powerful screening tool we currently have for seeking partners involved in such interactions.

ACKNOWLEDGMENTS

We would like to thank Delores Somerville and Christine Hatem for their technical support.

REFERENCES

1. FAMBROUGH, D. M. 1983. Cold Spring Harbor Symp. Quant. Biol. **48:** 297–304.
2. WOLITZKY, B. A. & D. M. FAMBROUGH. 1986. J. Biol. Chem. **261:** 9990–9999.
3. TAMKUN, M. M. & D. M. FAMBROUGH. 1986. J. Biol. Chem. **261:** 1009–1019.
4. TAKEYASU, K., M. M. TAMKUN, K. J. RENAUD & D. M. FAMBROUGH. 1988. J. Biol. Chem. **263:** 4347–4354.
5. TAKEYASU, K., V. LEMAS & D. M. FAMBROUGH. 1990. Am. J. Physiol. **259:** C619–630.
6. TAKEYASU, K., M. M. TAMKUN, N. R. SIEGEL & D. M. FAMBROUGH. 1987. J. Biol. Chem. **262:** 10733–10740.
7. LEMAS, M. V. & D. M. FAMBROUGH. 1993. Biochim. Biophys. Acta **1149:** 339–342.
8. LEMAS, M. V., K. TAKEYASU & D. M. FAMBROUGH. 1992. J. Biol. Chem. **267:** 20987–20991.
9. LEMAS, M. V., M. HAMRICK, K. TAKEYASU & D. M. FAMBROUGH. 1994. J. Biol. Chem. **269:** 8255–8259.

10. LEMAS, M. V., H. Y. YU, K. TAKEYASU, B. KONE & D. M. FAMBROUGH. 1994. J. Biol. Chem. **269:** 18651–18655.
11. RENAUD, K. J., E. M. INMAN & D. M. FAMBROUGH. 1991. J. Biol. Chem. **266:** 20491–20497.
12. HAMRICK, M., K. J. RENAUD & D. M. FAMBROUGH. 1993. J. Biol. Chem. **268:** 24367–24373.
13. FUERST, T. R., P. L. EARL & B. MOSS. 1987. Mol. Cell. Biol. **7:** 2538–2544.
14. KARIN, N. J., Z. KAPRIELIAN & D. M. FAMBROUGH. 1989. Mol. Cell. Biol. **9:** 1978–1986.
15. FAMBROUGH, D. M., M. V. LEMAS, M. HAMRICK, M. EMERICK, K. J. RENAUD, E. INMAN, B. HWANG & K. TAKEYASU. 1994. Am. J. Physiol. **266:** C579–C589.
16. FIELDS, S. & O. SONG. 1989. Nature **340:** 245–247.
17. KEEGAN, L., G. GILL & M. PTASHNE. 1986. Science **231:** 699–704.
18. BRENT, R. & M. PTASHNE. 1985. Cell **43:** 729–736.
19. FIELDS, S. & R. STERNGLANZ. 1994. Trends Genet. **10:** 286–292.
20. PHIZICKY E. M. & S. FIELDS. 1995. Microbiol. Rev. **59:** 94–123.
21. FIELDS, S. 1993. Methods (Orlando) **5:** 116–124.
22. LI, B. & S. FIELDS. 1993. FASEB J. **7:** 957–963.
23. LEGRAIN, P., M.-C. DOKHELAR & C. TRANSY. 1994. Nucleic Acids Res. **22:** 3241–3242.
24. YANG, M., Z. WU & S. FIELDS. 1995. Nucleic Acids Res. **23:** 1152–1156.
25. GLOOR, S., H. ANTONICEK, K. J. SWEADNER, S. PAGLIUSI, R. FRANK, M. MOOS & M. SCHACHNER. 1990. J. Cell Biol. **110:** 165–174.

Sorting of Ion Pumps in Polarized Epithelial Cells[a]

LISA A. DUNBAR, DENISE L. ROUSH,
NATHALIE COURTOIS-COUTRY, THEODORE R. MUTH,
C. J. GOTTARDI, VANATHY RAJENDRAN, JOHN GEIBEL,
MICHAEL KASHGARIAN, AND MICHAEL J. CAPLAN[b]

Department of Cellular and Molecular Physiology
Yale University School of Medicine
333 Cedar Street
New Haven, Connecticut 06520

The physiologic functions of a P-type ATPase are determined not only by its catalytic and regulatory properties but also by its distribution among a cell's various membranous compartments. With polarized epithelial cells that mediate vectorial ion fluxes, the restriction of P-type ion pumps to one or the other distinct domains of the plasmalemma in large measure determines the parent tissue's solute and fluid transport capacities. The biologic significance of these anisotropic distributions is well illustrated by the mechanisms through which the Na,K-ATPase drives the majority of active epithelial secretory and absorptive processes.[1] In most epithelial cells, the Na,K-ATPase is restricted to the basolateral plasmalemmal domain.[2] This membrane surface rests on the epithelial basement membrane, is in contact with the extracellular fluid compartment, and is involved in extensive contacts with neighboring epithelial cells. The basolateral membrane is separated by tight junctions from the apical plasmalemma, which generally confronts a compartment that is topologically continuous with the extracorporeal space. The nonequilibrium ion distributions generated by the sodium pump are exploited by secondary active transport systems to drive uphill secretory and absorptive fluxes. By expressing different classes of transport systems and restricting their distributions to one or the other surface compartment, epithelial cells can use the basolateral population of the Na,K-ATPase to catalyze a remarkably diverse array of unidirectional transport processes.

To achieve polarized distribution of P-type ATPase proteins, epithelial cells must be able to target newly synthesized ion pumps to the correct membrane surfaces and to retain them there following their delivery. To participate in these sorting functions, P-type ATPase subunit polypeptides must encode information within their structures that specify their sites of ultimate functional residence. Furthermore, machinery within the epithelial cell must be able to recognize this information and act on its messages.[3] Efforts to understand the nature of these sorting signals and of the cellular components that interpret them have largely relied on the extensive homology that relates the members of the P-type ATPase family. This high degree of structural

---

[a]This work was supported by National Institutes of Health grant GM-42136 and a National Young Investigator award from the National Science Foundation.

[b]Corresponding author. Tel: 203-785-7316; fax: 203-785-4951; e-mail: caplan@biomed.med.yale.edu

similarity has made it possible to use molecular techniques to create chimeric subunit polypeptides and to analyze the sorting behavior of these hybrid pumps to search for sorting signals.

Both the Na,K-ATPase and the gastric H,K-ATPase are heterodimers composed of α and β subunits.[4,5] The α subunits of both pumps are ~100-kD polypeptides that are predicted to span the membrane 10 times and do not carry any N-linked sugar residues. The H,K and Na,K-ATPase α subunits share ~65% amino acid sequence identity. Both β subunits span the membrane once in a type II orientation. These heavily glycosylated proteins exhibit apparent molecular weights of 55 kD, and their primary sequences are ~35% identical to one another. In almost every polarized epithelial cell type, the Na,K-ATPase is restricted to the basolateral domain of the plasma membrane. By contrast, the H,K-ATPase resides in a mobilizable pool of intracellular membranes that can fuse with and be retrieved from the gastric parietal cell's apical plasmalemma.[6–9] The fact that these pumps manifest dramatically different localizations when expressed endogenously[10] or by transfection[11] in the same cell type demonstrates that their constituent subunit polypeptides must transmit distinct sorting messages to the epithelial sorting machinery. We have generated chimeras composed of complementary portions of the Na,K-ATPase and the gastric H,K-ATPase subunit polypeptides. By analyzing the sorting of such chimeric subunits expressed by transfection in polarized kidney epithelial cell lines, we have found that both pump subunits can contain sorting information that plays a discrete role in orchestrating pump targeting.

SPECIFICITY OF SUBUNIT ASSEMBLY

Before we could examine the sorting properties of pump subunits and subunit chimeras, it was necessary to establish their requirements for holoenzyme assembly and surface delivery. Several investigators demonstrated that the newly synthesized α subunits of both the H,K and the Na,K-ATPases must assemble with a β subunit before they are released from their site of synthesis in the rough endoplasmic reticulum (RER) and allowed to proceed to the plasmalemma.[12] We expressed subunit polypeptides individually or in pairs by transient transfection in COS cells and examined the distributions of the exogenous proteins by immunofluorescence microscopy.[13] As would be expected, both α subunits were retained in the endoplasmic reticulum compartment when they were expressed alone. Coexpression of each α subunit with its appropriate β subunit resulted in the delivery of both subunits to the cell surface. Coexpression of the Na,K-ATPase α subunit with the β subunit of the H,K-ATPase did not permit exit from the endoplasmic reticulum. Similarly, no surface delivery of the H,K-ATPase α subunit was detected when this protein was coexpressed with the sodium pump's β polypeptide. The surface delivery of the Na,K-ATPase β subunit was similarly dependent upon interaction with its own α subunit. By contrast, the H,K-ATPase β subunit could reach the cell surface in the absence of any apparent α-subunit associations.

These results suggest that the pump subunits manifest fairly strict selectivity in choosing their assembly partners. It must be noted that our findings are somewhat different from those obtained in studies in which pump expression was examined in *Xenopus* oocytes,[14] *Sf* 9 insect cells,[15] and yeast.[16] In these experiments it was determined that both α subunits could form functional complexes with either β subunit. The relative lack of selectivity observed under these conditions is likely attributable to the comparatively low temperature at which each of these cell types is maintained. Whereas yeast, oocytes, and Sf9 cells are cultured at 20°C, COS cells are of mammalian origin and are grown at 37°C. Correct folding and subunit assembly at this higher temperature almost certainly demands stronger and more specific protein-protein interactions. Consequently, subunit assembly requirements are probably more stringent in mammalian cells than they are in the other expression systems. Results from the Fambrough laboratory indicate that assembly of the Na,K-ATPase α subunit and the H,K-ATPase β subunit, assayed by coimmunoprecipitation, can occur in transfected mammalian cells.[17] However, this assembly appears insufficient to support surface delivery of the complex. It is also extremely interesting that the assembly independence of the H,K-ATPase β subunit is observed in COS cells as well as in nonmammalian expression systems. Apparently, this protein is not recognized or retained by the endoplasmic reticulum's quality control machinery, and it is thus able to exit this compartment without participating in subunit oligomerization. Elegant studies from the laboratory of Kaethi Geering have identified portions of the H,K-ATPase polypeptide that endow it with this unique property.[18] Although the physiologic utility of unassembled H,K-ATPase remains unclear, it has been possible to exploit this behavior to examine β-subunit sorting signals in isolation from those present in the α subunits, as will be discussed.

The structures of the α-subunit chimeras that we have generated for our sorting studies are depicted in FIGURE 1. To determine their β-subunit assembly preferences, several chimeras were coexpressed in COS cells with the H,K or Na,K-ATPase β polypeptide and subjected to the same surface delivery assay just described.[13] As summarized in FIGURE 1, the COOH-terminal half of the α subunit appears to dictate specificity in β-subunit interactions. Those chimeras in which the COOH-terminal ~500 amino acids are derived from the Na,K-ATPase α subunit complex with the Na,K-ATPase β subunit, whereas the COOH-terminal ~500 amino acids of the H,K-ATPase α-subunit favor association with the H,K-ATPase β protein. These observations are consistent with a much more extensive analysis conducted in Fambrough's laboratory which has localized the α-subunit residues that interact with the β subunit to the ectodomain loop that connects transmembrane helices TM7 and TM8.[17,19]

α-SUBUNIT SORTING SIGNALS

The chimeric α subunits and their appropriate β subunits were coexpressed by stable transfection in the polarized pig kidney epithelial cell line LLC-PK$_1$. The distributions of the chimeric pump complexes were determined by immunofluorescence confocal microscopy and are summarized in FIGURE 1. We first determined that the NH$_2$-terminal ~500 amino acids of the H,K-ATPase possess dominant apical sorting information, whereas the corresponding portion of the Na,K-ATPase α subunit encodes a dominant basolateral sorting signal.[11,20] On further subdividing this region, it became clear that those chimeras that incorporated the H,K-ATPase TM4 were accumulated at the apical surface, whereas those in which TM4 derived from the Na,K-ATPase sequence were restricted to the basolateral surface. It would appear, therefore, that residues encompassed within the fourth predicted transmembrane domains of the H,K and Na,K-ATPase α subunits are necessary and sufficient to determine these pumps' subcellular localizations.[21]

A comparison of the sequences of the H,K and Na,K-ATPase TM4 is presented in FIGURE 2. There are only eight nonidentical amino acids. It is interesting that seven of these are clustered in the region of TM4 which might be expected to traverse the outer leaflet of the membrane's lipid bilayer. Previous site-directed mutagenesis studies

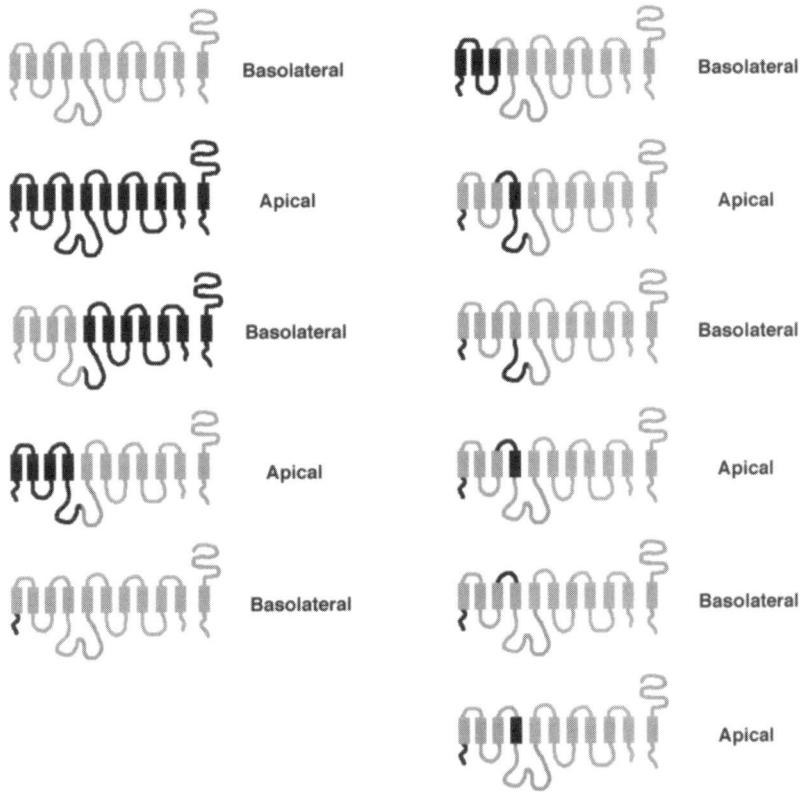

FIGURE 1. Assembly and sorting of Na,K-ATPase/H,K-ATPase α-subunit chimeras. Chimeric α subunits were created by standard molecular techniques and expressed by transfection in COS and LLC-PK$_1$ cells. *Solid bars* represent predicted transmembrane domains. Portions of the molecules colored in *black* represent sequence derived from the H,K-ATPase subunits, whereas *light gray* corresponds to the Na,K-ATPase sequence. As can be seen, all of the chimeras in which the COOH-terminal transmembrane helices are derived from the Na,K-ATPase assemble preferentially with the Na,K-ATPase β subunit. Chimeras that include the fourth transmembrane segment of the H,K-ATPase α subunit accumulate at the apical membrane.

demonstrate that at least one of these residues, Asn 329, may participate in interactions with transported cations.[22] Presumably, therefore, Asn 329 is directed towards the cation translocation pathway and away from the hydrophobic core of the membrane. Examination of a helical wheel representation of TM4 (not shown) suggests that two other residues, Leu 322 and Thr 343, share the orientation of Asn 329. It is likely, therefore, that the information required for sorting is encompassed in the five remaining residues which are predicted to face the membrane interior.

Before discussing the mechanism through which a signal embedded within a transmembrane domain might mediate pump sorting, it is necessary to examine the pathways that pump polypeptides pursue en route to the plasmalemma. Newly synthesized Na,K-ATPase subunit polypeptides are retained in the endoplasmic reticulum until they assemble to form holoenzyme, which is allowed to transit to the cell surface.[12] In epithelial cells, the Na,K-ATPase appears to be tightly bound to the cytoskeleton underlying the basolateral plasmalemma.[23–25] This interaction has been suggested to play an important role in establishing or stabilizing the basolateral localization of the Na,K-ATPase. Most newly synthesized basolateral proteins travel directly from the Golgi complex to the basolateral domain.[3] The subcellular site of sodium pump sorting, however, was the subject of recent controversy. Previous work from our laboratory used a photoactivatable derivative of ouabain (NAB-ouabain) to examine the surface arrival of Na,K-ATPase in MDCK (strain II) cells.[26] We found that this probe only detected newly synthesized enzyme at the basolateral surface, and hence we concluded that the sodium pump is delivered directly to its site of ultimate functional residence. Hammerton et al.[27] used a surface biotinylation technique to

```
Na,K-ATPase    A V I F L I G I I V A N V P E G L L A T V T V C L T L T A
H,K-ATPase     A M V F F M A I V V A Y V P E G L L A T V T V C L S L T A
```

FIGURE 2. Comparison of the fourth transmembrane segments of the Na,K and H,K-ATPase α subunits. Of the 28 amino acids that are predicted to comprise the fourth transmembrane helix, 8 are not conserved between the two pumps (*bold face*). Three of these (Na,K-ATPase Asn 329, Leu 322, and Thr 343) are predicted to face the hydrophilic cation translocation pathway, while the remaining 5 may face the membrane's lipid core.

monitor the initial appearance of newly synthesized Na,K-ATPase at the two cell surface domains. Employing a different clone of MDCK cells (clone J) from that used in the NAB-ouabain studies, these investigators found that the sodium pump seems to be delivered in roughly equal proportions to both sectors of the plasmalemma. They concluded that the pump's steady-state distribution is due to the relative stabilities of the enzyme inserted at the two surfaces (which their measurements suggested are: basolateral $t_{1/2} > 30$ hours and apical $t_{1/2} < 1$ hour). This difference was attributed to cytoskeletal interactions available at the basolateral but not the apical domain. On the basis of these findings Hammerton et al.[27] proposed that the machinery that interprets the sorting signals associated with the Na,K-ATPase resides not in the Golgi complex but in the cytoskeletal components apposed to the basolateral surface.

We have repeated these sodium pump targeting studies using the MDCK cell strain (strain II) employed in the NAB-ouabain studies under conditions that we have developed to optimize the efficiency of surface biotinylation. We found that the Na,K-ATPase was delivered directly and exclusively to the basolateral plasmalemma.[28] Mays et al.[29] examined Na,K-ATPase sorting in the two different strains of MDCK cells just discussed. Using the surface biotinylation technique, their results with strain II MDCK cells corroborated our own. Newly synthesized Na,K-ATPase was vectorially sorted to the basolateral surface. By contrast, sodium pump appeared once again to be delivered to both plasmalemmal domains of clone J cells. Further analysis revealed that clone J cells exhibit other anomalous sorting behaviors which may relate to their unique solution to the problem of sodium pump targeting.

Most strains of MDCK cell lines, including the widely used strain II, are able to develop polarized distributions of membrane lipids. Essentially all of the plasmalemmal glycosphingolipids (GSLs) are restricted to the outer leaflet of the apical plasma membrane in MDCK cells.[30] In clone J, however, this lipid polarity was not detected.[29] Mays *et al* proposed that the absence of GSL polarity may correlate with or cause the absence of direct sodium pump targeting observed in these cells.[29] To test this hypothesis, they treated strain II cells with fumonisin, a compound that interferes with GSL synthesis. Cells treated with this drug delivered newly synthesized Na,K-ATPase equally to both cell surface domains. The effect of the compound was reversible, and its removal restored vectorial basolateral sorting. These results are consistent with the possibility that GSL synthesis and sorting play a role in the initial targeting of the Na,K-ATPase.

Identification of sorting information within TM4 suggests a possible connection between the lipid and ion pump sorting processes. Recent evidence suggests that newly synthesized GSLs co-cluster during their passage through the Golgi complex to form "rafts" of detergent-insoluble lipid.[31] Furthermore, several apically directed membrane proteins appear to associate with these rafts and become detergent-inextractable en route to the cell surface. This behavior has been documented for apically sorted glycosylphosphatidyl inositol (GPI)-linked proteins[31] as well as for the influenza HA protein[32] and several intestinal brush border sugar hydrolases.[33] Analysis of the sequences that specify the apical sorting of the influenza neuraminidase protein demonstrate that its transmembrane domain is responsible for both apical targeting as well as the acquisition of detergent inextractability.[34]

It has been suggested that at least some membrane proteins destined for sorting to the apical surface are segregated from basolaterally directed polypeptides based on their ability to partition into the GSL-rich rafts as they transit the final stacks of the Golgi complex.[35] According to this model, the chemical properties of transmembrane domain residues in contact with the bilayer's outer leaflet determine a protein's capacity to enter the GSL-rich membrane subdomains which may, in turn, ultimately give rise to apically directed carrier vesicles. It is tempting to hypothesize that the TM4 sorting signals function to permit or deny their parent ion pumps access to the GSL rafts. The sodium pump's basolateral targeting would thus be attributable to its TM4-mediated exclusion from the Golgi's GSL-enriched apical staging area. Apical sorting of the H,K-ATPase would reflect the solubility of this pump's TM4 in the GSL rafts. This scheme would account for the affect of fumonisin on the sodium pump's initial targeting. Although this proposed mechanism is certainly attractive, preliminary experiments indicate that apically sorted chimeras do not exhibit the detergent inextractability that might be expected of proteins incorporated within GSL-rich domains.[21] It is possible that the detergent extraction assay is too insensitive to detect the partioning properties of a single membrane-spanning sequence when it is presented in the context of a protein with 10 transmembrane helices. Further experiments will be required to ascertain whether TM4's sorting message is interpreted through interactions with lipids or via associations with other, as yet unidentified, transmembrane polypeptides.

β-SUBUNIT SORTING SIGNALS

When expressed alone in LLC-PK$_1$ cells, the H,K-ATPase β subunit accumulates at the apical membrane as well as in subapical endosomes.[11] Examination of the sequence of the H,K-ATPase β subunit suggests a possible explanation for the protein's presence in the endosomal compartment. The primary structure of this protein's cytoplasmic NH$_2$-terminal tail includes the sequence Phe-Arg-Gln-Tyr (FRQY), which is highly homologous to the motif responsible for the rapid internalization of the transferrin receptor,[36] Tyr-Thr-Arg-Phe (YTRF) (FIG. 3). Mutagenesis studies performed on the transferrin receptor reveal that sequences of the form YXRF (where X can be any residue) permit this protein to be efficiently endocytosed. The YXRF sequence functions with ~30% effectiveness when presented in reverse orientation[37] (i.e., FRXY). Any substitution of the Tyr residue completely abolishes the motif's capacity to mediate rapid internalization. The sequence of the corresponding position of the Na,K-ATPase β subunit is Trp-Lys-Lys-Phe (WKKF).

Transferrin Receptor	NH$_2$- MMDQARSAFSNLFGGEPLS YTRF SLARQ---
Rat H,K-beta	NH$_2$- MAALQEKKSCSQRMAE FRQY CWN---
Rabbit H,K-beta	NH$_2$- MAALQEKKSCSQRMEE FRHY CWN---
Human H,K-beta	NH$_2$- MAALQEKKTCGQRMEE FQRY CWN---
Rat Na,K-beta	NH$_2$- MARGKAKEEGS WKKF IWN---
Chicken Na,K-beta	NH$_2$- MARGKANDGDGN WKKF IWN---
Human Na,K-beta	NH$_2$- MARGKAKEEGS WKKF IWN---

FIGURE 3. Comparison of the cytoplasmic NH$_2$-terminal tail of the transferrin receptor with those of the H,K and Na,K-ATPase β subunits. The transferrin receptor undergoes rapid endocytosis and recycling. The boxed sequence (YTRF) is necessary and sufficient to initiate this internalization.[36,37] Inversion of this sequence does not prevent it from mediating endocytosis.[37] All of the H,K-ATPase β subunits cloned to date possess a sequence that is highly homologous to the transferrin receptor's internalization motif, whereas this sequence is absent from Na,K-ATPase β subunits.

It is interesting to consider the physiologic significance of an endocytosis signal associated with the H,K but not the Na,K-ATPase β subunit. The H,K-ATPase is not a static component of the apical plasmalemma. In resting gastric parietal cells, the H,K-ATPase is associated with an intracellular population of membranous vesicles known as tubulovesicular elements[38] (TVEs). Secretagogue stimulation of acid production results in the fusion of these vesicles with the apical surface, leading to the formation of deeply invaginated secretory canaliculi as well as to the insertion of the H,K pump into the apical membrane. The cessation of acid secretion is associated with the internalization of the H,K-ATPase and its attendant membrane to regenerate the TVE compartment. The presence in the H,K-ATPase β protein of a Tyr-based motif resembling the transferrin receptor's endocytosis signal suggests the possibility that this sequence is responsible for the regulated internalization of the H,K pump.

If this model is correct, H,K-ATPase holoenzymes which incorporate β subunits whose critical tyrosine residues (Y20) have been mutagenized to alanine (Y20A) should not be substrates for endocytosis and thus should remain continuously at the parietal cell apical membrane. Consequently, expression of such mutant β subunits in parietal cells might produce a genetically dominant phenotype associated with the extended residence of H,K-ATPase at the cell surface. This, in turn, might lead to hypersecretion of gastric acid which persists after the suspension of secretagogue stimulation. To test this proposal, we prepared transgenic mice which express H,K-ATPase β subunits mutagenized in this manner.[39]

To assess acid secretion, gastric glands were hand dissected from the stomachs of normal mice as well as from the stomachs of mice transgenic for either the wild-type or the Y20A mutagenized H,K-ATPase β subunit. Confocal microscopy was used to monitor the accumulation of the pH-sensitive dye acridine orange before and after stimulation with the histamine.[40] We found that the secretagogue induced a marked increase in acid production in glands from nontransgenic and wild-type transgenic animals. By contrast, glands from animals that expressed Y20A produced acid at the maximally stimulated level in the absence of histamine. Application of the secretagogue did not induce further elevation of acid output. Ultrastructural examination of the gastric epithelium revealed that parietal cells in the Y20A-expressing mice exhibited the amplified apical membrane and secretory canaliculi associated with acid production despite the fact that the animals had been treated with an H_2-receptor blocking antisecretory drug prior to sacrifice. This pattern was observed with significantly lower frequency in parietal cells from nontransgenic mice. It would appear, therefore, that the tyrosine-based motif present in the H,K-ATPase β-subunit's cytoplasmic tail plays a critical role in the regulation of gastric acid secretion. Disruption of this signal leads to the constitutive presence of the H,K pump at the apical surface and consequently produces continuous gastric acid secretion.

Not surprisingly, mice that express Y20A exhibit the pathologic changes that might be expected to result from the hypersecretion of gastric acid. All of the mice examined to date have had gastric lesions that ranged in severity from mild erosions to severe ulcer disease. The stomachs of the nontransgenic and the wild-type transgenic mice were free of pathology. Mutation of Y20 is thus sufficient to produce a genetically dominant condition characterized by elevated basal acid secretion and peptic ulcer disease. It will be interesting to determine if the pathogenesis of a subset of human peptic ulcer disease can be referenced to a similar genetic peturbation.

A large body of research suggests that tyrosine-based motifs function in polarized sorting as well as in endocytosis.[41] Proteins bearing such signals are targeted to the basolateral surface when they are expressed in MDCK cells. It is perhaps surprising, therefore, that the H,K-ATPase β subunit behaves as an apical protein when it is expressed alone in LLC-PK$_1$ cells. We wondered if this phenomenon reflected differential interpretation of the tyrosine-based sorting motif by these two renal epithelial cell lines. To test this hypothesis, the H,K-ATPase β subunit was expressed alone in MDCK cells.[42] Immunocytochemical and biochemical localization techniques revealed that this protein accumulated at the basolateral surface in this cell type. When Y20A was expressed in MDCK cells, protein was detected in roughly equal quantities at both cell surfaces, suggesting that the tyrosine-based motif is responsible for the wild-type protein's basolateral localization. When expressed in LLC-PK$_1$ cells, Y20A was present only at the apical surface. It would appear, therefore, that LLC-PK$_1$ cells are incapable of interpreting the polarized sorting information encoded in the H,K-ATPase β protein's cytoplasmic tail. Furthermore, the apical distribution of this polypeptide in LLC-PK$_1$ cells hints at the possible presence of additional apical sorting information which may be embedded in some other domain of the H,K-ATPase β subunit.

CONCLUSIONS

In order to subserve their physiologic functions, the Na,K- and H,K-ATPase must be sorted to the correct subcellular locations. Multiple signals appear to mediate these sorting processes and to play important roles in regulating pump function. Both the α and the β subunits of the H,K-ATPase are endowed with information which specifies the pump's localization and determines its dynamic properties. Although the α subunit appears to direct the pump's apical localization, the β-subunit's signal is responsible for governing the duration of the pump's presence at the apical surface. Characterization of these signals at the molecular level has created tools that can be applied to studies of the mechanisms through which they determine and modulate ion pump distribution.

ACKNOWLEDGMENTS

Drs. G. Shull, G. Sachs, M. Reuben, D. Chow, and J. Forte generously provided reagents without which these studies would not have been possible. We wish to thank all of the past and present members of the Caplan laboratory for their insights and assistance.

REFERENCES

1. BERRIDGE, M. J. & J. L. OSCHMAN. 1972. Transporting Epithelia. Academic Press. New York.
2. CAPLAN, M. J. Am. J. Phys., in press.
3. CAPLAN, M. J. & K. S. MATLIN. 1989. *In* Functional Epithelial Cells in Culture. K. S. Matlin & J. D. Valentich, Eds.: 71–127. Alan R. Liss. New York.
4. HORISBERGER, J. D., V. LEMAS, J. P. KRAEHENBUHL & B. C. ROSSIER. 1991. Ann. Rev. Phys. **53:** 565–584.
5. RABON, E. C. & M. A. REUBEN. 1990. Ann. Rev. Phys. **52:** 321–344.
6. HIRST, B. H. & J. G. FORTE. 1985. Biochem. J. **231:** 641–649.
7. URUSHIDANI, T. & J. G. FORTE. 1987. Am. J. Physiol. **252:** G458–G465.
8. SMOLKA, A., H. F. HELANDER & G. SACHS. 1983. Am. J. Physiol. **245:** G589–G596.
9. SMOLKA, A. & W. A. WEINSTEIN. 1986. Gastroenterology **90:** 532–539.
10. SOROKA, C. J., C. S. CHEW, D. K. HANZEL, A. SMOLKA, I. M. MODLIN & J. M. GOLDENRING. 1993. Eur. J. Cell Biol. **60:** 76–87.
11. GOTTARDI, C. J. & M. J. CAPLAN. 1993. J. Cell Biol. **121:** 283–293.
12. GEERING, K., I. THEULAZ, F. VERRAY, M. T. HAUPTLE & B. C. ROSSIER. 1989. Am. J. Physiol. **257:** C851–C858.
13. GOTTARDI, C. J. & M. J. CAPLAN. 1993. J. Biol. Chem. **268:** 14342–14347.
14. HORISBERGER, J. D., P. JAUNIN, M. A. REUBEN, L. S. LASATER, D. C. CHOW, J. G. FORTE, G. SACHS, B. C. ROSSIER & K. GEERING. 1991. J. Biol. Chem. **266:** 19131–19134.
15. DE TOMASO, A. W., G. BLANCO & R. W. MERCER. 1994. J. Cell Biol. **127:** 55–69.
16. EAKLE, K. A., K. S. KIM, M. A. KABALIN & R. A. FARLEY. 1992. Proc. Natl. Acad. Sci. **89:** 2834–2838.
17. LEMAS, M. V., H. Y. YU, K. TAKEYASU, B. KONE & D. M. FAMBROUGH. 1994. J. Biol. Chem. **269:** 18651–18655.
18. JAUNIN, P., F. JAISSER, A. T. BEGGAH, K. TAKEYASU, P. MANGEAT, B. C. ROSSIER, J. D. HORISBERGER & K. GEERING. 1993. J. Cell Biol. **123:** 1751–1759.
19. LEMAS, M. V., M. HAMRICK, K. TAKEYASU & D. M. FAMBROUGH. 1994. J. Biol. Chem. **269:** 8255–8259.

20. MUTH, T. R., C. J. GOTTARDI, D. L. ROUSH & M. J. CAPLAN. Submitted.
21. DUNBAR, L. A. & M. J. CAPLAN. Submitted.
22. VILSEN, B. 1995. FEBS Lett. **363:** 179–183.
23. NELSON, W. J. & P. J. VESHNOCK. 1987. Nature **328:** 533–536.
24. NELSON, W. J. & R. W. HAMMERTON. 1989. J. Cell Biol. **108:** 893–902.
25. MORROW, J. S., C. D. CIANCI, T. ARDITO, A. S. MANN & M. KASHGARIAN. 1989. J. Cell Biol. **108:** 455–465.
26. CAPLAN, M. J., H. C. ANDERSON, G. E. PALADE & J. D. JAMIESON. 1986. Cell **46:** 623–631.
27. HAMMERTON, R. W., K. A. KRZEMINSKI, R. W. MAYS, T. A. RYAN, D. A. WOLLNER & W. J. NELSON. 1991. Science **254:** 847–850.
28. GOTTARDI, C. J. & M. J. CAPLAN. 1993. Science **260:** 552–554.
29. MAYS, R. W., K. A. SIEMERS, B. A. FRITZ, A. L. LOWE, G. VAN MEER & W. J. NELSON. 1995. J. Cell Biol. **130:** 1105–1115.
30. SIMONS, K. & G. VAN MEER. 1988. Biochemistry **27:** 6197–6202.
31. BROWN, D. & J. K. 1992. Cell **68:** 533–544.
32. SKIBBENS, J. E., M. G. ROTH & K. S. MATLIN. 1989. J. Cell Biol. **108:** 821–832.
33. DANIELSEN, E. M. 1995. Biochemistry **34:** 1596–1605.
34. KUNDU, A., R. T. AVALOS, C. M. SANDERSON & D. P. NAYAK. 1996. J. Virol. **70:** 6508–6515.
35. SIMONS, K. & A. WANDINGER-NESS. 1990. Cell **62:** 207–210.
36. COLLAWN, J. F., M. STANGEL, L. A. KUHN, V. ESEKOGWU, S. JING, I. S. TROWBRIDGE & J. A. TAINER. 1990. Cell **63:** 1061–1072.
37. GIRONES, N., E. ALVAREZ, A. SETH, I. M. LIN, D. A. LATOUR & R. J. DAVIS. 1991. J. Biol. Chem. **266:** 19006–19012.
38. HERSEY, S. J. & G. SACHS. 1995. Phys. Rev. **75:** 155–189.
39. COURTOIS-COUTRY, N., D. ROUSH, V. RAJENDRAN, J. B. MCCARTHY, M. KASHGARIAN, J. GEIBEL & M. J. CAPLAN. Cell, in press.
40. BERGLINDH, T., D. R. DIBONA, S. ITO & G. SACHS. 1980. Am. J. Physiol. **238:** G165–G176.
41. THOMAS, D. C. & M. G. ROTH. 1994. J. Biol. Chem. **269:** 15732–15739.
42. ROUSH, D. L., C. J. GOTTARDI, H. Y. NAIM, M. G. ROTH & M. J. CAPLAN. Submitted.

Distinct Distribution of Different Na^+ Pump α Subunit Isoforms in Plasmalemma

Physiological Implications[a]

MAGDALENA JUHASZOVA[b,d]
AND MORDECAI P. BLAUSTEIN[b,c]

[b]*Departments of Physiology and* [c]*Medicine
University of Maryland School of Medicine
Baltimore, Maryland 21201*

The Na^+ pump is a dimeric (α, β) transmembrane Na^+, K^+-ATPase complex that is expressed in virtually all animal cells. This transport system exports Na^+ and imports K^+ and consequently plays a critical role in Na^+ homeostasis by regulating the Na^+ electrochemical gradient across the plasma membrane. Inhibition of the Na^+ pump by ouabain increases the cytosolic Na^+ concentration ($[Na^+]_{cyt}$) and thereby decreases the plasma membrane Na^+ gradient. In most types of cells (including astrocytes and arterial myocytes), this enhances Ca^{2+} entry and reduces Ca^{2+} exit mediated by the plasma membrane Na^+/Ca^{2+} exchanger. The net Ca^{2+} gained by the cells is then efficiently sequestered in the sarcoplasmic/endoplasmic reticulum (S/ER) via the S/ER Ca^{2+} (SERCA) pump, so that the cytosolic free Ca^{2+} concentration ($[Ca^{2+}]_{cyt}$) rises only minimally.[1] For effective functional linkage of the Na^+ pump, the Na^+/Ca^{2+} exchanger, and the SERCA pump, we might expect these three transporters to be located in close proximity to one another.

The α subunit of the Na^+ pump contains the sites for catalytic activity (ATP hydrolysis and pump phosphorylation), for Na^+ and K^+ binding (and occlusion), and for ouabain binding.[2,3] At least three Na^+ pump α-subunit isoforms, α1, α2, and α3, have been identified in the tissues of mammals and birds.[4–6] In the rat, the α1 isoform has an unusually low affinity for ouabain (IC_{50} > 10,000 nM), whereas the α2 and α3 isoforms have high affinities for ouabain ($IC_{50}s$ ≈ 20–500 nM).[7] The Na^+ pump α2 and α3 isoforms in most other mammals also have high affinities for ouabain, but the α1 affinity for ouabain is usually moderately high, albeit somewhat lower than those of α2 and α3.[4,6]

Both low (i.e., α1) and high (i.e., α2 and/or α3) ouabain affinity isoforms are found in the plasma membrane of most rat cells. These isoforms are up- and downregulated independently, and they have different sensitivities to $[Na^+]_{cyt}$, $[Ca^{2+}]_{cyt}$,

---

[a]This work was supported by National Institutes of Health grants HL-32276 and NS-16106 (to MPB), by a Grant-in-Aid from the American Heart Association-Maryland Affiliate (to MJ), and by funds from the University of Maryland at Baltimore School of Medicine and Graduate School.

[d]Address for correspondence: Dr. Magdalena Juhaszova, Department of Physiology, University of Maryland School of Medicine, 655 West Baltimore Street, Baltimore, MD 21201 (tel: 410 706–2661; fax: 410 706–8341; e-mail: mjuhaszo@umabnet.ab.umd.edu).

and $[K^+]_o$ (the extracellular K^+ concentration) as well as to ouabain.[8,9] The different regulation of the α isoforms under various physiological or pathophysiological conditions[10–15] and their different kinetic properties imply that these isoforms subserve different physiological functions. Isoform-specific sequence regions that have been highly conserved among phylogenetically distant species[5,6] are likely to have an important functional role.

Recently, Moore and colleagues[16] reported that the Na^+ pumps in toad stomach smooth muscle (in which α-subunit isoforms have not been identified) were closely apposed to elements of the sarcoplasmic reticulum (SR). Na^+/Ca^{2+} exchanger molecules were localized to the same plasma membrane regions. In contrast, McDonough and colleagues[17] reported that α1 and α2 were distributed uniformly in, respectively, guinea pig and rat cardiac myocytes, and they concluded that there is *not* "a physiologically significant colocalization of Na^+ pump isoforms with Na^+/Ca^{2+} exchangers in heart."

In this study we examined the distribution of low ouabain affinity α1 and high ouabain affinity α2/α3 isoforms in the plasma membrane of rat astrocytes and arterial myocytes. These findings were compared to observations on the location of the Na^+/Ca^{2+} exchanger in the same preparations.[18,19] The results demonstrate that localization of the Na^+ pump α-subunit isoforms is organized and is isoform specific; the high ouabain affinity isoforms colocalize with the exchanger. We suggest that this isoform-specific localization may relate to the specific physiological functions of these different isoforms.[20]

MATERIALS AND METHODS

Isolation of Arterial Smooth Muscle Cells

Rat mesenteric artery or aorta was dissected in Ca^{2+}-free physiologic salt solution to prevent contraction. After removing the adipose tissue and adventitia, the arteries were incubated with collagenase and washed in physiologic salt solution. Arterial segments were agitated in a small volume of physiologic salt solution to separate individual myocytes, which were plated on 25-mm diameter coverslips coated with Cell-Tak adhesive (Collaborative Biomedical Products, Bedford, Massachusetts). The coverslips were immobilized for 45 minutes to permit the cells to settle and stick. The myocytes were then fixed in 2% formaldehyde, permeabilized with 0.5% Brij 58, and labeled with antibodies or treated with the lipophilic, cationic dicarbocyanine fluorochrome $DiOC_6(3)$ or "DiOC" (Molecular Probes, Eugene, Oregon), an agent that stains S/ER and mitochondria.[21] In the latter case, the cells were exposed to phosphate-buffered saline solution containing 0.5 µg/ml DiOC for 5 minutes; the extracellular DiOC was then washed away with fresh phosphate-buffered saline solution before examination in the microscope.

Primary Culture of Rat Arterial Smooth Muscle Cells and Astrocytes

Arterial myocytes were dissociated from the media of mesenteric arteries from adult rats. Cells were cultured as described.[22] Astrocytes were prepared from the brains of 1-day-old rats by mechanical disruption and filtration of the cell suspension without enzymatic digestion. Astrocytes were cultured using published methods.[23,24]

Immunofluorescence Microscopy

Primary cultured mesenteric artery cells and astrocytes were grown on 25-mm diameter coverslips and were studied 7–10 days after plating. Freshly isolated cells were labeled and studied immediately after immobilization. Coverslips were washed with phosphate-buffered saline solution and immunolabeled.[25] When monoclonal antibodies were used, the cells were fixed with periodate-lysine-paraformaldehyde (50 mM, 100 mM, 1% w/v) fixative.[26]

Na^+ pump $\alpha 1$ subunits were labeled with polyclonal antibodies PcSynA1[20,27] or one of two monoclonal antibodies (McK1,[28,29] C464-6B[30]). Monoclonal antibody McB2[28,29] was used to label $\alpha 2$ subunits. Polyclonal antibodies PcSynA3[20,27] (and TED[6]) were used to label $\alpha 3$ subunits. Rabbit polyclonal antibodies were raised against the purified Na^+/Ca^{2+} exchanger (NCX) from dog heart sarcolemma.[31] (Antibodies were generously provided by R. W. Mercer and M. J. Caplan [PcSynA1,

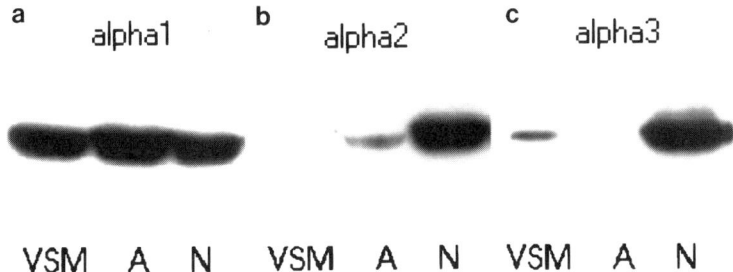

FIGURE 1. Western blot analysis of some of the Na^+ pump α-subunit isoform-specific monoclonal and polyclonal antibodies used in this study. The three gels were probed with (*top to bottom*): monoclonal (C464-6B) antibodies raised against $\alpha 1$ (**a**), monoclonal (McB2) antibody raised against $\alpha 2$ (**b**), and polyclonal antibodies (TED) raised against $\alpha 3$ (**c**). The three lanes in each gel each contained 10 µg membrane protein per lane (100 µg for the VSM lane probed for $\alpha 3$); lanes (*left to right*): vascular smooth muscle (VSM), astrocytes (A), and neurons (N). VSM = vascular smooth muscle.

C464-6B, and PcSynA3], K. J. Sweadner [McK1 and McB2], T. A. Pressley, and G. E. Lindenmayer [NCX]). All of the antibodies were previously well characterized; for this study, the specificity was reconfirmed by immunoblotting (FIG. 1). The figure legends indicate which antibody was used for each experiment illustrated. Rabbit preimmune serum was used as a control for primary rabbit polyclonal antisera, and MOPC-21 (Sigma, St. Louis, Missouri) was used as a control for the primary mouse monoclonal antibodies. In all cases, fluorescence of the specific antibody labeling was (with the same gain and exposure time) at least 8–10 times more intense than that of the corresponding controls.

Cy3- or FITC-conjugated donkey anti-rabbit or anti-mouse IgG (Jackson ImmunoResearch Laboratories, West Grove, Pennsylvania) was used to visualize the primary antibodies. Specimens were examined with a Nikon Diaphot fluorescence microscope (40× or 100× N.A. 1.3 Nikon UV-Flour objectives or an Olympus 60× N.A. 1.4 objective).

The CELLscan (Scanalytics, Billerica, Massachusetts) high resolution fluorescence imaging system was employed to reassign out-of-focus fluorescence. Images from multiple focal planes were collected. CELLscan uses an experimentally generated point spread function with an Exhaustive Photon Reassignment (EPR) "deblurring" algorithm to vector out-of-focus fluorescence back to its point of origin in each specimen image plane.[32] Each individual two-dimensional image generated by this system corresponds to a single deconvolved and restored optical image (i.e., the in-focus fluorescence in that image plane). An Image-1/MetaMorph Imaging System (Universal Imaging Corp., West Chester, Pennsylvania) was employed for statistical analysis of the colocalization of the plasma membrane ion transporters with the SR.

RESULTS

Sodium Pump α-Subunit Isoform Composition in Arterial Myocytes and Astrocytes

Standard immunochemical techniques were used to determine which Na^+ pump α isoforms are expressed in cultured rat mesenteric artery myocytes and cortical astrocytes. Membrane fractions were prepared from these cells, and their membrane proteins were characterized by immunoblot analyses using a panel of α-subunit isoform-specific antibodies. Both cell types express the low ouabain affinity α1 isoform. This is the prevalent isoform in both arterial myocytes (FIG. 1)[35] and astrocytes (FIG. 1).[33,34] Both cell types also express a high ouabain affinity α-subunit isoform, but the mesenteric artery myocytes express a small amount of α3,[35] whereas the astrocytes express a moderate amount of α2 (FIG. 1). Thus, as with many other cell types, including neurons and cardiac and skeletal muscle myocytes,[2,4,11,34] arterial myocytes and astrocytes express the α1 isoform and at least one other isoform of the Na^+ pump α subunit. However, whether cells express α2 and/or α3 is apparently specific for each cell type and may vary with development, cell activity, hormonal stimulation, pathological influences, and many other factors.[10–15,36]

Expression of Ca^{2+} Transporters in Arterial Myocytes and Astrocytes

Previously, we demonstrated with Western blot analyses as well as with immunocytochemistry that arterial myocytes and astrocytes also express three Ca^{2+} transport proteins.[18,19,37] Two of these are plasma membrane transporters: the plasma membrane Ca^{2+} pump and the Na^+/Ca^{2+} exchanger; the third is the SERCA pump located in the S/ER. Because the Na^+/Ca^{2+} exchanger is functionally coupled to the Na^+ pump, it seemed especially important to determine if any correlation exists between the distribution of the exchanger and the distribution of specific isoforms of the Na^+ pump.

Localization of Na^+ Pump α Isoforms and Na^+/Ca^{2+} Exchanger in Freshly Isolated Arterial Myocytes

Antibody labeling was studied in freshly isolated mesenteric artery myocytes to demonstrate that the Na^+ pump α subunits and the Na^+/Ca^{2+} exchanger are present primarily in the plasma membrane and not in the membranes of intracellular organelles such as the S/ER. The high resolution CELLscan imaging system was employed for these experiments.

FIURE 2 shows data from experiments conducted on freshly isolated mesenteric artery myocytes. These cells express both low ouabain affinity $\alpha 1$ and high ouabain affinity $\alpha 3$ Na$^+$ pump α subunits. The mesenteric artery myocyte in FIGURE 2Aa was

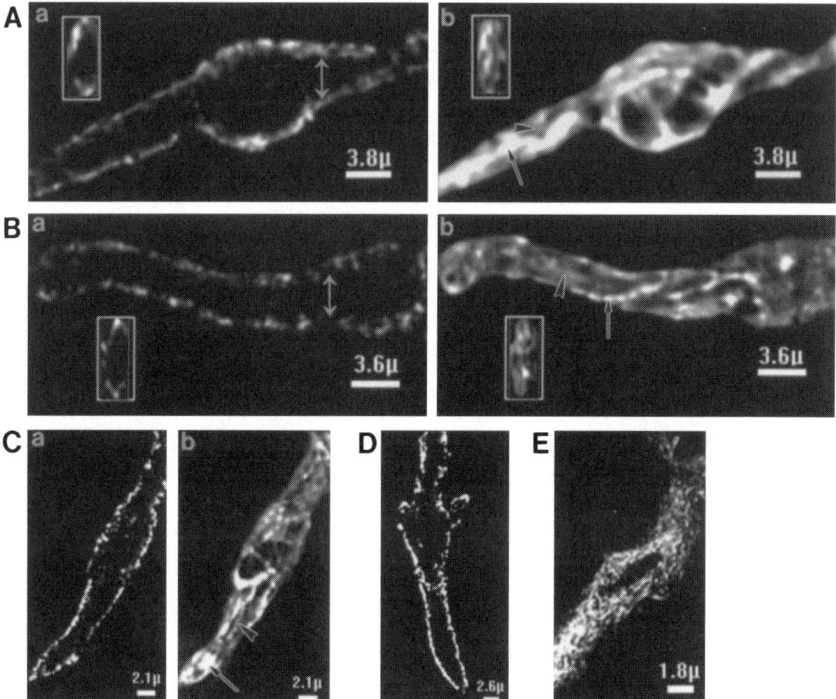

FIGURE 2. Distribution of Na$^+$ pump α-subunit isoforms, the plasma membrane and SR (SERCA) Ca^{2+} pumps, and the Na$^+$/Ca^{2+} exchanger in freshly isolated, fixed mesenteric artery myocytes. (**Aa**) Cell labeled with PcSynA1 directed against $\alpha 1$. (**Ba**) Cell labeled with PcSynA3 raised against $\alpha 3$. (**A,B insets**) Cell cross-sections, perpendicular to the plane of the coverslip, at the positions indicated by the *double arrows* in **Aa** and **Ba**. (**Ca**) Cell labeled with antibodies raised against the Na$^+$/Ca^{2+} exchanger. (**D**) Cell labeled with antibodies raised against the plasma membrane Ca^{2+} pump. After these images were captured, the cells in **A–C** were treated with DiOC (**Ab, Bb,** and **Cb**) to stain the SR (*arrowheads*) and mitochondria (*arrows*). (**E**) Cell labeled with anti-SERCA2b antibodies to identify the SR. The "single focal plane" images were restored with CELLscan.

immunolabeled with antibodies specific for the $\alpha 1$ isoform; note that antibody fluorescence is limited to the cell surface. The same cell was later stained with DiOC (FIG. 2Ab) to label the S/ER and mitochondria.[21] This dye reveals the extensive SR network; indeed, brightly fluorescent mitochondria lie on the SR (FIG. 2),[19] thereby helping us to visualize the organization of the more dimly fluorescent SR.

Another freshly isolated mesenteric artery myocyte was labeled with antibodies raised against α3 (FIG. 2Ba). The α3 immunofluorescence, like α1 immunolabeling (FIG. 2Aa), is confined to the cell surface (FIG. 2Ba). In cell cross-sections, however, the distribution of both α subunit labels appears to be more complex (FIG. 2Aa and 2Ba insets), because the surfaces of these freshly isolated cells are invaginated with folds, caveolae, and the like (see refs. 38 and 39). Nevertheless, the DiOC staining pattern in these cells is very different from that of either Na^+ pump antibody: DiOC stains structures that are distributed throughout the cytoplasm but excluded from the nuclei (FIGS. 2Ab, 2Bb, and 2Cb).

The Na^+/Ca^{2+} exchanger immunoreactive sites are also confined to the surfaces of the freshly isolated mesenteric artery myocytes (FIG. 2Ca), as are the plasma membrane Ca^{2+} pump immunoreactive sites (FIG. 2D). Indeed, none of the antibodies raised against these plasma membrane transport proteins labels the SR in these myocytes, even though the cells are fixed and permeabilized. This labeling contrasts with the DiOC staining (FIGS. 2Ab, 2Bb, and 2Cb) and with labeling by antibodies raised against the SR membrane Ca^{2+} transporter, the SERCA pump (FIG. 2E). In the latter case, immunofluorescent labeling is observed only in the cell interior, as is the DiOC staining of SR and mitochondria.

Localization of the Na^+ Pump Isoforms, the Na^+/Ca^{2+} Exchanger, and the Plasma Membrane Ca^{2+} Pump in Cultured Mesenteric Artery Myocytes

The preceding experiments on freshly isolated mesenteric artery myocytes demonstrate that the Na^+ pump, plasma membrane Ca^{2+} pump, and the Na^+/Ca^{2+} exchanger are all located on the cell surface, whereas the SERCA pump is located in the cell interior. It is noteworthy that the plasma membrane transporters are not detected in the Golgi/reticulum, where these proteins are synthesized and from whence they are transported to the plasma membrane.

Detailed analysis of the distribution of the various plasma membrane transporters in freshly isolated myocytes is hampered by the fact that the surfaces of these cells contain many folds, invaginations, and caveolae (FIG. 2Aa and 2Ba insets and see above). To circumvent problems in determining the precise antibody localization that arises from this surface complexity, we employed cultured cells that are flat and that have relatively smooth surfaces. Localization and distribution of the Na^+ pump α-subunit isoforms and the Ca^{2+} transporters were therefore studied in primary cultured rat mesenteric artery myocytes and astrocytes.

FIGURE 3 illustrates the distribution of the Na^+ pump α-subunit isoforms in cultured mesenteric artery myocytes. FIGURE 3Aa shows a myocyte labeled with antibodies raised against α1: the label is homogeneously distributed over the entire surface of the cell (3Aa). The same cell was later stained with DiOC: the reticular labeling of the SR by DiOC (FIG. 3Ab) is distinctly different from the α1 immunofluorescent pattern. FIGURE 3Ba shows mesenteric artery myocyte labeled with antibodies raised against the Na^+ pump α3 isoform. In this case, the pattern of immunofluorescent labeling is distinctly reticular and is clearly comparable to the underlying SR staining pattern obtained with DiOC (FIG. 3Bb). A very similar reticular labeling pattern was also observed when mesenteric artery myocytes were immunolabeled with antibodies raised against the Na^+/Ca^{2+} exchanger (FIG. 3Ca). The immunofluorescent patterns obtained with both the α3 and the Na^+/Ca^{2+} exchanger antibodies resemble strings of beads that appear to be organized into reticular networks. This reticular distribution is remarkably similar to the pattern observed with DiOC staining of the S/ER structures in these cells (FIG. 3Cb).

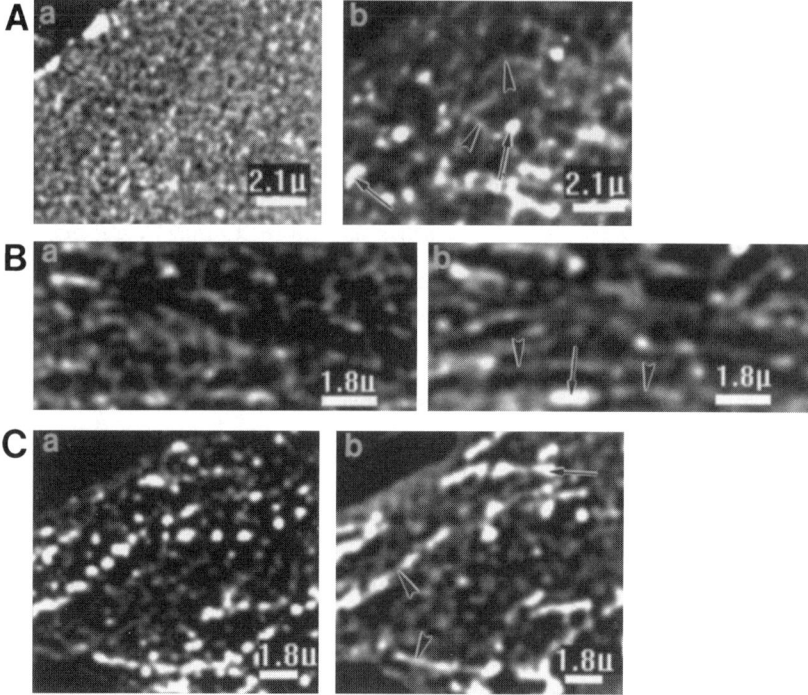

FIGURE 3. Localization of the Na^+ pump $\alpha 1$ and $\alpha 3$ isoforms and the Na^+/Ca^{2+} exchanger in primary cultured mesenteric artery myocytes. FIGURES 3–5 show high magnification, single focal plane images of portions of cells (astrocytes or myocytes) that were restored with CELLscan. (**Aa**) Cell was cross-reacted with polyclonal PcSynA1 antibodies raised against $\alpha 1$. (**Ba**) Cell was cross-reacted with polyclonal PcSynA3 antibodies raised against $\alpha 3$. (**Ca**) Cell was labeled with affinity-purified anti-Na^+/Ca^{2+} exchanger antibodies. All cells were later treated with DiOC to stain the SR (*arrowheads*) and mitochondria (*arrows*) (**Ab, Bb,** and **Cb**).

FIGURE 4 shows that antibodies raised against the plasma membrane Ca^{2+} pump (FIG. 4Aa) diffusely label the entire surface of mesenteric artery myocytes. This labeling is clearly different from the reticular labeling pattern observed with antibodies raised against the SERCA pump (FIG. 4Ba) and with DiOC staining (FIG. 4Ab and 4Bb). The SERCA antibodies (FIG. 4Ba) and DiOC (FIG. 4Bb) both label the same reticular structures (i.e., the SR) in cultured mesenteric artery myocytes. Thus, DiOC is a convenient and appropriate S/ER label.[21]

Localization of the Na^+ Pump Isoforms and the Na^+/Ca^{2+} Exchanger in Cultured Astrocytes

A similar comparison of Na^+ pump α-subunit isoform and Na^+/Ca^{2+} exchanger distribution was carried out on primary cultured rat cortical astrocytes (FIG. 5). As in the mesenteric artery cells, immunolabeling of $\alpha 1$ is distributed uniformly over the cell surface; labeling is intense at the cell edges, where the plasma membrane folds over (FIG. 5Aa). This pattern is distinctly different from the DiOC staining of the S/ER (and mitochondria) in the same cells (FIG. 5Ab). By contrast, immunofluorescent labeling of $\alpha 2$ in astrocytes is distributed in a reticular pattern (FIG. 5Ba) that corresponds to the reticular pattern of the underlying S/ER structures that are labeled with antibodies raised against the SERCA pump (FIG. 5Bb) or stained with DiOC (FIG. 5Bc).

A similar reticular pattern of immunofluorescent labeling is also observed in astrocytes labeled with anti-Na^+/Ca^{2+} exchanger antibodies (FIG. 5Ca). As in mesenteric artery myocytes (FIG. 3C), this resembles the DiOC staining pattern of the underlying S/ER (FIG. 5Cb).

DISCUSSION AND CONCLUSIONS

Immunocytochemical labeling of freshly isolated rat mesenteric artery myocytes provides evidence that the epitopes recognized by antibodies raised against the Na^+ pump $\alpha 1$- and $\alpha 3$-subunit isoforms (they do not express $\alpha 2$), the Na^+/Ca^{2+} exchanger, and the plasma membrane ATP-driven Ca^{2+} pump are all located primarily in the plasma membrane. These antibodies did not label the underlying SR that is stained with DiOC or labeled with antibodies raised against the SERCA pump. Because the cells were fixed and permeabilized, it seems likely that the SR is accessible to the plasma membrane transporter antibodies. Thus, the absence of label in the SR suggests that only the "mature" transporters in the plasma membrane are labeled.

Primary cultured rat mesenteric artery myocytes and astrocytes both express the

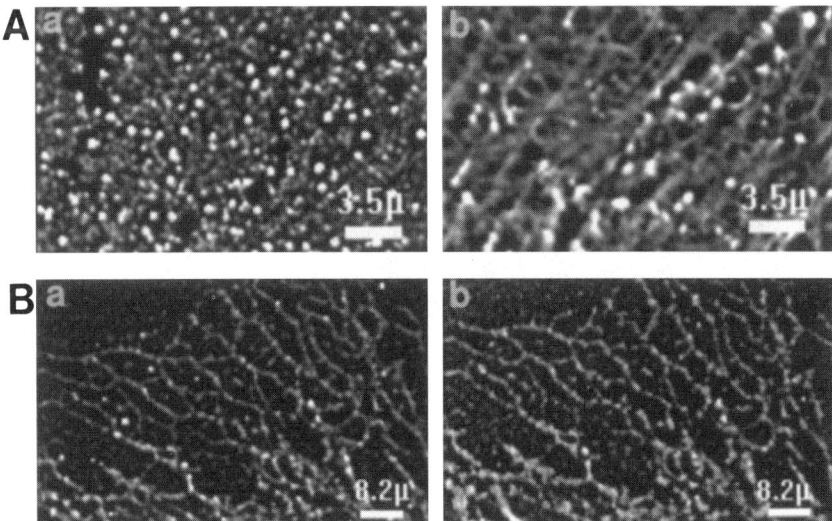

FIGURE 4. Immunolabeling of the plasmalemmal and SR Ca^{2+}-ATPase. High magnification, restored images of cultured mesenteric artery myocytes. The cell in **Aa** was immunolabeled with antibodies raised against plasmalemmal Ca^2-ATPase; the myocyte in **Ba** was labeled with anti-SERCA2b antibodies. Both cells were subsequently stained with DiOC to visualize the SR and mitochondria (**Ab** and **Bb**).

Na$^+$ pump α1 isoform, but the astrocytes also express α2 (but not α3), whereas myocytes express α3. Why cells should express more than one Na$^+$ pump α-subunit isoform, with their different kinetics and different sensitivities to ouabain, has not previously been explained. Our observations provide novel clues that suggest a

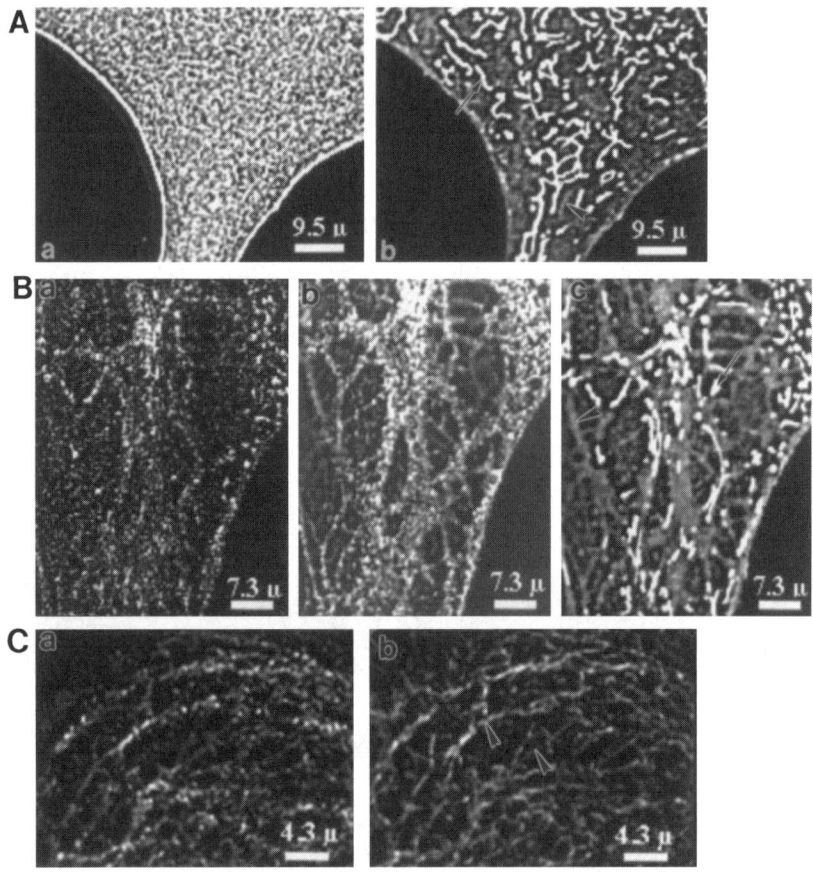

FIGURE 5. Localization of Na$^+$ pump α1 and α2 isoforms in primary cultured astrocytes. High magnification, restored images of an astrocyte (**Aa**) labeled with anti-α1 antibodies (PcSynA1), and the other (**Ba**) with anti-α2 antibodies (McB2). Both cells were later treated with DiOC to stain endoplasmic reticulum (*arrowheads*) and mitochondria (*arrows*) (**Ab and Bc**). The cell in **B** was also cross-reacted with anti-SERCA2b antibodies (**Bb**). All three labels in **B** show similar reticular distributions. (**C**) Another astrocyte was immunolabeled with affinity-purified anti-Na$^+$-Ca^{2+} exchanger antibody (**Ca**) and colabeled with DiOC (**Cb**).

solution to this dilemma. We found that the different Na$^+$ pump α-subunit isoforms are variously distributed in the cells. α1 is ubiquitously distributed in the plasma membrane of both astrocytes and arterial myocytes. By contrast, the high ouabain affinity isoforms, α2 in astrocytes and α3 in myocytes, are distributed in the plasma membrane in reticular patterns that parallel the underlying S/ER; data from the freshly isolated myocytes indicate that these α subunits are located in the plasma membrane and not the S/ER. Moreover, in physiologic experiments on intact, cultured myocytes and astrocytes, the high ouabain affinity isoforms respond to low dose (<1 μM) ouabain,[36,40] which indicates that these transport proteins are present in the plasma membrane. These different distributions therefore imply that the α1 and the α2/α3 isoforms have different functions.

Previously, we reported that the plasma membrane Na^+/Ca^{2+} exchanger is also distributed in reticular patterns that overlie S/ER in both astrocytes (refs. 19 and 37 and FIG. 5C) and arterial myocytes (refs. 18 and 19 and FIG. 3C). By contrast, the plasma membrane ATP-driven Ca^{2+} pump is ubiquitously distributed in the plasma membrane (ref. 19 and FIG. 4A).

The close proximity of the Na^+ pump α2/α3-subunit isoforms and Na^+/Ca^{2+} exchanger in the plasma membrane and the underlying "junctional" S/ER may imply that these elements are functionally related. Indeed, a recent report indicates that expression of the Na^+/Ca^{2+} exchanger and α2 (but not α1) are reciprocally regulated in rat heart under a variety of conditions.[14] We have suggested that the Na^+/Ca^{2+} exchanger helps to modulate the Ca^{2+} stores in the S/ER.[1,41] The current results now provide structural information as to why the Na^+ pump α2/α3 isoforms can be expected to act in concert with the Na^+/Ca^{2+} exchanger. Together, these transporters apparently control subplasmalemmal $[Na^+]_{cyt}$ and $[Ca^{2+}]_{cyt}$ in restricted microdomains between the plasma membrane and the S/ER that is closely apposed to the plasma membrane (i.e., the "junctional S/ER"). Indeed, this structurally distinct region of plasma membrane, S/ER, and intervening restricted subplasmalemmal cytosol acts as a functional unit which we shall refer to as the "plasmerosome." The plasma membrane Na^+ pump α2/α3 isoforms, the Na^+/Ca^{2+} exchanger, and S/ER Ca^{2+} pumps are all situated in the plasmerosome, where they can function in a coordinated manner. In this way, the Na^+ pump α2/α3 isoforms indirectly help to regulate the S/ER Ca^{2+} stores and thus influence the numerous processes that depend on mobilization of Ca^{2+} from these stores. In contrast, the α1 isoform of the Na^+ pump may help to regulate bulk $[Na^+]_{cyt}$. Indeed, there is evidence for spatial as well as functional organization of specific transporters and for subplasmalemmal microdomains of Na^+ and Ca^{2+}.[42–44] Furthermore, these ideas help to explain how low dose ouabain, which inhibits only a small fraction of the Na^+ pumps in arterial smooth muscle cells, can exert readily detectable effects on contraction.[45] This also provides a basis for understanding the selective up- and downregulation of specific Na^+ pump α-subunit isoforms and the action of an endogenous ouabain-like compound[1,46] that likely only regulates the α2 and α3 (high ouabain affinity) isoforms.

ACKNOWLEDGMENTS

We thank Dr. E. Santiago for preparations of freshly dissociated and cultured arterial myocytes and Drs. A. Ambesi, E. Carafoli, M. J. Caplan, G. Lindenmayer, R. W. Mercer, T. A. Pressley, K. J. Sweadner, and F. Wuytack for generous supplies of antibodies.

REFERENCES

1. BLAUSTEIN, M. P. 1993. The pathophysiological effects of endogenous ouabain: Control of stored Ca^{2+} and cell responsiveness. Am. J. Physiol. **264:** C1367–C1387.

2. SHULL, G. E., J. GREEB, & J. B LINGREL. 1986. Molecular cloning of three distinct forms of the Na$^+$, K$^+$-ATPase α-subunit from rat brain. Biochemistry **25:** 8125–8132.
3. LINGREL, J. B., J. ORLOWSKI, M. M. SHULL & E. M. PRICE. 1990. Molecular genetics of Na,K-ATPase. Prog. Nucleic Acid Res. Mol. Biol. **38:** 37–89.
4. SWEADNER, K. J. 1989. Isozymes of the Na$^+$/K$^+$-ATPase. Biochim. Biophys. Acta **988:** 185–220.
5. TAKEYASU, K., V. LEMAS & D. M. FAMBROUGH. 1990. Stability of Na$^+$-K$^+$-ATPase α-subunit isoforms in evolution. Am. J. Physiol. **259:** C619–C630.
6. PRESSLEY, T. A. 1992. Phylogenetic conservation of isoform-specific regions within α-subunit of Na$^+$-K$^+$-ATPase. Am. J. Physiol. **262:** C743–C751.
7. BLANCO, G., G. BERBERIAN & L. BEAUGÉ. 1990. Detection of a highly ouabain sensitive isoform of rat brainstem Na,K-ATPase. Biochim. Biophys. Acta **1027:** 1–7.
8. TURI, A., J. SOMOGYI & N. MULLNER. 1991. The effect of micromolar Ca^{2+} on the activities of the different Na$^+$/K$^+$-ATPase isozymes in the rat myometrium. Biochem. Biophys. Res. Commun. **174:** 969–974.
9. MUNZER, J. S., S. E. DALY, E. A. JEWELL-MOTZ, J. B LINGREL & R. BLOSTEIN. 1994. Tissue- and isoform-specific kinetic behavior of the Na,K-ATPase. J. Biol. Chem. **269:** 16668–16676.
10. ORLOWSKI, J. & J. B LINGREL. 1990. Thyroid and glucocorticoid hormones regulate the expression of multiple Na,K-ATPase genes in cultured neonatal rat cardiac myocytes. J. Biol. Chem. **265:** 3462–3470.
11. LUCCHESI, P. A. & K. J. SWEADNER. 1991. Postnatal changes in Na,K-ATPase isoform expression in rat cardiac ventricle. J. Biol. Chem. **266:** 9327–9331.
12. AZUMA, K. K., C. B. HENSLEY, M.-J. TANG & A. A. MCDONOUGH. 1993. Thyroid hormone specifically regulates skeletal muscle Na$^+$-K$^+$-ATPase α2- and β2-isoforms. Am. J. Physiol. **265:** C680–C687.
13. WELLING, P. A., M. CAPLAN, M. SUTTERS & G. GIEBISCH. 1993. Aldosterone-mediated Na/K-ATPase expression is α1 isoform specific in the renal cortical collecting duct. J. Biol. Chem. **268:** 23469–23476.
14. MAGYAR, C. E., J. WANG, K. K. AZUMA & A. A. MCDONOUGH. 1995. Reciprocal regulation of cardiac Na-K-ATPase and Na/Ca exchanger: Hypertension, thyroid hormone, development. Am. J. Physiol. **269:** C675–C682.
15. EWART, H. S. & A. KLIP. 1995. Hormonal regulation of the Na$^+$-K$^+$-ATPase: Mechanisms underlying rapid and sustained changes in pump activity. Am. J. Physiol. **269:** C295–C311.
16. MOORE, E. D. W., E. F. ETTER, K. D. PHILIPSON, W. A. CARRINGTON, K. E. FOGARTY, L. M. LIFSHITZ & F. S. FAY. 1993. Coupling of the Na$^+$/Ca^{2+} exchanger, Na$^+$/K$^+$ pump and sarcoplasmic reticulum in smooth muscle. Nature **365:** 657–660.
17. MCDONOUGH, A. A., Y. ZHANG, V. SHIN & J. S. FRANK. 1996. Subcellular distribution of sodium pump isoform subunits in mammalian cardiac myocytes. Am. J. Physiol. **270:** C1221–C1227.
18. JUHASZOVA, M., A. AMBESI, G. E. LINDENMAYER, R. J. BLOCH & M. P. BLAUSTEIN. 1994. Na-Ca exchange in arteries: Identification by immunoblotting and immunofluorescence microscopy. Am. J. Physiol. **266:** C234–C242.
19. JUHASZOVA, M., H. SHIMIZU, M. L. BORIN, R. K. YIP, E. M. SANTIAGO, G. L. LINDENMAYER & M. P. BLAUSTEIN. 1996. Localization of the Na$^+$-Ca^{2+} exchanger in vascular smooth muscle, and in neurons and astrocytes. Ann. N.Y. Acad. Sci. **779:** 318–335.
20. JUHASZOVA, M. & M. P. BLAUSTEIN. 1997. Distinct distribution of different Na$^+$ pump α subunit isoforms in plasmalemma: Physiological implications. Proc. Natl. Acad. Sci. USA **94:** 1800–1805.
21. TERASAKI, M. 1989. Fluorescent labeling of endoplasmic reticulum. Methods Cell Biol. **29:** 125–135.

22. GUNTHER, S., R. W. ALEXANDER, W. J. ATKINSON & M. A. GIMBRONE, JR. 1982. Functional angiotensin II receptors in cultured vascular smooth muscle cells. Cell. Biol. **92:** 289–298.
23. BOOHER, J. & M. SENSENBRENNER. 1972. Growth and cultivation of dissociated neurons and glial cells from embryonic chick, rat and human brain in flask cultures. Neurobiology **2:** 97–105.
24. GOLDMAN, W. F., P. J. YAROWSKY, M. JUHASZOVA, B. K. KRUEGER & M. P. BLAUSTEIN. 1994. Sodium/calcium exchange in rat cortical astrocytes. J. Neurosci. **14:** 5834–5843.
25. LUTHER, P. W. & R. J. BLOCH. 1989. Formaldehyde-amine fixatives for immunocytochemistry of cultured *Xenopus* myocytes. J. Histochem. Cytochem. **37:** 75–82.
26. MCLEAN, I. W. & P. K. NAKANE. 1974. Periodate-lysine-paraformaldehyde fixative: A new fixative for immunoelectron microscopy. J. Histochem. Cytochem. **22:** 1077–1083.
27. BLANCO, G., J. C. KOSTER & R. W. MERCER. 1994. The alpha subunit of the Na,K-ATPase specifically and stably associates into oligomers. Proc. Natl. Acad. Sci. USA **91:** 8542–8556.
28. FELSENFELD, D. P. & K. J. SWEADNER. 1988. Fine specificity mapping and topography of an isozyme specific epitope of the Na,K-ATPase catalytic subunit. J. Biol. Chem. **263:** 10932–10942.
29. URAYMA, O., H. SHUTT & K. J. SWEADNER. 1989. Identification of three isozyme proteins of the catalytic subunit of the Na,K-ATPase in rat brain. J. Biol. Chem. **264:** 8271–8280.
30. PIETRINI, G., M. MATTEOLI, G. BANKER & M. J. CAPLAN. 1992. Isoforms of the Na,K-ATPase are present in both axons and dendrites of hippocampal neurons in culture. Proc. Natl. Acad. Sci. USA **89:** 8414–8418.
31. AMBESI, A., E. E. BAGWELL Z & G. E. LINDENMAYER. 1991. Purification and identification of the cardiac sarcolemmal Na/Ca exchanger (abstr). Biophys. J. **59:** 138a.
32. CARRINGTON, W. A., K. E. FOGARTY & F. S. FAY. 1990. 3D Fluorescence imaging of single cells using image restoration. *In* Noninvasive Techniques in Cell Biology. K. Foster & S. Grinstein, Eds.: 53–72. Wiley-Liss, Inc. New York.
33. CAMERON, R., L. KLEIN, A. W. SHYJAN, P. RAKIC & R. LEVENSON. 1994. Neurons and astroglia express distinct subsets of Na,K-ATPase alpha and beta subunits. Brain Res. **21:** 333–343.
34. SWEADNER, K. J. 1992. Overlapping and diverse distribution of Na-K ATPase isozymes in neurons and glia. Can. J. Physiol. Pharmacol. **70:** S255–S259.
35. JUHASZOVA, M., R. W. MERCER, D. N. WEISS, D. J. PODBERSKY, J. HEIDRICH & M. P. BLAUSTEIN. 1994. Na^+/K^+-pump $\alpha 3$ isoform in rat vascular smooth muscle and its physiological role. *In* The Sodium Pump. Structure Mechanism, Hormonal Control and Its Role in Disease. E. Bamberg & W. Schoner, Eds.: 852–855. Steinkopff. Darmstadt.
36. MATSUDA, T., Y. MURATA, N. KAWAMURA, M. HAYASHI, K. TAMADA, K. TAKUMA, S. MAEDA & A. BABA. 1993. Selective induction of $\alpha 1$ isoform of $(Na^+ + K^+)$-ATPase by insulin/insulin-like growth factor-1 in cultured rat astrocytes. Arch. Biochem. Biophys. **307:** 175–182.
37. GOLDMAN, W. F., P. J. YAROWSKY, M. JUHASZOVA, B. K. KRUEGER & M. P. BLAUSTEIN. 1994. Sodium/calcium exchange in rat cortical astrocytes. J. Neurosci. **14:** 5834–5843.
38. FAY, F. & K. FOGARTY. 1984. The organization of the contractile apparatus in single isolated smooth muscle cells. *In* Smooth Muscle Contraction. N. L. Stephens, Ed.: 47–74. M. Dekker, Inc. New York.
39. SOMLYO, A. P. & A. V. SOMLYO. 1992. Smooth muscle structure and function. *In* The Heart and Cardiovascular System. A. H. Fozzard *et al.*, Eds.: 1295–1394. Raven Press, Ltd. New York.
40. ZHU, Z., M. TEPEL, M. NEUSSER & W. ZIDEK. 1996. Low concentrations of ouabain increase cytosolic free calcium concentration in rat vascular smooth muscle cells. Clin. Sci. (Lond.) **90:** 9–12.

41. BORIN, M. L., R. M. TRIBE & M. P. BLAUSTEIN. 1994. Increased intracellular Na^+ augments mobilization of Ca^{2+} from SR in vascular smooth muscle cells. Am. J. Physiol. **266:** C311–C317.
42. NELSON, M. T., H. CHENG, M. RUBART, L. F. SANTANA, A. D. BONEV, H. J. KNOT & W. J. LEDERER. 1995. Relaxation of arterial smooth muscle by calcium sparks. Science **270:** 633–637.
43. YOSHIKAWA, A., C. VAN BREEMEN & G. ISENBERG. 1996. Buffering of plasmalemmal Ca^{2+} current by sarcoplasmic reticulum of guinea pig urinary bladder myocytes. Am. J. Physiol. **271:** C833–C841.
44. SEMB, S. O. & O. M. SEJERSTED. 1996. Fuzzy space and the control of Na^+,K^+-pump rate in heart and skeletal muscle. Acta Physiol. Scand. **156:** 213–225.
45. WEISS, D. N., D. J. PODBERESKY, J. HEIDRICH & M. P. BLAUSTEIN. 1993. Nanomolar ouabain augments caffeine-evoked contractions in rat arteries. Am. J. Physiol. **265:** C1443–C1448.
46. HAMLYN, J. M., M. P. BLAUSTEIN, S. BOVA, D. W. DUCHARME, D. W. HARRIS, F. MANDEL, W. R. MATHEWS & J. H. LUDENS. 1991. Identification and characterization of ouabain-like compound from human plasma. Proc. Natl. Acad. Sci. USA **88:** 6259–6263.

α and β Subunits of Na,K-ATPase Interact with BiP and Calnexin[a]

AHMED T. BEGGAH AND KÄTHI GEERING[b]

Institute of Pharmacology and Toxicology
University of Lausanne
rue du Bugnon 27
1005 CH-Lausanne, Switzerland

Subunit assembly at the level of the endoplasmic reticulum (ER) is a prerequisite for the structural and functional maturation of oligomeric proteins. In oligomeric P-type ATPases, only assembled α-β complexes can leave the ER and become functional pumps. To be able to associate, individual subunits of oligomeric proteins must undergo an extensive folding process which is partly mediated by cotranslational modifications such as disulfide bond formation, core-glycosylation, and interaction with molecular chaperones. (For review see ref. 1.) In this study, we investigated whether molecular chaperones of the ER, namely, immunoglobulin binding protein (BiP) and calnexin, interact with α and β subunits of Na,K-ATPase and play a role in their initial folding.

RESULTS AND DISCUSSION

To study the role of BiP and calnexin in the early processing of α and β subunits of Na,K-ATPase, we first cloned the two genes from *Xenopus*. On the one hand, the cDNAs were used to produce fusion proteins and specific BiP and calnexin antibodies. On the other hand, the cDNAs were transcribed to yield cRNA which were injected into *Xenopus* oocytes together with *Xenopus* Na,K-ATPase α and/or βcRNA. Interaction of the exogenous or endogenous oocyte chaperones with the Na,K-ATPase subunits was followed after pulse-chase labeling of the newly synthesized proteins by nondenaturing immunoprecipitations with the specific chaperone antibodies.

After a 6-hour pulse, the wild-type β subunit expressed in the oocyte coprecipitated with both endogenous BiP and calnexin. The interaction time with the two chaperones closely paralleled the degradation rate of the β subunit. Interestingly, in opposition to the suggestion that calnexin is a glycoprotein-specific chaperone, a nonglycosylated β mutant also interacted with calnexin as long as it was detectable in the ER.

To study the interaction of the catalytic α subunit with BiP and calnexin, we expressed the α subunit alone or together with β subunits in the oocyte and performed nondenaturing immunoprecipitations after a pulse labeling or a chase period (FIG. 1). BiP antibodies coprecipitated newly synthesized α subunits with BiP (FIG. 1A, lanes 1

---

[a]This study was supported by the Swiss National Fund for Scientific Research (grant 31-42954.95).

[b]To whom correspondence should be addressed. Tel: 41 (21) 692 53 50/54 10; fax: 41 (21) 692 53 55; e-mail: kgeering@ulys.unil.ch

and 2) until the individual αs were degraded (FIG. 1C, lanes 1 and 2). A similar interaction was observed between α subunits and calnexin (FIG. 1B). The interaction of the two chaperones was abolished as soon as the α subunit associated with the β subunit. Indeed, in oocytes expressing α and β subunits, the α subunits became stabilized because of the formation of α-β complexes (FIG. 1C, lanes 3 and 4), but these α subunits were no longer coprecipitated with BiP (FIG. 1A, lanes 3 and 4) or calnexin (FIG. 1B, lanes 3 and 4).

In conclusion, both the catalytic α and the glycosylated or nonglycosylated β subunit of Na,K-ATPase transiently interacted with molecular chaperones such as BiP and calnexin during their synthesis and early processing in the ER. The wild-type β subunit behaved in this respect like many other glycoproteins. On the other hand, we showed for the first time that a large polytopic protein such as the α subunit, which exposes only about 10% of its mass to the ER lumen, can also associate with the

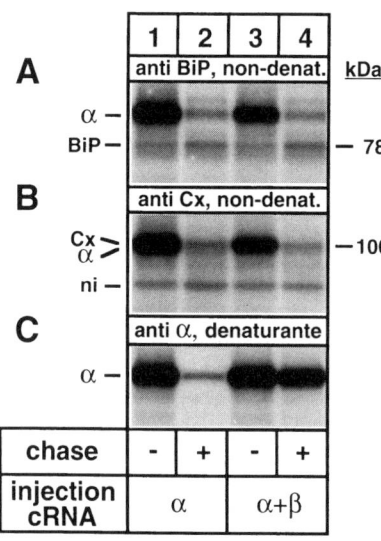

FIGURE 1. Interaction of BiP and calnexin with the α subunit and their dissociation after subunit assembly. The oocytes were injected with α or with α plus βcRNA, incubated with brefeldin A to retain newly synthesized proteins in the ER and labeled with ^{35}S-methionine during a 6-hour pulse or a 6-hour pulse and a 24-hour chase. Microsomes were prepared and nondenaturing (A and B) or denaturing (C) immunoprecipitations were performed with the indicated antibodies. The positions of the α subunit, of BiP, and of calnexin (Cx) are shown. ni = nonidentified band.

soluble BiP. Furthermore, our observations that the nonglycosylated α and the β mutant can interact with calnexin suggest that this previously assumed glycoprotein-specific chaperone can also associate with nonglycosylated proteins probably through its transmembrane domain. What might be the functional role of the chaperone interaction with Na,K-ATPase subunits? Our data show a coordinate, transient interaction of BiP and calnexin with α and β subunits which is abolished when subunits associate. These results strongly suggest that the two chaperones act in concert to facilitate the initial folding of the subunits and render them assembly competent. Since the interaction time with the two chaperones correlates with the subunits degradation rate, an additional function of chaperone interaction might be to retain misfolded or unassembled, overexpressed subunits in the ER until they are recognized by the pre-Golgi degradation system.

REFERENCE

1. GEERING, K. 1996. Oligomerization and maturation of eukaryotic membrane proteins. *In* Membrane Assembly, Molecular Biology Intelligence Unit Series, R. G. Landes Company. Gunnar von Heijne, Ed.: 173–188. New York.

Structural Domains Implicated in ER Degradation of α Subunits of Na,K-ATPase[a]

PASCAL BEGUIN AND KÄTHI GEERING[b]

Institute of Pharmacology and Toxicology
University of Lausanne
rue du Bugnon 27
1005 Lausanne, Switzerland

To avoid accumulation of misfolded, unfunctional proteins, cells exert a quality control which involves different degradative pathways. The best characterized proteolytic systems are the lysosomal and the ubiquitin-mediated pathways. Misfolded or unassembled subunits of secreted or membrane proteins are degraded in or close to the endoplasmic reticulum (ER), but little is known about the proteases involved or the structural determinants in proteins recognized by them. (For review see ref. 1.) To better define the ER degradation of large polytopic membrane proteins, we have chosen the α subunit of Na,K-ATPase as a model protein. Expressed in *Xenopus* oocyte without a β subunit, the α subunit is degraded in or close to the ER with a half-life of 4 hours and only becomes stable when assembled with the β subunit.

RESULTS AND DISCUSSION

To define the structural determinants in the α subunit which are involved in recognition by cellular proteases, we produced deletion or site-specific α mutants of *Xenopus* Na,K-ATPase and expressed them either alone or together with β subunits in *Xenopus* oocytes. Pulse-chase experiments combined with denaturing or nondenaturing immunoprecipitations permitted us to follow the assembly competence and the degradation rate of the mutant α subunits. The membrane topology of the mutants was checked by following the glycosylation of chimeric constructs containing the ectodomain of the β subunit of *Bufo* Na,K-ATPase. Endoglycosidase H (Endo H) sensitivity permitted us to distinguish the glycosylated mutants that expose the β-ectodomain to the ER lumen from the nonglycosylated mutants that expose the β ectodomain to the cytoplasm.

In contrast to individual, wild-type α subunits which were degraded after a 48-hour chase period (FIG. 1A, lanes 15 and 16), the deletion mutants M2 stop (for description of mutants see FIG. 1B) and M4 loop stop were stable (lanes 1–4). This result indicates that the transmembrane domains M1–M4 are stably inserted in the ER

---

[a]This study was supported by the Swiss National Fund for Scientific Research (grant 31-42954.95).
[b]To whom correspondence should be addressed. Tel: 41 (21) 692 53 50/54 10; fax: 41 (21) 692 53 55; e-mail: kgeering@ulys.unil.ch

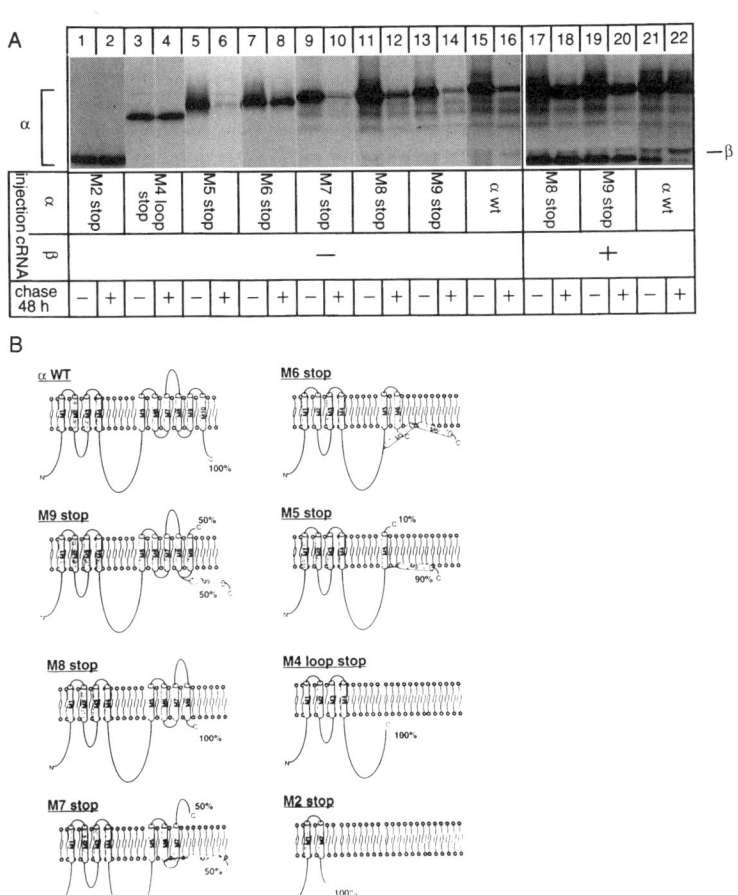

FIGURE 1. Degradation and membrane topology of wild-type and truncated α subunits expressed in the *Xenopus* oocyte in the presence or absence of β subunits. **(A)** Degradation of α subunits. Oocytes were injected with wild-type or mutant *Xenopus* α or α plus β cRNA and labeled with 35S-methionine during a 24-hour pulse followed or not by a 48-hour chase. Microsomes were prepared and immunoprecipitations were performed with α antibodies under denaturing (lanes 1–16) or nondenaturing (lanes 17–22) conditions. The position of wild-type and truncated α subunits and of the β subunit is shown. **(B)** Putative membrane topology of truncated α subunits as determined by following the glycosylation of chimeric α subunits containing the ectodomain of the Bufo β subunit.

membrane and that the large cytoplasmic domain of the α subunit is not degraded from the cytoplasmic side.

Mutants deleted after the putative M5 became unstable (FIG. 1A, lanes 5 and 6), but those deleted after M6 were again partially stabilized (lanes 7 and 8). Furthermore, if the putative M5 was added to a stable mutant such as M2 stop, it provoked its degradation (data not shown). This suggests that the putative M5 has a degradation signal than can be partially masked by M6. Topology studies of chimeric α-Bufo β constructs showed that only 10% of the putative M5 was transferred across the ER membrane (FIG. 1B). A double proline mutation in M5 (P801L/P803L), which permitted a complete membrane transfer, rendered the M5 stop mutant stable (data not shown), suggesting a relation between partial membrane insertion and degradation of the M5.

The mutants M7 stop (lanes 9 and 10), M8 stop (lanes 11 and 12), and M9 stop (lanes 13 and 14) were degraded. Only about 50% of M7 and M9 were transferred across the ER membrane (FIG. 1B). Mutants containing the pair M7/M8 linked by a β association domain[2] could be stabilized after coexpression with β subunits (FIG. 1A, lanes 17–20) similar to wild-type α subunits (lanes 21 and 22). Added to a stable protein M3 stop, the β association domain did not provoke its degradation (data not shown), indicating that the amino acids that are directly involved in β association are not themselves targets for proteolytic attack. Finally, ER exit of α-β complexes reflected by the full glycosylation of the β subunit necessitates an integral α subunit (lanes 17–22).

In conclusion, we have revealed several structural determinants that might be responsible for ER degradation of individual, unassembled α subunits of Na,K-ATPase. The results suggest that the degradation of this polytopic, membrane protein which exposes most of its mass to the cytoplasm is not mediated by cytosolic but rather by lumenal and/or intramembranous proteolytic attack of membrane segments such as M5. The assembly with the β subunit does not mask a degradation signal but favors the correct packing of M7 to M10 and in consequence the stabilization of the α subunit.

REFERENCES

1. GEERING, K. 1996. Oligomerization and maturation of eukaryotic membrane proteins. IN Membrane Assembly, Molecular Biology Intelligence Unit Series. Gunnar von Heijne, Ed.: 173–188. R. G. Landes Company. New York.
2. FAMBROUGH, D. M., M. V. LEMAS, M. HAMRICK, M. EMERICK, K. J. RENAUD, E. M. INMAN, B. HWANG & K. TAKEYASU. 1994. Analysis of subunit assembly of the Na-K-ATPase. Am. J. Physiol. **266:** C579–C589.

Posttranslational Processing of the Catalytic Subunit from Na^+/K^+-ATPase[a]

THOMAS A. PRESSLEY[b] AND SUSAN A. PETROSIAN

Department of Physiology
Texas Tech University Health Sciences Center
3601 4th Street
Lubbock, Texas 79430

The catalytic α subunit of the Na,K-ATPase exists as three distinct isoforms (α1, α2, and α3) whose primary structures are nearly identical. One exception is the amino terminus, and this divergence in sequence may be related to differences in posttranslational processing. The first five amino acids of the α1 isoform predicted from its cDNA are not found in enzyme purified from enriched tissues such as renal outer medulla.[1] This implies that α1 is cleaved enzymatically at some point during or after translation. By contrast, α3 is not thought to undergo this proteolysis. To facilitate evaluation of amino terminal structure and posttranslational cleavage, we developed site-directed antibodies specific for the amino termini of nascent α isoforms from rat.[2] These reagents were used to test explicitly the extent of posttranslational processing undergone by α1 and α3.

ANTIBODIES

We developed antibodies designed to distinguish processed and nonprocessed α subunits. The oligopeptide "VGR" was derived from the first nine residues of nascent rat α1 (MGKGVGRDK). The oligopeptide "DDK" was derived from the first eight residues of nascent rat α3 (MGDKKDDK). Both oligopeptides were linked covalently to hemocyanin and then used to immunize rabbits. The resulting antisera were purified by affinity chromatography, using target oligopeptide coupled to a solid support.

THE α1 SUBUNIT UNDERGOES POSTTRANSLATIONAL PROCESSING

Using anti-VGR, the α1-specific antibody, we tested explicitly the expectation that the α1 subunit is cleaved *in vivo*. In immunoblots of noncleaved polypeptide generated by *in vitro* translation, anti-VGR detected a prominent band with a mobility appropriate for the α1 subunit (100 kD). Immunoblots of total protein from various rat organs, however, revealed no significant binding to a 100-kD band. Similarly, immunoblots of isolated membranes from green monkey kidney cells transfected with rat α1 revealed barely detectable binding of anti-VGR. The absence of significant binding to anti-VGR in both intact tissues and transfected mammalian cells strongly

[a]This work was supported by National Institutes of Health grant RR-19799.
[b]Tel: 806-743-2521; fax: 806-743-1512; e-mail: phytap@ttuhsc.edu

suggests that most nascent α1 polypeptide undergoes posttranslational cleavage of the amino terminus.

THE α3 SUBUNIT DOES NOT UNDERGO POSTTRANSLATIONAL PROCESSING

Using anti-DDK, the α3-specific antibody, we tested explicitly the expectation that the α3 subunit is not cleaved *in vivo*. Of the organs surveyed, only brain revealed significant binding of anti-DDK to a band of the appropriate mobility (100 kD), consistent with the known distribution of the isoform. The presence of binding by anti-DDK in brain suggests that the amino termini of at least some α3 polypeptide were not processed.

CONCLUSIONS

As predicted from comparisons of the cDNA and protein sequences, α1 appears to undergo posttranslational cleavage of its amino terminus, whereas α3 does not. Additional studies will be necessary to determine the overall importance of this difference in subunit processing.

ACKNOWLEDGMENTS

We thank Dr. Jerry B Lingrel for providing the full-length α1 cDNA. The contributions of numerous colleagues are also appreciated.

REFERENCES

1. SHULL, G. E., J. GREEB & J. B LINGREL. 1986. Molecular cloning of three distinct forms of the Na^+,K^+-ATPase α-subunit from rat brain. Biochemistry **25:** 8125–8132.
2. PRESSLEY, T. A., J. C. ALLEN, C. H. CLARKE, T. ODEBUNMI & S. C. HIGHAM. 1996. Amino-terminal processing of the catalytic subunit from Na^+-K^+-ATPase. Am. J. Physiol. **271:** C825–C832.

Functional Control of Na$^+$, K$^+$-ATPase by Vasopressin and Aldosterone in the Cortical Collecting Duct

Role of Protein Phosphatase

N. COUTRY, N. FARMAN, J. P. BONVALET, AND M. BLOT-CHABAUD[a]

Inserm U 246
Faculté de Médecine X. Bichat
BP 416
75870 Paris, Cedex 18, France

The mammalian cortical collecting duct is an important site for control of renal sodium reabsorption and potassium secretion. Previous experiments have shown that aldosterone and vasopressin act synergistically to increase the apical sodium entry. At the basolateral side, Na$^+$, K$^+$-ATPase is currently assumed to rapidly adapt its turnover rate to this increase. Recently, we showed that aldosterone is responsible for the synthesis of a latent pool of Na$^+$, K$^+$-ATPase that can be activated and/or recruited in the basolateral membrane in response to an enhanced Na$_i$.[1] Because vasopressin rapidly increases apical sodium entry, we investigated whether this vasopressin-dependent increase could, in turn, provoke the recruitment or activation at the basolateral membrane of the aldosterone-dependent pool of latent pumps. In addition, we checked whether protein serine/theonine phosphatases (PP) such as PP1 or PP2A could be implied in the phenomenon, because recent studies had shown that Na$^+$, K$^+$-ATPase activity could be regulated by phosphorylation/dephosphorylation processes.

Experiments were performed on the isolated cortical collecting duct of female Swiss mice bilaterally adrenalectomized and treated with either 1 µg per 100 g body wt^{-1}/day^{-1} dexamethasone (ADX) or both 1 µg per 100 g body wt^{-1}/day^{-1} dexamethasone and 30 µg per 100 g body wt^{-1}/day^{-1} aldosterone (ALDO). The number of pumps present in the basolateral membrane of the cortical collecting duct was determined on microdissected cortical collecting duct using the ^{3}H-ouabain binding technique. The functional activity of pumps in the membrane was examined using the ^{86}Rb uptake technique, and protein phosphatase (PP) activity was measured using a kit (Life Technologies, France).[2,3]

The effect of vasopressin 10^{-8} M was determined on specific ^{3}H-ouabain binding and ouabain-sensitive ^{86}Rb uptake in the cortical collecting duct from ADX and ALDO mice. Results are given in TABLE 1. In ALDO animals, vasopressin induced the recruitment and/or activation of the aldosterone-dependent pool of Na$^+$, K$^+$-ATPase. By contrast, in ADX animals, no vasopressin-dependent recruitment could be ob-

---

[a] Author for correspondence. Tel: 331-44.85.63.25; fax: 331-42.29.16.44; e-mail: u246@bichat.inserm.fr

TABLE 1. Effect of Vasopressin (AVP) on Specific ^{3}H-Ouabain Binding and Ouabain-Sensitive ^{86}Rb Uptake in Cortical Collecting Duct from Adrenalectomized (ADX) and Aldosterone-Repleted (ALDO) Mice[a]

	Specific ^{3}H-Ouabain Binding (fmol/nl tub vol)	Ouabain-Sensitive ^{86}Rb Uptake (pmol/nl tub vol/min)
ADX; −AVP	7.08 ± 0.82; n = 46	7.33 ± 0.66; n = 14
ADX; +AVP	7.07 ± 0.93; n = 42	8.74 ± 1.33; n = 18
ALDO; −AVP	7.56 ± 0.78; n = 44	13.97 ± 2.38; n = 27†
ALDO; +AVP	10.65 ± 1.28; n = 46*	22.55 ± 3.85; n = 18*

[a]AVP increased specific ^{3}H-ouabain binding in ALDO animals. Ouabain-sensitive ^{86}Rb uptake was higher in ALDO than in ADX animals and was increased by AVP in ALDO animals.
*p < 0.05 ALDO; +AVP versus ALDO; −AVP.
†p < 0.05 ALDO; −AVP versus ADX; −AVP.

served. The effect of vasopressin was dose-dependent with a $K_{1/2}$ of about 10^{-9} M. The effect was rapid (present at 5 minutes) and transient (<20 minutes). It was reproduced by dDAVP 10^{-8} M, forskolin 10^{-5} M, or ^{8}Br-AMP 2.10^{-4} M and blocked by amiloride 10^{-5} M. This suggested that the vasopressin effect on basolateral pumps may be mediated, through a V_2 pathway, by an increase in apical sodium entry. Two PP inhibitors, okadaic acid (1 μM) and calyculin A (50 nM), prevented the vasopressin-induced increase in specific ^{3}H-ouabain binding and in ouabain-sensitive ^{86}Rb uptake. In the presence of nystatin 120 μg/ml the effect of vasopressin was also abolished, suggesting that the desphosphorylation process affected Na^+, K^+-ATPase rather than the apical sodium channel. The $K_{1/2}$ of inhibition by okadaic acid and calyculin A was about 2 nM and <6 nM, respectively, suggesting that PP2A rather than PP1 could be involved. This was confirmed by experiments designed to test the effect of 10^{-8} M vasopressin on PP1 or PP2A activity. Results are shown in TABLE 2. Both PP1 and PP2A activities are present in the cortical collecting duct, representing, respectively, 60% and 40% of the total activity. Vasopressin significantly increased PP2A activity in the cortical collecting duct (experiments in the presence of 3 nM tautomycin which

TABLE 2. Serine/Threonine Protein Phosphatase (PP) Activity and Effect of 10^{-8} M Vasopressin (AVP) on PP Activity in the Absence or Presence of 1 nM Okadaic Acid (OK) or 3 nM Tautomycin (TAUT)[a]

	C	OK	TAUT	OK + TAUT
PP activity (×10^{-3} nmol/nl tub vol/min)	1.38 ± 0.08 (n = 5)	0.84 ± 0.07* (n = 7)	0.49 ± 0.06† (n = 9)	0.03 ± 0.02‡ (n = 5)
AVP-induced increase in PP activity (% of control without AVP)	16.83 ± 5.94§ (n = 8)	0.01 ± 10.71 (n = 6)	28.57 ± 10.2¶ (n = 7)	

[a]Okadaic acid decreased PP activity by about 40%, whereas TAUT decreased it by about 60%. When both OK and TAUT were added, PP activity was about zero. PP activity was increased by AVP either in the absence of drug or in the presence of 3 nM TAUT, but it was not modified in the presence of 1 nM OK, indicating an effect of AVP on PP2A activity.
*p < 0.001 OK versus C; †p < 0.001 TAUT versus OK; ‡p < 0.001 OK + TAUT versus TAUT; §p < 0.05 AVP versus control; ¶p < 0.05 AVP versus control in the presence of 3 nM TAUT.

blocks PP1), whereas no effect could be observed on PP1 activity (experiments in the presence of 1 nM okadaic acid which blocks PP2A).

Taken together, these results show that aldosterone and vasopressin act synergistically not only on apical but also on basolateral membranes of cortical collecting duct to promote sodium reabsorption. Vasopressin increases the number of functional Na^+, K^+-ATPases present in the basolateral membrane by recruiting and/or activating an aldosterone-dependent pool of latent pumps. This effect is rapid and mediated by a vasopressin-dependent increase in PP2A activity which could, in turn, dephosphorylate latent Na^+, K^+-ATPases.

REFERENCES

1. BLOT-CHABAUD, M., F. WANSTOK, J. P. BONVALET & N. FARMAN. 1990. Cell sodium-induced recruitment of Na^+-K^+-ATPase pumps in rabbit cortical collecting duct is aldosterone dependent. J. Biol. Chem. **265:** 11676–11681.
2. COUTRY, N., N. FARMAN, J. P. BONVALET & M. BLOT-CHABAUD. 1995. Synergistic action of vasopressin and aldosterone on basolateral Na^+-K^+-ATPase in the cortical collecting duct. J. Membr. Biol. **145:** 99–106.
3. BLOT-CHABAUD, M., N. COUTRY, M. LAPLACE, J. P. BONVALET & N. FARMAN. 1996. Role of protein phosphatase in the regulation of Na^+-K^+-ATPase by vasopressin in the cortical collecting duct. J. Membr. Biol. **153:** 233–239.

Effect of Neurotensin on Synaptosomal Membrane ATPase and *p*-NPPase Activities[a]

M. G. LÓPEZ ORDIERES AND
G. RODRÍGUEZ DE LORES ARNAIZ[b]

*Instituto de Biología Celular y Neurociencias
"Prof. Eduardo De Robertis"
Facultad de Medicina
Universidad de Buenos Aires
Paraguay 2155
(1121) Buenos Aires, Argentina*

Neurotensin is a tridecapeptide first isolated from bovine hypothalamus[1] and widely distributed in the central nervous system (CNS).[2] The distribution of neurotensin in brain cell bodies and nerve terminals suggests that it may play a major role in neurotransmission or neuromodulation, subserving diverse physiological CNS functions.[3] The administration of neurotensin into the CNS induces a variety of effects, including potentiation of barbiturate and ethanol-induced sedation, reduction of food intake, antinociception, hypothermia, catalepsy, and alteration in locomotor activity. Several lines of evidence have demonstrated the coexistence and co-release of neurotensin and dopamine in diverse brain areas as well as the induction by neurotensin of membrane depolarization of dopaminergic neurons.[4,5]

Previous studies from this laboratory have shown that neuronal Na^+,K^+-ATPase activity is modified by several neurotransmitters including dopamine which inhibits the enzyme.[6] Taken jointly, these findings have awakened interest in the study of neurotensin's effect on ATPase and *p*-nitrophenylphosphatase (*p*-NPPase) activities.

METHODS

Wistar rats weighing 150–200 g were used. Reagents were analytical grade. Ouabain, neurotensin, disodium adenosine triphosphatase (ATP, grade I), and *p*-NPP were from Sigma Chemical Co., St. Louis, Missouri. Synaptosomal membranes from rat cerebral cortex were isolated by differential centrifugation and sucrose density gradient.[7] ATPase and *p*-NPPase activities were determined spectrophotometrically measuring, respectively, Pi and *p*-nitrophenol production in both the presence and the absence of ouabain.[8] Protein was determined by the method of Lowry *et al.*[9] using bovine serum albumin as the standard. Statistical analysis was performed using Student's *t* test.

---

[a]This study was supported by grants from the Consejo Nacional de Investigaciones Científicas y Técnicas and Universidad de Buenos Aires, Argentina.
[b]Corresponding author. Tel: 54-1-961-5010; fax: 54-1-962-5341 or 964–8274.

RESULTS AND CONCLUSION

In the absence of neurotensin, the specific activities of Na^+,K^+-ATPase and Mg^{2+}-ATPase in synaptosomal membranes were 49.2 ± 6.0 and 12.1 ± 2.6 µmol P_i released per milligram protein per hour (means $\pm$ SD; $n = 4$–6). Neurotensin within the 3×10^{-8} to 3×10^{-6} M range inhibited Na^+,K^+-ATPase activity 25–46%, but failed to change Mg^{2+}-ATPase activity (FIG. 1).

The peptide effect on Na^+,K^+-ATPase activity was also studied by assaying ATPase in the presence of variable concentrations of NaCl (3.1–200 mM), KCl (2.5–40 mM), and ATP (1–8 mM). Results indicated a noncompetitive type interaction (data not shown).

Alternatively, ATPase activity was tested by NPPase assay; basal K^+-p-NPPase

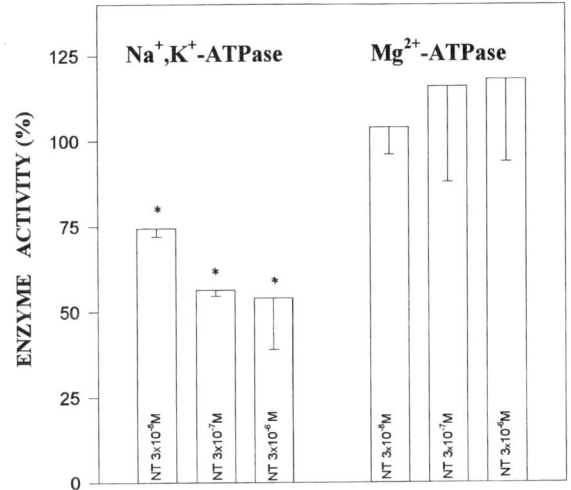

FIGURE 1. Effect of neurotensin on ATPase activities of synaptosomal membranes from rat cerebral cortex. Results are expressed as percentage enzyme activity in the presence of added neurotensin taking control values as 100% (mean $\pm$ SD; $n = 4$–6). *$p < 0.05$ using Student's t test.

and Mg^{2+}-p-NPPase activities in synaptosomal membranes were 7.9 ± 1.53 and 1.2 ± 0.13 µmol p-nitrophenol released per milligram protein per hour (means $\pm$ SD; $n = 8$). It was observed that neurotensin in the range of 8.6×10^{-8} to 8.6×10^{-6} M concentration had no effect on NPPase activities. However, 8.6×10^{-6} M neurotensin was found to inhibit K^+-p-NPPase 24% when ATP plus NaCl (0.6:45 mM) were added to preincubation and incubation enzyme media (FIG. 2).

Our results showed that neurotensin inhibits Na^+, K^+-ATPase activity but fails to alter basal K^+-p-NPPase activity (without added ATP + NaCl) of synaptosomal membranes. Because neurotensin inhibits K^+-p-NPPase when preincubation and incubation media are supplemented with ATP + NaCl, it seems that such effect takes place during enzyme phosphorylation.

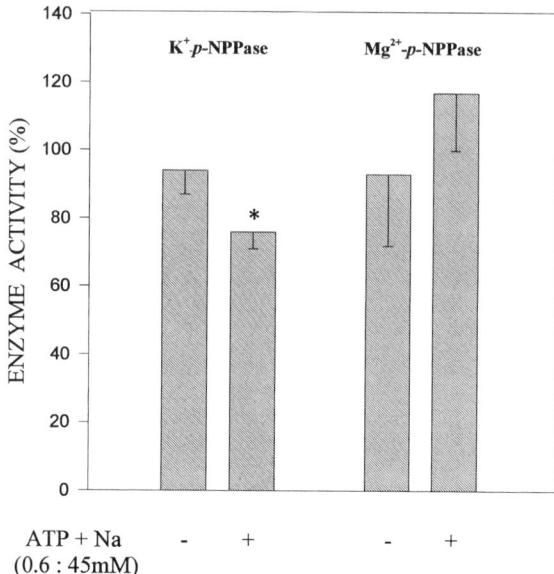

FIGURE 2. Effect of neurotensin and ATP + NaCl on NPPase activities of synaptosomal membranes from rat cerebral cortex. Enzyme activities were assayed in the presence and absence of 8.6×10^{-6} M neurotensin and ATP + NaCl (0.6 + 45 mM). Each column represents percentage enzyme activity taking as 100% values obtained in the absence of the peptide (mean ± SD; $n = 4$–8). *$p < 0.05$ using Student's t test.

It is postulated that Na^+, K^+-ATPase inhibition by neurotensin at the synaptic region may be involved in the reported effects of this neuropeptide in the nervous system.

REFERENCES

1. CARRAWAY, R. & S. LEEMAN. 1973. The isolation of a new hypotensive peptide, neurotensin, from bovine hypothalami. J. Biol. Chem. **248:** 6854–6861.
2. EMSON, P. C., M. GOEDERT, P. HORSFIELD, F. RIOUX & S. ST PIERRE. 1982. The regional distribution and chromatographic characterisation of neurotensin-like immunoreactivity in the rat central nervous system. J. Neurochem. **38:** 992–999.
3. ANGULO, J. A. & B. S. MCEWEN. 1994. Molecular aspects of neuropeptide regulation and function in the corpus striatum and nucleus accumbens. Brain Res. Rev. **19:** 1–28.
4. NEMEROF, C. B. & S. T. CAIN. 1985. Neurotensin-dopamine interactions in the CNS. Trends Pharmacol. Sci. **6:** 201–205.
5. BEAN, A. J. & R. H. ROTH. 1992. Dopamine-neurotensin interactions in mesocortical neurons. Evidence from microdialysis studies. *In* The Neurobiology of Neurotensin. P. Kitabgi & C. B. Nemeroff, Eds. Ann. N.Y. Acad. Sci. **668:** 43–53.
6. RODRÍGUEZ DE LORES ARNAIZ, G. 1983. Neuronal Na^+, K^+-ATPase and its regulation by catecholamines. *In* Neural Transmission, Learning and Memory. R. Caputto & C. Ajmone Marsand, Eds. Raven Press. New York.
7. RODRÍGUEZ DE LORES ARNAIZ, G., M. ALBERICI & E. DE ROBERTIS. 1965. Ultrastructural and

enzymic studies of cholinergic and non-cholinergic synaptic membranes isolated from brain cortex. J. Neurochem. **14:** 215–225.
8. ALBERS, R. W., G. RODRÍGUEZ DE LORES ARNAIZ & E. DE ROBERTIS. 1965. Sodium-potasium activated ATPase and potassium-activated p-nitrophenyl-phosphatase: A comparison of their subcellular localizations in rat brain. Proc. Natl. Acad. Sci. USA **53:** 557–564.
9. LOWRY, O. H., N. J. ROSEBROUGH, A. L. FARR & R. J. RANDALL. 1951. Protein measurement with the Folin phenol reagent. J. Biol. Chem. **193:** 265–275.

Na⁺/K⁺-ATPase Density Is Sexually Dimorphic in the Adult Rat Kidney[a]

LUIS EDUARDO M. QUINTAS,[b]
LUCIANE BARREIRO LOPEZ,[b] CADEN SOUCCAR,[c]
AND FRANÇOIS NOËL[b,d]

[b]Departamento de Farmacologia Básica e Clínica
Instituto de Ciências Biomédicas
Universidade Federal do Rio de Janeiro
Ilha do Fundão
21941-590, Rio de Janeiro, Brazil

[c]Departamento de Farmacologia
Escola Paulista de Medicina
Universidade Federal do Estado de São Paulo
Caixa Postal 20207
04034-900, São Paulo, Brazil

Sexual dimorphism refers to any difference between sexes manifested at a morphological, functional, and molecular level.[1] Although sex determination of lower classes of vertebrates may be partially influenced by environmental factors, in mammals sexual hormones are uniquely responsible for the sex differences of reproductive and nonreproductive organs.[2] Their actions are not restricted to the critical neonatal period, being observed in adulthood as well ("organizational" and "activational" action, respectively).[1,2] Several hormones affect the activity and differential expression of Na⁺/K⁺-ATPase isozymes. Endogenous steroids, peptides, and other factors have been reported to stimulate or inhibit the function of the Na⁺/K⁺-pump, and most often this is due to the specific action of such compounds on different enzyme isoforms.[3] Female sex steroids are known to modulate the activity of Na⁺/K⁺-ATPase, even though their positive or negative action on the activity is still controversial. Testosterone and congeners have been studied less.

In the present work we investigate whether expression of the Na⁺/K⁺-ATPase of kidneys from adult rats is sexually dimorphic. Ten male and ten female adult Wistar rats (4 months old; body weight 313 ± 7 g and 190 ± 5 g, respectively) of the 2BAW colony were separated into two groups each and decapitated, and their kidneys were rapidly excised and stored in N_2. After thawing, they were cut into small pieces and homogenized in a Potter homogenizer with a motor-driven teflon pestle at 4°C in 3 volumes 0.25 M buffered sucrose (pH 7.4) containing 100 μM PMSF/g organ. After filtration through four layers of gauze, the homogenate was centrifuged at 100,000 × g_{av} for 1 hour, and the pellet was resuspended and stored in N_2 until use. The flexibility to normalize data and the smaller variability in yield of required membranes were the reasons for choosing the crude homogenate preparation. Protein content was measured according to the method of Lowry *et al.*[4] Na⁺/K⁺-ATPase activity was evaluated according to Fiske and Subbarow.[5] The total number of kidney Na⁺/K⁺-

---

[a]This paper was supported by FAPERJ, CNPq, and FINEP.
[d]Corresponding author. Tel: +55-21-5909522 (ext. 244); fax: +55-21-5901841; e-mail: fnoel@pharma.ufrj.br

ATPases was estimated by [^{3}H]ouabain binding experiments carried out in a Mg-ATP-Na medium at equilibrium. (See details in TABLE 1.) For Western blotting, samples were run on 6% polyacrylamide gel (Laemmli's SDS-PAGE[6]) and were transferred to nitrocellulose sheets that were immersed for 1 hour in nonfat dry milk, washed, and incubated overnight with mouse anti-rabbit Na$^+$/K$^+$-ATPase α_1 isoform monoclonal IgG (Upstate Biotechnology Inc.). Blots were rinsed and treated for 1–1.5 hours with rabbit anti-mouse secondary IgG. Immunoreactivity was detected by enhanced chemiluminescence.

Na$^+$/K$^+$-ATPase activity was substantially higher (about 50% if expressed either per milligram protein or per gram kidney) in female kidneys than in male ones (TABLE

TABLE 1. Values of Protein Recovery, Na$^+$/K$^+$-ATPase Activity, and Binding Parameters of Female and Male Rat Kidneys[a]

	Female		Male	
	1st Group	2nd Group	1st Group	2nd Group
Protein recovery				
(mg · g kidney^{-1})	82.2	66.8	87.5	66.7
Na$^+$/K$^+$-ATPase activity[b]				
(μmol P$_i$ · mg protein^{-1} · h^{-1})	19.5	27.0	14.9	16.3
(μmol P$_i$ · g kidney^{-1} · h^{-1})	1600	1804	1300	1087
[^{3}H]Ouabain binding[c]				
K_d (μM)	9 ± 1	5 ± 1	8 ± 1	3 ± 1
B_{max} (pmol · mg protein^{-1})	9.5 ± 0.6	9.2 ± 1.6	5.3 ± 0.6	6.1 ± 1.2
(pmol · g kidney^{-1})	781 ± 49	615 ± 107	464 ± 52	407 ± 80

[a]K_d and B_{max} values are expressed as mean ± standard deviation, representing the "goodness of fit."
[b]The enzyme was incubated at 37°C for 1 hour in 20 mM maleate-Tris buffer, pH 7.4, containing 94 mM NaCl, 3 mM MgCl$_2$, 1.2 mM ATPNa$_2$, 10 mM NaN$_3$, 1 mM EGTA in the presence of 3 mM KCl (total activity), or 1 mM ouabain (ouabain-resistant activity).
[c]Around 250 µg protein were incubated at 37°C for 15 minutes in 20 mM maleate-Tris buffer, pH 7.4, containing 50 nM [^{3}H]ouabain, 3 mM MgCl$_2$, 3 mM ATPNa$_2$, 94 mM NaCl, 10 mM NaN$_3$, 1 mM EGTA, and various concentrations of unlabeled ouabain (0–3000 nM). Nonspecific binding was estimated in the presence of 1 mM unlabeled ouabain. After incubation, samples were rapidly diluted with 5 ml of ice-cold 5 mM Tris-HCl buffer, pH 7.4, and instantaneously filtered on Whatman glass fiber filters (GF/C) under vacuum, and test tubes were washed with 5 ml of Tris-HCl buffer. Filters were then washed twice with 10 ml of the same buffer, dried, and immersed in a scintillation cocktail, and the radioactivity was counted.

1). Scatchard plots derived from binding assays were linear and pointed to the presence of one class of low-affinity ouabain binding sites (FIG. 1) that were 1.6-fold more abundant in female than in male rats, with no differences among values of K_d (TABLE 1). A higher number of α1 isoforms in female rat kidneys was confirmed by Western blotting analysis using anti-Na$^+$/K$^+$-ATPase antibodies. Thus, a higher turnover of female Na$^+$/K$^+$-ATPases which might also explain the high ATPase activity *in vitro* was discarded.

Female rats exhibit a higher secretion rate, plasma titer, and greater diurnal variation in plasma corticosterone and a greater magnitude and duration of the adrenal cortex response to stress. This could be explained by the lower affinity for type I

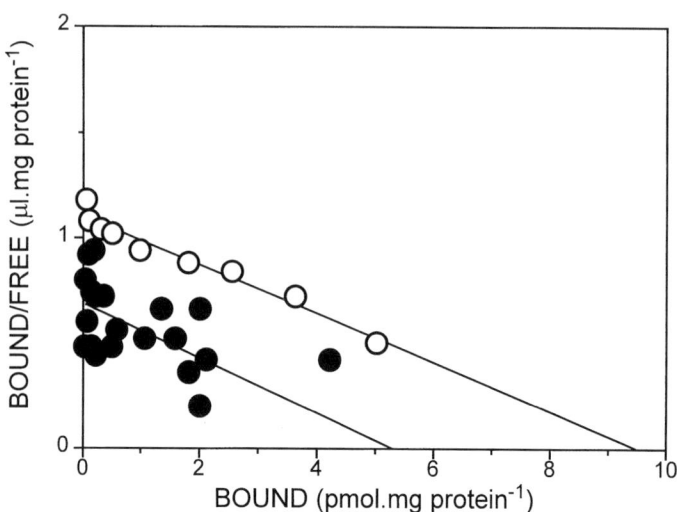

FIGURE 1. Scatchard plot for [^{3}H]ouabain binding to rat kidney crude preparation. About 250 µg protein of female (○) and male (●) preparation were incubated for 15 minutes in a Mg-ATP-Na medium. Each point is the mean of triplicate determinations. Lines were drawn using the parameters fitted by nonlinear regression analysis (TABLE 1, 1st Group). Bound = ouabain specifically bound; free = free ouabain concentration.

corticosteroid receptors in female rat hippocampus,[7] a limbic region implicated in corticosteroid negative feedback. Because basal ACTH levels are likely regulated by type I receptors, any reduction in affinity would raise the basal levels of circulating ACTH as well as corticosteroids. Interestingly, relative high secretion and excretion rates and circulating aldosterone titers are detected in women during the luteal phase of the menstrual cycle, possibly due to synergism between estrogens and progesterone in this phase.[8]

Thus, the sexual dimorphism here presented for the first time might be induced by adrenocorticoids, classical regulators of Na$^+$/K$^+$-ATPase gene expression, as a result of the action of estrogens and/or progesterone on peripheral (renin-angiotensin-aldosterone system) and/or central (hypothalamus-pituitary-adrenal axis) structures of gonadally intact females. At the renal level, if extrapolated to humans, this could explain the water retention that can be observed in the luteal phase of the menstrual cycle and during pregnancy.

REFERENCES

1. BARDIN, C. W. & J. F. CATTERALL. 1981. Science **211:** 1285–1293.
2. DÖHLER, K. D. 1991. Int. Rev. Cytol. **131:** 1–57.
3. EWART, H. S. & A. KLIP. 1995. Am. J. Physiol. **269:** C295–C331.
4. LOWRY, O. H., N. J. ROSEBROUGH, A. L. FARR & J. RANDALL. 1951. J. Biol. Chem. **193:** 265–275.
5. FISKE, C. H. & Y. SUBBAROW. 1925. J. Biol. Chem. **66:** 375–392.
6. LAEMMLI, U. K. 1970. Nature **227:** 680–685.
7. TURNER, B. B. 1992. Brain Res. **581:** 229–236.
8. SEALEY, J. E., J. ITSKOVITZ-ELDOR, S. RUBATTU, G. D. JAMES, P. AUGUST, I. THALER, J. LEVRON & J. H. LARAGH. 1994. J. Clin. Endocrinol. & Metab. **79:** 258–264.

Effect of Haloperidol on the Sarcoplasmic Reticulum Ca-ATPase

DELIA TAKARA[a] AND GUILLERMO L. ALONSO

Cátedra de Biofísica
Facultad de Odontología
Universidad de Buenos Aires
M. T. de Alvear 2142
1122 Buenos Aires, Argentina

Several hydrophobic drugs, such as propranolol, some local anesthetics, and phenothiazines, have a direct effect on the sarcoplasmic reticulum (SR) Ca-ATPase activity and the Ca^{2+} permeability of SR membranes.[1,2] Haloperidol is the neuroleptic agent most widely used in clinical medicine. This study was undertaken to determine the possibility of an eventual link between SR function and the neuroleptic malignant syndrome, an infrequent occurrence caused by neuroleptic agents in genetically predisposed patients[3] with increased myoplasmic $[Ca^{2+}]$.[4]

Sarcoplasmic reticulum membrane fractions were obtained from rabbit skeletal muscles.[5] Determinations of Ca-ATPase activity, phosphorylation of the Ca-ATPase with P_i, Ca uptake by SR vesicles, and Ca^{2+} permeability of the SR membranes were made as reported elsewhere.[6] The partition of haloperidol between the aqueous medium and the SR membrane phase was measured with (^{3}H)haloperidol.

Haloperidol inhibits Ca-ATPase activity (FIG. 1A). The apparent K_i is 0.4 mM. The figure also shows the effect of preexposure of SR membranes to the drug. The concentration of the drug during preincubations and incubations affects the ATPase activity differently. The inhibitory effect of haloperidol depends on preincubation conditions: inhibition increases with preincubation time; Ca and Mg protect the enzyme against the effect of the drug: the inhibitory effect of haloperidol decreases upon increasing $[Ca^{2+}]$, at constant [Mg], and disappears at 20 mM [Mg] for any $[Ca^{2+}]$, and at 0.5 mM $[Ca^{2+}]$ for any $[Mg^{2+}]$. Haloperidol also inhibits ATP-dependent Ca uptake (FIG. 1B), with apparent K_i of 0.15 mM, and increases the rate of Ca efflux from preloaded vesicles with apparent affinity of 0.12 mM.

Haloperidol inhibits the phosphorylation of the Ca-ATPase by P_i, as shown in FIGURE 2. The results were analyzed by simulation of a reaction model (inset). The concentration of the drug reducing phosphorylation to one half ranges within 0.10–0.12 mM at the different $[P_i]$.

The partition of (^{3}H)haloperidol between the SR and the aqueous phase largely favors the membrane phase. A 5,000/1 relationship in the concentration of the drug between the membrane lipids and the bulk water was estimated. This value is reduced to one half by mM Mg or Ca. A 1/1 molar relation between haloperidol and the Ca pump protein in the membrane is reached with less than 1 µM haloperidol in the medium.

Results suggest that haloperidol affects the environment of the catalytic site. A

---

[a]Tel: 54 1 961 0350; fax: 54 1 962 0176; e-mail: delia@biofis.odon.uba.ar; alonso@biofis.odon.uba.ar

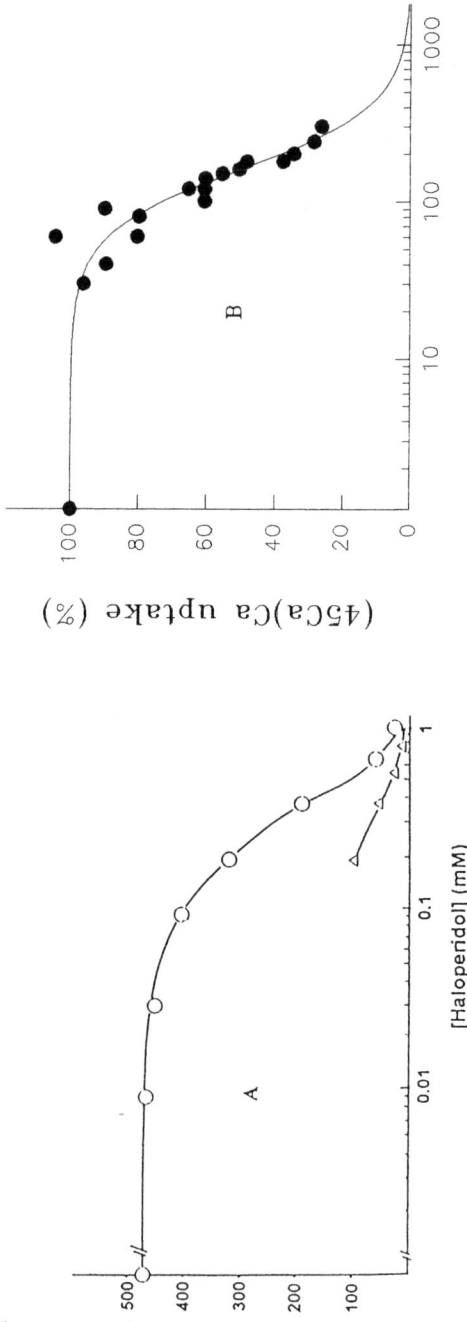

FIGURE 1. Haloperidol inhibits Ca-ATPase activity (**A**) and Ca uptake by sarcoplasmic reticulum (SR) vesicles (**B**). (**A**) P_i production was measured after incubation for 2 minutes at 37°C in 50 mM MOPS-Tris (pH 7.0), 3 mM ATP, 3 mM $MgCl_2$, 100 mM KCl, 0.1 mM $CaCl_2$, 0.1 mM EGTA, 0.01 mM calcimycin, and haloperidol as indicated. Some data (Δ) were obtained with SR membranes preexposed to 1 mM haloperidol for 20 minutes. (**B**) Incubations were in 0.05 mM (^{45}Ca) Ca, 0.05 mM EGTA, and other reactants as in FIGURE 1A, and radioactivity retained by the SR vesicles separated by filtration was measured. Further experimental details are given in ref. 6. (Reprinted with permission from Takara & Alonso.[6])

reaction model in which haloperidol interacts with the E_2 conformers of the enzyme, with lower affinity for the phosphoenzyme than for the dephosphoenzyme species is consistent with the results of the phosphorylations with P_i (FIG. 2). Inhibition of Ca uptake by SR vesicles is ascribed to increased Ca permeability rather than to inhibition of the Ca-ATPase, which requires higher concentrations of the drug. Impairment of the

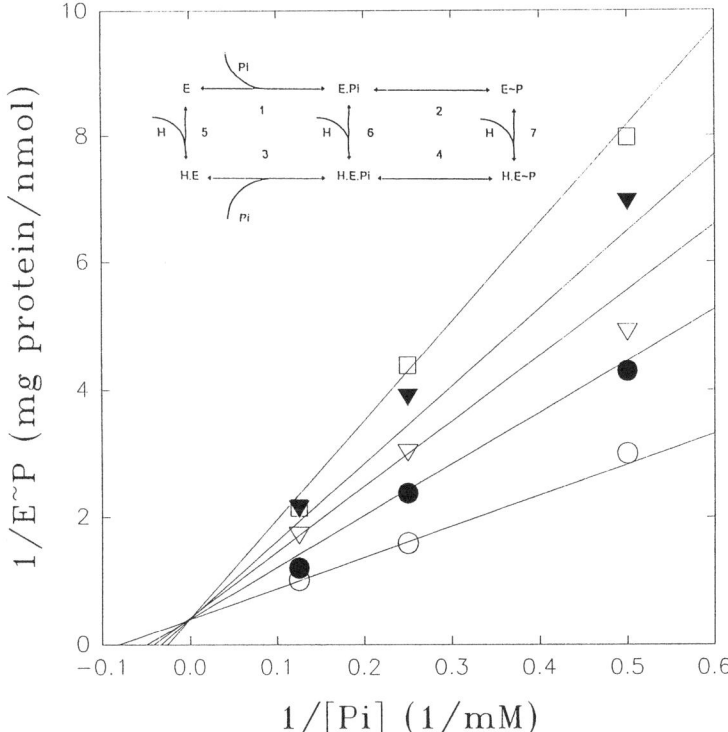

FIGURE 2. Double reciprocal of the phosphoenzyme formed (E ~ P) against [P_i]. Incubations were at 22°C and pH 7.0 in 50 mM MOPS-Tris, 10 mM $MgCl_2$, 1 mM EGTA, (^{32}P)P_i as indicated, and 0 (○), 0.1 (●), 0.2 (▽), 0.3 (▼), or 0.6 (□) mM haloperidol. *Inset:* Reaction model of the phosphorylation reaction. The equilibrium was simulated with the following constants: K_2 = [E.P_i]/[E ~ P] = 0.025, K_4 = [H.E.P_i]/H.E ~ P] = 0.125, and the dissociation constants $K_1 = K_3 = 0.5$ M, $K_5 = K_6 = 1.10^{-4}$ M, and $K_7 = 5.10^{-4}$ M. The *straight lines* (main figure) are drawn through the theoretically calculated results. Further details are given in ref. 6. (Reprinted with permission from Takara & Alonso.[6])

effects of the drug on the Ca-ATPase by the divalent cations occurs either because the cations, acting as cosubstrates at the catalytic site, affect the interaction of the drug or because the cations affect the partition of haloperidol in the membrane phase. The effects of haloperidol on Ca transport in SR membranes are similar to those of the mentioned hydrophobic reactants.[7]

REFERENCES

1. ESCUDERO, B. & C. GUTIERREZ-MERINO. 1987. Biochim. Biophys. Acta **902:** 374–384.
2. DE MEIS, L., M. T. GOMEZ-PUYOU & A. GOMEZ-PUYOU. 1988. Eur. J. Biochem. **171:** 343–349.
3. PEARLMAN, C. A. 1986. J. Clin. Psychopharmacol. **6:** 257–273.
4. LÓPEZ, J. R., V. SÁNCHEZ & M. J. LÓPEZ. 1989. Cell Calcium **10:** 223–233.
5. CHAMPEIL, P., F. GUILLAIN, C. VENIEN & P. M. GINGOLD. 1985. Biochemistry **24:** 69–81.
6. TAKARA, D. & G. L. ALONSO. 1996. Biochim. Biophys. Acta **1314:** 57–65.
7. WOLOSKER, H., A. G. F. PACHECO & L. DE MEIS. 1992. J. Biol. Chem. **267:** 5785–5789.

Changes in Actin Filament Organization Regulate Na^+,K^+-ATPase Activity

Role of Actin Phosphorylation[a]

HORACIO F. CANTIELLO

Renal Unit
Massachusetts General Hospital East
Charlestown, Massachusetts 02129

The actin cytoskeleton frames the specific localization of transmembrane proteins such as the Na^+,K^+-ATPase within the plasma membrane. Ankyrin, for example, links actin and spectrin to the Na^+,K^+-ATPase, thus helping localize the enzyme to the basolateral pole of epithelial cells.[1,2] The molecular mechanisms associated with functional regulation of the Na^+,K^+-ATPase by the actin cytoskeleton, however, are largely unknown. Previous studies have established that the colocalization of actin filaments with epithelial Na^+ channels,[3] for example, is responsible for a novel regulatory mechanism of ion channel function.[3,4] Actin filament organization may also be required for proper protein kinase A activation of both epithelial Na^+ channels[5] and the anion channel CFTR.[6] In this study, the functional role of actin filament organization was assessed on epithelial Na^+,K^+-ATPase.

EFFECT OF ACTIN ON THE PURIFIED Na^+,K^+-ATPase

The Na^+- and K^+-dependent, ouabain-sensitive ATP hydrolysis mediated by purified rat kidney cortex Na^+,K^+-ATPase increased 74% by the addition of actin (24 µM), 2.06 ± 0.03 ($n = 13$) vs 3.58 ± 0.13 µmol $P_i \cdot$ mg prot$^{-1} \cdot$ min^{-1} ($n = 15$, $p < 0.001$) for control and actin-stimulated, respectively. DNAse I, which prevents actin filament formation, returned actin stimulation of the Na^+,K^+-ATPase to control levels. Addition of prepolymerized actin was without effect on Na^+,K^+-ATPase activity. To further assess the nature of the actin-induced regulation of the enzyme, the effect of actin was tested on the various substrates for the Na^+,K^+-ATPase. Actin increased the affinity of the Na^+,K^+-ATPase for its intracellular substrate Na^+ (11.6 ± 0.81 mM vs 7.82 ± 0.62, $n = 5$, $p < 0.01$) but not K^+, Mg^{2+}, or ATP. The stimulatory effect of actin on Na^+,K^+-ATPase activity depended on both the conformation of the enzyme and the actin:Na^+,K^+-ATPase ratio. (For details, see ref. 7.) Actin stimulation of Na^+,K^+-ATPase as high as 296% was observed at a ratio Na^+,K^+-ATPase:actin of 1:50,000. Approximately 4.47 ± 1.08 ($n = 6$) actin monomers per enzyme were required for activation of the enzyme. The regulatory role of actin on the Na^+,K^+-ATPase seems to be elicited by the direct binding of actin to the enzyme, as determined by immunoblotting, and further supported by the finding of putative actin-binding domains in the α subunit of the Na^+,K^+-ATPase.[7]

[a]This study was funded by NIDDK R01-DK48040.

EFFECT OF ACTIN PHOSPHORYLATION ON Na^+,K^+-ATPase ACTIVITY

Conventional second messengers regulate the Na^+,K^+-ATPase. However, the molecular mechanisms associated with regulation of the enzyme by specific phosphorylation events are largely ill-defined. Protein kinase A (PKA) activation associated

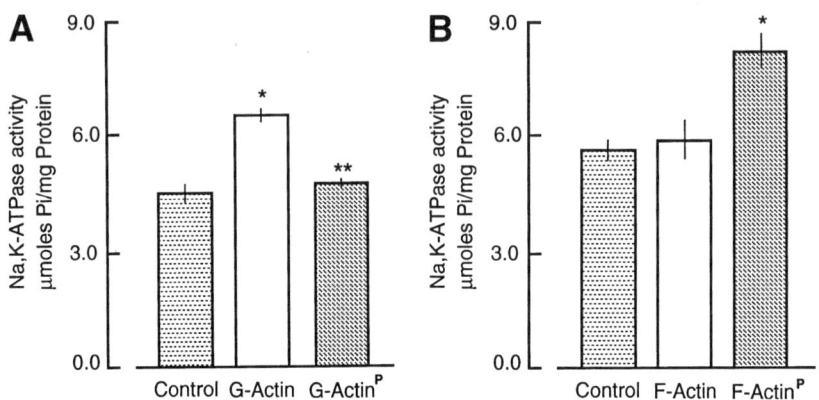

FIGURE 1. Effect of protein kinase A on the actin-mediated regulation of the Na^+,K^+-ATPase. (**A**) Na^+,K^+-ATPase activity was assessed in the absence (*left*) or presence of unpolymerized actin (24 µM) and in the absence (*center*) or presence of PKA (5 µg, *right*, superscripted P). PKA had no effect on Na^+,K^+-ATPase activity in the absence of actin (data not shown). Values are the mean ± SEM of 15, 19, and 20 experiments, respectively. One asterisk (*) indicates statistical significance ($p < 0.05$) with respect to control. Two asterisks (**) indicate statistical difference between samples in the presence of actin with and without PKA. (**B**) Na^+,K^+-ATPase activity was assessed in the absence (*left*) or presence of prepolymerized actin (24 µM) and in the absence (*center*) or presence of PKA (*right*, superscripted P). Values are the mean ± SEM of 16, 18, and 17 experiments, respectively. One asterisk (*) indicates statistical significance ($p < 0.05$) with respect to control. Actin was polymerized for 2 hours at room temperature with KCl, 50 mM, and $MgCl_2$, 1 mM, in the absence or presence of PKA (5 µg) and ATP (1 mM) prior to the Na^+,K^+-ATPase assay. *Na^+,K^+-ATPase activity.* The epithelial Na^+,K^+-ATPase was purified from rat renal cortex as previously described.[12] Briefly, Na^+,K^+-ATPase activity was determined as the linear (15 minutes at 37°C) ^{32}P-ATP hydrolysis. The assay solution (100 µl) contained (in mM): KCl 5.0, $MgCl_2$ 10, EGTA 1.0, Tris-HCl 50, Na_2-ATP 3.0 (Sigma Chem. Co., St. Louis, Missouri), and ^{32}P-ATP (NEN, Billerica, Massachusetts) in tracer amounts (for details, see ref. 7). The ouabain-insensitive ATPase activity was assayed in the presence of ouabain (1 mM, Sigma) in a medium devoid of Na^+ and K^+. Purified rabbit muscle actin (Sigma) was kept unpolymerized in a solution containing (in mM): 10 imidazole-HCl, 0.2 $CaCl_2$, 0.1 ATP, pH 8.0. Whenever indicated, actin was allowed to polymerize for 2 hours in a buffer containing (in mM): KCl 5.0, NaCl 50, $MgCl_2$ 10, ATP 3.0, Tris-HCl 50, pH 7.4.

with cAMP stimulation has been linked to both activation[8] and inhibition[9] of epithelial Na^+,K^+-ATPase. Thus, despite the fact that the Na^+,K^+-ATPase is a target for phosphorylation by kinases, including PKA,[10,11] a molecular role for Na^+,K^+-ATPase regulation by PKA is still lacking. To determine regulatory mechanisms linking changes in actin filament organization with Na^+,K^+-ATPase activity, the effect of

PKA, which reduces the rate of actin polymerization,[5] was assessed by the effect of actin on the enzyme. The actin-stimulated Na^+,K^+-ATPase activity (4.57 ± 0.42 µmol $P_i \cdot$ mg prot$^{-1} \cdot$ min^{-1}, $n = 15$, vs 6.58 ± 0.35, $n = 19$, $p < 0.005$, FIG. 1A) was brought back to control values in the presence of PKA (4.81 ± 0.19 µmol $P_i \cdot$ mg prot$^{-1} \pm$ min^{-1}, $n = 20$, NS, FIG. 1A). By contrast, prepolymerized actin, which does not stimulate Na^+,K^+-ATPase activity (5.73 ± 0.53 µmol $P_i \cdot$ mg prot$^{-1} \cdot$ min^{-1}, $n = 16$, vs 5.98 ± 0.98, $n = 18$, NS, FIG. 1B), was still capable of stimulating Na^+,K^+-ATPase activity by 44% (8.27 ± 0.87 µmol $P_i \cdot$ mg prot$^{-1} \cdot$ min^{-1}, $n = 17$, $p < 0.02$) in the presence of PKA.

In conclusion, the data are consistent with a functional role of actin filament organization, length in particular, on the Na^+,K^+-ATPase. Binding of actin to the Na^+,K^+-ATPase modifies the conformational state of the enzyme, also changing the affinity for the intracellular substrate Na^+. Changes in actin cytoskeleton organization, including those associated with direct phosphorylation of actin by PKA, may play a regulatory role on the Na^+,K^+-ATPase by modifying the rate of actin binding to the enzyme. Actin may be an important target for the regulation of Na^+,K^+-ATPase activity, which is explained in part by modifying actin cytoskeleton dynamics.

ACKNOWLEDGMENTS

The author gratefully thanks Dr. Anita Aperia (Karolinska Institutet, Sweden) for providing the purified enzyme used in the study and Dr. Adriana G. Prat and Mr. George R. Jackson, Jr. for superb technical help.

REFERENCES

1. NELSON, W. J. & P. L. VESHNOCK. 1987. Nature **328**: 533–536.
2. BENNETT, V. 1990. Curr. Opin. Cell Biol. **2**: 51–56.
3. CANTIELLO, H. F., J. STOW, A. G. PRAT & D. A. AUSIELLO. 1991. Am. J. Physiol. **261**: C882–C888.
4. BERDIEV, B., A. PRAT, H. CANTIELLO, D. AUSIELLO, C. FULLER, B. JOVOV, D. BENOS & I. ISMAILOV. 1996. J. Biol. Chem. **271**: 17704–17710.
5. PRAT, A. G., A. M. BERTORELLO, D. A. AUSIELLO & H. F. CANTIELLO. 1993. Am. J. Physiol. **265**: C224–C233.
6. PRAT, A., Y.-F. XIAO, D. AUSIELLO & H. CANTIELLO. 1995. Am. J. Physiol. **268**: C1552–C1561.
7. CANTIELLO, H. F. 1995. Am. J. Physiol. **269**: F637–F643.
8. SILVA, P., J. STOFF & F. EPSTEIN. 1982. Activation of Na^+,K^+-ATPase by cAMP in Shark Rectal Gland.: 1-215–221. Elsevier Science Publishing. New York.
9. APERIA, A., A. M. BERTORELLO & I. SERI. 1987. Am. J. Physiol. **252**: F39–F45.
10. MARDH, S. 1983. Curr. Top. Membr. Transp. **19**: 999–1004.
11. BEGUIN, P., A. T. BEGGAH, A. V. CHIBALIN, P. BURGENER-KAIRUZ, F. JAISSER, P. M. MATHEWS, B. C. ROSSIER, S. COTECCHIA & K. GEERING. 1994. J. Biol. Chem. **269**: 24437–24445.
12. JORGENSEN, P. L. 1974. Biochim. Biophys. Acta **356**: 36–52.

Specific Expression and Regulation of CHIF in Kidney and Colon

CLAUDIA CAPURRO,[a,d] NATHALIE COUTRY,[a]
JEAN-PIERRE BONVALET,[a] BRIGITTE ESCOUBET,[b]
HAIM GARTY,[c] AND NICOLETTE FARMAN[a]

Institut National de la Santé et de la Recherche Médicale
[a]*U246 and* [b]*U426*
Institut Fédératif de Recherches "Cellules épithéliales"
Faculté de Médecine Xavier Bichat
BP 416
75870 Paris Cedex 18, France

[c]*Department of Membrane Research and Biophysics*
The Weizmann Institute of Science
Rehovot, 76100, Israel

Channel-inducing factor (CHIF) is a novel cDNA recently cloned from a rat distal colon cDNA library of dexamethasone-treated rats.[1] CHIF primary structure is close ($>50\%$ similarity) to that of putative regulatory proteins, such as phospholemman, γ subunit of Na^+-K^+-ATPase, or Mat-8. Expression of CHIF in *Xenopus laevis* oocytes sometimes evokes a K^+-specific channel activity similar to that induced by the K^+ channel Isk (minK). However, CHIF and Isk show no homology. We examined the cellular localization of CHIF mRNA in kidney, intestine, and other organs and whether its expression could be modulated.[2] Both *in situ* hybridization and RNase protection assay were used.

CHIF mRNA expression is restricted to limited cell types. It was found only in epithelial cells of the distal colon, where its expression increased towards the tip of the villi and in the terminal portions of the nephron, including cortical, medullary, and papillary collecting duct (FIG. 1). No specific signal was detected in proximal colon, glomerulus, proximal tubule, loop of Henle, and distal tubule. Also negative were the small intestine, lung, choroid plexus, salivary glands, skin, and brain. This is a unique pattern of expression, not reported before for any other epithelial mRNA, except and only partially for the colonic H^+-K^+-ATPase.[3]

As the renal medullary collecting duct and the distal colon are involved in the reabsorption of large amounts of ions and water and as corticosteroid hormones modulate the functions of these epithelia, we investigated if CHIF expression could be modulated by corticosteroids and low sodium diet, by potassium depletion, and by acidosis. In the distal colon, a lower signal was observed after adrenalectomy as compared to control rats. A clear increase in colonic CHIF mRNA was seen in animals treated for 7 days with 30 μg · 100 g body wt^{-1} · day^{-1} aldosterone or 2 days with 1.2 mg · 100 g body wt^{-1} · day^{-1} dexamethasone (DEX). By contrast, in the kidney, no change in CHIF mRNA was present between adrenalectomy- and corticosteroid-

---

[d]Tel: 33-1-44.85.63.23; fax: 33-1-42.29.16.44; e-mail: u246@bichat.inserm.fr

treated rats. Sodium restriction (10 days), which raises endogenous aldosterone levels, induced a clear increase in CHIF expression in the epithelial cells of the colon, while its expression did not change in the renal medullary and papillary collecting duct. Low potassium diet (low K, 4 days) resulted in a significant increase in colonic CHIF mRNA, whereas no change was apparent in the kidney. Metabolic acidosis (6 days) resulted in increased expression of CHIF mRNA in both kidney and colon. On the whole, all these changes alter colonic expression of CHIF, whereas renal expression is very stable, except after metabolic acidosis. Because CHIF is induced by metabolic

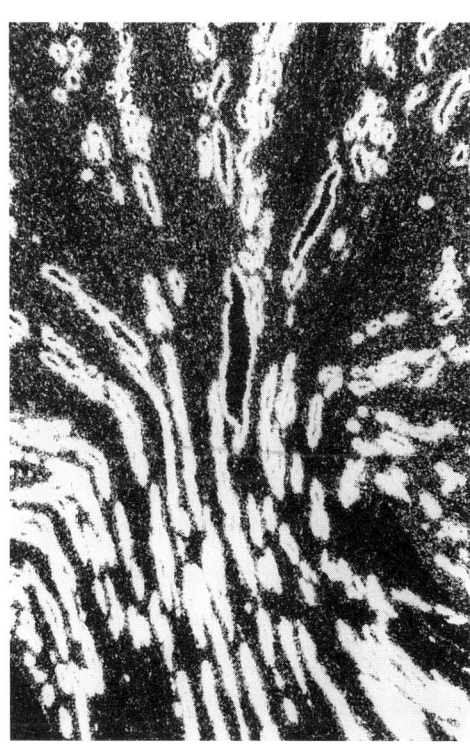

FIGURE 1. Channel-inducing factor (CHIF) mRNA expression in the kidney of control rats. *In situ* hybridization was performed on kidney sections from control rats using an antisense probe. CHIF mRNA (dark field photograph) is expressed all along the terminal parts of the collecting duct, while distal and connecting tubules, as well as loop of Henle, proximal tubules, and glomerulus are negative.

acidosis in both distal colon and kidney, proton transporters are good candidates to consider as putatively CHIF-regulated proteins. It was found that metabolic acidosis also regulates the colonic putative H^+-K^+-ATPase mRNA in both kidney and distal colon. TABLE 1 compares the observed variations of both CHIF and colonic H^+-K^+-ATPase mRNAs; only partial parallelism is apparent. Therefore, if CHIF cannot be excluded as a regulator of H^+-K^+-ATPase(s), it is not possible to relate CHIF expression solely to this function.

In conclusion, the current study demonstrates that CHIF is selectively expressed in distal parts of the nephron and the gastrointestinal tract. It is regulated by several factors known to modulate epithelial ion transport in a tissue-specific manner. The cellular role of this epithelial transcript remains to be identified.

TABLE 1. Comparison of Expression of Channel-Inducing Factor (CHIF) and Colonic Putative H^+-K^+-ATPase mRNAs

	Distal Colon		Kidney	
	CHIF	H^+-K^+-ATPase	CHIF	H^+-K^+-ATPase
Adrenalectomy	↘	↘ [a]	→	→ [a]
Aldosterone	↗	↗ [a]	→	→ [a]
Dexamethasone	↗	↗ [a]	→	→ [a]
Low K	↗	→ [a]	→	↗ [a]
Acidosis	↗	↗	↗	↗

[a]Data from ref. 3.

REFERENCES

1. ATTALI, B., H. LATTER, N. RACHAMIM & H. GARTY. 1995. A corticosteroid-induced gene expressing an "IsK-like" K^+ channel activity in *Xenopus* oocytes. Proc. Natl. Acad. Sci. USA **92:** 6092–6096.
2. CAPURRO, C., N. COUTRY, J. P. BONVALET, B. ESCOUBET, H. GARTY & N. FARMAN. 1996. Cellular localization and regulation of CHIF in kidney and colon. Am. J. Physiol. **271** (Cell Physiol. 40): C753–C762.
3. JAISSER, F., B. ESCOUBET, N. COUTRY, E. EUGENE, J. P. BONVALET & N. FARMAN. 1996. Differential regulation of putative K-ATPase by low-K diet and corticosteroids in rat distal colon and kidney. Am. J. Physiol. **270** (Cell Physiol. **39**): C679–C687.

Effects of Extracellular Sodium Concentration on the Activity of Na,K-ATPase in Dogfish Rectal Gland Epithelial Cells[a]

J. EDWARDS, S. MACKENZIE, C. P. CUTLER, AND G. CRAMB[b]

School of Biological and Medical Sciences
Bute Medical Buildings
University of St Andrews
Fife, UK KY16 9TS

The rectal gland of the European dogfish (*Scyliorhinus canicula*), in combination with the kidney and gill, works to preserve plasma electrolyte balance in the presence of a constant uptake of NaCl from sea water through permeable body surfaces (mainly the gill) and the intermittent uptake of NaCl from food and ingested sea water by the gut. This gland is composed of a network of secretory tubules, veins, and arteries and has a central region consisting of a major vein and a central canal that terminates at the drainage duct of the rectum. The secretory tubules actively transport NaCl from the blood into the tubular lumen at concentrations of approximately 0.5 M (twice that of the plasma), and this fluid then drains from the ducts into the central canal before being excreted from the fish. The cells that make up the secretory tubules possess numerous mitochondria and a greatly expanded basolateral plasma membrane that typifies cells engaged in transepithelial ion transport. These epithelial cells exhibit high levels of expression of the Na,K-ATPase (sodium pump) and can easily be isolated and cultured from the dogfish rectal gland using a collagenase perfusion technique. The cell cultures provide a suitable model for studying the hormonal and ionic regulation of Na,K-ATPase activity and expression. In this report we describe how raising the extracellular sodium concentration of the growth medium induces a transient but marked increase in Na,K-ATPase activity in cell homogenates.

METHODS

Rectal gland epithelial cells were isolated by collagenase digestion. The rectal gland was perfused at room temperature with 5 ml calcium and magnesium-free Shark Ringer (SR, 240 mM NaCl, 7 mM KCl, 23 mM NaHCO$_3$, 0.5 mM Na$_2$HPO$_4$, 0.5 mM Na$_2$SO$_4$, 360 mM urea, 60 mM trimethylamine oxide and 1% (w/v) glucose, pH 7.4, by gassing with 95% air/5% CO$_2$) containing 1 mM EGTA followed by 5 ml SR

---

[a]This work was supported by the Natural Environment Research Council and the Wellcome Trust. J. E. and S. M. are recipients of NERC studentships.

[b]Corresponding author. Tel: (+) 1334 463530; fax: (+) 1334 463600; e-mail: gc@st-and.ac.uk

containing 100 µM $CaCl_2$ and 2 mg/ml collagenase (Boehringer type D). The outer capsule of the gland was then removed, and the digested tissue was suspended in SR using a wide bore needle and syringe. Epithelial cells in the suspension were isolated by centrifugation (200 × g_{max}) for 90 seconds at 4°C, resuspended in growth medium (Hams F-12/ 95 mM NaCl, 360 mM urea, 60 mM trimethylamine oxide, 5% (v/v) Nu-serum, 1% (v/v) ITS$^+$, 100 units/ml penicillin/streptomycin), and plated on collagen-coated plates. Cell cultures were grown at 20°C for 5–7 days in an atmosphere of 95% air/5% CO_2. Confluent monolayers of rectal gland cells were incubated for various times with medium containing increased sodium chloride concentrations and other effectors, before being homogenized and assayed for Na,K-ATPase activity. Maximal Na,K-ATPase activity was determined in the presence of 0.001% sodium deoxycholate at 20°C essentially as reported by Esmann.[1]

RESULTS

Na,K-ATPase activity in homogenates of rectal gland epithelial cell cultures increased approximately 3.5-fold in response to a 12-hour incubation of cells in a 'high salt' medium containing a 50% increase in NaCl concentration (FIG. 1). This was a delayed and transient response with the first significant increase in Na,K-ATPase activity found in homogenates of cells incubated for 6 hours in the high salt medium, but within 24 hours activity returned to control levels. The increase in Na,K-ATPase

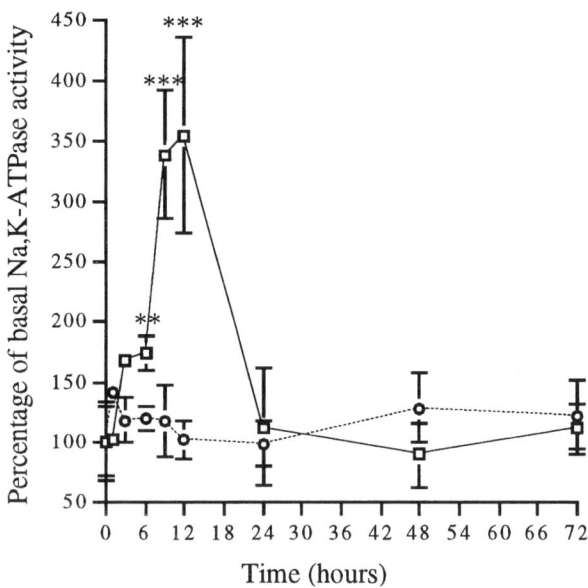

FIGURE 1. Cells were incubated for the times indicated in medium containing either 240 mM NaCl (normal growth medium) (○) or 360 mM NaCl (□). Cells were then homogenized and Na,K-ATPase activities were determined. Each point represents the mean ± SEM for four separate experiments. **$p < 0.01$; ***$p < 0.001$.

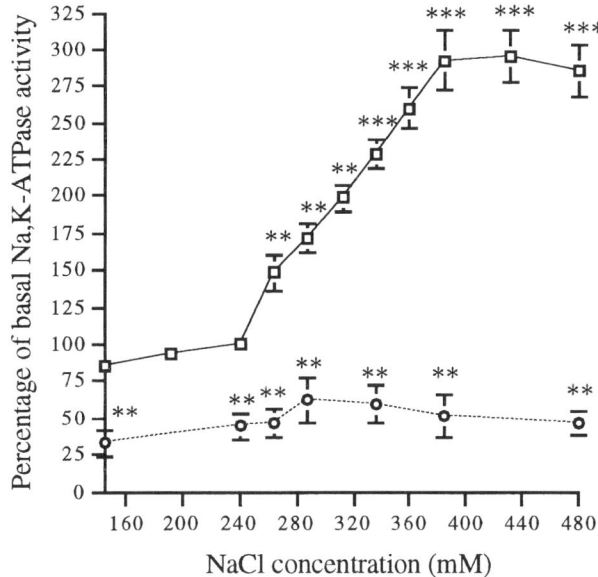

FIGURE 2. Cells were incubated for 12 hours in growth medium containing the NaCl concentrations indicated, in either the absence (□) or the presence (○) of 0.1 mM bumetanide. Cells were homogenized and Na,K-ATPase activities were determined. Each point represents the mean ± SEM for four separate experiments. **$p < 0.01$; ***$p < 0.001$.

activity was not due to an increase in osmolality, as the addition of equivalent osmotic concentrations of mannitol, choline chloride, or Na_2SO_4 to the medium had no effect on basal Na,K-ATPase activity (results not shown). This increased activity was therefore dependent on increases in the concentrations of both Na and Cl ions in the medium.

Inclusion of the sodium ionophore monensin to the growth medium also stimulated Na,K-ATPase activity (results not shown). Although the rate of increase in activity was similar to that found with elevated NaCl, the effect was not transient and elevated activities were maintained for the duration of the experiment (72 hours).

Cells incubated for 12 hours in the presence of various sodium chloride concentrations exhibited characteristic dose-dependent increases in Na,K-ATPase activity (FIG. 2). Small increases (10%) in extracellular [NaCl] resulted in relatively large increases (50%) in Na,K-ATPase activity. Maximal stimulation of activity was found when NaCl concentrations were increased by 60%.

The NaCl-mediated increases in activity were completely inhibited by the addition of the loop diuretic bumetanide to the growth medium (FIG. 2). In addition, basal Na,K-ATPase activities in rectal gland epithelial cell homogenates were also reduced by over 50% (FIG. 2) by incubating the cells in bumetanide with maximum reductions found within 30 minutes of the addition of the loop diuretic (results not shown). This effect was reversible with activities returning to control values within 30 minutes of bumetanide removal. The extent of recovery from this inhibition, however, was dependent on the length of time the cells were in the presence of bumetanide incubation with the $t_{1/2}$ for loss of pumps being approximately 4 hours (results not shown).

CONCLUSIONS

Increasing the NaCl concentration of the growth medium resulted in a transient three- to fourfold increase in Na,K-ATPase activity in cell homogenates, peaking approximately 12 hours after the medium change. The response depended on both sodium and chloride ions and was also inhibited by the loop diuretic bumetanide, indicating that entry of the ions into the cell occurred via the Na/K/Cl cotransporter. Incubation of cells in normal medium in the presence of the sodium ionophore monensin also resulted in a sustained increase in Na,K-ATPase activity. These results suggest that intracellular [Na] is an important regulator of the *de novo* synthesis and/or cycling of Na,K-ATPase units between plasma membrane and some intracellular vesicular compartment. Increases in $[Na]_i$ result in increased synthesis and recruitment of active Na,K-ATPase units to the plasma membrane. Decreasing $[Na]_i$ results in the rapid uptake of pumps into some intracellular store. The Na,K-ATPase units appear to be only functionally active when in the plasma membrane, and prolonged increases in intracellular [Na] may result in a delayed inactivation of the Na/K/Cl cotransporter.

REFERENCE

1. ESMANN, M. 1988. Methods Enzymol. **156:** 105–115.

Phosphorylation Site-Independent Downregulation of Na-Pump Current in A6 Epithelia by Protein Kinase C

Decrease in Na,K-ATPase Cell-Surface Expression

JÖRG BERON AND FRANÇOIS VERREY[a]

Institute of Physiology
University of Zürich
Winterthurerstrasse 190
CH-8057 Zürich, Switzerland

The role of protein kinase C (PKC) in regulating the function of Na pumps in the context of epithelia formed by A6 cells (*Xenopus laevis* distal nephron) was investigated after permeabilization of the apical membrane with amphotericin B. The function of the Na pumps was measured as Na-pump current (I_p) at controlled intracellular Na concentrations.[1]

The stimulation of PKC by phorbol myristate acetate (PMA), which was prevented by the PKC inhibitor bisindolylmaleimide GF 109203X (BIM), decreased the I_p by 28–35% within 25 minutes independent of the intracellular Na concentration. In addition to this major effect of PMA, a further BIM-resistant decrease in I_p was observed at low Na concentrations (total reduction ~50% at 5 mM Na). To determine if phosphorylation of the Na,K-ATPase by PKC was playing a role in downregulation of Na-pump function by PMA, we established cell lines expressing wild-type or PKC site-mutant[2] ouabain-resistant α1 subunits from *Bufo marinus*. Residues Thr-15 and Ser-16 which are replaced by two Ala in this mutant α1 subunit had been shown to be the unique phosphorylation sites for PKC *in vitro* and in intact cells.[2] PMA produced a (BIM-sensitive and BIM-resistant) downregulation of the I_p carried by Na pumps containing exogenous wild-type or mutant α1 subunit equal to that observed for endogenous Na pumps (FIG. 1).

To measure the effect of PKC stimulation on cell-surface expression of Na,K-ATPase, independent of its functional state, we labeled the basolateral surface proteins with sulfosuccinimidobiotin and visualized the biotinylated immunoprecipitated β subunit by streptavidin blotting. The amount of β subunits labeled at the cell surface was reduced by PKC stimulation by 20% within 25 minutes. The same treatment was shown to stimulate endocytosis sevenfold measured as uptake of the fluid phase marker horseradish peroxidase. Furthermore, using capacitance measurements to estimate the basolateral surface area of apically amphotericin-permeabilized epithelia, we could demonstrate that PMA treatment produced 16% reduction of the basolateral cell surface area.

In conclusion, PKC stimulation produces withdrawal of Na pumps and a corre-

---

[a]Address for correspondence: François Verrey, Institute of Physiology, University of Zürich, Winterthurerstrasse 190, CH-8057 Zürich, Switzerland (Tel: +41 1 635-5044/37; fax: +41 1 635-6814; e-mail: Verrey@physiol.unizh.ch).

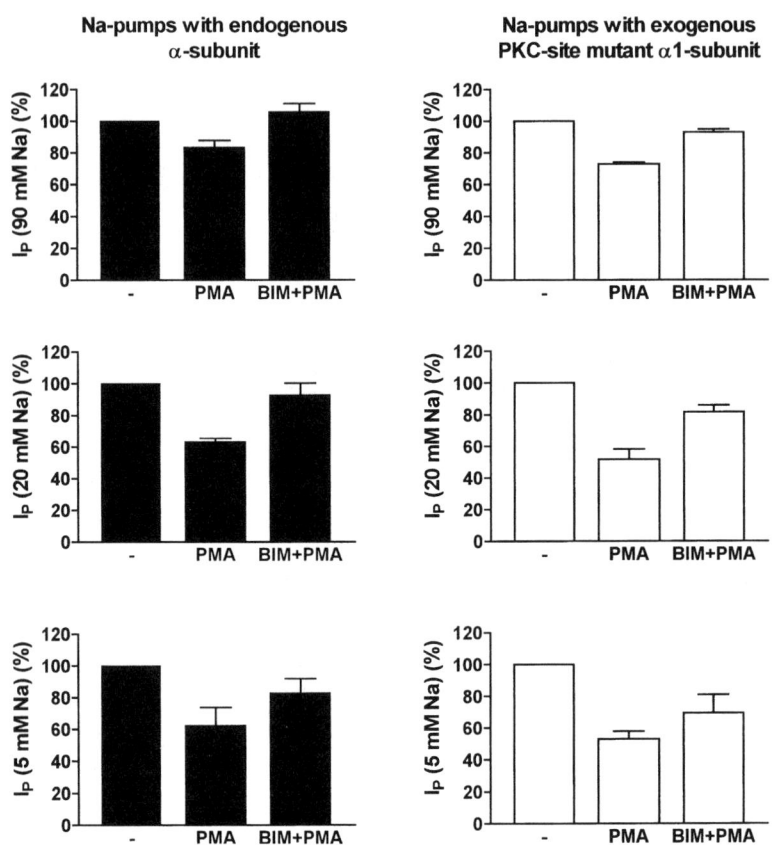

FIGURE 1. Phorbol myristate acetate (PMA) decreases the pump current (I_p) carried by endogenous Na pumps and that carried by Na pumps containing an exogenous wild-type or protein kinase C (PKC)-site mutant α1 subunit to the same extent. A6 epithelia were apically permeabilized for 25 minutes with 20 μg/ml amphotericin B and concomitantly treated with 100 nM PMA ± 5 μM BIM or vehicle. The Na pumps were then activated with 90, 20, or 5 mM Na. The current inhibited by 20 μM strophanthidin was taken as I_p of endogenous pumps and the remaining I_p blocked by 2 mM ouabain as that carried by the pumps containing an exogenous α1 subunit. Bars represent mean fractional changes. The SE is indicated for 90 and 20 mM Na ($n = 3$), $n = 2$ for 5 mM Na.

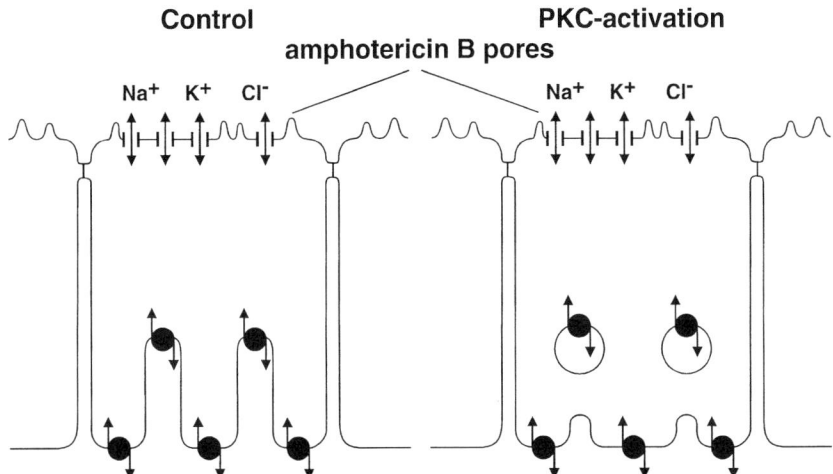

FIGURE 2. The phorbol myristate acetate-induced decrease in Na pump current across A6 epithelia can be explained by withdrawal of Na pumps from the basolateral cell surface which appears to be due to an increase in basolateral endocytosis and a consequent reduction of the basolateral cell surface. The Na-pump surface expression was tested by cell-surface biotinylation followed by immunoprecipitation, endocytosis was measured as uptake of the fluid phase marker horseradish peroxidase, and the basolateral surface was measured by electrical capacitance measurements.

sponding reduction in basolateral membrane area in A6 epithelia which can account for the observed reduction in Na-pump current (FIG. 2). This effect is independent of PKC phosphorylation sites located in the NH_2-terminal domain of the $\alpha1$ subunit.

REFERENCES

1. BERON, J., L. MASTROBERARDINO, A. SPILLMANN & F. VERREY. 1995. Mol. Biol. Cell **6:** 261–271.
2. BEGUIN, P., A. T. BEGGAH, A. V. CHIBALIN, P. BURGENER-KAIRUZ, F. JAISSER, P. M. MATHEWS, B. C. ROSSIER, S. COTECCHIA & K. GEERING. 1994. J. Biol. Chem. **269:** 24437–24445.

Regulation of the α2β1 and α3β1 Isozymes of the Na,K-ATPase by Ca^{2+}, PKA, and PKC[a]

GUSTAVO BLANCO[b] AND ROBERT W. MERCER

Department of Cell Biology and Physiology
Washington University School of Medicine
St. Louis, Missouri 63110

The molecular mechanisms underlying the regulation of the Na,K-ATPase are poorly understood. Of the various isoforms of the catalytic α (α1, α2, α3, and α4) and glycosylated β (β1 and β2) subunits that comprise the Na,K-ATPase of mammalian cells,[1] the α1β1 isozyme has generally been used to study the modulation of activity by intracellular Ca^{2+} and regulatory phosphorylation/dephosphorylation.[2,3] On the other hand, there is relatively little evidence for the regulation of the α2 and α3 isoforms by second messengers. Here we show the effects of Ca^{2+} and activators of protein kinase A and C (PKA, PKC) on the activity of the Na,K-ATPase isozymes of the rat, expressed in *Sf*-9 insect cells using recombinant baculoviruses.

METHODS

Sf-9 cells grown in 150-mm petri dishes were infected with recombinant baculoviruses containing the rat α1, α2, α3, and β1 Na,K-ATPase cDNAs.[4] At 40 hours postinfection, cells were treated with the PKA and PKC modulators. For PKA activation, 2 mM of $N^6,2'$-O-dibutyryladenosine 3′:5′-cyclic monophosphate (dibutyryl cAMP) was used for 1 hour. As an antagonist of PKA, 8-bromoadenosine 3′,5′ cyclic monophosphothioate, Rp-isomer was added at a concentration of 50 µM 1 hour before the addition of the dibutyryl cAMP. PKC activation was obtained with phorbol 12-myristate 13-acetate (PMA) at 1 µM for 1 hour. Staurosporine (200 nM) or 50 µM 1-(5-isoquinolinesulfonyl)-2-methylpiperazine (H7) was used as PKC inhibitors 1 hour before and during treatment with PMA. After treatment, cells were harvested, centrifuged at 1,500 × g for 10 minutes, and resuspended in 10 mM imidazole hydrochloride (pH 7.5), 1 mM EGTA, with 100 nM tautomycin to inhibit protein phosphatases.

Enzyme activity from permeabilized intact cells was determined by the initial rate of release of $^{32}P_i$ from [γ-^{32}P]ATP.[4] For Ca^{2+} dose-response curves, a cellular membrane fraction was prepared.[4] Maximal binding of [^{3}H]ouabain was performed as

---

[a]This work was supported by National Institutes of Health grant GM 39746 and the George M. O'Brien Kidney and Urological Diseases Center at Washington University School of Medicine.

[b]Address for correspondence: Dr. Gustavo Blanco, Department of Cell Biology and Physiology, Washington University School of Medicine, 660 S. Euclid Avenue, St. Louis, MO 63110 (tel: 314-362-6922; fax: 314-362-7463; e-mail: gblanco@cellbio.wustl.edu).

before[4] and was calculated from Scatchard plots of the data, which in all cases showed a single population of binding sites. Curve fitting of the experimental data was carried out using the equations described previously.[5] Free Ca^{2+} concentrations were calculated with a program written by M. Kurzmak (University of Maryland), which was based on the formulations of Fabiato and Fabiato.[5]

RESULTS AND DISCUSSION

To investigate the effect of Ca^{2+} on Na,K-ATPase isozyme activity, the activity of the isozymes in the presence of various concentrations of Ca^{2+} was determined. The Ca^{2+} dose-response curves indicated that the individual isozymes are differentially inhibited by Ca^{2+}. For example, the calculated inhibition constants of the $\alpha1\beta1$, $\alpha2\beta1$, and $\alpha3\beta1$ isozymes were: $1.0 \pm 0.2 \times 10^{-4}$ M, $7.3 \pm 4.6 \times 10^{-6}$ M, and $1.9 \pm 1.0 \times 10^{-5}$ M, respectively. This suggests that in excitable cells after depolarization, when the intracellular Ca^{2+} concentration approaches 5–10 µM, the $\alpha1$ isozyme remains active, while the $\alpha2$ and $\alpha3$ are functioning at approximately half their maximal capability. Thus, intracellular Ca^{2+} may regulate cellular contractility and excitability by selectively inhibiting specific Na,K-ATPase isozymes. This may be particularly relevant in the heart, where the rise in intracellular Ca^{2+} elicited by the cardiotonic steroids may be enhanced by further inhibition of the ouabain-sensitive $\alpha2$ and $\alpha3$ isoforms.

Phosphorylation of discrete Ser/Thr residues of the $\alpha1$ subunit influences the activity of the enzyme. Although the effect is cell-type dependent, phosphorylation has primarily been associated with inhibition of the Na,K-ATPase.[3] To determine if protein kinases also influence the function of the other isoforms, the effects of activators of PKA and PKC on the specific activity of Na-pump isozymes were investigated. FIGURE 1 shows that activation of PKC with the phorbol ester, phorbol 12-myristate 13-acetate (PMA), results in a decrease in activity of the $\alpha1\beta1$, $\alpha2\beta1$, and $\alpha3\beta1$ isozymes expressed in *Sf*-9 insect cells. The effect of PMA could partially be reversed by treating the cells with the specific inhibitors of PKC, staurosporin and H7. Moreover, Na,K-ATPase activity could be restored by treatment with protein phosphatase 2A. To assess whether the effect of PMA was a result of complete inactivation of a fraction of the Na,K-ATPase or partial inactivation of the total Na,K-ATPase, the turnover numbers of the $\alpha2$ and $\alpha3$ isoforms were determined. The molecular activity of the isozymes was estimated from the maximal Na,K-ATPase activity and the amount of equilibrium [^{3}H]ouabain binding. These values indicated that PMA treatment of the $\alpha2$ and $\alpha3$ isoforms reduced molecular activity by $71 \pm 4\%$ and $55 \pm 2\%$ of the control, respectively. This decrease in molecular activity accounts for the similar reduction in ATP hydrolysis. In contrast to these results, PKA activation by dibutyryl cAMP resulted in an increase in $\alpha3\beta1$ activity and a decrease in $\alpha1\beta1$ and $\alpha2\beta1$. This effect was blocked by the PKA competitive inhibitor 8-bromoadenosine cyclic monophosphothioate Rp-isomer (FIG. 2). Once more, the variation in activity paralleled changes in the molecular activity of the treated isozymes, with values of $88 \pm 6\%$ and $123 \pm 4\%$ for the $\alpha2$ and $\alpha3$ isoforms, respectively. The mechanisms by which PKA and PKC regulate the Na,K-ATPase isozymes are difficult to explain. For the rat $\alpha1$ isoform, phosphorylation by PKA has been mapped to Ser943, while PKC phosphorylates the polypeptide on Ser16 and Ser23. Ser943 is conserved in the $\alpha2$ and $\alpha3$ polypeptides; however, Ser16 is not and Ser23 is present in α1 and α3. Thus, the effect of PKA and PKC activators may depend on phosphorylation of additional groups on the α polypeptides, on differential modifications of the structure and hence function of the proteins, or on the involvement of another second messenger(s). In any case, because the Na,K-ATPase isoforms have unique kinetic characteristics,[4] the differential modulation of their activity may be important in adapting Na-pump function to specific cellular requirements.

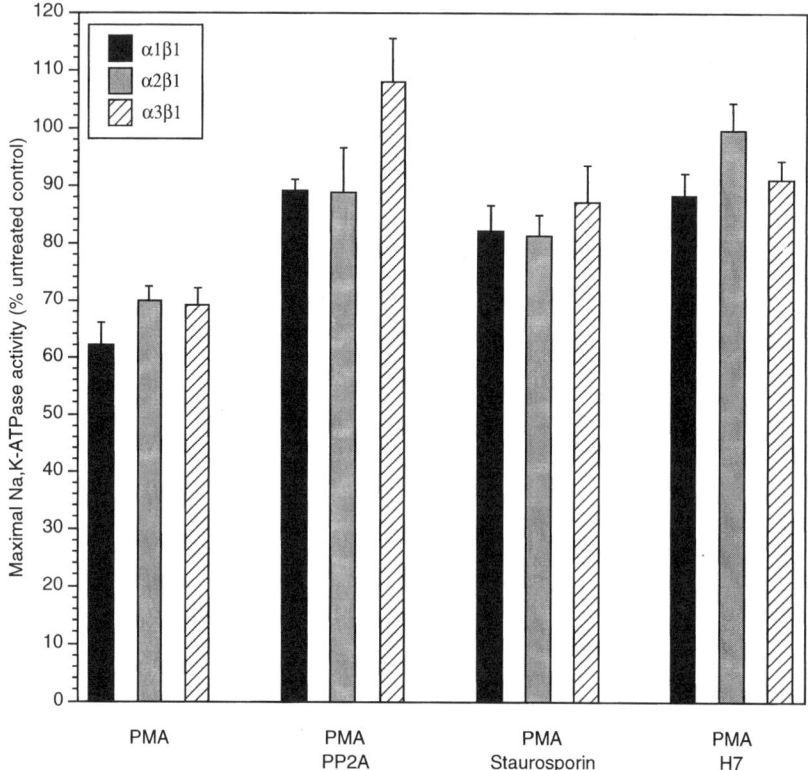

FIGURE 1. Effect of phorbol 12-myristate 13-acetate (PMA) on the Na,K-ATPase activity of the rat α1β1, α2β1, and α3β1 Na,K-ATPase isozymes. *Sf*-9 cells expressing the α1β1, α2β1, or α3β1 isoforms were treated 40 hours after infection with PMA alone (1 μM for 1 hour) or with the addition of either staurosporine (200 nM) or H7 (50 μM). Cells were then processed for Na,K-ATPase activity. Aliquots of the cells treated with only PMA were homogenized and incubated for 1 hour at 25°C in the absence and presence of 2 U/mg protein of protein phosphatase 2A (PP2A). Na,K-ATPase activity was assayed in a reaction medium containing: 120 mM NaCl, 30 mM KCl, 3 mM $MgCl_2$, 0.2 mM EGTA, 3 mM [γ-^{32}P]ATP-cold ATP, 30 mM Tris-HCl (pH 7.4) in the absence or presence of 1 mM ouabain. Values are expressed as a percentage of the untreated cells, taken as 100%. Each value is the mean, and error bars represent the standard errors of the mean of three to seven experiments performed in triplicate on samples obtained from different infections.

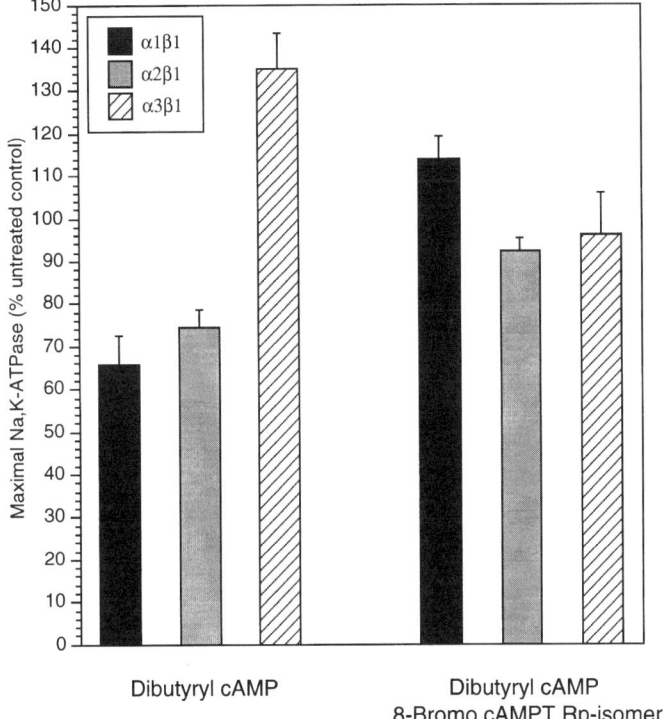

FIGURE 2. Effect of $N^6,2'$-O-dibutyryladenosine 3′:5′-cyclic monophosphate (dibutyryl cAMP) on the Na,K-ATPase activity of the rat $\alpha1\beta1$, $\alpha2\beta1$, and $\alpha3\beta1$ Na,K-ATPase isozymes. *Sf*-9 cells expressing the $\alpha1\beta1$, $\alpha2\beta1$, or $\alpha3\beta1$ isoforms were treated 40 hours after infection with dibutyryl cAMP alone (2 mM for 1 hour) or with the addition of 50 μM of 8-bromo adenosine cyclic monophosphothioate, Rp-isomer (8-bromo cAMPT). Na,K-ATPase activity was determined as described in Figure 1. Values are expressed as percentage of the untreated cells, taken as 100%. Each value is the mean, and error bars represent the standard errors of the mean of three to seven experiments performed in triplicate on samples obtained from different infections.

REFERENCES

1. SWEADNER, K. J. 1989. Isozymes of the Na,K-ATPase. Biochim. Biophys. Acta **988:** 185–220.
2. BEAUGÉ, L. & M. A. CAMPOS. 1983. Calcium inhibition of the ATPase and phosphatase activities of Na,K-ATPase. Biochim. Biophys. Acta **729:** 137–149.
3. BERTORELLO, A. M. & A. I. KATZ. 1993. Short term regulation of renal Na,K-ATPase activity: Physiological relevance and cellular mechanisms. Am. J. Physiol. **265:** F743–F755.
4. BLANCO, G., G. SÁNCHEZ & R. W. MERCER. 1995. Comparison of the enzymatic properties of the Na,K-ATPase $\alpha3\beta1$ and $\alpha3\beta2$ isozymes. Biochemistry **34:** 9897–9903.
5. FABIATO, A. & F. FABIATO. 1979. Calculator programs for computing the composition of the solutions containing multiple metals and ligands used for experiments in skinned muscle cells. J. Physiol. **75:** 463–505.

Roles of PKA and PKC in Regulation of Na$^+$ Pump Activity in Vascular Smooth Muscle Cells

MIKHAIL L. BORIN[a]

*Department of Physiology and
Center for Vascular Biology and Hypertension
University of Maryland School of Medicine
Baltimore, Maryland 21201*

Recent studies demonstrated that activity of the Na$^+$ pump can be modulated by phosphorylation/dephosphorylation processes.[1,2] The goal of this study was to determine if and how protein kinase C (PKC), protein kinase A (PKA), and phosphatases regulate the Na$^+$ pump activity in vascular smooth muscle cells. Also, the objectives of the study were to monitor the activity of the Na$^+$ pump in live cells and to be able to discern a direct effect of agents on the pump from changes in the pump's activity secondary to changes in intracellular Na$^+$.

Intracellular free Na$^+$ concentration, [Na$^+$]$_i$, was measured in primary cultured rat aortic myocytes using Na$^+$-sensitive fluorescent indicator SBFI and digital imaging microscopy as described.[3,4] Cells were loaded with Na$^+$ up to 30–40 mM by temporary inactivation of the Na$^+$ pump in K$^+$-free medium. Thereafter, cells were exposed to Na$^+$-free, K$^+$-containing medium in the presence or absence of test substances (FIG. 1). The resulting decline in [Na$^+$]$_i$ reflected Na$^+$ efflux. The rate of Na$^+$ efflux was assessed as time constant of [Na$^+$]$_i$ decline. Preliminary experiments demonstrated that agents used in this study affected only the Na$^+$ pump-mediated component of Na$^+$ efflux, because they had no effect on changes in [Na$^+$]$_i$ evoked in the presence of ouabain.

The experimental protocol is illustrated in FIGURE 1. The results of the experiments are summarized in FIGURE 2 and TABLE 1. Reduction in the rate of Na$^+$ pump-mediated decrease in [Na$^+$]$_i$ connotes inhibition of the Na$^+$ pump, and conversely augmentation in the rate of decrease in [Na$^+$]$_i$ connotes stimulation of the pump.

In addition to the experiments first described, a possible role of dephosphorylation processes was studied using the phosphatase inhibitor caliculin. Caliculin (20 nM) reduced the rate of Na$^+$ pump-mediated Na$^+$ efflux, which suggested activation of the Na$^+$ pump by dephosphorylation. Furthermore, caliculin augmented inhibition of the Na$^+$ pump by PKA activators (8-Br cAMP, forskolin); this effect is consistent with activatory effect of dephosphorylation on the Na$^+$ pump.

---

[a]Address for correspondence: Mikhail L. Borin, Ph.D., 9 Empire Ct., Reisterstown, MD 21136 (tel: 410 526-7863).

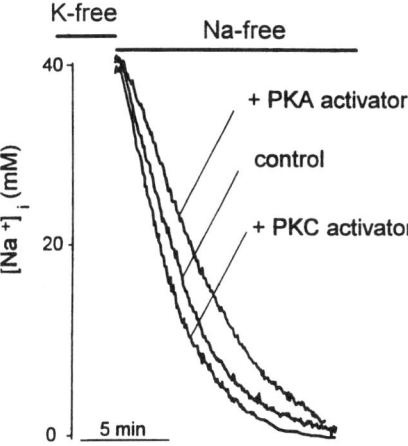

FIGURE 1. Experimental protocol.

Taken together, these results suggest that in rat vascular smooth muscle cells: (1) PKA mediates inhibition of the Na^+ pump; (2) PKC mediates activation of the Na^+ pump; and (3) dephosphorylation mediates activation of the Na^+ pump. Apparently, the sequence of events underlying the observed effects is more complicated. These mechanisms require further investigation.

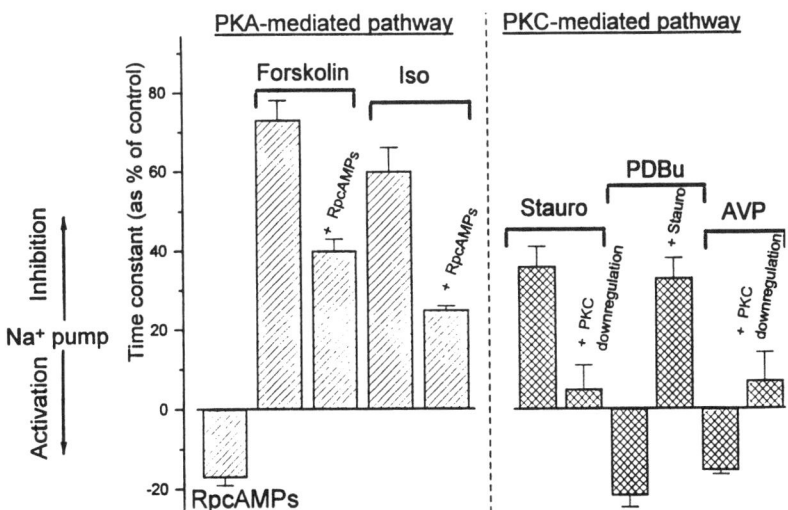

FIGURE 2. Results of experiments.

TABLE 1. Results of Experiments

PKA-Mediated Events	PKC-Mediated Events
1. PKA inhibitor Rp-cAMPS (10 µM) stimulated the Na^+ pump	1. PKC inhibitor, staurosporine (Stauro; 100 nM), inhibited Na^+ pump. The effect of staurosporine was absent in the cells with downregulated PKC (24-hour treatment with 2 µM PDBu).
2. PKA activators, 8-Br cAMP (0.1–5 mM) or forskolin + IBMX (10 µM and 0.1 mM, respectively) inhibited the Na^+ pump. This effect was prevented by the PKA inhibitor Rp-cAMPS.	2. PKC activators (phorbol esters, PDBu, or PMA, 0.2–5 µM) stimulated the Na^+ pump. The effect of phorbol esters was prevented by staurosporine.
3. Isoproterenol (Iso, 10–100 µM), a physiological agonist that triggers PKA-mediated processes, inhibited the Na^+ pump. The effect of Iso was prevented by PKA inhibitor Rp-cAMPS.	3. Arginine vasopressin (AVP, 2 nM), a physiological agonist that triggers PKC-dependent processes, stimulated the Na^+ pump. This effect was absent in the cells with downregulated PKC (24-hour treatment with 2 µM PDBu).

REFERENCES

1. BERTORELLO, A. M. & A. I. KATZ. 1993. Short-term regulation of renal Na-K-ATPase activity: Physiological relevance and cellular mechanisms. Am. J. Physiol. **265:** F743–F755.
2. BERTORELLO, A. M. & A. I. KATZ. 1995. Regulation of Na/K pump activity: Pathways between receptors and effectors. NIPS **10:** 253–259.
3. BORIN, M. L. 1995. cAMP evokes a rise in intracellular Na^+ mediated by Na^+ pump inhibition in rat aortic smooth muscle cells. Am. J. Physiol. **269:** C884–C891.
4. BORIN, M. L., W. F. GOLDMAN & M. P. BLAUSTEIN. 1993. Intracellular free Na^+ in resting and activated A7r5 vascular smooth muscle cells. Am. J. Physiol. **264:** C1513–C1524.

Effects of Protein Kinase Modulators on the Sodium Pump Activities of HeLa Cells Transfected with Distinct Alpha Isoforms of Na,K-ATPase[a]

NESTOR B. NESTOR,[b] LOIS K. LANE,[c]
AND RHODA BLOSTEIN[b]

[b]Department of Medicine
McGill University
Montreal, Canada

[c]Department of Pharmacology and Cell Biophysics
University of Cincinnati College of Medicine
Cincinnati, Ohio 45267–0575

Stimulation of protein kinases *in vivo* has been reported to activate as well as to inhibit pump activity in various systems, and increased phosphorylation of the Na,K-ATPase α subunit has been demonstrated in conjunction with both of these effects,[1,2] as well as in situations in which no change in activity has been detected.[3] This study addresses the question of how the individual α isoforms of the Na,K-ATPase are affected *in vivo* following activation of the endogenous protein kinases A (PKA) and C (PKC). Effects were evaluated by measuring ouabain-sensitive $K^+(Rb^+)$ influx into isoform-transfected cells. For PKC, a deletion mutant, α1M32, which lacks the NH_2-terminal sites of PKC phosphorylation,[3,4] was also evaluated.

The effect of the PKC activator PMA was tested on wild-type HeLa cells and HeLa cells transfected with the ouabain-resistant rat α1, α2*, and α3* isoforms (gifts of Dr. J. B Lingrel) as well as on α1M32. Cells were first incubated overnight in serum-free DMEM; the phorbol ester PMA, used to activate PKC, was added 10 minutes prior to the flux assay. As shown in TABLE 1, PMA pretreatment results in significant pump inhibition of wild-type and of all four transfectants.

PKA was activated by adding forskolin, an activator of adenylate cyclase, and IBMX, an inhibitor of 3′ 5′-cyclic nucleotide phosphodiesterase. Decreases in activity were consistently observed with each transfected cell line, that is, 22 ± 1.4%, 22 ± 1.7%, and 26 ± 1.6% for α1, α2*, and α3*, respectively. With the inactive forskolin analog 1,9-dideoxyforskolin, no inhibition was observed. In other experiments (not shown), there was evidence that PKA activation causes a small, but significant decrease in apparent affinity of α2 for extracellular K^+. With α3 as well, we have rudimentary evidence that PKA activation decreases the apparent affinity for cytoplasmic Na^+; a greater decrease in activity was observed at low (12 mM) compared to

---

[a]This work was supported by grants from the Medical Research Council of Canada, the Quebec Heart and Stroke Foundation (to R.B.), and the National Institutes of Health (to L.K.L.).

[b]Address for correspondence: Dr. R. Blostein, Montreal General Hospital, 1650 Cedar Avenue, Montreal, Que. Canada, H3G 1A4.

relatively high (40 mM) Na^+ concentration, that is, $33 \pm 8.7\%$ (5 experiments) versus $14 \pm 4.3\%$ (4 experiments), respectively.

The result of primary importance is the observation that activation of PKC by PMA inhibited pump activity of the wild-type (human) $\alpha 1$, the transfected rat $\alpha 1$, $\alpha 2^*$, and $\alpha 3^*$ isoforms as well as the truncated rat $\alpha 1M32$ mutant. These results are interesting from the following points of view. First, Feschenko and Sweadner[3] localized phosphorylation of the rat kidney ($\alpha 1$) enzyme to residues in the amino terminus, particularly serines 11 and 18, and Beguin et al.[4] reported that threonine 15 and serine 16 were phosphorylated in the *Bufo marinus* ($\alpha 1$) enzyme. If phosphorylation of the amino terminus is responsible for the observed inhibition of the $\alpha 1$ isoform, inhibition should not have been detected in the $\alpha 1M32$ mutant. Furthermore, with rat $\alpha 1$, 75% of the phosphorylation was located on serine 18,[3] which is not present in either the human $\alpha 1$ isoform or the rat $\alpha 2$ isoform.

The most straightforward explanation for these inhibitory effects of protein kinase activation is that phosphorylation of residue(s) in the amino terminal region of the α subunit is not responsible for inhibition of sodium pump activity in both isoform-

TABLE 1. Inhibition of Pump Activity of Transfected HeLa Cells after Activation of PKC with Phorbol Ester[a]

Transfected Cells	Control (4-α-PMA)	PMA Added	% Change
Untransfected	$1,242 \pm 56$	$1,024 \pm 54$	-18
$\alpha 1$	697 ± 58	512 ± 23	-27
$\alpha 2$	$1,341 \pm 27$	$1,007 \pm 92$	-25
$\alpha 3$	590 ± 48	484 ± 1	-18
$\alpha 1M32$	648 ± 67	495 ± 22	-24

[a]Cells were cultured overnight in serum-free medium, with 1 μM ouabain present with the transfected cells to inhibit endogenous Na,K-ATPase. All cells were treated with 1 μM PMA for 10 minutes prior to measurement of ouabain-sensitive $^{86}Rb^+(K^+)$ influx in buffer containing 40 mM NaCl, with monensin added to maintain intracellular $Na^+ \approx 50$ mM and with saturating (4 mM) KCl, conditions that approach V_{max} for $\alpha 1$, $\alpha 2$, and $\alpha 1M32$, as described elsewhere.[6]

transfected and untransfected HeLa cells. Accordingly, inhibition of pump activity by protein kinase activation may be indirect in certain systems and due presumably to phosphorylation/dephosphorylation of another distinct cell constituent that regulates pump activity and/or expression at the cell surface.

ACKNOWLEDGMENT

We thank Dr. Jerry B Lingrel for generously providing us with a2*- and a3*-transfected HeLa cells.

REFERENCES

1. LYNCH, C. J., A. C. MADER, K. M. MCCALL, Y.-C. NG & S. A. HAZEN. 1994. FEBS Lett. **355:** 157–162.

2. MIDDLETON, J. P., W. A. KHAN, G. COLLINSWORTH, Y. A. HANNUM & R. M. MEDFORD. 1993. J. Biol. Chem. **268:** 15958–15964.
3. FESCHENKO, M. S. & K. J. SWEADNER. 1995. J. Biol. Chem. **270:** 14072–14077.
4. BEGUIN, P., A. T. BEGGAH, A. V. CHIBALIN, P. BURGENER-KAIRUZ, F. JAISSER, P. MATHEWS, B. C. ROSSIER, S. COTECCHIA & K. GEERING. 1994. J. Biol. Chem. **269:** 24437–24445.
5. PFEFFER, L. M., B. STRULOVICI & A. R. SALTIEL. 1990. Proc. Natl. Acad. Sci. USA **87:** 6537–6541.
6. MUNZER, J. S., S. E. DALY, E. A. JEWELL-MOTZ, J. B LINGREL & R. BLOSTEIN. 1994. J. Biol. Chem. **269:** 16668–16676.

Phosphorylation of Tyr⁷, Tyr¹⁰, and Ser²⁷ of α-Chain in H⁺,K⁺-ATPase by Intrinsic and Extrinsic Kinases[a]

KATSUHIKO TOGAWA,[b] SHUNJI KAYA,[b]
MASANOBU MORI,[b] AKIRA SHIMADA,[b]
TOSHIAKI IMAGAWA,[b] KAZUYA TANIGUCHI,[b,f]
SVEN MÅRDH,[c] JACKIE CORBIN,[d] AND USHIO KIKKAWA[e]

[b]*Biological Chemistry
Graduate School of Science
Hokkaido University
Sapporo 060, Japan*

[c]*Department of Cell Physiology
Linkoping University
Sweden*

[d]*Department of Molecular Physiology and Biophysics
Vanderbilt University
Nashville, Tennessee 37232*

[e]*Biosignal Research Center
Kobe University
Kobe, 657, Japan*

Acid secretion is regulated by second messenger pathways; histamine-stimulated acid secretion appears to be mediated by an increase in cellular cAMP and cholinergic and gastrin stimulation to be mediated by an increase in the intracellular calcium concentration.[1–5] Activation of cAMP-dependent acid stimulation required cAMP and an additional increase of cytosolic Ca^{2+} for maximal stimulation.[6] Protein tyrosine kinase inhibitors reverse the inhibition of parietal cell secretion by transforming growth factor-α.[7]

Recently, sequential phosphorylation of Tyr¹⁰ and Tyr⁷ [8] in the α-chain[9] of pig stomach H^+,K^+-ATPase of the G_1 fraction[10] by endogenous membrane-bound kinase and dephosphorylation by endogenous phosphatase were shown unequivocally.[8] However, the absence of vanadate reduced the phosphorylation from 0.5 to 0.05 mol/mol α-chain.[8] Acid hydrolysis of the α-chain produced Tyr(P) with a slight amount of Ser(P), which indicated that the G_1 fraction contained not only tyrosine kinase but also some kinase which phosphorylated Ser residues.[8] Data suggest that the

---

[a]This work was supported in part by Grants-in-Aid for Scientific Research (06454648, 07558220) and for International Scientific Research Program (07044049, 08044047) from the Ministry of Education, Science and Culture of Japan and Swedish Medical Research Council (4X-4965).

[f]To whom correspondence should be addressed. Tel: 81-11-706-2698; fax: 81-11-736-2074; e-mail: KTAN@hucc.hokudai.ac.jp

extent of phosphorylation was dependent on both activities of protein kinase and phosphatase in the G_1 fraction.

To further investigate Tyr and Ser phosphorylation in the α-chain, the G_1 fraction was incubated with [γ-^{32}P]ATP and Mg^{2+}; the addition of Ca^{2+} biphasically increased ($K_{0.5}$ = 6 and 200 μM) the phosphorylation of both Tyr and Ser residues in the NH_2-terminal α-chain of H^+,K^+-ATPase without any detectable phosphorylation in other regions of the α-chain. This phosphorylation was reversible. Mild tosylphenylalanyl chloromethyl ketone-trypsin treatment followed by reverse-phase column chromatography gave three radioactive peptide peaks. The first peak contained both

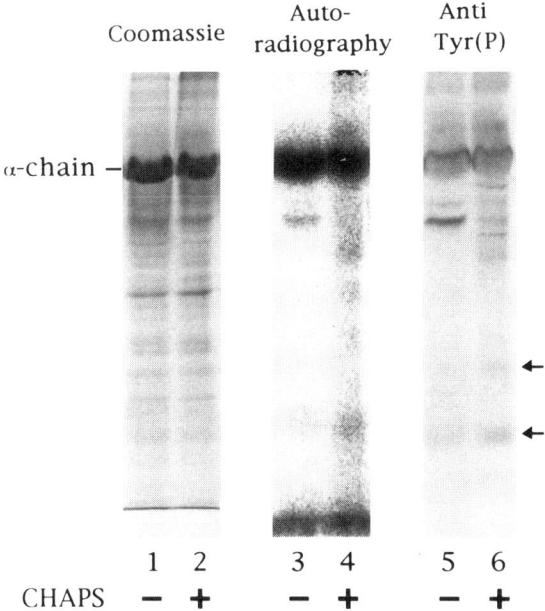

FIGURE 1. Phosphorylation of the α-chain of H^+,K^+-ATPase in the presence of CHAPS. The G_1 fraction phosphorylated without (*lanes 1, 3, and 5*) or with (*lanes 2, 4, and 6*) CHAPS was subjected to SDS-gel electrophoresis, stained with Coomassie blue (*lanes 1 and 2*), and examined by enzyme-linked immunoassay to detect Tyr(p) (*lanes 5 and 6*), and radioactivities were detected (*lanes 3 and 4*).[8]

$Tyr^{10}(^{32}P)$ and $Tyr^{7}(^{32}P)$, and the second peak contained $Tyr^{10}(^{32}P)$.[8] The third peak contained $Ser^{27}(^{32}P)$ which was also obtained after trypsin treatment of partially purified H^+,K^+-ATPase preparations phosphorylated with protein kinase C + Ca^{2+} or protein kinase A. The addition of CHAPS to the G_1 fraction seemed to reduce the substrate specificity of kinases present (FIG. 1).

Ca^{2+}-dependent phosphorylation of the α-chain of H^+,K^+-ATPase by intrinsic kinases and others suggest that the NH_2-terminal domain of the α-chain of H^+,K^+-ATPase is under possible control of various protein kinase and phosphatase activities and imply the participation of Ca^{2+} and cAMP in these phosphorylation reactions.

REFERENCES

1. SOLL, A. H. & A. WOLLIN. 1979. Am. J. Physiol. **237:** E444–E450.
2. CHEW, C. S., S. J. HERSEY, G. SACHS & T. BERGLINDH. 1980. Am. J. Physiol. **238:** G312–320.
3. CHEW, C. S. 1986. Biochim. Biophys. Acta **888:** 116–125.
4. HERSEY, S. J. & J. SACHS. 1995. Physiol. Rev. **75:** 155–189.
5. LI, Z.-Q., J. L. CABERO & S. MÅRDH. 1995. Am. J. Physiol. **268:** G82–89.
6. LI, Z.-Q. & S. MÅRDH. 1996. Biochim. Biophys. Acta **1311:** 133–142.
7. TSUNODA, Y., I. M. MODLIN & J. R. GOLDENRING. 1993. Am. J. Physiol. **264:** G351–G356.
8. TOGAWA, K., T. ISHIGURO, K. KAYA, A. SHIMADA, T. IMAGAWA & K. TANIGUCHI. 1995. J. Biol. Chem. **270:** 15475–15478.
9. MAEDA, M., J. ISHIZAKI & M. FUTAI. 1988. Biochem. Biophys. Res. Commun. **157:** 203–209.
10. CHANG, H., G. SACCOMANI, E. RABON, R. SCHACKMANN & G. SACHS. 1977. Biochim. Biophys. Acta **464:** 313–327.

Regulatory Phosphorylation of the Na$^+$/K$^+$-ATPase from Mammalian Kidneys and *Xenopus* Oocytes by Protein Kinases

Characterization of the Phosphorylation Site for PKC[a]

LARISA A. VASILETS,[b,d] HEIKE FOTIS,[b]
AND EVA-MARIA GÄRTNER[b]

[b]*Max-Planck-Institute for Biophysics*
Frankfurt/M, Germany

[c]*Institute of Chemical Physics*
Russian Academy of Science
Chernogolovka
Moscow region, Russia

Protein kinase-mediated phosphorylation of the Na$^+$/K$^+$-ATPase has been studied in enzymes purified from pig, dog, sheep, and rat kidneys and in *Xenopus* oocytes. None of the α subunits from mammalian kidney ATPases is phosphorylated by casein kinase II or Ca^{2+}/calmodulin-dependent protein kinase. For the purified enzymes, rat protein kinase C (PKC) phosphorylates only the α subunit of the Na$^+$/K$^+$-ATPase of rat kidney. Selective proteolytic digestion of the rat α1 enzyme under conditions of mild trypsinolysis demonstrates that phosphorylation occurs at the NH$_2$-terminus before the T2 cleavage site. There are Ser-16 and Ser-23 within the cleaved fragment of the rat ATPase that can potentially be phosphorylated by PKC. Only Ser-16 is present in the other mammalian kidney isoforms that are poorly phosphorylated by PKC. Mutated rat α1 subunits with substitution of Ser-23 (S23A) or Ser-16 (S16A) by Ala were coexpressed in *Xenopus* oocytes together with β subunits, and modulation of transport activity (TABLE 1) as well as phosphorylation of pumps by PKC in yolk-free oocyte homogenates (FIG. 1) was investigated. PKC produces incorporation of ^{32}P into α subunits of wild-type pumps and inhibits ouabain-sensitive ^{86}Rb uptake by 80 ± 3%. By contrast, in the S23A mutant, much weaker phosphorylation and no inhibition of ^{86}Rb uptake by rat PKC were observed. In the S16A mutant, the α subunit still can be phosphorylated by PKC at the Ser-23 site, but the sensitivity of transport activity to PKC is lost. The data demonstrate that Ser-23 is the actual site of regulatory phosphorylation for rat brain PKC in rat kidney α1 Na$^+$/K$^+$-ATPase, but Ser-16 is also important for PKC-mediated changes in pump function. Effects of modulation of

---

[a]This work was supported by DFG grant Schw 446/2-1 and HHMI grant 75195–547101.

[d]Address for correspondence: Larisa Vasilets, Max-Planck-Institute for Biophysics, Kennedyallee 70, D-60596 Frankfurt/M, FRG (tel: +49 69 6303 338; fax: +49 69 6303 340; e-mail: 100304.1621@compuserve.com).

TABLE 1. Relative Changes in ^{86}Rb Uptake and (^{3}H) Ouabain Binding after Microinjection into the Oocyte of Rat Brain Protein Kinase C (PKC) or Activation of *Xenopus* PKC with PMA[a]

	Rat PKC		PMA	
	^{86}Rb Uptake	(^{3}H)Ouabain Binding	^{86}Rb Uptake	(^{3}H)Ouabain Binding
Rα(WT)	20 ± 3	138 ± 11	32 ± 3	87 ± 6
Rα(S16A)	104 ± 4	104 ± 10	16 ± 4	70 ± 32
Rα(S23A)	111 ± 5	109 ± 6	20 ± 4	70 ± 9
Xenopus	141 ± 8	97 ± 7	59 ± 3	87 ± 7

[a]Before experiment, oocytes with only endogenous pumps or with expressed rat wild-type Rα (WT) or mutated Rα(S16A, S23A) pumps were loaded with Na$^+$. Measurements were done in Na-free solution (Na substituted with TMA)[4] 30 minutes after microinjections of rat brain PKC or application of PMA. ^{86}Rb uptake represents ouabain-sensitive Rb uptake calculated as a difference of the uptake in the absence and presence of 10 mM ouabain. Values are normalized: for cells injected with PKC to the values obtained from cells injected with the buffer for PKC, for cells incubated with PMA to the values in nontreated cells. In the batches with expressed rat pumps, (^{3}H)ouabain binding represents the contribution of endogenous ouabain binding sites and can be used as a marker of cellular endocytosis.

transport activity of expressed pumps by stimulation of endogenous *Xenopus* PKC with PMA differ from those of rat brain PKC. Incubation with PMA leads to considerable inhibition of ^{86}Rb uptake in wild-type pumps as well as in S16A or S23A mutants (TABLE 1), and all α subunits can be phosphorylated (lanes 2, 3, and 4 in FIG. 1). However, ^{32}P-incorporation into α subunits of S16A and S23A mutants was less than that for the wild type. This suggests that activation of *Xenopus* PKC may lead to

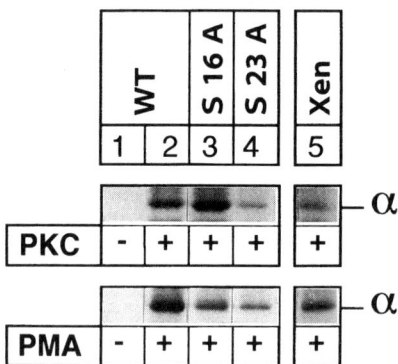

FIGURE 1. Phosphorylation of expressed and endogenous α subunits in yolk-free oocyte homogenates by different types of PKC. One microliter of commercial stock solution of PKC (Calbiochem) or 50 nM of PMA was added to the yolk-free oocyte homogenates.[3] After 30 minutes, phosphorylation reactions were stopped and α subunits were coprecipitated under nondenaturing conditions from digitonin extracts with anti-β antibody (lanes 1–4) or were precipitated from triton extracts with anti-α antibody (lane 5). Proteins were separated by SDS-PAGE, and ^{32}P incorporation was visualized by autoradiography.

phosphorylation of sites that are different from those phosphorylated by rat brain PKC.

Effects of different PKCs on modulation of transport activity and phosphorylation of the α subunit were also studied for endogenous *Xenopus* pumps. Activation of *Xenopus* PKC with PMA produces inhibition of ^{86}Rb uptake; microinjection into the oocytes of rat PKC leads to the opposite effect (TABLE 1). Both PKCs produce incorporation of ^{32}P into the *Xenopus* α subunit with somewhat stronger incorporation by stimulation with PMA (lane 5, FIG. 1). This work demonstrates a diversity of effects of regulatory phosphorylation depending on the type of pump and of PKC. Activity of the *Xenopus* pump is inhibited by endogenous *Xenopus* PKC and is stimulated by rat PKC. In complementary systems (rat α subunit/rat PKC) and *Xenopus* α subunit/ *Xenopus* PKC phosphorylation always leads to inhibition of transport activity. Rat PKC inhibits transport activity of expressed pumps and stimulates that of endogenous *Xenopus* pumps. There are 10–11 motifs within α1 subunit isoforms[1] that may be theoretically phosphorylated by about 9 isoforms of PKC.[2] This may form the basis for specificity of regulatory phosphorylation with respect to pump and PKC isoforms and may account for the variability in the effects of PKC on transport activity of the Na$^+$/K$^+$-ATPase in our work and in that reported in the literature.

ACKNOWLEDGMENTS

The authors are thankful to Dr. W. Schwarz for support of this work and critical discussions and to Rolf Postina for help with the mutations.

REFERENCES

1. VASILETS, L. A. & W. SCHWARZ. 1993. Biochim. Biophys. Acta **1154:** 201–222.
2. HUG, H. & T. F. SARRE. 1993. Biochem. J. **291:** 329–343.
3. CHIBALIN, A. V., L. A. VASILETS, H. HENNEKES, D. PRALONG & K. GEERING. 1992. J. Biol. Chem. **267:** 22378–22384.
4. VASILETS, L. A., H. OMAY, T. OHTA, S. NOGUCHI, M. KAWAMURA & W. SCHWARZ. 1991. J. Biol. Chem. **266:** 16285–16288.

Ion-Sensitive Domains of the SERCA- and the Na^+/K^+-ATPases Identified by Chimeric Recombination

SHIGE H. YOSHIMURA, TOSHIAKI ISHII,[a]
JIRO C. YASUHARA, MASA H. SATO,
AND KUNIO TAKEYASU[b]

*Department of Natural Environment Sciences
Faculty of Integrated Human Studies
Kyoto University
Sakyo-ku, Kyoto 606-01, Japan*

We have identified functional domains specific to the calcium and the sodium pump, using the chicken sarcoplasmic/endoplasmic reticulum Ca^{2+}-ATPase (SERCA1) as a parental molecule and replacing various portions with corresponding portions of the chicken Na^+/K^+-ATPase $\alpha 1$ subunit.[1] We found that the SERCA-ATPase possesses two different populations of K^+ binding sites, one of which could also bind Na^+. This population of K^+-/Na^+ binding sites possesses a higher affinity for K^+ than for Na^+. Historically, although the effects of Na^+ and K^+ on the ouabain-sensitive Na^+/K^+-ATPase have been well characterized, the effects of these ions on the SERCA-ATPase activity have been ignored except in a very few studies (e.g., ref. 2). We demonstrated that the thapsigargin-sensitive SERCA-ATPase does not depend on Na^+ and K^+ for its basal activity, but it can be further activated either by K^+ in a two-step fashion with high ($ED_{50} \sim 20$ mM) and low affinity ($ED_{50} \sim 70$ mM) or by Na^+ in a single-step fashion with an ED_{50} value of ~ 50 mM.[1] When high concentrations of K^+ (e.g., 100 mM) were present in the SERCA-ATPase assay system, the regulatory effect of Na^+ was abolished. On the other hand, when saturating concentrations of Na^+ were present, only the high-affinity effect of K^+ was abolished and the low-affinity activation by K^+ was still observed.[1] Use of the chimeric SERCA-ATPase (FIG. 1) led us to identify the domains sensitive to Na^+ and K^+.

1. The SERCA-ATPase activity of the chimera CC[c/n/n] lost the low sensitivity to K^+ and displayed only a high sensitivity to K^+, whereas those of the chimeras CNC and [n/c]CC retained both high and low sensitivity to K^+. Thus, the carboxy-terminal region of the SERCA1 is responsible for the low sensitivity to K^+. The corresponding region (Leu861-Tyr1021) of the Na^+/K^+-ATPase $\alpha 1$ subunit is possibly responsible for the high-affinity K^+ binding, as suggested by Karlish *et al.*[3,4]

2. The SERCA-ATPase activities of all the chimeras ([n/c]CC, CCC, CNC, and CC[c/n/n]) were regulated in a single-step fashion by Na^+ in the absence of K^+ and became insensitive to Na^+ in the presence of K^+, indicating that the Na^+ site in the

---

[a]Department of Veterinary Pharmacology and Research Institute for Advanced Science and Technology, University of Osaka Prefecture, Sakai, Osaka 591, Japan.

[b]Address for correspondence: K. Takeyasu, Department of Natural Environment Sciences, Faculty of Integrated Human Studies, Kyoto University, Sakyo-ku, Kyoto 606-01, Japan (tel: 81-75-753-6852; fax: 81-75-753-6549; e-mail: takeyasu@gaia.h.kyoto-u.ac.jp).

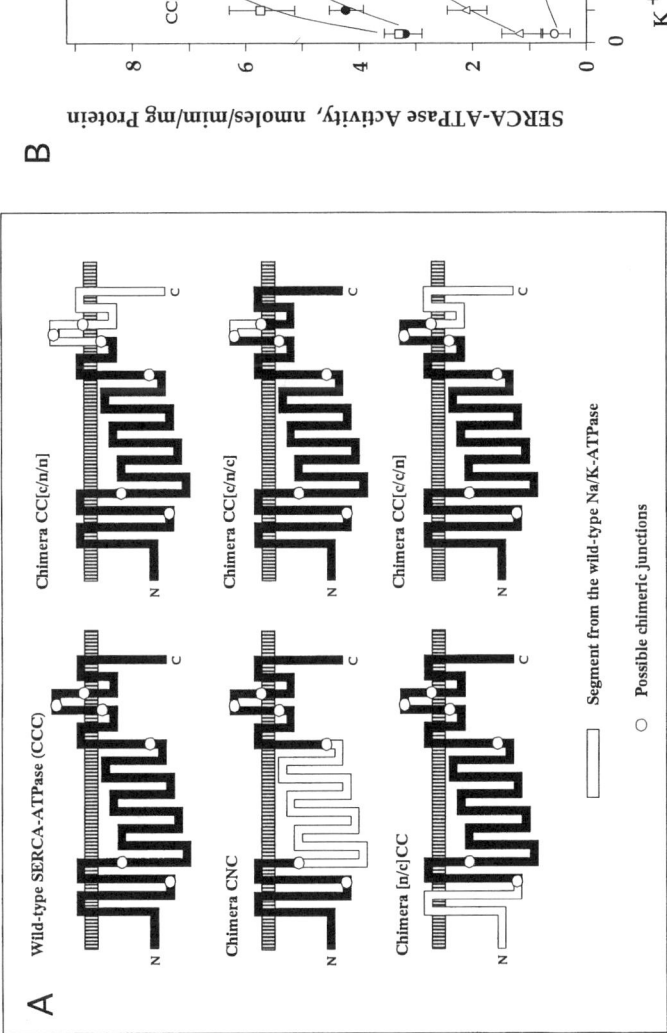

FIGURE 1. (A) Schematic representation of chimeric molecules. The methods for construction of the recombinant molecules have been described elsewhere.[1,5] Replacement of the cytoplasmic loop (Gly354-Lys712) and the amino-terminal regions (Met1-Asp162) of the SERCA1 with the corresponding portions of the Na$^+$/K$^+$-ATPase α1 subunit yield CNC and [n/c]CC, respectively. (B) Existence of a distinct K$^+$-binding site at the carboxy-terminal region of the Na$^+$/K$^+$- and the SERCA-ATPase. Stable cell lines that express CCC, CC[c/n/n], CC[c/n/c], and CC[c/c/n] were used for to measure their thapsigargin- and Ca^{2+}-sensitive (SERCA) ATPase activities as described.[5] The SERCA-ATPase activities were measured at 5 μM free Ca^{2+} and different concentrations of K$^+$ in the absence of Na$^+$. Total ionic strength was kept at 180 mM with KCl plus choline-chloride.

wild-type and chimeric SERCA-ATPase was blocked by K^+. Thus, regions Ile163-Gly354 and/or Lys712-Ser30 of the SERCA1 are sufficient for activation by Ca^{2+} and inhibition by thapsigargin[5] and also for further stimulation by Na^+ and K^+. The existence of the corresponding Na^+/K^+-sensitive regions (referred to as a core domain

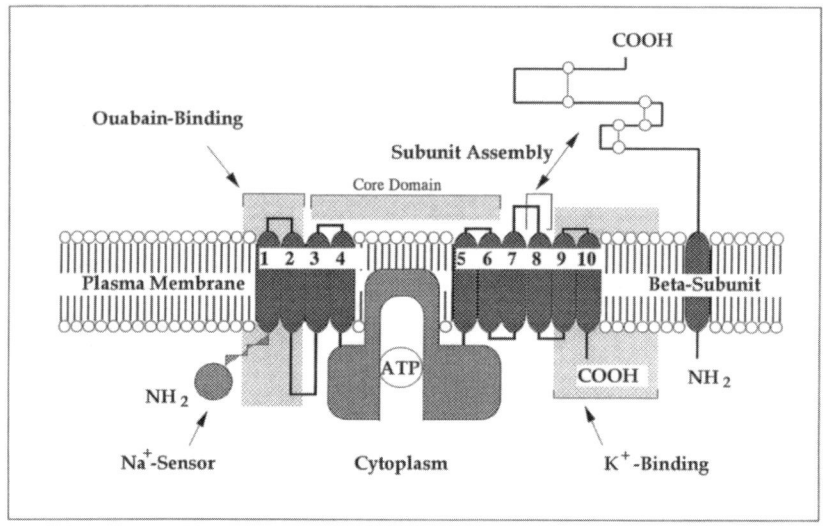

FIGURE 2. Functional domain model for structure and function of the Na^+/K^+-ATPase that accommodates all the results from functional analyses on the SERCA-ATPase, the Na^+/K^+-ATPase, and their chimeric derivatives.[1,5,7] The transmembrane segments are numbered (M1–M10). The model includes Na^+ sensor, ouabain-binding and subunit-assembly domains, and K^+-sensitive region. A major ouabain binding domain is located within the amino-terminal 131 amino acids (Ala70-Asp200), a minimum subunit-assembly domain at the carboxyl-terminus (Asn894–Ala919). The amino-terminal 69 amino acids (Met1-Leu69, "Na^+ sensor") of the Na^+/K^+-ATPase are required to maintain Na^+ sensitivity, whereas the carboxy-terminal 102 amino acids (Phe920-Thr1021) constitute a K^+-sensitive domain. The Na^+ sensor together with the ouabain-binding domain can confer Na^+ and ouabain sensitivity to the SERCA-ATPase in a chimeric molecule [n/c]CC. A region similar to the Na^+ sensor is present in all isoforms of the Na^+/K^+-ATPase α subunits, but not in the SERCA-ATPase, that is, the amino terminus of the SERCA-ATPase is ~40 amino acids shorter than that of the Na^+/K^+-ATPase α subunit. The addition of the isolated Na^+-sensor fragments to the assay system for the SERCA-ATPase activity resulted in stimulation of the ATPase activity in a dose-dependent manner. These results suggest that the isolated Na^+ sensor can directly interact with the SERCA-ATPase molecule. Two additional ion-binding sites for K^+ and Na^+ are possibly localized in the core domain based on the analogy to the minimum structural requirement for SERCA-ATPase activity. (o-o) = S-S bond in the β subunit.

in FIG. 2) within the Na^+/K^+-ATPase α subunit is consistent with recent results by site-directed mutagenesis of Glu327 in M4 and Ser775 in M5.[6]

3. Analysis of chimeras (CC[c/c/n] and CC[c/n/c]) demonstrated that the carboxy-terminal 101 amino acids (Phe920-Tyr1021) of the Na^+/K^+-ATPase α1 subunit are sufficient to shift the K^+ affinity ($ED_{50} < 10$ mM, similar to that of the Na^+/K^+-

ATPase) without the β subunit (FIG. 1B). No change in the two-step activation of SERCA-ATPase by K^+ was seen when residues Thr871–Thr898 of the SERCA1 were replaced with residues Asn894-Ala919 of the Na^+/K^+-ATPase α1 subunit, a region known to bind the Na^+/K^+-ATPase β subunit.[7] Thus, the Na^+/K^+-ATPase subunit assembly domain and the K^+-sensitive region are distinct within the carboxy-terminal 161 amino acids of the Na^+/K^+-ATPase α1 subunit.

From these results together with our previous findings on the other regulatory regions, we propose the functional domain model (FIG. 2) for the Na^+/K^+-ATPase. The membranous regions of the P-type ATPase include several negatively charged residues[8] that are essential for Ca^{2+} [8,9] or K^+ binding.[6] These membranous regions may be critical for ion dependence of all P-type ATPases. Indeed, a chimeric Na^+/K^+-ATPase (NCN in which the cytoplasmic loop [Gly354–Lys712] of the SERCA1 was incorporated) transported ^{86}Rb, indicating that the cytoplasmic loop is not essential for K^+ uptake.

REFERENCES

1. ISHII, T., F. HATA, M. V. LEMAS, D. M. FAMBROUGH & K. TAKEYASU. 1997. Biochemistry **36:** 442–451.
2. SHIGAKAWA, M. & L. J. PEARL. 1976. J. Biol. Chem. **251:** 6947–6952.
3. GOLDSHLEGER, R., D. M. TAL, N. MOORMAN, W. D. STEIN & S. J. D. KARLISH. 1992. Proc. Natl. Acad. Sci. USA **89:** 6911–6915.
4. CAPASSO, J. M., S. HOVING, D. M. TAL, R. GOLDSHLEGER & S. J. D. KARLISH. 1992. J. Biol. Chem. **267:** 1150–1158.
5. ISHII, T., M. V. LEMAS & K. TAKEYASU. 1994. Proc. Natl. Acad. Sci. USA **91:** 6103–6107.
6. LINGREL, J. B. & T. A. KUNTZWEILER. 1994. J. Biol. Chem. **269:** 19659–19662.
7. LEMAS, M. V., M. HAMRICK, K. TAKEYASU & D. M. FAMBROUGH. 1994. J. Biol. Chem. **269:** 8255–8259.
8. CLARKE, D. M., T. W. LOO, G. INESI & D. H. MACLENNAN. 1989. Nature **339:** 476–478.
9. ANDERSEN, J. P. & B. VILSEN. 1995. FEBS Lett. **359:** 101–106.

Selective Inhibition of the Gastric H^+,K^+-ATPase by Omeprazole and Related Compounds

P. LORENTZON, A. BAYATI, H. LEE, AND K. ANDERSSON

AB Astra Hässle
S-431 83 Mölndal, Sweden

Acid-related diseases of the gastrointestinal tract have long been treated with pharmacological agents that not only neutralize the acid (antacids), but also inhibit the production and secretion of acid from the parietal cell in the stomach (antisecretory drugs). Receptor antagonists/agonists such as anticholinergics, prostaglandin analogs, and especially histamine H_2-antagonists were the first pharmacological tools clinically used to reduce acid secretion. Gastric H^+,K^+-ATPase inhibitors were developed during the late 1970s and 1980s, omeprazole[1] being the first compound available for clinical use. Omeprazole is now registered for use in duodenal and gastric ulcer disease, reflux esophagitis, the Zollinger-Ellison syndrome, and in combination with antibiotics to eradicate *Helicobacter pylori*. Omeprazole is a so-called substituted benzimidazole which reacts covalently with the H^+,K^+-ATPase, the terminal step of acid formation. Similar compounds have since followed, and currently lansoprazole[2] and pantoprazole[3] are also registered antisecretory drugs and still others such as rabeprazole/pariprazole,[4] are being developed. (See FIG. 1 for their respective structure.) All of these compounds share the same core structure but differ with respect to substituents, features that result in a common mechanism of action but somewhat different chemical characteristics. A second class of H^+,K^+-ATPase inhibitors, the so-called K^+ competitive inhibitors, is presently under development. These compounds act as reversible inhibitors of the potassium-binding site at the luminal face of the H^+,K^+-ATPase.[5]

MECHANISM OF ACTION OF OMEPRAZOLE-LIKE COMPOUNDS

Omeprazole, lansoprazole, and pantoprazole together with rabeprazole/pariprazole and other compounds in this class all share a common mechanism of action, which will be described with omeprazole as example.

The events underlying the selective inhibition of the gastric proton pump by omeprazole are represented schematically in FIGURE 2.[6] Omeprazole is a weak base with a pKa of 4, which means that it will be concentrated in acid compartments such as the secretory canaliculi of the parietal cell where the pH is below 4. At physiological pH it can readily cross biological membranes, whereas in the acid spaces of the parietal cell it will gain a proton and thus become less permeable. The secretory canaliculi of the parietal cell have a pH of about 1, and omeprazole can thus be concentrated by a factor of 1,000 into this compartment. In this highly acidic environment of the canaliculus, an acid-catalyzed intramolecular rearrangement of omeprazole into a sulfenamide form takes place. This compound subsequently reacts rapidly with luminally accessible sulfhydryl groups on the H^+,K^+-ATPase.[7] A disulfide is formed between the inhibitor and the enzyme, resulting in long-lasting inhibition of the proton pump.

That acid is necessary for the inhibitory action of omeprazole has been illustrated in various models, both *in vitro* and *in vivo*.[7–10] At the enzyme level, it was demonstrated in vesicular preparations of the H^+,K^+-ATPase that omeprazole had an inhibitory effect only under conditions in which acid was allowed to accumulate in the vesicular lumen.[7] As illustrated in FIGURE 3, the use of K^+ and NH_4^+-ions has allowed

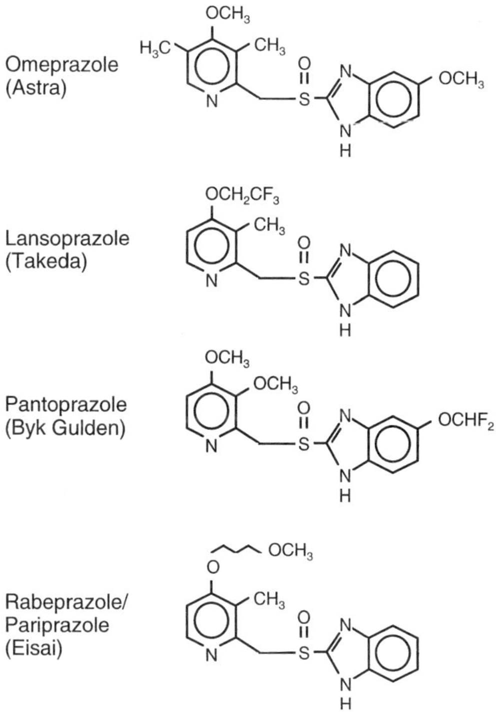

FIGURE 1. Structures of covalent H^+,K^+-ATPase inhibitors.

for the generation or dissipation of acidic intravesicular spaces. In a concentration of 10 μM, omeprazole inhibited the ATPase activity to about 90% under acid-accumulating conditions. However, the same concentration of omeprazole had no effect on the NH_4^+-stimulated ATPase activity, that is, under neutral conditions.

Two important factors provide the basis for the selective inhibition of gastric acid secretion by omeprazole and related compounds: (1) their intrinsic physicochemical properties, which result in a concentration within the acidic compartment of the parietal cell and the subsequent formation of the active inhibitor; (2) the restricted localization of the target enzyme, the H^+,K^+-ATPase, to the apical membrane of the parietal cell, which is in close proximity to the site of formation of the active inhibitor.

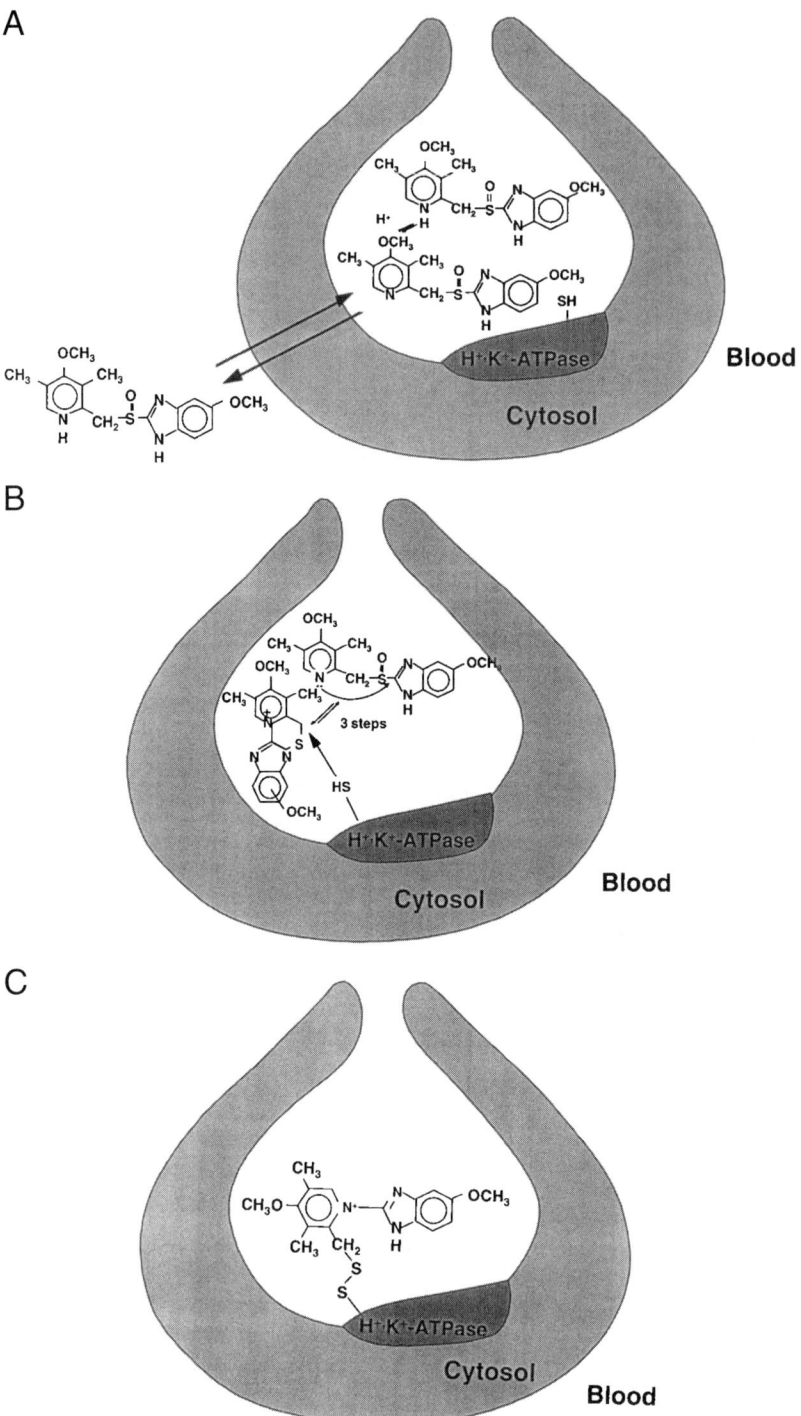

FIGURE 2. Mechanisms of action of omeprazole. **(A)** Concentration of the protonated omeprazole form in the secretory canaliculi; **(B)** transformation of omeprazole into the sulfenamide form; and **(C)** structure of the enzyme inhibitor complex.

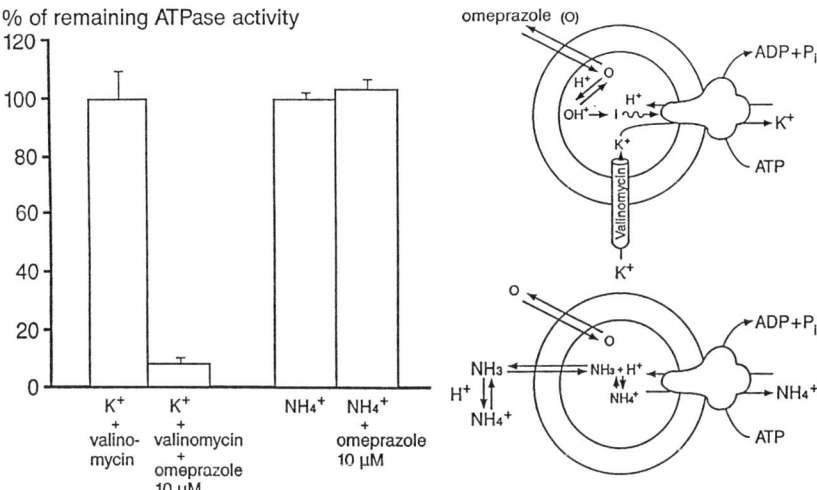

FIGURE 3. Effect of omeprazole on ATPase activity in isolated gastric vesicles. The reaction mixture contained 2 mmol/l MgCl$_2$, 2 mmol/l Hepes/NaOH pH 7.4, 100 µmol/l glutathione, 175 mmol/l KCl plus 1 µg valinomycin/µg protein or 200 mmol/l NH$_4$Cl. The reaction was started by the addition of 2 mmol/l ATP. After 5 minutes of preincubation at 37°C, omeprazole was added in a concentration of 10 µmol/l. At t = 30 minutes, the reaction was stopped and the ATPase activity was determined.

Although the absolutely most important function and localization of the H$^+$,K$^+$-ATPase reside in the gastric mucosa, experimental evidence provided by polymerase chain reaction amplification of rat renal mRNA indicates the presence of the gastric H$^+$,K$^+$-ATPase in the kidney.[11,12] Various functional *in vitro* studies have indeed indicated the presence of an H$^+$,K$^+$-ATPase in the kidney. (For a review see ref. 13.) However, *in vivo* experiments in acid-loaded rats with induced levels of renal H$^+$,K$^+$-ATPase using high intravenous doses of omeprazole failed to show any effect of the drug on urinary excretion of potassium or hydrogen ions.[11] Furthermore, an isoform of the H$^+$,K$^+$-ATPase, with approximately 65% homology to the α subunits of both the H$^+$,K$^+$-ATPase and the Na$^+$,K$^+$-ATPase, has been detected in the colon, kidney, and uterus.[12,14]

TABLE 1. Binding Sites in the H$^+$,K$^+$-ATPase for Omeprazole, Pantoprazole, and Lansoprazole (data from refs. 16–18)

	Omeprazole	Pantoprazole	Lansoprazole
Cys 813 (M5/M6)	X[a]	X	X[a]
Cys 822 (M5/M6)	X[a]	X	X[a]
Cys 892 (M7/M8)	X		X
Cys 321 (M3/M4)			X

[a]Compound binds at either Cys 813 or Cys 822.

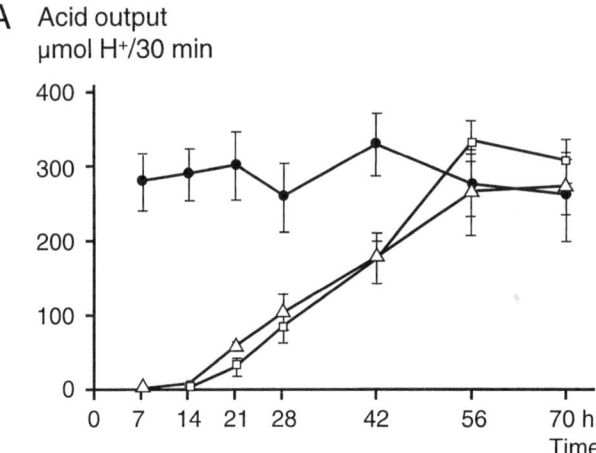

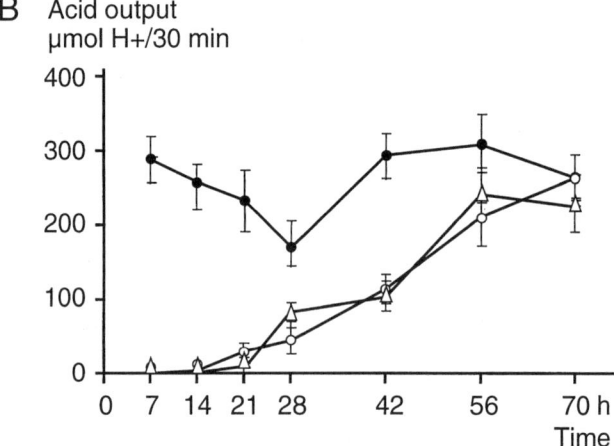

FIGURE 4. *See legend on facing page.*

REVERSAL OF INHIBITION *IN VITRO* AND MECHANISM OF RECOVERY *IN VIVO*

The disulfide formed between the inhibitor and the H^+,K^+-ATPase can be cleaved by sulfhydryl reagents such as dithiotreitol, β-mercaptoethanol, and glutathione as shown in various *in vitro* models.[7,8,10] These sulfhydryl-containing reagents can also act as scavengers of the sulfenamide and thus protect against the establishment of inhibition.[10,15] This scavenging action of glutathione, for example, might be a contributing factor to the safety and selectivity of the substituted benzimidazoles. Any sulfenamide formed in cytosolic compartments, for examples, would immediately be scavenged by the glutathione that is present in high concentrations. Although difficult to demonstrate, reversal, that is, breakage of the covalent inhibitory bond, most likely

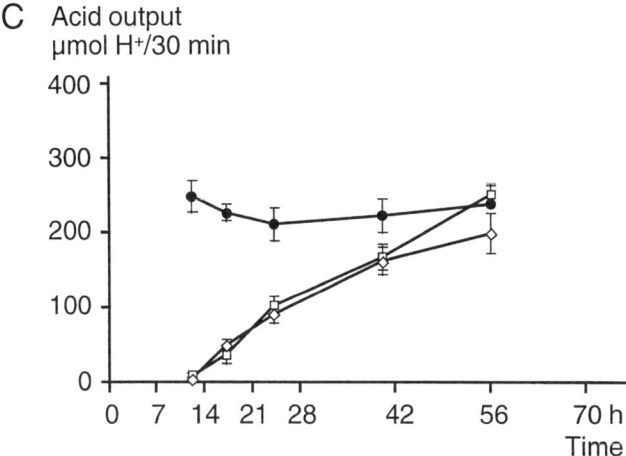

FIGURE 4. Duration of the inhibitory action of omeprazole, pantoprazole, lanzoprazole, and rabeprazole/pariprazole in the rat. Rats were treated orally for 3 days, and acid secretion was measured at intervals after the last dose. Acid secretion was stimulated by a mixture of pentagastrin (20 nmol/kg, h) and carbachol (110 nmol/kg, h) known to induce maximal secretion. **(A)** ● = vehicle; △ = omeprazole 270 µmol/kg; □ = lansoprazole 135 µmol/kg. **(B)** ● = vehicle; ○ = omeprazole 400 µmol/kg; △ = pantoprazole 200 µmol/kg. **(C)** ● = vehicle; □ = omeprazole 135 µmol/kg; ◇ = rabe/pariprazole 400 µmol/kg.

also takes place in the physiological *in vivo* situation. In a comparative study in the rat, both turnover of the H^+,K^+-ATPase and recovery of the H^+,K^+-ATPase activity was measured after 7 days' administration of omeprazole.[16] The half-life of the enzyme was ~50 hours, while the H^+,K^+-ATPase activity recovered with a half-life of ~15 hours. Hence, *de novo* synthesis of the H^+,K^+-ATPase cannot be the sole factor responsible for the recovery of ATPase activity after omeprazole inhibition. Cleavage of the inhibitory bond must also be a contributing factor. This study also showed that the turnover rate of the H^+,K^+-ATPase was unaffected by omeprazole treatment, implying that omeprazole inhibition does not interfere with regulation of the enzyme.

DIFFERENCES IN BINDING SITES FOR OMEPRAZOLE, LANSOPRAZOLE, PANTOPRAZOLE, AND RABEPRAZOLE/PARIPRAZOLE ON THE H^+,K^+-ATPase AND ITS POSSIBLE IMPLICATIONS

The α subunit of the H^+,K^+-ATPase contains 28 cysteine residues, of which only a few are predicted to be accessible from the luminal surface of the enzyme. The particular cysteines that react with the three compounds have been identified in experiments using radioactive compound to label them under acid-transporting conditions in isolated gastric vesicles, followed by tryptic digestion of the enzyme and sequencing of the fragments.[17–19] Despite their common mechanism of action, the binding pattern for these three compounds was somewhat different (TABLE 1). The reason for these differences has not yet been clarified, but involvement of various factors, such as differences in rates of transformation into the active inhibitor, steric factors in the compounds, and the structure of the pump in the binding region, has been discussed. Whether one or more of these binding sites are essential for inhibition has not yet been established. Although similar labeling studies with radioactive rabeprazole/pariprazole have not been presented, experiments in which the bound compound was made fluorescent indicated that Cys322 was the site of interaction.[20]

Functionally or clinically relevant implications of these differences in binding patterns, however, remain to be found. The most evident effects would be possible differences in the rate of inhibition or the rate of recovery. Evidence for differences in the onset of inhibition is absent for the physiological *in vivo* situation, and the differences that have been obtained in *in vitro* studies correlate with the known differences in pH-stability profiles between the compounds.[21] Differences in recovery rates and mechanisms, however, have been both proposed and denied. For example, it has been claimed that rabeprazole/pariprazole has a shorter duration of action than does lansoprazole in the rat,[22] but it has also been stated that rabeprazole/pariprazole is as long lasting as omeprazole and lansoprazole in man.[23] We performed a series of comparative experiments in which chronic gastric fistula rats were treated orally once daily for 3 days, and the recovery of acid secretion was subsequently followed. In order to compare the duration of action of the different compounds, we carefully titrated out doses that resulted in that acid secretion started to return at the same time point as after the last dose. We thereby were able to compare the compounds in pairs, and we could not detect any differences in the duration of action between the four compounds (FIG. 4). The results indicate that the rates and thus possibly the mechanisms of recovery are very similar for these four compounds. The reported differences in binding patterns have thus most probably no relevance whatsoever to the action and effects of these drugs.

REFERENCES

1. LINDBERG, P., A. BRÄNDSTRÖM, B. WALLMARK, H. MATTSSON, L. RIKNER & K. J. HOFFMAN. 1990. Med. Rev. Res. **10:** 1–54.
2. BARRADELL, L. B., D. FAULDS & D. MCTAVISH. 1992. Drugs **44:** 225–250.
3. HUBER, R., B. KOHL, G. SACHS, J. SENN-BILFINGER, W. A. SIMON & E. STORM. 1995. Aliment. Pharmacol. Ther. **9:** 363–378.
4. YASUDA, S., A. OHNISHI, T. OGAWA, Y. TOMONO, J. HASEGAWA, H. NAKAI, Y. SHIMAMURA & N. MORISHITA. 1994. Int. J. Pharmacol. Ther. **32:** 466–473.
5. POPE, A. J. & M. E. PARSONS. 1993. Trends Pharmacol. Sci. **14:** 323–325.
6. LINDBERG, P., P. NORDBERG, T. ALMINGER, A. BRÄNDSTRÖM & B. WALLMARK. 1986. J. Med. Chem. **29:** 1327–1329.
7. LORENTZON, P., R. JACKSON, B. WALLMARK & G. SACHS. 1987. Biochim. Biophys. Acta **897:** 41–51.
8. FRYKLUND, J., K. GEDDA & B. WALLMARK. 1988. Biochem. Pharmacol. **37:** 2543–2549.
9. DE GRAEF, J. & M. C. WOUSEN-COLLE. 1986. Gastroenterology **91:** 333–337.
10. WALLMARK, B., A. BRÄNDSTRÖM & H. LARSSON. 1984. Biochim. Biophys. Acta **778:** 549–558.
11. NILSSON, A., A. TORVÉN & M. SOHTELL. 1995. Gastroenterology **108:** A179.
12. KRAUT, J., F. STARR, G. SACHS & M. REUBEN. 1995. Am. J. Physiol. **268:** F581–F587.
13. KLEINMAN, J. G. 1994. J. Am. Soc. Nephrol. **5:** S6–S11.
14. CROWSON, S. M. & G. E. SHULL. 1992. J. Biol. Chem. **267:** 13740–13748.
15. LORENTZON, P., B. EKLUNDH, A. BRÄNDSTRÖM & B. WALLMARK. 1985. Biochim. Biophys. Acta **817:** 25–32.

16. GEDDA, K., D. SCOTT, M. BESANCON, P. LORENTZON & G. SACHS. 1995. Gastroenterology **109:** 1134–1141.
17. BESANCON, M., J. M. SHIN, F. MERCIER, K. MUNSON, M. MILLER, S. HERSEY & G. SACHS. 1993. Biochemistry **32:** 2345–2355.
18. SACHS, G., J. M. SHIN, M. BESANCON & C. PRINZ. 1993. Aliment. Pharmacol. Ther. **7:** 4–12.
19. SHIN, J. M., M. BESANCON, W. A. SIMON & G. SACHS. 1993. Biochim. Biophys. Acta **1148:** 223–233.
20. MORII, M., K. HAMATANI & N. TAKEGUCHI. 1995. Biochem. Pharmacol. **49:** 1729–1734.
21. SACHS, G., J. M. SHIN, C. BRIVING, B. WALLMARK & S. HERSEY. 1995. Annu. Rev. Pharmacol. Toxicol. **35:** 277–305.
22. TOMIYAMA, Y., M. MORII & N. TAKEGUCHI. 1994. Biochem. Pharmacol. **48:** 2049–2055.
23. KOVACS, T. O. G., B. SYTNIK, T. J. HUMPHRIES & J. H. WALSH. 1996. Gastroenterology **110:** A161.

Vacuolar H$^+$-ATPases

Targets for Drug Discovery?

D. J. KEELING,[a] M. HERSLÖF, B. RYBERG, S. SJÖGREN,
AND L. SÖLVELL

*Preclinical R&D
Astra Hässle AB
43183 Mölndal, Sweden*

Vacuolar H$^+$-ATPases are a class of multi-subunit ATP-driven proton pumps found in all eukaryotic cells. The precise number of subunits constituting these proteins is not yet resolved, and whether individual peptides constitute true subunits or accessory/regulatory subunits is a matter of debate. Nevertheless, the following overall picture is emerging.[1–3] Functionally the subunits are divided into two subcomplexes (FIG. 1): a cytosolic catalytic complex (Vc) which is responsible for hydrolysis of ATP, and an integral membrane subcomplex (Vb) responsible for proton conduction across the membrane. The Vc subcomplex consists of six subunits, named A-F, with apparent molecular masses of approximately: A (70 kD), B (58 kD), C (40 kD), D (34 kD), E (33 kD), and F (16 kD). The ATP binding sites are situated on a hexamer comprised of three copies of each of the A and B subunits. The Vb complex consists of six copies of a 17-kD subunit which are believed to form the proton channel along with one copy of each of a 100–116 kD and 39 kD subunit. In addition, a 20-kD subunit is claimed to be present in some systems.[2] Recently, an additional (G) subunit has been reported that may link the Vc and Vb halves of the pump.[4,5] Additionally, a 54-kD peptide in yeast[6] and a corresponding 50-57 kD heterodimer in clathrin-coated vesicles[7] have been suggested to play a role in the activity and regulation of proton pumping.

The Vc/Vb structural design of eukaryotic vacuolar H$^+$-ATPases is shared by the ATP-synthases of eubacteria, chloroplasts, and mitochondria (the F1/FO-ATPases) and by ATP-synthases from archaebacteria (A-ATPases). Sequence homologies between several of the subunits suggest that all three classes of proton pumps derive from a common ancestor that existed before the divergence of eubacteria, archaebacteria, and eukaryotes.[8] Thus, this overall structure represents one of nature's fundamental designs for active membrane transport. In contrast to the F1/FO-ATPases and A-ATPases, which can both synthesize ATP from an existing protonmotive gradient and hydrolyze ATP to form such a gradient, vacuolar H$^+$-ATPases can only function in the proton pumping mode.

Proton transport by vacuolar H$^+$-ATPases is an electrogenic process and results in the formation of an inside-positive membrane potential. In many cases a chloride channel is also present in parallel with the acid pump, allowing the membrane potential to be converted into a ΔpH[9].

Characterization of the functional role of vacuolar H$^+$-ATPases has been greatly assisted by the discovery of the bafilomycins (FIG. 2). These bacterially produced

---

[a]Address for correspondence: Department of Cell Biology, Astra Hässle AB, 431 83 Mölndal, Sweden (tel: +46 31 7761528; fax: +46 31 7763761; e-mail: david.keeling@hassle.se.astra.com).

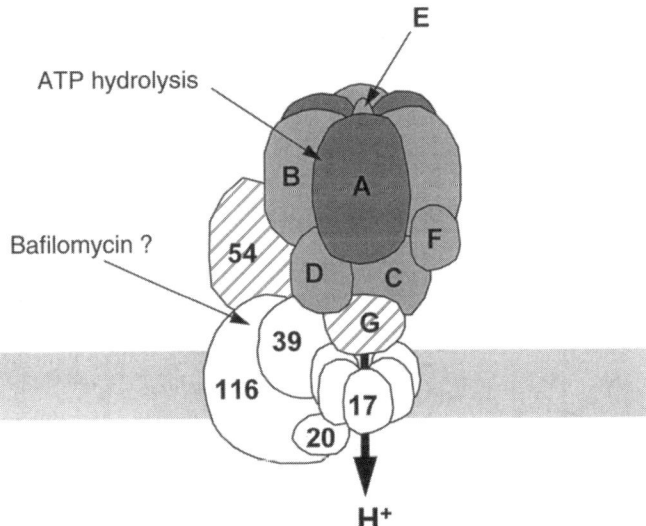

FIGURE 1. Putative structure of the vacuolar H^+-ATPase. The pump is divided into two parts, a soluble, cytosolic catalytic sector (Vc), shown as *shaded subunits,* and a membrane-spanning proton channel sector, shown as *unshaded subunits.* Additional subunits believed to link the two halves of the pump and regulate proton transport are shown with *diagonal shading.* The relative positions of the subunits within each subcomplex are speculative. Vc has been drawn to resemble the structure of the F1 sector of the distantly related FO/F1-ATPase synthase.

macrolides[10] have been shown to be potent and selective inhibitors of vacuolar H^+-ATPases.[11] The precise manner in which bafilomycin A1 inhibits vacuolar H^+-ATPase activity is not known. However, bafilomycin A1 has been shown to block the proton conductance of the isolated membrane channel part of the vacuolar H^+-ATPase,[12] suggesting a binding site somewhere within this subcomplex (hence the name Vb for "bafilomycin binding"). Direct binding of bafilomycin to the Vb complex has been demonstrated.[13] A further study has claimed that the 100–116 kD subunit of the Vb complex contains the bafilomycin binding site.[14]

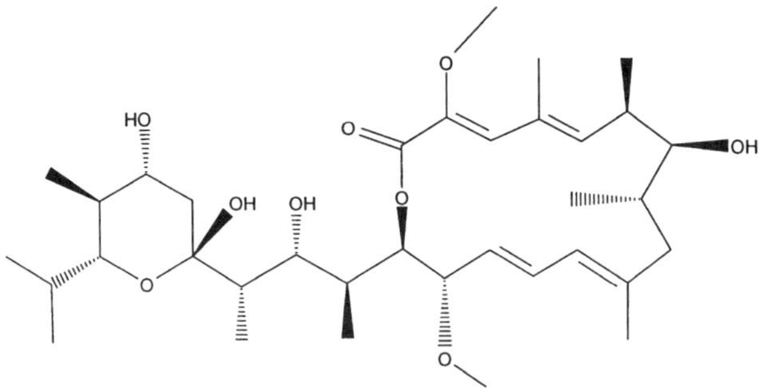

FIGURE 2. Structure of bafilomycin A1.

THE ROLE OF VACUOLAR H$^+$-ATPases IN EUKARYOTIC CELLS

Intracellular Vacuolar H$^+$-ATPases

In eukaryotic cells, vacuolar H$^+$-ATPases are responsible for the acidification of many intracellular compartments and as such play important roles in a number of cell processes. Acidification of the trans-Golgi network is important in the targeting of newly synthesized proteins to their appropriate destinations.[15] Acidification of immature secretory granules is required for the activity of proteases necessary for the processing of pro-proteins to their mature, secreted products.[16] In the endocytotic pathway, the dissociation of internalized receptors from their ligands is favored by low pH.[17] The uptake of neurotransmitters into synaptic vesicles is driven by a protonmotive force created by vacuolar H$^+$-ATPase activity.[18] The low pH within the lysosome favors activity of the hydrolases present in this organelle.[19] Furthermore, vacuolar H$^+$-ATPase is present in the vacuolar membranes of plants and fungi where the protonmotive force generated by this acid pump drives secondary active transport of solutes into the vacuole.[20,21]

Plasma Membrane Vacuolar H$^+$-ATPase

In addition to being found in intracellular membranes, vacuolar H$^+$-ATPases have been identified in the plasma membranes of a number of specialized eukaryotic cells. In insects the active uptake of nutrients in the gut and ion homeostasis in the Malpighian tubule are driven by vacuolar H$^+$-ATPases in the apical membranes of these epithelia.[22] In the vertebrate kidney, a vacuolar H$^+$-ATPase in the apical membrane of intercalated cells of the collecting tubule secretes excess metabolic acid (from the catabolism of protein) and ensures that overall acid/base balance is maintained.[23] Furthermore, in the proximal tubule, acid secretion is important for the reabsorption of bicarbonate from the glomerular filtrate, where it is estimated that a third of bicarbonate uptake depends on a vacuolar H$^+$-ATPase in the apical membrane.[23] In the male reproductive tract a low pH environment is believed to be important for the maturation and storage of viable sperm. The luminal membranes of certain cells in the epididymus have been shown to be rich in vacuolar H$^+$-ATPase,[24] and vacuolar H$^+$-ATPase-mediated acid transport has been demonstrated in vas deferens.[25] A vacuolar H$^+$-ATPase has also been identified in the plasma membrane of the macrophages and neutrophils where it is believed to assist in the maintainance of cytosolic pH under conditions of acid load.[26,27]

The focus of this report, however, is the osteoclast, the cell responsible for the breakdown (resorption) of bone. Evidence has accumulated that bone resorption requires the action of a vacuolar H$^+$-ATPase situated in the plasma membrane of the osteoclast adjacent to the site of resorption (the ruffled membrane). Because certain pathological conditions, including osteoporosis, result from an imbalance in the formation and breakdown of bone, the inhibition of bone resorption by an inhibitor of vacuolar H$^+$-ATPase could be of therapeutic value.

BONE RESORPTION AND OSTEOPOROSIS

Bone is a dynamic tissue in a continual state of renewal and adaptation to the load placed upon it. Old bone is broken down (resorbed) and new bone is built in its place. In the young adult skeleton these processes act in balance and the overall mass of the skeleton is maintained. However, in the aging skeleton the net rate of resorption exceeds that of formation and bone is progressively lost. This imbalance is further exacerbated in postmenopausal women when reduced levels of circulating estrogens cause an acceleration of bone turnover. With time, the skeleton can be sufficiently weakened that even slight trauma results in fracture. This is the condition known as osteoporosis which is estimated to affect 200 million women world-wide and results in a cost to society approaching $10 billion per year in the United States alone.[28]

Role of Vacuolar H^+-ATPase in Bone Resorption

The cell responsible for bone resorption is the osteoclast. In order to resorb bone the osteoclast must first remove the mineral (hydroxyapatite) component of the bone, thus exposing the proteinaceous bone matrix (mostly collagen) for degradation. It has been suggested that removal of mineral could depend on the secretion of acid onto the bone surface by the osteoclast.[29] The detection of vacuolar H^+-ATPase activity in membrane vesicles derived from osteoclast-rich sources along with staining of the osteclast ruffled membrane with antibodies to vacuolar H^+-ATPase subunits[30,31] supported this view. Functional evidence for the role of vacuolar H^+-ATPase was provided by the observation that the selective vacuolar H^+-ATPase inhibitor bafilomycin A1 was able to inhibit bone resorption in isolated rat osteoclast preparations and isolated mouse calvariae (skull bones) with potencies in the range of 1 nM.[32,33] We have extended these results to assess the effect of vacuolar H^+-ATPase inhibition in the whole animal.

Effect of Bafilomycin A1 on Bone Resorption in Vivo

The skeletons of young growing rats can be conveniently labeled with ^{3}H-tetracycline. Bone resorption causes a release of the bound tetracycline which is rapidly excreted by the kidneys without further change. By following the excretion of radioactivity in urine the rate of bone resorption can be indirectly followed. After the first few days during which ^{3}H-tetracycline is still being cleared from the soft tissues, the excretion of radioactivity follows an approximately exponential decline which can be used to calculate the rate of basal bone resorption. Twice daily, intravenous administration of bafilomycin A1 during this period induced a dose-dependent inhibition of the resorption rate (FIG. 3). This inhibition was reversible as indicated by the return of the resorption rate to control levels upon the cessation of dosing. The maximal extent of inhibition achievable by vacuolar H^+-ATPase blockade could not be determined, as doses of bafilomycin A1 higher than 0.3 µmol/kg i.v. caused acute toxicity effects typically manifest as disturbances in locomotor control and piloerection (5 minutes after dosing) and, with doses of 1 µmol/kg i.v. and above, signs of cyanosis and convulsions (10–15 minutes after dosing). Somewhat higher doses (1.4

µmol/kg) could be administered once daily, subcutaneously without toxic symptoms. This dose was therefore chosen to assess the effect of vacuolar H^+-ATPase inhibition on bone density.

Bone turnover in young growing rats is high. This is particularly so at the ends of the long bones (e.g., distal femur) where template bone and cartilage is broken down and rebuilt as part of the growth process. Inhibition of resorption leads to an increase in bone density at these sites. Subcutaneous administration of bafilomycin A1 for 14 days, at doses of 0.7 and 1.4 µmol/kg, caused a dose-dependent increase in bone density in the distal femur (FIG. 4).

These results indicate that vacuolar H^+-ATPase plays a central role in the process of bone resorption. They furthermore suggest that selective blockade of the osteoclast vacuolar H^+-ATPase might be effective in reducing the rate of bone resorption in pathological conditions such as osteoporosis.

SELECTIVITY: ISOFORMS OF VACUOLAR H^+-ATPases

The key issue regarding the therapeutic usefulness of vacuolar H^+-ATPase inhibitors is that of selectivity. Vacuolar H^+-ATPases are present in all eukaryotic cells where they are involved in many important cell processes. Nonspecific inhibition of vacuolar H^+-ATPases can be expected to cause a multitude of effects, most of which

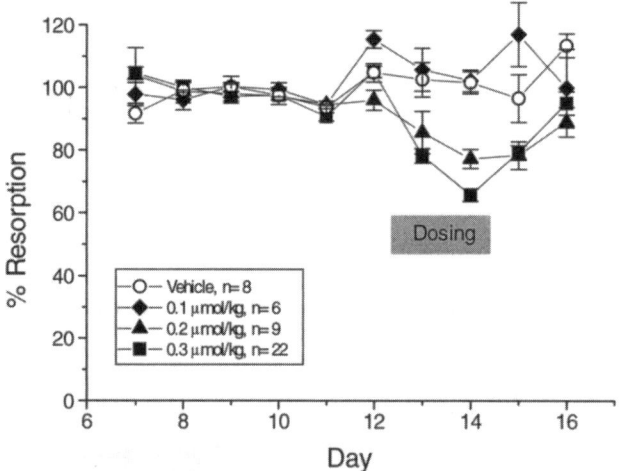

FIGURE 3. Inhibition of basal bone resorption in the rat by bafilomycin A1. Bone resorption was measured as the urinary excretion of ^3H-tetracycline in rats whose skeletons had previously been labeled with ^3H-tetracycline. Excretion of ^3H-tetracycline followed an exponential pattern from which a control rate of resorption could be calculated (100%) that was stable between 6 and 16 days following ^3H-tetracycline labeling. Bafilomycin A1 was dosed intravenously, twice daily, at the doses shown for 2 days (days 12 and 13), inducing inhibition in the corresponding 24-hour urine collection periods (days 13 and 14). Following cessation of dosing the resorption rate returned to normal. Vehicle was 5% solutol/5% ethanol/90% physiological saline solution. The numbers of animals in each group are indicated. *Error bars* represent standard error of the mean.

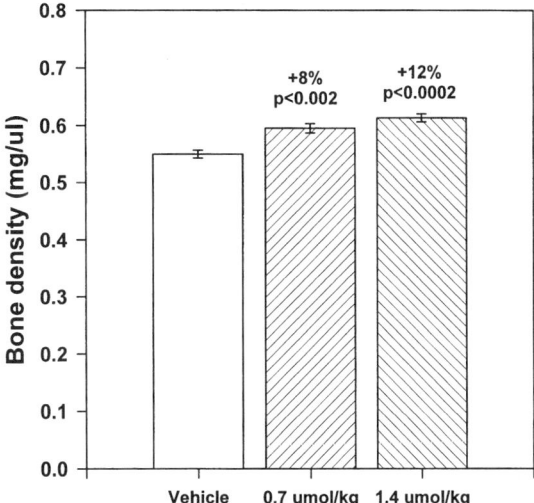

FIGURE 4. Effect of bafilomycin A1 on bone density in young growing rats. Young female Sprague-Dawley rats (170 g) were dosed subcutaneously with bafilomycin A1 once daily for 14 days. The distal femur was removed and bone density was determined as ash weight/volume. Vehicle was 3% solutol/3% ethanol/94% physiological saline solution. Values shown are means and standard error of the mean from six rats in each group.

are undesirable. Indeed, bafilomycin A1 is such a nonspecific agent, inhibiting acid transport in microsomal membrane vesicles from avian medullary bone and avian brain similar high affinities (FIG. 5). It is most likely that the signs of toxicity observed in *in vivo* experiments at high doses of bafilomycin A1 were due to the inhibition of vacuolar H^+-ATPase in tissues other than bone.

Evidence is accumulating, however, for the existence of different isoforms of the vacuolar H^+-ATPase. A novel form of the A subunit was reported in human osteoclastoma tissue,[34] but subsequent work failed to detect this form in tissues other than the original tumor.[35] Another group has reported alternative-spliced variants of the A subunit in chicken tissues,[36] although both variants were expressed in all tissues studied. Two isoforms of the B subunit have been reported,[37,38] one expressed in all tissues studied, including osteoclasts, and the other in kidney and placenta.[39–41] Thus, the isoform in osteoclasts appears to be the same as that in other cells. Greater isoform diversity may exist for the 100–116 kD (VBA) subunit. In yeast, two isoforms of this subunit are coded for by the VPH1 and STV1 genes.[42] The VPH1 isoform appears to be localized to the vacuolar membrane, whereas the STV1 isoform was found in cytoplasmic organelles. Thus, different isoforms can exist within the same cell. In mammalian systems, RNA for two alternatively spliced forms of the 116-kD subunit have been identified, one of which was widely expressed whereas the other was only detected in brain.[43] A further isoform has been identified by sequence homology comparisons. This protein was originally reported as an immunoregulatory protein, TJ6, from mouse thymus,[44] but based on homology, it is most probably an isoform of the 116-kD subunit. More recently, a putative human osteoclast-specific isoform of the 116-kD subunit was isolated from human osteoclastoma tissue.[45] mRNA coding for this isoform was detected only in osteoclast-like cells of the tumor and not in a range of other tissues and cell lines; however, its presence in native human osteoclasts has yet to be demonstrated. Interestingly, it has been claimed that bafilomycin A1 interacts with the 116-kD subunit,[14] although there is no evidence to date of tissue-selective inhibition of vacuolar H^+-ATPase using this compound.

The presence of distinct isoforms of the vacuolar H^+-ATPase is encouraging, although, as yet, definitive evidence for an osteoclast-specific isoform is lacking. Furthermore, identification of isoforms does not automatically confer therapeutic utility. This will require novel vacuolar H^+-ATPase inhibitors that can distinguish between such isoforms. To date there is little conclusive evidence in support of such pharmacological differences. Claims that the osteoclast vacuolar H^+-ATPase was

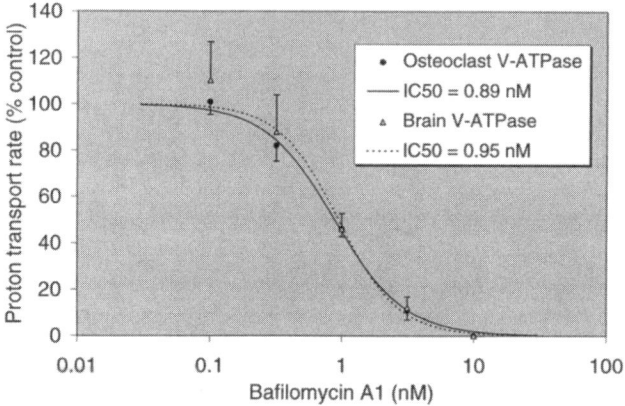

FIGURE 5. Bafilomycin A1 inhibits osteoclast and brain vacuolar H^+-ATPase with similar potency. Microsomal membrane fractions were prepared from chicken medullary bone (osteoclast-rich) and chicken brain. Proton transport was assessed by the initial rate of quenching of the dye acridine orange (5 μM) in the presence of 5 mM Hepes/Tris, 125 mM KCl, 3 mM MgSO4, 1 μM valinomycin, and 1 mM ATP (pH 7.4). Each assay contained either 30 μg protein (osteoclasts) or 50 μg protein (brain). Control acidification rates were (in milliabsorbance units per minute, mOD/min): osteoclast 15 ± 3 and brain 14 ± 3. Data shown are means with standard error of the mean from three experiments for each tissue source.

selectively inhibited by vanadate[46] or by the bisphosphonate tiludronate[47] were not repeatable in our laboratory. Nonetheless, biochemical preparations of vacuolar H^+-ATPase from different parts of the kidney have been reported to have distinct pH optima and sensitivities to copper,[48] suggesting that functional properties can vary between different vacuolar H^+-ATPases.

CONCLUSIONS

Bafilomycin A1, an inhibitor of vacuolar H^+-ATPase, inhibits bone resorption in a reversible manner and increases bone density in the young growing rat. This indicates a central role for the osteoclast vacuolar H^+-ATPase in the process of bone resorption. The development of a useful therapeutic agent with this mechanism of action requires selective inhibition of the osteoclast vacuolar H^+-ATPase relative to vacuolar H^+-ATPases in other cell types. Growing evidence for multiple isoforms of the vacuolar H^+-ATPase increases the probability that such an agent can be identified.

REFERENCES

1. STONE, D. K., B. P. CRIDER, T. C. SUDHOF & X. S. XIE. 1989. J. Bioenerg. Biomembr. **21:** 605–620.
2. FORGAC, M. 1992. J. Bioenerg. Biomembr. **24:** 341–350.
3. NELSON, H., S. MANDIYAN & N. NELSON. 1994. J. Biol. Chem. **269:** 24150–24155.
4. SUPEKOVA, L., F. SUPEK & N. NELSON. 1995. J. Biol. Chem. **270:** 13726–13732.
5. LEPIER, A., R. GRAF, M. AZUMA, H. MERZENDORFER, W. R. HARVEY & H. WIECZOREK. 1996. J. Biol. Chem. **271:** 8502–8508.
6. HO, M. N., R. HIRATA, N. UMEMOTO, Y. OHYA, A. TAKATSUKI, T. H. STEVENS & Y. ANRAKU. 1993. J. Biol. Chem. **268:** 18286–18292.
7. XIE, X. S., B. P. CRIDER, Y. M. MA & D. K. STONE. 1994. J. Biol. Chem. **269:** 25809–25815.
8. NELSON, N. 1992. Curr. Opin. Cell Biol. **4:** 654–660.
9. AL-AWQATI, Q., J. BARASCH & D. LANDRY. 1992. J. Exp. Biol. **172:** 245–266.
10. WERNER, G., H. HAGENMAIER, H. DRAUTZ, A. BAUMGARTNER & H. ZAHNER. 1984. J. Antibiot. **37:** 110–117.
11. BOWMAN, E. J., A. SIEBERS & K. ALTENDORF. 1988. Proc. Natl. Acad. Sci. USA **85:** 7972–7976.
12. CRIDER, B. P., X. S. XIE & D. K. STONE. 1994. J. Biol. Chem. **269:** 17379–17381.
13. MATTSSON, J. P. & D. J. KEELING. 1996. Biochim. Biophys. Acta **1280:** 98–106.
14. ZHANG, J., Y. FENG & M. FORGAC. 1994. J. Biol. Chem. **269:** 23518–23523.
15. ROTHMAN, J. H., C. T. YAMASHIRO, P. M. KANE & T. H. STEVENS. 1989. Trends Biochem. Sci. **14:** 347–350.
16. HAASS, C., A. CAPELL, M. CITRON, D. B. TEPLOW & D. J. SELKOE. 1995. J. Biol. Chem. **270:** 6186–6192.
17. DIAZ, R., T. E. WILEMAN, S. J. ANDERSON & P. STAHL. 1989. Biochem. J. **260:** 127–134.
18. HELL, J. W., P. R. MAYCOX, H. STADLER & R. JAHN. 1988. EMBO J. **7:** 3023–3029.
19. YOSHIMORI, T., A. YAMAMOTO, Y. MORIYAMA, M. FUTAI & Y. TASHIRO. 1991. J. Biol. Chem. **266:** 17707–17712.
20. THOM, M. & E. KOMOR. 1984. Eur. J. Biochem. **138:** 93–99.
21. STEVENS, T. H. 1992. J. Exp. Biol. **172:** 47–55.
22. ZEISKE, W. 1992. J. Exp. Biol. **172:** 323–334.
23. GLUCK, S. & R. NELSON. 1992. J. Exp. Biol. **172:** 205–218.
24. BROWN, D., B. LUI, S. GLUCK & I. SABOLIC. 1992. Am. J. Physiol. **263:** C913–916.
25. BRETON, S., P. J. S. SMITH, B. LUI & D. BROWN. 1996. Nat. Med. **2:** 470–472.
26. SWALLOW, C. J., S. GRINSTEIN, R. A. SUBSBURY & O. D. ROTSTEIN. 1990. Surgery **108:** 363–368.
27. NANDA, A., A. GUKOVSKAYA, J. TSENG & S. GRINSTEIN. 1992. J. Biol. Chem. **267:** 22740–22746.
28. RIGGS, B. L. & L. MELTON, JR. 1995. Bone **17:** 505S–511S.
29. BARON, R., L. NEFF, D. LOUVARD & P. J. COURTOY. 1985. J. Cell Biol. **101:** 2210–2222.
30. BLAIR, H. C., S. L. TEITELBAUM, R. GHISELLI & S. GLUCK. 1989. Science **245:** 855–857.
31. VAANANEN, H. K., E. K. KARHUKORPI, K. SUNDQUIST, B. WALLMARK, I. ROININEN, T. HENTUNEN, J. TUUKKANEN & P. LAKKAKORPI. 1990. J. Cell Biol. **111:** 1305–1311.
32. MATTSSON, J. P., K. VAANANEN, B. WALLMARK & P. LORENTZON. 1991. Biochim. Biophys. Acta **1065:** 261–268.

33. SUNDQUIST, K., P. LAKKAKORPI, B. WALLMARK & K. VAANANEN. 1990. Biochem. Biophys. Res. Comm. **168:** 309–313.
34. VAN HILLE, B., H. RICHENER, D. B. EVANS, J. R. GREEN & G. BILBE. 1993. J. Biol. Chem. **268:** 7075–7080.
35. VAN HILLE, B., H. RICHENER, J. R. GREEN & G. BILBE. 1995. Biochem. Biophys. Res. Comm. **214:** 1108–1113.
36. HERNANDO, N., M. BARTKIEWICZ, P. COLLINOSDOBY, P. OSDOBY & R. BARON. 1995. Proc. Natl. Acad. Sci. USA **92:** 6087–6091.
37. SUDHOF, T. C., V. A. FRIED, D. K. STONE, P. A. JOHNSTON & X. S. XIE. 1989. Proc. Natl. Acad. Sci. USA **86:** 6067–6071.
38. BERNASCONI, P., T. RAUSCH, I. STRUVE, L. MORGAN & L. TAIZ. 1990. J. Biol. Chem. **265:** 17428–17431.
39. NELSON, R. D., X. L. GUO, K. MASOOD, D. BROWN, M. KALKBRENNER & S. GLUCK. 1992. Proc. Natl. Acad. Sci. USA **89:** 3541–3545.
40. VAN HILLE, B., H. RICHENER, P. SCHMID, I. PUETTNER, J. R. GREEN & G. BILBE. 1994. Biochem. J. **303:** 191–198.
41. LEE, B. S., S. HOLLIDAY, B. OJIKUTU, I. KRITS & S. L. GLUCK. 1996. Am. J. Physiol. **39:** C382–C388.
42. MANOLSON, M. F., D. WU, D. PROTEAU, B. E. TAILLON, B. T. ROBERTS, M. A. HOYT & E. W. JONES. 1994. J. Biol. Chem. **269:** 14064–14074.
43. PENG, S. B., B. P. CRIDER, X. S. XIE & D. K. STONE. 1994. J. Biol. Chem. **269:** 17262–17266.
44. LEE, C., K. GHOSHAL & K. D. BEAMAN. 1990. Mol. Immunol. **27:** 1137–1144.
45. LI, Y. P., W. CHEN & P. STASHENKO. 1996. Biochem. Biophys. Res. Comm. **218:** 813–821.
46. CHATTERJEE, D., M. CHAKRABORTY, M. LEIT, L. NEFF, S. JAMSA-KELLOKUMPU, R. FUCHS, M. BARTKIEWICZ, N. HERNANDO & R. BARON. 1992. J. Exp. Biol. **172:** 193–204.
47. DAVID, P., H. NGUYEN, A. BARBIER & R. BARON. 1995. Bone **16:** 166S.
48. WANG, Z. Q. & S. GLUCK. 1990. J. Biol. Chem. **265:** 21957–21965.

The Plasma Membrane H^+-ATPase of Fungi

A Candidate Drug Target?[a]

DAVID S. PERLIN,[b,d] DONNA SETO-YOUNG,[b]
AND BRIAN C. MONK[c]

[b]Public Health Research Institute
455 First Avenue
New York, New York 10016

[c]Department of Oral Biology and Pathology
Faculty of Dentistry
University of Otago
Dunedin, New Zealand

Opportunistic fungal infections are widespread in HIV and other immunosuppressed individuals and are a growing problem for the management of such patients. The commensal *Candida albicans,* which causes both topical and invasive disease, is the predominant organism associated with fungal disease, although other fungal pathogens such as *Aspergillus fumigatus, Cryptococcus neoformans, Pneumocystis carinii, Histoplasma capsulatum,* and other Candida species are important. These organisms have become major causes of morbidity and mortality in the immunocompromised[1,2] and are principal causes of nosocomial disease.[3] Mucosal Candida infections, which are frequently observed in AIDS and other immunocompromised patients, can usually be treated with existing azole-based antifungal drugs, such as fluconazole and itraconazole. However, a major concern is the increasing numbers of treatment failures associated with resistance to azole-based drugs caused by target mutations, multidrug efflux pumps,[4] and colonization by less susceptible fungi.[5] Invasive fungal disease involving one or more organs is intrinsically more difficult to treat. Therapeutic agents that are effective against mucosal infections are largely ineffective against invasive disease. The treatment options for invasive infections are extremely limited and almost always involve the use of the nephrotoxic agent amphotericin B.[6] There is an urgent clinical need to develop more effective therapeutics to deal with invasive disease, and the establishment of new targets is critical to this process.

THE FUNGAL H^+-ATPase AS A NOVEL ANTIFUNGAL TARGET

The fungal plasma membrane H^+-ATPase is an important new target for therapeutic intervention.[7] It is an essential enzyme that plays a critical role in fungal cell physiology by maintaining the large transmembrane electrochemical proton gradient

---

[a]This work was generously supported by grants (to D.S.P.) from the National Institutes of Health (AI 35411) and Astra Hässle.

[d]To whom correspondence should be addressed.

necessary for nutrient uptake and by helping to regulate intracellular pH. It is one of the few antifungal targets that have been shown to be essential by gene disruption.[8] In addition to its role in cell growth, the H^+-ATPase has been implicated in fungal pathogenicity through its effects on dimorphism, nutrient uptake, and medium acidification.[9] These properties, along with the availability of numerous *in vivo* and *in vitro* screens that facilitate high through-put screening of compound libraries, make the fungal H^+-ATPase a highly desirable drug discovery target.

The P-type ATPase family of ion pumps already serves as a valuable source of cellular targets for the clinically important therapeutics. These include the cardiac glycosides and antiulcer drugs, which block the Na^+,K^+-ATPase and H^+,K^+-ATPase, respectively. Each of these therapeutics acts in a highly enzyme-specific manner and interacts with the extracellular surface of its respective target enzyme.[10,11] The specificity and mechanism of action of these well-established therapeutics suggest that it should be possible to develop similar types of antagonists of the fungal H^+-ATPase. The high degree of sequence similarity among diverse fungal *PMA* genes suggests that a selective H^+-ATPase antagonist could have broad-spectrum activity. By contrast, the relative low identity (<30%) between the fungal *PMA* genes and their mammalian counterparts suggests that it should be possible to identify antagonists that distinguish between the two types of related enzymes.

Recently, we demonstrated the importance of the H^+-ATPase as an antifungal target by showing that inhibition of the enzyme by the sulfhydryl-reactive reagent omeprazole was closely correlated with inhibition of cell growth by *C. albicans* and *Saccharomyces cerevisiae* and that omeprazole-induced inhibition of the proton pump was fungicidal.[12] Omeprazole is not suitable as an antifungal candidate. However, its well defined and controlled properties make it a valuable probe to selectively examine the role of the H^+-ATPase as an antifungal target.

OMEPRAZOLE AS A PROBE OF THE H^+-ATPase

Omeprazole was previously shown to be an effective antagonist of fungal cell growth.[12] The cell growth antagonism was attributed to the formation of a highly reactive sulfenamide species, inasmuch as the effects were low pH dependent (pH <4.0) and were protected by sulfhydryl reagents, such as reduced glutathione and mercaptoethanol. A close correlation was established between inhibition of the H^+-ATPase, loss of cell growth, and ultimately cell death and inhibition of H^+-ATPase-mediated proton transport. In these studies, inhibition of the H^+-ATPase was proposed to occur from outside the cell because the charged sulfenamide species is weakly permeant across the membrane.[10] This result is important for the drug discovery process because it implies that a target region on or near the extracellular face of the enzyme could be exploited for drug discovery. An antagonist operating at this site would be highly desirable because it would not need to enter the cell and would not be subject to the action of high capacity drug efflux pumps.

The well-established genetics and biochemistry of the H^+-ATPase have been exploited to better understand the mechanistic basis of the interaction between omeprazole and the H^+-ATPase. It was previously demonstrated that omeprazole-induced antagonism of the plasma membrane H^+-ATPase was altered by mutations in transmembrane segments 1 and 2 (M1, M2). The role of this region in omeprazole-induced antagonism was assessed by screening a collection of well-characterized mutations in this transmembrane loop of the H^+-ATPase from *S. cerevisiae*.[13,14] The *pma1* mutants were screened for changes in omeprazole-dependent growth inhibition profiles by inoculating log-phase cells in YPD medium at pH 3.5 containing increasing amounts of omeprazole (0–150 μM). FIGURE 1 summarizes these results in diagrammatic form. Several of the mutations in M1 and the turn region enhanced the sensitivity of cells to omeprazole. By contrast, two mutations, A130G and a double mutant G158D,G156C, significantly decreased the sensitivity of cells for omeprazole. Most other mutations were without significant effect, including C148 in transmembrane segment 2, which was a likely candidate for omeprazole interaction.[12] These

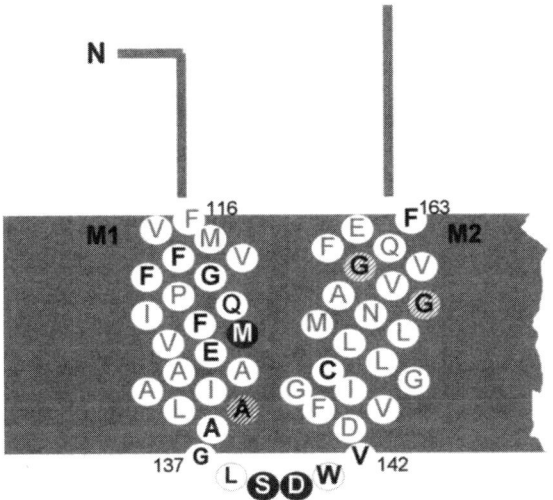

FIGURE 1. Effect of mutations in M1 and M2 on omeprazole sensitivity. A collection of *pma1* mutants with well-characterized mutations in the M1, M2 region[13,14] were screened for omeprazole sensitivity in a 24-hour end-point growth assay. Log-phase cells were added to YPD medium, pH 3.5, containing 0–150 μM omeprazole, and the cells were grown for 24 hours at 22°C in duplicate. The amino acids in *bold letters* represent mutated residues, as follows: F119A, F120A, G122A, Q125A, F126A, M128C, E129A, A138V, G138A, L139A, S139C, D140A, W141A, V142A, C148A, G158D-G156C, and F163A. Amino acid positions conferring enhanced (>fourfold) omeprazole sensitivity are shown by the *solid circles;* those showing growth resistance to omeprazole are indicated by the *cross-hatch circles,* and those without effect are indicated by the *open circles.*

whole-cell results confirm that modification of amino acid side groups in transmembrane segments 1 and 2 can modulate omeprazole-induced inhibition of the H^+-ATPase.

A biochemical demonstration that acid-activated omeprazole inhibited the enzyme from its extracellularly exposed surface was achieved using a novel *in vitro* assay system. This assay, illustrated in FIGURE 2A, uses the light-driven proton pump bacteriorhodopsin (BR) to acidify the intravesicular space of liposomes coreconstituted with BR and the H^+-ATPase. Only the omeprazole molecules that partition into the internal space are activated by the localized acid conditions. While reconstituted

A.

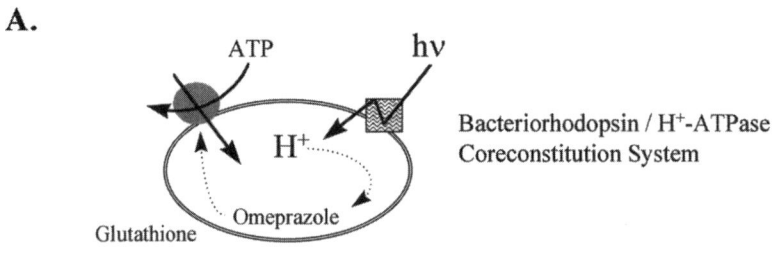

Bacteriorhodopsin / H⁺-ATPase Coreconstitution System

B.

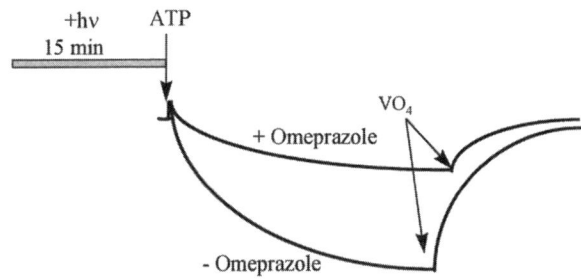

FIGURE 2. *In vitro* assay system for omeprazole-induced inhibition of the H^+-ATPase. **(A)** Schematic diagram showing liposomes co-reconstituted with bacteriorhodopsin and the H^+-ATPase. Reconstituted vesicles are loaded with omeprazole, and light is used to initiate proton pumping by bacteriorhodopsin, which lowers intravesicular pH and converts omeprazole to its reactive species. Glutathionine-conjugated agarose in the external buffer serves as a chemical trap to eliminate a stray reaction at the catalytic ATP hydrolysis site. **(B)** In a typical experiment, reconstituted liposomes, with and without 100 µg/ml omeprazole, were pretreated with actinic light for 15 minutes in a buffer consisting of 5 mM Hepes-KOH, pH 7.0, 1 M NaCl, 5 mM ATP, 4 µM acridine orange, and 100 µg membrane protein. 5 mM $MgSO_4$ was added to initiate proton transport by the H^+-ATPase, which was followed by the quenching of acridine orange fluorescence. The reaction was stopped by the addition of 10 µM vanadate, as indicated.

H^+-ATPase molecules are inserted in both possible orientations, only those enzymes with their catalytic ATP hydrolysis domain facing the medium will be active. The quenching of acridine orange fluorescence is used as a qualitative indicator of interior acid pH gradient formation. A 15-minute actinic light pulse is used to activate omeprazole in the intravesicular space of the liposome. Reduced glutathione conjugated to agarose is present in the external buffered medium to act as a chemical trap for the small amount of charged sulfenamide that might diffuse across the bilayer. MgATP is then added to activate the H^+-ATPase, and proton pumping by the enzyme is assessed. The intravesicular compartment is routinely acidified down to pH 4.0. Although this pH is not optimal for omeprazole activation, it is sufficient to convert about 5–10% of the molecules to the activated sulfenamide. FIGURE 2B shows that proton transport by the H^+-ATPase was strongly inhibited by 100 µg/ml omeprazole following a 15-minute actinic light pretreatment. The inhibition of H^+-ATPase function was fully dependent on acid conversion of omeprazole catalyzed by bacteriorhodopsin. These results support the previous suggestion that omeprazole can block the H^+-ATPase from its extracellular surface.

EXPLORING THE ROLE OF THE H^+-ATPase IN CELLULAR DIMORPHISM

A significant portion of the pathogenicity of *C. albicans* has been attributed to the formation of germ tubes that mediate tissue penetration through the action of acid-activated proteases and lipases.[15] The H^+-ATPase plays a pivotal role in this process by extruding protons and ensuring that the extracellular environment remains acidic. In addition, the H^+-ATPase has been suggested to induce a transient cytoplasmic alkalinization prior to germ tube emergence that may be a critical feature of the morphogenic switch.[16,17] The putative role of the H^+-ATPase in the dimorphic transformation by *C. albicans* was explored by blocking the enzyme with omeprazole under enzyme-selective conditions.[12] Preincubation with omeprazole effectively blocked germ tube formation in a manner consistent with the inhibitory properties of the reagent on the H^+-ATPase (FIG. 3A). A concentration of 100 μg/ml was required to block germ tube formation 90%, which is somewhat less than the dosage required to fully inhibit the H^+-ATPase and completely block cell growth.[12] This result suggests that partial inhibition of the H^+-ATPase may be sufficient to block dimorphism.

A partially inhibited H^+-ATPase should have a lower capacity for proton transport, which would result in a depolarized membrane potential and a reduced ability to efficiently regulate intracellular pH. The requirement for a highly hyperpolarized membrane potential in germ tube formation would best be explored by directly measuring membrane potential in these cells. However, such measurements are not possible. To indirectly assess the role of membrane potential in dimorphism, cells were incubated during the germ tube induction period with increasing amounts of

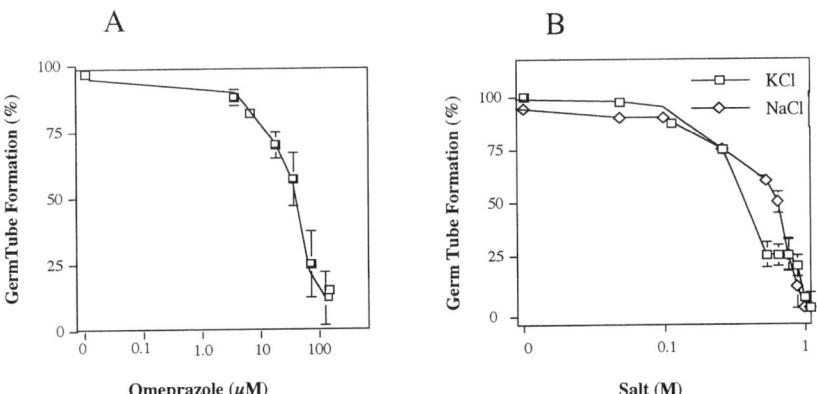

FIGURE 3. Blocking germ tube formation by *Candida albicans*. Dimorphism in *C. albicans* can be readily examined by the induction of germ tubes, as described by Monk *et al.*[17] Germ tube formation was assessed following an omeprazole pretreatment at pH 3.5 during the preinduction period (**A**) or in the presence of depolarizing amounts of KCl or NaCl (**B**). Control levels were determined in the absence of omeprazole or added salt.

monovalent cations (K^+, NH_4^+, Na^+), up to 1 M salt. FIGURE 3B shows that at concentrations sufficient to collapse the membrane potential (>750 mM), all cations tested fully blocked germ tube formation. The relative efficacy of inhibition by the various cations correlated with their known permeability across the cell membrane with NH_4^+ (not shown) and K^+ requiring a lower concentration than Na^+ to abolish germ tube formation. The concentrations of monovalent cations required to eliminate germ tube formation had no effect on either the rate or final extent of growth by *C. albicans*, indicating that the effects of monovalent cations were not simply due to a generalized growth arrest. These results are highly suggestive of an additional role for membrane potential, generated by the electrogenic action of the H^+-ATPase, in dimorphism. A greater understanding of the role of the H^+-ATPase in cellular function and its relationship to microbial pathogenicity is an important goal.

HIGH THROUGH-PUT SCREENS

A primary advantage of the fungal H^+-ATPase system is the diversity of screens available to assess the functionality of the proton pump both *in vitro* and *in vivo*, as shown in TABLE 1. Most of these screens can be performed in 96-well microplate format to facilitate high through-put screening of compound libraries. Typically, natural product or combinatorial libraries (>10,000 members) can be readily screened for lead compound generation by the simultaneous application of two primary screens that assess cell growth and ATPase activity. The growth screen is nonselective for the H^+-ATPase but is an obvious essential prerequisite for any lead compound. It consists of a 24-hour end-point determination in which a ~10^4 cell inoculum of log-phase *C. albicans* (ATCC10261) is challenged with a fixed amount of a test compound, usually present in a concentration range of 1–10 µM. By contrast, the ATPase assay examines inhibition of ATP hydrolysis by purified H^+-ATPase from either *S. cerevisiae* or *C. albicans* under cytoplasmic conditions of pH (pH 7.1) and ionic strength (100 mM KCl). Generally, the ease of purification and stability of the *Saccharomyces* enzyme favors its use for generalized screens. The high degree of amino acid sequence identity and similarity of biochemical parameters permits this substitution.[18] The assay can be performed by inhibiting the ATPase under steady-state hydrolysis conditions, in which the test compounds are present during hydrolysis. In the case of slow-reacting covalent modifiers, the purified enzyme can be preincubated with compounds for a

TABLE 1. Summary of Screens Available for Drug Discovery

Screen	Priority	Cellular Level (Organism)	H^+-ATPase Function
Growth	Primary	*In vivo* (*Candida*)	Nonspecific
ATPase assay	Primary	*In vitro*	ATP hydrolysis
Medium acidification	Secondary	*In vivo* (*Candida*)	Proton pumping
Acetate-induced acid loading	Secondary	*In vivo* (*Candida*)	Intracellular pH regulation
Hygromycin breakthrough	Secondary	*In vivo* (*Saccharomyces*)	Membrane potential
Reconstitution	Tertiary	*In vitro*	Proton pumping/membrane potential

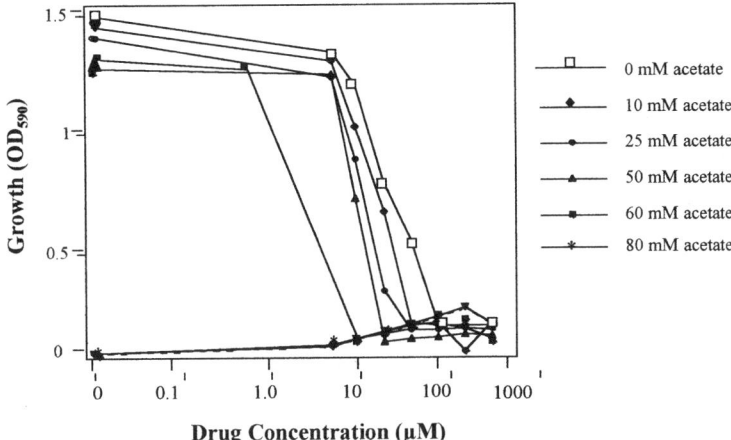

FIGURE 4. Acetate-induced shift in MIC for growth sensitivity to omeprazole. *Candida albicans* was inoculated into 200 µl of YPD medium, pH 3.5, containing increasing amounts of omeprazole in the presence of fixed amounts of acetic acid (0–80 mM), as indicated. The cells were grown for 24 hours, and final growth yields were determined by observing cell densities at $A_{590\,nm}$.

short duration, usually 10–15 minutes, and then assayed for residual activity. Once active compounds are identified in the primary screens, inhibitory properties are confirmed and explored in more detail by obtaining MIC_{90} values for growth inhibition and IC_{50} values for inhibition of ATP hydrolysis. An evaluation of compounds in secondary screens is then feasible.

Glucose-induced medium acidification is a secondary *in vivo* assay that evaluates proton transport linked to the proton pump. In this assay, which was adapted from Serrano and colleagues,[19] cells are grown to log-phase and then starved of carbon source to inactivate the H^+-ATPase. The cells are transferred to the wells of a 96-well plate containing weakly buffered KCl medium at pH 5.0 and a pH-sensitive dye bromophenol blue. Upon addition of glucose, the cells are energized and rapid medium acidification follows as the H^+-ATPase is activated. The assay is performed in the presence of potential lead compounds to assess whether they are active antagonists of whole cell proton transport.

The hygromycin B break-through assay is a secondary screen that exploits the ability of the H^+-ATPase to form a highly hyperpolarized membrane potential. In the presence of hygromycin B, which is a positively charged aminoglycoside antibiotic, yeast cells are strongly inhibited in a standard end-point growth assay. However, when the H^+-ATPase is partially attenuated due either to mutation[20] or to the action of an antagonist, the membrane potential is depolarized and the driving force for hygromycin B uptake is diminished, leading to cell resistance. Thus, compounds that antagonize the H^+-ATPase promote hygromycin B resistance at sublethal concentrations and prevent cell growth at higher concentrations.

The final secondary screen is an acetate-induced acid-loading assay, which assesses the relative ability of the proton pump to regulate intracellular pH in response to an acid load on the cytoplasm. In this assay, a weak acid acetate is added to cell growth medium at pH 5.0 in the presence of increasing amounts of a test compound. Inhibition of the proton pump at sublethal doses of antagonist yields an enzyme that is less kinetically competent to maintain intracellular pH in the presence of an acid load. Thus, there is an acetate-induced shift in the MIC90 for growth inhibition by a given test compound, as illustrated in FIGURE 4.

Finally, reconstitution of the purified H^+-ATPase in asolectin liposomes provides a convenient *in vitro* assay in which the effects of a given antagonist on proton transport and membrane potential formation can be explored.

Using these screens, it has been possible to identify diverse classes of compounds that block cell growth of *Candida, Aspergillus,* and *Cryptococcus* at concentrations below 5 µM and inhibit *in vitro* H^+-ATPase activity at less than 1 µM.

SUMMARY AND CONCLUSIONS

The fungal plasma membrane H^+-ATPase possesses important attributes that make it desirable as a target for antifungal drug discovery. First, the enzyme is essential to fungal cell physiology, being required for the formation of a large electrochemical proton gradient and the maintenance of intracellular pH. While complete inhibition of the proton pump will certainly be lethal, partial inhibition can also be lethal depending on the environment of the cell (gastrointestinal tract, etc.). Thus, an effective antagonist of the proton pump will be fungicidal, which is an important attribute for a drug being developed to treat opportunistic infections in the severely immunocompromised. Secondly, the well-characterized biochemistry and genetics of the H^+-ATPase (encoded by the *PMA*1 gene) facilitate detailed analysis of interaction of lead or model compounds with the enzyme. Studies with omeprazole, which is not suitable as an antifungal but can be used under selective conditions to target the H^+-ATPase, indicate that the enzyme can be inhibited from its extramembrane surface. Detailed genetic analysis suggests that modification of amino acids in transmembrane segments 1 and 2 can either enhance or diminish the omeprazole sensitivity of the H^+-ATPase, depending on the nature and location of the amino acid substitution. This region in mammalian P-type enzymes has been implicated in the interaction of cardiac glycosides and reversible gastric pump inhibitors. Our results suggest that this region in the H^+-ATPase may be valuable as a potential interaction domain for antifungal agents. Finally, a number of primary and secondary screens are available to identify compounds that are targeted to the H^+-ATPase and affect one or more functional properties. These screens assess enzyme functionality in the cell as well as *in vitro* and can be used in 96-well microplate format to facilitate high through-put screening. These screens have already yielded promising H^+-ATPase-directed antagonists.

In conclusion, the plasma membrane H^+-ATPase is a highly desirable target for the development of novel antifungal therapeutics.

REFERENCES

1. WEY, S. B., M. MOTOMI & M. A. PFALLER. 1988. Arch. Intern. Med. **148:** 2642.
2. STERNBERG, S. 1994. Science **266:** 1632–1635.
3. BANERJEE, S. N., T. G. EMORI, D. H. CULVER, R. P. GAYNES, W. R. JARVIS, T. HORAN, J. R. EDWARDS, J. TOLSON, T. HENDERSON, W. J. MARONE & T. N. N. I. S. SYSTEM. 1991. Am. J. Med. **91:** 86S–89S.

4. Sanglard, D., K. Kuchler, F. Ischer, J.-L. Pagani, M. Monod & J. Bille. 1995. Antimicrob. Agents Chemother. **39:** 2378–2386.
5. Rex, J. H., M. G. Rinaldi & M. A. Pfaller. 1995. Antimicrob. Agents Chemother. **39:** 1–8.
6. Armstrong, D. 1993. Clin. Infect. Dis. **16:** 1–9.
7. Monk, B. C. & D. S. Perlin. 1994. Crit. Rev. Microbiol. **20:** 209–223.
8. Serrano, R., M. C. Kielland-Brandt & G. R. Fink. 1986. Nature **319:** 689–693.
9. Prasad, R. 1991. The plasma membrane of *Candida albicans:* Its relevance to transport phenomena. *In Candida albicans.* :108–127. Springer-Verlag. Berlin.
10. Sachs, G., J. M. Shin, C. Briving, B. Wallmark & S. Hersey. 1995. Annu. Rev. Pharmacol. Toxicol. **35:** 277–305.
11. Forbush, B., III. 1983. Curr. Top. Membr. Transp. **19:** 167–201.
12. Monk, B. C., B. Mason, G. Abramochkin, J. E. Haber, D. Seto-Young & D. S. Perlin. 1995. Biochim. Biophys. Acta **1239:** 81–90.
13. Seto-Young, D., S. Na, B. C. Monk, J. E. Haber & D. S. Perlin. 1994. J. Biol. Chem. **269:** 23988–23995.
14. Seto-Young, D., M. J. Hall, S. Na, J. E. Haber & D. S. Perlin. 1996. J. Biol. Chem. **271:** 581–587.
15. Ghannoum, M. A. & K. H. Abu-Elteen. 1990. Mycoses **33:** 265–282.
16. Gow, N. A. R. & G. W. Gooday. 1987. CRC Crit. Rev. Microbiol. **15:** 73–85.
17. Monk, B. C., M. Niimi & M. G. Shepherd. 1993. J. Bacteriol. **175:** 5566–5573.
18. Monk, B. C., M. B. Kurtz, Marrinan & D. S. Perlin. 1991. J. Bacteriol. **173:** 6826–6836.
19. Serrano, R. 1983. FEBS Lett. **156:** 11–14.
20. Perlin, D. S., C. L. Brown & J. E. Haber. 1988. J. Biol. Chem. **263:** 18118–18122.

Demonstration of a Specific Transport Protein for Cardiac Glycosides in Bovine Blood[a]

ROBERTO ANTOLOVIC,[b,e] HOLGER KOST,[b]
DIETMAR LINDER,[c] MONICA LINDER,[c]
DETLEF THÖNGES,[b] DAVID LICHTSTEIN,[d]
AND WILHELM SCHONER[b]

[b]*Institut für Biochemie & Endokrinologie*
Fachbereich Veterinärmedizin
Justus-Liebig-Universität Giessen
Frankfurter Str. 100
D-35392 Giessen, Germany

[c]*Institut für Biochemie*
Fachbereich Humanmedizin
Justus-Liebig-Universität Giessen
Friedrich Str. 24
D-35392 Giessen, Germany

[d]*Department of Physiology*
Hadassah Medical School
Hebrew University-Hadassah Medical School
Jerusalen, Israel

Cardiac glycosides are specific inhibitors of Na^+/K^+-ATPase, the biochemical equivalent of the Na^+ pump. They inhibit the Na^+ pump in the range of 10^{-8}–10^{-9} M, where the therapeutic concentrations of cardiac glycosides are also found. In plasma, hydrophobic cardiac glycosides are considered to be transported unspecifically bound to serum albumin.[1] In their studies on the distribution of a digitalis-like compound (DLC) in the toad *Bufo viridis*, Lichtstein *et al.*[2] reported that about 30% of DLC was bound to plasma proteins of a molecular mass of 48–53 kD. This finding led to the postulation of the existence of a specific cardiac glycoside transport protein in vertebrates. This communication shows that a cardiac glycoside transport globulin (CGTG) of 58–72 kD and of high affinity for cardiac glycosides exists in the globulin fraction of bovine plasma.

---

[a]This work was supported by Grant Scho 139/21-1 from the Deutsche Forschungsgemeinschaft, Bonn-Bad Godesberg, and the Fonds der Chemischen Industrie, Frankfurt/M.
[b]Tel: +49-641-99-38171; fax: +49-641-99-38179; e-mail: Antolovic@vetmed.uni-giessen.de

METHODS

The globulin fraction from 10 L of bovine blood was further purified on a CM-Sephadex C 50 column (5.5 × 50 cm). The proteins of interest were precipitated with 100% $(NH_4)_2SO_4$ and dialyzed against TBS buffer (50 mM Tris/HCl, 150 mM NaCl, pH 7.5). After equilibration with 20 mM Na-acetate buffer pH 5.5, the protein was loaded to a Resource-S column (Pharmacia). Bound proteins were eluted with a linear gradient of up to 0.5 M NaCl. The CGTG was then labeled with the digoxigenin-derivative HDMA.[3] Further impurities of the CGTG-protein fraction were removed by passage through Protein-A Sepharose (Pharmacia) followed by chromatography on a Ni-NTA column (Qiagen). The affinity of CGTG for cardiac glycosides was measured by the increase in fluorescence of anthroyl-ouabain (AO) upon binding in 0.1 M imidazole buffer pH 7.2 and at a protein concentration of 0.9

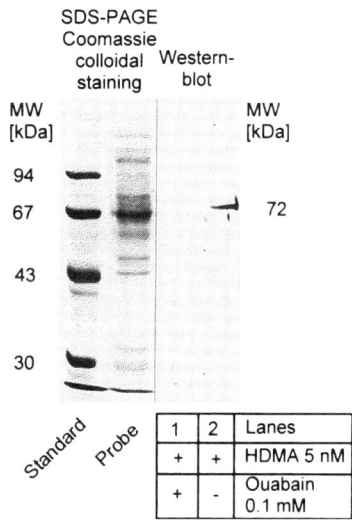

FIGURE 1. A digoxigenin binding protein in bovine serum. Isolated and HDMA-labeled CGTG after the Red-A column purification step was subjected to 10% SDS-PAGE, and proteins were visualized by Coomassie colloidal staining. HDMA-labeled proteins were detected on western blot after transfer to a PVDF membrane by anti-digoxigenin-specific polyclonal antibodies (Boehringer-Mannheim) (*lane 2*). Pretreatment with 0.1 mM ouabain suppressed HDMA labeling (*lane 1*).

mg/ml with a Hitachi F-3000 Fluorescence Spectrophotometer (Ex. 365 nm, Em. 495 nm). This fluorescent derivative of ouabain is known to bind specifically to the cardiac glycoside receptor of the Na^+/K^+-ATPase.[4]

RESULTS AND DISCUSSION

Incubation of the purified globulin fraction with 5×10^{-9} M HDMA at pH 7.2 revealed that a protein of molecular mass 58–72 kD was specifically labeled by the digoxigenin derivative. Many proteins were still detectable on SDS-PAGE after Red-A column chromatography, but only one protein was specifically labeled by HDMA as evident from western blot using a polyclonal anti-digoxigenin-specific antibody (FIG. 1, lane 2). A 20,000-fold molar excess of ouabain almost completely

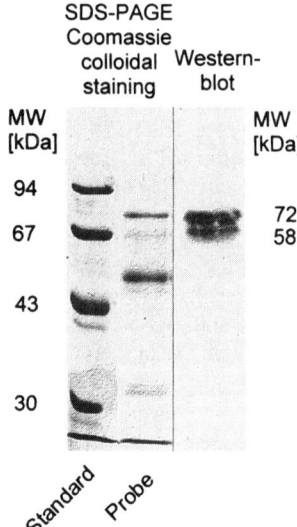

FIGURE 2. Heterogeneity of pure CGTG. The HDMA-labeled protein used in FIGURE 1 was further purified on Protein-A Sepharose and Ni-NTA columns (see Methods). Coomassie colloidal staining of the gel shows two major protein bands of molecular weight 58 and 72 kD which also reacted on western blots with anti-digoxigenin antibodies.

suppressed labeling with HDMA (FIG. 1, lane 1), but other steroid hormones such as estrogen, testosterone, progesterone, and cortisone did not (not shown). The fact that ouabain as a specific inhibitor of the Na^+ pump protected binding and labeling of a 72-kD protein by HDMA indicates the existence of a specific transport globulin for cardiac glycosides that may transport cardiac glycosides in blood. Binding of AO to the CGTG produced an increase in fluorescence that could be suppressed by excess ouabain. The apparent $K_{0.5}$ of the transport globulin thus obtained was $K_{0.5} = 0.54 \pm 0.02$ μM. This affinity of CGTG to AO is about 1,000-fold higher than that of albumin for cardiac glycosides. When compared to the Na^+/K^+-ATPase from pig kidney, its $K_{0.5}$ (0.27 ± 0.07 μM) for AO was lower than that of CGTG from bovine serum globulin ($K_{0.5} = 0.54 \pm 0.02$ μM). Contrary to its binding to the Na^+/K^+-ATPase, the binding of AO to purified CGTG showed positive cooperativity. Whether this is important in the release of the cardiac glycoside at the extracellular surface of the sodium pump is speculative. Positive cooperativity is indicative of the existence of two interacting binding sites for AO in CGTG. Consistent therewith we found two HDMA-labeled proteins of 58 and 72 kD in very pure preparations of CGTG (FIG. 2). We may not currently exclude the possibility that the 58-kD band is a degradation product of the 72-kD protein. Consistent therewith is the observation that an amino acid sequence analysis of both proteins showed no sequence homology between both NH_2-termini. Moreover, no homology of the NH_2-terminus of the 72-kD protein was found in nonredundant Gen-Bank CDS translations + PDB + SwissProt + SPupdate PIR 207.

REFERENCES

1. KRIEGLSTEIN, J. 1981. Handb. Exp. Pharmacol. **56/II:** 95–104.
2. LICHTSTEIN, D. *et al.* 1993. J. Cardiac Pharmacol. **22:** S102–S105.
3. ANTOLOVIC, R. *et al.* 1995. Eur. J. Biochem. **227:** 61–67.
4. FORTES, G. P. A. 1977. Biochemistry **16:** 531–540.

Functional Characterization of an Endogenous Digitalis-Like Factor in Human Newborn Plasma

Effects on Rat (Na^+/K^+)-ATPase Isoforms and on Binding to Placenta

G. CRAMBERT,[a] S. BALZAN,[b] A. PACI,[b] S. DECOLLOGNE,[a]
U. MONTALI,[c] S. GHIONE,[b] AND L. G. LELIÈVRE[a]

[a] *University Paris VII. Pharmacology and Ion Transports*
Paris, France

[b] *CNR Institute of Clinical Physiology*
Pisa, Italy

[c] *Institute of Biological Chemistry*
University of Pisa
Pisa, Italy

Several endogenous digitalis-like factors (EDLF) from various mammalian organs have been at least partially isolated and characterized.[1,2] The demonstration of multiple isoenzymes of the (Na^+/K^+)-ATPase which differ in their sensitivity to cardiac glycosides[3] raises the possibility that these isoforms may be differentially modulated by EDLF. We here report on two aspects of the functional characterization of an EDLF extracted by us from the human umbilical cord plasma[4,5]: (1) its activity on (Na^+/K^+)-ATPase isoforms in the rat (in which these isoforms are more extensively characterized than in the human[6,7]); (2) its competition with labeled digitalis glycosides for human placental binding sites.

MATERIALS AND METHODS

The methods of purification of neonatal EDLF,[4] the preparation of membrane fractions enriched in (Na^+/K^+)-ATPase from rat brains and kidneys[6] and from hearts,[8] the measurement of (Na^+/K^+)-ATPase activities employing an enzymatic coupled assay,[9] the preparation of human placental membranes,[10] and binding experiments with ^{125}I-digoxin[11] and ^{3}H-ouabain[10] have previously been described. The sensitivity of (Na^+/K^+)-ATPase to EDLF was measured after preincubation for 30 minutes at 37°C.

RESULTS

All control experiments demonstrated that inhibitory potencies of our EDLF on rat (Na^+/K^+)-ATPases could only be found in the fractions inhibiting erythrocyte ^{86}Rb uptake (data not shown). EDLF had to be preincubated with rat membranes for at least 15 minutes at 37°C in a medium devoid of ATP in order to inhibit the (Na^+/K^+)-ATPase activity. By contrast, the presence of potassium ions (5–10 mM) in the preincubation medium devoid of ATP did not affect the inhibitory effects of EDLF. The addition of ATP (1 or 4 mM) in the preincubation medium almost completely prevented EDLF-induced enzyme inhibition, even at EDLF concentrations leading to complete inhibition in the absence of ATP during preincubation. EDLF inhibited rat (Na^+/K^+)-ATPase in a dose-dependent manner. Amounts of EDLF that completely inhibited rat cardiac and renal (Na^+/K^+)-ATPase activities had no detectable effects on cardiac Ca^{2+}-ATPases or ouabain-insensitive Mg^{2+}-ATPases assayed in the same membrane preparations.

With rat *kidney* membranes (which contain only $\alpha 1$ isoforms), monophasic and complete inhibition was obtained for about a 10-fold increase in the volume of original plasma assayed (typical curve shown in FIG. 1, left). With the rat *heart* (Na^+/K^+)-ATPase (a third of whose activity is accounted for by the $\alpha 1$ isoform and two thirds by the $\alpha 2$ isoform[12]), all six EDLF extracts tested led to complete inhibition. In four of them, the dose-response curves were very steep (FIG. 1, middle; curve A), full inhibition occurring with an amount of extract corresponding to 4 ml of original plasma per tube. For the other two lots, complete inhibition was reached for a 100-fold change in the amounts of EDLF (FIG. 1, middle; curve B). These different patterns presumably reflect differences in the amounts of EDLF in the original plasma and/or in the extracts. With rat *brain* microsomes ($\alpha 1$, $\alpha 2$, and $\alpha 3$ isoforms), inhibition of the (Na^+/K^+)-ATPase never exceeded 70% (FIG. 1, right) even at concentrations that induced full inhibition of rat cardiac and kidney preparations. Complete inhibition of the rat brain (Na^+/K^+)-ATPase was achieved when digitoxigenin at a dose known to affect only the $\alpha 3$ isoform (30 nM) was added to high doses of EDLF (20 ml of original plasma per assay). The inhibition curves of ^{125}I-digoxin binding to placental membranes (FIG. 2, right) showed that the EDLF displacement curve was steeper than that of ouabain when binding inhibition was assessed at equilibrium, whereas ouabain and EDLF appeared to be parallel when analyzed by a sequential approach. Lineweaver-Burk plots obtained from (^{3}H)ouabain-binding studies to human placental membranes resulted in straight lines passing through a common point on the ordinate, indicating that the inhibition was competitive (FIG. 2, left).

DISCUSSION AND CONCLUSIONS

Our results confirm that EDLF purified from human neonate plasma shares many biological properties with cardiac glycosides. In fact, in addition to inhibiting ion transport in erythrocytes and competing with digoxin and ouabain for placental digitalis-binding sites, neonate EDLF extracts specifically inhibited membrane-bound (Na^+/K^+)-ATPase activity in rat kidney, heart, and brain without affecting other membrane-bound ATPases.

These data are also consistent with the idea that human neonate EDLF is functionally similar but not identical to ouabain. In fact, (1) at variance with ouabain, which inhibits (Na^+/K^+)-ATPase activity within 10 minutes, EDLF required 15–30 minutes of preincubation at 37°C; (2) EDLF and ouabain differ in their ligand requirements: (2a) K^+, which antagonizes the binding of ouabain, had no effect on EDLF-induced inhibition of (Na^+/K^+)-ATPase; (2b) ATP, which favors ouabain

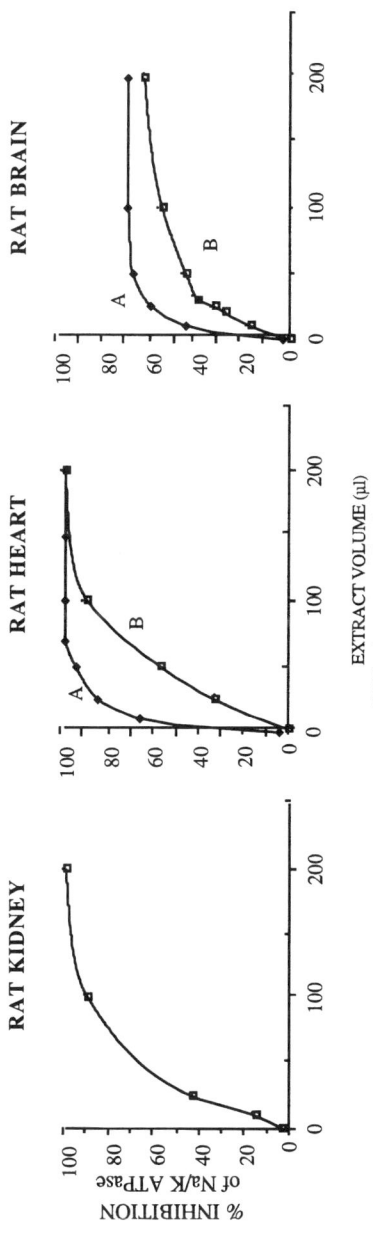

FIGURE 1. Dose-response curves of purified newborn EDLF in (Na^+/K^+)-ATPase membranes obtained (from *left* to *right*) from kidney, heart, and brain. A and B refer to lots of EDLF preparations having different amounts of EDLF.

binding, prevented EDLF action when present during preincubation; (3) EDLF also differs from ouabain (and digoxin) in the slope of the dose-response curves of enzymatic activity: for EDLF, with both å1 and å2 isoforms, complete inhibition occurred in most cases within a 4- to 10-fold increase in concentration (curves A, FIG. 1), whereas for each individual rat (Na^+/K^+)-ATPase isoforms, ouabain and digoxin a completely inhibited enzymatic activity over a 100-fold range in concentration[6,9,13]; (4) EDLF differs from ouabain in its relative potency on rat (Na^+/K^+)-ATPase isoforms. In fact, whereas the various (Na^+/K^+)-ATPase isoforms differ in sensitivity to ouabain in decreasing order,[6,9] our EDLF inhibited with apparently similar potency α1 (in kidney) and α2 (in brain and heart) rat isoforms and appeared to have no effect on rat brain α3 isoform, which is known to have the highest affinity for ouabain[6]; (5) inhibition binding curves to human placental digitalis receptors showed that EDLF is a competitive inhibitor; however, EDLF displacement curves at equilibrium were steeper than those of ouabain.

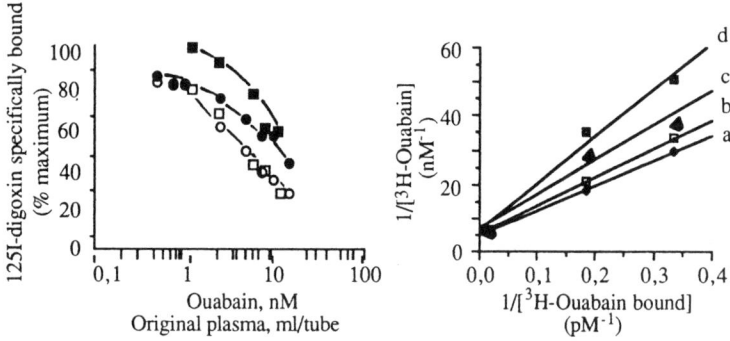

FIGURE 2. (Left) Inhibition curves of ^{125}I-digoxin binding to human placental membranes by ouabain (*circles*), and newborn EDLF (*squares*) at equilibrium (*full symbols*) and under sequential conditions (*open symbols*). (**Right**) Lineweaver-Burk plots of inhibition of (^{3}H) ouabain binding to human placental membranes, respectively: (a) without EDLF, (b) with 5 ml PO/tube, (c) 15 ml PO/tube, and (d) 30 ml PO/tube.

REFERENCES

1. GOTO, A., K. YAMADA, N. YAGI, M. YOSHIOKA & T. SUGIMOTO. 1992. Physiology and pharmacology of endogenous digitalis-like factors. Pharmacol. Rev. **44:** 377–399.
2. HAMLYN, J. M. 1994. Ouabain: An endogenous mammalian ligand for the sodium pump. DuPont NEN Biotech Update. **9:** 16–18.
3. SWEADNER, K. J. 1989. Isozymes of the Na^+/K^+-ATPase. Biochim. Biophys. Acta **988:** 185–220.
4. BALZAN, S., S. GHIONE, P. BIVER, P. GAZZETTI & U. MONTALI. 1991. Partial purification of endogenous digitalis-like compound(s) in cord blood. Clin. Chem. **37:** 277–281.
5. MONTALI, U., S. BALZAN & S. GHIONE. 1991. Purification of endogenous digitalis-like factor(s) from cord blood of neonate by immunoaffinity chromatography. Biochem. Int. **25:** 853–859.
6. BERREBI-BERTRAND, I., J. M. MAIXENT, G. CHRISTE & L. G. LELIÈVRE. 1990. Two active N/K-ATPase of high affinity for ouabain in adult rat brain membranes. Biochim. Biophys. Acta **1021:** 148–156.

7. SWEADNER, K. J. 1979. Two molecular forms of (Na + K) stimulated ATPase in the brain: Separation and difference in affinity for strophanthidin. J. Biol. Chem. **254:** 6060–6067.
8. MANSIER, P., D. CHARLEMAGNE, B. ROSSI, M. PRETESEILLE, B. SWYNGHEDAW & L. LELIÈVRE. 1983. Isolation of impermeable inside-out vesicles from an enriched sarcolemma fraction of rat heart. J. Biol. Chem. **258:** 6628–6635.
9. LELIÈVRE, L. G., P. LORENTE, J. M. MAIXENT, C. MOUAS & B. SWYNGHEDAUW. 1986. Prolonged responsiveness to ouabain in hypertrophied rat heart. Physiological and biochemical studies. Am. J. Physiol. **251:** H923–H931.
10. PACI, A., F. COCCI, F. PIRAS, G. CIARIMBOLI & A. CLERICO. 1989. Specific binding of cardiac glycoside drugs and endogenous digitalis-like substances to particulate membrane fractions from human placenta. Clin. Chem. **35:** 2093–2097.
11. PACI, A., G. CIARIMBOLI & P. BIVER. 1996. Human placenta radioreceptor assay with digoxin and ouabain to detect endogenous digitalis-like factor(s) in human plasma and urine. Clin. Chem. **42:** 270–278.
12. CHARLEMAGNE, D., E. MAYOUX, M. POYARD, P. OLIVIEIRO & K. GEERING. 1987. Identification of two isoforms of the catalytic subunit of $(Na^+ K^+)$ATPase in myocytes from adult rat heart. J. Biol. Chem. **262:** 8941–8943.
13. MANSIER, P. & L. LELIÈVRE. 1982. Ca^{2+}-free perfusion of rat heart reveals a $(Na^+ K^+)$-ATPase activity highly sensitive to ouabain. Nature **300:** 535–537.

Evidence of an Endogenous Ouabain-Like Immunoreactive Compound with Digitalis-Like Properties in the Human

SILVANA BALZAN,[a] UMBERTO MONTALI,[b]
AND SERGIO GHIONE[a]

[a]CNR Institute of Clinical Physiology and
[b]Institute of Biological Chemistry
University of Pisa
Pisa, Italy

A major advance in the search for an endogenous digitalis-like factor (EDLF) was made by Hamlyn and coworkers who showed that a factor indistinguishable from ouabain is present in human plasma[1] and that anti-ouabain antiserum could be used for assay of this factor.[2] These observations led these investigators to conclude that ouabain is an endogenous circulating agent.[3] More recently, Haupert and coworkers reported that an isomer of ouabain, rather than ouabain per se, is present in the bovine hypothalamus[4] and that this compound may be identical to the one isolated in human plasma by Hamlyn's group.[5] The existence of a factor closely similar or identical to ouabain in the hypothalamus of the rat was confirmed by immunohistochemical studies[6] and biochemical characterization studies on partially purified extracts.[7]

The original report that ouabain may be present in human plasma prompted several confirmatory studies by other investigators who employed either a commercially available radioimmunoassay (RIA) kit by Dupont-NEN or self-produced anti-ouabain antisera. Unfortunately, the results appeared more confusing than clarifying. In fact, the commercially available RIA kit has been criticized on methodological grounds,[8,9] and discordant results with the other antisera have been obtained both in studies on ouabain-like immunoreactivity measured in less purified plasma extracts and in studies that employed additional HPLC purification. In normal human subjects, levels in samples after a single extraction (usually on C_{18} cartridges) ranged widely from undetectable,[10] to 30–50[11,12] and 100–500 pmol/L,[2,13–15] up to 1 nmol/L.[3] Even more disturbing are three recent reports that were unable to confirm the existence of plasma endogenous ouabain-like immunoreactivity in HPLC fractions in which exogenous ouabain was eluted.[10,12,13] These results contrasted with the findings of Hamlyn et al.[16] and with our observations. We observed[17] that newborn (umbilical cord) plasma contains a factor that inhibits human erythrocyte ^{86}Rb uptake, is eluted in our HPLC system in the same fraction as ouabain, and cross-reacts with anti-ouabain antibodies (see also FIG. 1). In addition, these antibodies neutralized the inhibitory effect of the purified EDLF on erythrocyte ^{86}Rb uptake[17] (FIG. 2).

How can our results be reconciled with the negative findings of Doris,[10] Gomez,[12] Lewis,[13] and their coworkers? It is unlikely that these differences are due to the fact that we employed newborn plasma, because Lewis et al., in their thorough study, also investigated newborn plasma.[13] In addition, numerous independent evidence supports the existence of an EDLF in newborn plasma (see ref. 18 for a review). Another, perhaps more likely explanation is that EDLF is a compound structurally very similar but not identical to ouabain (possibly an isomer of ouabain, in accordance with the reports of Haupert's group[4,5]). As pointed out by Pidgeon et al.,[9] this could explain why different antisera raised against ouabain may vary in their capacity to detect EDLF. Moreover, our EDLF seems to differ from ouabain in its effect on the

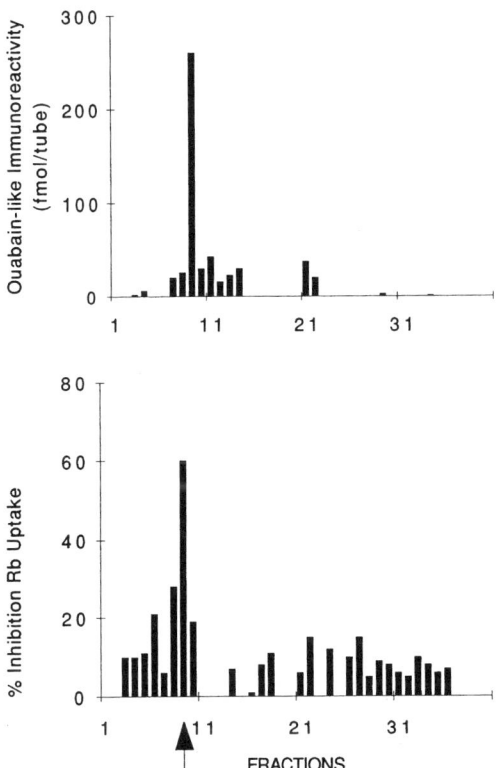

FIGURE 1. HPLC elution profile of newborn plasma pre-purified on SepPakC$_{18}$, assayed as: (**top**) ouabain-like immunoreactivity and (**bottom**) inhibition of erythrocyte ^{86}Rb uptake. HPLC conditions: 300 × 7.8 mm C$_{18}$ μBondapaK column, Waters; mobile phase: acetonitrile/methanol/water: 14/14/72 by vol.; flow 1.5 ml/min, fractions 2 ml. Authentic ouabain eluted in fraction nine (*arrow*).

Na$^+$/K$^+$-ATPase isoforms of the rat,[19] and our HPLC system is probably not selective enough to discriminate between ouabain and other similar compounds.

The existence of an endogenous factor with digitalis-like immunoreactive properties is further supported by studies employing anti-digoxin Fab fragments, which are commercially available as an antidote against digoxin intoxication, but which also bind bufodienolides[20] and ouabain.[21] We observed that these antibodies can be used to purify newborn EDLF by immunoaffinity chromatography[22] and can neutralize its effect on erythrocyte ^{86}Rb uptake.[21] Moreover, Huang and Leenen[23] demonstrated that the cerebrospinal fluid of the rat contains a factor with hypertensive properties which is specifically inhibited by anti-digoxin Fab fragments.

A conclusive answer will probably be available only after the definitive structural characterization of EDLF, independently confirmed by different investigators. In the meantime, we believe that the possibility of a factor in the human body that has strong biological and structural similarities to ouabain (or other cardiac glycosides) should not be dismissed.

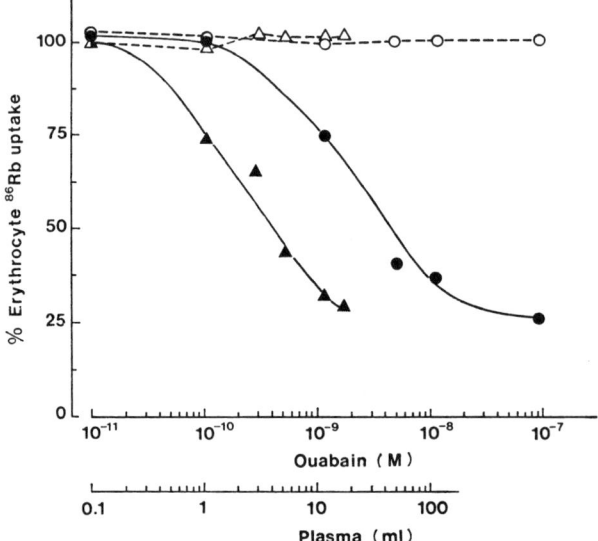

FIGURE 2. Neutralizing effect of anti-ouabain antibodies on the inhibitory activity of ouabain and newborn endogenous digitalis-like factor (EDLF) on erythrocyte ^{86}Rb uptake. Ouabain (*circles*) and purified EDLF (*triangles*) were incubated at various concentrations without (*full symbols*) and with (*open symbols*) anti-ouabain antiserum. The antiserum fully neutralizes the effect of ouabain and EDLF on erythrocyte ^{86}Rb uptake. (Reproduced from ref. 17 with permission.)

REFERENCES

1. LUDENS, J. H., M. A. CLARK, D. W. DUCHARME, D. W. HARRIS, B. S. LUTZKE, F. MANDEL, W. R. MATHEWS, D. M. SUTTER & J. M. HAMLYN. 1991. Purification from human plasma of an endogenous digitalis-like factor for structural analysis. Hypertension **17:** 923–929.
2. HARRIS, D. W., M. A. CLARK, J. F. FISHER, J. M. HAMLYN, K. P. KOLBASA, J. H. LUDENS & D. W. DULHARME. 1991. Development of an immunoassay for endogenous digitalis-like factor. Hypertension **17:** 936–943.
3. HAMLYN, J. M. & P. MANUNTA. 1992. Ouabain, digitalis-like factors and hypertension. J. Hypertens. Suppl. **10:** 952–1178.

4. TYMIAK, A. A., J. A. NORMAN, M. BOLGAR, G. C. DIDONATO, H. LEE, W. L. PARKER, L. C. LO, N. BEROVA, K. NAKANISHI, E. HABER et al. 1993. Physicochemical characterization of a ouabain isomer isolated from bovine hypothalamus. Proc. Natl. Acad. Sci. USA **90:** 8189–8193.
5. ZHAO, N., L. LO, N. BEROVA, K. NAKANISHI, A. TYMIAK, J. LUDENS & G. HAUPERT. 1995. Na,K-ATPase inhibitors from bovine hypothalamus and human plasma are different from ouabain: Nanogram scale CD structural analysis. Biochemistry **34:** 9893–9896.
6. YAMADA, H., M. NARUSE, K. NARUSE, H. DEMURA, H. TAKAHASHI, M. YOSHIMURA & J. OCHI. 1992. Histological study on ouabain immunoreactivities in the mammalian hypothalamus. Neurosci. Lett. **141:** 143–146.
7. FERRANDI, M., E. MINOTTI, S. SALARDI, M. FLORIO, G. BIANCHI & P. FERRARI. 1992. Ouabainlike factor in Milan hypertensive rats. Am. J. Physiol. **263:** F739–F748.
8. PACI, A., A. LEDDA, G. CIARIMBOLI, P. BIVER & G. BERNINI. 1996. Commercial enzyme immunoassay reagent pack for ouabain compared with human placenta radioreceptor assay. Clin. Chem. **42:** 648–650.
9. PIDGEON, G., L. LEWIS, T. YANDLE, A. RICHARDS & M. NICHOLLS. 1996. Endogenous ouabain, sodium balance and blood pressure. J. Hypertens. **14:** 169–171.
10. DORIS, P. A., L. A. JENKINS & D. M. STOCCO. 1994. Is ouabain an authentic endogenous mammalian substance derived from the adrenal? Hypertension **23:** 632–638.
11. NARUSE, K., M. NARUSE, A. TANABE, T. YOSHIMOTO, Y. WATANABE, F. KURIMOTO, N. HORIBA, M. TAMURA, T. INAGAMI & H. DEMURA. 1994. Does plasma immunoreactive ouabain originate from the adrenal gland? Hypertension **23:** 0194–0911.
12. GOMEZ-SANCHEZ, E., M. FOECKING, D. SELLERS, M. BLANKERSHIP & C. GOMEZ-SANCHEZ. 1994. Is the circulating ouabain-like compound ouabain? Am. J. Hypertens. **7:** 647–650.
13. LEWIS, L., T. YANDLE, J. LEWIS, A. RICHARDS, G. PIDGEON, R. KAAJA & M. NICHOLLS. 1994. Ouabain is not detectable in human plasma. Hypertension **24:** 549–555.
14. GOTTLIEB, S. S., A. C. ROGOWSKI, M. WEINBERG, C. M. KRICHTEN, B. P. HAMILTON & J. M. HAMLYN. 1992. Elevated concentrations of endogenous ouabain in patients with congestive heart failure. Circulation **86:** 420–425.
15. MASUGI, F., T. OGIHARA, T. HASEGAWA & Y. KUMAHARA. 1987. Ouabain-like and non-ouabain-like factors in plasma of patients with essential hypertension. Clin. Exp. Hypertens. **A9:** 1233–1242.
16. HAMLYN, J. M., M. P. BLAUSTEIN, S. BOVA, D. W. DUCHARME, D. W. HARRIS, F. MANDEL, W. R. MATHEWS & J. H. LUDENS. 1991. Identification and characterization of a ouabain-like compound from human plasma. Proc. Natl. Acad. Sci. USA **88:** 6259–6263.
17. DI BARTOLO, V., S. BALZAN, L. PIERACCINI, S. GHIONE, S. PEGORARO, P. BIVER, R. REVOLTELLA & U. MONTALI. 1995. Evidence for an endogenous ouabain-like immunoreactive factor in human plasma coeluted with ouabain on HPLC. Life Sci. **57:** 1417–1425.
18. GHIONE, S., S. BALZAN, S. DECOLLOGNE, A. PACI, L. PIERACCINI & U. MONTALI. 1993. Endogenous digitalis-like activity in the newborn. J. Cardiovasc. Pharmacol. **22**(Suppl. 2): S25–S28.
19. CRAMBERT, G., S. BALZAN, A. PACI, S. DECOLLOGNE, U. MONTALI, S. GHIONE & L. G. LELIEVRE. 1997. Functional characterization of an endogenous digitalis-like factor in human newborn plasma: Effects on rat (Na^+/K^+)-ATPase isoforms and on binding to placenta. Ann. N.Y. Acad. Sci. New York.
20. CHERN, M. S., C. Y. RAY & D. L. WU. 1991. Biologic intoxication due to digitalis-like substance after ingestion of cooked toad soup. Am. J. Cardiol. **67:** 443–444.
21. BALZAN, S., U. MONTALI, P. BIVER & S. GHIONE. 1991. Antidigoxin antibodies neutralize the effect of newborn endogenous digitalis-like factor on erythrocyte ^{86}Rb uptake. J. Nucl. Biol. Med. **35:** 38–40.

22. MONTALI, U., S. BALZAN & S. GHIONE. 1991. Purification of endogenous digitalis-like factor(s) from cord blood of neonate by immunoaffinity chromatography. Biochem. Int. **25:** 853–859.
23. HUANG, B. S. & F. H. LEENEN. 1994. Brain "ouabain" mediates the sympathoexcitatory and hypertensive effects of high sodium intake in Dahl salt-sensitive rats. Circ. Res. **74:** 586–595.

Effect of a Low Molecular Weight Factor from Na^+,K^+-ATPase Preparations on Ouabain Binding[a]

CARLOS F. L. FONTES,[b] FÁBIO E. V. LOPES,[b]
HELENA M. SCOFANO, HÉCTOR BARRABIN,[b]
AND JENS G. NØRBY[c,e]

[b]*Departamento de Bioquímica Médica*
Instituto de Ciências Biomédicas
CCS, Bloco H2
Universidadé Federal do Rio de Janeiro
Cidade Universitária
21941-590, Rio de Janeiro, Brazil

[c]*Department of Biophysics*
University of Aarhus
DK-8000 Aarhus C, Denmark

We previously reported[1] that the reactions of the Na^+,K^+-ATPase are profoundly influenced by a reduction of water activity with dimethylsulfoxide (DMSO). It is especially relevant for this communication that in 40% DMSO the phosphorylated intermediate EP is formed with inorganic phosphate, P_i, at high affinity and that this process *is inhibited by ouabain,* which binds preferentially to the unphosphorylated enzyme, all in drastic contrast to what happens in an aqueous milieu.

METHODS AND RESULTS

In subsequent experiments with similar Na^+,K^+-ATPase preparations from pig kidney,[2] we discovered that inhibition of EP formation by ouabain, which is thus proportional to ouabain binding, was significantly dependent on the enzyme concentration in the assay (FIG. 1, open bars). The *standard medium* for these experiments was: 40% DMSO, 5 mM $MgCl_2$, 0.5 mM KCl, 1 mM EDTA, pH 7, t = 27°C. Ouabain binding was assessed by preincubation of the enzyme with 20 µM ouabain for 5 minutes, the addition of 10 µM $^{32}P_i$, and after 1 minute, the amount of acid-stable EP was measured by precipitation with $HClO_4$ and filtering as described.[1]

From the results in FIGURE 1 (open bars), we suspected that a ouabain binding-promoting factor was diluted out in the medium and was present in subsaturating concentrations when the enzyme concentration was low (e.g., 12 µg/ml). Therefore, it

---

[a]This work was supported by Conselho Nacional de Desenvolvimento Científico e Tecnológico (CNPq), Financiadora de Estudos e Projetos (FINEP), and Biomembrane Research Center, Aarhus, Denmark.
[d]Tel: +55-21-5904548; fax: +55-21-2708647; e-mail: CFONTES@Server.Bioqmed.UFRJ.BR.
[e]Tel: (+45) 89422935; fax: (+45) 86129599; e-mail: JGN@MIL.AAU.DK

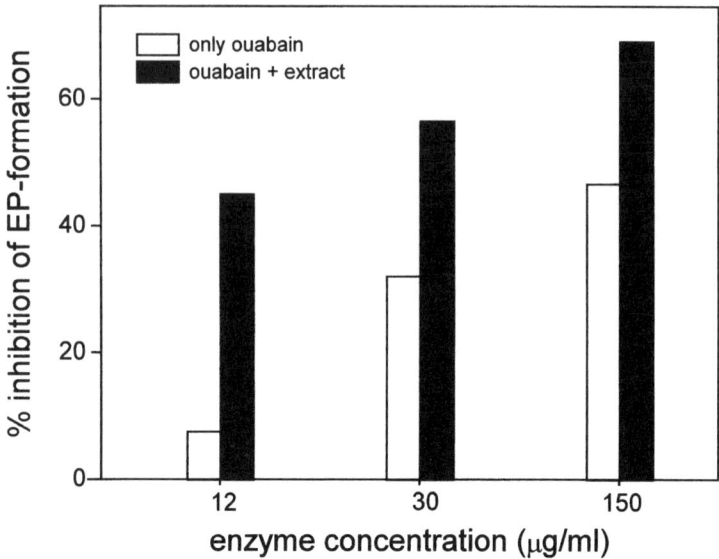

FIGURE 1. Inhibition of EP formation, which is proportional to ouabain binding, with enzyme in the standard medium (*open bars*) and enzyme in the standard medium + 0.1 volume of "DMSO extract" (*closed bars*). For further details see text.

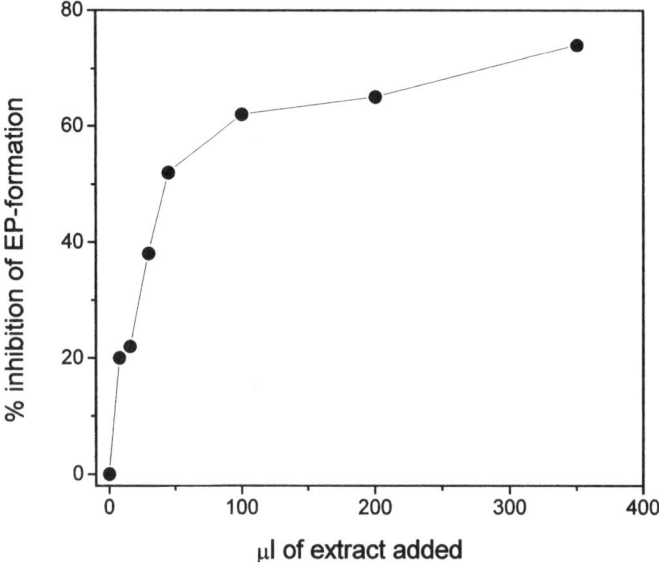

FIGURE 2. Ouabain binding (see Fig. 1) to the enzyme, 12 µg/ml, as a function of the amount of "DMSO extract" added to 0.5 ml assay solution.

could be possible to make an extract from the enzyme preparation, the addition of which would increase the binding of ouabain. Accordingly, Na^+,K^+-ATPase stock solution was diluted with two volumes to make the composition of final suspension equal to the 40% DMSO standard medium, and a $148,000 \times g$ supernatant, the DMSO "extract," was prepared. When an aliquot of this DMSO extract was added to preincubation medium with ouabain, a considerable increase in ouabain binding was observed (FIG. 1, closed bars), and in accordance with the foregoing hypothesis the increase was largest at the lowest enzyme concentrations. As in FIGURE 2, stimulation of ouabain binding is clearly dose dependent in a saturable manner. An estimate of the size of the active component was obtained by 24-hour dialysis of DMSO extract against standard medium with a "cut-off" membrane for 3.5 kD or 12–14 kD. The active factor was retained by the 3.5 kD but not by the 12–14 kD membrane. This factor is removed when membranes are diluted in the DMSO medium, leaving the ATPase less sensitive to ouabain.

CONCLUSION

We demonstrated that the Na^+,K^+-ATPase preparation contains a low molecular weight factor that enhances ouabain binding in 40% DMSO.

REFERENCES

1. FONTES, C. F. L., H. M. SCOFANO, H. BARRABIN & J. G. NØRBY. 1995. Biochim. Biophys. Acta **1235:** 43–51.
2. JØRGENSEN, P. L. 1974. Biochim. Biophys. Acta **356:** 36–52.

High Affinity Anti-Digoxin Antibodies as Model Receptors for Cardiac Glycosides

Comparisons with Na$^+$,K$^+$-ATPase[a]

R. KASTURI,[b] L. R. McLEAN,[c] M. N. MARGOLIES,[d]
AND W. J. BALL, JR.[b]

[b]*University of Cincinnati*
Pharmacology & Cell Biophysics
Cincinnati, Ohio 45267

[c]*Hoechst Marion Roussel*
Cincinnati, Ohio 45215

[d]*Department of Surgery*
Massachusetts General Hospital
Boston, Massachusetts 02114

Although Na$^+$,K$^+$-ATPase (NKA), a transmembrane pump responsible for maintaining Na$^+$ and K$^+$ gradients across cell membranes, has long been known as the receptor for cardiac glycosides such as digoxin, neither its structure nor the site of cardiac glycoside binding has been determined. The binding of glycosides to Na$^+$,K$^+$-ATPase occurs extracellularly[1] and is regulated by the H1–H2 and H7–H8 extracellular loops of the α subunit of the pump.[2] High affinity anti-digoxin monoclonal antibodies (mAbs)[3] with known structures potentially provide a model system for characterization of the relationship of receptor structure to glycoside binding. The objective of this study was to use the fluorescent ouabain derivative anthroylouabain[4] to determine whether anti-digoxin mAbs can serve as models of the digitalis receptor.

Na$^+$,K$^+$-ATPase was purified from lamb kidney according to the method of Lane *et al.*[5] Antibody-producing cell lines 26–10 (γ_{2a},κ), 45–20 (γ_{2a},λ_1), and 40–50 (γ_{2b},κ) were obtained as reported by Mudgett-Hunter *et al.*[3] SLM/Aminco SPF-500 and ISS (Industria Strumentazioni Scientifiche, Champaign, Illinois) Greg K2 spectrofluorometers were used for steady-state fluorescence and polarization, and lifetime measurements, respectively.

The apparent affinities of the mAbs for anthroylouabain relative to their apparent affinities for digoxin and ouabain were determined using an antibody capture ELISA. The order of affinities based on the IC$_{50}$ values was digoxin > anthroylouabain > ouabain for mAb 26–10 (FIG. 1). Anthroylouabain binding to Na$^+$,K$^+$-ATPase and 26–10 resulted in 3.0- and 3.5-fold increases in fluorescence intensity, respectively (FIG. 2). The emission maxima of the fluorescence spectra were blue-shifted for both anthroylouabain-protein complexes, suggesting that anthroylouabain binding results in the transfer of the anthroyl group to a hydrophobic environment. Consistent with this, average fluorescence decay lifetimes ($\langle\tau\rangle$) of 11.30 and 5.74 ns were obtained for

---

[a]This work was supported by National Institutes of Health Training Grant (5 T32 HLO7382-19), an American Heart Association, Ohio affiliate grant to W.J.B., and a National Institutes of Health RO1 grant (HL47415) to M.N.M.

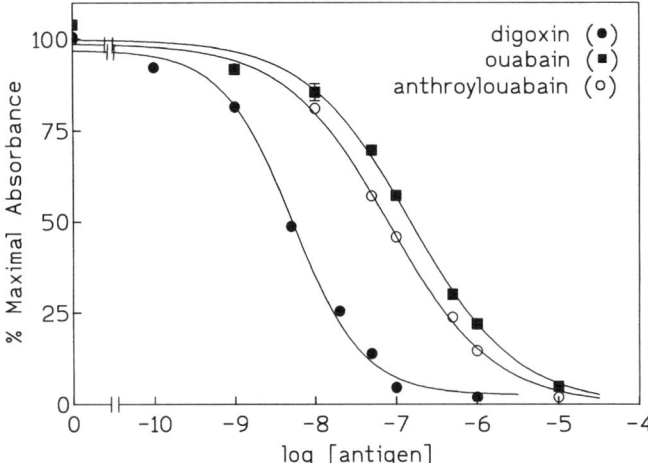

FIGURE 1. Determination of relative binding affinities of anti-digoxin mAb 26–10 for digoxin, ouabain, and anthroylouabain using ELISA. The curves show the inhibition of binding of digoxin-alkaline phosphatase (dig-AP) to mAb 26–10 by increasing concentrations of digoxin (●), ouabain (■), and anthroylouabain (○). 26–10 (2 µg/ml) was captured by 5.0 µg/ml goat anti-mouse Fc-specific antibody adsorbed to a microtiter plate and antibody bound digoxin-AP was monitored at 405 nm using a Molecular Devices microplate reader.

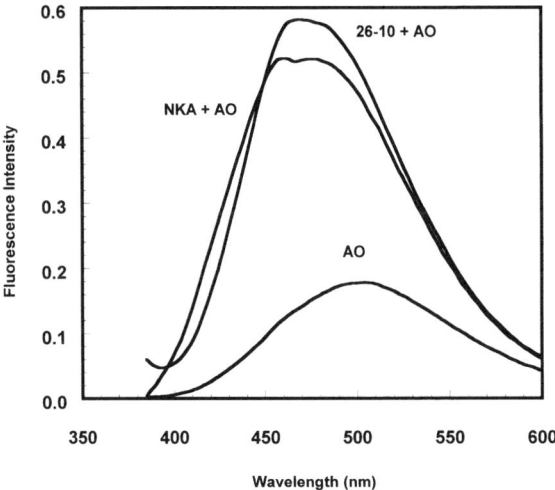

FIGURE 2. Emission spectra of free and protein-bound anthroylouabain. Fluorescence emission spectra (λ_{ex} = 364 nm) were obtained using 0.3 µM anthroylouabain and anthroylouabain saturated with either 0.6 µM Na^+,K^+-ATPase (NKA) or mAb 26–10. Buffer was 50 mM Tris, 2 mM EGTA, pH 7.4, and 5 mM Mg^{2+}/P_i added for measurements with Na^+,K^+-ATPase.

anthroylouabain complexes with 26–10 and Na$^+$,K$^+$-ATPase, respectively, compared with an $\langle\tau\rangle$ of 0.6 ns for free probe. The increased anthroylouabain lifetimes are attributable to a hydrophobic environment in the vicinity of the anthroyl group. The measured anisotropy (r) values for anthroylouabain complexed to 26–10 and Na$^+$,K$^+$-ATPase were 0.173 and 0.158, respectively, compared with an r value of 0.078 for free anthroylouabain. This implies that anthroylouabain is not rigidly bound, as the anthroyl group is relatively free to undergo rotational motion and is consistent with binding domains that are located near the surface of the proteins. In addition, Förster resonance energy transfer, previously shown to occur from tryptophan(s) in Na$^+$,K$^+$-ATPase to anthroylouabain,[4] was also found to occur from mAb 26–10 to anthroylouabain, indicating that in both cases one or more tryptophans is located in the vicinity of the binding site of anthroylouabain. Our initial model of the binding of anthroylouabain to 26–10 Fab, generated using the crystal structure of the 26–10 Fab-digoxin complex,[6] shows that tryptophan H100 is in close proximity to the anthroyl moiety of anthroylouabain (<3Å). This residue, in the heavy chain CDR 3 of Fab 26–10, has been shown to be in contact with the lactone and steroid backbone of digoxin.[6] These data suggest that the digoxin-binding site of mAb 26–10 resembles that of Na$^+$,K$^+$-ATPase and is a good model of the digitalis receptor. By contrast, mAbs 45–20 and 40–50 do not appear to resemble Na$^+$,K$^+$-ATPase in their binding of anthroylouabain.

REFERENCES

1. HOFFMAN, J. F. 1966. Am. J. Med. **41**: 666.
2. WALLICK. 1995. Cell Physiol. Sourcebook. Sperelakis, Ed.: 148. Academic Press. New York.
3. MUDGETT-HUNTER, M., W. ANDERSON, E. HABER & M. MARGOLIES. 1985. Mol. Immunol. **22**: 477.
4. FORTES, P. A. G. 1977. Biochemistry **16**: 531.
5. LANE, L., J. POTTER & J. COLLINS. 1979. Prep. Biochem. **9**: 157.
6. JEFFREY, P., R. STRONG, L. SIEKER, C. CHANG et al. 1993. Proc. Natl. Acad. Sci. USA **90**: 10310.

Synthetic Candidates for EDLF
Activity on Human Placenta Digitalis Receptors

ANNA PACI,[b] MASAYUKI SAKAKIBARA,[a]
PAOLA DEL BENE, AND AKI OGAWA UCHIDA[a]

CNR Institute of Clinical Physiology
Pisa, Italy

Istituto Patologia Medica I
University of Pisa
Pisa, Italy

[a]*Pharmaceutical Research Laboratories*
Kirin Brewery Co., Ltd.
Gunma, Japan

The existence of a mammalian analog of plant-derived cardiac glycosides (endogenous digitalis-like factor, EDLF) has long been postulated (for review see ref. 1), but despite intensive studies no definitive substance has been found. Compounds indistinguishable from plant digoxin and ouabain have been reported by mass spectrometry from human urine[2] and plasma.[3] A ouabain-like substance with a molecular weight of 336 has been isolated by Tamura and Inagami[4] from bovine adrenals. An isomer of ouabain has been found in bovine hypothalamus[5] and human plasma,[6] and 19-norbufalin has been isolated from human lens.[7] Efforts to isolate and structurally characterize mammalian EDLF have been limited by the small amount of substance typically isolated. Sakakibara and colleagues[8,9] decided to approach the structure of the mammalian EDLF by synthesizing and analyzing some plausible compounds to confirm the probability of the proposed undefined structure. Among the several plausible compounds deduced to account for the molecular weight of the Inagami-Tamura EDLF[4] and EDLF properties reported by others,[8] the same authors hypothesized that the 14β-hydroxy group of digitalis compounds might be more significant than the lactonic moiety to elicit their activity[8] and found that two polyoxygenated estrogen compounds (14β,15β,16β,17α)- and (14β,15α,16α,17α)-1,3,5(10)-estratriene-2,3,14,15,16,17-hexaols (DHE1 and DHE2A, FIG. 1) induced a contractile response in isolated rat aorta and in isolated guinea pig left atrium.[8]

Using (^{3}H)ouabain and (^{125}I)digoxin as tracers, we demonstrated that human placenta contains large amounts of specific, high-affinity digitalis receptors.[10] Human placenta membranes were then used to develop a radioreceptor assay (RRA) to detect EDLF in human plasma and urine.[11]

The aim of this study was to analyze the reactivity of some synthetic oxygenated estrogen derivatives and several derivatives of digitoxin to the digitalis receptors from

---

[b]Address for correspondence: Dr. Anna Paci, CNR Institute of Clinical Physiology, Via Savi 8, 56100 Pisa, Italy (tel: 0039-50-583248/111; fax: 0039-50-553461; e-mail: paci@nsifc.ifc.pi.cnr.it).

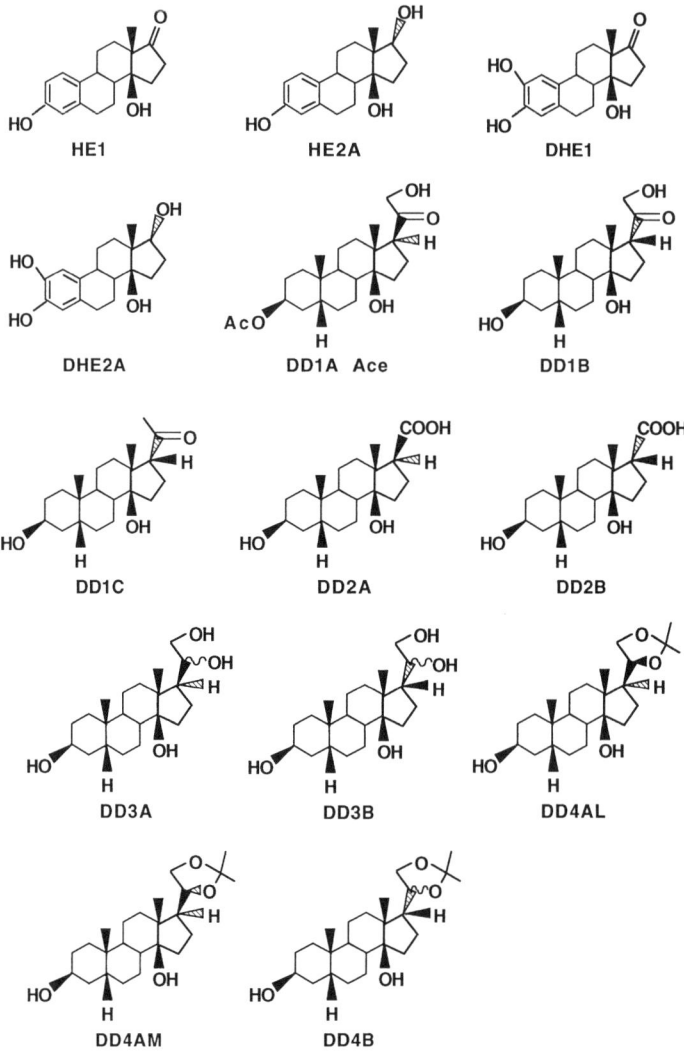

FIGURE 1. Structures of the synthetic compounds tested. HE, DHE = oxygenated estrogen derivatives; DD = digitoxin derivatives; M = more polar; L = less polar.

human placenta. Receptor interaction was analyzed at equilibrium and by a sequential technique that was demonstrated to increase the sensitivity of the RRA.[11] The question of whether synthetic compounds inhibiting (^{125}I)digoxin binding acted via the digitalis binding site (competitive or noncompetitive inhibitor?) was analyzed by measuring (^{3}H)ouabain binding to human placental membranes at various fixed concentrations of the inhibitor.

MATERIALS AND METHODS

Four oxygenated estrogen derivatives (HE1, HE2A, DHE1, and DHE2A, FIG. 1) and 10 digitoxin derivatives lacking the sugar moiety and the lactone ring of digitoxin (DDxx, FIG. 1) were synthesized[8,12] and weighed in glass vials. Stock solutions (10^{-1} M or 10^{-2} M) were prepared in pyridine and stored at 4°C. DHE1 and DHE2A pyridine solutions were taken and stored under nitrogen. Working solutions were then freshly prepared in assay buffer (5 mM Na_2HPO_4, 5 mM $MgCl_2$, 50 mM Tris-HCl, pH 7.4 at 37°C, BSA 0.1%). For cross-reactivity experiments at equilibrium, 0.3 mg/ml of human placental membranes prepared as previously described[10] were incubated for 5 hours at 37°C with increasing amounts of the synthetic compounds and/or 40 pM (^{125}I)digoxin (DuPont-New England Nuclear, Boston, Massachusetts). Bound and free radioactivity was separated by filtration through Whatman GF/C glass fiber filters, and retained radioactivity was counted in a γ-spectrometer. Nonspecific binding was assessed with 10 µM ouabain and was subtracted from total binding to calculate specific binding. Protein concentration was quantified with Pierce BCA protein assay reagent using BSA as a standard. Cross-reactivity experiments by the sequential technique were performed as previously described[11] using identical dilutions of the synthetic compounds. (^{3}H)ouabain binding was measured as described previously[10] using filtration as above to separate bound from free ligand. To construct Lineweaver-Burk plots, (^{3}H)ouabain (DuPont-New England Nuclear) at concentrations of 70, 20, 10, and 5 nM was incubated with four concentrations (from 5×10^{-7} to 10^{-4} M) of the inhibiting synthetic compounds.

RESULTS AND DISCUSSION

Analysis of the apparent cross-reactivities of the 14 synthetic compounds tested at concentrations of 10^{-10}, 10^{-9}, 10^{-8}, 10^{-7}, 10^{-6}, 10^{-5}, and 10^{-4} M showed that only three of the compounds, namely, DHE1, DHE2A, and (3β,5β,14β,17β,20R)-20,21-dimethylmethylenedioxy-pregnane-3,14-diol (DD4AM, FIG. 1), produced inhibition of (^{125}I)digoxin binding to human placental digitalis receptors (FIG. 2, left panels). For all the other compounds, specific (^{125}I)digoxin binding, as the percentage of (^{125}I)digoxin binding in assay buffer was higher than 90% at the highest concentration tested (not shown). Similar results were obtained by analyzing cross-reactivities at equilibrium and by the sequential technique (not shown). Higher concentrations of the synthetic compounds were not tested because they would have involved the use of an unacceptable amount of pyridine in the assay. At equilibrium, IC_{50} values (i.e., 50% inhibitory concentrations) for DD4AM, DHE1, and DHE2A were 5.5×10^{-5}, 7×10^{-5}, and 10^{-4} M, respectively (FIG. 2, left panels). By the sequential technique, IC_{50} of 4×10^{-5} and 7.5×10^{-5} and 10^{-4} M were obtained. Interestingly, the light decrease in IC_{50} values obtained by the sequential technique with DD4AM is similar to the effects found, in the same binding system, with ouabain and digoxin.[11]

The dose dependencies of the inhibition of (^{125}I)digoxin binding by the three synthetic inhibiting compounds span two logarithmic units as observed with known cardiac glycosides.[10,11] The concentration range, however, is well above ($\approx$10,000-fold) the typical range of the plant digitalis compounds.[10,11] These effects are completely different from those obtained by fatty acids[11] in which sharp dose-response curves were obtained.

Lineweaver-Burk plots for DD4AM and DHE2A are reported in the right panels of FIGURE 2. The competition of DHE1 was not assessed. For DD4AM, straight lines passing through a common point on the ordinate were obtained, indicating that the inhibition was competitive. The resulting inhibition constant was 1.51×10^{-5} M. For DHE2A, inhibition appeared to be noncompetitive.

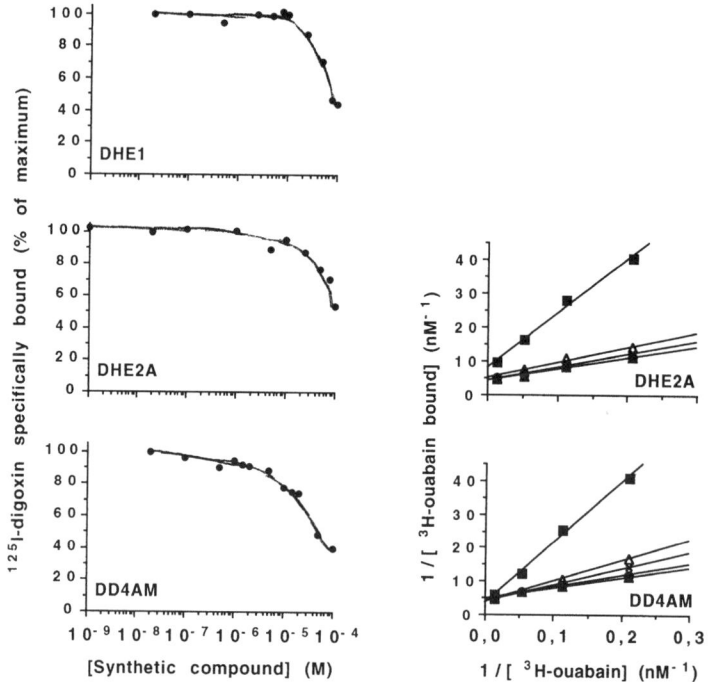

FIGURE 2. Displacement at equilibrium of (^{125}I)digoxin specifically bound to human placenta digitalis receptors by some of the synthetic compounds tested (*left panels*) and Lineweaver-Burk plots obtained from specific (^{3}H)ouabain binding inhibition data (*right panels*) (○ = no inhibitor; ■ = 10^{-4} M inhibitor; △ = 2×10^{-5} M inhibitor; X = 10^{-5} M inhibitor; ▲ = 5×10^{-6} M inhibitor).

The observation of binding inhibition by DHE1 and DHE2A confirms previous results with isolated tissues.[8] No contractile response of isolated rat aorta had previously been demonstrated for digitoxigenin derivatives,[13] but also no effect was observed with plant ouabain under used conditions.

In conclusion, these findings may support the proposed steroidal structure for mammalian EDLF and prompt us to persist in the synthetic approach to EDLF. Clearly, more effort is necessary to elucidate the structure of this factor. Moreover, identification of the Na^+,K^+-ATPase isoform(s) of the human placenta appears to be essential for the functional interpretation of the results of the present study. Experiments with specific antibodies against the isoforms of the Na^+,K^+-ATPase are in progress.

ACKNOWLEDGMENT

We thank Dr. Giuseppina Nicolini, Istituto di Patologia Medica I e Metodologia Clinica, Università di Pisa, Pisa, Italy, for helpful collaboration.

REFERENCES

1. GOTO, A., K. YAMADA, N. YAGI, M. YOSHIOKA & T. SUGIMOTO. 1992. Physiology and pharmacology of endogenous digitalis-like factors. Pharmacol. Rev. **44:** 377–399.
2. GOTO, A., T. ISHIGURO, K. YAMADA, M. ISHII, M. YOSHIOKA, C. EGUCHI, M. SHIMORA & T. SUGIMOTO. 1990. Isolation of a urinary digitalis-like factor indistinguishable from digoxin. Biochem. Biophys. Res. Commun. **173:** 1093–1101.
3. MATHEWS, W. R., D. W. DUCHARME, J. M. HAMLYN, F. MANDEL, M. A. CLARK & J. H. LUDENS. 1991. Mass spectral characterization of an endogenous digitalis-like factor from human plasma. Hypertension **17:** 930–935.
4. TAMURA, M., T.-T. LAM & T. INAGAMI. 1987. Specific endogenous Na,K-ATPase inhibitor purified from bovine adrenal. Biochem. Biophys. Res. Commun. **149:** 468–474.
5. TYMIAK, A. A., J. A. NORMAN, M. BOLGAR, G. C. DIDONATO, H. LEE, W. L. PARKER, L.-C. LO, N. BEROVA, K. NAKANISHI, E. HABER & G. T. HAUPERT, JR. 1993. Physicochemical characterization of a ouabain isomer isolated from bovine hypothalamus. Proc. Natl. Acad. Sci. USA **90:** 8189–8193.
6. ZHAO, N., L.-C. LO, N. BEROVA, K. NAKANISHI, A. A. TYMIAK, J. H. LUDENS & G. T. HAUPERT, JR. 1995. Na,K-ATPase inhibitors from bovine hypothalamus and human plasma are different from ouabain: Nanogram scale CD structural analysis. Biochemistry **34:** 9893–9896.
7. LICHTSTEIN, D., I. GATI, S. SAMUELOV, D. BERSON, Y. ROZENMAN, L. LANDAU & J. DEUTSCH. 1993. Identification of digitalis-like compounds in human cataractous lenses. Eur. J. Biochem. **216:** 261–268.
8. SAKAKIBARA, M. & A. OGAWA UCHIDA. 1996. Syntheses of $(14\beta,15\beta,16\beta,17\alpha)$- and $(14\beta,15\alpha,16\alpha,17\alpha)$-1,3,5(10)-estratriene-2,3,14,15,16,17-hexaols, possible candidates for the Inagami-Tamura endogenous digitalis-like factor, and their activity. Biosci. Biotech. Biochem. **60:** 405–410.
9. SAKAKIBARA, M. & A. OGAWA UCHIDA. 1996. Syntheses of $(14\beta,17\alpha)$-14-hydroxy- and $(14\beta,17\alpha)$-2,14-dihydroxyestradiols and their activities. Biosci. Biotech. Biochem. **60:** 411–414.
10. PACI, A., F. COCCI, F. PIRAS, G. CIARIMBOLI & A. CLERICO. 1989. Specific binding of cardiac glycoside drugs and endogenous digitalis-like substances to particulate membrane fractions from human placenta. Clin. Chem. **35:** 2093–2097.
11. PACI, A., G. CIARIMBOLI & P. BIVER. 1996. Human placenta radioreceptor assay with digoxin and ouabain to detect endogenous digitalis-like factor(s) in human plasma and urine. Clin. Chem. **42:** 1–10.
12. SAKAKIBARA, M., M. TAMURA, F. KONISHI & T. INAGAMI. 1991. XVIII International Symposium on Natural Product Chemistry, Monterrey.
13. KASHIWABARA, T., Y. INAGAKI, H. OHTA, A. IWMATSU, M. NOMIZU, A. MORITA & K. NISHIKORI. 1989. FEBS Lett. **247:** 73–76.

Kinetics of Na$^+$,K$^+$-ATPase Inhibition by Brain Endobains[a]

G. RODRÍGUEZ DE LORES ARNAIZ,[b,d] A. REINÉS,[b]
T. HERBIN,[b] AND C. PEÑA[c]

[b]*Instituto de Biología Celular y Neurociencias*
"Prof. Eduardo De Robertis"
Facultad de Medicina
Universidad de Buenos Aires,
Paraguay 2155
(1121) Buenos Aires, Argentina

[c]*IQUIFIB*
Facultad de Farmacia y Bioquímica
Universidad de Buenos Aires
Junín 956
(1113) Buenos Aires, Argentina

Na$^+$,K$^+$-ATPase, the enzymatic version of the sodium pump, is involved in the restoration and maintenance of sodium and potassium equilibrium through neuronal membranes both at rest and after passage of a nervous impulse.[1,2] Na$^+$,K$^+$-ATPase is highly concentrated in synaptic nerve endings, and regulatory mechanisms that control this enzymatic activity are essential in maintaining metabolic activity of the synaptic region and in processes directly related to neurotransmission. Inhibition of Na$^+$,K$^+$-ATPase activity by ouabain blocks the various functions occurring in isolated synaptosomes such as the synthesis of low molecular weight substances (ATP, creatine phosphate), macromolecules, and neurotransmitters as well as Na$^+$ extrusion and K$^+$ entry.[3]

Previous findings from this laboratory indicated that catecholamines and brain endogenous factors modulate neuronal Na$^+$,K$^+$-ATPase activity.[4] Isolation of brain-soluble fractions able to stimulate (peak I) and inhibit (peak II) synaptosomal membrane Na$^+$,K$^+$-ATPase activity has already been described.[5] Na$^+$,K$^+$-ATPase inhibition by peak II seems to be enzyme specific, because other enzymes bound to synaptosomal membranes such as Mg^{2+}-ATPase, acetylcholinesterase, and 5'-nucleotidase are not affected by this fraction.[6] The inhibitory material from peak II shares with ouabain the ability not only to inhibit Na$^+$,K$^+$-ATPase activity, but also to enhance natriuresis and diuresis, to induce neurotransmitter release, and to block high affinity (^{3}H)ouabain binding, which justify the suggested term endobain.[7] Further

---

[a]This study was supported by grants from the Consejo Nacional de Investigaciones Científicas y Técnicas and Universidad de Buenos Aires, Argentina.
[d]Tel: 54-1-961-5010; fax: 54-1-962-5341 or 964-8274.

peak II fractionation rendered eight fractions (A-H), two of which (II-A and II-E) presented inhibitory activity.[8]

This study evaluates the kinetics of synaptosomal membrane Na^+,K^+-ATPase inhibition by endobains II-A and II-E isolated from rat cerebral cortex.

METHODS

Wistar rats (body weight 100–150 g) were used. Pooled cerebral cortices from five rats were homogenized in 0.32 M sucrose and processed by differential centrifugation

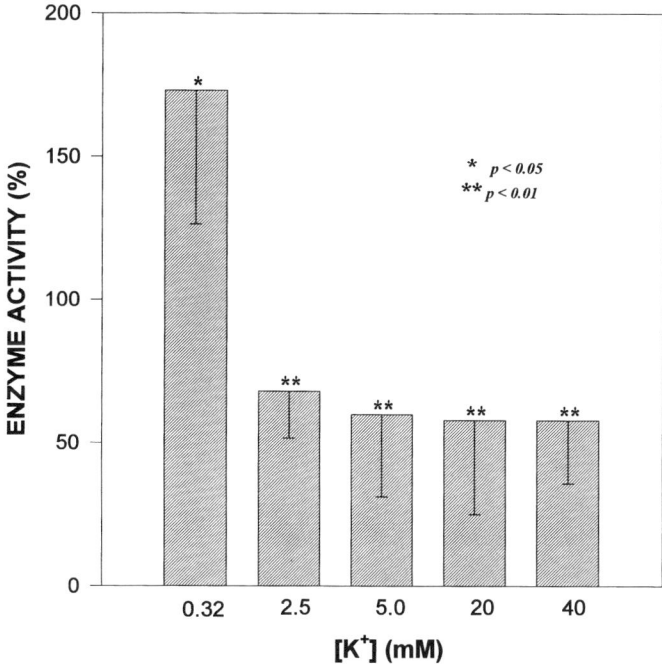

FIGURE 1. K^+ dependence of II-A endobain effect on synaptosomal membrane Na^+,K^+-ATPase activity. Results are expressed as percentage of enzymatic activity (mean ± SD; $n = 7$–10) in the presence of added II-A, taking control values as 100%. Statistical significance was calculated by the Student t test.

and sucrose gradient centrifugation to isolate synaptosomal membranes.[9] Separately, cerebral cortices from five rats were homogenized in bidistilled water, spun down at 100,000 g for 30 minutes, and the supernatant filtered in turn through Sephadex G-10 and G-50 to separate peaks I and II.[5] Peak II was further fractionated by anionic exchange HPLC performed on a Synchropak AX-300 column to obtain II-A and II-E

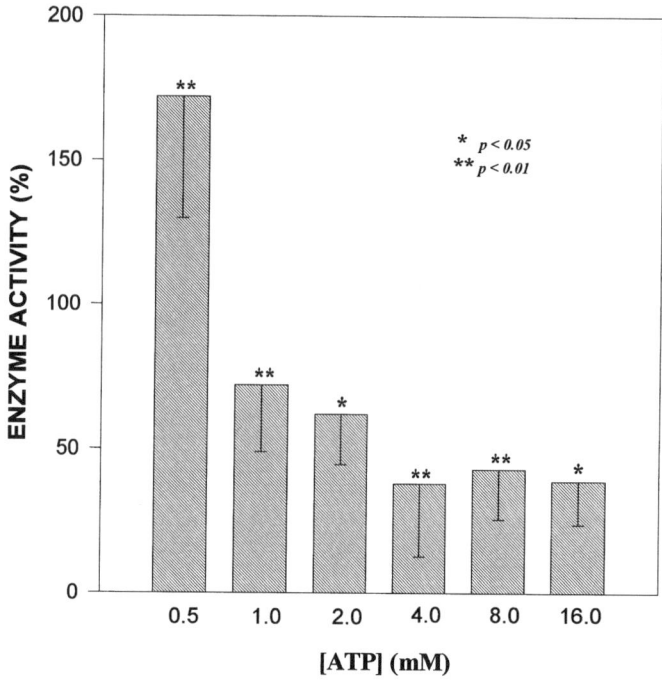

FIGURE 2. ATP dependence of II-E endobain effect on synaptosomal membrane Na^+,K^+-ATPase activity. Results are expressed as percentage of enzymatic activity (mean ± SD; $n = 4$–10) in the presence of added II-E, taking control values as 100%. Statistical significance was calculated by the Student t test.

fractions.[8] Na^+,K^+-ATPase activity in the membranes was determined spectrophotometrically following the release of orthophosphate.[10]

RESULTS AND CONCLUSION

Control synaptosomal membrane Na^+,K^+-ATPase activity ranged from 15 to 40 μmol P_i released per milligram protein per hour. Enzymatic activity was inhibited by II-A, an effect that proved noncompetitive for Na^+ within the 3.1–200 mM range and for ATP within the 2–8 mM range. In the presence of 2.5–40 mM K^+, II-A noncompetitively inhibited the enzyme, whereas at lower K^+ concentration (0.32 mM), it stimulated Na^+,K^+-ATPase activity (FIG. 1).

Na^+,K^+-ATPase activity was inhibited by II-E, an effect that proved noncompetitive for Na^+ within the 0.16–100 mM range and for K^+ within the 0.13–40 mM range. In the presence of 1–16 mM ATP, II-E noncompetitively inhibited the enzyme, whereas at lower ATP concentration (0.5 mM), it stimulated Na^+,K^+-ATPase activity (FIG. 2).

Endobains II-A and II-E decreased V_{max} values 30–60%, whereas they failed to modify K_m values. Our results indicate that endobains II-A and II-E behaved as noncompetitive inhibitors of synaptosomal membrane Na^+,K^+-ATPase activity versus Na^+,K^+ or ATP concentration. However, the enzymatic activity was stimulated by II-A at suboptimal K^+ concentration and by II-E at suboptimal ATP concentration.

Another difference between these endobains is related to their chemical nature. II-E has been characterized as a low molecular component, heat labile, neither lipidic nor peptidic in nature, sensitive to acid hydrolysis, and highly unstable to alkaline pH, whereas II-A presents properties indicative of a peptidic molecule.[8]

Jointly, these findings led us to postulate that endobains II-A and II-E may act diversely as sodium pump modulators at the synaptic region.

REFERENCES

1. STAHL, W. L. 1986. The Na^+,K^+-ATPase of nervous tissue. Neurochem. Int. **4:** 449–476.
2. WU, P. H. 1986. Na^+,K^+-ATPase in nervous tissue. *In* Neuromethods, Enzymes. A. A. Boulton, G. B. Baker & P. H. Wu, Eds.: 451–502. Humana. Clifton, NJ.
3. RODRÍGUEZ DE LORES ARNAIZ, G. & E. DE ROBERTIS. 1972. Properties of isolated nerve endings. Current Topics in Membranes and Transport, Vol. 3. F. Bronner & A. Kleinzeller, Eds.: 237–272. Academic Press. New York.
4. RODRÍGUEZ DE LORES ARNAIZ, G. & M. MISTRORIGO DE PACHECO. 1978. Regulation of (Na^+, K^+) adenosine triphosphatase of nerve ending membranes: Action of norepinephrine and a soluble factor. Neurochem. Res. **3:** 733–744.
5. RODRÍGUEZ DE LORES ARNAIZ, G. & M. ANTONELLI DE GÓMEZ DE LIMA. 1986. Partial characterization of an endogenous factor which modulates the effect of catecholamines on synaptosomal Na^+,K^+-ATPase. Neurochem. Res. **11:** 933–947.
6. RODRÍGUEZ DE LORES ARNAIZ, G., M. ANTONELLI DE GÓMEZ DE LIMA & E. GIRARDI. 1988. Different properties of two brain fractions separated in Sephadex G-50 that modify synaptosomal ATPase activities. Neurochem. Res. **3:** 229–235.
7. RODRÍGUEZ DE LORES ARNAIZ, G. 1993. An endogenous factor which interacts with synaptosomal membrane Na^+,K^+-ATPase activation by K^+. Neurochem. Res. **18:** 655–661.
8. RODRÍGUEZ DE LORES ARNAIZ, G. & C. PEÑA. 1995. Characterization of synaptosomal membrane Na^+,K^+-ATPase inhibitors. Neurochem. Int. **27:** 319–327.
9. RODRÍGUEZ DE LORES ARNAIZ, G., M. ALBERICI & E. DE ROBERTIS. 1965. Ultrastructural and enzymic studies of cholinergic and non-cholinergic synaptic membranes isolated from brain cortex. J. Neurochem. **14:** 215–225.
10. ALBERS, R., G. RODRÍGUEZ DE LORES ARNAIZ & E. DE ROBERTIS. 1965. Sodium-potasium activated ATPase and potassium-activated *p*-nitrophenylphosphatase: a comparison of their subcellular localizations in rat brain. Proc. Natl. Acad. Sci. USA **53:** 557–564.

Proscillaridin A Immunoreactivity: A New Endogenous Cardiac Glycoside?[a]

SU-QIN LI,[b,c] CHRISTIAN EIM,[b] RALF SCHNEIDER,[b]
BEATE SICH,[b] ULRIKE KIRCH,[b]
AND WILHELM SCHONER[b,d]

[b]Institut für Biochemie & Endokrinologie
Fachbereich Veterinärmedizin
Justus-Liebig-Universität Giessen
Frankfurter Str. 100
D-35392 Giessen, Germany

[c]Department of Biochemistry
Institute of Experimental Medicine
Hebei Academy of Medical Sciences
Shijiazhuang 050021, China

An isomer of ouabain circulates in increased concentrations in blood in essential hypertension and under various conditions of volume expansion.[1] It cross-reacts with antibodies against ouabain. Some vertebrates such as the toad *Bufo bufo* synthesize bufadienolides. Therefore, the possibility exists that mammals also synthesize bufadienolides. In fact, 19-norbufalin was extracted from human cataractous lenses,[2] and a material cross-reacting with antibodies against marinobufagenin elevates with the pressure response as a result of hypoventilatory breathing.[3] To further substantiate the suggestion that bufadienolide-like material may circulate in mammalian blood, we raised antibodies against proscillaridin A and looked for its distribution in the tissues of humans, rats, and cattle.

MATERIAL AND METHODS

Antibodies against proscillaridin A not cross-reacting with ouabain[4] and those against ouabain not cross-reacting with bufadienolides were used for an ELISA. Methanol extracts from sera and tissues were chromatographed on Lichrospher RP18 with a propanol/isopropanol gradient in 0.1% HFBA.[4] A fraction with a retention time like that of ouabain was used in these studies. This fraction of bovine adrenals was further purified by affinity chromatography on an anti-proscillaridin A IgG column, eluted by a pH jump, and subsequently processed by several RP18 column chromatographies.

RESULTS AND DISCUSSION

Blood of humans, rats, and cattle contains material with a retention time similar to that of ouabain which cross-reacts with anti-proscillaridin A IgG.[4] Although its

---

[a]This work was supported by grant Scho 139/20-1 of the Deutsche Forschungsgemeinschaft, Bonn-Bad Godesberg, and the Fonds der Chemischen Industrie, Frankfurt/Main.
[d]Tel: +49-641-99-38170; fax: +49-641-99-38179; e-mail: Schoner@vetmed.uni-giessen.de

relative concentrations vary with the species (TABLE 1), ouabain immunoreactivity seems to be higher than proscillaridin A immunoreactivity. We previously demonstrated in humans that proscillaridin A immunoreactivity correlated with pulse pressure ($p < 0.0001$) and mean arterial blood arterial blood pressure ($p < 0.05$).[4] This behavior resembles the observations of Hamlyn et al.[1] for ouabain immunoreactivity, but the inability of anti-proscillaridin A IgG to recognize ouabain seems to exclude the identity of both substances. Therefore, we tried to get additional information on the identity/nonidentity of both immunoreactivities. Extraction of bovine tissues with methanol followed by HPLC chromatography resulted in a similar distribution of immunoreactivity for proscillaridin A and ouabain in all tissues tested so far. The concentration of proscillaridin A immunoreactivity in adrenals was 1.1×10^{-9} mol equivalents/kg wet weight, in hypothalamus 7.1×10^{-14} mol equivalents/kg wet weight, and in muscle 1.3×10^{-14} mol equivalents/kg wet weight in muscle tissue. The data are consistent with the assumption that in cattle both immunoreactivities are secreted by the adrenals. To determine if the proscillaridin A immunoreactivity is a different compound, it was purified by affinity chromatography on an antibody column and purified further by HPLC reversed chromatography. The pure material,

TABLE 1. Plasma Concentrations of Proscillaridin A and Ouabain Immunoreactivities in Various Species

Species	Ouabain Immunoreactivity (nmol/L)	Proscillaridin A Immunoreactivity (nmol/L)	Ratio
Human	0.138 ± 43^a ($n = 11$)[5]	$0.102 \pm$ DF 8.8^b ($n = 18$)	1.35
Rat	0.700 ± 0.07^a ($n = 37$)[6]	$0.026 \pm$ DF 5.9^b ($n = 12$)	26.9
Cattle	$0.788 \pm$ DF 0.02^b ($n = 10$)	$0.0489 \pm$ DF 0.04 ($n = 5$)	16.1

[a]Mean value $\pm$ standard deviation.
[b]Geometric mean and dispersion factor (DF).

which inhibited the sodium pump at 2×10^{-8} mol equivalents/L (ouabain = 4×10^{-8} mol/L), was more polar than ouabain in RPC18 HPLC under identical conditions. It showed an UV maximum neither at 220 nm like ouabain nor at 300 nm like a bufadienolide but at 250 nm. Mass spectrometry gave a mass 16 Da larger than that of ouabain. We conclude, therefore, that the proscillaridin A immunoreactive material is most likely a new inhibitor of the sodium pump and not identical with endogenous ouabain nor with a known bufadienolide.

REFERENCES

1. HAMLYN, J. M. et al. 1996. J. Hypertens. **14:** 151–167.
2. LICHTSTEIN, D. et al. 1993. Eur. J. Biochem. **216:** 261–268.
3. BAGROV, A. Y. et al. 1995. Hypertension **26:** 781–788.
4. SICH, B. et al. 1996. Hypertension **27:** 1073–1078.
5. HARRIS, D. W. et al. 1991. Hypertension **17:** 936–943.
6. HAMILTON, B. P. et al. 1994. Endocrin. Society Abstr. 1087, p. 472.

Skeletal Muscle Na,K-ATPase Concentration Changes and Intramuscular and Extrarenal K Homeostasis in Animals and Humans[a]

H. BUNDGAARD,[b] T. A. SCHMIDT, AND K. KJELDSEN

Department of Medicine B 2142
The Heart Centre, Rigshospitalet
National University Hospital
Blegdamsvej 9
DK-2100 Copenhagen Ø, Denmark

Na,K-pump handling of K leakage during muscular activity is of importance for intramuscular as well as for extrarenal K homeostasis. Physiologically, K handling has implications for skeletal muscle function.[1,2] Clinically, changes in plasma K fluctuations due to modulations of skeletal muscular Na,K-pumps may be of importance.

The relationship between skeletal muscle Na,K-ATPase concentration and active K uptake and the influence thereof on intramuscular as well as on extrarenal K homeostasis has been assessed in various ways during the last decade (TABLE 1): *in vitro* by Na loading[3] and electrical stimulation[4] of skeletal muscle fibers. These studies showed good agreement between Na,K-ATPase concentration and maximal capacity for active K uptake, and they demonstrated that skeletal muscle Na,K-ATPase concentration changes by age, K depletion, or hypo- or hyperthyroidism modified the maximal capacity for active K uptake accordingly. *In vivo* the relationship has been recognized as hypokalemic attacks in patients with increased skeletal muscle Na,K-ATPase concentration due to hyperthyroidism[5,6] and by plasma-K increase in response to functional skeletal muscle Na,K-pump concentration reductions due to cardiac glycoside intoxication in humans[7] and by ouabain administration to rats.[8] Furthermore, in a clinical setting a 9% reduction in the functional skeletal muscle Na,K-pump concentration by digitalization caused an increased plasma-K rise during exhausting exercise in patients with heart failure. Correspondingly, digitalization was followed by an attenuated plasma-K decline during the recovery period after exercise.[9] These findings are in accord with other recent human studies demonstrating decreased plasma-K rise during exercise after a training period, shown to cause significant upregulation of skeletal muscle Na,K-pump concentration.[10,11]

On this basis, we have developed a new simple model using small experimental animals to further assess the relation between skeletal muscle Na,K-ATPase concentration changes and intramuscular and extrarenal K homeostasis. The neck vessels in anesthetized rats were catheterized, and during i.v. infusion of 0.75 mmol KCl/100 g body weight per hour delivered by an electric infusion pump, arterial plasma-K concentration was closely monitored by blood sampling at 5–15-minute intervals (FIG.

---

[a]This study was supported in part by the Danish Heart Foundation.
[b]Corresponding author.

TABLE 1. Effects of Na,K-ATPase Activity and Concentration Changes as Evaluated by Active K Uptake in Skeletal Muscle Fibers and by Changes in Plasma-K

Condition	Species	Effect	Refs.
Na loading	Rat	90% of maximum active K uptake in muscles	3
Electrical stimulation	Rat	60–100% of maximum active K uptake in muscles	4
Ouabain injection	Rat	100% plasma-K increase	8
Digitalization	Human	18% higher increase in exercise plasma-K	9
Hyperthyroidism	Human	Hypokalemic attacks	5, 6
Training	Human	19% lower increase in exercise plasma-K	11
Training	Human	15% lower increase in exercise plasma-K	10

1). Plasma-K concentrations were determined using a K-sensitive electrode. To focus on extrarenal K homeostasis, measurements were performed on functionally nephrectomized animals and on animals with preserved renal function. From the measurements in the nephrectomized animals, it could be calculated that a total of 0.38 mmol KCl had been infused and ECV K content had increased by 0.25 mmol K. This indicates that 0.13 mmol K may have been taken up into the skeletal muscle pool, inducing a mean increase in skeletal muscle K content of about 3 µmol/g wet weight. Indeed, this is in good accord with the measured tendency to an increase in

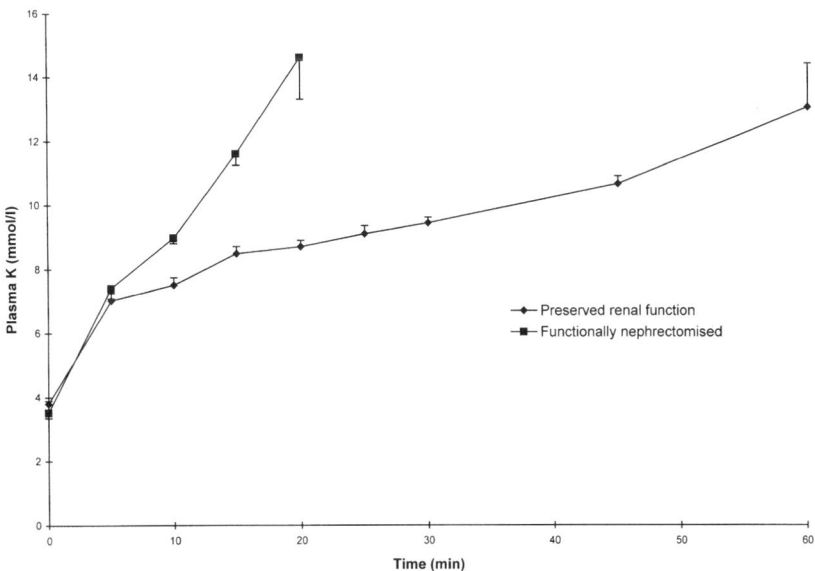

FIGURE 1. Arterial plasma-K concentrations measured in response to continous i.v. KCl infusion of 0.75 mmol KCl/100 g body weight per hour in female Wistar rats (body weight range 140–160 g) after functional nephrectomy or in animals with preserved renal function, respectively. The end point of each curve indicates mean duration (5-minute intervals) of KCl infusion before respiratory and cardiac arrest occurred. Values are given as means. Bars denote SEM. Each point represents measurements on 4–5 animals.

gastrocnemius muscle K content from 96 ± 4 to 101 ± 3 µmol/g wet weight ($n = 4$–5, $p > 0.2$) in response to KCl infusion.

Results obtained by the present *in vivo* model indicate that increased skeletal muscle Na,K-ATPase concentration is associated with correspondingly improved intramuscular and plasma-K handling.[12] In conclusion, knowledge of the relations between skeletal muscle Na,K-ATPase concentration changes and intramuscular as well as extrarenal K homeostasis is increasing, and the described model seems suitable to further address various aspects of these relations.

ACKNOWLEDGMENT

We thank Stig Haunsø for valuable discussions.

REFERENCES

1. NIELSEN, O. B. & T. CLAUSEN. 1996. The significance of active Na^+,K^+ transport in the maintenance of contractility in rat skeletal muscle. Acta Physiol. Scand. **157:** 199–209.
2. SJØGAARD, G. 1990. Exercise induced muscle fatigue: The significance of potassium. Acta Physiol. Scand. Suppl. **593:** 1–63.
3. CLAUSEN, T., M. E. EVERTS & K. KJELDSEN. 1987. Quantification of the maximum capacity for active sodium-potassium transport in rat skeletal muscle. J. Physiol. Lond. **388:** 163–181.
4. EVERTS, M. E. & T. CLAUSEN. 1994. Excitation-induced activation of the Na^+-K^+ pump in rat skeletal muscle. Am. J. Physiol. **266:** 925–934.
5. CONWAY, M. J., J. A. SEIBEL & R. P. EATON. 1974. Thyrotoxicosis and periodic paralysis: improvement with beta blockage. Ann. Intern. Med. **81:** 332–336.
6. KJELDSEN, K., A. NØRGAARD, C. O. GÖTZSCHE, A. THOMASSEN & T. CLAUSEN. 1984. Effect of thyroid function on number of Na-K pumps in human skeletal muscle. Lancet **2:** 8–10.
7. SMITH, T. W., V. P. BUTLER, E. HABER, H. FOZZARD, F. I. MARCUS, W. F. BREMNER, I. C. SCHULMAN & A. PHILLIPS. 1982. Treatment of life-threatening digitalis intoxication with digoxin-specific Fab antibody fragments. N. Engl. J. Med. **307:** 1357–1362.
8. CLAUSEN, T., O. HANSEN, K. KJELDSEN & A. NØRGAARD. 1982. Effect of age, potassium depletion and denervation on specific displaceable [^{3}H]ouabain binding in rat skeletal muscle *in vivo*. J. Physiol. Lond. **333:** 367–381.
9. SCHMIDT, T. A., H. BUNDGAARD, H. L. OLESEN, N. H. SECHER & K. KJELDSEN. 1995. Digoxin affects potassium homeostasis during exercise in patients with heart failure. Cardiovasc. Res. **29:** 506–511.
10. GREEN, H. J., E. R. CHIN, M. BALL-BURNETT & D. RANNEY. 1993. Increases in human skeletal muscle Na^+-K^+-ATPase concentration with short-term training. Am. J. Physiol. **264:** 1538–1541.
11. MCKENNA, M. J., T. A. SCHMIDT, M. HARGREAVES, L. CAMERON, S. L. SKINNER & K. KJELDSEN. 1993. Sprint training increases human skeletal muscle Na(+)-K(+)-ATPase concentration and improves K+ regulation. J. Appl. Physiol. **75:** 173–180.
12. BUNDGAARD, H., T. A. SCHMIDT, J. S. LARSEN & K. KJELDSEN. 1997. K^+ supplementation increases [Na,K-ATPase] and improves extrarenal K homeostasis in rats. J. Appl. Physiol. **82:** 1136–1144.

Alpha-2 Na,K-ATPase Contributes to Lung Liquid Clearance[a]

K. RIDGE, W. OLIVERA, D. H. RUTSCHMAN,
R. W. MERCER, B. UHAL, S. HOROWITZ, F. HUGHES,
P. FACTOR, M. L. BARNARD, AND J. I. SZNAJDER[b]

Pulmonary and Critical Care Medicine Division
Michael Reese Hospital and
University of Illinois
Chicago, Illinois 60680

Winthrop Hospital
New York, New York 11501

Washington University
St. Louis, Missouri 63110

Rush Presbyterian-St. Luke's Medical Center and
Northeastern Illinois University
Chicago, Illinois 60637

An important function of alveolar epithelial cells is to keep the airspace free of liquid and preserve gas exchange. Previous studies have suggested that Na,K-ATPase may play a role in alveolar epithelial vectorial sodium transport and thus lung liquid clearance.[1–3] In the lungs the $\alpha 1$ and $\beta 1$ isoforms have been localized to alveolar type 2 (AT2) cells[4,5]; the $\alpha 2$ isoform has been described in whole lung homogenates from adult rats[6,7] but not to a specific cell type. The alveolar surface contains two types of epithelial cells. AT2 cells are cuboidal and cover ~5% of the surface area. Their major distinguishing characteristic is the capacity for synthesis and processing of pulmonary surfactant.[8] Alveolar type 1 (AT1) cells are large flat cells that cover 95% of the alveolar surface and lack the capacity for surfactant synthesis.[8] During lung injury, AT1 cells are sloughed off and the establishment of normal epithelial structure and function is dependent on the ability of AT2 cells to proliferate and differentiate into AT1 cells.

Expression of Na,K-ATPase $\alpha 1$, $\alpha 2$, and $\beta 1$ subunits was investigated in rat AT2 cells cultured for 7 days, a period during which they lose their phenotypic markers and differentiate to an alveolar type 1-like cell phenotype. Differentiation of AT2 cells to an AT1-like phenotype resulted in a decrease of $\alpha 1$ and an increase of $\alpha 2$ mRNA and protein abundance without changes in the $\beta 1$ subunit. The existence of the distinct functional classes of Na,K-ATPase in AT2 and AT1-like cells was confirmed by ouabain inhibition of Na,K-ATPase activity. Ouabain inhibition of AT2 cells was

---

[a]This work was supported by HL-48129, the American Lung Association, and Michael Reese Hospital.

[b]Address for correspondence: Jacob Iasha Sznajder, MD, Pulmonary and Critical Care Medicine, Michael Reese Hospital, 2929 S. Ellis, RC-216, Chicago, IL 60616 (tel: 312-791-5776; fax: 312-791-2349.

consistent with expression of the α1 isozyme ($IC_{50} = 4 \times 10^{-5}$ M), whereas in AT1-like cells it was consistent with the presence of both α1 and α2 isozymes ($IC_{50} = 9.0 \times 10^{-5}$ and 1.5×10^{-7} M, respectively). (^{3}H)ouabain binding studies corroborated these findings.

To test whether the Na,K-ATPase α2 subunit was localized to the alveolar epithelium in the lung, *in situ* hybridzations were performed using isoform-specific, digoxigenin-label riboprobes. A strong hybridization signal for the α2 isoform mRNA was detected throughout the alveolar epithelium. No signal was detected in sections hybridized with Na,K-ATPase α2 sense strands. Confocal microscopy localized the Na,K-ATPase α2 isoform protein to most of the normal rat alveolar epithelial surface. Definitive localization of the Na,K-ATPase α2 isoform within the alveolar epithelium was established by immunoelectron microscopy. Isoform-specific monoclonal antibody to the Na,K-ATPase α2 subunit was used in conjunction with a 5-nm gold-labeled secondary antibody, localizing the α2 protein in AT1 cells.

We then studied whether the α2 isozyme contributes to vectorial Na$^+$ transport by inhibiting the Na,K-ATPase α2 isozyme in the isolated rat lung model. Using the isolated perfused rat lung model we perfused the pulmonary circulation with ouabain at 10^{-7} M and 10^{-5} M (which inhibits mostly the α2 isozyme in rats) which decreased active Na$^+$ transport and lung liquid clearance by 28% and 59%, respectively, as compared to those of controls and lungs perfused with 10^{-9} M ouabain. These results demonstrate for the first time the presence of α2 Na,K-ATPase isozyme in rat alveolar epithelial cells *in situ* and suggest a role for the α2 isozyme in the vectorial Na$^+$ flux across the lung epithelium.

REFERENCES

1. OLIVERA, W., K. RIDGE, L. D. H. WOOD & J. I. SZNAJDER. 1994. Active sodium transport and alveolar epithelial Na,K-ATPase increase during subacute hyperoxia in rats. Am. J. Physiol. **266:** L577–L584.
2. MATTHAY, M. A. & J. P. WIENER-KRONISH. 1990. Intact epithelial barrier function is critical for the resolution of alveolar edema in humans. Am. Rev. Respir. Dis. **142:** 1250–1257.
3. SZNAJDER, J. I., W. G. OLIVERA, K. M. RIDGE & D. H. RUTSCHMAN. 1995. Mechanisms of lung liquid clearance during hyperoxia in isolated rat lungs. Am. J. Respir. Crit. Med. **151:** 1519–1525.
4. SCHNEEBERGER, E. E. & K. M. MCCARTHY. 1986. Cytochemical localization of Na$^+$-K$^+$-ATPase in rat type II pneumocytes. J. Appl. Physiol. **60:** 1584–1589.
5. SUZUKI, S., D. ZUEGE & Y. BERTHIAUME. 1995. Sodium-independent modulation of Na,K-ATPase activity by β-adrenergic agonist in alveolar type II cells. Am. J. Physiol. **268:** L983–L990.
6. SHYJAN, A. W. & R. LEVENSON. 1989. Antisera specific for the α1, α2, α3, and β subunits in rat tissue membranes. Biochemistry **28:** 4531–4535.
7. ORLOWSKI, J. & J. B. LINGREL. 1988. Tissue specific and developmental regulation of rat Na,K-ATPase catalytic α and β subunit mRNAs. J. Biol. Chem. **263:** 10436–10442.
8. ADAMSON, I. Y. R. & D. H. BOWDEN. 1974. The type 2 cell as progenitor of alveolar epithelial regeneration. Lab. Invest. **30:** 35–42.

Opposite Expression Pattern of the Human Na,K-ATPase β1 Isoform in Stomach and Colon Adenocarcinomas[a]

JULIO AVILA,[b] EMILIA LECUONA,[b] MANUEL MORALES,[c]
ARTURO SORIANO,[d] TERESA ALONSO,[b]
AND PABLO MARTÍN-VASALLO[b,e]

[b]Laboratorio de Biología del Desarrollo
Departamento de Bioquímica y Biología Molecular
Universidad de La Laguna
Avda Astrofísico Sánchez s/n
38206 La Laguna, Tenerife, Spain

[c]Sección de Oncología Médica
Departamento de Medicina Interna and
[d]Servicio de Cirugía Digestiva
Hospital Nuestra Señora de la Candelaria
Tenerife, Spain

Stomach and colon cancers are among the most common and fatal cancers throughout the world. We studied the expression patterns of β-subunit isoforms in both types of adenocarcinoma using isoform-specific antibodies against the human Na,K-ATPase β1 and β2 subunits.[1]

Microsomes were prepared from surgical samples of 16 colon and 6 stomach adenocarcinomas collected from patients in the Oncology Division, following the guidelines of the Helsinki Convention. All of them were classified according to the TNM/pTNM-Classification of the International Union Against Cancer (UICC) for Malignant Tumours (TNM).[2] None of them underwent any chemotherapy treatment prior to surgery. The sample consisted of the adenocarcinoma in addition to a portion of healthy tissue distant from the tumor for use as the control.

Expression of both β isoforms in gastric and colonic adenocarcinomas and their respective controls was determined by western blot analysis and quantitated by densitometric scanning. Two different findings were noted: (1) As shown in FIGURE 1 A, glycosylation of the β1 isoform in colon adenocarcinoma is less than that in normal colon, in which the amount of glycosyl residues is between that of brain and kidney. Glycosylation of this isoform in gastric adenocarcinoma is similar or identical to that of control stomach. Na,K-ATPase β2 subunit was detected neither in stomach or colon adenocarcinomas nor in controls. (2) Both types of adenocarcinoma showed opposite patterns of β1-isoform glycoprotein expression. Stomach adenocarcinomas showed lower expression levels of Na,K-ATPase β1 subunit than did control tissue, and colon

---

[a]This work was supported by grants 93/0831 and 96/0453 from FIS (Spain) to P.M.-V.
[e]To whom correspondence should be addressed. Tel: 34 22 603728; fax: 34 22 063724; e-mail: PMARTIN@ULL.ES

adenocarcinomas showed higher expression of the same isoform than of normal surrounding tissue (FIG. 1B and C, respectively, and TABLE 1).

It is plausible that our findings on the changes in glycosylation of the Na,K-ATPase β1-subunit isoform (FIG. 1 and TABLE 2) are due to changes in the Golgi and/or rough endoplasmic reticulum glycosylation systems. The final product, in this case the Na,K-ATPase β1 subunit as a membrane glycoprotein,[3] might participate in the cell identification, recognition, interaction, or adhesion processes, which are typically altered in cancer cells.

These differences in the glycosylation patterns of Na,K-ATPase β1 subunit could be used to develop new diagnostic tests on biopsy specimens that could uncover early changes in the colon before advancement to the malignant state.

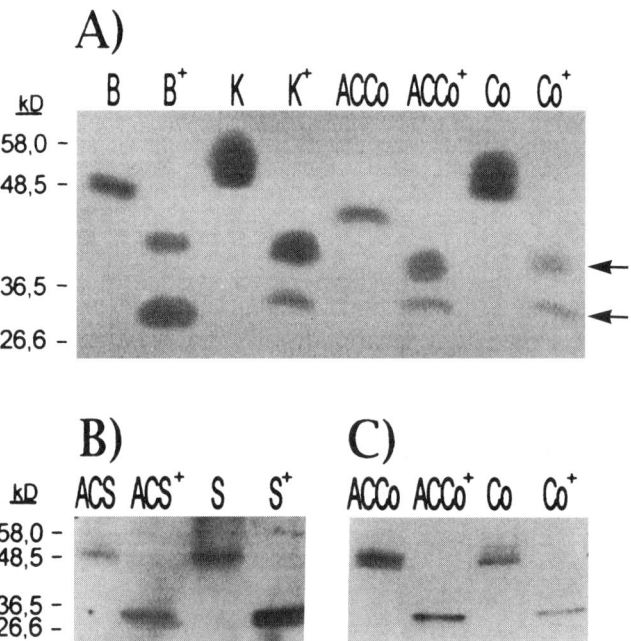

FIGURE 1. (A) Comparison of expression of Na,K-ATPase β1 isoform in brain, kidney, normal colon, and colon adenocarcinoma. Microsomes containing 20 μg of protein were fractionated by SDS-PAGE, transferred to Immobilon, and probed with specific antisera. Molecular weight markers run in a parallel lane of the gel are indicated. B = brain; K = kidney; Co = normal colon; ACCo = colon adenocarcinoma; + = treated with N-glycosidase F. *Arrows* point to the partially deglycosylated forms and core proteins. (**B** and **C**) Examples of protein expression of Na,K-ATPase β1 isoform in control stomach, gastric adenocarcinoma (**B**), normal colon, and colon adenocarcinoma (**C**), both deglycosylated to completion and ready for quantitative scanning. Microsomes containing 20 μg of protein were fractionated by SDS-PAGE, 10% polyacrylamide, transferred to Immobilon, and probed with specific antisera. Molecular weight markers run in a parallel lane of the gel are indicated. S = stomach; ACS = gastric adenocarcinoma; Co = normal colon; ACCo = colon adenocarcinoma; + = treated with N-glycosidase F.

TABLE 1. Quantitation of Densitometric Scanning of Western Blots of Adenocarcinomas and Their Controls[a]

	Total Density	SD	Core Density	SD
Normal colon	8.7	3.1	9.4	2.9*
Colon cancer	16.5	6.4	16.6	5.3*
Normal stomach	40.1	18.1	54.5	8.5†
Gastric cancer	4.1	2.3	12.9	5.5†

[a]Total density column shows the mean and standard deviation (SD) in densitometric units of the glycosylated native Na,K-ATPase β1 subunit. Core density column refers to the mean and SD in densitometric units of the completely deglycosylated core peptide of the same protein. Densitometric scanning was done in a BioRad GS-270 Imaging Densitometer (Hercules, California). Statistical parameters for pair marked by an *asterisk* are $n = 16$, $t = 3.673$, $p < 0.05$; statistical parameters for pair marked by a *dagger* are $n = 7$, $t = 0.528$, $p < 0.05$.

REFERENCES

1. GONZÁLEZ MARTÍNEZ, L. M., J. AVILA, E. MARTÍ, E. LECUONA & P. MARTÍN-VASALLO. 1994. Biol. Cell **81:** 215–222.
2. INTERNATIONAL UNION AGAINST CANCER (UICC). 1989. TNM Atlas. Illustrated Guide to the TNM/pTNM-Classification of Malignant Tumours. B. Spiessl, O. H. Beahrs, P. Hemaneck, R. V. P. Hutter, O. Scheibe, L. H. Sobin & G. Wagner, Eds.: 71–89. Springer-Verlag. Berlin, Heidelberg, New York, London, Paris, Tokyo.
3. BENALLAL, M. & B. M. ANNER. 1994. Experientia **50:** 664–668.

Immunological Identification of Na,K-ATPase Isoforms in Nonfailing and Failing Myocardium[a]

ODILE BARBEY, ALAIN GERBI, KARINE ROBERT,
VINCENT MAYOL, SANDRINE PIERRE,
FRANCK PAGANELLI, AND JEAN-MICHEL MAIXENT[b]

Cardiac Research Laboratories
School of Medicine
University Aix-Marseille
15 Bld Dramard
F-13326 Marseille, Cedex 15, France

Digitalis therapies are widely used in the treatment of congestive heart failure. The myocardial effect of digitalis to improve contractility results from specific binding and inhibition of the Na,K-ATPase. Failing hearts respond differently to digitalis than do healthy myocardiums,[1] which may be related to a change in expression of the cardiac isoform of Na,K-ATPase. The Na,K-ATPase consists of two subunits, a catalytic α subunit bearing the digitalis binding site and a glycosylated β subunit. We investigated the relative levels of expression of α- and β-subunit isoforms in patients with end-stage heart failure and whether these changes in humans could be predicted from experimental failing heart models.

METHODS

Five different types of failing myocardiums were used and originated from (1) genetically cardiomyopathic Syrian hamsters, (2) diabetic rats, (3) dogs by rapid ventricular pacing, (4) rabbits by double volume and pressure overload, and (5) pieces of transplanted hearts from patients with end-stage heart failure. Sham-operated animals and hearts from subjects without heart disease served as controls. Protein levels of digitalis receptors were determined by Western blot analysis using specific poly- and monoclonal α_1, α_2, α_3, β_1, and β_2 antibodies (purchased from U.B.I., Lake Placid, New York or provided by R. W. Mercer, K. S. Sweadner, and P. Martín-Vasallo) as we previously described.[2]

RESULTS AND DISCUSSION

Western blot analysis of tissue microsomal preparations with antisera specific for each subunit is the most direct approach to demonstrate differential expression of α

---

[a] This work was supported by grants from Procter & Gamble Pharmaceuticals France F91165 Longjumeau Cedex, France and the Assistance Publique à Marseille, Marseille, France.
[b] To whom correspondence should be addressed. Tel: (33) 91.69.88.81; fax: (33) 91.09.05.06.

TABLE 1. Immunoblotting of α and β Subunits of the Na,K-ATPase in Hamster, Rat, Dog, Rabbit, and Human Left Ventricular Myocardium Effects of Heart Failure[a]

	Hamster	Rat	Dog	Rabbit	Human
Failing Effects	Cardiac Dilatation	Ischemia + 30% Hypertrophy	Cardiac Dilatation	Dilatation + Hypertrophy	Ischemia NYHA Class 4
α1	↘	↗	→	→	↗
α2	(−)	↘	ND	ND	↘
α3	ND	ND	↗	↗	→
β1	↘	↘	↗	↘	↘
β2	↘	ND	(−)	ND	↗

[a]Protein extracts prepared from nonfailing and failing left ventricular myocardium were subjected to SDS-PAGE, electrophoretic transfer to nitrocellulose membranes, and incubation with anti α and β subunits of the Na,K-ATPase as described in methods. ND = not detected; (−) not investigated.

and β subunits of Na,K-ATPase isoforms. Five polypeptides corresponding to the five isoforms were detected in human hearts. Only three to four were detected in all the animal models. Heart failure induced distinct and opposite changes (reductions and increases) in the α and β contents of myocardial membranes (TABLE 1). Discrepancies between the findings in animal models and changes in the isoform levels of the Na,K-ATPase in failing human myocardium suggest that animal models are limited in their applicability to heart failure in human. Of interest is that changes in diabetic rats resemble those in humans with severe heart failure. However, two isoforms are not detected in the rat heart. These findings may provide a biochemical basis for an explanation of the contrasting results obtained in digitalis receptors in various heart failure models. Widespread dysfunction of digitalis receptors cannot be extrapolated from all of these studies, but we can speculate that this may be related, to some extent, to animal species and isoform patterns, to the pathophysiologic state, and to the progressive nature of myocardial disease that leads to severe heart failure.

REFERENCES

1. EZZAHER, A., R. MOUGENOT, A. BAGGIONI, T. E. OUAZZANI, B. CROZATIER & J. M. MAIXENT. 1995. C.R. Acad. Sci. Paris **318:** 1–7.
2. BARBEY, O., A. GERBI, F. PAGANELLI, K. ROBERT, S. LEVY & J-M. MAIXENT. 1997. J. Rec. Res. Signal Transduction **17:** 447–458.

Regulation of Na$^+$,K$^+$-ATPase α Subunit Isoforms in Mouse Cortex during Focal Ischemia

I. JAMME,[a] P. TROUVÉ,[b] J. M. MAIXENT,[c] A. GERBI,[c]
D. CHARLEMAGNE,[b] AND A. NOUVELOT[a]

[a]Université de Caen
CNRS-UMR 6551
Laboratoire de Neurosciences
Bd H. Becquerel
BP 5229
14074 Caen cedex, France

[b]Inserm U127
Biologie et Physiopathologie du Système Cardiovasculaire
Hôpital Lariboisière
41, Bd de la Chapelle
75010 Paris, France

[c]Laboratoire de Recherche Cardiologique
Faculté de Médecine
IFR J. Roche
Bd P. Dramard
13015 Marseille, France

Reduction of cerebral blood flow results in several acute metabolic disturbances, including a reduction in Na$^+$,K$^+$-ATPase activity.[1] Functional reduction in sodium pump capacity may be an important factor in hyperexcitability, homeostasis dysregulation, and neuronal death. Nonetheless, little evidence of eventual modulation of cerebral Na$^+$,K$^+$-ATPase activity via its three isoforms in pathology such as ischemic injury is available today.[2,3] To explore the incidence of ischemic brain damage on the Na$^+$,K$^+$-ATPase and its isoforms, the present study of adult mice examines the effect of ischemia on the affinity of the binding sites for ouabain of the Na$^+$,K$^+$-ATPase and on the regulation of mRNA expression of the α isoforms of the Na$^+$,K$^+$-ATPase.

METHODS

Ischemia was produced by middle cerebral artery occlusion[4] in groups of three mice (B6D2F1) sacrificed at 30 minutes, 1 hour, 3 hours, or 6 hours postocclusion. Enzyme activity on cortical microsomes was measured by a coupled enzymatic assay, under saturating substrate levels[5]; α isoforms were distinguished by their different affinities for ouabain; curves were fitted to experimental data by a nonlinear regression model[6] using MK Model® Software (Biosoft, Cambridge, England). mRNA expression of each isoform was studied by Northern blot and slot blot analysis and subsequently hybridized with α isoform Na$^+$,K$^+$-ATPase-specific cDNAs. Normalization for mRNA was performed by 18S quantification.

RESULTS

Time course analysis after ischemia revealed significant decreases in total Na$^+$,K$^+$-ATPase activity (38, 40.5, 68.6, and 52.8% of inhibition, respectively, $p < 0.001$) (FIG. 1). In sham-operated mice and in the contralateral cortices of ischemic mice, ouabain sensitivity disclosed three binding sites corresponding to α1, α2, and α3 isozymes (20.2 ± 5.5%, IC$_{50}$ of 0.18 ± 0.03 mM; 44.8 ± 5.5%, IC$_{50}$ of 7.1 ± 0.3 μM; 35.0%, IC$_{50}$ of 4.2 ± 0.1 nM, respectively) (FIG. 2).

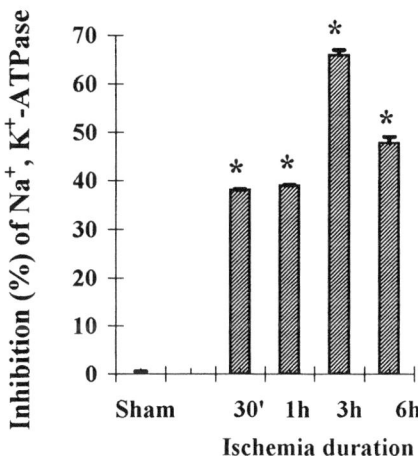

FIGURE 1. Percentage of inhibition of Na$^+$,K$^+$-ATPase in function of ischemia duration. Values represent the mean ± SEM of more than 40 tests. Sham significantly different from each time of ischemia, $p < 0.001$*.

After ischemia, we observed only two binding sites for ouabain: (1) the affinity of the first binding site (low affinity site) at 30 minutes (IC$_{50}$ of 0.15 ± 0.01 mM) was slightly increased at 1 and 3 hours (IC$_{50}$ of 0.03 ± 0.001 mM and IC$_{50}$ of 0.039 ± 0.001 mM, respectively) as compared to that of α1 isoform; (2) the second site (high affinity site) exhibited intermediate affinity (IC$_{50}$ of 0.20 ± 0.01 μM) between those of α2 and α3 isoforms at 30 minutes which evolved towards that of α3 at 6 hours postocclusion (IC$_{50}$ of 2.9 ± 0.1 nM). Its relative contribution increased from 51.4% at 30 minutes to 68.3% at 6 hours (FIG. 2).

mRNA analysis demonstrated that all three catalytic isoforms expression were expressed in mouse cortex (data not shown). Relative quantification of mRNA revealed no major difference in mRNA between the ipsilateral or contralateral cortices of ischemic mice. However, mRNA levels of α1, α2, and α3 isoforms presented a slight tendency to increase, respectively, in the ipsilateral cortex of the 30 minutes and 3 hours ischemic cortex compared to the contralateral cortex. A temporal evolution

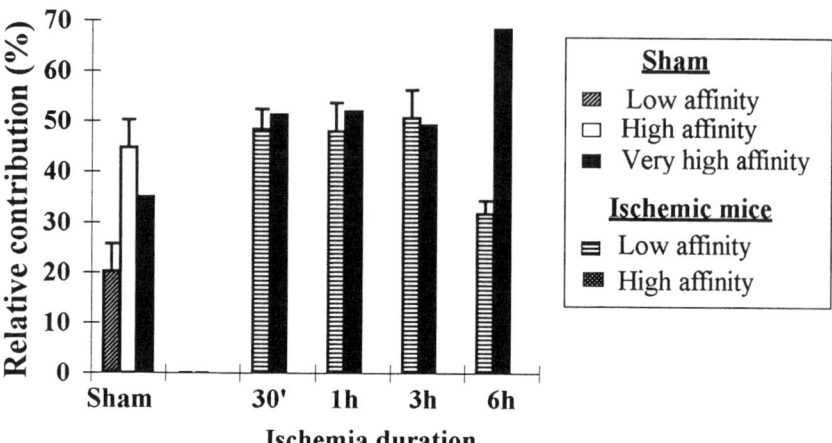

FIGURE 2. Relative contribution of each ouabain site of Na^+,K^+-ATPase to the function of ischemia duration. Each relative contribution was determined by using MK Model Software. Values are means ± SD.

was observed corresponding mainly to an increase in the levels of mRNA of α3 isoform following 30 minutes of ischemia in both contralateral and ipsilateral cortices (data not shown).

CONCLUSION

In response to ischemia, decreases in total ouabain-sensitive Na^+,K^+-ATPase in association with a shift in ouabain affinity for the second site (high affinity site) towards the affinity of α3 isoform were observed. Even if mRNA expression of α1, α2, and α3 isoforms presents a slight tendency towards an increase, this could not explain the shift in the second ouabain binding site towards the affinity of α3 isoform. Conformational changes in the α subunits, probably due to changes in lipid environment, may explain in part the total decrease in activity. However, further studies are needed to determine whether these changes reflect deleterious alterations that impair ionic homeostasis or are adaptive, potentially beneficial modifications produced in areas partially depleted of energy following focal cerebral ischemia.

REFERENCES

1. YANG, G. Y., S. F. CHEN, H. KINOUCHI, P. H. CHAN & P. R. WEINSTEIN. 1992. Stroke **23:** 1331–1336.
2. MAHADIK, S. P., V. A. BHARLICHA, A. STADLIN, A. ORTIZ & S. E. KARPIAK. 1992. J. Neurosci. Res. **32:** 209–220.
3. GRAHAM, E., O. P. MISHRA & M. DELIVORIA-PAPADOPOULOS. 1995. Neurosci. Lett. **185:** 159–162.
4. WELSH, F. A., T. SAKAMOTO, A. E. MCKEE & R. E. SIMS. 1987. J. Neurochem. **49:** 846–851.
5. JAMME, I., E. PETIT, D. DIVOUX, A. GERBI, J. M. MAIXENT & A. NOUVELOT. 1995. Neuroreport **7:** 333–337.
6. GERBI, A., M. DEBRAY, J. M. MAIXENT, C. CHANEZ & J. M. BOURRE. 1993. J. Neurochem. **60:** 246–252.

Kinetic Parameters of Na/K-ATPase Modified by Free Radicals *in Vitro* and *in Vivo*[a]

E. KURELLA,[b,e] M. KUKLEY,[b] O. TYULINA,[c] D. DOBROTA,[d]
M. MATEJOVICOVA,[d] V. MEZESOVA,[d] AND A. BOLDYREV[c]

[b]*Laboratory of Clinical Neurochemistry*
Institute of Neurology
Russian Academy of Medical Sciences
123367 Moscow, Russia

[c]*Department of Biochemistry*
International Biotechnology Center
Moscow State University
119899 Moscow, Russia

[d]*Department of Biochemistry*
Komenius University
Martin, Slovakia

Na/K-ATPase is characterized by complex kinetic behavior reflected in abnormal substrate dependence and described as a curve with an intermediary plateau.[1] This feature does not depend on the source of the enzyme studied,[2] but is closely connected to interprotomer interaction of the enzyme and the modulating effect of ATP on the activity. Solubilization of the enzyme into a monomeric state by nonionic detergents results in transformation of the complex substrate dependence curve into one resembling a hyperbola.[3,4] From these observations we concluded that ATP stimulates enzyme activity modifying interprotomer interactions within oligomeric ensembles of Na/K-ATPase.

Na/K-ATPase is a key enzyme regulating ionic homeostasis of the cell. Unfavorable conditions such as oxidative stress accompanied by increased generation of reactive oxygen species result in inhibition of Na/K-ATPase *in vivo*.[5] The same result can be achieved using different oxidants *in vitro*.[6] In our experiments we compared the kinetic properties of Na/K-ATPase after oxidative modification by hydrogen peroxide or hypochlorous anion *in vitro* and after experimental brain ischemia *in vivo*.

Hydrogen peroxide inhibited Na/K-ATPase very slowly; 5 mM H_2O_2 led to 25–30% inhibition after 30 minutes of preincubation. Hypochlorous anion provided the same inhibiting effect much faster and at as low a concentration as 5 µM. Experimental brain ishemia for 15 minutes in rats or gerbils was also accompanied by pronounced (24–28%) inhibition of brain Na/K-ATPase.

To elucidate the mechanism of inhibition, we compared the kinetic properties of highly purified membrane-bound enzyme[2] before and after oxidative attack. Kinetic

---

[a]This work was supported by the Russian Foundation for Basic Research, grant 96-04-49078 (1996, Russia).
[e]Tel/Fax: +7 (095)490–2408; e-mail: kurella@inevro.msk.su.

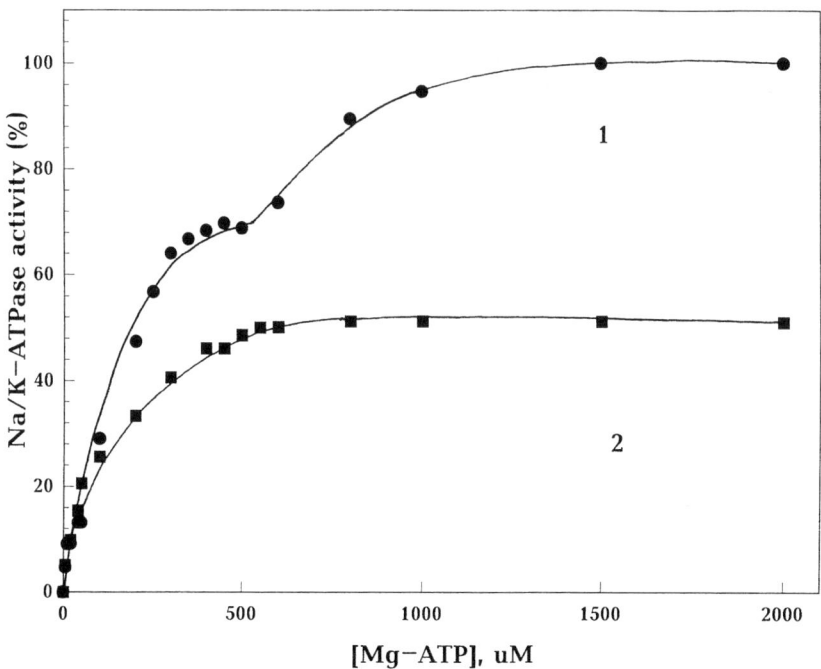

FIGURE 1. Dependence of the activity of purified dog kidney Na/K-ATPase on MgATP concentration. (1) Control; (2) after 20 minutes of oxidation in the presence of 20 mM H_2O_2.

analysis was done as recently described.[7] Whereas the control enzyme sample had substrate dependence being described as a "curve with intermediary plateau" (sum of hyperbola and sygmoid), partial oxidative modification of enzyme transformed the substrate dependence curve into one close to a hyperbola (FIG. 1). Michaelis constant for ATP was slightly increased, and kinetic cooperativity was lost, with the Hill coefficient being decreased from 7.5 to 1. Correspondingly, the maximal rate of ATP hydrolysis was also decreased (TABLE 1). On the whole, the kinetic properties of the

TABLE 1. Kinetic Parameters of Dog Kidney Na/K-ATPase under Control Conditions and after Oxidation by 20 mM H_2O_2 (20 minutes, 37°C)[a]

Parameter	Control	H_2O_2
V_M (μmol/mg protein per hour)	170	125
V_H (μmol/mg protein per hour)	65	—
$V_{max} = V_M + V_H$	235	125
K_M (μM)	80.5	135
K_H (μM)	800	—
n	7.5	1
Correlation coefficient	0.957	0.992

[a]Parameters obtained from computer simulation of hydrolysis; model is the sum of hyperbola and sygmoid.[1,8]

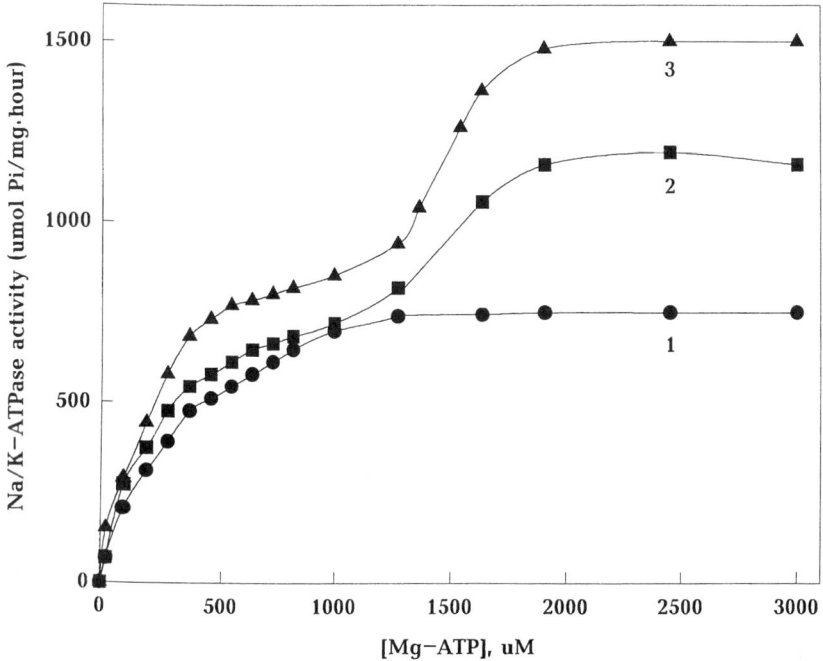

FIGURE 3. Dependence of the activity of purified duck salt gland Na/K-ATPase on MgATP concentration at different pH: (1) 6.5; (2) 8.5; and (3) 7.4.

However, the substrate dependence curve of native rat brain enzyme could be described by both (1) a hyperbola + sygmoid equation, and (2) a two hyperbola equation, with a close correlation coefficient (TABLE 2). The substrate dependence of Na/K-ATPase after its *in vivo* ischemic modification was presented by a hyperbola-like curve (FIG. 2).

When complex enzyme kinetics are described as the sum of a hyperbola and sygmoid, removal of the sygmoid by oxidative modification can be explained by the disappearance of the modulating effect of ATP on oligomer formation.[7] With two hyperbolic curves, a possible explanation is the difference in isozyme pattern of kidney and brain enzymes if they possess different sensitivity to oxidation. As a matter of fact, evidence indicates that brain Na/K-ATPase is more vulnerable to oxidative attack than is to kidney enzyme.[9,10]

Characteristically, similar transformation of substrate dependence[11] was observed under acidification of the reaction medium with purified enzyme (FIG. 3). These data show that several unfavorable conditions might be expressed in the same result, that is, removal of the apparent regulating effect of ATP on enzyme activity.

REFERENCES

1. BOLDYREV, A. *et al.* 1991. Biomed. Sci. **2:** 450–454.
2. LOPINA, O. *et al.* 1995. Arch. Biochem. Biophys. **321:** 429–433.

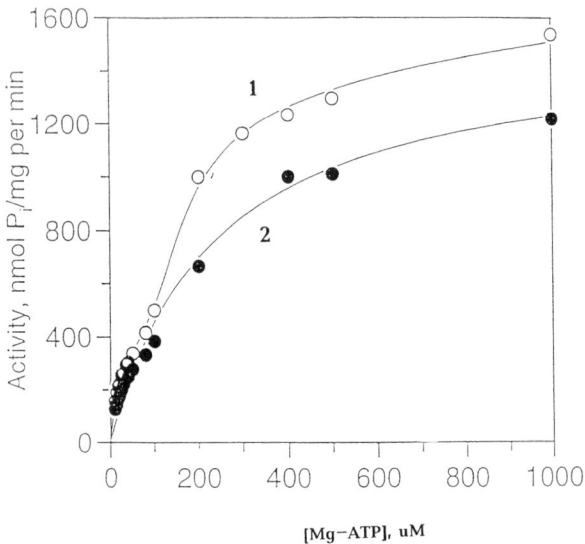

FIGURE 2. Dependence of the activity of rat brain Na/K-ATPase on MgATP concentration. (1) Control (sham-operated); (2) after 15 minutes of ischemia followed by 60 minutes of reperfusion.

enzyme after oxidative modification *in vitro* were close to those described earlier for detergent-solubilized Na/K-ATPase.[8]

Na/K-ATPase from rat brain microsomes[9] demonstrated complex kinetics (FIG. 2) and also changed the substrate dependence curve after an *in vivo* oxidative attack.

TABLE 2. Kinetic Parameters of Rat Brain Na/K-ATPase under Control Conditions and after 15 Minutes of Ischemia followed by 60 Minutes of Reperfusion[a]

Parameter	Control	Ischemia
Model I. Sum of Hyperbola and Sigmoid		
V_M (nmol/mg protein per min)	1,514	1,610
V_H (nmol/mg protein per min)	330	—
$V_{max} = V_M + V_H$	1,814	1,610
K_M (µM)	180	250
K_H (µM)	241	—
n	8	1
Correlation coefficient	0.992	0.992
Model II. Sum of Two Hyperbolas		
V_{max} 1 (nmol/mg protein per min)	1,171	1,610
V_{max} 2 (nmol/mg protein per min)	816	—
K_M 1 (µM)	236	250
K_M 2 (µM)	236	—
Correlation coefficient	0.989	0.992

[a]Parameters obtained from computer simulation of hydrolysis.[1,8]

3. HUANG, W.-H. & A. ASKARI. 1980. Molec. Pharmacol. **18:** 53–56.
4. BOLDYREV, A. *et al.* 1997. Ann. N.Y. Acad. Sci., this volume.
5. MINTOROVITCH, J. *et al.* 1994. J. Cerebr. Blood Flow Metab. **14:** 332–336.
6. BOLDYREV, A. *et al.* 1995. Biochemistry (Moscow) **60:** 1293–1298.
7. BOLDYREV, A. & E. KURELLA. 1996. Biochem. Biophys. Res. Commun. **202:** 483–487.
8. BOLDYREV, A. & N. FEDOSOVA. 1990. Biochim. Int. **22:** 903–911.
9. XIE, Z. *et al.* 1995. Biochem. Biophys. Res. Commun. **207:** 155–159.
10. DOBROTA, D. *et al.* 1996. Cell Molec. Neurobiol. **24:**.
11. FEDOSOVA, N. *et al.* 1992. Biochemistry (Moscow) **58:** 1077–1084.

Na/K-ATPase and Oxidative Stress[a]

ALEXANDER A. BOLDYREV[b] AND ELENA R. BULYGINA

Department of Biochemistry
International Biotechnology Center
Moscow State University
119899 Moscow, Russia

Oxidative stress is accompanied by the generation of a number of reactive oxygen species that attack, among other biomacromolecules, membrane-bound proteins such as Na/K-ATPase.[1–3] In our *in vivo* experiments on brain soon after ischemic injury, inhibition of Na/K-ATPase was not accompanied by significant changes in the products of lipid peroxidation, the LPO level (TABLE 1). In the *in vitro* experiments, Na/K-ATPase was also inhibited by a number of oxidants, hypochlorous anion, hydroxyl, peroxyl, and nitroxyl radicals being the most effective.

After oxidative modification of Na/K-ATPase samples, we measured: (1) hydrolytic activity (tested under optimal conditions); (2) accessibility of SH groups to Ellman reagent; and (3) ordering of the lipid environment being reflected in excimerization of pyrene fluorescent probe embedded into the hydrophobic bilayer.[4,5] When enzyme was oxidized at as low a temperature as +4°C, the lipid environment was not sufficiently changed, whereas inhibition of hydrolytic activity was very pronounced. A decrease in the rate of ATP hydrolysis correlated perfectly with a decrease in the amount of SH groups in enzyme preparation (TABLE 2). Moreover, protection of SH groups by 100 μM ascorbic acid (in the case of peroxyl radical attack[5]) or restoration of their initial amount by 1 mM cysteine or 100 μM dithiotreitol (in the case of NO-Fe compounds) was accompanied by protection or restoration of enzyme activity.

After oxidative modification of the enzyme, its kinetic properties were also modified. Oxidation suppresses Na-dependent ATP hydrolysis and $^{86}Rb^+$ occlusion[6] as well as affinity to ATP.[5] The substrate–velocity curve, which for native enzyme was described as a curve with an intermediary plateau, was transformed into a nearly hyperbolic shape.[7] Such a curve is typical for protomeric (detergent-solubilized) enzyme or for enzyme operating with GTP or UTP instead of ATP.[8,9] Therefore, we conclude that oxidative modification of Na/K-ATPase results in disordering of interprotein interactions characteristic of native enzyme.[10,11] In agreement with this finding, cooperativity for ATP of Na/K-ATPase after its oxidation disappeared.

---

[a]This work was supported by National Institutes of Health Fogarty Grant TW 00080 (1993–1995) and by Russian Foundation for Basic Research grant 96-04-49078 (1996–1998).
[b]Tel: +7 (095)9391398; fax: +7 (095)9393955; e-mail: aab@atpase.bio.msu.su

TABLE 1. Na/K-ATPase and LPO Products (measured as thiobarbituric acid-reactive substances, TBARS) in the Hemispheric Cortex of Animals with Experimental Brain Ischemia

Conditions	Na/K-ATPase Activity (% to control level)	TBARS (% to control level)	n
Wistar Rats			
Ischemia (15 min)	79 ± 4	86 ± 9	10
Sham operation	109 ± 4	105 ± 4	3
Mongolian Gerbils			
Ischemia (15 min)	88 ± 17	98 ± 4	8
Ischemia (15 min) + reperfusion (60 min)	76 ± 11	Not measured	10
Sham operation	Not measured	100 ± 3	3

Our data demonstrate that oxidative modification of Na/K-ATPase results in oxidation of SH groups of protein and in simultaneous disappearance of the regulating effect of ATP resulting from disordering of the interprotomer ensembles of the membrane-bound enzyme. The further aim of the study is to determine if oxidation of thiol groups of the enzyme directly induces the loss of interprotomer interactions or if oligomers formed are characterized by lower affinity to ATP and poor activity.

TABLE 2. Effect of Oxidants on Na/K-ATPase Activity (I, % of inhibition), SH-groups Content (II, nmol/mg protein), and Lipid Bilayer Ordering (III, pyrene excimerization ratio, F_{465}/F_{395})

Conditions	I	II	III
Dog Kidney Na/K-ATPase			
Control	100	33	Not determined
ROO$^\bullet$ generating system[5]			Not determined
10 min	56	24.9	Not determined
30 min	68	12.4	Not determined
60 min	78	11.5	Not determined
Dinitrosyl-Fe-Cys			
20 μM	58	19.2	Not determined
60 μM	69	10.0	Not determined
H_2O_2			
20 mM 20 min	51	22.6	Not determined
Ox Brain Na/K-ATPase			
Control	100	23.0	0.303
H_2O_2			
5 mM 10 min	45	13.0	Not determined
15 mM 10 min	72	5.4	0.291
NaOCl			
25 μM 10 min	26	17.0	0.303

ACKNOWLEDGMENT

We thank Professor Amir Askari for helpful discussion.

REFERENCES

1. HUANG, W.-H. *et al.* 1992. Int. J. Biochem. **24:** 621–626.
2. MINTOROVICH, J. *et al.* 1994. J. Cerebr. Blood Flow Metab. **14:** 332–336.
3. BOLDYREV, A. A. 1995. Biochemistry (Moscow) **60:** 1173–1177.
4. BOLDYREV, A. *et al.* 1995. Biochemistry (Moscow) **60:** 1293–1298.
5. KURELLA, E. *et al.* 1995. Biochim. Biophys. Acta **1232:** 52–58.
6. WANG, Y. *et al.* 1994. *In* The Sodium Pump. W. Schoner *et al.,* Eds.: 880–883. Darmstadt Press. Darmstadt, Germany.
7. BOLDYREV, A. & E. KURELLA. 1996. Biochim. Biophys. Res. Commun. **222:** 483–487.
8. BOLDYREV, A. & N. FEDOSOVA. 1990. Biochim. Int. **22:** 903–911.
9. BOLDYREV, A. & I. SVINUKHOVA. 1982. Biochim. Biophys. Acta **707:** 167–172.
10. ASKARI, A. 1987. J. Biomembr. Bioenerg. **19:** 10362–10367.
11. BOLDYREV, A. & P. QUINN. 1994. Int. J. Biochem. **26:** 1323–1331.

A Hypothetic to Explain the Non-Michaelis Substrate Dependence Curve of Na/K-ATPase[a]

ALEXANDER BOLDYREV,[b,e] ANATOLY KOTLOBAY,[b]
EKATERINA KURELLA,[c] OLGA LOPINA,[b]
AND NUNE SARVAZYAN[d]

[b]*Department of Biochemistry*
International Biotechnology Center
Moscow State University
119899 Moscow, Russia

[c]*Laboratory of Clinical Neurochemistry*
Institute of Neurology
Russian Academy of Medical Sciences
123367 Moscow, Russia

[d]*Department of Pharmacology*
Medical College of Ohio
Toledo, Ohio 4369–0008

Careful analysis of substrate dependence of Na/K-ATPase measured under optimal conditions demonstrates a deviation from that which might be described by the Michaelis-Menten equation (FIG. 1). Such dependence is typical for enzyme from different sources, such as ox brain,[1] dog heart,[2] dog kidney, or duck salt gland,[3] independent of the purity of enzyme samples or their isozyme pattern.

The substrate dependence curve in FIGURE 1 can be represented as the sum of two curves: a hyperbola (or "negative cooperativity curve" which is close to it in shape) in the range of low ATP concentrations and a sygmoid in the higher range starting from 300–500 µM ATP. The sigmoidal part of the dependence reflects the modulating effect of ATP resulting from its interaction with low affinity sites. To elucidate the possible mechanism of such modulation, the following facts have to be taken into account.

Na/K-ATPase is disposed to form oligomers,[4] although the functional significance is not fully understood. Estimation of the molecular size of the enzyme performing the overall cycle at high ATP levels suggests that they are functionally active oligomers, whereas certain partial reactions (Na-dependent ATPase, Rb$^+$ occlusion, and K-phosphatase) are apparently performed by enzyme in a monomeric (protomeric) state.[5,6] The same low molecular size, characteristic for the protomeric state, was measured when the Na/K-ATPase was operating with ATP concentrations as low as 10–30 µM or after substitution of 3 mM ATP by 3 mM GTP, the latter failing to show modulating activity.[6,7] These data clearly demonstrate that oligomeric structure is not

---

[a]This work was supported by TW-00080 grant from the Fogarty International Center, 1993–1995 (USA), and by Russian Foundation for Basic Research grant 96-04-49078, 1996–1998 (Russia).

[e]Tel: +7 (095)9391398; fax: +7 (095)9393955; e-mail: aab@atpase.bio.msu.su

a characteristic of the entire hydrolytic cycle, but oligomers are transiently formed at some step between E_2 and E_1 when the enzyme is preparing to begin a new cycle (FIG. 2).

Further analysis shows that realization of the modulating effect of high concentrations of ATP coincides with complex substrate dependence, and unfavorable factors result in the disappearance of this sigmoidal part of substrate dependence. Thus, acidification of the medium, a decrease in the temperature of incubation to 10–15°C, or substitution of ATP by GTP, all result in simplification of the substrate dependence curve.[7-10] Moreover, inhibition of Na/K-ATPase by hydrogen peroxide results in a change of enzyme kinetics[11,12] and, in particular, in the disappearance of the sigmoidal part of substrate dependence.[13] At the same time, the fraction of SH groups of enzyme available for DTNB is decreased in proportion with the suppression of hydrolytic activity.

Heart and brain enzymes, both consisting of a number of isoforms ($\alpha 1 + \alpha 2 + \alpha 3$), are more vulnerable to oxidation than kidney enzyme which consists of only the $\alpha 1$ isoform.[3,11,14] This fact correlates well with the different amounts of SH groups in the isozymes, because $\alpha 3$ contains 4 thiols, and $\alpha 2$ contains 1 thiol in addition to those groups (23) determined in $\alpha 1$ primary structure (for data on primary structure of rat isoenzymes, see ref. 15.)

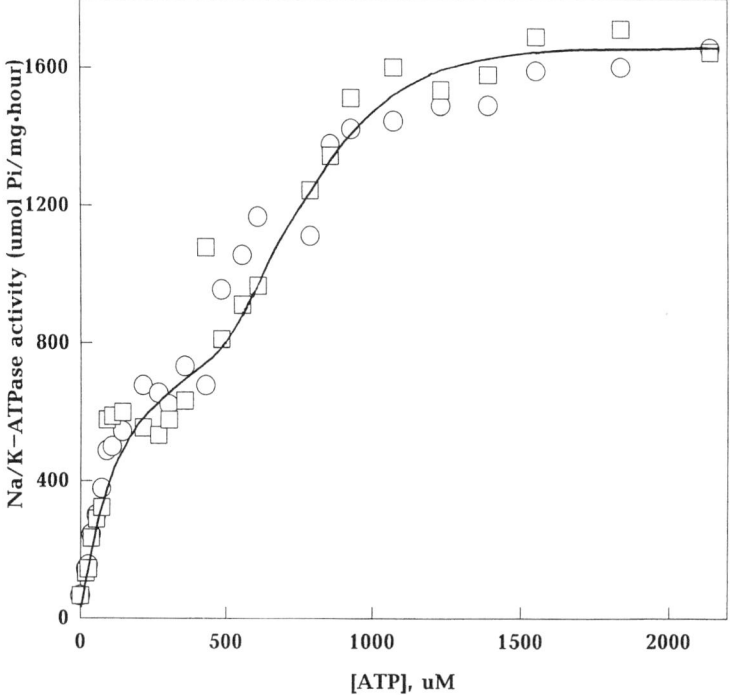

FIGURE 1. Dependence of Na/K-ATPase activity on concentration of ATP under optimal conditions (NaCl 130 mM, KCl 20 mM, $MgCl_2$ 3 mM, PIPES buffer 30 mM, pH 7.4, 37°C). □ = dog kidney enzyme; ○ = duck salt gland enzyme).

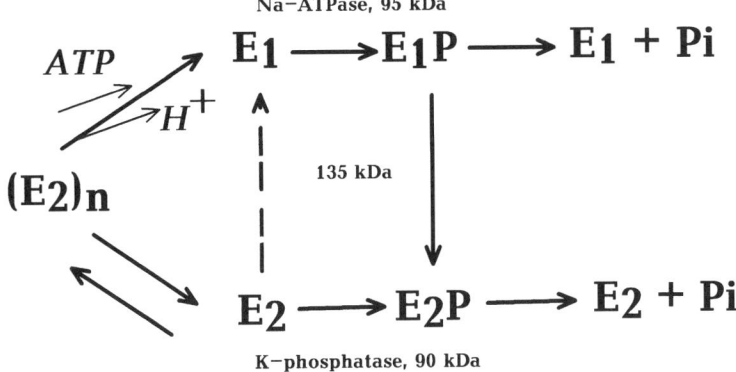

FIGURE 2. Schematic presentation of overall Na/K-ATPase cycle, indicating apparent molecular size of enzyme expressing activity under appropriate conditions.[5–7,17] Molecular size of the enzyme corresponding to the α subunit (90–95 kD) is measured under both Na-ATPase or K-pNPPase conditions but to the αβ protomer (135 kD), or (αβ)$_2$ dimer (240 kD), when the overall cycle takes place with low, or high, concentrations of ATP, respectively.

Most of the 23 thiols of α1 (16–17 groups) are located in the hydrophilic head of the globular protein and about half of them lie to the COOH-terminal side of Ala439, which was demonstrated to be important in interprotomer interaction during the phosphorylation-dephosphorylation reaction.[16] It seems reasonable to suggest that these are the thiols that take part in interprotomer interactions, because modification of these groups leads to disappearance of the modulating effect of ATP on enzyme activity.

It is important to understand why ATP should accelerate the overall cycle by stimulating E_2 oligomer decomposition rather than E_2 monomer oligomerization (FIG. 2). First, a decrease in temperature or an increase in H$^+$ concentration shifts the $E_1 \rightarrow E_2$ equilibrium towards the E_2 conformer,[17] thus increasing the portion of oligomers, while leading to a simplification of the substrate dependence curve.[6,9] Second, ATP dissociates oligomeric ensembles in detergent-solubilized enzyme.[18] These findings suggest that in stimulating $E_2 \rightarrow E_1$ transition, ATP actually induces ionization of some groups of protein, allowing it to switch its high affinity from the more "hydrophobic" ion (potassium) to affinity for the more hydrophilic sodium ion,[19] simultaneously decreasing the ability of the enzyme to be oligomerized.

In conclusion, we propose that the complex substrate dependence curve, which is characteristic for native Na/K-ATPase, is a result of the transient formation and decomposition of E_2 oligomers, with the rate of their dissociation being regulated by ATP (FIG. 2).

REFERENCES

1. BOLDYREV, A. *et al.* 1981. Biochem. Int. **2:** 137–144.
2. SHCHEGLOVA, M. *et al.* 1984. Neurochemistry (Moscow) **3:** 34–40.

3. LOPINA, O. *et al.* 1995. Arch. Biochim. Biophys. **321:** 429–433.
4. ASKARI, A. 1987. J. Biomembr. Bioenerg. **19:** 359–374.
5. JENSEN, J. & J. NORBY. 1988. J. Biol. Chem. **263:** 18063–18070.
6. BOLDYREV A. *et al.* 1990. Biochim. Int. **21:** 45–52.
7. BOLDYREV, A. & N. FEDOSOVA. 1990. Biochem. Int. **22:** 903–911.
8. BOLDYREV, A. *et al.* 1995. Biochemistry (Moscow) **60:** 1032–1039.
9. FEDOSOVA, N. *et al.* 1993. Biochemistry (Moscow) **58:** 1077–1084.
10. BOLDYREV, A. & I. SVINUKHOVA. 1982. Biochim. Biophys. Acta **707:** 167–172.
11. HUANG, W.-H. *et al.* 1992. Int. J. Biochem. **24:** 621–626.
12. WANG, Y. *et al.* 1994. *In* The Sodium Pump. W. Schoner *et al.*, Eds.: 880–883. Darmstadt Press. Darmstadt, Germany.
13. BOLDYREV, A. & E. KURELLA. 1996. Biochem. Biophys. Res. Commun. **222:** 483–487.
14. XIE, Z. *et al.* 1995. Biochem. Biophys. Res. Commun. **207:** 155–159.
15. SWEADNER, K. 1989. Biocheim. Biophys. Acta. **988:** 185–220.
16. GANJIEZADEH, M. *et al.* 1995. J. Biol. Chem. **270:** 15707–15710.
17. BOLDYREV, A. *et al.* 1991. *In* The Sodium Pump: Recent Developments. J. Kaplan *et al.*, Eds.: 483–487. The Rockefeller University Press.
18. HAYASHI, Y. *et al.* 1989. Biochim. Biophys. Acta **983:** 217–229.
19. SKOU, J. C. 1988. Methods Enzymol. **156:** 1–25.

The Plasma Membrane Ca^{2+}-ATPase in Spontaneously Hypertensive Rats[a]

BASIL D. ROUFOGALIS,[b,d] SHI CHEN,[b]
ELEANOR P. W. KABLE,[a] TUAN H. KUO,[c]
AND G. R. MONTEITH[b]

[b]*Department of Pharmacy*
University of Sydney
NSW 2006, Australia

[c]*Department of Pathology*
Wayne State University
Detroit, Michigan 48201

Intracellular Ca^{2+} regulation has been one of the factors examined in the possible etiology of hypertension over many years. These studies stem from the knowledge that $[Ca^{2+}]_i$ is an important regulator of vascular tone.[1] The plasma membrane Ca^{2+}-ATPase (PM Ca^{2+}-ATPase) is a Ca^{2+} translocating enzyme which, along with the Na^+/Ca^{2+} exchanger,[2] is responsible for the removal of Ca^{2+} across the plasma membrane. The relative importance of these two transporters is cell-type dependent.[2] It should be noted, however, that the high Ca^{2+} affinity of the PM Ca^{2+}-ATPase means that this is probably the principal pathway in the regulation of basal $[Ca^{2+}]_i$ and the removal of at least small Ca^{2+} loads.[2]

Our laboratory has focused on the possible role of the PM Ca^{2+}-ATPase in hypertension. We compared various aspects of Ca^{2+} homeostasis in cultured aortic smooth muscle cells isolated from the spontaneously hypertensive rat (SHR) and its normotensive control, the Wistar-Kyoto rat (WKY). Two basic hypotheses have been proposed for the role of the PM Ca^{2+}-ATPase in hypertension. Firstly, the activity or expression of the PM Ca^{2+}-ATPase may be diminished in hypertension, elevating resting $[Ca^{2+}]_i$ and augmenting responses to vasoconstrictive agents by diminishing the ability of the cell to lower $[Ca^{2+}]_i$ after agonist stimulation. Alternatively, the activity or expression of the PM Ca^{2+}-ATPase may be upregulated in hypertension as an attempt by the cell to lower $[Ca^{2+}]_i$ and offset the effects of elevated Ca^{2+} influx.

We have found evidence for elevated Ca^{2+} influx in SHR. Cultured aortic smooth muscle cells from SHR were associated with elevated ^{45}Ca uptake[3] and Mn^{2+} influx.[7] Hence, our studies are consistent with others which indicate that hypertension, at least in the SHR model, is associated with elevated Ca^{2+} influx.

To probe the activity of the PM Ca^{2+}-ATPase we examined $^{45}Ca^{2+}$ efflux in resting, angiotensin II (Ang II)-, and ionomycin-stimulated cells from both strains. PM Ca^{2+}-ATPase-mediated Ca^{2+} efflux was measured in Na^+-free buffer to inhibit Ca^{2+} efflux via the Na^+/Ca^{2+} exchanger.[3] These studies indicate that resting, Ang II-,

---

[a]This work was supported by the National Health and Medical Research Council and the National Heart Foundation of Australia.

[d]To whom reprint requests should be addressed. Fax: 61-2-9351-4447; e-mail: basilr@pharm.usyd.edu.au

and ionomycin-stimulated $^{45}Ca^{2+}$ efflux values are elevated in SHR compared to WKY.[3,4] This finding is not simply a consequence of the elevated $^{45}Ca^{2+}$ uptake in the hypertensive strain, because the fraction of $^{45}Ca^{2+}$ lost after Ang II stimulation is also elevated in SHR.[3] Indeed, the results in FIGURE 1 show that after Ang II (100 nM) stimulation, the peak Ca^{2+} efflux rate is elevated in SHR compared to WKY despite no significant difference between the two strains in peak $[Ca^{2+}]_i$ (FIG. 1). Furthermore, Ang II-stimulated PM Ca^{2+}-ATPase-mediated $^{45}Ca^{2+}$ efflux is elevated in SHR vs WKY when determined at the same $[Ca^{2+}]_i$ levels, strongly suggesting that the PM Ca^{2+}-ATPase-mediated Ca^{2+} efflux pathway is upregulated in SHR.[5] This may be due to an increase in total PM Ca^{2+}-ATPase expression, an alteration in the types of PM Ca^{2+}-ATPase isoforms expressed, or increased stimulation of the PM Ca^{2+}-ATPase in SHR by its regulators, such as protein kinase C and calmodulin.[2] Alternatively, fura-2 may not accurately assess the pool of Ca^{2+} which is the source of activation of the PM Ca^{2+}-ATPase.[5]

To determine the expression of the PM Ca^{2+}-ATPase in SHR and WKY, we compared mRNA for the PMCA1 isoform of the PM Ca^{2+}-ATPase using Northern blotting. FIGURE 2 gives initial data showing increased PMCA1 mRNA in SHR compared to WKY. No change could be detected in total PM Ca^{2+}-ATPase protein using Western blotting with 5F10 anti-PM Ca^{2+}-ATPase antibodies (FIG. 2, inset). This may indicate that the increased PMCA1 mRNA in SHR does not express greater Ca^{2+} pump protein or that the difference in protein is beyond the limit of sensitivity of this technique. Alternatively, an increase in PMCA1 protein could be masked by alterations in the expression of other PM Ca^{2+}-ATPase isoforms, because the 5F10 anti-PM Ca^{2+}-ATPase antibody used recognizes all isoforms. Further studies will allow these issues to be addressed directly. It should be noted that the increase in PMCA1 mRNA may be a consequence of altered Ca^{2+} homeostasis in SHR. Recently, it was proposed that the expression of the PM Ca^{2+}-ATPase is regulatorily linked to the expression of other proteins associated with Ca^{2+} regulation.[6] Hence, our

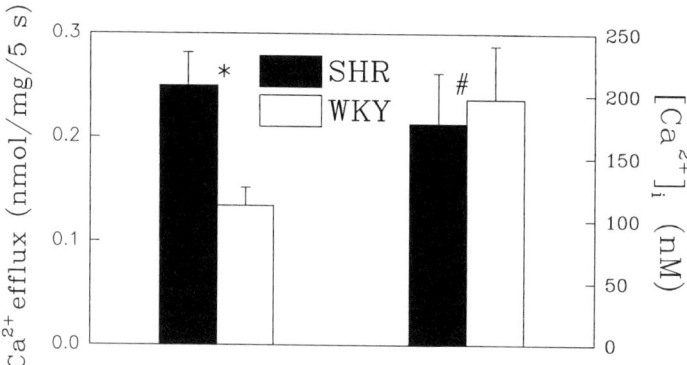

FIGURE 1. Peak PM Ca^{2+}-ATPase-mediated Ca^{2+} efflux and peak $[Ca^{2+}]_i$ after 100 nM angiotensin II stimulation. (**Left**) Angiotensin II-stimulated peak PM Ca^{2+}-ATPase-mediated Ca^{2+} efflux in the spontaneously hypertensive rat (SHR) (*filled bars*) and the Wistar-Kyoto rat (WKY) (*unfilled bars*) (*$p < 0.05$). However, no significant difference was observed between SHR (*filled bars*) and WKY (*unfilled bars*) in angiotensin II-stimulated peak $[Ca^{2+}]_i$ (**right**, #$p > 0.05$).

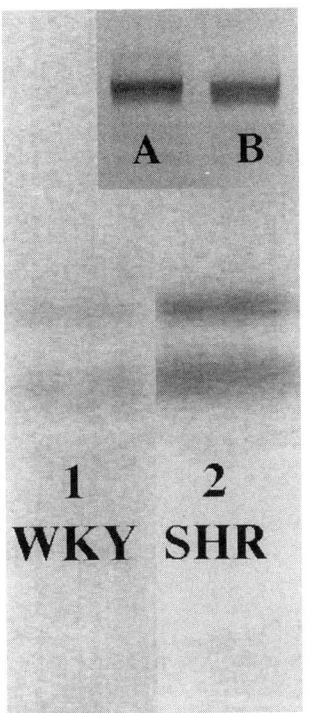

FIGURE 2. PMCA1 mRNA and total PM Ca^{2+} pump protein levels in SHR and WKY. PMCA1 mRNA is greater in the spontaneously hypertensive rat (SHR) (*lane 2*) than in the Wistar Kyoto rat (WKY) (*lane 1*), as measured by Northern blotting. However, no significant difference was observed in total PM Ca^{2+}-ATPase protein as measured by Western blotting between WKY (*lane A*) and SHR (*lane B*).

observations of increased PM Ca^{2+} pump activity and PMCA1 mRNA in SHR may be the result of increased Ca^{2+} influx. Upregulation of Ca^{2+} efflux pathways in conditions in which Ca^{2+} influx is increased may be a compensatory response of the cell to maintain Ca^{2+} homeostasis close to optimal, so that the effects of Ca^{2+} overload are at least minimized.

REFERENCES

1. REMBOLD, C. M. 1992. Hypertension **20**: 129–137.
2. MONTEITH, G. R. & B. D. ROUFOGALIS. 1995. Cell Calcium **18**: 459–470.
3. CHEN, S. & B. D. ROUFOGALIS. 1994. Am. J. Hypertens. **7**: 597–602.
4. CHEN, S., G. R. MONTEITH & B. D. ROUFOGALIS. 1995. Am. J. Hypertens. **8**: 1015–1022.
5. MONTEITH, G. R., E. P. W. KABLE, S. CHEN & B. D. ROUFOGALIS. 1996. J. Hypertens. **14**: 435–442.
6. LIU, B-F., X. XU, R. FRIDMAN, S. MUALLEM & T. J. KUO. 1996. Biol. Chem. **271**: 5536–5544.
7. MONTEITH, G. R., E. P. W. KABLE & B. D. ROUFOGALIS. 1997. Clin. Exp. Hypertens. **19**: 431–443.

Regulation of Myocardial Na,K-ATPase Concentration in Experimental and Human Heart Disease[a]

T. A. SCHMIDT, H. BUNDGAARD,[b] AND K. KJELDSEN

Department of Medicine B 2142
The Heart Centre, Rigshospitalet
National University Hospital
Blegdamsvej 9
DK-2100 Copenhagen Ø, Denmark

Because Na,K-ATPase is of importance for Na, K, and Ca handling as well as for digoxin treatment, changes in myocardial Na,K-ATPase capacity may be of pathogenetic and pharmacologic importance in heart disease. Studies of myocardial Na,K-ATPase, however, may be misleading due to low and inconsistent enzyme recovery.[1,2] Vanadate-facilitated (^{3}H)ouabain binding to intact necropsy or biopsy specimens of human heart as well as K-activated 3-*O*-methyl-fluorescein phosphatase (3-*O*-MFPase) and *para*-nitrophenyl phosphatase (*p*-NPPase) measurements in crude homogenates from rodent hearts, on the other hand, have proven applicable in quantitative studies.[3–5] Furthermore, with extensive washing with digoxin antibodies of necropsy or biopsy specimens of human heart, it is possible to remove previously bound digoxin before Na,K-ATPase quantification by (^{3}H)ouabain binding.[6]

FIGURE 1 gives an overview of the conditions that have the potential or indeed have been found to change myocardial Na,K-ATPase concentration in animals and humans, as evaluated by quantitative methods. In both animal models as well as humans, heart failure is associated with reduced left ventricular Na,K-ATPase concentration of 15–40%. Indeed, a positive linear correlation between myocardial Na,K-ATPase concentration and heart performance as evaluated by left ventricular ejection fraction has been observed.[7] Whereas short-term ischemia and reperfusion in dogs[8] did not cause an acute reduction in Na,K-ATPase concentration, chronic ischemic heart disease (IHD) did.[9,10] Although myocardial hypertrophy may initially be associated with minor upregulation of myocardial Na,K-ATPase concentration,[11,12] later stages with progression into heart failure are associated with downregulation in animals as well as humans.[11] Hence, in myocardial samples from heart failure patients undergoing aortic valve surgery for combined aortic stenosis and regurgitation, endomyocardial biopsies from the left ventricle showed a reduction in the Na,K-ATPase of as much as 60% ($n = 5, p < 0.01$).[11]

Digoxin is often used in the clinical management of heart failure patients. Thus, it is of interest that cardiac glycoside treatment has been associated with 24–34% inhibition of myocardial Na,K-pumps in these patients.[9,10] These values seem to be in

---

[a] This study was supported in part by the Danish Heart Foundation.
[b] Address for correspondence: Henning Bundgaard, Department of Medicine B 2142, The Heart Centre, Rigshospitalet, National University Hospital, Blegdamsvej 9, DK-2100 Copenhagen Ø, Denmark.

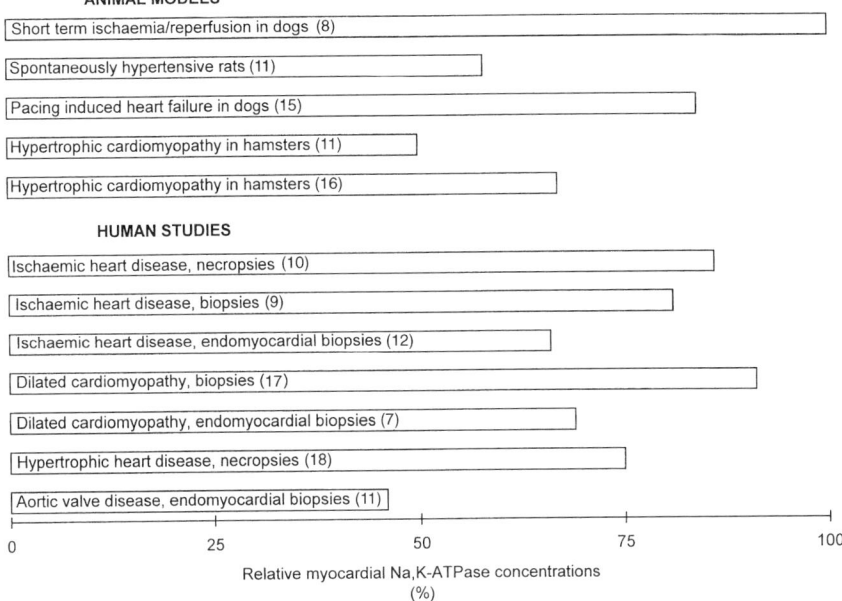

FIGURE 1. Relative myocardial Na,K-ATPase concentration changes in various heart diseases in animals and human beings as compared to control values. Figures in parentheses refer to reference numbers.

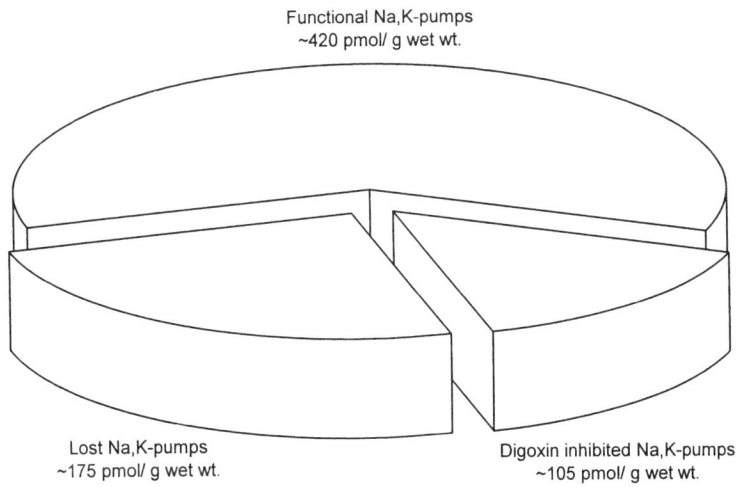

FIGURE 2. Reductions in functional human myocardial Na,K-pump concentration per gram wet weight of tissue due to heart failure and inhibition by cardiac glycoside, respectively. Calculations are based on a myocardial Na,K-ATPase concentration of 700 pmol/g wet weight in normal human subjects.[9,14]

accord with the observed 33% reduction in maximal hyperpolarization of resting membrane potential in human atrial tissue as a result of digitalization.[13] Thus, not less than 20% of myocardial Na,K-pumps are functionally lost during digitalization. This corresponds to around 15% of the number of Na,K-pumps present in the normal human myocardium. It is furthermore of interest that whereas studies of cultured cells as well as peripheral blood cells have indicated that long-term digitalization should be associated with upregulation of myocardial Na,K-ATPase, such an effect was not found in the human heart.[9,10] On the contrary, the myocardial Na,K-ATPase concentration in the human heart of digitalized patients was reduced even after removal of bound digoxin from the Na,K-ATPase. This is probably due to the underlying heart disease just discussed.

In conclusion, it has become possible to quantify myocardial Na,K-ATPase in animals as well as in human subjects. In the normal human myocardium, the Na,K-ATPase concentration is 700 pmol/g wet weight.[9,14] In heart disease, around 25% of these Na,K-pumps may be lost. By digitalization, around 15% may be functionally lost. Thus, only about 60% corresponding to around 400 pmol/g wet weight may remain functional (FIG. 2). This reduced capacity for maintaining myocardial electrolyte homeostasis is of importance in the further progression of heart failure and for generation of arrhythmias and should probably be taken into account in digoxin therapy.

ACKNOWLEDGMENT

We thank Stig Haunsø for valuable discussions.

REFERENCES

1. HANSEN, O. & T. CLAUSEN. 1996. Studies on sarcolemma components may be misleading due to inadequate recovery. FEBS Lett. **384:** 203.
2. JONES, L. R. & J. H. R. BESCH. 1984. Isolation of canine cardiac sarcolemmal vesicles. Meth. Pharmacol. **5:** 1–12.
3. LARSEN, J. S. & K. KJELDSEN. 1995. Quantification in crude homogenates of rat myocardial Na^+,K^+- and Ca^{2+}-ATPase by K^+ and Ca^{2+}-dependent pNPPase. Age-dependent changes. Basic Res. Cardiol. **90:** 323–331.
4. NØRGAARD, A., K. KJELDSEN, O. HANSEN, T. CLAUSEN, C. G. LARSEN & F. G. LARSEN. 1986. Quantification of the ^{3}H-ouabain binding site concentration in human myocardium: A postmortem study. Cardiovasc. Res. **20:** 428–435.
5. NØRGAARD, A., K. KJELDSEN & O. HANSEN. 1985. K^+-dependent 3-O-methylfluorescein phosphatase activity in crude homogenate of rodent heart ventricle: Effect of K^+ depletion and changes in thyroid status. Eur. J. Pharmacol. **113:** 373–382.
6. SCHMIDT, T. A. & K. KJELDSEN. 1991. Enhanced clearance of specifically bound digoxin from human myocardial and skeletal muscle samples by specific digoxin antibody fragments: Subsequent complete digitalis glycoside receptor (Na,K-ATPase) quantification. J. Cardiovasc. Pharmacol. **17:** 670–677.
7. NØRGAARD, A., J. P. BAGGER, P. BJERREGAARD, U. BAANDRUP, K. KJELDSEN & P. E. THOMSEN. 1988. Relation of left ventricular function and Na,K-pump concentration in suspected idiopathic dilated cardiomyopathy. Am. J. Cardiol. **61:** 1312–1315.
8. SCHMIDT, T. A., J. H. SVENDSEN, S. HAUNSØ & K. KJELDSEN. 1990. Quantification of the total Na,K-ATPase concentration in atria and ventricles from mammalian species by

measuring ³H-ouabain binding to intact myocardial samples. Stability to short term ischemia reperfusion. Basic Res. Cardiol. **85:** 411–427.

9. SCHMIDT, T. A., P. D. ALLEN, W. S. COLUCCI, J. D. MARSH & K. KJELDSEN. 1993. No adaptation to digitalization as evaluated by digitalis receptor (Na,K-ATPase) quantification in explanted hearts from donors without heart disease and from digitalized recipients with end-stage heart failure. Am. J. Cardiol. **71:** 110–114.

10. SCHMIDT, T. A., P. HOLM NIELSEN & K. KJELDSEN. 1991. No upregulation of digitalis glycoside receptor (Na,K-ATPase) concentration in human heart left ventricle samples obtained at necropsy after long term digitalisation. Cardiovasc. Res. **25:** 684–691.

11. LARSEN, J. S., T. A. SCHMIDT, H. BUNDGAARD & K. KJELDSEN. 1997. Reduced concentrations of myocardial Na^+,K^+-ATPase in human aortic valve disease as well as of Na^+,K^+- and Ca^{2+}-ATPase in rodents with hypertrophy. Moll. Cell. Biochem. **169:** 85–93.

12. NØRGAARD, A. & K. KJELDSEN. 1989. Human myocardial Na,K-pumps in relation to heart disease. J. Appl. Cardiol. **4:** 239–245.

13. RASMUSSEN, H. H., G. T. OKITA, R. S. HARTZ & R. E. TEN EICK. 1990. Inhibition of electrogenic Na^+-pumping in isolated atrial tissue from patients treated with digoxin. J. Pharmacol. Exp. Ther. **252:** 60–64.

14. BUNDGAARD, H. & K. KJELDSEN. 1996. Human myocardial Na,K-ATPase concentration in heart failure. Moll. Cell. Biochem. **163/164:** 277–283.

15. SCHMIDT, T. A., J. S. LARSEN, R. P. SHANNON, K. KOMAMURA, D. E. VATNER & K. KJELDSEN. 1993. Reduced ³H-ouabain binding site (Na,K-ATPase) concentration in ventricular myocardium of dogs with tachycardia induced heart failure. Basic. Res. Cardiol. **88:** 607–620.

16. NØRGAARD, A., U. BAANDRUP, J. S. LARSEN & K. KJELDSEN. 1987. Heart Na,K-ATPase activity in cardiomyopathic hamsters as estimated from K-dependent 3-O-MFPase activity in crude homogenates. J. Mol. Cell Cardiol. **19:** 589–594.

17. SCHWINGER, R. H. G., M. BÖHM & E. ERDMANN. 1990. Effectiveness of cardiac glycosides in human myocardium with and without "downregulated" β-adrenoceptors. J. Cardiovasc. Pharmacol. **15:** 692–697.

18. ELLINGSEN, Ø., R. HOLTHE, A. SVINDLAND, G. AKSNES, O. M. SEJERSTED & A. ILEBEKK. 1994. Na,K-pump concentration in hypertrophied human hearts. Eur. Heart J. **15:** 1184–1190.

Oxygen-Free Radicals Directly Attack the ATP Binding Site of the Cardiac Na^+,K^+-ATPase[a]

KAI Y. XU,[b] JAY L. ZWEIER, AND LEWIS C. BECKER

Department of Medicine
Cardiology Division
The Johns Hopkins Medical Institutions
Baltimore, Maryland 21224

Highly reactive oxygen-free radicals play a crucial role in sarcolemmal injury following ischemia/reperfusion and cause irreversible inhibition of Na^+,K^+-ATPase activity.[1,2] However, the free radical targeting sites and the derangements responsible for the altered function of the enzyme are not yet fully understood. To investigate whether the active site of the cardiac Na^+,K^+-ATPase is directly involved in free radical inhibition, ATP was utilized to specifically protect enzymatic activity against hydroxyl free radical (·OH)-induced inactivation. We reasoned that occupation of the active site would protect enzymatic function if the ATP binding site were a critical target of ·OH attack. The hydroxyl free radical was chosen for our experiments because it is generated in the postischemic heart and results in contractile dysfunction.[3]

METHODS

Cardiac sarcolemmal Na^+,K^+-ATPase was purified from Sprague-Dawley rat heart muscle using a protocol based on a modification of the methods of Jones[4] and Watanabe *et al.*[5] Enzymatic activity is defined as the strophanthidin-sensitive hydrolysis of MgATP in the presence of Na^+ and K^+.[6] The hydroxyl radicals were generated from a system consisting of hydrogen peroxide (H_2O_2) and the ferric iron chelate Fe^{3+}/nitrilotriacetic acid (NTA). A final concentration of 1 mM H_2O_2 was used in all experiments. Electron paramagnetic resonance (EPR) spin-trapping measurements were performed in the presence of 5,5-dimethyl-1-pyrroline-*N*-oxide (DMPO) and recorded at room temperature for 30 minutes using an IBM-Bruker ER 300 spectrometer operating at X-band with a TM_{110} cavity and flat cell. Spectral simulations were performed on a personal computer and directly matched with the experimental data to obtain the spectral parameters, as described previously.[7]

---

[a]This work was supported by National Institutes of Health grants Hl-33360, Hl-52315, and Hl-17655 and Johns Hopkins Institutional Research Grant S07 RR05378.

[b]Address for correspondence: The Johns Hopkins Medical Institutions, Department of Medicine, Division of Cardiology, 5501 Hopkins Bayview Circle, 1A2, Baltimore, MD 21224 (tel: 410-550-2021; fax: 410-550-2448; e-mail: kxu@welchlink.welch.jhu.edu).

RESULTS AND CONCLUSIONS

When purified rat cardiac Na^+,K^+-ATPase was exposed to the hydroxyl radicals generated from 80 μM Fe^{3+}-NTA and 1 mM H_2O_2, there was a 79% inhibition of Na^+,K^+-ATPase activity, as shown in FIGURE 1. By contrast, when Na^+,K^+-ATPase was premixed with 1 mM ATP before exposure to ·OH, complete protection was observed; there was no loss of enzymatic activity. No significant preservation of enzymatic activity occurred when the enzyme was premixed with adenosine, sucrose, AMP, or ADP (1 mM each), respectively, before exposure of the enzyme to the same ·OH free radical generating system. Preincubation of Na^+,K^+-ATPase with 1 mM 5′-adenylylimidodiphosphate (AMP-PNP, a derivative of ATP) before exposure to ·OH resulted in 90% protection, suggesting that the terminal phosphate group of ATP may play an important role in protecting enzymatic function. No inhibition was seen when purified Na^+,K^+-ATPase was exposed to 80 μM Fe^{3+}-NTA or 1 mM H_2O_2 by itself. To investigate whether the concentration of ·OH used in our experiments damaged the polypeptide chain of the enzyme, purified cardiac Na^+,K^+-ATPase was incubated with 80 μM Fe^{3+}/NTA + 1 mM H_2O_2. Polyacrylamide gel electrophoresis demonstrated no change in the apparent molecular weight of the Na^+,K^+-ATPase polypeptide, and no tattered peptide fragments were observed, indicating that ·OH did not damage the primary structure of the enzyme (data not shown). To investigate whether ATP, at the concentration used in this study, acts as a direct scavenger of ·OH, EPR measurements were performed in the presence of the spin trap DMPO (50 mM) and quantitated by double integration. No significant signal was detected with buffer

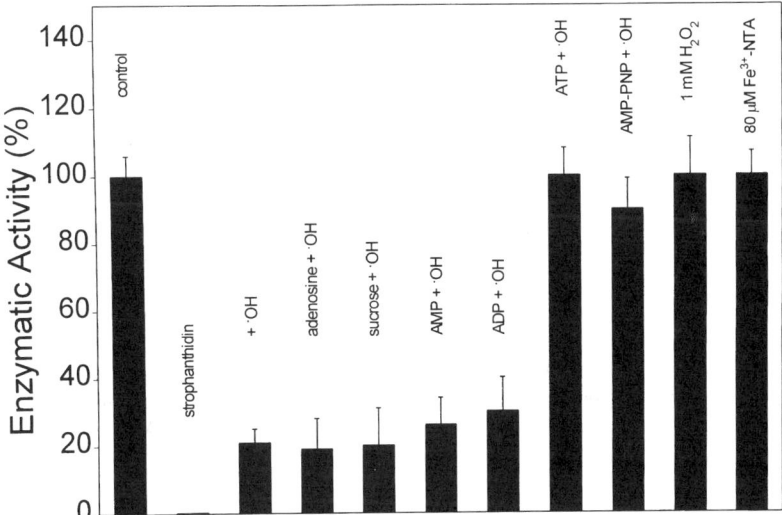

FIGURE 1. Effect of ATP, ADP, AMP, AMP-PNP, adenosine, and sucrose on the inactivation of cardiac Na^+,K^+-ATPase by ·OH free radical. Data represent the mean of three experiments. ATPase activity was completely protected by 1 mM ATP, but not by ADP, AMP, adenosine, or sucrose (1 mM each). Control = no ·OH. Protection is also provided by AMP-PNP, a derivative of ATP which has high affinity at the ATP binding site of the enzyme.

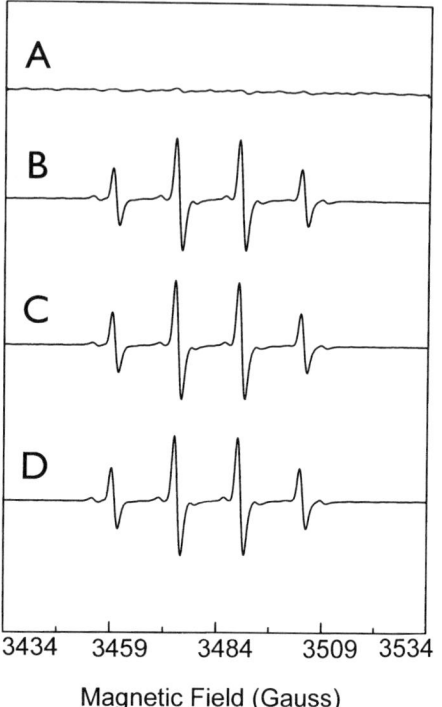

FIGURE 2. Electron paramagnetic resonance (EPR) spectra of the hydroxyl free radical generating system in the presence of 50 mM DMPO under various conditions. (**A**) 25 mM imidazole/HCL buffer; (**B**) H_2O_2 + Fe^{3+}/NTA; (**C**) H_2O_2 + Fe^{3+}/NTA + AMP-PNP; (**D**) H_2O_2 + Fe^{3+}/NTA + ATP. EPR results indicate that ·OH was the only type of free radical present and ATP did not scavenge ·OH.

alone, but in the presence of H_2O_2 + Fe^{3+} NTA, a prominent EPR signal consisting of a quartet 1:2:2:1 signal with hyperfine coupling constants $a_N = a_H = 14.9$ G was seen indicative of DMPO-OH. This signal was not altered by ATP and AMP-PNP, demonstrating that ATP and AMP-PNP did not scavenge ·OH (FIG. 2). In addition, the time course and maximal level of ·OH generation were not altered by ATP, ADP, AMP, or AMP-PNP (data not shown). These results suggest that: (1) ·OH denatures the cardiac sarcolemmal Na^+,K^+-ATPase by directly attacking the ATP binding site; (2) occupation of the ATP binding site of the enzyme protects against ·OH-induced loss of enzymatic activity; and (3) the depletion of ATP that occurs during ischemia may enhance the toxic effect of ·OH formed at the time of reperfusion.

ACKNOWLEDGMENT

We thank Dr. P. Kuppusamy for supervision of the technical aspects of the electron paramagnetic resonance measurements.

REFERENCES

1. KUKREJA, R. C. & M. L. HESS. 1992. The oxygen free radical system: From equations through membrane protein interactions to cardiovascular injury and protection. Cardiovasc Res. **26:** 641–655.
2. KIM, M. S. & T. AKERA. 1987. O_2 free radicals: Cause of ischemia-reperfusion injury to cardiac Na^+,K^+-ATPase. Am. J. Physiol. **252:** H252–257.
3. ZWEIER, J. L. 1988. Measurement of superoxide-derived free radicals in reperfused heart. J. Biol. Chem. **263:** 1353–1357.
4. JONES, L. R. 1988. Rapid preparation of canine cardiac sarcolemmal vesicles by sucrose flotation. Methods Enzymol. **157:** 85–91.
5. WATANABE, T., Y. TAWADA & M. SHIGEKAWA. 1988. Purification of cardiac (Na^+,K^+)-activated adenosine triphosphatase from rat. Anal. Biochem. **175:** 284–288.
6. KYTE, J. 1971. Purification of the sodium- and potassium-dependent triphosphatase from canine renal medulla. J. Biol. Chem. **246:** 4157–4165.
7. JOSEPHSON, R. A., H. S. SILVERMAN, E. G. LAKATTA, M. D. STERN & J. L. ZWEIER. 1991. Study of the mechanisms of hydrogen peroxide and hydroxyl free radical-induced cellular injury and calcium overload in cardiac myocytes. J. Biol. Chem. **266:** 2354–2361.

Effect of *Leptospira interrogans* Endotoxin on Renal Tubular Na,K-ATPase and H,K-ATPase Activities

M. YOUNES-IBRAHIM,[a,c] P. BURTH,[b] M. CASTRO-FARIA,[b]
L. CHEVAL,[a] B. BUFFIN-MEYER,[a] S. MARSY,[a]
AND A. DOUCET[a]

[a]*Laboratoire de Biologie Intégrée des Cellules Rénales*
CNRS URA 1859
Commissariat à l'Energie Atomique
Saclay, France

[b]*Laboratório de Biologia Celular*
Universidade do Estado do Rio de Janeiro
Rio de Janeiro, Brazil

We recently demonstrated that glycolipoprotein (GLP) isolated from *Leptospira interrogans* significantly inhibits Na,K-ATPase activity.[1] These inhibitory effects were verified on both purified enzyme from kidney medulla and renal epithelial cells from different segments of the nephron. Inhibition was also verified on intact cells by Rb uptake assay. The GLP did not change the apparent affinity of Na,K-ATPase for potassium, whereas it increased that for sodium, revealing a mechanism of inhibition different from that of ouabain. Indeed, we also described ouabain- and SCH28080-sensitive H,K-ATPase activity in single segments of the rat nephron.[2] Because it is now established that ouabain inhibits not only Na,K-ATPase but also some specific types of H,K-ATPase, such as that recently described in the rat nephron,[2] we investigated whether GLP would be more specific than ouabain as a Na,K-ATPase inhibitor.

METHODS

Animals. Microdissection of single nephron segments from male Wistar rats weighing 170–200 g was carried out after treatment of the kidney with collagenase.[3]

ATPase Assay. ATPase activities were determined on permeabilized individual nephron segments by measuring ^{32}P formed from exogenous $[\gamma^{32}P]$-ATP, with slight modifications of the method described previously.[2,3] ATP hydrolysis by the enzymes was determined after a 5-minute preincubation in the absence or the presence of GLP.

[c]Address for correspondence: Dr. M. Younes-Ibrahim, Rua Carlos Oswald, 230 b13-804, 22793-120 Rio de Janeiro, Brazil (tel: 55 21 5876227; fax: 55 21 5877377).

The distinction between the different ATPases present in the same segment of nephron was based on their cation-specific stimulation and their sensitivity to specific inhibitors. For each structure, 10–14 samples were distributed randomly into two groups, one for measuring basal ATPase activities and the other for measuring stimulated Na,K-ATPase or H,K-ATPase activities.

Na-free Assay Conditions. Using micro-flame photometry in tubular samples treated exactly as for ATPase measurement, except for [γ^{32}P]-ATP, which was omitted, we determined Na$^+$ concentrations in the assay medium after tubular permeabilization. According to the sodium activation curves of Na,K-ATPase previously established in rat nephron segments in our lab, the low concentrations of sodium we found (40–70 µmol) would induce Na,K-ATPase activities below 1 pmol · mm^{-1} · h^{-1} in the corresponding segments of nephron.

Adenylate Cyclase Activity. This was determined on single segments of nephron as previously described.[4]

Glycolipoprotein (GLP). Glycolipoprotein was extracted from *L. interrogans* according to the method of Vinh et al.[5]

TABLE 1. Effect of GLP Endotoxin on Enzymatic Activities in Single Segments of Rat Nephron (% of control)

	PCT	MTAL	CTAL	CCD	OMCD
Na,K-ATPase[a]	34 ± 8[c]	11 ± 3[c]	3.0 ± 3[c]	18 ± 9[c]	6.0 ± 3[c]
Mg-ATPase[a]	92 ± 5	113 ± 10	90 ± 7	112 ± 20	108 ± 18
Adenylate Cyclase[b]					
Basal	—	80 ± 10	—	133 ± 2	—
Hormone-stimulated (vasopressin 10^{-6} M)	—	96 ± 9	—	97 ± 11	—
H,K-ATPase					
Ouabain-sensitive[a]	—	103 ± 10	—	—	—
Ouabain-insensitive[a]	—	—	—	—	96 ± 12

Note: Enzymatic activities were determined in rat nephron segments. Proximal convoluted tubules (PCT); medullary and cortical thick ascending limb of Henle's loop (MTAL and CTAL); cortical and outer medullary collecting tubule (CCD and OMCD). Data are expressed as percentage of the respective control enzymatic activities (measured in the absence of GLP) in the corresponding segments of nephron and are means ± SE from 4–7 animals. Segments were preincubated for 5 minutes at 37°C in the presence of: [a]190 µg Prot/ml GLP or [b]175 µg Prot/ml GLP.
[c]$p < 0.001$, values statistically different from controls as determined by Student's *t* test.

RESULTS

Results presented in TABLE 1 show that single nephron segments preincubated with GLP display lower inhibition of Na,K-ATPase activity. GLP did not alter significantly either basal Mg-ATPase activity or hormone-sensitive adenylate cyclase activity. Finally, the presence of leptospiral endotoxin does not modify either ouabain-sensitive or ouabain-insensitive H,K-ATPase activity in distinct single segments from rat nephron.

CONCLUSION

The inhibitory effect of GLP seems to be specific for the Na,-K-ATPase enzyme, because GLP did not modify other cellular enzymatic activities including adenylate cyclase which, as with Na,K-ATPase, is present in the same basolateral cell membrane. GLP also had no effect on renal H,K-ATPase activity, even in the ouabain-sensitive form, indicating that the active principle of GLP is more specific for Na,K-ATPase than is ouabain itself. Although cellular effects verified *in vitro* with leptospiral endotoxin remain to be demonstrated *in vivo*, a specific interaction between GLP and the ubiquitous Na,K-ATPase supports the hypothesis that inhibition of Na,K-ATPase activity by GLP might be a primary cellular defect in the physiopathology of leptospirosis.

REFERENCES

1. YOUNES-IBRAHIM, M., P. BURTH, M. V. CASTRO FARIA, B. BUFFIN-MEYER, S. I. MARSY, C. BARLET-BAS, L. CHEVAL & A. DOUCET. 1995. Inhibition of Na,K-ATPase by an endotoxin extracted from *Leptospira interrogans:* A possible mechanism for the physiopathology of leptospirosis. C. R. Acad. Sci. Paris, Life Sci. **318:** 619–625.
2. YOUNES-IBRAHIM, M., C. BARLET-BAS, B. BUFFIN-MEYER, L. CHEVAL, R. RAJERISON & A. DOUCET. 1995. Ouabain-sensitive and -insensitive K-ATPases in rat nephron: Effect of K-depletion. Am. J. Physiol. **268**(37): F1141–1147.
3. DOUCET, A., A. I. KATZ & F. MOREL. 1979. Determination of Na,K-ATPase activity in single segments of mammalian nephron. Am. J. Physiol. **237:** F105–F113.
4. MOREL, F., D. CHABARDES & M. IMBERT-TEBUL. 1978. Methodology for enzymatic studies of isolated tubular segments: Adenylate cyclase. In Methods in Pharmacology. M. Martinez-Maldonado, Ed., 297–323. Plenum. New York.
5. VINH, T., B. ADLER & S. FAINE. 1986. Glycolipoprotein cytotoxin from *Leptospira interrogans:* Serovar *copenhageni.* J. Gen. Microbiol. **132:** 111–123.

Transgenic Mice Expressing Human α3 Na,K-ATPase Isoform in Heart

RAPHAEL ZAHLER,[a] MARK LUFBURROW, MIRA MANOR,
RADHA SHENOY, DIEGO FORNASARI, MARC ROMANA,
AND WEI SUN

*Yale University School of Medicine
New Haven, Connecticut 06520*

The Na,K-ATPase is an enzyme crucial for normal mechanical and electrical function of the heart, because it supports cell volume regulation, membrane potential, and ion gradients for transport of other solutes. The α3 pump isoform is associated with the sites of conduction of the cardiac impulse; expression of α3 in adult rat heart is confined to the cardiac conduction system and the junctional complex.[1] Also, our laboratory and others have shown that the α3 isoform differs functionally from α1, having higher ouabain affinity, lower Na^+ affinity, and greater maximum turnover rate.[2,3] Isoform differences could thus underlie the well known discrepancy between the inotropic and conduction-system effects of cardiac glycosides. Also, a major consequence of myocardial ischemia is cell swelling caused by Na influx. Because hypertrophied hearts have fewer Na^+ pumps and a higher percentage of α1 pumps, their increased vulnerability to ischemia could be offset by restoring high-capacity α3 pumps. Thus, to investigate the consequences of manipulating pump isoforms on transport and contractility in heart cells, we constructed transgenic mice (TGM) with cardiac-specific overexpression of human α3 Na,K-ATPase.

We verified that normal mouse heart does not express detectable amounts of α3 mRNA and protein via Northern and Western blotting with isoform-specific reagents. We then prepared a construct in which a 640-bp fragment of α-myosin heavy chain 5′-upstream region drove expression of the full-length human α3 cDNA, terminated by the SV40 small t antigen splice and polyadenylation signal. An Sph I-Pvu I fragment of this plasmid (containing all the aforementioned elements but excluding almost all plasmid sequence) was isolated, purified, and injected into mouse oocyte pronuclei. A line of mice was developed which was hemizygous for this transgene and whose hearts reproducibly expressed a 97-kD membrane protein immunoreactive with isoform-specific anti-α3 antibody by both Western blotting and immunocytochemistry. Such immunoreactivity was not found in normal mouse heart or in any organ of TGM except brain and (at low levels) lung. Expression of immunoreactive α1 protein was reduced in TGM heart compared to normal heart, suggesting that a compensatory decrease in α1 could exaggerate the effect of the transgenic isoform.

TGM hearts were morphologically normal, and heart weight/body weight did not differ from that of controls. However, quantitative competitive (^{3}H)ouabain binding on microsomes from TGM heart showed that the high-affinity component in TGM

---

[a]Address for correspondence: Cardiology/Fitkin 3, Yale University School of Medicine, PO Box 208017, New Haven, CT 06510-8017 (tel: 203 785-4102; fax: 203 785-7144; e-mail: raphael.zahler@yale.edu).

heart had a threefold increase in affinity; TGM heart K_D for ouabain was 38.8 nM compared with 144 nM for normal mouse heart. In fact, the high-affinity ouabain K_D measured in TGM heart was similar to the K_D of human α3 measured independently in SY5Y neuroblastoma cells.[4] Furthermore, the ouabain-sensitive fraction of Na,K-ATPase activity was 15% in TGM heart compared to <5% in controls, a statistically significant difference (FIG. 1). Surface electrocardiograms were obtained from six TGM and six control mice. Heart rates were similar; however, TGM had significant prolongation of the QRS complex (65 vs 39 ms, $p = 0.014$) and corrected QT interval QT_c (291 vs 184 ms, $p = 0.02$) compared to those of the controls (FIG. 2).

Thus, we have constructed transgenic mice with cardiac-specific overexpression of human α3; data suggest that our mice express physiologically significant amounts

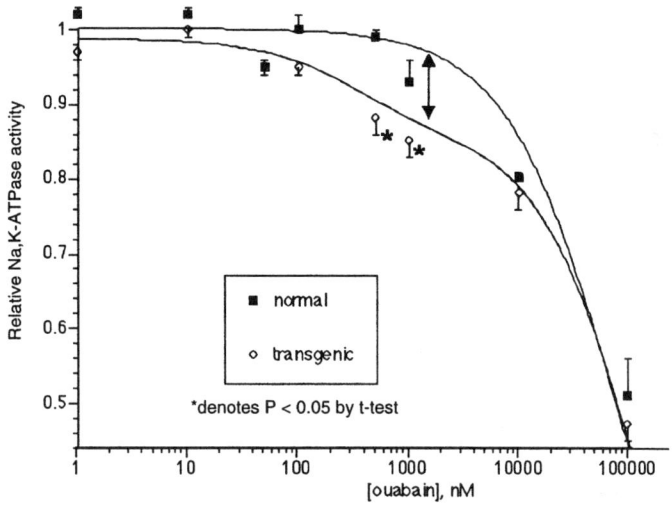

FIGURE 1. Ouabain inhibition profiles of Na,K-ATPase activity in cardiac microsomes from normal and α3-transgenic mice expressed as a fraction of activity in the absence of ouabain. Significant inhibition in normal mice does not occur with ouabain concentrations of less than 1,000 nM, whereas approximately 15% of Na,K-ATPase in the transgenic heart (*arrow*) is inhibited by nanomolar ouabain concentrations.

of functional membrane-inserted Na,K-ATPase α3 isoform in heart. The data also suggest that TGM hearts may have a wider therapeutic window for glycosides. Specifically, in normal mice a ouabain concentration sufficient to increase contractility by partially inhibiting α1 pumps in ventricular muscle will completely inhibit the more sensitive α3 pumps in the conduction system, causing toxic effects. In transgenic mice, however, the ouabain-sensitivity curve of cardiac muscle should be shifted closer to that of the conduction system; thus, low concentrations of ouabain should increase contractility without toxicity. Studies of transport and contractility in perfused TGM hearts and isolated myocytes are underway to test the hypotheses that an increased fraction of α3 pumps will lead to increased basal $[Na^+]_i$, improve the ability to clear an Na^+ load, and increase the contractility response to low-dose ouabain.

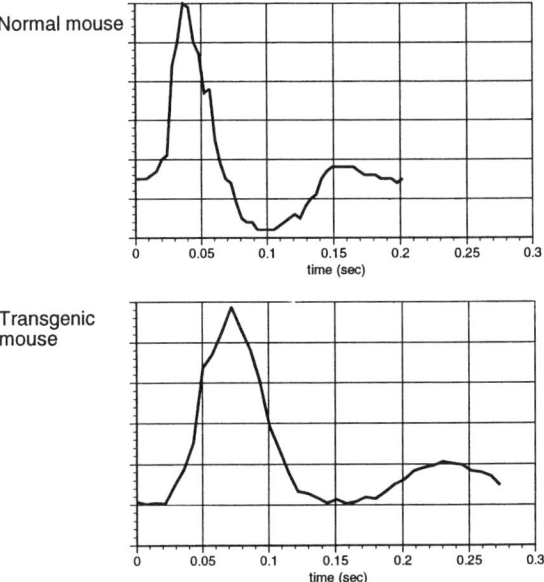

FIGURE 2. Representative signal-averaged surface ECGs from a normal mouse and a transgenic mouse. For all mice studied, QRS complex duration and heart rate-corrected QT interval were longer ($p < 0.02$ for each comparison) in transgenic mice.

REFERENCES

1. ZAHLER, R., W. SUN, T. ARDITO, J. KOCSIS & M. KASHGARIAN. 1996. The α3 isoform protein of the Na,K-ATPase is associated with the sites of neuromuscular and cardiac impulse transmission. Circ. Res. **78:** 870–879.
2. ZAHLER, R., Z.-T. ZHANG, M. MANOR & W. BORON. 1997. Sodium kinetics of Na,K-ATPase α isoforms measured separately in intact transfected cells. J. Gen. Physiol. In press.
3. MUNZER, J. S., S. E. DALY, E. A. JEWELL-MOTZ, J. B LINGREL & R. BLOSTEIN. 1994. Tissue- and isoform-specific kinetic behavior of the Na,K-ATPase. J. Biol. Chem **269:** 16668–16676.
4. ZAHLER, R., W. SUN, D. FORNASARI, M. BRINES & M. ROMANA. 1994. Antisense oligodeoxynucleotide selectively inhibits expression of endogenous α1 sodium pump isoform. Circulation **90:** 1–147.

Index of Contributors

Adamo, H. P., 56–64, 449–451, 452–453
Albers, R. W., 129–131, 280–296, 376–377
Allen, J. C., 457–458
Alonso, G. L., 400–403, 555–558
Alonso, T., 110–114, 653–655
Alvarez de la Rosa, D., 110–114
Andersen, J. P., 297–309, 333–338
Andersson, K., 592–599
Anner, B. M., 367–371
Antolovic, R., 618–620
Apell, H.-J., 221–230, 364–366, 420–423
Argüello, J. M., 194–206, 339–342
Arnadottir, J., 424–425
Arulanantham, P. R., 376–377
Askerlund, P., 77–87
Avila, J., 110–114, 653–655
Axelsen, K. B., 77–87

Ball, W. J., Jr., 634–636
Balzan, S., 621–625, 626–630
Bamberg, E., 270–279
Bamberg, K., 65–76
Barakeh, H., 475–478
Barbey, O., 656–657
Barnard, M. L., 651–652
Barrabin, H., 631–633
Barreiro Lopez, L., 552–554
Baunsgaard, L., 77–87
Bayati, A., 592–599
Beaugé, L., xv, 378–380, 416–419
Becker, L. C., 680–683
Beggah, A. T., 537–539
Beguin, P., 540–542
Berlin, J. R., 251–259, 339–342
Beron, J., 569–571
Besancon, M., 65–76
Bezanilla, F., 231–243
Bharadwaj, A., 424–425
Blanco, G., 88–96, 104–106, 135–138, 572–575
Blaustein, M. P., 524–536
Blostein, R., 489–499, 579–581
Blot-Chabaud, M., 545–547
Boldyrev, A., 661–665, 666–668, 669–672
Bonvalet, J. P., 545–547, 562–564
Borin, M. L., 576–578
Brumfeld, V., 146–148
Buffin-Meyer, B., 684–686
Buhagiar, K. A., 347–349
Bulygina, E. R., 666–668
Bundgaard, H., 648–650, 676–679

Burth, P., 684–686
Buxbaum, E., 381–385

Campos, M., 378–380
Cantiello, H. F., 559–561
Caplan, M. J., 514–523
Capurro, C., 562–564
Caricati-Neto, A., 115–118
Caride, A. J., 459–461
Castello, P. R., 126–128
Castro-Faria, M., 684–686
Cavieres, J. D., 381–385, 432–434
Charlemagne, D., 658–660
Chen, S., 673–675
Cheval, L., 684–686
Colonna, T., 498–513
Corbin, J., 582–584
Cornelius, F., 386–389, 390–393, 394–396
Cortas, N., 475–478
Courtois-Coutry, N., 514–523
Coutry, N., 545–547, 562–564
Cramb, G., 123–125, 565–568
Crambert, G., 97–100, 621–625
Cutler, C. P., 123–125, 565–568

Daly, S. E., 489–499
Daoud, S., 45–55
De Pont, J. J. H. H. M., 101–103, 472–474
De Weer, P., 231–243
Decollogne, S., 621–625
Del Bene, P., 637–641
DeTomaso, A. W., 88–96
Dobrota, D., 661–665
Domaszewicz, W., 420–423
Donnet, C., 459–461
Doucet, A., 684–686
Dunbar, L. A., 514–523

Edmonds, Z. V., 376–377
Edwards, J., 565–568
Efendiyev, R. E., 153–154
Eim, C., 646–647
Enyedi, A., 56–64
Escoubet, B., 562–564
Esmann, M., 310–321, 410–411

Factor, P., 104–106, 651–652
Faller, L. D. 442–444
Falson, P., 142–145
Fambrough, D. M., 498–513
Farman, N., 545–547, 562–564

691

Fedosova, N. U., 310–321, 386–389, 390–393, 394–396
Fendler, K., 270–279, 280–296, 361–363
Feschenko, M. S., 479–488
Filoteo, A. G., 56–64
Fontes, C. F. L., 631–633
Forbush, B., III, 386–389
Fornasari, D., 687–689
Fotis, H., 585–587
Franz, A., 97–100
Friedrich, T., 270–279, 435–438
Froehlich, J. P., 129–131, 280–296
Fuglsang, A. T., 77–87
Fukushima, Y., 462–465
Futai, M., 149–152

Gadsby, D. C., xv, 231–243, 426–431
Gagliardino, J. J., 126–128
Gao, D., 158–160
García, M. P., 397–399
García-Segura, L. M., 110–114
Garrahan, P. J., xv, 327–332
Gärtner, E.-M., 585–587
Garty, H., 562–564
Gatto, C., 45–55
Geering, K., 537–539, 540–542
Geibel, J., 514–523
Gerbi, A., 656–657, 658–660
Gevondyan, N. M., 364–366, 466–468
Gevondyan, V. S., 466–468
Ghione, S., 621–625, 626–630
Gilliam, T. C., 155–157
Glitsch, H. G., 354–356
González, D. A., 400–403
González Flecha, F. L., 126–128
Gottardi, C. J., 514–523
Gray, D. F., 347–349
Grimaldi, M. E., 452–453
Grinberg, A. V., 466–468
Gu, Q., 372–375

Hagiwara, E., 19–29, 132–134
Hamrick, M., 498–513
Hansen, O., 404–406
Hansen, P. S., 347–349
Hatfield, W. R., 88–96, 135–138
Hayashi, Y., 19–29, 132–134
Hellen, E. H., 439–441, 445–448
Herbin, T., 642–645
Herscher, C. J., 407–409
Herslöf, M., 600–608
Hilgemann, D. W., 260–269
Hoffman, J. F., 119–122
Holmgren, M., 231–243

Horisberger, J.-D., 244–250, 343–346
Horowitz, S., 651–652
Hughes, F., 651–652
Hwang, B., 498–513

Imagawa, T., 129–131, 582–584
Inesi, G., 207–220
Ishii, T., 372–375, 588–591
Iwamoto-Kihara, A., 149–152

Jaffe, H. A., 104–106
Jamme, I., 658–660
Jemelka, S., 457–458
Jensen, J., 404–406
Jørgensen, P. L., 161–174, 454–456, 469–471
Juhaszova, M., 524–536
Jurkiewicz, A., 115–118
Juul, B., 142–145

Kable, E. P. W., 673–675
Kameyama, K., 19–29
Kaplan, J. H., 45–55, 155–157
Karlish, S. J. D., 30–44, 146–148
Kashgarian, M., 514–523
Kasho, V. N., 442–444
Kasturi, R., 634–636
Kaufman, S. B., 327–332, 397–399
Kaya, S., 129–131, 186–193, 582–584
Keeling, D. J., 600–608
Kenney, L. J., 45–55
Khater, K. A., 350–353
Kikkawa, U., 582–584
Kirch, U., 646–647
Kirtley, M., 207–220
Kjeldsen, K., 648–650, 676–679
Klaassen, C. H. W., 101–103, 472–474
Klodos, I., 386–389, 390–393, 394–396
Kobayashi, T., 19–29, 132–134
Kockskämper, J., 354–356
Kost, H., 618–620
Koster, J. C., 88–96, 135–138
Kostich, M., 498–513
Kotlobay, A., 669–672
Krishna, S., 158–160
Kukley, M., 661–665
Kuntzweiler, T. A., 194–206
Kuo, T. H., 673–675
Kurella, E., 661–665, 669–672

Lacapère, J.-J., 9–18, 400–403
Lafayette, S. S. L., 115–118
Lane, L. K., 489–499, 579–581
Lanfermeijer, F. C., 77–87, 139–141

INDEX OF CONTRIBUTORS

Le Maire, M., 142–145
Leblanc, G., 9–18
Lecuona, E., 110–114, 653–655
Lee, H., 592–599
Lelièvre, L. G., 97–100, 621–625
Lewis, D., 207–220
Li, S.-Q., 646–647
Lichtstein, D., 618–620
Lin, S.-H., 442–444
Linder, D., 618–620
Linder, M., 618–620
Lingrel, J. B, 194–206, 339–342
Linnertz, H., 322–326
Lopes, F. E. V., 631–633
López Ordieres, M. G., 548–551
Lopina, O., 669–672
Lorentzon, P., 592–599
Lücking, K., 107–109
Lufburrow, M., 687–689
Luquín, S., 110–114
Lutsenko, S., 45–55, 155–157

Mackenzie, S., 565–568
MacLennan, D. H., 175–185
Mahaney, J. E., 280–296
Mahfouz, H., 475–478
Maixent, J.-M., 656–657, 658–660
Malkov, D. Y., 357–360
Malmström, S., 77–87
Manor, M., 687–689
Mårdh, S., 582–584
Margolies, M. N., 634–636
Marsy, S., 684–686
Martín-Vasallo, P., 110–114, 653–655
Matejovicova, M., 661–665
Mayol, V., 656–657
McLean, L. R., 634–636
Mercer, R. W., 88–96, 104–106, 135–138, 457–458, 572–575, 651–652
Mezesova, V., 661–665
Minor, N. T., 88–96
Møller, J. V., 142–145
Monk, B. C., 609–617
Montali, U., 621–625, 626–630
Monteith, G. R., 673–675
Morales, M., 653–655
Mori, M., 582–584
Mosser, G., 9–18
Muth, T. R., 514 523

Nagel, G., 270–279, 435–438
Nandi, A., 207–220
Nestor, N. B., 579–581
Nettikadan, S., 149–152

Nielsen, J. M., 107–109, 161–174
Noël, F., 115–118, 552–554
Nørby, J. G., 327–332, 410–411, 631–633
Nouvelot, A., 658–660

Odebunmi, T., 457–458
Odermatt, A., 175–185
Okkels, F. T., 77–87
Olivera, W., 651–652
Omote, H., 149–152
Ordahl, C. P., 207–220

Paci, A., 621–625, 637–641
Paganelli, F., 656–657
Palit, A., 445–448
Palmgren, M. G., 77–87, 139–141
Pavlov, K. V., 357–360
Pedersen, P. A., 161–174, 454–456, 469–471
Peluffo, R. D., 251–259, 339–342
Penniston, J. T., 56–64, 449–451, 452–453
Penny, J., 158–160
Peña, C., 642–645
Perlin, D. S., 609–617
Petrosian, S. A., 543–544
Petrukhin, K., 155–157
Pierre, S., 656–657
Pintschovius, J., 361–363
Platoshkina, E. A., 153–154
Plesner, I. W., 412–415
Potapenko, N. A., 153–154
Pratap, P. R., 439–441, 445–448
Pressley, T. A., 457–458, 543–544

Quintas, E. M., 552–554
Quintas, L. E. M., 115–118

Rajendran, V., 514–523
Rakowski, R. F., 231–243, 350–353
Ramlov, D., 297–309
Ranck, J.-L., 9–18
Rasmussen, H. H., 347–349
Rasmussen, J. H., 161–174, 454–456, 469–471
Rawn, J. D., 498–513
Rega, A. F., 407–409, 449–451, 452–453
Reinés, A., 642–645
Rice, W. J., 175–185
Ridge, K., 104–106, 651–652
Rigaud, J.-L., 9–18
Robert, K., 656–657
Roberts, G., 416–419
Rodríguez de Lores Arnaiz, G., 548–551, 642–645

Romana, M., 687–689
Rossi, J. P. F. C., 126–128, 459–461
Rossi, R. C., 327–332, 397–399
Roufogalis, B. D., 673–675
Roush, D. L., 514–523
Rutschman, D. H., 651–652
Ryberg, B., 600–608

Sachs, G., 65–76
Sakakibara, M., 637–641
Sánchez, G., 88–96
Sanders, I. L., 123–125
Sarvazyan, N., 669–672
Sato, M. H., 588–591
Scarborough, G. A., 1–8
Schmidt, T. A., 648–650, 676–679
Schneeberger, A., 420–423
Schneider, R., 646–647
Schoner, W., 322–326, 432–434, 618–620, 646–647
Schwarz, W., 372–375
Schwarzbaum, P. J., 327–332, 397–399
Scofano, H. M., 631–633
Scriven, D. R. L., 424–425
Seifert, K., 361–363
Senne, C., 104–106
Seto-Young, D., 609–617
Shainskaya, A., 146–148
Shenoy, R., 687–689
Shimada, A., 582–584
Shin, J. M., 65–76
Shinji, N., 19–29, 132–134
Shinohara, Y., 462–465
Sich, B., 646–647
Sjögren, S., 600–608
Smirnova, I. N., 442–444
Sokolov, V. S., 357–360, 364–366
Sölvell, L., 600–608
Sørensen, T., 333–338
Soriano, A., 653–655
Souccar, C., 552–554
Stangeland, B., 77–87
Stengelin, M., 119–122
Stokes, D. L., 9–18
Stukolov, S. M., 364–366
Sumbilla, C., 207–220
Sun, W., 687–689
Swarts, H., 101–103, 472–474
Sweadner, K. J., 479–488
Sznajder, J. I., 104–106, 651–652

Takagi, T., 19–29
Takara, D., 555–558
Takeyasu, K., 149–152, 158–160, 372–375, 588–591
Talgham, S. A., 459–461
Taniguchi, K., 129–131, 186–194, 280–296, 582–584
Thoenges, D., 322–326, 618–620
Thomas, D. D., 280–296
Togawa, K., 582–584
Tokumasu, F., 149–152
Tostesson, D. C., 424–425
Tostesson, M. T., 424–425
Trouvé, P., 658–660
Tsuda, T., 186–193
Tyulina, O., 661–665

Uchida, A. O., 637–641
Uhal, B., 651–652
Ushimaru, M., 462–465
Usta, J., 475–478

Van Huysse, J., 194–206
Vasilets, L. A., 372–375, 585–587
Venema, K., 77–87, 139–141
Verma, A. K., 56–64
Verrey, F., 569–571
Vilsen, B., 297–309, 333–338
Vladimirova, N. M., 153–154

Wagg, J., 231–243, 426–431
Walton, T. J. H., 381–385
Wang, S., 158–160
Wang, X., 244–250, 343–346
Ward, D. G., 381–385, 432–434
Wetzel, R. K., 479–488

Xu, K. Y., 680–683

Yasuhara, J. C., 588–591
Yokoyama, A., 129–131
Yokoyama, T., 186–193
Yoshimura, S. H., 588–591
Younes-Ibrahim, M., 684–686

Zahler, R., 687–689
Zhao, J., 372–375
Zhao, X., 457–458
Zweier, J. L., 680–683

OHIO UNIVERSITY LIBRARY
Please return this book as soon as you have